JN191047

Basics of Physics

by

HARA Yasuo

Translated by

KAWAIZUMI Fumio

UEMATSU Tsuneo

Gakujutsu Tosho Shuppan-sha Co., Ltd.

第 5 版　物理学基礎

by

©HARA Yasuo 2016

Translated by

KAWAIZUMI Fumio

UEMATSU Tsuneo

Published 2019 in Japan by

Gakujutsu Tosho Shuppan-sha Co., Ltd.

Preface

This book is designed as a textbook or a reference book in physics in the basic education of science and engineering students.

Recently, concerning the education at the science and engineering departments, what has been required is the cultivation of the student abilities which can be applied to the discovery and the solution of the practical problems.

The construction of physics has been performed through the processes where the natural phenomena are investigated by introducing the physical quantities (concepts) such as velocity, acceleration, mass, work, energy, temperature, charge, current, electric field, magnetic field etc. together with the physical entities like molecules, atoms, electrons etc. to discover physical laws which are mathematical relationships among physical quantities. Studying physics in this viewpoint is not only acquiring the knowledge about physics but also learning physical points of view as well as physical ways of thinking.

Mathematics is an academic discipline which was born by abstracting the mathematical aspects of various phenomena and has later been developed so far. In order to understand abstract concepts in mathematics it would be effective to learn mathematical concepts in connection with concrete examples. While, in order to train applied skills of mathematics it would be important to learn physical contents combining the mathematical methods used there. These two points are clear if we look back the history of the development of physics and mathematics.

This book was written based on the above two viewpoints, expecting the readers to acquire the basic knowledge of physics and to cultivate the applied skills, and at the same time to become able to understand the mathematical methods in physics.

In writing the present book, the author keeps in mind that the contents are well above the international standards of the basic education in physics at the science and engineering departments, while these contents can be well understood by those who studied physics insufficiently at high-school but have strong motivation to learn physics at university. To this end, the author tried to express both logics and mathematical equations as easy to understand as possible. Explanations are given by making connection with concrete phenomena and trying to let the readers be able to realize the effectiveness of physics.

The characteristics of the present book are as follows :

(1) Important concepts (physical quantities) and laws are explained in a careful way.

(2) Using plenty of cases and examples we try to help readers to understand laws and ways of their applications.

(3) At the end of each chapter, a number of problems for exercise (A is relatively easy, B is relatively difficult) are given. Their detailed answers/solutions are attached at the end of this book to promote the better understanding of each chapter.

(4) Derivation and calculation of equations are made in a careful manner, and the middle of the calculation including calculation of units is not omitted.

(5) Unnecessary use of the higher mathematics is avoided and the necessary mathematical matters are explained in the text. Proper care has been taken not to hinder the understanding of physics due to the difficulties in mathematics.

(6) We try to make readers understand systematically an overview of physics in an organized form.

(7) Since there might be the case where the contents of this book are too many for the lectures, we try to place the basic matters in the former half and the optional matters in the latter half in each chapter (We have carefully written the present book so that the item indicated with the symbol ✣ marked on the right shoulder of the heading such as sections etc. can be skipped without causing difficulties in the learning of later parts of the book).

The present author expects that you students will deeply understand fundamental physics and acquire applied skills by reading this book. For those students who want to learn more, my book with the title "Butsurigaku Tsuron I, II (Gakujutsu Tosho Shuppansha Co., Ltd)" is recommended.

The first time I gave the course in physics was 50 years ago when I took up the position of associate professor at Tokyo University of Education (Now Tsukuba University). I well remember that around that time while we had a chat at the laboratory, Dr. Shin-ichiro Tomonaga told me "Hara-kun (friendly call of Mr. Hara), since laws in physics are not what we derive, you can present to the students basic laws in a way easy to understand for yourself." Since then I have been keeping this word in my mind at any time when I give a lecture or write a textbook. Since I believe that this word should be very useful in learning of physics, I introduce it to the readers of this book.

The first edition of "Butsurigaku Kiso (Basics of Physics, original edition of this book in Japanese)" was published in 1986 and the revised edition in 1993. Fortunately, the book was adopted as a textbook in many universities. During this period, I had opportunities to have experiences at the sites of education and to have had communication with educators so that I have improved my understanding of physics education. Making use of these experiences, I have revised this book for a number of times.

In the third edition of "Butsurigaku Kiso" published in 2004, in addition to the contents, we renewed the style and fully colorized the text incorporating photos to a large extent. Fortunately, the present book has gained a reputation that for the students who are familiar with the color printing of the textbook until high-school, the full-color text is easy to get familiar with as well as easy to access.

This time we revised the book for the fourth time and delivered the fifth edition of "Butsurigaku Kiso" to the readers. The chapter organization of the fifth edition is the same as that of the fourth edition, but I have received valuable instructions and useful advises from several specialists at this occasion of the revision. Thanks to these instructions and advises, better revision was possible. So, I deeply thank from the heart those people who gave me instructions and advises. In particular, I am very much indebted to Dr. Shuji Ukon, from whom I got the opinion about physics education at various occasions and received appropriate instructions. In addition, the column in Chapter 1 "Similarity rule" borrows a passage written by Dr. Ukon in the book co-authored by Dr. Ukon and myself "Unravelling the questions in daily life by physics" (Science-i paperback). Here let me thank him for his kindness concerning this matter.

In order to make the contents of this book easy to understand, the colors are used as follows :

(1) As shown in the following table, position, force, velocity, acceleration etc. are indicated using arrows with different color.

Color of arrows	
Position	→
Force	→
Velocity	→
Acceleration	→
Lines of electric force	→
Electric filed	→
Lines of magnetic force	→
Magnetic field	→
Electric current	→

(2) Important conclusions and definitions are printed in blue color.

(3) Important equations/formulas are printed in yellow color.

I would be very happy if the use of these colors could make the book clear and easy to learn.

Moreover, we have included a lot of color photographs so that the readers can feel familiar with the book. The sources of the photos are indicated in Photo Credits. Here I express my deep gratitude to the universities, institutes, foundations, companies, researchers and others that provide us with the photographs.

I have asked Mr. Takao Hotta, Mr. Shuji Takahashi and Ms. Mika Sugimura at the Gakujutsu Tosho Shuppan Co., Ltd. for collecting the photographs. These three people also took care of editing works. I am deeply grateful to these three people.

If you have any question concerning the contents of this book, please feel free to contact the editorial office to the following e-mail address :

info@gakujutsu.co.jp

October, 2016

HARA Yasuo

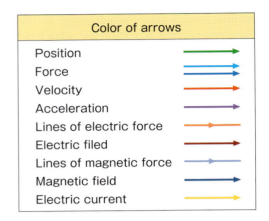

Contents

Column

Description of the photo on the right page

RIKEN heavy ion linear accelerator (RILAC) which linearly accelerates heavy ions using a high-frequency electric field. In July 2004, using this accelerator, Dr. Kosuke Morita (currently professor at RIKEN, Kyushu University and director of Riken group) successfully synthesized a new element with atomic number 113 (see page 343). In November 2016, the International Union of Pure and Applied Chemistry (IUPAC) decided that the name of this 113th element is nihonium.

Basics of Physics

HARA Yasuo

Introduction

What is physics?

Physics is a human attempt to understand natural phenomena from empirical as well as logical points of view. Physics originated from the efforts to comprehend the phenomena that humans observed with the naked eye. As a result of these efforts, physics could show in a clear fashion the existence of atom, and electric and magnetic fields. Study of physics permitted humans to understand more profoundly the nature around us. In addition, the fruitful results obtained therefrom provided human race with new energy resources and communication means.

In this introductory chapter, the following topics will be treated : 1) a brief explanation of the relations between space and time found in various natural phenomena, 2) an explanation of the role played by physical quantities and physical laws, both of which support empirical and mathematical foundations of physics as a field of science, and 3) the introduction to the International System of Units employed nowadays in physics.

A photo of Crab Nebula taken by the Subaru Telescope. At the center of the nebula lies a neutron star which emits X-rays and γ-rays with a period of 0.33 seconds while rotating at high speed and it is called "Crab Pulsar".

0.1 What is physics?

Physics as an empirical science and a demonstrative science

What kind of science is physics? Physics has changed with the times but characteristic points of modern physics are to be empirical and demonstrative. In other words, the goal of physics is to create a system in which we look into **various laws** ruling the phenomena happened in nature around us **by means of observation and experiments,** and then express the **obtained laws** in mathematical form and finally derive these laws from more fundamental laws. This is the goal of physics.

Physics aims to be logical and systematic but its feature is different from that of geometry. Geometry is founded on the assumption of several axioms regarded as reasonable and commonly accepted and other theorems are derived logically from them. Development of experimental technology has constantly led us to find new phenomena and to create new materials. This indicates that in physics there exists in general a limit in applying the fundamental laws corresponding to the axioms in geometry. Once we encounter the application limit of the conventional laws, the search for a new law system that can surpass the previous one is explored. This is the reality of physics

Overview of development of physics Modern physics to explore the natural law on the basis of the observed facts started to clarify the phenomena we see with the naked eye and notice by touching with hands. Fall motion of an object we see with the naked eye, celestial motion, heat we feel in our hands, light visible to us, sound we hear, and stimulus due to the frictional electricity, all of these phenomena were objects of interest of physicists.

But the advance in physical study has made it clear that in order to have a deep understanding of the world we see and touch in our daily life, such as light, thermal phenomena, electromagnetic phenomena, and physicochemical properties of materials, we must know the space filled with lines of electric force and those of magnetic force as well as the atomic world. All of these are undetectable with the naked eye.

The electromagnetic theory developed in the 19th century revealed that the light visible to our eyes is a kind of electromagnetic waves and it travels through space as a wave with light speed. In addition, application of laws of electromagnetics provoked the invention of electric motor, generator, and transformer. Results of electromagnetics provide humans with new sources of power and energy, lightning in the night, and information technology and have had a great impact on our social life.

In the 20th century, structure of solid, atomic structure consisting of atomic nucleus and electrons, structure of atomic nucleus, and structures of proton and neutron both of which consist of quarks were unveiled successively. Electron, proton, neutron, atomic nucleus, atom and molecule which are the constituent elements of materials are found to have, similarly as the case of light and contrary to our sense in daily life, the dual property of light and wave and not to obey the Newtonian mechanics but to obey the quantum mechanics. With the advance in the study of nucleus, origins of energy of solar radiation,

Fig. 0-1 Subaru telescope. A large scale optical-infrared telescope at the summit (height 4200 m) of Mauna Kea, Hawaii. Open cluster Subaru and Jupiter are twinkling in the night sky.

volcanic activity, and earthquake became clear.

In the recent information society, semiconductors and lasers play an important role. They are the applied results of quantum mechanics and the knowledge of quantum mechanics is indispensable to conduct research in these fields.

In addition to the study of microscopic world, the study of structure of the universe and the history of its development has advanced and the observed data of the Hubble Space Telescope on the artificial satellite and of the Subaru Telescope on the ground revealed that the universe was born about 13.8 billion years ago and since then the universe has continued to expand.

Fig. 0-2 The Hubble Space Telescope. An optical telescope launched into Earth orbit in 1990 by the US. Length 13.1 m and weight 11 t. Diameter of the mirror 2.4 m.

The advance in physics has contributed to the clarification of the laws and structures of materials lying on the basic layer of natural phenomena. But at the same time, it has also become clear that among the constituent materials we know, atoms (proton, neutron, electron etc.) occupy only ca. 20% and the rest of the materials in the universe is the unknown material called dark matter. As mentioned up to this point, a vast unknown area is still left in nature. In future, studies of physics will surely advance more and new scientific results will have to be produced, which should deepen the understanding of nature, and also the discovery of a new unknown world is expected.

For us living in a modern society, energy-supply system, traffic system, information-communication system, and industrial products are indispensable and all these rely on the fruitful results of physics. For every human, therefore, the physical view and way of thinking as well as the knowledge of physics are needed, although there may be some differences depending on his/her personal or social conditions. Moreover, the society needs presence of personnel standing on the basis of physics and capable of making discoveries and doing inventions in the field of science and technology.

0.2 Space and time

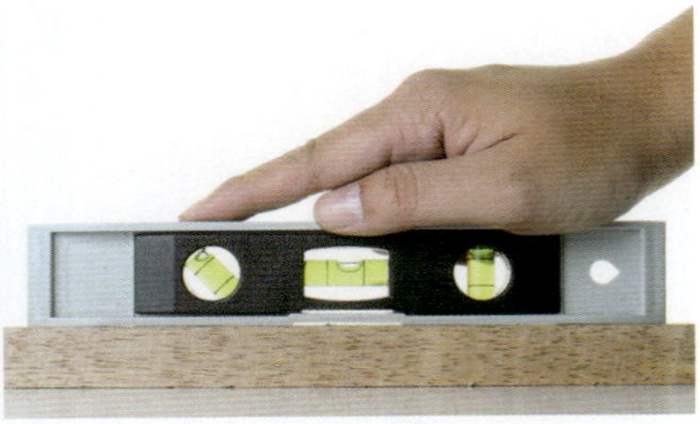

Fig. 0-3 A level showing the horizontal direction

Expression of "phenomena that occur in nature around us" means the presence of "space" around us and "time" in which the phenomena occur. The fact that the three directions of front-back, right-left, and up-and-down exist in space is referred to as "Space has three dimensions". Since the up-and-down direction is the direction of the suspended plomb, it is called the vertical [in Japanese 鉛 (plomb) 直 (linear)] direction. The term four directions is used to express the four directions (north, south, east and west) on the still-water surface and refers to the horizontal direction (Fig. 0-3).

Mathematics of space is the geometry Concerning the phenomena which occur on Earth, the vertical direction and the horizontal directions are entirely different. In contrast to this, physics established to be applicable not only to the phenomena near the Earth surface but also to those of the universe including the solar system is the Newtonian mechanics. In Newtonian mechanics, no special direction is considered in the discussion of the universe.

Accumulated conventional understanding of the space through daily life is geometry you learned in junior-high school and high school. Newton established his mechanics assuming that the space should obey the Euclidian geometry.

Coordinate axis and coordinate To specify the position of an object in space, we introduce coordinate axes. As illustrated in Fig. 0-4 for the vertical motion of the spring-hung weight, in the case of linear motion of an object along the straight line, we first choose this line as the axis (x-axis), and then determine the origin O, the positive direction of the x-axis, and the unit of length. After these preparations, the position of the object can be expressed by the coordinate, for example, $x = 2.0$ m.

In the case of motion in three-dimensional space, we introduce three mutually orthogonal coordinate axes as shown in Fig. 0-5. Then, the coordinates of the objects are expressed using this coordinate system as $x = 1$ m, $y = 2$ m, and $z = 2$ m. Since each numerical value is assumed to correspond to any real number from $-\infty$ (infinite) to $+\infty$, the space discussed in Newtonian mechanics is the space extending infinitely in every direction.

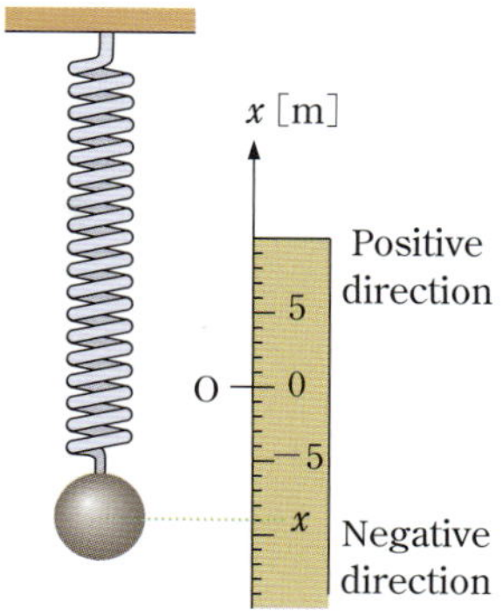

Fig. 0-4[*1]

*1 In this book, x [m] means the numerical part of x in units of m. If $x = -9$ m, then x [m] $= -9$. Namely, x [m] $= x/\text{m}$.

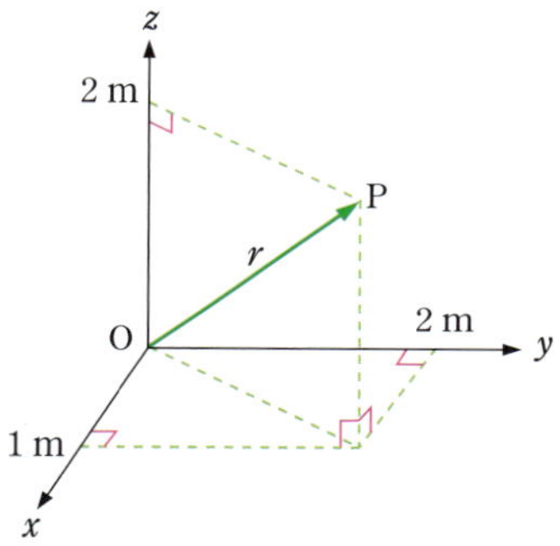

Fig. 0-5

Unit of length, meter [m] The international unit of length is meter [m]. Historically, 1 m was regulated to be $1/10^7$ of the meridian length from the equator to the north pole of the Earth[*2]. International prototype meter was constructed based on the above-mentioned regulation and the length of 1 m was defined by this prototype.

However, with the advance in precision, the definition based on the prototype became inadequate. As a result of lots of precise experiments, values of the speed of light in vacuum determined by experiments were found to be constant. For this reason, since 1983 the speed of light in vacuum was not an experimental value but was defined to be 299 792 458 m/s. Using this defined value and the time precisely measured with the atomic clock, the length of 1 m is defined as follows : "The meter is the length of the path travelled by light in vacuum during a time interval of 1/299 792 458 of a second".

*2 One round of the Earth is about 4×10^7 m $= 4\times10^4$ km.

Time An instrument to measure the time is called a clock. The principle of clock lies on the use of time unit based on the duration of periodic motion in which the same motion is repeated at a fixed time interval (period). For precise measurement of time, it is required that the period should be exactly constant and short. Quartz clocks we use today in our daily life are based on the vibration generating from the artificial quartz to which the voltage is applied. The period of this vibration is extremely constant and varies little with the change in temperature. But since the period of light emitted from atoms is far more accurate and constant, nowadays the atomic clock is adopted as the reference of time.

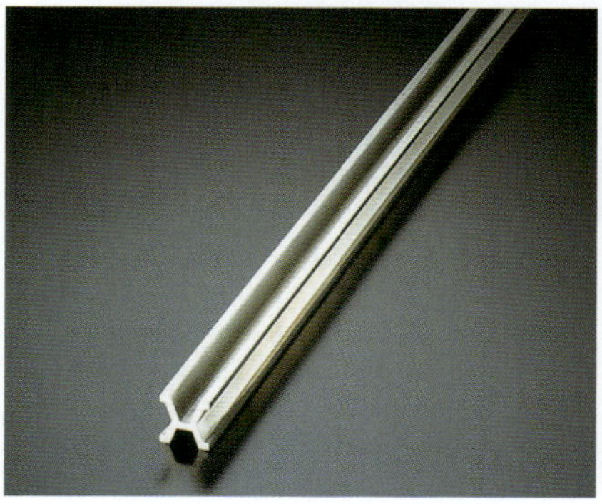

Fig. 0-6 In the first International Association of Weights and Measures held in 1889, of the 30 prototypes of 1 m, the No. 6 prototype was approved to be the International prototype meter. The No. 22 prototype was distributed to Japan.

In Newtonian mechanics founded by Newton, when an object travels between two points A and B, it passes through any point continuously between the two points (Fig. 0-7). Similarly like this, the time is considered to be able to take any instance continuously between the initial starting time and the final arriving time. In Newtonian mechanics

Fig. 0-7

time flies uniformly and there is neither beginning nor end.

But we only can measure and record the position of moving objects intermittently. For example, the movie shooting is a record of discontinuous measurements performed at a rate of 24 times per second. Advances in technology can shorten the time interval of measurements but cannot make it zero.

Arrow of time It seems that the flow of time cannot be stopped nor reversed. This impression is based on the fact that the flow of time of natural phenomena has a direction. For example, heat is transferred from high temperature bodies to low temperature bodies but heat transfer from low temperature bodies to high temperature bodies does not occur. Also, water spilled from a container does not return to the container. From these facts the term "arrow of time" was created.

Is there a beginning in the flow of time? Does the time flow from the infinite past to the infinite future? Currently it is thought that the universe, namely time and space, started about 13.8 billion years ago. This fact was discovered through the measurement on the universe and the study based on Einstein's general theory of relativity.

Unit of time, second [s] The international unit of time is second [s]. Originally, one second was defined to be $\frac{1}{24}\times\frac{1}{60}\times\frac{1}{60}=\frac{1}{86\,400}$ of a day which is equal to the interval between two successive transits of the Sun across the meridian*. However, the rotation speed of the Earth is not constant and it decreases, although only slight, gradually with time. For this reason, currently 1 second is defined to be exactly the duration of 9 192 631 770 periods of the specified electromagnetic radiation of the cesium-133 (^{133}Cs) atom.

* The orbit of the Earth's revolution around the Sun is not circular but elliptical, so the length of day changes. Accordingly, as the length of day the averaged value of mean solar day is used.

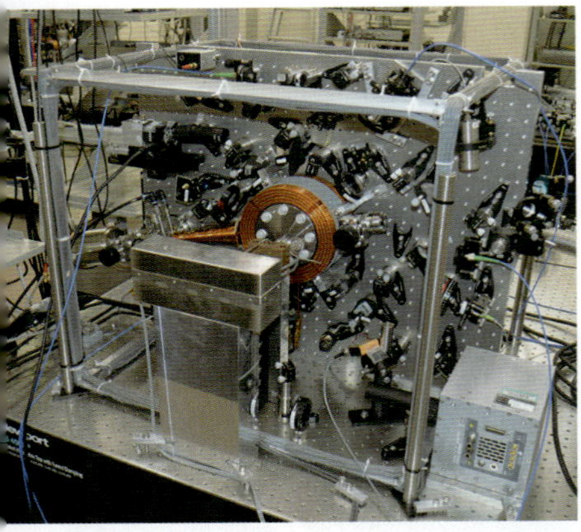

Fig. 0-8 An optical lattice clock. By trapping sophisticatedly one million atoms spatially with laser beams, it is probable to improve the accuracy of 15 significant figures given by the currently defined cesium atomic clock to the accuracy of 18 significant figures.

0.3 Physical quantities and physical laws

Physical quantities which serve as the key to understand natural phenomena In order to pursue physical laws based on observed and/or experimental facts, the first thing to do is to look for concepts which serve as the key to understand natural phenomena. As a result of such efforts, concepts of mass, velocity, acceleration, force, work, energy, etc. were found. "Concept" is the extracted common property without paying any consideration on the features of individual materials or phenomena

Concepts in physics, in discussing individual materials and phenomena, are expressed in the form of comparison with the unit indicating the reference magnitude and are given quantitatively by the formula of "numerical value" × "unit", such as 1 kg [kilogram], 5 m [meter] and 1 s [second], so they are referred to as **physical quantities**. In simple terms, the key to understand the nature lies in physical quantities.

Laws of physics are expressions of mathematical relationships that physical quantities follow Laws of physics are expressions of mathematical relationships followed by physical quantities. As Galileo expressed "The book called nature is written with the word called mathematics", laws of physics that the natural phenomena follow are expressed as mathematical expressions such as proportional relationship, reciprocal relationship, etc., depending on how physical quantities and their time rates of change behave.

For example, Newton observed that mass, acceleration and force are important in every motion and found the law of motion

$$\text{mass} \times \text{acceleration} = \text{force}. \quad (0.1)$$

This is a mathematical relationship (mathematical formula) of the three physical quantities (see Section **2.1**). Equation (0.1) expresses that if the mass is constant the acceleration is proportional to the force and if the force is constant the acceleration is inversely proportional to the mass. Therefore, the law of motion is expressed in words as "When force acts on an object, the acceleration of the object is proportional to the force and inversely proportional to the mass of the object".

In physics, terms commonly used in daily life such as power and work are also used as physical terms of physical quantities. The terms power and work used in our daily life are words to express something qualitatively, but the power and work used as terms of physical quantity are quantitative words whose magnitudes are expressed using units in accordance with the laws of physics. Therefore, when readers of this book want to understand physical quantities, they are advised to understand by combining the concrete phenomena for which the physical laws expressed by physical quantities holds with the definition of the physical quantities.

In physics, physical quantities which are treated as numerical values are represented by symbols such as mass m, acceleration a, and force F, etc., and the physical law is expressed using these symbols like

$$Ma = F. \quad (0.2)$$

However, when the readers encounter such an equation like equation (0.2) in which symbols are used, they are asked first to read equation (0.2) as an equation in Japanese terms like equation (0.1), and then, to translate it into the law in Japanese sentence, namely "When the force acts, the acceleration is proportional to the force and is inversely proportional to the mass*".

* m, a, and F are the initial letters of mass, acceleration, and force, respectively.

0.4 Units

The physical quantity has the form of "numerical value" × "unit"

In physics, experimental and observation results are first understood in terms of physical concepts called **physical quantities** such as speed, force, mass, energy, electric field, etc., and then, describe these results. Physics is a discipline to explore physical quantities and their relationships, and these relationships are **physical laws**. When a physical quantity is expressed, it shows how many times it is larger than the reference **unit** of the amount used in measuring that physical quantity. For example, the height of a tower is expressed as 50 m or 60 m as compared to the ruler of 1 m length of the reference length. In other words, the physical quantity has the form of "**numerical value**" × "**unit**". Accordingly, units should be well understood in order to quantitatively handle the physical problems.

In physics, physical quantities are represented by symbols in Roman

Fig. 0-9 Japanese Kilogram Prototype (National Institute of Advanced Industrial Science and Technology)

alphabet or Greek letters. These symbols representing physical quantities also stand for "numerical value"×"unit".

Unit of length m
Unit of mass kg
Unit of time s
Unit of electric current A

The International System of Units The metrology law of Japan is based on the International System of Units (abbreviated as SI). The International System of Units is a metric system built on seven base units which are the length in metre (meter) (m), mass in kilogram (kg), time in second (s), electric current in ampere (A), temperature in kelvin (K), light intensity in candela (cd), and amount of substance in mole (mol)*1.

Units of physical quantities other than base units are derived from base units using definitions and physical laws, and are called **derived units**. For example, since the unit of length is m and the unit of time is s,

"speed" = "travel distance" ÷ "travel time", and its international unit is the unit of length divided by the unit of time m/s,

"acceleration" = "change in speed"÷"duration of time", and its international unit is m/s^2 obtained by dividing the unit of speed by the unit of time.

A/B represents $A \div B$. As will be learned in Chapter 2, since

"force" = "mass"×"acceleration", its international unit is the unit of mass of kg multiplied by the unit of acceleration m/s^2, or $kg \cdot m/s^2$ *2.

In honor of Newton, the founder of mechanics, we call this unit of $kg \cdot m/s^2$ as newton and use the symbol N. Even so, the international unit of force Newton is not the base unit. Table 0-1 shows the SI

*1 1 kg of the international unit of mass was historically stipulated as the mass of 1000 cm^3 (1 L) of water at the temperature at which the density of water becomes maximum (about 4 °C). Since 1889 the magnitude of the mass of 1 kg has been defined by the mass of the International Prototype Kilogram made of platinum-iridium alloy stored in the International Bureau of Weights and Measures located at Sèvres in France. One copy of this "International Prototype Kilogram" is the "Japanese Kilogram Prototype". The stability of "International Kilogram prototype" became questioned and the definition of kilogram that does not rely on artifacts such as weights was decided in 2018 by the International Committee of Weights and Measures.

*2 $A \cdot B$ represents $A \times B$.

Table 0-1 SI derived units with special names and used in this book

Quantity	Unit	Symbol	Expression in other SI units	Expression in SI base units	page in this book
Frequency	hertz	Hz		s^{-1}	60, 146, 313
Force	newton	N	J/m	$m \cdot kg \cdot s^{-2}$	29
Pressure, stress	pascal	Pa	N/m^2	$m^{-1} \cdot kg \cdot s^{-2}$	128
Energy, work	joule	J	N·m, C·V	$m^2 \cdot kg \cdot s^{-2}$	69, 176
Power, electric power	watt	W	J/s	$m^2 \cdot kg \cdot s^{-3}$	70, 264
Electric charge	coulomb	C	A·s	s·A	216
Electric potential, voltage	volt	V	J/C	$m^2 \cdot kg \cdot s^{-3} \cdot A^{-1}$	228
Electric capacity	farad	F	C/V	$m^{-2} \cdot kg^{-1} \cdot s^4 \cdot A^2$	242
Electric resistance	ohm	Ω	V/A	$m^2 \cdot kg \cdot s^{-3} \cdot A^{-2}$	258
Magnetic flux	weber	Wb	V·s, $T \cdot m^2$	$m^2 \cdot kg \cdot s^{-2} \cdot A^{-1}$	272
Magnetic field (magnetic flux density)	tesla	T	N/(m·A), Wb/m^2	$kg \cdot s^{-2} \cdot A^{-1}$	271, 279
Inductance	henry	H	Wb/A	$m^2 \cdot kg \cdot s^{-2} \cdot A^{-2}$	309
Radioactivity	becquerel	Bq		s^{-1}	372
Absorbed dose	gray	Gy	J/kg	$m^2 \cdot s^{-2}$	372
Equivalent dose	sievert	Sv	J/kg	$m^2 \cdot s^{-2}$	372

derived units with special names and used in this book.

In addition, in this book, from the view of importance in educating students, the author uses in some cases units called the practical units which are not the ones of SI.

We are in the transition from the period using prototypes of standard for units to the period using physical constant criteria The international unit of mass 1 kg was historically stipulated as the mass of 1000 cm^3 (1 L) of water at the temperature at which the density of water becomes maximum (about 4 °C). Since 1889 the magnitude of the mass of 1 kg was defined by the mass of the International Prototype Kilogram made of platinum-iridium alloy stored in the International Bureau of Weights and Measures in France. However, the limit came in using an artifact of prototype as the unit. Therefore, from the year 2019 on, it follows that "1 kg is set by accurately defining the Planck constant as $6.626\,070\,15\times10^{-34}$ J·s*".

* The unit of current, ampere A, is set by acurately defining the elementary charge e as $1.602\,176\,634\times10^{-34}$ C.

The unit of thermodynamic temperature, kelvin K, is set by acurately defining the Boltzmann constant k as $1.380\,649\times10^{-23}$ J/K.

The unit of amount of substance, mol, is defined to contain acurately $6.022\,140\,76\times10^{23}$ constitutive particles.

How to express large quantities and small quantities (exponent, prefix) There are two ways to express physical quantities when the magnitude of physical quantities appearing in the phenomenon under consideration is much larger or smaller than the magnitude of the base unit or the derived unit.

One way is to express the quantity in the form of power of 10 and to express 1 000 000 as 10^6 and 0.000 001 as 10^{-6}. In other words, this is a method to express a large quantity as $a\times10^n$ (n is a positive integer) and a small quantity as $a\times10^{-n}$. The n of 10^n and $-n$ of 10^{-n} are called the index. As an example, the equatorial radius of the Earth 6 378 000 m is expressed as 6.378×10^6 m.

The other is to express the unit with prefix specified in the SI and shown on the back side of back cover of this book. Examples are

$$10^6\ \mathrm{Hz} = 1\ \mathrm{MHz},\quad 1000\ \mathrm{m} = 1\ \mathrm{km},\quad 100\ \mathrm{Pa} = 1\ \mathrm{hPa},$$
$$10^{-3}\ \mathrm{m} = 1\ \mathrm{mm},\quad 10^{-3}\ \mathrm{kg} = 1\ \mathrm{g},\quad 10^{-15}\ \mathrm{m} = 1\ \mathrm{fm}.$$

The unit of frequency MHz is read as megahertz, unit of pressure hPa as hectopascal, and unit of length fm as femtometer.

One more comment to add is that since the base unit of mass, kilogram (kg), includes the prefix "kilo", the name of the unit of integer multiplication of 10 in units of mass is constructed by fixing the proper prefix to the word "gram".

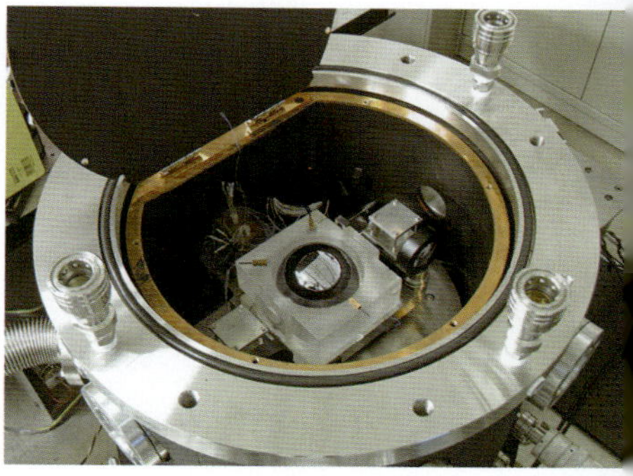

Fig. 0-10 A laser interferometer that measures the shape of a silicon sphere with nanometer precision by precise control of the wavelength of light. In combination with other studies, Avogadro's constant can be determined with high accuracy. By relying on highly accurate basic physical constants, the redefinition of kilograms that do not rely on artifacts was attained in 2019.

Dimension As a concept closely related to the unit, dimension is discussed here. The units of all the physical quantities appearing in mechanics can be expressed in three base units, that is length unit m, mass unit kg, and time unit s. Dimension shows from what kind of combination of the base units the derived units such as speed and force are made. For example, when the unit of a physical quantity Y is $\mathrm{m}^a\,\mathrm{kg}^b\,\mathrm{s}^c$, $\mathrm{L}^a\,\mathrm{M}^b\,\mathrm{T}^c$ is called the **dimension** of the physical quantity Y. Here, L, M, and T are the initial letters of length, mass, and time, respectively. For example, the dimension of speed is LT^{-1}, and the dimension of force is LMT^{-2}.

Fig. 0-11 The quantity mc^2 has the dimension of energy.

In the expression of $A = B$ which appears in the middle of calculation or in the results, the dimensions of the left-hand side A and the right-hand side B must always be the same. For this reason, to examine whether the dimensions of both sides of the expression of the calculated results are the same or not is one way to check whether the calculated results are correct or not. If the readers do not know the unit in a calculation containing a derived unit with a special name, consult the column "Expression in other SI units" or "Expression in SI base units" in Table 0-1 and check the results.

Always remember that multiplication and division of two quantities with different dimensions are possible but addition and subtraction are no way possible.

0.5 Uncertainty in measured values and significant figures

Significant figures When a certain physical quantity is repeatedly measured under the same condition, variations are observed in measured values obtained by the measurements. The average value of these measurements is the best estimate of this physical quantity. According to the invention and improvement of the measuring devices, it is possible to reduce variations in measured values, but it is impossible to make the variation zero.

In many cases, as shown in Fig. 0-12, the measured values have a distribution called the **normal distribution** having a bell-shaped curve around the mean value m. In statistics, σ in Fig. 0-12 is called the **standard deviation** of the measurement results of this physical quantity. The standard deviation means that the measured value of the magnitude between $m-\sigma$ and $m+\sigma$ is 68.3% of the total [Fig. 0-12 (a)], and the measured value of the magnitude between $m-2\sigma$ and $m+2\sigma$ is 95.4% of the total [Fig. 0-12 (b)]. The measurement results are expressed as $m\pm\sigma$, and σ is referred to as the **standard uncertainty**.

Since there is uncertainty, it is meaningless to express the results with unnecessarily increasing the number of digits of the mean value m. For example, if the average value of the height measurements of a person is 161.414 cm and the standard deviation is 0.1 cm, the mean value of the height measurement results is 161.4 cm. In this case, a meaningful four-digit number 1614 is called a **significant figure**. When the measurement value is expressed as $a\times 10^n$, we use the significant figure that makes $1 \leqq |a| < 10$ as a (for example, 1.614×10^2 cm).

Since this book is mainly aimed at understanding the physical meanings of physical phenomena and laws, much care will not be paid on significant figures nor uncertainties in the solution of problems.

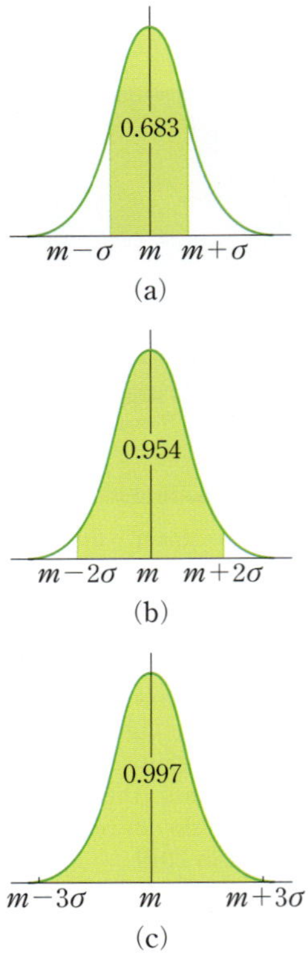

Fig. 0-12 Normal distribution

Motion

Mechanics is a field of science to study "force and motion". Motion of an object is the position change of the object with time. Therefore, what is first required to describe the motion is to describe the position of the object. Quantities which describe the motion state of the object are **velocity** and **acceleration**. Through many times of experience of taking a car or a train, we are familiar with the concept of velocity and acceleration

In this chapter, as a preparation for study of mechanics, you will learn the notion of position, displacement, velocity, and acceleration. In addition, the relationships between velocity, acceleration and differentiation (derivatives) will be discussed.

It is also important for the readers of this book first to draw an appropriate graph and then to deduce the features of velocity and acceleration of the object from the graph.

1.1 Definition of velocity and acceleration in linear motion, and differentiation

Learning objective To understand the definition of velocity and acceleration of an object in linear motion and their expressions using differentiation (derivatives). To be able to deduce information from x-t graphs and v-t graphs.

Mass point Although an object has a certain volume, up to Chapter 7 in this book, discussion is limited to the case where the object volume is negligible and a whole mass is concentrated to one point. The object which can be described as a point having the whole mass of the object but no volume is called the **mass point**.

Position Consider a mass point moving along a straight line. We can choose the straight line as the coordinate axis (x-axis) and set up the origin O and the positive direction of the x-axis and then determine the unit length. After these processes the position of mass point is expressed by the coordinate x as $x = 3$ m (Fig. 1-1).

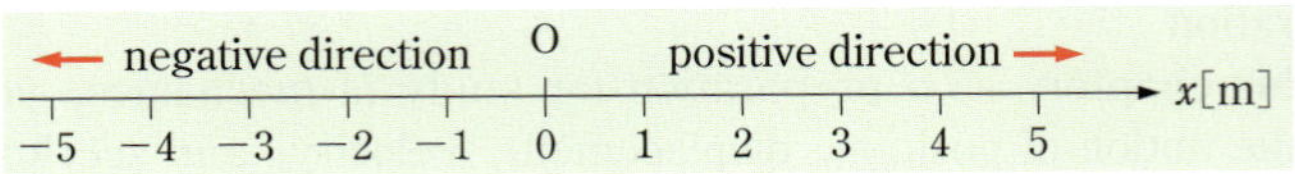

Fig. 1-1 Coordinate axis (x-axis)[*1]

The position of a mass point varies with time. We, therefore, express the position coordinate of the object at time t as $x(t)$[*2]. The variation of the position of mass point with time can be expressed by the function

$$x = x(t) \tag{1.1}$$

and by the ***x*-*t* graph** (Fig. 1-2), where the horizontal axis is time t and the vertical axis is the position x of mass point.

Displacement and mean velocity of the linear motion The speed (mean speed) of the mass point is "moving distance" divided by "time duration". In the case of linear motion, to distinguish the speed of the object moving in the positive direction from that of the object in the negative direction, the term velocity which has a positive or negative sign corresponding to the moving direction is used. For this purpose, in place of moving distance the term displacement is used for which a positive or negative sign is assigned.

When a mass point is at P of its position $x = x(t)$ at time t moves to P′ of its position $x = x(t+\Delta t)$ during time interval Δx, the position change in Δt is $\Delta x = x(t+\Delta t)$-$x(t)$. This change in position Δx is called the **displacement** of the object for interval between t and $t+\Delta t$ (see Fig. 1-3)[*3]. We define the mean velocity during interval Δt as the quantity given by "displacement" "divided by "time interval"

$$\bar{v} = \frac{\Delta x}{\Delta t} = \frac{x(t+\Delta t)-x(t)}{\Delta t}. \tag{1.2}$$ [*4]

The mean velocity $\bar{v}$ is equal to the slope of the directed line segment

*1 x [m] is the numeral part of x measured in units of m. $x = 3$ m shows x [m] $= 3$. And x [m] $= x/\text{m}$. See *1 on p. 5.

*2 The expression $x = x(t)$ does not mean the multiplication $x \times t$ but a function of the variable t. For details, see the description given in the margin of p. 16.

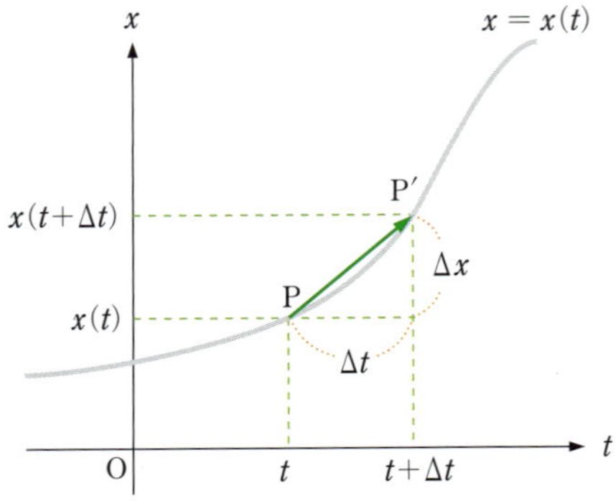

Fig. 1-2 x-t graph. The slope $\dfrac{\Delta x}{\Delta t}$ of the directed line segment $\overrightarrow{PP'}$ is the mean velocity during time interval Δt.

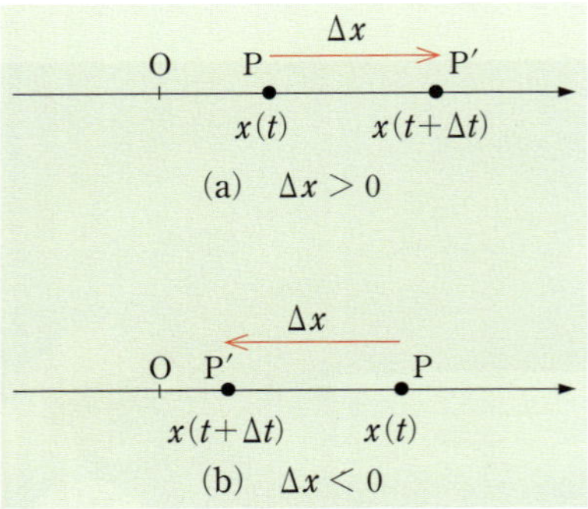

Fig. 1-3 Displacement $\Delta x = x(t+\Delta t)-x(t)$

*3 Pay attention to the fact that Δx is a single quantity to express the displacement, and not the product of Δ (read as delta) and x. Similarly, Δt is a single quantity to express the interval between two times and not the product of Δ and t.

*4 The superscript bar $^{-}$ of $\bar{v}$ attached to v indicates the mean quantity.

$\overrightarrow{PP'}$ shown in Fig. 1-2. If the mass point moves toward the positive direction of x-axis, then $\Delta x > 0$. Therefore, the average velocity is positive ($\bar{v} > 0$) and the directed line segment $\overrightarrow{PP'}$ is positively sloped. If the mass point moves toward the negative direction of x-axis, then $\Delta x < 0$, so the average velocity is negative ($\bar{v} < 0$) and the directed line segment $\overrightarrow{PP'}$ is negatively sloped. The steeper the directed line segment $\overrightarrow{PP'}$ is, the higher the velocity is.

Velocity (instantaneous velocity) of linear motion and derivatives

When the velocity changes with time, the limiting value obtained by letting Δt of the mean velocity $\frac{\Delta x}{\Delta t}$ in equation (1.2) be infinitely small

$$v(t) = \lim_{\Delta t \to 0} \frac{\Delta x}{\Delta t} = \lim_{\Delta t \to 0} \frac{x(t+\Delta t)-x(t)}{\Delta t} = \frac{dx}{dt} \tag{1.3}$$

is the **velocity** at time t, or the **instantaneous velocity**. The third equality in the above equation is the definitional identity of the derivative $\frac{dx}{dt}$ of the function $x(t)$. Therefore, the velocity $v(t)$ of a mass point is the derivative obtained by differentiating the position-describing function $x(t)$ with respect to t.

Let time interval Δt get shorter and shorter. Then, in the limit $\Delta t \to 0$, the direction of the directed line segment $\overrightarrow{PP'}$ in the x-t graph falls on that of the tangent of the x-t graph at point P (Fig. 1-4). In other words, the velocity $v(t)$ is equal to the slope of the tangent in the x-t graph at time t. If the tangent is positively sloped, then $v(t) > 0$ and the motion is toward the positive direction. If it is negatively sloped, then $v(t) < 0$ and the motion is toward the negative direction. If the tangent is horizontal, the instantaneous velocity $v(t)$ is zero (Fig. 1-5).

Unit of velocity m/s

The variation of velocity v of a mass point with time can be expressed by the ***v*-*t* graph** showing the function $v = v(t)$ (Fig. 1-6), where the horizontal axis refers to time t, and the vertical axis refers to velocity v.

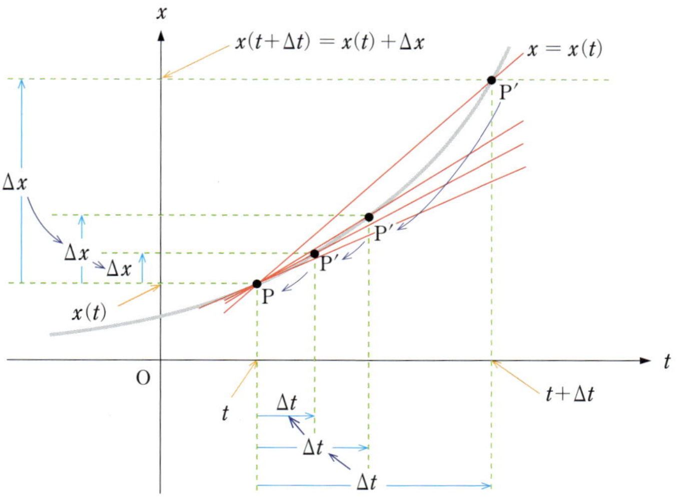

Fig. 1-4 x-t graph and velocity. The slope $\frac{\Delta x}{\Delta t}$ of the directed line segment $\overrightarrow{PP'}$ is the mean velocity during time interval Δt. The limit of the slope of the directed line segment $\overrightarrow{PP'}$ at $\Delta t \to 0$ is equal to the slope of the tangent of x-t graph at time t. This slope of the tangent is the velocity (instantaneous velocity) at time t.

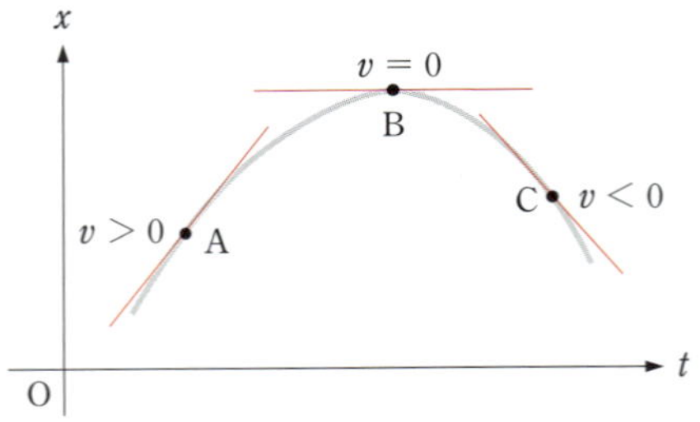

Fig. 1-5 Slope and velocity in x–t graph. At point A, the tangent is positively sloped, so $v>0$. At point B, the tangent is horizontal, so $v=0$. At point C, the tangent is negatively sloped, so $v<0$.

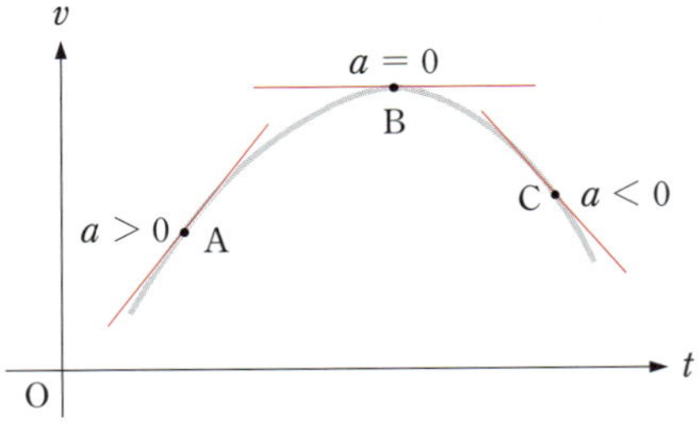

Fig. 1-6 v–t graph. The slope of the tangent in v–t graph is equal to the acceleration.

Question 1 Compare the six x–t graphs of Fig. 1-7 (a) with the six v–t graphs of Fig. 1-7 (b) to find the 1 : 1 correspondence.

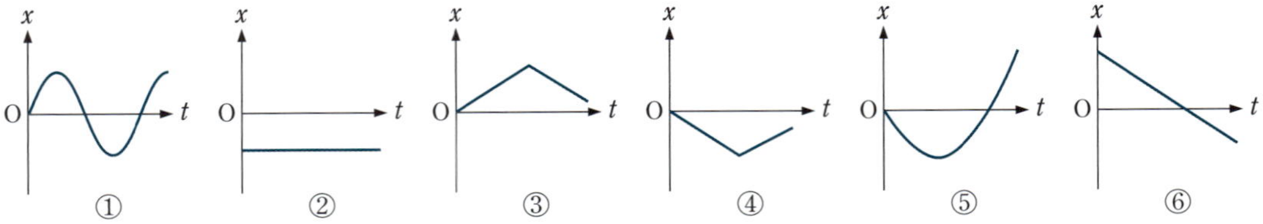

(a) x–t graph

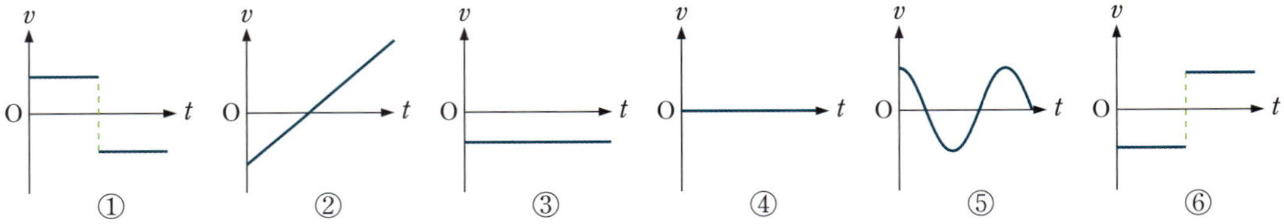

(b) v–t graph

Fig. 1-7

Acceleration of the linear motion When you drive a car and step on the accelerator pedal or apply brakes to the car, its velocity changes. The changing rate of velocity, or the change in velocity in a unit time is defined as **acceleration**.

The **mean acceleration** $\bar{a}$ during time interval Δt is defined as "change in velocity Δv" divided by "time interval Δt",

$$\bar{a} = \frac{\Delta v}{\Delta t}. \tag{1.4}$$

If the acceleration is negative, it means that the change in velocity is negative.

When a car at rest is accelerated to the velocity 20 km/s in 5 seconds, the mean acceleration $\bar{a}$ is given as

$$\bar{a} = \frac{(20\ \mathrm{m/s})-(0\ \mathrm{m/s})}{5\ \mathrm{s}} = 4\ \mathrm{m/s^2}.$$

Fig. 1-8 Shinkansen

The unit of acceleration in the SI units is m/s^2.

Unit of acceleration m/s^2

The **acceleration** at time t (instantaneous acceleration) $a(t)$ is equal to the limiting value obtained by letting Δt be infinitely small in equation (1.4) for the mean acceleration.

$$a(t) = \lim_{\Delta t \to 0} \frac{\Delta v}{\Delta t} = \lim_{\Delta t \to 0} \frac{v(t+\Delta t)-v(t)}{\Delta t} = \frac{dv}{dt}. \quad (1.5)$$

Thus, the acceleration $a(t)$ is equal to the slope of the tangent of a v–t graph at time t (Fig. 1-6). The acceleration $a(t)$ is the derivative of the velocity $v(t)$, and the velocity $v(t)$ is the derivative of position $x(t)$. Therefore, the following relation holds.

$$a(t) = \frac{dv}{dt} = \frac{d}{dt}\left(\frac{dx}{dt}\right) = \frac{d^2x}{dt^2} \quad (1.6)$$

Derivatives obtained by applying the differentiation twice to a function are called second derivatives, namely the acceleration is the second derivative of position.

Example 1 Fig. 1-9 is the x–t graph for two cars A and B running on a straight road with two lanes on one side. Answer whether the following each sentence is correct or not.

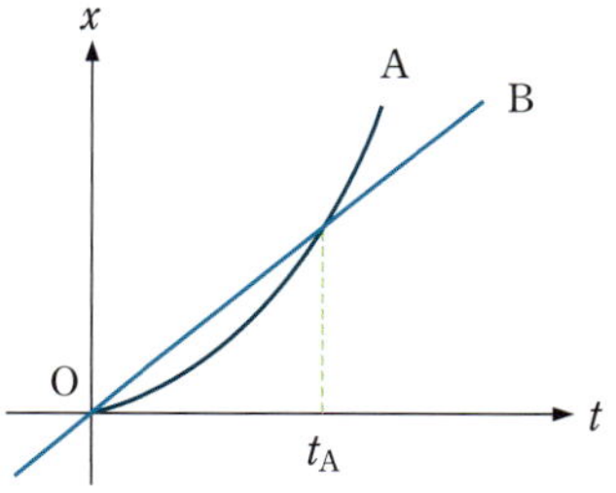

Fig. 1-9

(1) At time t_A, velocities of the two cars are equal.

(2) At time t_A, positions of the two cars are equal.

(3) The two cars continue accelerating.

(4) At a certain time before t_A velocities of the two cars become equal.

(5) At a certain time before t_A, accelerations of the two cars become equal.

Solution (1) × (From the comparison of the slopes of tangents in the x–t graph at time t_A, we know $v_A > v_B$.)

(2) ○

(3) × (The x–t graph for B is linear, indicating that B is moving at a constant speed.)

(4) ○ (In the x–t graph for car A we can find the time at which the slope of the tangent is equal to the corresponding slope for car B.)

(5) × (The acceleration for car A is constantly positive, while the acceleration for car B is always zero.)

Gravitational acceleration Experiments have shown that when an object falls in the air near the Earth surface, provided that the air resistance is negligible, any body is accelerated in a vertical and downward direction and the acceleration is constant. That is to say the acceleration in the falling motion is, irrespective of size and/or type of falling objects, constant and does not change during the fall. This acceleration is called **gravitational acceleration** and its magnitude is denoted by the symbol g. Depending on the location on Earth, slight differences are observed among the numerical values of g but we can assume

Gravitational acceleration $g \approx 9.8\ m/s^2$

$$g \approx 9.8\ m/s^2. \quad (1.7)^*$$

As in the case of falling motion without air resistance, the linear motion at constant acceleration is referred to as the uniformly accelerated linear motion.

* $A \approx B$ means "A and B are approximately equal, or numerically equal".

Comment on the equation $x = x(t)$

The equation $x = x(t)$ means that the position x of an object at t is determined and its value is $x(t)$. The variable t can represent the general time as well as the specific time. The t at the right edge of the horizontal axis of Figs. 1-2 and 1-4 means that the horizontal axis of each figure represents the variable t corresponding to the general time, while the t shown below the horizontal axis represents a specified time t, and the $t+\Delta t$ on the right of t means "the value at a specific time duration" $+\Delta t$. If we change equation (1.8) into the two-equation form

$$x = x(t), \quad x(t) = \frac{1}{2}gt^2 \qquad (1.8')$$

then, the second equation shows that the calculation rule at time t is $x(t) = \frac{1}{2}gt^2$. Even if we know the value of t we cannot calculate the position x unless the calculation rule of $x(t)$ is given. In place of $x = x(t)$ we can write $x = f(t)$, but use of the expression $x(t)$ shows more clearly the calculation rule of the position x.

Fig. 1-10

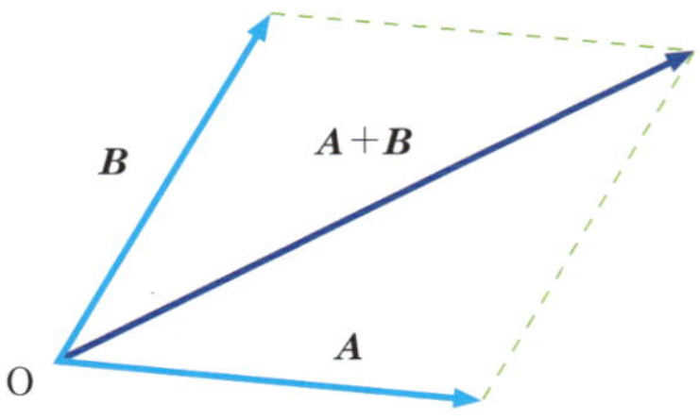

Fig. 1-11 $\boldsymbol{A}+\boldsymbol{B}$

Case 1 Galileo found that in the free-fall, which is the falling motion without initial velocity, the fall length x is proportional to the square of the fall time. In other words, he found

$$x = \frac{1}{2}gt^2 \qquad (g \text{ is a constant}) \qquad (1.8)$$

(Refer to the column on p. 55 and Problems 3-A3). From the above relation, two equations, one for velocity v and the other for acceleration a, are derived.

$$v(t) = \frac{dx}{dt} = \frac{d}{dt}\left(\frac{1}{2}gt^2\right) = gt \qquad (1.9)$$

and

$$a(t) = \frac{dv}{dt} = \frac{d}{dt}(gt) = g. \qquad (1.10)$$

Therefore, the free-fall is the uniformly accelerated linear motion in which an object falls under the condition of constant acceleration (gravitational acceleration) g.

The equation $x(t) = \frac{1}{2}gt^2$ indicates that the position (distance) x at time (fall time) t is $\frac{1}{2}gt^2$. To emphasize that the x on the left-hand side of equation (1.8) is the position at time t, we can also write as $x(t) = \frac{1}{2}gt^2$.

Question 2 For an object in free-fall, calculate its velocities and fall lengths at 1, 2, and 3 seconds after it starts to fall. For simplicity, set $g = 10\ \mathrm{m/s^2}$.

From Equation (1.8) the following equation

$$t = \sqrt{\frac{2x}{g}} = \sqrt{\frac{2x}{9.8\ \mathrm{m/s^2}}} = \sqrt{\frac{x}{4.9\ \mathrm{m}}}\ \mathrm{s} \qquad (1.11)$$

is derived. This equation gives the fall time when the fall length is x.

Question 3 Reaction time of nerves As illustrated in Fig. 1-10, a girl A is pinching the top edge of a thousand-yen bill and a boy B is spreading his forefinger and his thumb near the bottom edge of the thousand-yen bill. If the girl suddenly opens her fingers, the bill starts to fall. From the moment of the start of fall, a time is needed so that the boy perceives the fall of the bill and catches the falling bill. By measuring the fall length x of the bill, we can estimate the reaction time of nerves (reflex time) of the boy B. If the fall length x is 4.9 cm, how long (in second) does it take for the nerves to react to the fall?

For reference Vector

Physical quantities are divided into vector quantities and scalar quantities. A vector quantity has a magnitude and a direction, and the summation (addition) is defined according to the parallelogram law (Fig. 1-11). A scalar quantity has a magnitude, but not a direction. In this book, vectors are described by bold roman letters like $\boldsymbol{A}$ and their magnitudes are described like $|\boldsymbol{A}|$ or A.

In the rectangular coordinate system, the magnitude and direction of the vector $\boldsymbol{A}$ is specified by a set of coordinates (A_x, A_y, A_z) which describe the end point coordinates of the arrow indicating the vector $\boldsymbol{A}$ (Fig. 1-12). We express the vectors having the length 1 for $+x$-direction, $+y$-direction, and z-dirction (called unit vector) as $\boldsymbol{i}$, $\boldsymbol{j}$, and $\boldsymbol{k}$, respectively. Then, the vector $\boldsymbol{A}$ is described as

$$\boldsymbol{A} = A_x\boldsymbol{i}+A_y\boldsymbol{j}+A_z\boldsymbol{k} \qquad (1.12)$$

Applying twice the Pythagorean theorem $[A^2 = (A_x^2+A_y^2)+A_z^2]$, the magnitude $A = |\boldsymbol{A}|$ of the vector $\boldsymbol{A}$ is given as

$$A = |\boldsymbol{A}| = \sqrt{A_x^2+A_y^2+A_z^2} \qquad (1.13)$$

The sum of two vectors $\boldsymbol{A} = (A_x, A_y, A_z)$ and $\boldsymbol{B} = (B_x, B_y, B_z)$ is the vector $\boldsymbol{A}+\boldsymbol{B}$ and its each component is the sum of the corresponding components of $\boldsymbol{A}$ and $\boldsymbol{B}$.

$$\boldsymbol{A}+\boldsymbol{B} = (A_x+B_x, A_y+B_y, A_z+B_z) \qquad (1.14)$$

This can be seen from Fig. 1-13 showing the case of two-dimensional plane.

Multiplication of vector with scalar : Let k be an arbitrary scalar and let $\boldsymbol{A}$ be an arbitrary vector. Then, $k\boldsymbol{A}$ is a vector whose magnitude is $|k|$ times the magnitude $|\boldsymbol{A}|$ of the vector $\boldsymbol{A}$. When $k > 0$, then, the direction of vector $k\boldsymbol{A}$ is the same as $\boldsymbol{A}$, and when $k < 0$, it is a vector with its direction opposite to $\boldsymbol{A}$ (Fig. 1-14). $-\boldsymbol{A} = (-1)\boldsymbol{A}$ is a vector having the same magnitude as $\boldsymbol{A}$ but the direction opposite to $\boldsymbol{A}$. The vector having the magnitude 0 is called the **zero vector** and is denoted by $\boldsymbol{0}$ (Fig. 1-15). The vector $\boldsymbol{A}-\boldsymbol{B}$ is the difference obtained by subtracting the vector $\boldsymbol{B}$ from the vector $\boldsymbol{A}$. The vector $\boldsymbol{A}-\boldsymbol{B}$ is the sum of addition of the vector $\boldsymbol{A}$ and the vector $\boldsymbol{B}$ multiplied by -1 $[\boldsymbol{A}-\boldsymbol{B} = \boldsymbol{A}+(-\boldsymbol{B})]$ (Fig. 1-16).

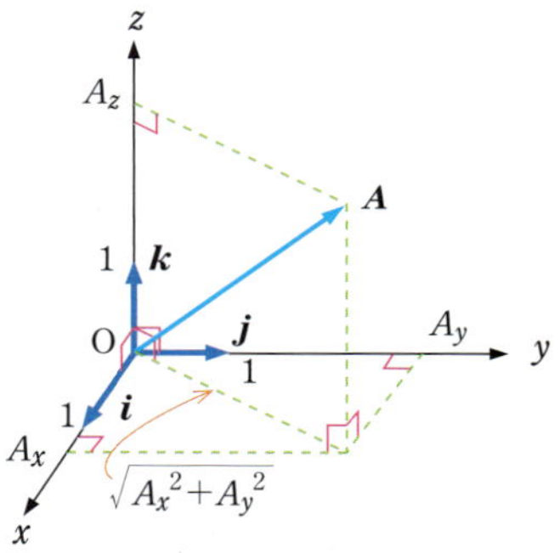

Fig. 1-12 Rectangular coordinate system and vector $\boldsymbol{A} = (A_x, A_y, A_z)$

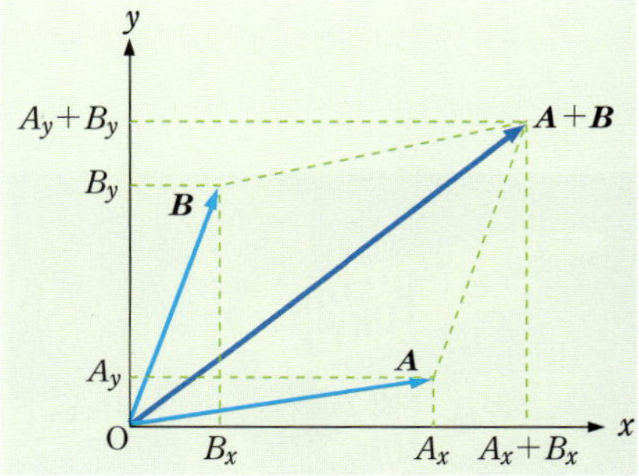

Fig. 1-13 $\boldsymbol{A}+\boldsymbol{B} = (A_x+B_x, A_yB_y)$ when vectors $\boldsymbol{A}$ and $\boldsymbol{B}$ are on the xy plane.

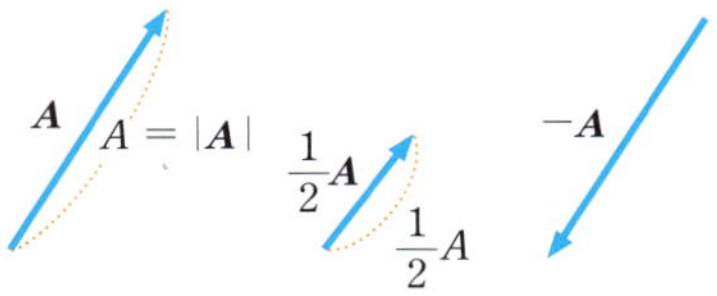

Fig. 1-14 $k\boldsymbol{A}$ obtained by scalar multiplication of vector $\boldsymbol{A}$

Fig. 1-15 Zero vector

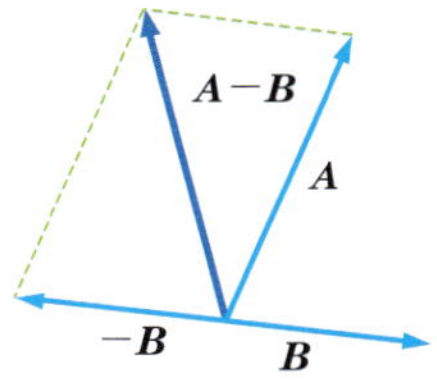

Fig. 1-16 $\boldsymbol{A}-\boldsymbol{B} = \boldsymbol{A}+(-\boldsymbol{B})$

1.2 Velocity and acceleration of the general motion

Learning objective To understand the definition of velocity and acceleration of the general motion which is outside the frame of linear motion. To be able to have clear images of various types of acceleration arising from the representative velocity changes.

In the following sections in this chapter, we will discuss the generalized motion which is outside the linear motion. One method to describe the position of a mass point in these cases is to use the vector whose initial point is the origin O and the end point is P. The vector $\overrightarrow{\mathrm{OP}}$ is called the **position vector** of mass point and is described by $\boldsymbol{r}$ (Fig. 1-

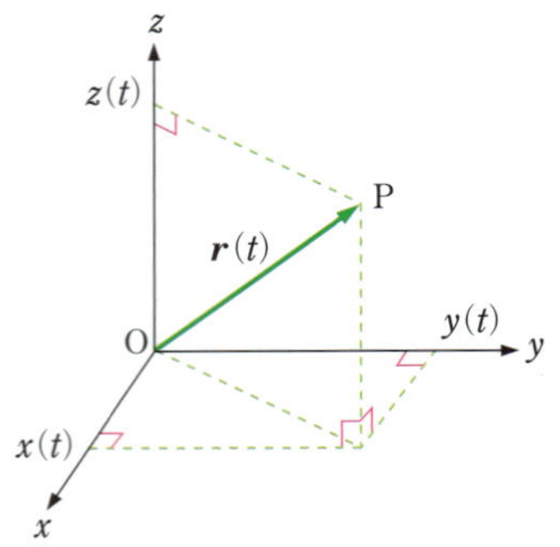

Fig. 1-17 Position vector and rectangular coordinate system

17). Since the position of a moving mass point varies with time, the position vector at time t is described as $\boldsymbol{r}(t)$. The path through which the mass point travels is called the orbit.

Let the rectangular coordinates of a mass point be $[x(t), y(t), z(t)]$ at time t. Then, the position vector expressed in terms of unit vectors is

$$\boldsymbol{r}(t) = x(t)\boldsymbol{i} + y(t)\boldsymbol{j} + z(t)\boldsymbol{k}. \tag{1.15}$$

Distance between the origin and the mass point is

$$r = |\boldsymbol{r}| = \sqrt{x^2+y^2+z^2}. \tag{1.16}$$

Fig. 1-18 Position vectors (direction and distance) representing various places when the current location is taken as the origin.

Displacement The change of the position vector of a mass point during the time interval Δt between t and $t+\Delta t$ is the vector $\Delta\boldsymbol{r}$ whose initial point at t is P and end point at $t+\Delta t$ is P′

$$\Delta\boldsymbol{r} = \boldsymbol{r}(t+\Delta t) - \boldsymbol{r}(t). \tag{1.17}$$

$\Delta\boldsymbol{r}$ is referred to as the **displacement** of the mass point during time interval Δt. If $\Delta\boldsymbol{r}$ is expressed in terms of the components, it follows

$$\begin{aligned}\Delta\boldsymbol{r} &= [\Delta x, \Delta y, \Delta z] \\ &= [x(t+\Delta t)-x(t), y(t+\Delta t)-y(t), z(t+\Delta t)-z(t)]\end{aligned} \tag{1.18}$$

(Fig. 1-19).

Velocity and speed Similarly as in the case of circular motion, when the direction of a mass point changes, two states with the same speed but different direction are the separate states in motion. Here we introduce the vector $\boldsymbol{v}$ which has the same direction as that of motion and the length equal to the speed v. This vector $\boldsymbol{v}$ is called the **velocity**. Since the velocity $\boldsymbol{v}$ is a vector, it has x-component v_x, y-componet v_y, and z-componet v_z.

Mean velocity The mean velocity $\bar{\boldsymbol{v}}$ of a mass point during time interval Δt is defined to be "displacement" divided by "time interval"

$$\bar{\boldsymbol{v}} = \frac{\Delta\boldsymbol{r}}{\Delta t} = \left(\frac{\Delta x}{\Delta t}, \frac{\Delta y}{\Delta t}, \frac{\Delta z}{\Delta t}\right) \tag{1.19}$$

(Fig. 1-19). $\dfrac{\Delta\boldsymbol{r}}{\Delta t}$ is a vector which has the same direction as the displacement $\Delta\boldsymbol{r}$ or the direction of motion and has the magnitude $\dfrac{|\Delta\boldsymbol{r}|}{\Delta t}$.

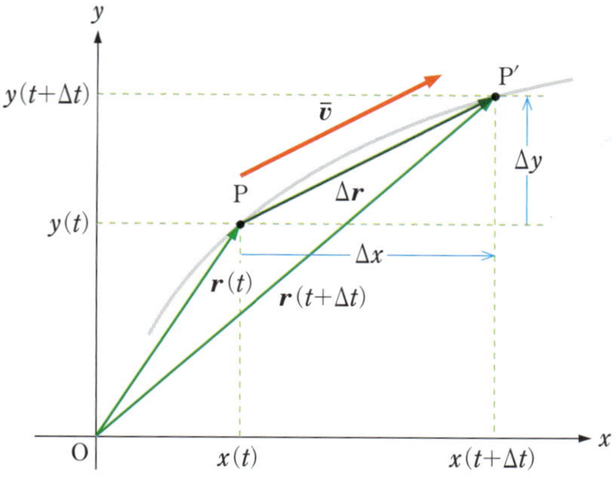

Fig. 1-19 Displacement $\Delta \boldsymbol{r}$ during the time interval from time t to time $t+\Delta t$. Mean velocity is $\bar{\boldsymbol{v}} = \dfrac{\Delta \boldsymbol{r}}{\Delta t}$.

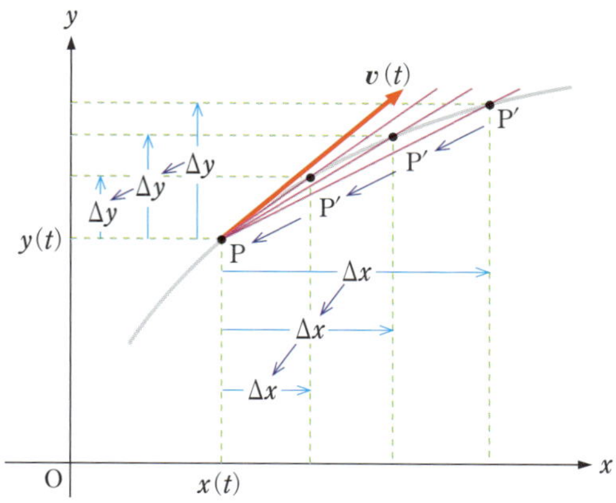

Fig. 1-20 Instantaneous velocity $\boldsymbol{v}(t)$ at time t takes the same direction as the tangent of the motion path.

Velocity (instantaneous velocity) The instantaneous velocity $\boldsymbol{v}(t)$ at time t is the value in the limit $\Delta t \to 0$ of the mean velocity given by equation (1.19)

$$\begin{aligned}\boldsymbol{v}(t) &= [v_x(t), v_y(t), v_z(t)] \\ &= \lim_{\Delta t \to 0} \frac{\Delta \boldsymbol{r}}{\Delta t} = \frac{\mathrm{d}\boldsymbol{r}}{\mathrm{d}t} = \left(\frac{\mathrm{d}x}{\mathrm{d}t}, \frac{\mathrm{d}y}{\mathrm{d}t}, \frac{\mathrm{d}z}{\mathrm{d}t}\right).\end{aligned} \tag{1.20}$$

Fig. 1-20 shows that the velocity $\boldsymbol{v}(t)$ is a vector and its direction is the direction of the tangent to the curve (orbit) drawn in space by the moving mass point, namely the direction of motion, and the magnitude is the speed at time t.

$$v(t) = |\boldsymbol{v}(t)| = \sqrt{v_x(t)^2 + v_y(t)^2 + v_z(t)^2}. \tag{1.21}$$

Mean acceleration In the generalized motion, the velocity changes with flow of time. Let us consider the case in which the velocity changes from $\boldsymbol{v}(t)$ to $\boldsymbol{v}(t+\Delta t)$ in time interval Δt from t to $t+\Delta t$.

$$\Delta \boldsymbol{v} = \boldsymbol{v}(t+\Delta t) - \boldsymbol{v}(t).$$

Dividing this velocity change $\Delta \boldsymbol{v} = (\Delta v_x, \Delta v_y, \Delta v_z)$ by time interval Δt gives

Fig. 1-21 Car speed meter

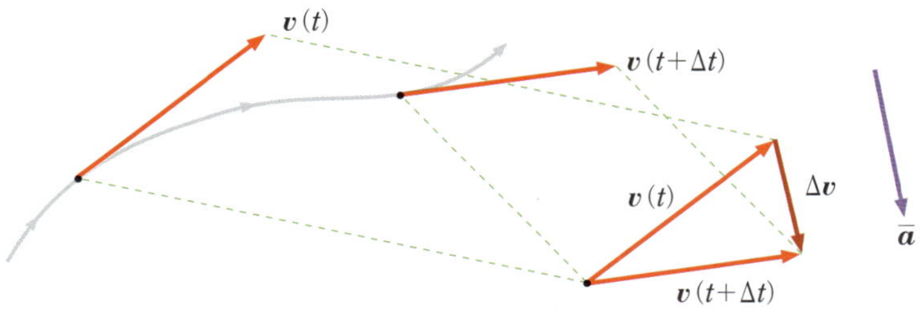

Fig. 1-22 Mean acceleration $\bar{\boldsymbol{a}} = \dfrac{\Delta \boldsymbol{v}}{\Delta t}$

$$\bar{\boldsymbol{a}} = \frac{\Delta \boldsymbol{v}}{\Delta t} = \left(\frac{\Delta v_x}{\Delta t}, \frac{\Delta v_y}{\Delta t}, \frac{\Delta v_z}{\Delta t}\right) \tag{1.22}$$

and $\bar{\boldsymbol{a}}$ is called the **mean acceleration** during this time interval (Fig. 1-22). Since the change in velocity $\Delta \boldsymbol{v}$ is a vector, the mean acceleration $\bar{a} = \dfrac{\Delta \boldsymbol{v}}{\Delta t}$ is a vector with the same direction as that of change in velocity $\Delta \boldsymbol{v}$ and the magnitude $\dfrac{|\Delta \boldsymbol{v}|}{\Delta t}$.

Case 2 If the velocity at time 0 is $\boldsymbol{v}_0$ and the velocity at time t is $\boldsymbol{v}(t)$, then the mean acceleration $\bar{\boldsymbol{a}}$ at time t is given by

$$\bar{\boldsymbol{a}} = \frac{\boldsymbol{v}(t) - \boldsymbol{v}_0}{t}. \tag{1.23}$$

This equation is transformed into $\boldsymbol{v}(t) - \boldsymbol{v}_0 = \bar{\boldsymbol{a}}t$, so the following equation is obtained.

$$\boldsymbol{v}(t) = \bar{\boldsymbol{a}}t + \boldsymbol{v}_0. \tag{1.24}$$

Changes in velocity cause the acceleration. When the speed (magnitude of the velocity) of mass point changes with time, the acceleration is not **0**. In the case where the speed does not change but the direction of force changes, the acceleration is not **0**, too. For example, when you are driving a car on a straight road and you press on the accelerator pedal or apply the brakes to the car, your action causes acceleration of the car [Fig. 1-23 (a), (b)]. When you turn the steering wheel without doing any action on the accelerator or brake pedals, the acceleration is, in this time also, not **0** [Fig. 1-23 (c)].

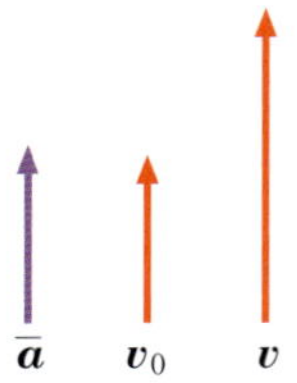

(a) When we press on the accelerator pedal.

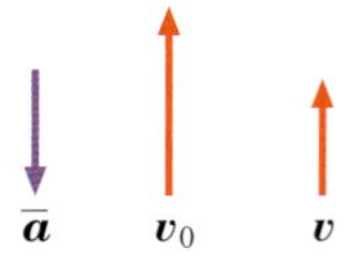

(b) When we apply the brakes.

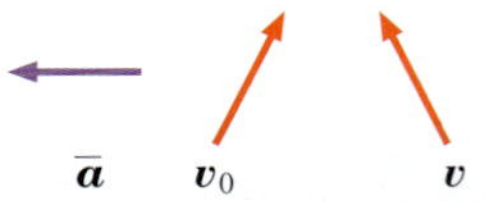

(c) When we turn the steering wheel.

Fig. 1-23 Mean acceleration $\bar{\boldsymbol{a}} = \dfrac{\boldsymbol{v} - \boldsymbol{v}_0}{t}$

Acceleration (instantaneous acceleration) The **acceleration** at time t (instantaneous acceleration) $\boldsymbol{a}(t)$ is $\dfrac{\mathrm{d}\boldsymbol{v}}{\mathrm{d}t}$, which is the value of the mean acceleration $\bar{\boldsymbol{a}} = \dfrac{\Delta \boldsymbol{v}}{\Delta t}$ [equation (1.22)] in the limit $\Delta t \to 0$. That is to say

$$\begin{aligned} \boldsymbol{a}(t) = (a_x, a_y, a_z) &= \frac{\mathrm{d}\boldsymbol{v}}{\mathrm{d}t} = \left(\frac{\mathrm{d}v_x}{\mathrm{d}t}, \frac{\mathrm{d}v_y}{\mathrm{d}t}, \frac{\mathrm{d}v_z}{\mathrm{d}t}\right) \\ &= \frac{\mathrm{d}^2\boldsymbol{r}}{\mathrm{d}t^2} = \left(\frac{\mathrm{d}^2x}{\mathrm{d}t^2}, \frac{\mathrm{d}^2y}{\mathrm{d}t^2}, \frac{\mathrm{d}^2z}{\mathrm{d}t^2}\right). \end{aligned} \tag{1.25}$$

Some readers of this book may be unfamiliar with derivatives of vectors such as velocity $\boldsymbol{v} = \dfrac{\mathrm{d}\boldsymbol{r}}{\mathrm{d}t}$ and acceleration $\boldsymbol{a}(t) = \dfrac{\mathrm{d}\boldsymbol{v}}{\mathrm{d}t} = \dfrac{\mathrm{d}^2\boldsymbol{r}}{\mathrm{d}t^2}$. Such readers will be encouraged to compare these symbols with physical images of velocity and acceleration that they encounter in their everyday life. It is also noted that components of velocity $\boldsymbol{v}$ and acceleration $\boldsymbol{a}$ (for example, x-components v_x and a_x) are respectively the velocity and the acceleration of the linear motion of the foot point of perpendiculars dropped from the mass point to the corresponding axes (x-axes) (Fig. 1-19).

Case 3 | Relative velocity Let the fall velocity of the rain drop in the windless air be $\boldsymbol{v}_1$. It is adequate for a man standing still to hold an umbrella straight up to the sky (Fig. 1-24). If he (object 2) is walking in the rain at velocity $\boldsymbol{v}_2$, the fall velocity of the rain (object 1) relative to the man is $\boldsymbol{v}_{12} = \boldsymbol{v}_1 - \boldsymbol{v}_2$. It is therefore adequate for him to walk with his umbrella inclined forward (in the direction of $-\boldsymbol{v}_{12}$).

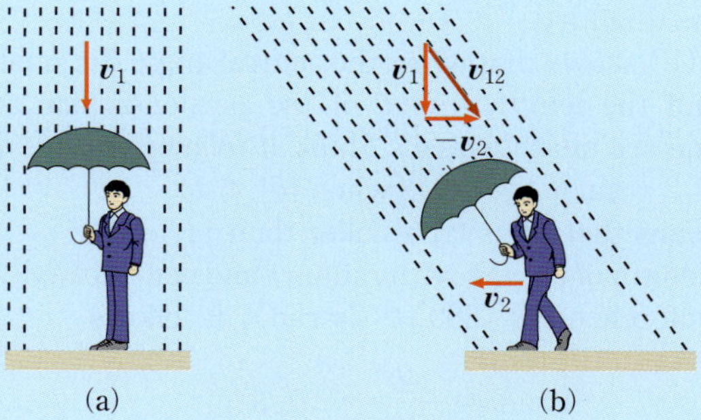

Fig. 1-24 Relative velocity

$$\boldsymbol{v}_{12} = \boldsymbol{v}_1 - \boldsymbol{v}_2 \tag{1.26}$$

is referred to as **relative velocity** of the object 1 relative to the object 2.

For example, in Fig. 1-25, the relative velocity of the car-1 to the car-2 is $\boldsymbol{v}_{12} = \boldsymbol{v}_1 - \boldsymbol{v}_2 = (-50\ \mathrm{m/s}, 0) - (0, 50\ \mathrm{m/s}) = (-50\ \mathrm{m/s}, -50\ \mathrm{m/s})$. It has the magnitude $50\sqrt{2} \approx 71\ \mathrm{m/s}$ and is directed toward southwest.

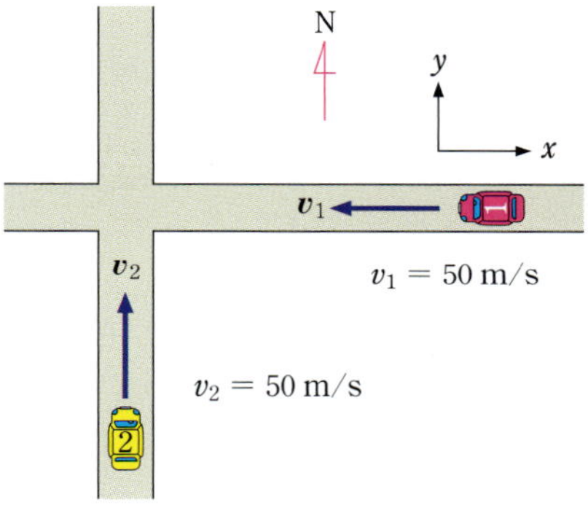

Fig. 1-25

1.3 Uniform circular motion

Learning objective To understand that the acceleration of mass point in uniform circular motion is directed to the center of circle, and its magnitude is proportional to the square of velocity and inversely proportional to the radius.

Polar coordinates To describe the motion of a mass point P on the plane, it is sometime convenient to use the two dimensional polar coordinates r, θ which are related to the rectangular coordinates x, y as follows:

$$x = r\cos\theta, \qquad y = r\sin\theta \tag{1.27}$$

(Fig. 1-26). r is the distance between the origin O and the mass point P. The angle θ indicating the direction (angle position) of mass point P seen from the origin O has + or − sign. When the $+x$-axis is taken to be the reference to measure the angle θ, the sign is decided as follows : for the mass point circulating anticlockwise on the circumference, the angle θ increases ($\Delta\theta > 0$), and for the point circulating clockwise, the angle θ decreases ($\Delta\theta < 0$).

Throughout this book unless otherwise noted, the angle is described in radian whose symbol is rad. When the length of an arc of circle with

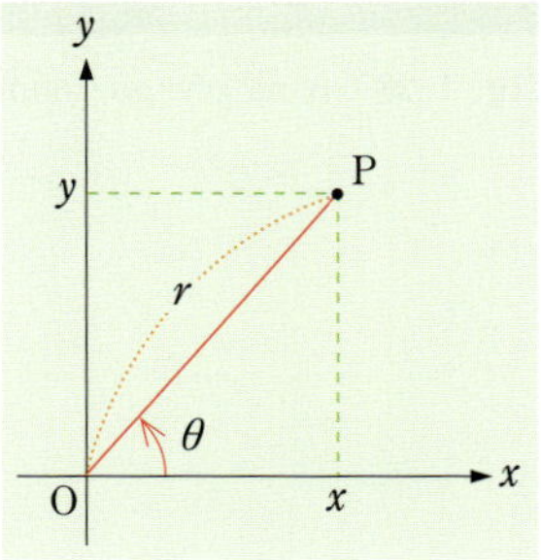

Fig. 1-26 Two-dimensional polar coordinates (r, θ). $x = r\cos\theta$ and $y = r\sin\theta$.

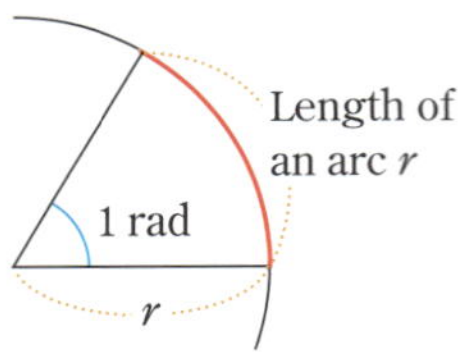

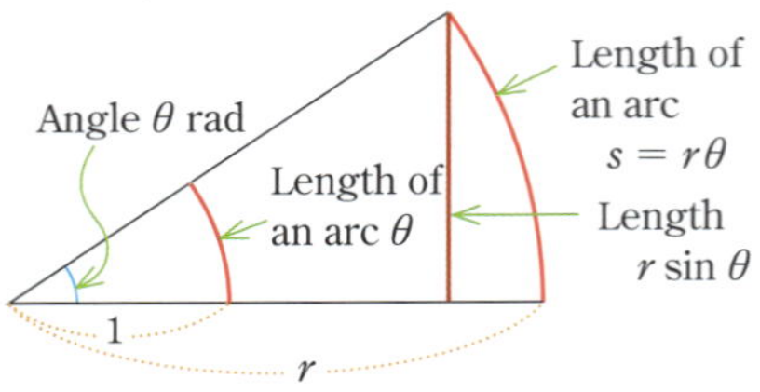

Fig. 1-27 (a) The length of the arc with its central angle 1 rad is equal to the radius. (b) The length s of the arc with its radius r and central angle θ rad is given by $s = r\theta$.

Unit of angle rad

its radius r is equal to r, the magnitude of the central angle subtended by the arc is defined as 1 rad [Fig. 1-27 (a)]. Since the length of arch is proportional to the central angle θ, the length s of an arc with radius r and central angle θ rad is given by

$$s = r\theta \tag{1.28}$$

[Fig. 1-27 (b)]). Since both s and r are lengths, the angle $\theta = \frac{s}{r}$ is a dimensionless quantity.

Fig. 1-27 (b) shows that when the central angle θ is small, the length $r\theta$ of arc and the length $r \sin\theta$ of the perpendicular drawn to the horizontal line are nearly equal. Thus, it follows

$$\sin\theta \approx \theta \qquad (\text{when } |\theta| \ll 1) \tag{1.29}$$

$|\theta| \ll 1$ means that $|\theta|$ is far smaller than 1.

Since the length of the arc with radius r and central angle 360° is equal to the circumference $2\pi r$ (360° = 2π rad), it follows

$$1\ \text{rad} = \frac{360°}{2\pi} \approx 57.3°. \tag{1.30}$$

Question 4 Explain the followings : $180° = \pi$, $90° = \frac{\pi}{2}$, $60° = \frac{\pi}{3}$, $45° = \frac{\pi}{4}$

Question 5 Explain that the area of a sector with radius r and central angle θ rad shown in Fig. 1-27 (b) is equal to $\frac{1}{2}r^2\theta$.

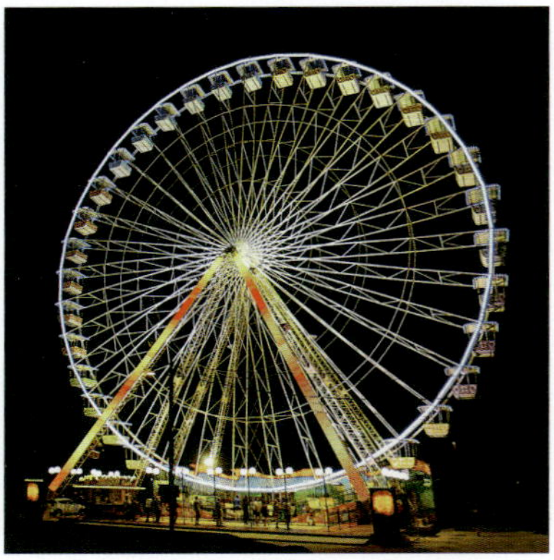

Fig. 1-28 A merry-go-round

Uniform circular motion When a mass point P moves at a constant speed v on the circumference of a circle with its origin O and radius r, that is to say when the mass point is in **uniform circular motion**, the angle position θ of the mass point P increases uniformly with time. Thus, if $\theta = \theta_0$ at $t = 0$, then, the angle position $\theta(t)$ at time t is given by

$$\theta(t) = \omega t + \theta_0. \tag{1.31}$$

The rate of change in angle position $\theta(t)$ with time t (rotation angle per unit time) $\omega = \frac{d\theta}{dt}$ is called the **angular velocity**. Insertion of equation (1.31) into equation (1.27) leads to the following equation describing the position of mass point at t which is in uniform circular motion around the circle with radius r.

$$x(t) = r\cos(\omega t + \theta_0), \quad y(t) = r\sin(\omega t + \theta_0) \tag{1.32}$$

Differentiating equation (1.32) with respect to t using differential formulae for trigonometric functions, we can derive the components of velocity $\boldsymbol{v}(t)$ as well as acceleration $\boldsymbol{a}(t)$

$$\begin{aligned} v_x &= \frac{dx}{dt} = \frac{d}{dt}[r\cos(\omega t + \theta_0)] = -\omega r \sin(\omega t + \theta_0) \\ v_y &= \frac{dy}{dt} = \frac{d}{dt}[r\sin(\omega t + \theta_0)] = \omega r \cos(\omega t + \theta_0) \end{aligned} \tag{1.33}$$

and

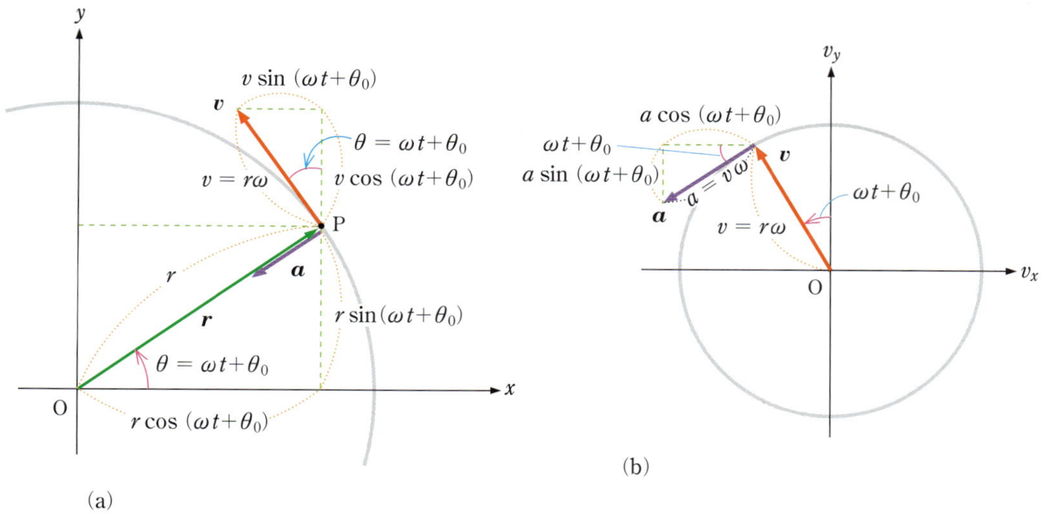

Fig. 1-29 Uniform circular motion. (a) velocity $\boldsymbol{v}$ of the uniform circular motion, (b) acceleration $\boldsymbol{a}$ in the case where the velocity $\boldsymbol{v}$ is equal to the acceleration given in the left figure.

$$a_x = \frac{\mathrm{d}v_x}{\mathrm{d}t} = -\omega^2 r\cos(\omega t+\theta_0) = -\omega^2 x$$
$$a_y = \frac{\mathrm{d}v_y}{\mathrm{d}t} = -\omega^2 r\sin(\omega t+\theta_0) = -\omega^2 y. \quad (1.34)$$

If readers do not remember the differential formulae for trigonometric functions, they are advised to accept equation (1.33) as the differential formula.

From the above calculated results and Fig. 1-29 illustrating them we can understand that the speed v is given by

$$v = \sqrt{v_x{}^2+v_y{}^2} = r\omega \quad (1.35)$$

and the velocity $\boldsymbol{v}$ is perpendicular to the radial (position vector) $\boldsymbol{r}$ (namely, the tangent of the circle is perpendicular to the radius)

$$\boldsymbol{v} \perp \boldsymbol{r}. \quad (1.36)$$

In addition, the magnitude a of acceleration is

$$a = \sqrt{a_x{}^2+a_y{}^2} = r\omega^2 = v\omega = \frac{v^2}{r} \quad (1.37)$$

and the acceleration $\boldsymbol{a}$ is perpendicular to the velocity $\boldsymbol{v}$ and is in the opposite direction to the position vector $\boldsymbol{r}$

$$\boldsymbol{a} = -\omega^2\boldsymbol{r}. \quad (1.38)$$

Acceleration of the uniform circular motion takes the direction toward the center of circle. For this reason, it is called **centripetal acceleration***.

Question 6 A car is running at a constant speed on a flat road as illustrated in Fig. 1-30. Of the four sections 1 → 2, 2 → 3, 3 → 4, and 4 → 1, (1) in which section does the magnitude of acceleration take the maximum, and (2) in which section does the magnitude of acceleration take the minimum?

* In the case of a mass point in non-uniform circular motion, the angular velocity $\omega = \frac{\mathrm{d}\theta}{\mathrm{d}t}$ changes and the angular acceleration $\alpha = \frac{\mathrm{d}\omega}{\mathrm{d}t} = \frac{\mathrm{d}^2\theta}{\mathrm{d}t^2} \neq 0$. In this case the acceleration $\boldsymbol{a}$ of mass point has not only the component directed to the radius but also the one a_t directed to the tangent and the relation $a_t = r\alpha$ holds.

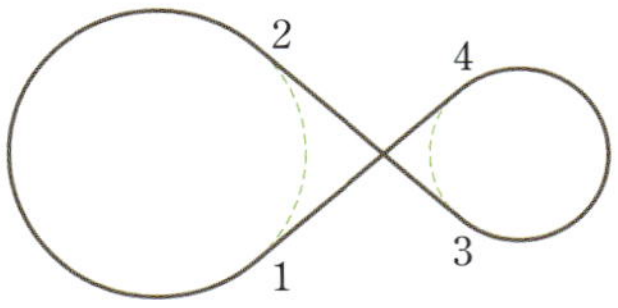

Fig. 1-30

The angular velocity ω is the rotation angle per unit time. The rotation angle is 2π rad for a mass point to rotate once around the circumference. If the number of rotation per unit time is f, the following relation holds :

$$\omega = 2\pi f. \tag{1.39}$$

Similarly, as in the case of uniform circular motion, motions which take the same state repeatedly in equal time interval are referred to as the periodic motion. This time interval is called the **period**. The period T of uniform circular motion is the time for the mass point to rotate once the circumference of circle. The period is the reciprocal of rotational number f per unit time :

$$T = \frac{1}{f} = \frac{2\pi}{\omega}. \tag{1.40}$$

The speed v of a mass point which rotates around the circle with circumference $2\pi r$ at frequency f per unit time is given by $v = 2\pi rf = r\omega$.

Problems for exercise 1

Problems for exercise given at the end of each chapter consist of A and B parts. Problems in B part are a little more difficult than in A.

A

1. (1) Using that 1 km = 1000 m and 1 h = 3600 s, derive that 1 km/h = $\frac{1}{3.6}$ m/s and 1 m/s = 3.6 km/h.
   (2) The equation 1 m/s = 3.6 km/h means that if we walk for 1 hour at the rate of 1 m per 1 second, we can go 3.6 km. Explain this description.
2. Among "Kodama" trains of Tokaido Shinkansen (Bullet Train Line), there are some which, stopping at every station, run between Tokyo and Shin-Osaka in 4 hours and 12 minutes. Assuming that the distance between Tokyo and Shin-Osaka is equal to the official working distance of 552.6 km, calculate the average speed of these "Kodama" trains.
3. A car is running on a flat straight road which stretches in the east-west direction. Taking the east as the +direction and the west as the −direction, draw examples of the v-t curves which correspond to the following conditions.
   (1) A car is running east at a constant speed.
   (2) A car running west changed the speed in a uniform fashion and it started running east.
   (3) A car running east is increasing its speed at a constant rate.
4. Positions of an object moving on the x-axis are given in Fig. 1 (a)-(f). Place one of your hands at a desk edge and display each case of Fig. 1 (a)-(f) by the motion of your hand.

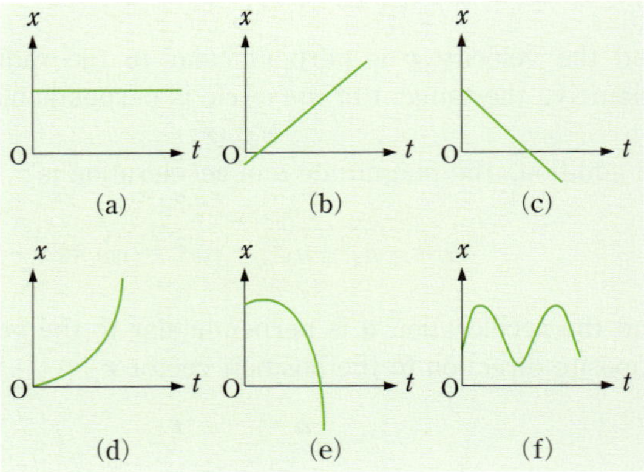

Fig. 1

5. The x-t graph for two objects A and B which move on the x-axis is shown in Fig. 2. Find the collision point and the collision time.

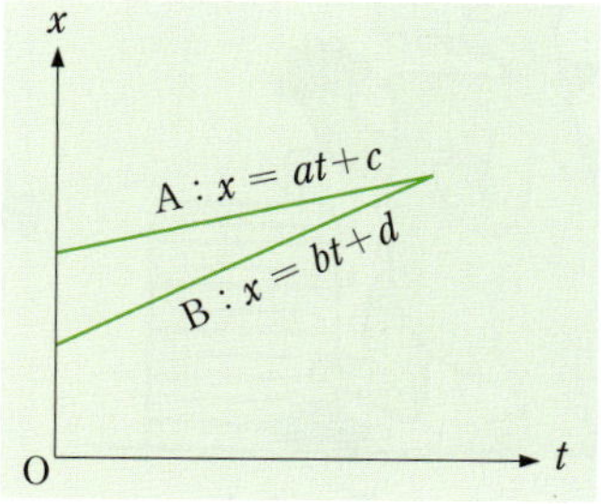

Fig. 2

6. A "Kodama" train, after leaving the station, is accelerated uniformly to the speed 198 km/h at the rate of 0.25 m/s in speed per second. That is to say the acceleration is constant and is equal to $a = 0.25$ m/s^2. Find the time till the speed reaches 198 km/h = 55 m/s.

7. Fig. 3 is a flash photo of a free-fall ball taken at every 1/30 s and the ruler is graduated in cm. Based on this flash photo, calculate the acceleration g of the free-fall motion (uniformly accelerated motion for the case of initial velocity zero) and confirm that $g \approx 9.8$ m/s^2. Use equation (1.8).

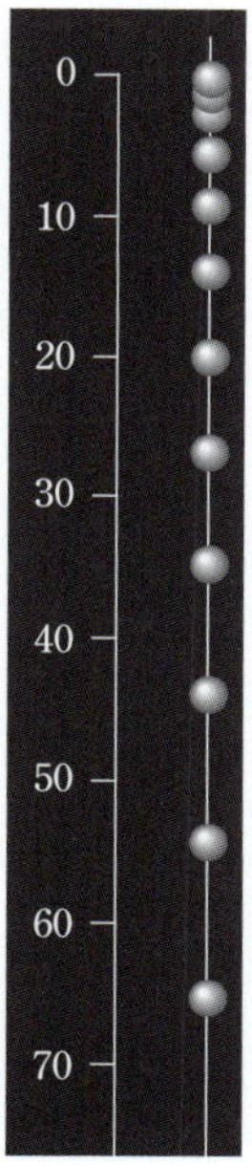

Fig. 3

8. Calculate the acceleration a when the velocity of a mass point moving on the x-axis is given by $v = V_0(1-bt)$ and also the case $v = V_0(1-bt)^2$, where V_0 and b are constants.

9. Differentiate the following functions with respect to t. And once again differentiate the obtained functions with respect to t. All parameters except t are constants.
   (1) $x = at^2+bt+c$, (2) $x = a\sin(bt+c)$,
   (3) $x = a\cos(bt+c)$, (4) $x = a\log(bt+c)$,
   (5) $x = ae^{mt}+be^{-nt}$

10. A ball was thrown at speed 20 m/s and at angle 30° relative to the horizontal direction. Find the horizontal and vertical components of the initial velocity of this ball.

11. Assuming that the Earth rotates once in 24 hours, find its angular velocity.

12. A toy car running the circular track will run off the track when the centripetal acceleration roughly exceeds the gravitational acceleration. Find the speed at which the car runs off when the radius of the circular track is 0.5 m.

13. A merry-go-round with radius 5 m rotates at period 10 s.
   (1) Find the number of rotations per second f.
   (2) Calculate the speed v of the dummy horse placed at 4 m away from the center.
   (3) Calculate the acceleration a of this dummy horse. How many times is this acceleration as large as the gravitational acceleration g?

B

1. A train is running on the curved rail tracks at speed 20 m/s and is changing its traveling direction at a rate of 0.01 rad per second. Find the acceleration applied on the train passengers.

2. The position $x(t)$ of an object which moves along the x-axis in uniformly accelerated motion with its initial velocity v_0 and acceleration a_0 can be described as shown in Fig. 4, where the position at $t = 0$ is set to be $x_0 = 0$. Explain this description.

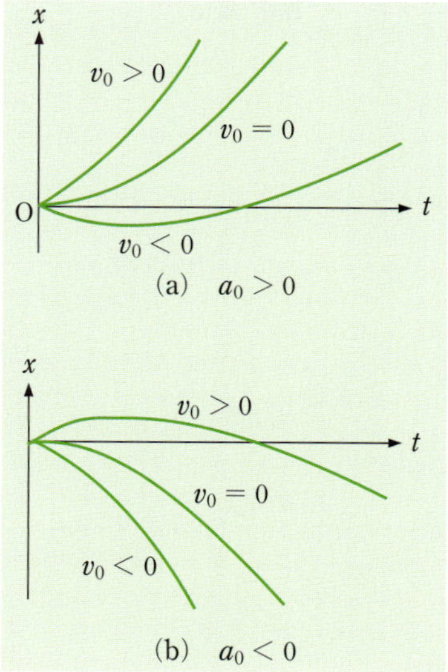

Fig. 4

3. Describe qualitatively the behavior of an object moving along the x-axis in linear motion with positive or negative sign in velocity v and acceleration a as given below.
   (1) $v > 0$, $a > 0$ (2) $v > 0$, $a < 0$
   (3) $v < 0$, $a > 0$ (4) $v < 0$, $a < 0$

Similarity rule

In the filming of "monster movies" in which a huge monster heavily destroys the city, a miniature model of the city is prepared in the studio and an actor wearing a monster-suit ramps around the city. However, when the audience watches the scene filmed as it is, such film can only give an impression that an actor wearing a monster-suit ramps around in the miniature city. For example, if, in the miniature city of scale 1/16, the monster lifts a model house to the height 2 m and releases it from his hand, for the audience of the film the fall scene of the house cannot look like the case from the height of 32 m. This is because the audience realizes well that the duration of fall should be much longer.

Therefore, in the filming of monster movies, the slow-motion technique is adopted. The following relation holds for every free-falling body irrespective of its mass.

$$\text{fall length} = \frac{1}{2}(\text{gravitational acceleration } g) \times (\text{fall time})^2.$$

Fig. 1-A

Since the fall length is proportional to the square of fall time, in order to make the fall length look like 16 times (= 4×4 times), it is appropriate to perform slow-motion shooting that extends the fall time by a factor of 4. Generally, to make the model reproducing the real thing at $1/N^2$ (or $1/N^2$ model) correspond to the real world, we must stretch the time by N times in shooting The above-mentioned technique is said to be based on the similarity rule.

Laws of Motion and Laws of Force

Various types of forces act on objects. Forces acting on the objects follow the laws of force such as gravity and electromagnetic force, while the laws of motion determine how the objects move by the action of force.

About 350 years ago, Newton overturned the conventional common sense of his era and showed that the motion of the celestial object and the movement of objects on the Earth surface follow the same laws. The reason why the spacecraft launched from the Earth surface can be soft-landed on the moon surface is that the laws of motion and the law of universal gravitation (gravity) discovered by Newton hold exactly everywhere from the Earth surface to the lunar surface. Not only the motion of objects around us but also the motion of celestial objects such as planets and the moon can be explained satisfactorily by Newton's theory. For this reason, the three rules for motion of objects proposed by Newton in "Philosophiae Naturalis Principia Mathematica "which was published in 1687 and simply referred to as" Principia" are called Newton's laws of motion.

2.1 Laws of motion

Learning objective To fully understand what the three Newton's laws of motion (law of inertia, law of motion, and law of action and reaction) are and to be able to explain their meanings by applying these laws to simple examples.

Fig. 2-1 A scene of cycling race

The first law of motion (law of inertia) The first law is a law concerning the motion of an object on which the resultant force of the acting forces is **0**.

When we push a book on the desk the book moves. If we push it with a stronger force, it moves faster, but when we stop pushing it, it stops moving. From such experiences in everyday life, we have an impression that objects move only while the force is acting on them and they stop moving immediately soon after the acting force vanishes. It is, therefore, likely for us to think that the velocity of an object is proportional to the force acting on it. However, there are phenomena which are contradictory to this hypothesis. For example, when pushing a light cart on a flat road, if we keep pushing with a force of the same strength, the speed of the cart will increase and even if we stop pushing and stop applying force, the cart keeps running for a while. If the speed is proportional to the strength of force, at the moment the force acting on an object vanishes, the speed should be 0.

Ancient people also noticed that there are cases where objects continue moving even if forces do not act on them. It is the case when we shoot an arrow with a bow. The arrow starts to move due to the elasticity of the string, but it continues moving even while the force is not acting on the arrow which is away from the string. Some scholars in old times who saw this scene thought that moving objects had the property to maintain their moving state, and they named this property the **inertia**.

Newton, considering all objects have own inertia, proposed the following rule called the **first law of motion** or the **law of inertia**.

The first law If the resultant force of the forces acting on an object is **0**, the object at rest remains stationary, and the moving object keeps its linear motion at a constant speed. Conversely, if the object is kept at rest or keeps the linear motion at a constant speed, the resultant force of the forces acting on the object is **0**[*].

* Since velocities of objects kept at rest with a velocity of **0** or objects in linear motion at constant velocity are constant, their accelerations are all **0**. It follows, therefore, an alternative expression of the first law is "Velocities are constant for objects on which acting resultant forces are **0**, and accelerations are **0**".

The reason why we must keep pushing the objects on the desk to move them is due to the action of friction force which acts in the direction hindering the motion of the objects. On account of the presence of gravity and friction force, near the Earth surface it is difficult to think of the state in which no force acts on moving objects. However, if we succeed in reducing the friction caused by the floor, then, we can ignore the forces acting horizontally. For example, if a thin disc of dry ice placed on a flat floor is flicked with fingers, the disc slides linearly at a constant speed. This is because the friction between the dry ice and the floor is significantly reduced by the sublimed CO_2 gas, and the horizontal force acting on dry ice becomes approximately zero.

Newton's laws of motion do not hold for every coordinate system. The coordinate systems in which the first and second laws hold are called inertial coordinate systems or inertial frames (refer to Chapter 9). When the effects of rotation and revolution of the Earth can be ignored, the Earth surface can be approximated as an inertial frame.

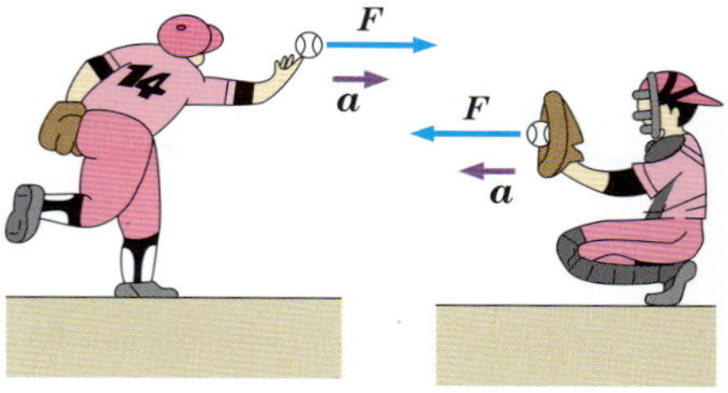

Fig. 2-2 The acceleration $\boldsymbol{a}$ of an object has the same direction of the force $\boldsymbol{F}$ acting on the object and its magnitude is proportional to $\boldsymbol{F}$.

The second law (law of motion) The first law of motion means that when the motion state of an object changes and acceleration occurs, a force is acting on the object.

To throw or catch a ball, the hand must exert the force $\boldsymbol{F}$ on the ball. The direction of the force $\boldsymbol{F}$ that the hand exerts on the ball is the same as the direction of acceleration $\boldsymbol{a}$ of the ball (Fig. 2-2).

When we try to stop a shot of shot-put or a ball of baseball both of which are rolling on the ground at the same speed, we must apply the stronger force to stop the shot which has a mass larger than that of the baseball ball. In other words, the magnitude of mass represents the magnitude of inertia.

Fig. 2-3 A baseball pitcher

These facts are expressed as a quantitative relationship among force, acceleration and mass by Newton's **second law of motion,** also called the **law of motion**.

The second law When forces act on an object, acceleration is generated in the object in the direction of force (resultant force). The magnitude of acceleration is proportional to the magnitude of force (resultant force) and inversely proportional to the mass of the object.

When expressed as a formula, $\boldsymbol{a} \propto \dfrac{\boldsymbol{F}}{m}$. In the International System of Units, since the proportionality constant is 1, the second law of motion is expressed as "mass m"×"acceleration $\boldsymbol{a}$" = "force $\boldsymbol{F}$", that is to say

$$m\frac{\mathrm{d}^2\boldsymbol{r}}{\mathrm{d}t^2} = \boldsymbol{F}. \tag{2.1}$$

This is called **Newton's equation of motion**. When dividing into components, we obtain the following relations :

$$m\frac{\mathrm{d}^2x}{\mathrm{d}t^2} = F_x, \quad m\frac{\mathrm{d}^2y}{\mathrm{d}t^2} = F_y, \quad m\frac{\mathrm{d}^2z}{\mathrm{d}t^2} = F_z, \tag{2.1$'$}$$

where F_x, F_y and F_z are the x-, y- and z-components of the force $\boldsymbol{F}$, respectively.

In the International System of Units, the equation of motion takes the form $m\boldsymbol{a} = \boldsymbol{F}$. This is because the units kg for mass, m/s^2 for acceleration, and kg·m/s^2 for force are adopted. In other words, the magnitude of force causing the acceleration of 1 m/s^2 on an object with a mass of 1 kg is used in the equation. This international unit of force is called Newton (symbol N).

$$\mathrm{N} = \mathrm{kg{\cdot}m/s^2} \tag{2.2}$$

Unit of force N = kg·m/s^2

1 N is roughly equal to the gravity acting on 100 g of water (a half cup).

In the case of an object spreading in space, equation (2.1) is the equation of motion for the center of gravity. Therefore, acceleration $\boldsymbol{a}$

stands for the acceleration of the center of gravity. The center of gravity will be discussed in more detail in Chapter 7*.

* For consideration of the rotation of an object spreading in space around its center of gravity, refer to Chapters 7 and 8.

Case 1 To stop in 3 seconds a 30 kg object moving on a straight line at a speed of 15 m/s, how much force on average should be applied on the object?

$$\text{average acceleration } \bar{a} = \frac{v(t)-v_0}{t} = \frac{(0\,\text{m/s})-(15\,\text{m/s})}{3\,\text{s}} = -5\,\text{m/s}^2$$

$$F = m\bar{a} = (30\,\text{kg})\times(-5\,\text{m/s}^2) = -150\,\text{kg}\cdot\text{m/s}^2 = -150\,\text{N}.$$

Accordingly, 150 N. The negative sign indicates that the direction of force and the direction of motion are opposite.

Case 2 When a force of $16\,\text{N} = 16\,\text{kg}\cdot\text{m/s}^2$ is applied to an object of mass 4 kg, acceleration a is

$$a = \frac{F}{m} = \frac{16\,\text{kg}\cdot\text{m/s}^2}{4\,\text{kg}} = 4\,\text{m/s}^2.$$

The direction of acceleration is the same as the force.

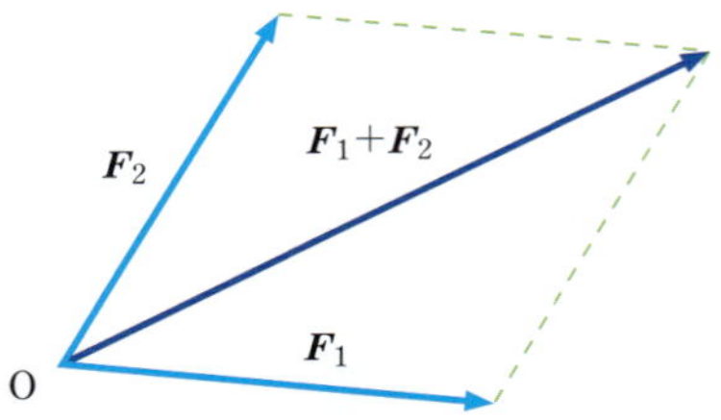

Fig. 2-4 $\boldsymbol{F} = \boldsymbol{F}_1+\boldsymbol{F}_2$

Resultant force When several forces are acting on one object, one force giving the same effect as these forces is called the **resultant force** of these forces. According to the experiment, the resultant force $\boldsymbol{F}$ of the two forces $\boldsymbol{F}_1$ and $\boldsymbol{F}_2$ whose lines of action of force* intersect is the force corresponding to the diagonal of the parallelogram with $\boldsymbol{F}_1$ and $\boldsymbol{F}_2$ as two adjacent sides (Fig. 2-4), so using the vector sum notation we can express as

* A straight line passing through the point where the force acts on the object and facing the direction of force.

$$\boldsymbol{F} = \boldsymbol{F}_1+\boldsymbol{F}_2. \tag{2.3}$$

Inversely, the two forces $\boldsymbol{F}_1$ and $\boldsymbol{F}_2$ that exert the same effect as the force $\boldsymbol{F}$ are called **components of the force $\boldsymbol{F}$**.

The resultant force $\boldsymbol{F}$ is the sum of the components of two forces $\boldsymbol{F}_1$ and $\boldsymbol{F}_2$,

$$\boldsymbol{F} = \boldsymbol{F}_1+\boldsymbol{F}_2 = (F_{1x}+F_{2x},\ F_{1y}+F_{2y},\ F_{1z}+F_{2z}). \tag{2.3'}$$

When more than three forces $\boldsymbol{F}_1$, $\boldsymbol{F}_2$, $\boldsymbol{F}_3$, $\cdots$ are acting, the resultant force is

$$\boldsymbol{F} = \boldsymbol{F}_1+\boldsymbol{F}_2+\boldsymbol{F}_3+\cdots \tag{2.4}$$

(Fig. 2-5).

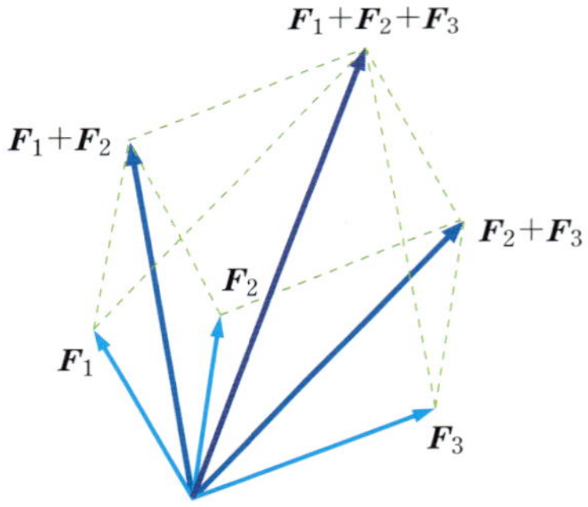

Fig. 2-5 Sum of three vectors $\boldsymbol{F}_1$, $\boldsymbol{F}_2$, and $\boldsymbol{F}_3$. $\boldsymbol{F}_1+\boldsymbol{F}_2+\boldsymbol{F}_3$

To find the sum of three vectors $\boldsymbol{F}_1$, $\boldsymbol{F}_2$, and $\boldsymbol{F}_3$, we should first find the sum of $\boldsymbol{F}_1$ and $\boldsymbol{F}_2$ using the rule of parallelogram, and then find the vector sum of $\boldsymbol{F}_1+\boldsymbol{F}_2$ and $\boldsymbol{F}_3$ again using the rule of parallelogram as $(\boldsymbol{F}_1+\boldsymbol{F}_2)+\boldsymbol{F}_3$. We can obtain the same result through the step first to find $\boldsymbol{F}_2+\boldsymbol{F}_3$ equal to the sum of two vectors $\boldsymbol{F}_2$ and $\boldsymbol{F}_3$, and then add $\boldsymbol{F}_1$ to the sum of $\boldsymbol{F}_2+\boldsymbol{F}_3$ as $\boldsymbol{F}_1+(\boldsymbol{F}_2+\boldsymbol{F}_3)$ to obtain the vector sum. The vector sum of the three vectors $\boldsymbol{F}_1$, $\boldsymbol{F}_2$, and $\boldsymbol{F}_3$ found as mentioned above is denoted as $\boldsymbol{F}_1+\boldsymbol{F}_2+\boldsymbol{F}_3$.

Centripetal acceleration and centripetal force As described in Section **1.3**, acceleration of a mass point moving on the circumference with radius r at speed v and angular velocity ω takes the direction to the center of circle and has the magnitude $a = \dfrac{v^2}{r} = r\omega^2$. It follows, therefore, according to the second law of motion, the force of magnitude

$$F = mr\omega^2 = \frac{mv^2}{r} \quad (\boldsymbol{F} = -m\omega^2\boldsymbol{r}) \tag{2.5}$$

directed to the center of circle is acting on the mass point. This force is

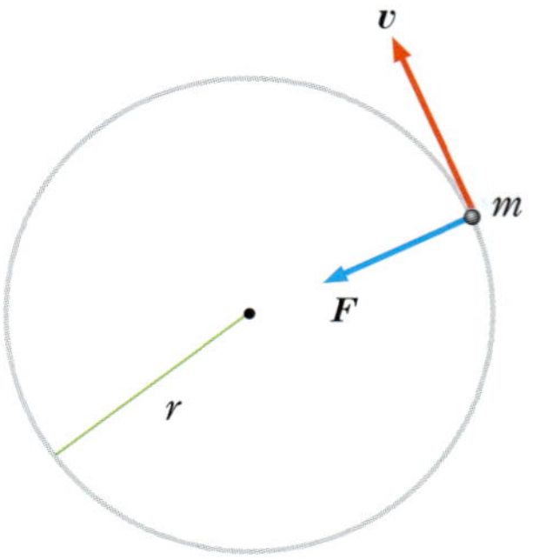

Fig. 2-6 Centripetal force $F = mr\omega^2 = \dfrac{mv^2}{r}$

called the **centripetal force** (Fig. 2-6).

Question 1 Using arrows, indicate the direction and the relative magnitude of the force which acts on a car running at a constant speed when the car passes the points A, B, and C on the curve depicted in Fig. 2-7.

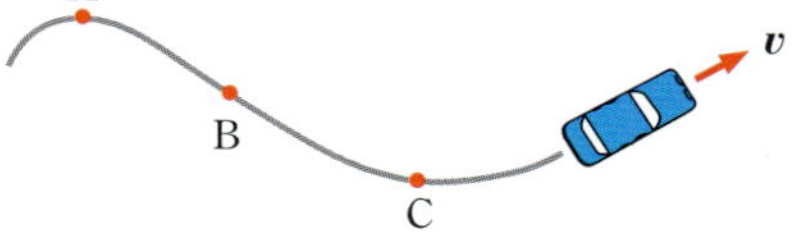

Fig. 2-7

The third law of motion (law of action and reaction) When we lift a baggage, our hands are pulled by the baggage. When we are on a boat and push the shore with an oar, the shore pushes back the oar. We can walk forward because the ground pushes our foot forward when our foot pushes the ground backwards. In this way, the force acts between two objects, and when an object A exerts the force on an object B, B also exerts the force on A. This is a phenomenon called the **action** and **reaction**. The relation between action and reaction is expressed by the **third law of motion,** also called the **law of action and reaction** (Fig. 2-8).

The third law When the object A exerts the force $\boldsymbol{F}_{B \leftarrow A}$ on the object B, the object B also exerts the force $\boldsymbol{F}_{A \leftarrow B}$ on the object A[*1]. The two forces mutually take the opposite directions and their magnitudes are the same[*2].

$$\boldsymbol{F}_{B \leftarrow A} = -\boldsymbol{F}_{A \leftarrow B} \tag{2.6}$$

*1 The reaction does not need time to occur after the action but occurs simultaneously in time. When the road surface is slippery, the reaction by the road surface does not occur, so the foot fails to act on the road surface.

*2 Some electromagnetic forces do not follow the law of action and reaction (see Problems for exercise 20, B6). The reason is that the electromagnetic field mediating the electromagnetic forces possesses momentum (see Section **22.3**).

(a) Force $\boldsymbol{F}_{B \leftarrow A}$ and Force $\boldsymbol{F}_{A \leftarrow B}$, $\boldsymbol{F}_{B \leftarrow A} = -\boldsymbol{F}_{A \leftarrow B}$
(b) Force $\boldsymbol{F}_{A \leftarrow B}$ that B acts on A
(c) Force $\boldsymbol{F}_{B \leftarrow A}$ that A acts on B

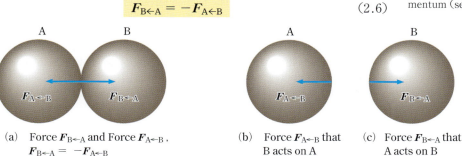

Fig. 2-8

Question 2 When an adult and an infant push each other, the adult moves forward. Explain that the law of action and reaction also holds in this case (Fig. 2-9).

Fig. 2-9

Example 1 | Internal force and external force

As shown in Fig. 2-10 (a), when we connect two carts A and B on a flat smooth floor, and pull the cart A with force $F = 40$ N, two carts start to move. Find the magnitude a of acceleration $\boldsymbol{a}$ common to A and B. We assume that the masses of carts A and B are $m_A = 10.0$ kg and $m_B = 6.0$ kg, respectively.

Solution For cart A the equation of motion in the horizontal direction is

$$m_A \boldsymbol{a} = \boldsymbol{F} + \boldsymbol{F}_{A \leftarrow B}. \quad \text{[Fig. 2-10 (c)]}$$

For cart B the equation of motion in the horizontal direction is

$$m_B \boldsymbol{a} = \boldsymbol{F}_{B \leftarrow A}. \quad \text{[Fig. 2-10 (b)]}$$

Adding both sides of the above two equations and using the law of action and reaction $\boldsymbol{F}_{A \leftarrow B} + \boldsymbol{F}_{B \leftarrow A} = \boldsymbol{0}$, we obtain the following relation:

$$m_A \boldsymbol{a} + m_B \boldsymbol{a} = (m_A + m_B) \boldsymbol{a} = \boldsymbol{F}.$$

It follows that the magnitude a of acceleration of the carts is

$$a = \frac{F}{m_A + m_B} = \frac{40\ \text{N}}{16.0\ \text{Kg}} = 2.5\ \text{m/s}^2.$$

The equation of motion in the horizontal direction $(m_A + m_B)\boldsymbol{a} = \boldsymbol{F}$ can be immediately derived from the fact that the horizontal force acting from the outside is only $\boldsymbol{F}$ after taking the two carts as a combined object with its mass $m_A + m_B$ [Fig. 2-10 (a)]. The forces $\boldsymbol{F}_{B \leftarrow A}$ and $\boldsymbol{F}_{A \leftarrow B}$ acting on the two carts A and B cancel each other. The force acting between constituents of the object system such as $\boldsymbol{F}_{B \leftarrow A}$ and $\boldsymbol{F}_{A \leftarrow B}$ in this case is called the **internal force** while the force acting on constituents of the object system from the outside of the object is called the **external force**. The motion of the object system as a whole is determined only by the external force, and the internal force is irrelevant.

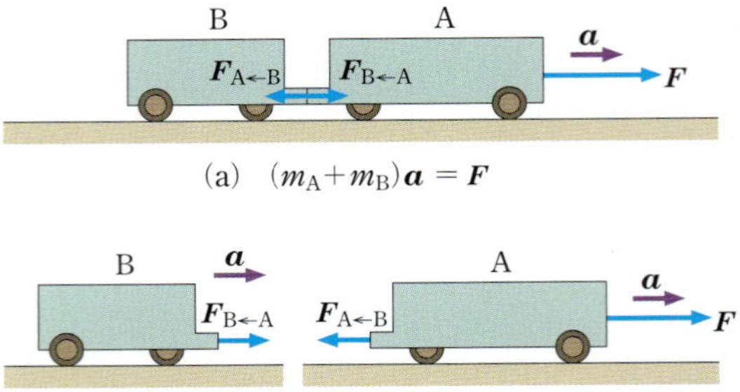

(a) $(m_A + m_B)\boldsymbol{a} = \boldsymbol{F}$

(b) $m_B \boldsymbol{a} = \boldsymbol{F}_{B \leftarrow A}$ (c) $m_A \boldsymbol{a} = \boldsymbol{F} + \boldsymbol{F}_{A \leftarrow B}$

Fig. 2-10

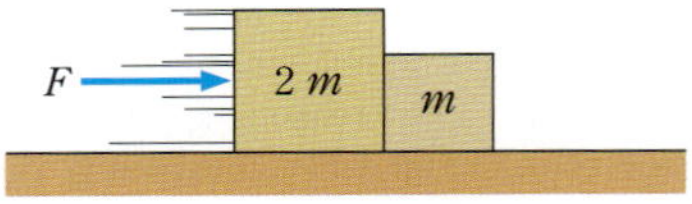

Fig. 2-11

Question 3 If a passenger in a car at rest pushes the front windshield, will the car move?

Question 4 By the action of horizontal force $\boldsymbol{F}$, the objects of masses m and $2m$ shown in Fig. 2-11 are moving. Find the force $F_{m \leftarrow 2m}$ acting between the two objects. The friction force between the objects and the floor can be neglected

Question 5 When two carts A and B of Example 1 are tied with a light string ($m_{\text{string}} \approx 0$), the magnitudes of the tension ($\boldsymbol{F}_{A \leftarrow \text{string}}$) that the string draws the cart A and the tension ($\boldsymbol{F}_{B \leftarrow \text{string}}$) that pulls the cart B are approximately equal ($F_{A \leftarrow \text{string}} \approx F_{B \leftarrow \text{string}}$). Show this using the equation of motion of the string $m_{\text{string}} \boldsymbol{a}_{\text{string}} = \boldsymbol{F}_{A \leftarrow \text{string}} + \boldsymbol{F}_{B \leftarrow \text{string}}$.

2.2 Various types of forces and laws of forces

Learning objective To understand that various forces act on objects, and near the Earth surface, on any object (with mass m) the gravity of magnitude mg acts vertically downward, and that the cause of gravity which the Earth exerts is the fundamental force of the universal gravitation.

To understand the properties of friction force and the relationship between friction force and normal force.

In order to apply Newton's laws of motion to the motion of objects, it is necessary to know the laws of force acting on the objects. The **force** is a cause to change the motion state of objects or to deform them.

Various types of forces act on objects[*1]. The Earth's gravity acts on any object. For the ball rolling in the schoolyard, the ground applies the normal force perpendicular to the contact surface and the friction force acts parallel to the contact surface. Resistance of air also acts on this ball. Resistance force and lift of the air act on a jet airplane in flight, and the emitted jet gives the forward propulsive force. Liquids and gases exert pressure on the objects placed in them. The electric force acts between charged objects, and the magnetic force acts between magnets and iron nails. Various other types of forces act on various other objects.

*1 In Japanese, expressions like "a force works, to exert force, to apply force, and to be subjected to ○○ force", are often used. On the other hand, in English, the word act is so frequently used that in this book, the author adopts mostly such expression as "The force acts on an object".

Fundamental forces and phenomenological forces Forces are divided into fundamental forces and phenomenological forces. Fundamental forces are those acting between electrons, protons, neutrons, etc. which are the basic constituent elements of matter. The fundamental forces which appear in mechanics are only three, namely gravity (universal gravitation), electric force, and magnetic force. Because the electric and magnetic forces are related to each other, in many cases they are called in a combined term the electromagnetic force. The electromagnetic force will be treated in electromagnetism

Universal gravitation From the study of revolution of the planets around the Sun and the gravity force of the Earth acting on the objects on the Earth, Newton discovered the **universal gravitation force** acting between any two objects[*2].

*2 These days, the universal gravitation is commonly referred to simply as the gravity.

Law of universal gravitation Between any two objects, an attractive force acts in every case in proportion to the product of the masses of the two objects and inversely proportional to the square of their distance (Fig. 2-12).

$$F = G\frac{m_1 m_2}{r^2} \tag{2.7}$$

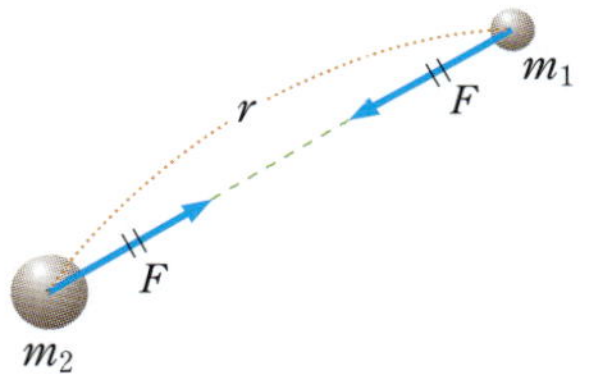

Fig. 2-12 Universal gravitation
$F = G\frac{m_1 m_2}{r^2}$

This proportionality constant is called the **gravitational constant**. Cavendish succeeded in its measurement first in 1798 by measuring the universal gravitation acting between two large lead balls placed in the ground laboratory. Recent measurements give the value

$$G = 6.674\times 10^{-11}\ \mathrm{m^3/(kg{\cdot}s^2)}. \tag{2.8}$$

Gravitational constant
$G = 6.674\times 10^{11}\ \mathrm{m^3/(kg{\cdot}s^2)}$

Seince this G value is small, the universal gravitation is significant only for objects with large mass. The universal gravitation is the force that binds celestial objects to make stars, the solar system, the galaxy, etc. It is the force to cause objects near the Earth surface to fall.

The universal gravitation acting between two objects spreading in space can be regarded as the resultant force of the universal gravitation acting between each part by considering the object as the sum of minute parts. Then, we can prove the followings : As in the case of the universal gravitational force acting between the Sun and the Earth, the

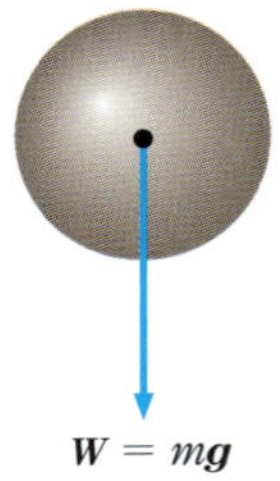

Fig. 2-13 Gravity of the Earth $\boldsymbol{W} = m\boldsymbol{g}$ acting on the object of mass m.

Practical unit of force
kilogram-force kgf
kilogram-weight kgw
Gravitational kilogram
1 kgf = 9.80665 N

* Newton's law of motion $m\boldsymbol{a} = \boldsymbol{F}$ is also an equation to find the force $\boldsymbol{F}$ from mass m and acceleration $\boldsymbol{a}$.

Fig. 2-14 An astronaut working on the lunar surface

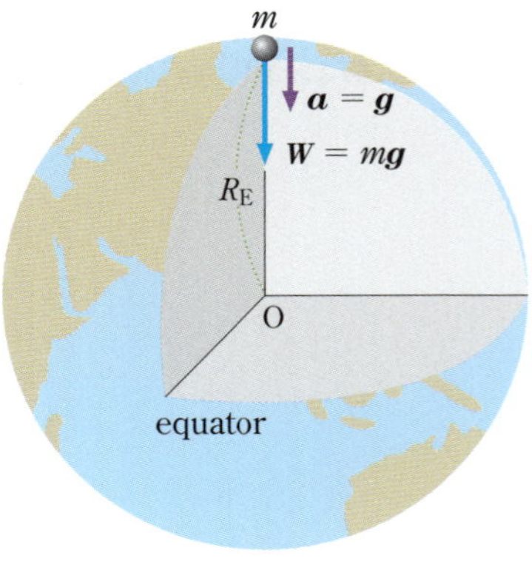

Fig. 2-15 Gravity that the Earth exerts.

universal gravitational force acting between two spherically symmetric objects A and B is equal to the case when the masses of A and B are gathered at their respective centers (see Problems for exercise 16, B 5).

Gravity of the Earth $\boldsymbol{W} = m\boldsymbol{g}$ Objects fall in the air near the Earth surface, because the Earth exerts the attractive force on them. This attractive force is called the **gravity**.

As discussed in Section **1.1**, when the air resistance is negligible, the gravitational acceleration, which is the acceleration of the falling motion due to the gravity, is constant regardless of the object

$$g \approx 9.8\ \mathrm{m/s^2}. \tag{2.9}$$

Then, according to Newton's law of motion (2.1), the gravitation $\boldsymbol{W}$ acting on an object is the product of the mass m of the object and the vertically downward gravitational acceleration $\boldsymbol{g}$, or $m\boldsymbol{g}$,

$$\boldsymbol{W} = m\boldsymbol{g} \tag{2.10}$$

(Fig. 2-13)*. That is, the magnitude of gravity acting on an object is proportional to the mass. **Mass** represents the magnitude of inertia, but it can cause gravity at the same time. The international unit of mass is kg.

An expression "a force of magnitude 1 N" is rather difficult to understand intuitively. For this reason, the magnitude of gravity acting on an object with a mass of 1 kg was called 1 **kilogram-force** (symbol kgf) or 1 **kilogram-weight** (symbol kgw) and was used as a practical unit of force. Since $g \approx 10\ \mathrm{m/s^2}$, 1 kgf is about 10 N. 1 N is about 0.1 kgf, that is about 100 gf. The unit of kilogram-force is easy to understand but the Earth's gravity has slight differences depending on places, so we cannot use this when high accuracy is required. Therefore, in the field of engineering, kilogram-force has been defined as

$$1\ \mathrm{kgf} = 9.80665\ \mathrm{N}. \tag{2.11}$$

The magnitude of gravity differs from place to place, but the magnitude of inertia is a quantity characteristic to each object and does not depend on location. For example, on the lunar surface, the gravitational acceleration is about 1/6 of that on the Earth, so the magnitude of gravity acting on an iron ball is about 1/6 of that on the Earth. However, the magnitude of force ($F = ma$) required to stop the iron ball (mass m) rolling on the moon at the same acceleration $\boldsymbol{a}$ as on the Earth is the same magnitude as on the Earth.

Gravity acts on each part of the object spreading in space. However, in the case of rigid object, the resultant force can be considered to act on a point called the center of gravity (will be discussed in Chapter 7).

The gravity mg of the Earth acting on an object of mass m placed near the Earth surface is the resultant force of gravity which each part of the earth with radius R_E and mass M_E acts on the object. Therefore, when the total mass M_E of the Earth with the radius R_E is in the center of the Earth, mg is equal to the gravity $G\dfrac{mM_E}{{R_E}^2}$ acting on objects placed near the Earth surface (Fig. 2-15). It follows that

$$mg = G\frac{mM_E}{{R_E}^2} \quad \therefore \quad g = G\frac{M_E}{{R_E}^2}. \tag{2.12}$$

Substituting the value of 6.37×10^6 m for the radius R_E of the Earth and $9.8\ \mathrm{m/s^2}$ for the value of gravitational acceleration g into the relation $M_E = g{R_E}^2/G$ derived from the above equation, we can find that the mass M_E of the Earth is 6.0×10^{24} kg.

Artificial satellite Newton expected the possibility of artificial satellites. Newton wrote "When an object is launched horizontally from the top of a high mountain, the object will draw a parabola and fall to the ground if the launching speed is not enough. However, as the launching speed increases, since the Earth is a circle, the orbit of the object deviates from the parabola and follows the curves like B, C, and D shown in Fig. 2-16. If the launching speed is further increased, the object will rotate in a circular orbit around the Earth".

In this case, the centripetal force acting on the artificial satellite is the Earth's gravity mg and the equation of motion (2.5) is

$$m\frac{v^2}{r} = mg \tag{2.13}$$

(Fig. 2-17). From this equation, we derive $v^2 = rg$ and substitute the radius $R_E = 6370$ km of the Earth for r. Then, the speed v of the artificial satellite rotating the circular orbital just above the ground is

$$v = \sqrt{R_E g} = \sqrt{(6.37\times10^6\mathrm{m})\times(9.8\ \mathrm{m/s^2})} = 7.9\times10^3\ \mathrm{m/s}. \tag{2.14}$$

It follows that this artificial satellite rotates around the Earth at a speed of 7.9 km/s. The period T of rotation is

$$T = \frac{2\pi R_E}{v} = \frac{2\pi\times6.37\times10^6\ \mathrm{m}}{7.9\times10^3\ \mathrm{m/s}} = 5.06\times10^3\ \mathrm{s} = 84\ \mathrm{min}. \tag{2.15}$$

The period is 84 minutes. However, the artificial satellite rotating the circular orbit just above the ground is decelerated by the air resistance and will fall soon after.

Case 3 | Dark matter There are galaxies called spiral galaxies because collections of stars are distributed in a disk shape and appear to rotate and to swirl (Fig. 2-18). By observing the Doppler effect of the light radiated from each part of the galaxy, we can know the rotation speed $v(r)$ of a star or gas away from the center of the galaxy (Fig. 2-19). Assuming that the mass distribution of the galaxy is spherically symmetric and that the total mass inside the sphere with

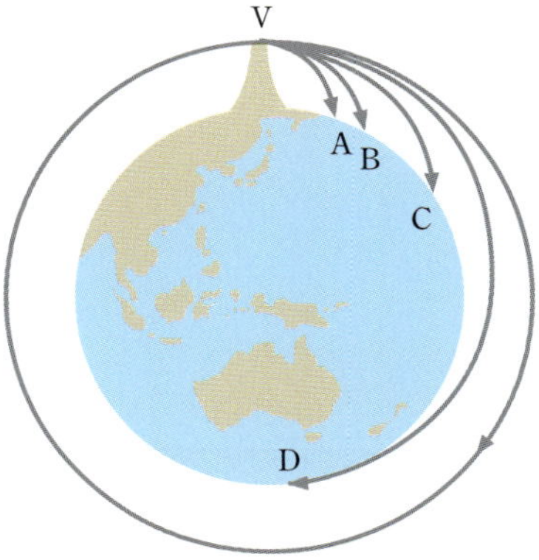

Fig. 2-16 Newton's expectations for the existence of artificial satellites

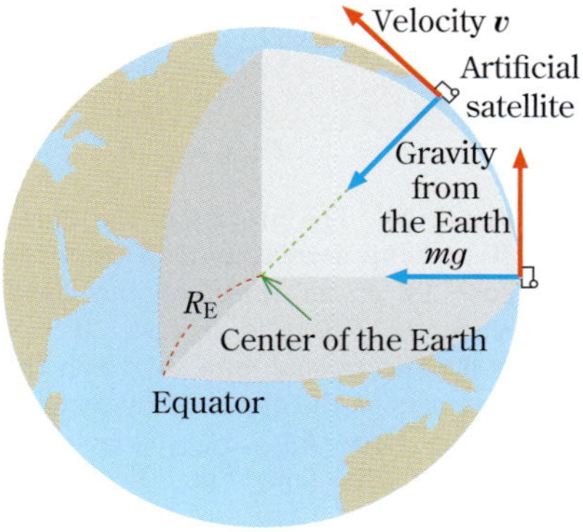

Fig. 2-17 The equation of motion for the artificial satellite rotating just above the Earth surface is $m\frac{v^2}{R_E} = mg$.

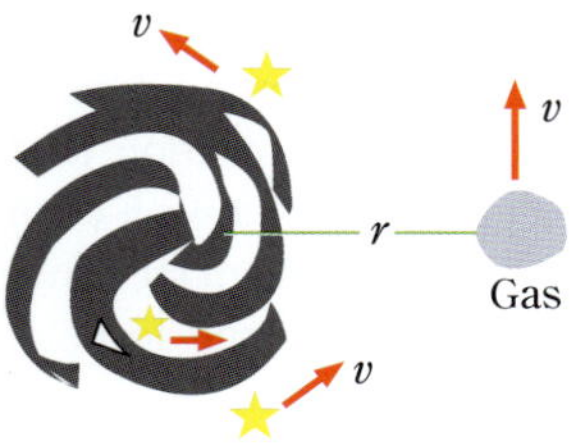

Fig. 2-18 The spiral galaxy

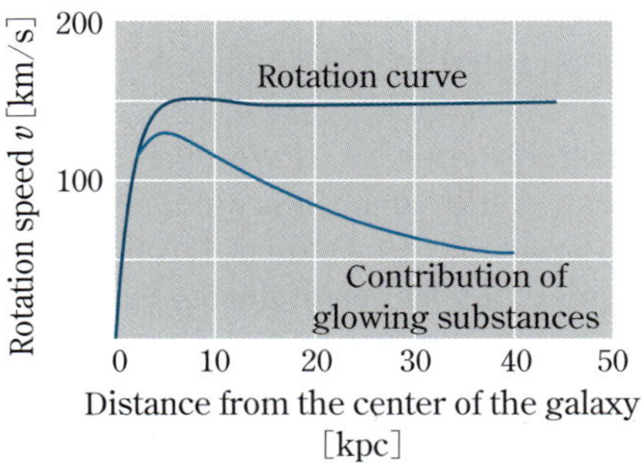

Fig. 2-19 Rotation curve of a spiral galaxy

its radius r from the center of galaxy is $M(r)$, the equation of motion of fixed stars and gas at distance r from the center of galaxy is

$$\frac{mv(r)^2}{r} = G\frac{mM(r)}{r^2}. \qquad (2.16)$$ *1

*1 See equation (16.21) on page 225.

Therefore, the rotation speed $v(r)$ of stars or gas is

$$v(r) = \sqrt{\frac{GM(r)}{r}}. \qquad (2.17)$$

According to observation results, however, $v(r)$ is almost constant, so that

$$M(r) = \frac{v(r)^2 r}{G} \propto r \qquad (2.18)$$

and it becomes clear that a large amount of substance having a mass also exists in the non-shiny part outside the spiral galaxy. This is called the **dark matter**. Currently it is estimated that ordinary substances such as atoms are responsible for 4 % of the energy of the universe, while the unknown substance of dark matter carries 24 %, and unknown energy of dark energy occupies 72 % of the energy of the universe.

Phenomenological force (force in macroscopic scale) Atoms which are the basic constituents of substance are composed of positively charged nucleus and negatively charged electrons. The force that acts between nucleus and electrons to create atoms, molecules, and crystals is the electric force. Therefore, the causes of friction force and normal force acting directly between two objects, elastic force of spring, adhesion of glue, resistance of air and water, etc. are basically electric forces acting between nucleus and electrons. That is, the force acting between objects which are aggregates of many atoms and molecules is the resultant force of electric forces acting between constituents of the object in a microscopic level. However, it is more convenient to treat the forces experienced in everyday life such as friction force, normal force, elasticity as macroscopic phenomenological forces. Study of the resistance of air and water will be treated in Chapter 3.

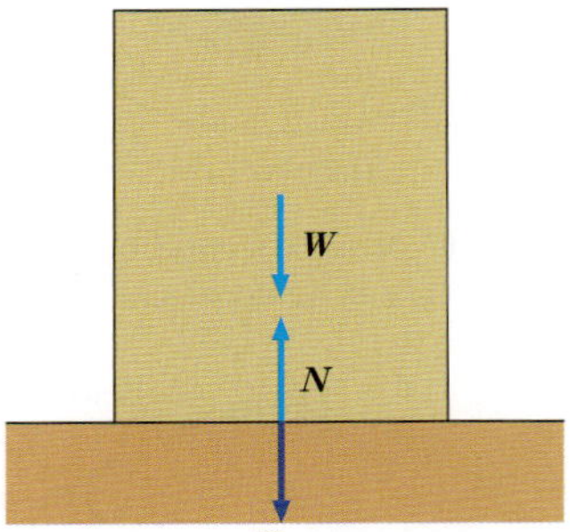

Fig. 2-20 The gravity of the Earth $\boldsymbol{W}$ and the normal force $\boldsymbol{N}$ from the floor act on an object on the floor. Since the object is at rest, $\boldsymbol{W}+\boldsymbol{N} = \boldsymbol{0}$. The normal force $\boldsymbol{N}$ which the floor acts on the object is the reaction of the force $\boldsymbol{F}_{floor\leftarrow object}$ which the object acts on the floor, so $\boldsymbol{N} = -\boldsymbol{F}_{floor\leftarrow object}$. Accordingly, $\boldsymbol{F}_{floor\leftarrow object} = -\boldsymbol{N} = \boldsymbol{W}$. Therefore, we see that the force $\boldsymbol{F}_{floor\leftarrow object}$ of the object to push the floor is equal to the gravity $\boldsymbol{W}$ of the Earth acting on the object.

Normal force We can stand on the ground, but we cannot stand on the surface of the water or the mire. The reason is that the ground can act to balance the gravity of the Earth acting on us, but the water surface and mire cannot. When two objects (solid) are in contact, the force acting on the other object vertically through the contact surface is called the **normal force** (Fig. 2-20). In this book, the normal force is represented by the symbol $\boldsymbol{N}$*2.

*2 Pay attention not to confuse $\boldsymbol{N}$ with the unit of force N described in roman letter.

Static friction If we push an object on the floor with a horizontal force $\boldsymbol{f}$, the object does not move while the force $\boldsymbol{f}$ is small. This is because the floor exerts the force $\boldsymbol{F}$ on the object in a direction that hinders the motion of the object. When two objects in contact with each other exert mutually the force in parallel with the contact surface in a direction that hinders the movement of the other object, the force is called the **friction force**. When there is no difference in the speed of two objects, such as in the case of the object kept at rest on the floor, the friction force is called the **static friction force**. Since the object remains at rest, from the condition of the balance of forces acting on the object in the horizontal direction, the force $\boldsymbol{f}$ which a person exerts on the object and the static friction force $\boldsymbol{F}$ which the floor exerts on the object are equal in magnitude and opposite in direction (Fig. 2-21). Therefore, $\boldsymbol{F} = -\boldsymbol{f}$.

When the force $\boldsymbol{f}$ pressing the object is increased to a certain limiting magnitude of F_{max} or more, the object starts to move and the static friction force changes into the kinetic friction force. The magnitude F_{max} of the static friction force at this limit is called the **maximum friction force**. The magnitude of maximum friction force F_{max} is approximately proportional to the magnitude of normal force N

$$F_{\text{max}} = \mu N. \tag{2.19}$$

The proportional constant μ is called the **coefficient of static friction**. μ is a constant depending on such factors as material properties, roughness of the surface, wet-dry state of the contacting two objects, and presence or absence of lubricants, but it hardly changes with the contacting area.

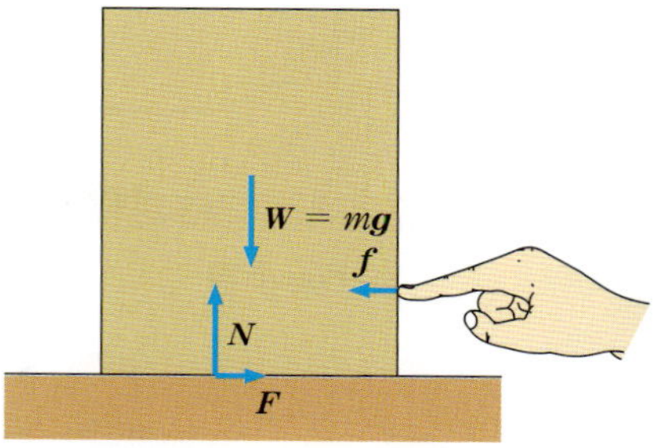

Fig. 2-21 Static friction $F \leqq \mu N$. Since the object is stationary, the magnitude of the force f pressed by the hand and the magnitude of the static friction F are equal. $f = F$. The floor exerts the normal force on the entire surface contacting the object, but in this case the normal force on the left side is larger than the normal force on the right side, so the arrow of the normal force $\boldsymbol{N}$ is drawn on the left side a little apart from the midpoint (refer to Section **8.1**).

Kinetic friction force As in the case between the object moving on the floor and the floor, between two objects moving in different speeds, a friction force acts along the contact surface to reduce the difference in speed. This frictional force is called the **kinetic friction force** (Fig. 2-22). Experiments have shown that the magnitude F of kinetic friction force is also approximately proportional to the magnitude N of normal force.

$$F = \mu' N \tag{2.20}$$

The proportional constant μ' is called the **coefficient of kinetic friction**. μ' is a constant that depends on the type of two objects in contact, properties of the contacting material, and roughness of the surface, the wet-dry state, presence or absence of lubricants, etc., but it is almost irrelevant to the area of the contact surface or the sliding speed. In general the coefficient of kinetic friction μ' is smaller than the coefficient of static friction μ.

$$\mu > \mu' > 0. \tag{2.21}$$

Examples of coefficients of friction are given in Table 2-1.

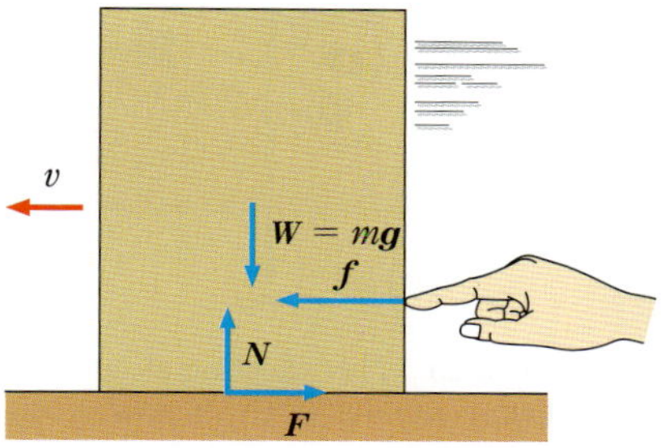

Fig. 2-22 Kinetic friction $F = \mu' N$

Table 2-1 Examples of coefficients of friction

Object I	Object II	Coefficients of static friction	Coefficients of kinetic friction
Steel	Steel	0.8	0.4
Wood	Wood	0.6	0.5
Wood	Wet wood	0.4	0.2
Rubber	Wood	0.7	0.5
Glass	Glass	0.9	0.4
Cupper	Glass	0.7	0.5

In the case where the object I is at rest or in motion on the object II.
Values obtained by rounding off to the nearest tenth the values listed in "Physics IB" published in 1995, Jikkyo Shuppan

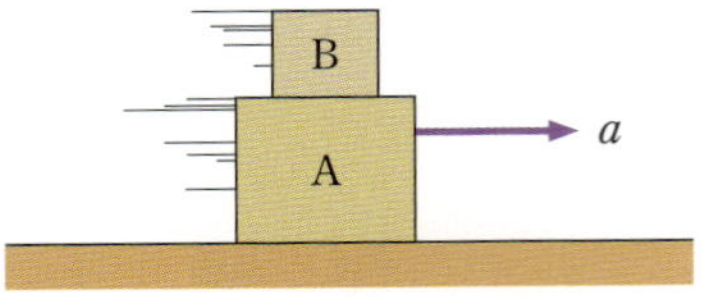

Fig. 2-23

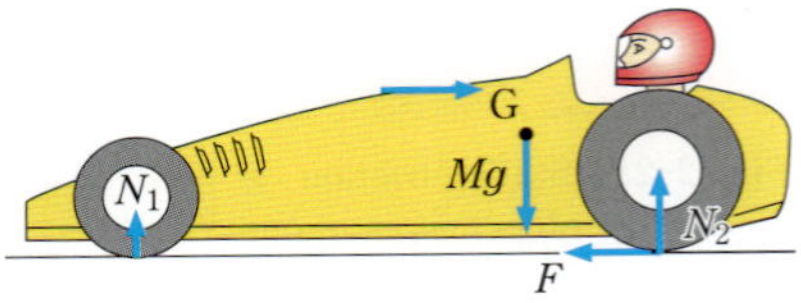

Fig. 2-24

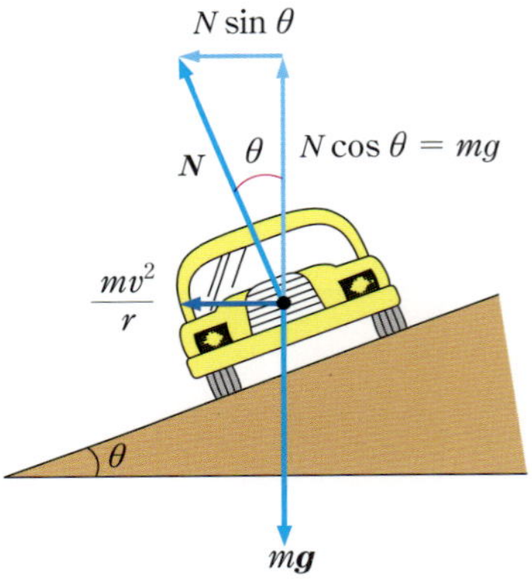

Fig. 2-25 The horizontal component of normal force $N\sin\theta$ is the centripetal force $\frac{mv^2}{r}$.

Fig. 2-26 Curve-turning cars.

Question 6 In Fig. 2-23, the mass of object A is 25 kg, and the mass of object B is 10 kg. These two objects are moving to the right under the condition of uniform acceleration $a = 2\ \mathrm{m/s^2}$. The coefficient of static friction between the two objects is $\mu_s = 0.8$, Find the static friction force between two objects.

Cars and friction force Driving force of cars to move forward comes from the power of engine. But, according to Newton's laws of motion, a car at rest can move forward when the force directed from the outside towards the car acts on it. This forward force is the friction force induced by the work of the engine and it is the force that the road exerts on the tires of driving wheels (Fig. 2-24). Evidence of this claim is as follows : The friction does not work between the frozen road surface and the car tires in a snow country. In this case even if we step on the accelerator pedal after starting the engine, the wheels will only spin and the car will not run forward.

When we apply the brakes to stop a moving car, the force to stop the car is the backward frictional force acting on the tires induced by the action of the brakes. When the car turns a curve, the centripetal force acts on it. This centripetal force is also a frictional force which the road surface exerts laterally on the tires. In the express way curves, the inner side road surface is made to be lower. This is because the normal force which the road surface exerts on the car has the horizontal component to reduce the necessary magnitude of the friction force facing the center direction for turning so that the risk of slipping in the lateral direction will be reduced.

Let us find the angle θ of inclination of the road surface in order that the friction force will be 0 for a car running at a speed of 72 km/h (20 m/s) on a curve with a radius of 100 m (Fig. 2-25). The centripetal acceleration $a = \frac{v^2}{r}$ is in this case

$$a = \frac{v^2}{r} = \frac{(20\ \mathrm{m/s})^2}{100\ \mathrm{m}} = 4\ \mathrm{m/s^2}. \tag{2.22}$$

From the condition of balance of the force in the vertical direction, the vertical component of gravity $\boldsymbol{mg}$ and that of normal force $\boldsymbol{N}$ should be balanced, namely the relation $N\cos\theta = mg$ holds. The condition that the horizontal component $N\sin\theta$ of the normal force $\boldsymbol{N}$ is equal to the centripetal force of the circular motion

$$m\frac{v^2}{r} = N\sin\theta = mg\tan\theta \tag{2.23}$$

is required so that the friction force becomes 0. Therefore, when the gravitational acceleration g multiplied by $\tan\theta$ is equal to the centripetal acceleration $\frac{v^2}{r}$, that is, when

$$\frac{v^2}{r} = g\tan\theta \qquad \therefore \quad \tan\theta = \frac{v^2}{rg} = 0.4 \tag{2.24}$$

the friction force becomes 0. The angle θ of inclination of the road surface is $\theta = 22°$.

Problems for exercise 2

A

1. A force acts on an object with a mass of 30 kg, and the object is moving at an acceleration of 4 m/s^2. Find the force acting on the object.
2. How much force should be added on average to stop in 6 seconds an object with a mass of 20 kg running at a speed of 30 m/s on a straight line?
3. What is the acceleration when a force of 12 N acts on a 2 kg object?
4. Find the speed of the object when a force of 20 N acts for 3 seconds on a stationary object with a mass of 2 kg.
5. A 1000 kg car running on a straight road was uniformly accelerated from 20 m/s to 30 m/s in 5 seconds.
   (1) What is the acceleration of the car while it is being accelerated?
   (2) Find the magnitude of force which was applied at this time.
6. Passengers feel a large impact in the lateral direction when the train departing from the platform ① shown in Fig. 1 passes the switching point P. Explain the reason.

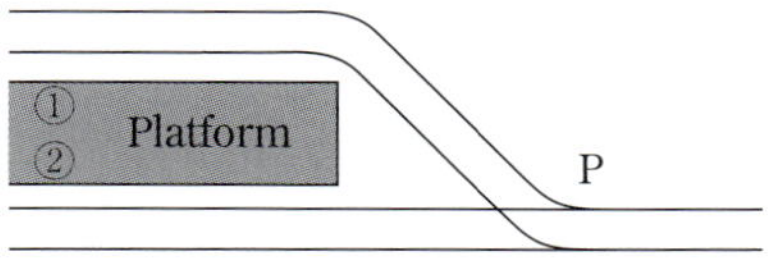

Fig. 1

7. In Figs. 2 (a) and (b), which cart is moving faster? In (a) you continue drawing a 400 g cart horizontally with a constant force so that the spring balance will indicates 100 g. In (b) you tie a 400 g cart to a 100 g weight with a string put on a pulley, and then gently release your hand which holds the weight.

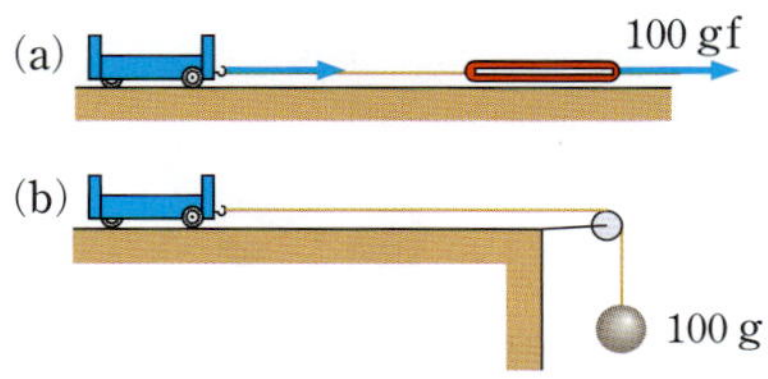

Fig. 2

8. An elevator of mass M is lifting a person with a mass m and receives a tension T from the rope. Find the acceleration.

B

1. Is the tension S acting on the string connecting the objects of mass m_A and m_B in Fig. 3 greater or less than the gravity $m_A g$ acting on the falling object A? Will the tension S become larger or smaller as m_B becomes larger?

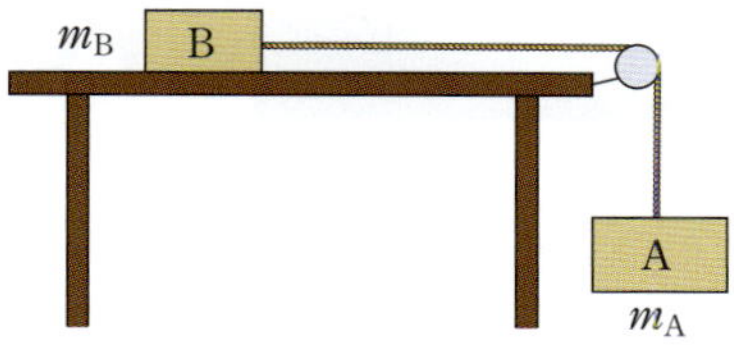

Fig. 3

2. Show that the magnitude F of the universal gravitation acting between two gold balls with each mass 1 kg is 2.7×10^{-8} N when the centers of two balls are separated by 5 cm. The radius of 1 kg of gold ball is 2.31 cm.
3. As shown in Fig. 4, the sled is pulled with a rope at an angle of 30° from the horizontal plane. The coefficient of static friction between the sled and the ground is 0.25 and the sum of the mass of the passenger and the sleigh is 60 kg. Find the magnitude F (in kgf) of the tension of the rope when the sled can start to move.

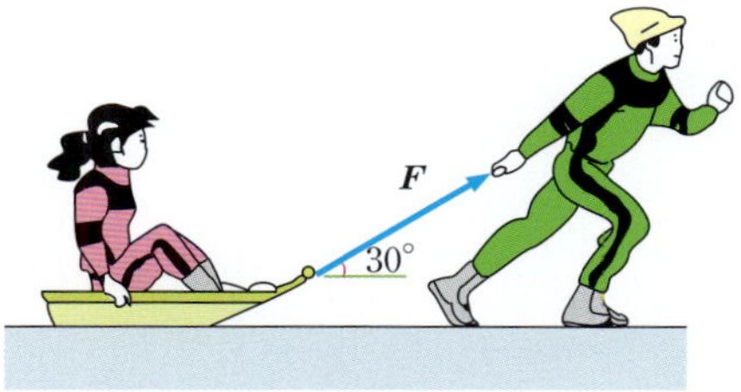

Fig. 4

4. In Fig. 5, find the magnitude F of the force required to make a 20 kg cuboid start to move.

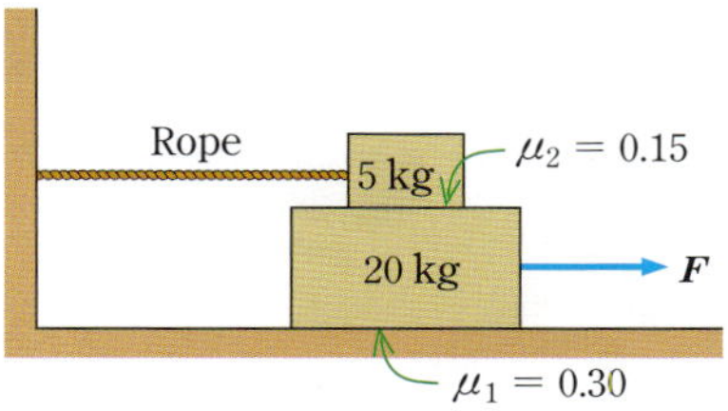

Fig. 5

Force and Motion

In order to apply the laws of motion to the motion of objects, it is necessary to know the laws of force acting on the objects. If the laws of force are discovered, substituting them into equations of motion and solving the equations of motion as differential equations, we can know the motion of the object of mass m on which the force $\boldsymbol{F}$ acts.

In this chapter, you will learn the method to solve simple differential equations and apply it to the parabolic motion and the falling motion in the presence of viscous drag. You will also learn the relationship between changes in momentum and impulses. The momentum is a quantity standing for the intensity of the motion of object.

3.1 Differential equation and integration

Learning objective To fully understand the definition of indefinite integrals and definite integrals in preparation for solving equations of motion which are differential equations. Using the fundamental theorem of calculus, to derive the impulse-momentum relation, relationship between velocity and displacement, and the relationship between acceleration and change in velocity

Differential equation An equation containing an unknown derived function (derivative) is called a **differential equation**. The order of the highest order derivative contained in the differential equation is called the rank of the differential equation. The equation of motion

$$m\frac{\mathrm{d}^2\boldsymbol{r}}{\mathrm{d}t^2} = \boldsymbol{F} \tag{3.1}$$

is a second order differential equation since it contains a second order derivative of the position vector $\boldsymbol{r}(t)$ with respect to time t. To find a function satisfying the differential equation is referred to as **solving** the differential equation, and the function found is called a **solution**. The solution $\boldsymbol{r}(t)$ of the differential equation (3.1) represents the motion of an object of mass m on which the force $\boldsymbol{F}$ acts.

Integration Before solving the differential equation (3.1), let us review the integration. Integration is an inverse operation of differentiation. The velocity $v(t)$ of an object which is moving linearly can be obtained by differentiating the position $x(t)$ of the object. On the contrary, if we integrate the velocity $v(t)$ of the object, we can find the position $x(t)$. The integral consists of indefinite integral and definite integral.

Indefinite integral The function that becomes $f(t)$ when differentiated is called the **primitive function** of $f(t)$. Therefore, if

$$\frac{\mathrm{d}}{\mathrm{d}t}F(t) = f(t) \tag{3.2}$$

then, the function $F(t)$ is a primitive function of $f(t)$. Since $F(t)+C$ obtained by adding an arbitrary constant C to $F(t)$ also leads to $f(t)$ through differentiation, $F(t)+C$ is also a primitive function of the function $f(t)$. Therefore, a myriad of primitive functions of the function $f(t)$ are collectively referred to as an **indefinite integral** of $f(t)$ and expressed as

$$\int f(t)\,\mathrm{d}t \qquad \text{or} \qquad \int \mathrm{d}t\, f(t). \tag{3.3}$$

Therefore,

$$\int f(t)\,\mathrm{d}t = F(t)+C \qquad (C \text{ is an arbitrary constant}). \tag{3.4}$$

The indefinite integral of the function used in this book is shown in Appendix (p. 38).

Case 1 $$\frac{\mathrm{d}}{\mathrm{d}t}\left(\frac{1}{2}at^2+bt+c\right) = at+b \qquad (a,\ b,\ \text{and } c \text{ are constants}) \tag{3.5}$$

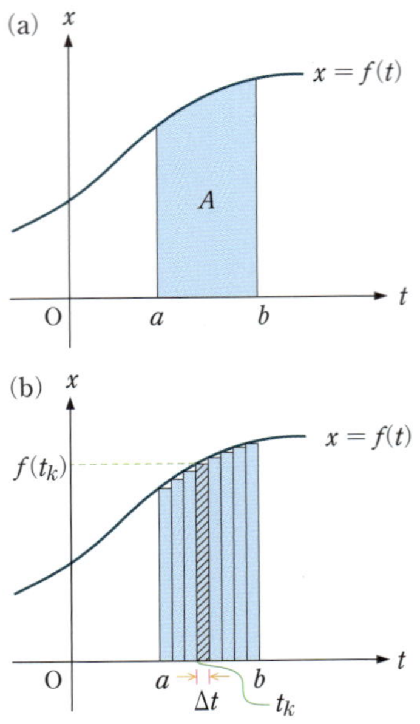

Fig. 3-1 (a) $A = \int_a^b f(t)\,dt$.
(b) The area of the shaded part is $f(t_k)\,\Delta t$. The area of the colored section is $\sum_{k=1}^{N} f(t_k)\Delta t$.

* $\sum_{k=1}^{N} A_k \equiv A_1+A_2+\cdots+A_N$
$A \equiv B$ means that A and B are equal by definition.

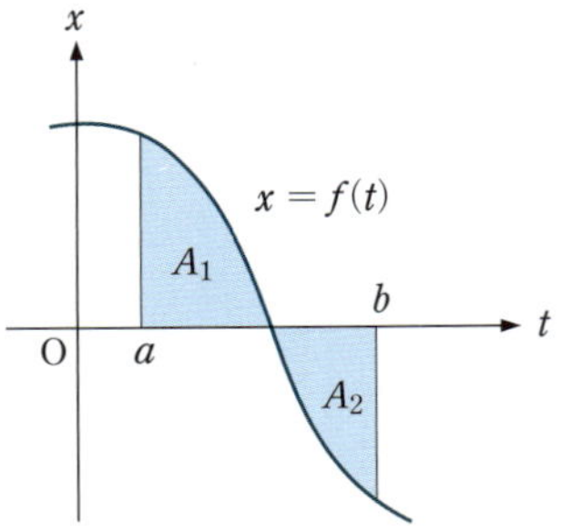

Fig. 3-2 $\int_a^b f(t)\,dt = A_1 - A_2$

Therefore,

$$\int (at+b)\mathrm{d}t = \frac{1}{2}at^2+bt+C \qquad (C \text{ is an arbitrary constant}). \quad (3.6)$$

Definite integral When we draw a graph of the function $x = f(t)$, the area A [the colored section in Fig. 3-1 (a)] surrounded by the curve $x = f(t)$ and the three straight lines $t = a$, $t = b$, and $x = 0$ (t-axis) is called the **definite integral** of the function $f(t)$ in the interval $[a, b]$. The definite integral is mathematically defined as follows :

The interval $[a, b]$ is divided into N equal parts, and this colored section is divided into N elongated parts of width $\Delta t = \frac{b-a}{N}$ [Fig. 3-1 (b)]. If the t coordinate of the left end of the k-th ($k = 1, 2, \cdots, N$, $t_1 = a$, $t_N+\Delta t = b$) part from the left is t_k, then the area of the k-th part from the left is approximated by the area of the rectangle with width Δt and height $f(t_k)$, that is $f(t_k)\,\Delta t$. When the number N of elongated rectangles is increased indefinitely and the width Δt is made as small as possible, the sum of the area of the elongated rectangle becomes as close as possible to the colored section in Fig. 3-1 (a). Therefore, the definite integral of the function $f(t)$ over the interval $[a, b]$ is defined as*

$$\int_a^b f(t)\,\mathrm{d}t \equiv \lim_{N\to\infty} \sum_{k=1}^{N} f(t_k)\,\Delta t. \qquad (3.7)$$

However, in this definition of definite integral, since the area of $f(t) < 0$ below the t-axis is negative, the definite integral in the case of Fig. 3-2 is $A_1 - A_2$.

Let one primitive function of $f(t)$ be $F(t)$, or $\frac{\mathrm{d}F}{\mathrm{d}t} = f(t)$. Then

$$\frac{F(t_k+\Delta t)-F(t_k)}{\Delta t} \approx f(t_k),$$

so

$$f(t_k)\,\Delta t \approx F(t_k+\Delta t)-F(t_k), \qquad (3.8)$$

Therefore,

$$\sum_{k=1}^{N} f(t_k)\,\Delta t \approx \sum_{k=1}^{N} [F(t_k+\Delta t)-F(t_k)] = F(b)-F(a) \qquad (3.9)$$

and the definite integral of the function $f(t)$ can be expressed using the indefinite integral $F(t)$ of $f(t)$ as

$$\int_a^b f(t)\,\mathrm{d}t = \int_a^b \frac{\mathrm{d}F}{\mathrm{d}t}\mathrm{d}t = F(b)-F(a) \equiv F(t)\big|_a^b. \qquad (3.10)$$

This is called the **fundamental theorem of calculus**. The following two relationships derived from this equation are important.

$$F(t) = \int_{t_0}^{t} \frac{\mathrm{d}F}{\mathrm{d}t}\mathrm{d}t + F(t_0) \qquad (3.11)$$

$$\text{If} \quad \frac{\mathrm{d}}{\mathrm{d}t}F(t) = 0, \qquad \text{then} \qquad F(t) = \text{constant}. \qquad (3.12)$$

Momentum As a quantity standing for the intensity of motion, the **momentum** (symbol $\boldsymbol{p}$) which is the product of mass m and velocity $\boldsymbol{v}$

$$\boldsymbol{p} = m\boldsymbol{v} \qquad (3.13)$$

is used. $\boldsymbol{p}$ is a vector having the direction of motion.

Since the mass m is constant, using the law of motion, we obtain

$$\frac{\mathrm{d}(m\boldsymbol{v})}{\mathrm{d}t} = m\frac{\mathrm{d}\boldsymbol{v}}{\mathrm{d}t} = \boldsymbol{F}.$$

$$\therefore \quad \frac{\mathrm{d}\boldsymbol{p}}{\mathrm{d}t} = \boldsymbol{F}. \tag{3.14}$$

Equation (3.14) shows "The time rate of change in momentum is equal to the force acting on the object". It follows that equation (3.14) is an alternative expression of the second law of motion.

Impulse-momentum relation Integration of equation (3.14) from time t_1 to t_2 leads to

$$\int_{t_1}^{t_2} \frac{\mathrm{d}\boldsymbol{p}}{\mathrm{d}t}\mathrm{d}t = \boldsymbol{p}(t_2)-\boldsymbol{p}(t_1) = \int_{t_1}^{t_2} \boldsymbol{F}\,\mathrm{d}t. \tag{3.15}$$

The integral on the right-hand side of this equation showing "product of force and duration of action of force" is called the **impulse**, so equation (3.15) which means

The change in momentum is equal to the impulse of force acting during that time,

is referred to as the **impulse-momentum relation**.

When a large force acts for a time $\Delta t = t_2 - t_1$ from time t_1 to t_2 and the momentum change $\Delta\boldsymbol{p} = m\cdot\Delta\boldsymbol{v}$ occurs, letting $\bar{F}$ be the average of the applied forces (Fig. 3-3), then we have the following relation :

$$\Delta\boldsymbol{p} = m\cdot\Delta\boldsymbol{v} = \bar{F}\Delta t. \tag{3.16}$$

It can be seen that the magnitude $\bar{F}$ of the force causing the same momentum change $\boldsymbol{p}(t_2)-\boldsymbol{p}(t_1) = \Delta\boldsymbol{p} = m\cdot\Delta\boldsymbol{v}$ is inversely proportional to the duration Δt during which the force acts.

Question 1 Show that when the constant force $\boldsymbol{F}$ acts during the time length $T = t_2 - t_1$, the impulse is given by

$$\int_{t_1}^{t_2} \boldsymbol{F}\,\mathrm{d}t = \boldsymbol{F}(t_2-t_1) = \boldsymbol{F}T. \tag{3.17}$$

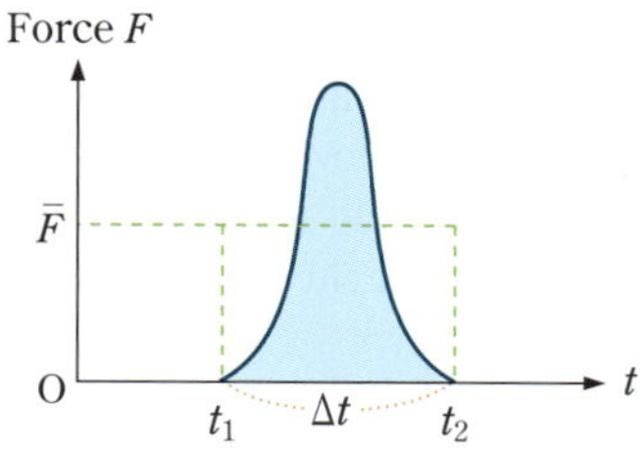

Fig. 3-3

A seat belt is a device that weakens the magnitude of applied force by lengthening the time during which the force acts to the human body.

Even in sports, the relationship between momentum change and impulse is utilized. Examples are found in a baseball game. It is necessary to keep adding as strong force as possible to a ball for the time length as long as possible, for a batter to fly the ball far and for a pitcher to throw a fast ball.

Question 2 When we jump off a tall platform or rack, if we land bending our knees, the impact on our body is reduced. Explain the reason for the decrease in impact on our body.

Fig. 3-4 A baseball batter

Example 1 A car running at 20 m/s speed hit the wall. A passenger in the front passenger seat not wearing a seat belt bumped his forehead against a front windshield and stopped after 0.0005 second.

(a) Assuming that the head mass m is 5 kg and the forehead and windshield contact area A is 4 cm^2, find the magnitude $\bar{F}$ of the mean force acting on the head, and the intensity $\bar{P}$ of force per unit contact area of the bumped forehead.

(b) When this 50 kg passenger is wearing a seat belt from the shoulder to the chest, he stops in 0.2 seconds. Find the average force $\bar{F}$ acting on the passenger in this case. Find the intensity of force $\bar{P}$ per unit area when the contact area between the seat belt and the body is 0.02 m^2.

Solution (a) Since the speed of the head immediately before bump is 20 m/s, and the speed immediately after bump is 0 m/s, we derive, from the impulse-momentum relation $\bar{F}\Delta t = \Delta p = m\cdot\Delta v$, the following results :

$$\bar{F} = \frac{m\cdot\Delta v}{\Delta t} = \frac{5\ \text{kg}\cdot 20\ \text{m/s}}{0.0005\ \text{s}} = 2\times10^5\ \text{N}$$

$$\bar{P} = \frac{\bar{F}}{A} = \frac{2\times10^5\ \text{N}}{4\ \text{cm}^2} = 5\times10^8\ \text{N/m}^2.$$

(b) $$\bar{F} = \frac{m\cdot\Delta v}{\Delta t} = \frac{50\ \text{kg}\cdot 20\ \text{m/s}}{0.2\ \text{s}} = 5000\ \text{N}$$

$$\bar{P} = \frac{\bar{F}}{A} = \frac{5000\ \text{N}}{0.02\ \text{m}^2} = 2.5\times10^5\ \text{N/m}^2.$$

As shown in the above results, when we wear the seat belt, the time during which momentum changes is prolonged and at the same time the area which the force acts on greatly increases, so the mean force acting on our body becomes weaker. All these effects lead to the drastic decrease in the magnitude of the force acting per unit area.

Example 2 A car with a mass of 1000 kg collided head on the wall at 72 km/h ($v_0 = 20$ m/s) and crashed, and was bounced back at a speed of $v = 3.0$ m/s (Fig. 3-5). Let the collision time length be 0.10 s. Find the mean $\bar{F}$ of the magnitude of the force the wall acted on the car for 0.1 s.

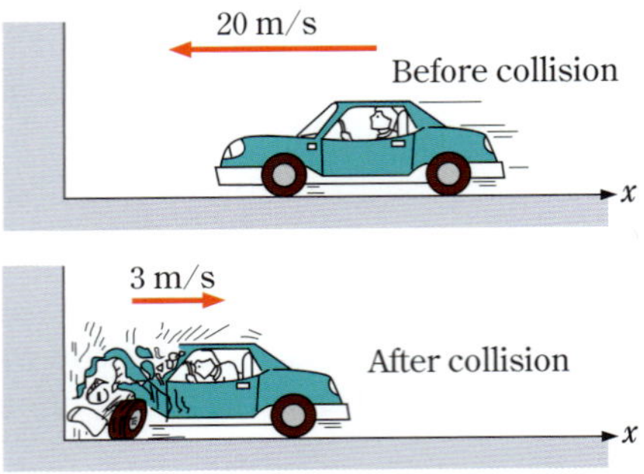

Fig. 3-5

Solution The momentum change Δp of the car is evaluated as follows :

just before collision

$$p_0 = mv_0 = (1000\ \text{kg})(-20\ \text{m/s}) = -2.0\times10^4\ \text{kg}\cdot\text{m/s}$$

immediately after collision

$$p = mv = (1000\ \text{kg})(3.0\ \text{m/s}) = 3.0\times10^3\ \text{kg}\cdot\text{m/s}.$$

Accordingly,

$$\Delta p = p - p_0 = [0.3\times10^4 - (-2.0\times10^4)]\ \text{kg}\cdot\text{m/s} = 2.3\times10^4\ \text{kg}\cdot\text{m/s}.$$

Therefore, the mean magnitude $\bar{F}$ of force that the wall acted on the car for 0.1 second is

$$\bar{F} = \frac{\Delta p}{\Delta t} = \frac{2.3\times10^4\ \text{kg}\cdot\text{m/s.}}{0.1\ \text{s}} = 2.3\times10^5\ \text{N}.$$

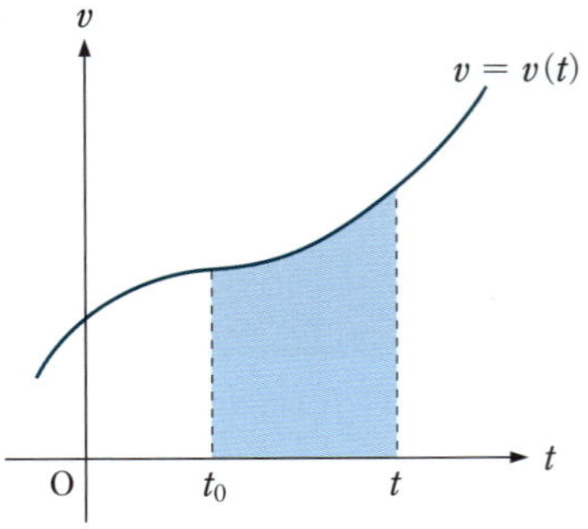

Fig. 3-6 $x(t)-x(t_0) = \int_{t_0}^{t} v(t)\,dt$

Relationship between velocity and displacement in linear motion

Let the position of an object moving on the x-axis be $x(t)$. Then, the velocity is $v(t) = \frac{dx(t)}{dt}$. When $x(t)$ is chosen as $F(t)$ in equation (3.11), the displacement $x(t)-x(t_0)$ of the object for time from t_0 to t is given as a definite integral of velocity $v(t)$

$$x(t)-x(t_0) = \int_{t_0}^{t} v(t)\,dt. \tag{3.18}$$

Therefore, the displacement is equal to the area of the colored section in the v-t graph of Fig. 3-6 with the vertical axis showing speed v and the horizontal one showing time t.

Displacement in the case of linear motion at a constant velocity

We express the constant velocity as v_0. The displacement of an object $x(t)-x(t_0)$ from t_0 to t can be obtained using equation (3.18) as follows

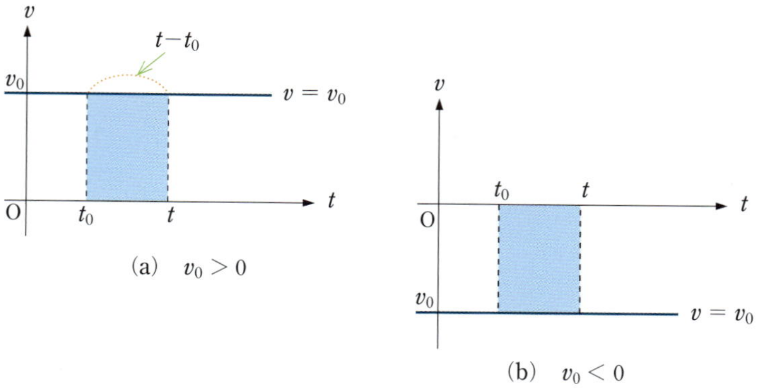

Fig. 3-7 v-t graph of linear motion at a constant speed.
(a) When $v_0 > 0$, the area of the ■ colored section is the displacement during the time t_0 to t.
$x(t) - x_0 = v_0(t - t_0) > 0$.
(b) When $v_0 < 0$,
$x(t) - x_0 = v_0(t - t_0) < 0$.
$|v_0|(t - t_0)$ is the displacement distance in $-x$ direction.

(Fig. 3-7):
$$x(t) - x(t_0) = \int_{t_0}^{t} v_0 \, dt = v_0 \big|_{t_0}^{t} = v_0(t - t_0), \quad (3.19)$$

In other words, "displacement" = "velocity" × "time". When $v_0 > 0$, it is a movement in the $+x$ direction and when $v_0 < 0$, it is a movement in the $-x$ direction.

> **Question 3** Using the relation between the definite integral and the area in the v-t graph, calculate the displacement from time $t = 0$ s to 150 s in the case shown in Fig. 3-8.

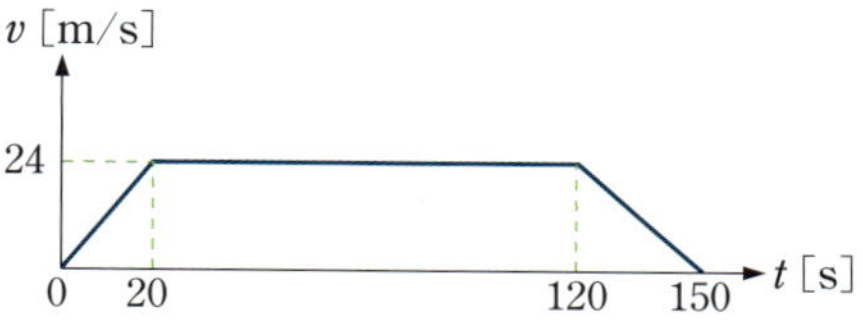

Fig. 3-8

Relation between acceleration and change in velocity in linear motion Let $v(t)$ be the velocity of an object moving on the x-axis. Then the acceleration is $a(t) = \dfrac{dv(t)}{dt}$. When $v(t)$ is chosen as $F(t)$ in equation (3.11), the change in velocity $v(t) - v(t_0)$ of the object for time from t_0 to t is given as a definite integral of acceleration $a(t)$.

$$v(t) - v(t_0) = \int_{t_0}^{t} a(t) \, dt. \quad (3.20)$$

The change in velocity is equal to the area of the ■ colored section in the a-t graph of Fig. 3-9.

3.2 Solving simple differential equations

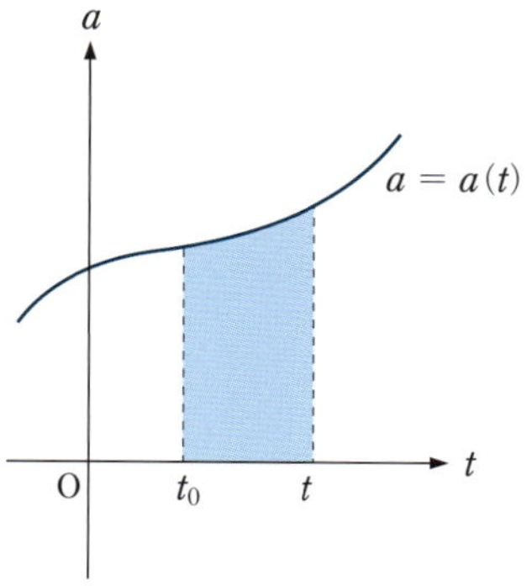

Fig. 3-9 $v(t) - v(t_0) = \int_{t_0}^{t} a(t) \, dt$

Learning objective Taking a parabolic motion and a falling motion in the presence of viscous drag as examples, to understand how to solve simple differential equations. In addition, to understand that given the initial condition, the motion of an object under the action of force is obtained as a solution to the equation of motion.

Uniformly accelerated linear motion When an object moves on the x-axis at a constant acceleration of a, that is when the relation

$$\frac{d^2x(t)}{dt^2} = \frac{dv(t)}{dt} = a = \text{constant} \quad (3.21)$$

holds, we, using equation (3.20), can derive from equation (3.21)

$$\frac{dx(t)}{dt} = v(t) = v_0 + \int_0^t a \, dt = at + v_0 \quad (v_0 = v(0)). \quad (3.22)$$

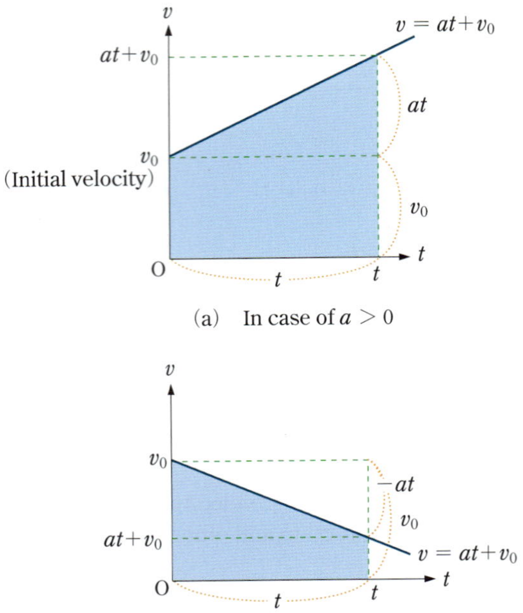

(a) In case of $a > 0$

(b) In case of $a < 0$

Fig. 3-10 Uniformly accelerated motion. The area of the colored section $v_0t+\frac{1}{2}at^2$ is the displacement $x(t)-x_0$.

Using this equation and equation (3.18), we obtain the relation

$$x(t) = \int_0^t (at+v_0)\,\mathrm{d}t+x_0 = \frac{1}{2}at^2+v_0t+x_0 \quad (x_0 = x(0)) \tag{3.23}$$

where v_0 and x_0 are the velocity $v(0)$ and position $x(0)$ of the object at $t = 0$, respectively.

Question 4 Using the relationship between the definite integral and the area of the v-t graph, derive the equation (3.23) for the displacement $x(t)-x_0$ in the case of $v(t) = at+v_0$. Refer to Fig. 3-10.

We transform equation (3.23) into the following form.

$$x(t)-x_0 = \frac{1}{2}t(at+2v_0) = \frac{1}{2}t\{v(t)+v_0\}.$$

We then insert $t = \dfrac{v(t)-v_0}{a}$ obtained from equation (3.22) into the third side of the above equation and obtain the relation

$$v(t)^2-{v_0}^2 = 2a\{x(t)-x_0\}. \tag{3.24}$$

When $v_0 = 0$ and $x_0 = 0$, equations (3.22)-(3.24) become

$$v(t) = at, \qquad x(t) = \frac{1}{2}at^2, \quad \text{and} \quad v(t)^2 = 2ax(t). \tag{3.25}$$

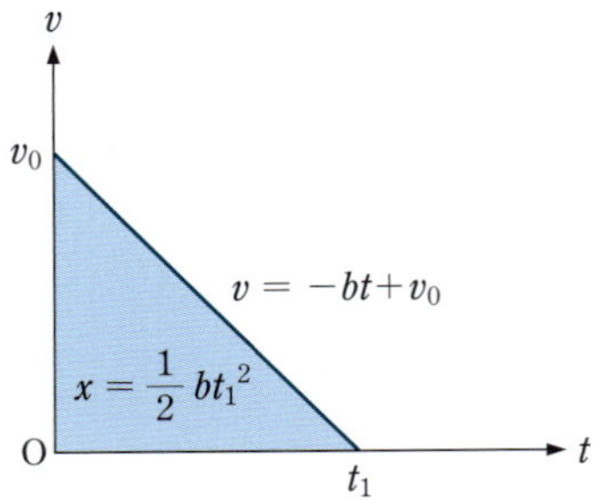

Fig. 3-11 Moving distance $x = \frac{1}{2}b{t_1}^2$

Question 5 The object with speed v_0 $(v_0 > 0)$ at $t = 0$ is decelerated uniformly at a constant acceleration of $-b$ $(b > 0)$ and stops at time t_1 after moving the distance x. Show that the following relationships hold in this case (Fig. 3-11).

$$v_0 = bt_1, \qquad x = \frac{1}{2}v_0 t_1 = \frac{1}{2}bt_1{}^2, \qquad v_0{}^2 = 2bx \qquad (3.26).$$

Question 6 As shown in left end diagram of Fig. 3-12, an object of mass m which is at rest at point A slides down the slope and then slides on the horizontal plane. Assuming that the section ABC is smooth and after passing the point C the uniform friction works, find the acceleration due to the force acting on the object and choose the most appropriate v-t graph for the object.

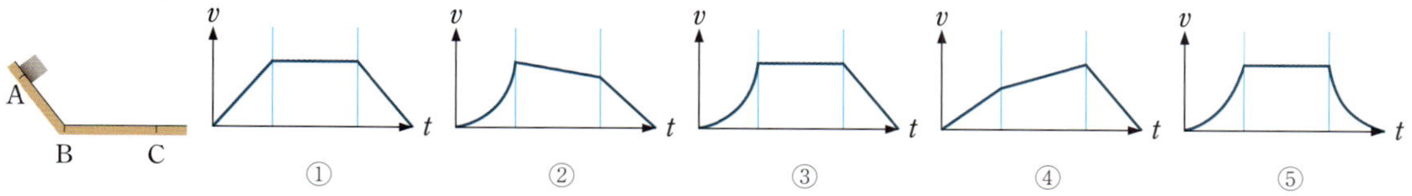

Fig. 3-12

Solutions of differential equations that contain the same number of independent arbitrary constants as the rank of differential equations are called **general solutions**. A function obtained by giving a specific value to an arbitrary constant of a general solution of a differential equation is a solution of the original differential equation and is called the **particular solution**. In physics, conditions that determine arbitrary constants of general solutions of differential equations are called **initial conditions** (or **boundary conditions**).

Equation (3.21) is a differential equation of rank two and given the object position $x(0) = x_0$ and velocity $v(0) = v_0$ at time $t = 0$, the later motion $x(t)$ of this object is completely determined. x_0 and v_0 are initial conditions. The relation that if initial conditions are given, the subsequent motion is completely determined is called the **causality** (causality is the principle that the result is always pre-determined when a cause is given).

Example 3 | Parabolic motion What kind of motion does the object do when we throw an object of mass m from a point on the ground in the direction of angle θ_0 ($0 < \theta_0 < 90°$) with respect to the horizontal plane at time $t = 0$ with the initial velocity v_0? The air resistance can be neglected.

Solution In order to solve the equation of motion, we first have to find the equation of motion. Therefore, it is necessary to select the coordinate axes. In selecting the coordinate axes, attention should be paid so that the coordinate system can make it easier to solve the equation of motion.

The horizontal plane is the xy plane and the $+z$-axis (the positive direction of the z-axis) is taken vertically upward, and the object is thrown in the $+x$-axis direction from the origin O (Fig. 3-13). The only force acting on this object is gravity $m\boldsymbol{g}$ directed in the $-z$ direction (the negative direction of the z-axis). When both sides of the equation of motion $m\boldsymbol{g} = m\boldsymbol{a}$ are divided by m, the acceleration of the object is

$$\frac{d^2x}{dt^2} = 0, \qquad \frac{d^2y}{dt^2} = 0, \qquad \frac{d^2z}{dt^2} = -g. \qquad (3.27)$$

In other words, the motion of this object is a superimposed motion of a constant velocity motion in the horizontal direction and a uniformly accelerated motion in the vertical direction.

Using equation (3.18), equation (3.20) and the relations

$$\int(-g)\,dt = -gt + \text{an arbitrary constant}$$

$$\int(-gt+v_{0z})\,dt = \int(-gt)\,dt + \int v_{0z}\,dt$$

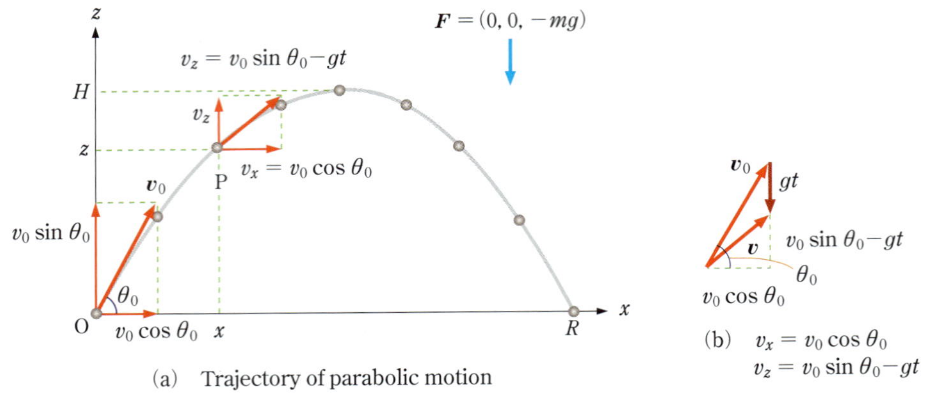

Fig. 3-13 Parabolic motion

$$= -\frac{1}{2}gt^2 + v_{0z}t$$

$+$an arbitrary constant,

we integrate both sides of equation (3.27) from $t=0$ to t and have the following set of equations :

$$v_x = \frac{dx}{dt} = v_{0x}, \qquad v_y = \frac{dy}{dt} = v_{0y},$$

$$v_z = \frac{dz}{dt} = -gt + v_{0z} \tag{3.28}$$

$$x = v_{0x}t + x_0, \qquad y = v_{0y}t + y_0$$

$$z = -\frac{1}{2}gt^2 + v_{0z}t + z_0. \tag{3.29}$$

The terms on the left hand side v_x, v_y, v_z, x, y, z are simplified expressions of the values of $v_x(t)$, $v_y(t)$, $v_z(t)$, $x(t)$, $y(t)$, $z(t)$ at time $t=0$ respectively. The arbitrary constants $\boldsymbol{r}_0 = (x_0, y_0, z_0)$, and $\boldsymbol{v}_0 = (v_{0x}, v_{0y}, v_{0z})$ are the position and velocity at $t=0$, so $\boldsymbol{r}_0 = (0, 0, 0)$, and $\boldsymbol{v}_0 = (v_0 \cos\theta_0, 0, v_0 \sin\theta_0)$. Thus, from equations (3.28) and (3.29), it follows that

$$v_x = v_0 \cos\theta_0, \qquad v_y = 0,$$

$$v_z = v_0 \sin\theta_0 - gt \tag{3.30}$$

$$x = v_0 t \cos\theta_0, \qquad y = 0,$$

$$z = -\frac{1}{2}gt^2 + v_0 t \sin\theta_0. \tag{3.31}$$

Substituting the relation $t = \dfrac{x}{v_0 \cos\theta_0}$ which is given by the first equation of equation (3.31) into the third equation, we obtain the equation describing the trajectory of the object

$$z = -\frac{gx^2}{2{v_0}^2 \cos^2\theta_0} + (\tan\theta_0)x. \tag{3.32}$$

This is an "upwardly convex parabola" in the xz plane [Fig. 3-13 (a)]. In this way, it has been demonstrated that the thrown object moves on a parabola.

For the flat ground we can find the falling point ($z=0$) by solving expression (3.32) at $z=0$ and selecting the solution different from $x=0$ (site from which the object is thrown). It follows

$$x = R = \frac{2{v_0}^2 \sin\theta_0 \cos\theta_0}{g} = \frac{{v_0}^2 \sin 2\theta_0}{g}. \tag{3.33}$$

In deriving equation (3.33), the addition theorem of trigonometric functions

$$2\sin\theta_0 \cos\theta_0 = \sin 2\theta_0$$

is used. In the case where we throw the object at the same initial speed v_0, the distance R to the point of fall is the largest when $\sin 2\theta_0 = 1$, or $\theta_0 = 45°$, and the reaching distance is

$$R = \frac{{v_0}^2}{g} \qquad (\text{when } \theta_0 = 45°). \tag{3.34}$$

At time t_1 when the object reaches the highest point, the vertical component of speed $v_z = v_0 \sin\theta_0 - gt$ becomes 0. Thus

$$t_1 = \frac{v_0 \sin\theta_0}{g} \tag{3.35}$$

and the height H of the highest point is obtained by substituting t_1 of equation (3.35) into the third equation of equation (3.31) and we find

$$H = \frac{(v_0 \sin\theta_0)^2}{2g}. \tag{3.36}$$

Question 7 Show that in Example 3, the residence time of the object in the air is $\dfrac{2v_0 \sin\theta_0}{g}$.

Question 8 We throw a ball on the flat ground at the same speed and in the direction of angle θ and also $90° - \theta$ with respect to the ground. Show that the ball reaches the same distance in both cases (we ignore the air resistance). Are the residence times of the ball in the air the same?

Question 9 In order that the ball hit by the batter becomes a home run with a reaching distance of 100 m, how much minimum initial speed (in m/s unit) is needed for the ball?

Question 10 Till how much distance will we be able to update in maximum the world record of long jump?

Resistance of air and water When considering the parabolic motion in Example 3, we ignore the air resistance. However, in discussing the motion occurring around us, the air resistance cannot be ignored in many cases. Fall of raindrops and skydiving are such examples. The resistance of air and water means the force that acts in the direction to hinder the motion of objects in air or in water.

Fig. 3-14 Skydiving

Resistance of objects (solids) moving in liquids and gases is complicated. So long as the speed is small, the magnitude F of the resistance is proportional to the speed v,

$$F = bv \qquad (b \text{ is a constant}). \tag{3.37}$$

Resistance proportional to the speed v is called the **viscous drag**. It is due to the viscous force which will be treated in Section **11.4**.

The viscous drag acting on a spherical object of radius R is

$$F = 6\pi\eta R v. \tag{3.38}$$

This is called **Stokes' law**. η is called the viscosity and is a constant determined for each gas or liquid (see Section **11.4**).

When the speed v of an object moving in a liquid or gas of density ρ increases and a vortex is formed behind the moving object, the magnitude F of the resistance force which the moving object receives is proportional to the square of speed v and expressed as

$$F = \frac{1}{2} C\rho A v^2 \tag{3.39}$$

where A is the cross-sectional area of the moving object and the resistance coefficient C is about 0.5 for the sphere, and it gets smaller if the object is streamlined. The resistance expressed by (3.39) is called the **inertial resistance** (or pressure resistance) (see Section **11.4**). When an automobile runs at high speed, the resistance from the air is the inertial resistance.

Example 4 | Fall of raindrop Write the equation of motion of a raindrop of mass m falling vertically downward while receiving a resistance force bv proportional to the speed v in windless air, and discuss qualitatively the motion of raindrops. Air buoyancy can be ignored.

Solution We take the $+x$ direction vertically downward. Since the force acting on the raindrop is the vertical downward gravity mg and the vertical upward viscous drag bv, the resultant force is $F = mg - bv$ (Fig. 3-15). Therefore, the equation of motion is

$$m\frac{\mathrm{d}^2x}{\mathrm{d}t^2} = m\frac{\mathrm{d}v}{\mathrm{d}t} = mg - bv. \tag{3.40}$$

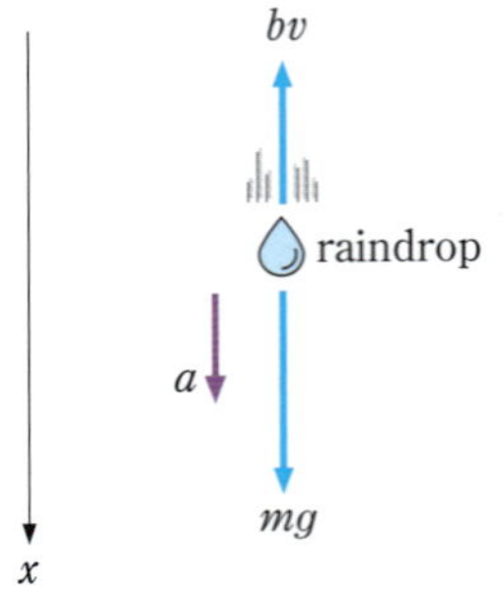

Fig. 3-15 Fall of raindrop $ma = mg - bv$

At the beginning of drop, the viscous drag can be neglected since the speed of raindrop is slow ($F = mg - bv \approx mg$). Therefore, the raindrop performs the uniformly accelerated linear motion due to the gravitational acceleration. Since the viscous drag increases as the raindrop speed v increases, the magnitude of the resultant force acting downward on the raindrop decreases and the acceleration also decreases. Finally when the speed v reaches

$$v_t = \frac{mv}{b} \tag{3.41}$$

the resultant force F acting on the raindrop becomes 0, so the raindrop will continue to fall at a constant speed v_t. This speed v_t is called the **terminal velocity**. The equation of motion (3.40) will be solved in the following Example 5. The reason why raindrops fall on the ground at a constant speed is because their falling speed has reached the terminal velocity.

Question 11 Several spherical objects are falling at the terminal velocity in water under the influence of viscous drag. If the sizes are the same, show that the terminal velocity is proportional to the mass of the object.

Question 12 Write the equation of motion of the object of a mass of m which falls vertically downward while receiving resistance force $\frac{1}{2}C\rho Av^2$ proportional to the square of speed v in windless air, and discuss qualitatively the motion of the object. Show that the terminal velocity of this object is given by

$$v_t = \sqrt{\frac{2mg}{C\rho A}}. \tag{3.42}$$

Fig. 3-16

Case 2 Using a lead weight and a set of paper cup used to put side dish into a lunch box, we can examine the relationship between the falling speed and the air resistance. We prepare four stages from one cup to four piled one (see Fig. 3-16) and drops these gently. We obtain from this experiment the v-t graph shown in Fig. 3-17.

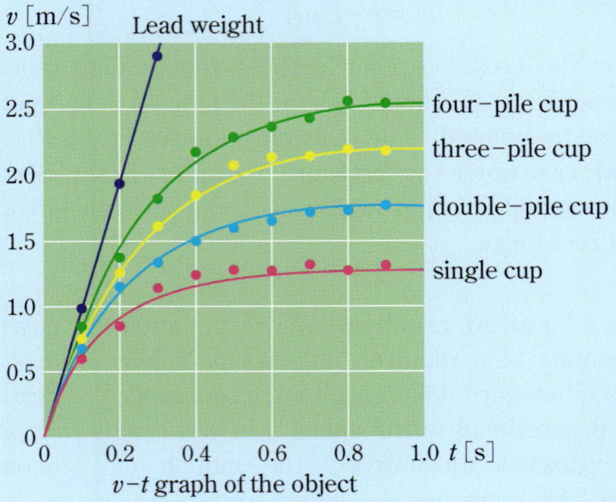

v-t graph of the object

Fig. 3-17*

* v [m/s] is the numerical part of velocity v expressed in m unit and t [s] is the numerical part of time t expressed in s unit. Refer to *1 on p. 5.

(1) Since the experimental data of lead weight is on the curve of $v = (9.8\,\mathrm{m/s^2})t = gt$, we can see that the weight performs the uniformly accelerated motion due to gravity.

(2) The cup also performs the uniformly accelerated motion due to gravity like the lead weight so long as the falling speed in the initial stage is slow enough and the air resistance is negligible, but it can be seen that with the increase in speed the resistance of air increases and the rate of increase decreases and eventually the air resistance and the gravity are balanced. In this state, the motion is performed at a constant terminal velocity

(3) The ratio of terminal velocity is $1 : \sqrt{2} : \sqrt{3} : \sqrt{4} = 1 : 1.41 : 1.73 : 2$, so it is roughly proportional to the square root of the number of sheets. Since the mass of the piled-cup is proportional to the number of sheets, this experimental result shows that the terminal velocity is proportional to the square root of the piled-cup mass. This fact indicates that the cup is subjected to inertial resistance [see (3.42)]. If the air resistance is the viscous drag, from the expression (3.41), the terminal speed is proportional to the piled-cup mass (number of sheets).

Separable differential equations Differential equations with the form

$$f(y)\frac{\mathrm{d}y}{\mathrm{d}x} = g(x) \tag{3.43}$$

are referred to as **separable differential equations**. The general solution of the above-type differential equation is obtained by transforming it into the form $f(y)\,\mathrm{d}y = g(x)\,\mathrm{d}x$ and then integrating both sides :

$$\int f(y)\,\mathrm{d}y = \int g(x)\,\mathrm{d}x. \tag{3.44}$$

($\frac{\mathrm{d}y}{\mathrm{d}x}$ can be taken as the quotient obtained by dividing $\mathrm{d}y$ by $\mathrm{d}x$).

Example 5 When a raindrop of mass m falls in air receiving the resistance force proportional to the speed, the speed v satisfies the differential equation

$$m\frac{\mathrm{d}v}{\mathrm{d}t} = mg - bv. \tag{3.40}$$

(Example 4). Assuming that the speed of raindrop at $t = 0$ is 0, solve equation (3.40). In addition, find the terminal velocity $v_\mathrm{t} = \lim_{t\to\infty} v(t)$.

Solution Since $\frac{\mathrm{d}v}{\mathrm{d}t}$ is Δv divided by Δt in the limit $\Delta t \to 0$, we set $\frac{\mathrm{d}v}{\mathrm{d}t} = \mathrm{d}v \div \mathrm{d}t$ and transform equation (3.40) into

$$\frac{\mathrm{d}v}{\frac{mg}{b} - v} = \frac{b}{m}\mathrm{d}t$$

and integrate the above equation

$$\int \frac{\mathrm{d}v}{\frac{mg}{b} - v} = \frac{b}{m}\int \mathrm{d}t.$$

Using that $-\log|A-v|$ is the primitive function of $\frac{1}{A-v}$, we can derive

$$-\log\left|\frac{mg}{b} - v\right| = \frac{b}{m}t + C$$

(C is an arbitrary constant) (3.45)

In this book, log mean the logarithm whose base is e (natural logarithm)*. The equalities $A = \mathrm{e}^B$ and $B = \log A$ are equivalent, so equation (3.45) is changed into

* In scientific electronic calculators and many Western text books of physics, the natural logarithm $\log_\mathrm{e} x$ is expressed as $\ln x$.

$$\left|\frac{mg}{b}-v(t)\right| = e^{-C}e^{-\frac{bt}{m}}. \tag{3.46}$$

Fall starts at $t = 0$. Accordingly $v(0) = 0$. At $t = 0$, equation (3.46) gives $e^{-C} = \frac{mg}{b}$, from which the constant C is determined. It follows that from equation (3.46) the speed $v(t)$ of the raindrop is shown to be

$$v(t) = \frac{dx}{dt} = \frac{mg}{b}(1-e^{-\frac{bt}{m}}) \tag{3.47}$$

(Fig. 3-18). At $t \to \infty$, $e^{-t} \to 0$, and we obtain

$$v_t = \lim_{t\to\infty} v(t) = \frac{mg}{b}. \tag{3.48}$$

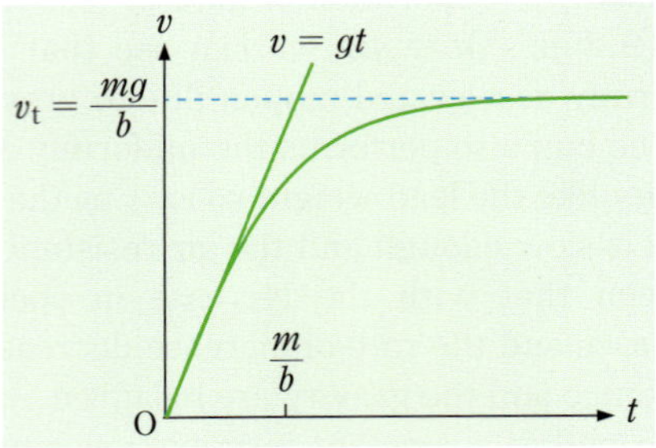

Fig. 3-18 Falling velocity v of raindrop and terminal velocity $v_t = \frac{mg}{b}$

Question 13 When $|bt| \ll 1$ holds, using the relation $e^{-bt} \approx 1-bt$, show $v \approx gt$.

Question 14 Using equation (3.47) and the relation between velocity and displacement given by equation (3.18), find the fall length $x(t)$.

Differential equations in the form of linear equation of an unknown function and its derivatives, such as

$$\frac{d^2x}{dt^2}+a\frac{dx}{dt}+bx = f(t) \qquad (a, b \text{ are constants}) \tag{3.49}$$

and

$$\frac{dv}{dt}+\frac{b}{m}v = g \qquad (b, m, \text{and } g \text{ are constants}) \tag{3.50}$$

are called the linear differential equations. The coefficients a, b, and $\frac{b}{m}$ are constants, and these differential equations are called differential equations at constant coefficients. Since these equations include $f(t)$ and g which are the terms that contain neither unknown functions nor their derivatives (inhomogeneous terms), these equations are called inhomogeneous equations.

The general solution of an inhomogeneous constant coefficient linear differential equation is the sum of one particular solution of the differential equation and the general solution of the homogeneous equation with the inhomogeneous terms being set to be zero. As an example, the procedure to obtain the general solution of inhomogeneous first order linear differential equation (3.50) is mentioned below : One particular solution of equation (3.50) is,

$$v = \frac{mg}{b} \tag{3.51}$$

(a solution expressing the uniform linear motion at terminal velocity). The equation in which inhomogeneous terms are set to be zero is

$$\frac{dv}{dt}+\frac{b}{m}v = 0 \tag{3.52}$$

and the general solution of equation (3.52) is

$$v = Ce^{-\frac{bt}{m}} \qquad (C \text{ is an arbitrary constant}). \tag{3.53}$$

Therefore, the general solution of equation (3.50) is the sum* of the above equation and equation (3.51), that is

$$v = \frac{mg}{b} + Ce^{-\frac{bt}{m}}. \qquad (3.54)$$

Equation (3.47) which is the solution of equation (3.40) is obtained by setting an arbitrary constant C of general solution (3.54) to be $-\frac{mg}{b}$.

* The relation $\frac{d}{dt}(Ce^{kt}) = k(Ce^{kt})$ is used.

Problems for exercise 3

A

1. Calculate the following indefinite integrals. Latin characters other than t are constants (let the arbitrary constant be C.)
   (1) $\int \left(at^2 + bt + c + \frac{d}{t} + \frac{e}{t^2}\right) dt$
   (2) $\int a \sin(\omega t + b)\, dt$
   (3) $\int a \cos(\omega t + b)\, dt$
2. An object was dropped from a height of 122.5 m. Find the time to reach the ground and the speed just before it reaches the ground. The air resistance can be ignored.
3. Galileo Galilei inferred that "In the uniformly accelerated motion with initial velocity 0, the fall at fixed time intervals increase in such a way as in the arithmetic progression of 1 : 3 : 5 : 7 : …". By checking carefully the arithmetic increase of the fall lengths of the ball falling down the slope, he confirmed the fact that the falling motion with the initial velocity of 0 is the uniformly accelerated motion. Explain his reasoning using Fig. 1.

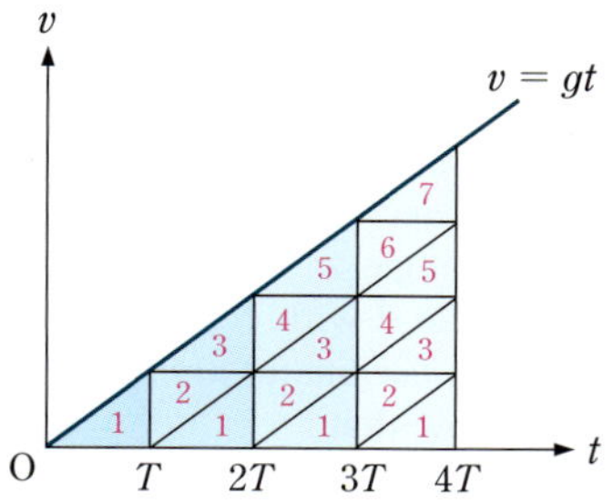

Fig. 1

4. (1) When we are driving a car, in such an emergency where a child jumps out forward, we must stop the car with a sudden braking. Assume that the time between when the driver of a car running at 50 km/h notices the danger and when he/she steps on the brakes is 0.5 s, and calculate the distance (free running distance) that the car runs in 0.5 s.
   (2) In some cars equipped with good performance brakes and tires, when we apply the brakes on such a running car, the car can decelerate at about 7 m/s^2. How long will the car move at 100 km/h run before it stops? The distance after stepping on the brakes is called the braking distance.
5. An object is moving in a uniformly accelerated linear motion of -10 m/s^2 in the x direction. The speed at time $t = 0$ was 20 m/s.
   (1) Find the equation expressing the speed at time t.
   (2) Find the travel distance and displacement from time $t = 0$ to $t = 5$ s.
6. When we performed a free-drop of metal ball from the roof onto the ground, the fall time was 3.0 seconds. Answer the following questions on the assumption that the air resistance is negligible.
   (1) The speed of the metal ball just before reaching the ground.
   (2) Roof height.
   (3) Mean speed of the falling metal ball.
7. An object moves parabolically in the trajectories a, b, and c shown in Fig. 2.

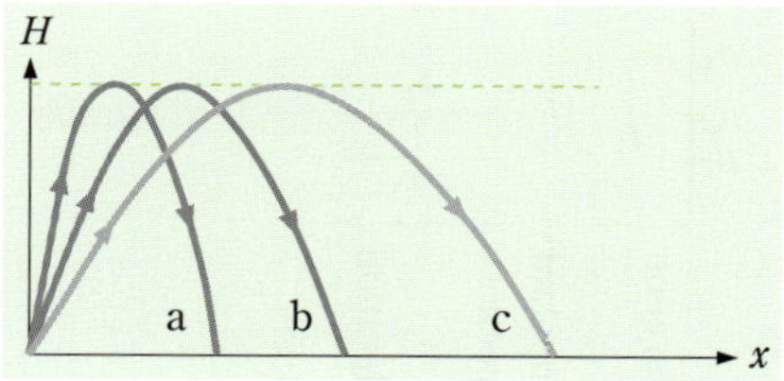

Fig. 2

(1) Compare the residence times in which the object is in the air.
(2) Compare the vertical components of the initial velocity.
(3) Compare the horizontal components of the initial velocity.
(4) Compare the magnitudes of the initial speed.

8. A person performed a service of tennis ball horizontally at a speed of 36 m/s from the height 2.5 m above the ground. The net is 12 meters apart from the servicing point and its height is 0.9 m. Does this ball go over the net? Find the distance to the point of fall of this ball.

B

1. The x-t, v-t, and a-t graphs of an object are given in Fig. 3.,
(1) Using v, t_1, t_2, and t_3, express the acceleration a_1 in the acceleration stage and the acceleration $-a_2$ in the deceleration stage.
(2) Using v, t_1, t_2, and t_3, express the positions x_1, x_2, and x_3.
(3) Show that when $t_2 < t < t_3$, the position x is given by

$$x = x_2+v(t-t_2)-\frac{a_2(t-t_2)^2}{2}.$$

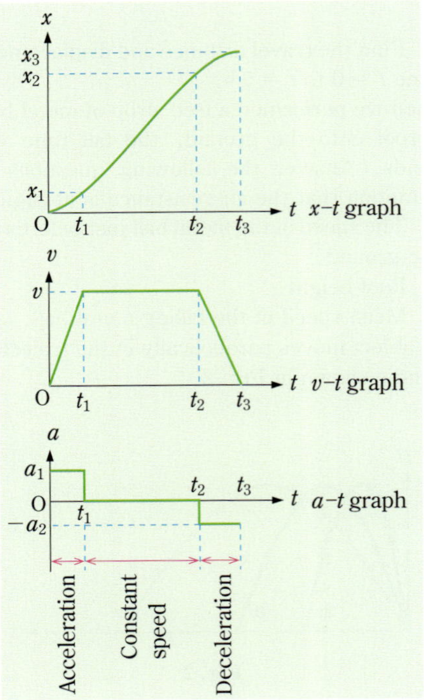

Fig. 3

2. An object with mass $m = 10$ kg is moving on the x-axis subject to a constant force F.
(1) Find the acceleration when the force of $F = 20$ N acts in the $+x$ direction.
(2) The force of $F = 10$ N acted from time $t = 0$ on the object which was at rest at the origin. Find the position x and velocity v at time $t = 10$ s.
(3) The force of $F = -20$ N is applied to the object whose position x_0 and velocity v_0 at time $t = 0$ are $x_0 = 0$ and $v_0 = 20$ m/s. Find the time such that the speed of the object becomes 0 and the distance x until that time.
(4) The speed at $t = 0$ was $v_0 = 20$ m/s and the speed at $t = 5$ s was $v = 40$ m/s. Find the magnitude F of the force acting on the object during this time.

3. The training vehicle for astronauts to withstand the large acceleration in the rocket, on which the astronauts boarded, was decelerated from a speed of 200 m/s and stopped after running the distance d. What is the minimum value of d in order that the acceleration exerted on the astronauts does not exceed 6 times the gravitational acceleration?

4. A man jumps down from the cliff of 3 m in height. Immediately after his feet touch the ground beneath the cliff, he tries to bend his knees so that the body as a whole is decelerated uniformly.
(1) What is the speed at the moment when his feet reach the ground?
(2) Find the force that the legs exert on the 40 kg body (including the arms and head) during deceleration. Let the length of the leg be 60 cm.

5. A person skydiving from the top of a high tower falls under the action of gravity and inertial resistance. Show that the falling speed is given by

$$v(t) = v_t\frac{1-\exp\left(-\dfrac{2gt}{v_t}\right)}{1+\exp\left(-\dfrac{2gt}{v_t}\right)}$$

where t is the time after the fall begins and v_t is the terminal velocity. Assume that $v(t = 0) = 0$.

6. We throw an object at an initial velocity v_0 at time $t = 0$ with an angle α with respect to the slope forming the angle θ with respect to the horizontal plane. If x and y coordinates are selected as shown in Fig. 4, then show that

$$x = -\frac{1}{2}gt^2 \sin\theta+v_0 t\cos\alpha$$

$$y = -\frac{1}{2}gt^2 \cos\theta+v_0 t\sin\alpha$$

In addition, show that the time T needed to reach the slope surface and the x coordinate X of the arrival point are given respectively by

$$T = \frac{2v_0 \sin \alpha}{g \cos \theta},$$

$$X = \frac{{v_0}^2}{g \cos^2 \theta}\{\sin(2\alpha+\theta) - \sin \theta\}.$$

Also, find the angle α at which the object reaches the furthest distance.

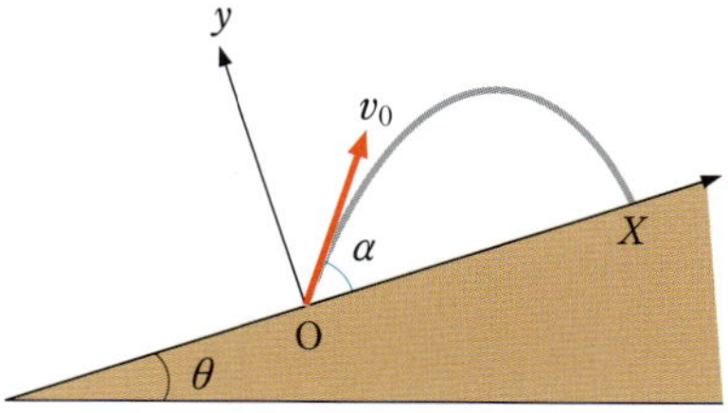

Fig. 4

7. The equation of motion of the object (mass m) receiving resistance $bv = m\beta v$ proportional to velocity is

$$m\frac{dv_x}{dt} = mg - bv_x$$

(+x-axis is vertically downward)

$$m\frac{dv_y}{dt} = -bv_x \qquad (y\text{-axis is horizontal})$$

(Fig. 5). Assume that at time $t = 0$, $x = y = 0$, $v_x = v_{x0}$, and $v_y = v_{y0}$. Then show that the solution of the equation of motion is

$$v_x = \left(v_{x0} - \frac{g}{\beta}\right)e^{-\beta t} + \frac{g}{\beta}, \qquad v_y = v_{y0}\,e^{-\beta t}$$

$$x = +\frac{g}{\beta}t + \frac{1}{\beta}\left(v_{x0} - \frac{g}{\beta}\right)\left(1 - e^{-\beta t}\right),$$

$$y = \frac{v_{y0}}{\beta}(1 - e^{-\beta t}).$$

Show that in the limit $t \to \infty$, $v_x \to v_{xt} = \dfrac{g}{\beta}$, and

$v_y \to v_t = \dfrac{v_{y0}}{\beta}$.

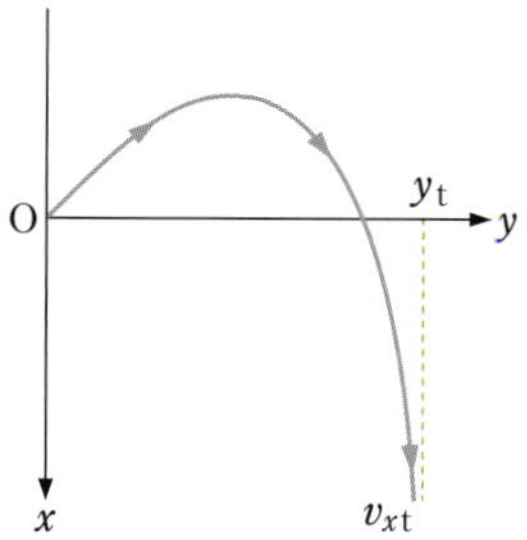

Fig. 5

Founder of physics Galileo (1564-1642)

The founder of physics is said to be Galileo. The reason is that Galileo was the first to conduct physics study based on the modern scientific research methods. Galileo created a telescope and discovered Jupiter's satellites and sunspots and observed their movement. He not only established the heliocentric theory but also conducted the research on the free fall with modern research procedures. That is to say, "first to establish a hypothesis by observing natural phenomena, and second, to make hypotheses, and then to make reasoning based on hypotheses to verify the results derived from the hypothesis through experiments and finally to find the natural laws", and he discovered "When the air resistance is negligible, any object falls at the same acceleration".

In the days of Galileo, ancient Aristotle's theory that "Every object falls at a constant speed (constant velocity) proportional to the weight immediately after it begins to fall" was believed. On the contrary, Galileo argued that "Free fall is a uniformly accelerated motion, and when iron balls of different weights are simultaneously dropped from the same height, the two iron balls reach the ground almost at the same time".

Galileo proved that the free fall of an iron ball is a uniformly accelerated motion, that is, the speed v of an iron ball increases in proportion to the fall time t, such as $v = gt$ (g is a proportional constant) based on the procedures as follows : The position of the iron ball at each time point can be measured with good precision from the stroboscopic photo-

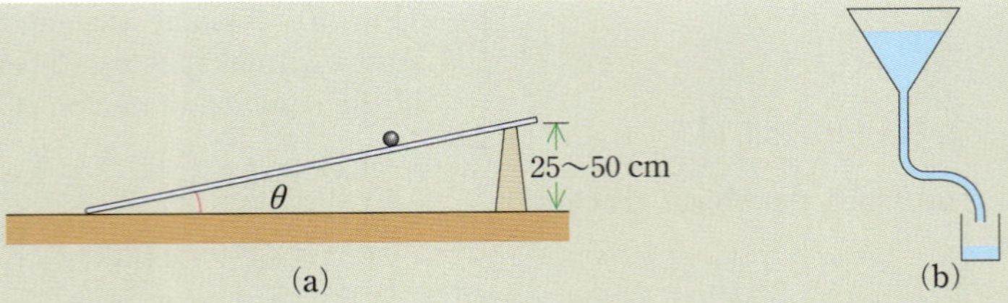

Fig. 3-A Galileo's experiments. (a) A groove was cut on a square timber having a length of about 6 m, and a metal ball was rolled on this groove. (b) In order to measure the time, he placed at a high place a large water-containing vessel and attached a thin tube to the bottom of this vessel so that the water flowing through the tube is ejected from the pipe to collecte this water in a cup. The weight of the water was measured with a precise balance.

graph of Fig. 3-A in Problems for exercise of Chapter 1, but the speed of the iron ball at each time cannot be estimated with good precision. When we try to examine the speed from the above-mentioned photographic picture, only the average speed for 1/30 second can be obtained from the position data at two time points. In other words. it is difficult to show based on the data of free fall of iron ball that the falling speed v follows the equation $v = gt$.

Meanwhile, when the fall speed is $v = gt$, since the average speed in free fall in time t is $\frac{1}{2}gt$, the relation "fall length" = "average speed" × "fall time" gives $\frac{1}{2}gt^2$ for fall length. Therefore, the fall lengths d's at time $t = T, 2T, 3T, 4T, \cdots$ are $d = \frac{1}{2}gT^2, 2gT^2, \frac{9}{2}gT^2, 8gT^2, \cdots$. Thus the fall lengths for each time interval T after the fall starts are $\frac{1}{2}gT^2$, $2gT^2 - \frac{1}{2}gT^2 = \frac{3}{2}gT^2$, $\frac{9}{2}gT^2 - 2gT^2 = \frac{5}{2}gT^2$, $8gT^2 - \frac{9}{2}gT^2 = \frac{7}{2}gT^2, \cdots$, so the ratio of magnitude becomes an arithmetic progression of $1:3:5:7:\cdots$. Galileo conducted his experiments based on the inference like this and found that the ratio of fall lengths at every fixed time becomes an arithmetic progression and confirmed the validity of his inference.

Free falls were too fast to observe at the time where stroboscopic photographs were unavailable. Thus, he carried out experiments in which metal balls fell down on sloped grooves on the squared timber (Fig. 3A). When the angle θ between the slope and the horizontal plane is small, the ball slowly falls down and it is easy to carry out experiments. Galileo confirmed that even if the inclination of the slope is changed, the ratio of the fall length at a certain fall time is an arithmetic progression. Galileo inferred that this result holds even when the angle θ is 90°.

He also confirmed that when balls of different weight roll down the same slope they fall at the same acceleration.

Although Galileo measured the speeds of the fall motion at each time, he failed to directly demonstrate that the fall motion is the uniformly accelerated motion. However, by confirming experimentally that the ratio of the fall length at a specified fall time is an arithmetic progression, he succeeded indirectly in proving that the fall motion is the uniformly accelerated motion.

Galileo thought that all objects would fall to the ground at the same time if the air resistance could be ignored. A vacuum pump was invented soon after the death of Galileo, and Boyle showed in 1660 that wing feathers and coins fall at the same speed in the vacuum vessel, demonstrating the correctness of Galileo's idea.

Oscillation

The word oscillation (vibration) is reminiscent of vibrations of spreading objects, such as strings of string instruments, membranes of drums and tuning forks. In this chapter, you will mainly learn the periodic motion called the simple harmonic oscillation which is caused by the action of restoring force proportional to the displacement (deviation) from the balancing location of the object (mass point) whose volume in space can be neglected like the case of the pendulum weight.

Unless energy is externally supplied, the amplitude of the pendulum will decrease. This is called damped oscillation. To make the pendulum continue oscillating, it is sufficient to apply the periodically oscillating force to the pendulum. When such an oscillating force is applied to the pendulum, the oscillation generated at the same frequency as the force is called forced vibration.

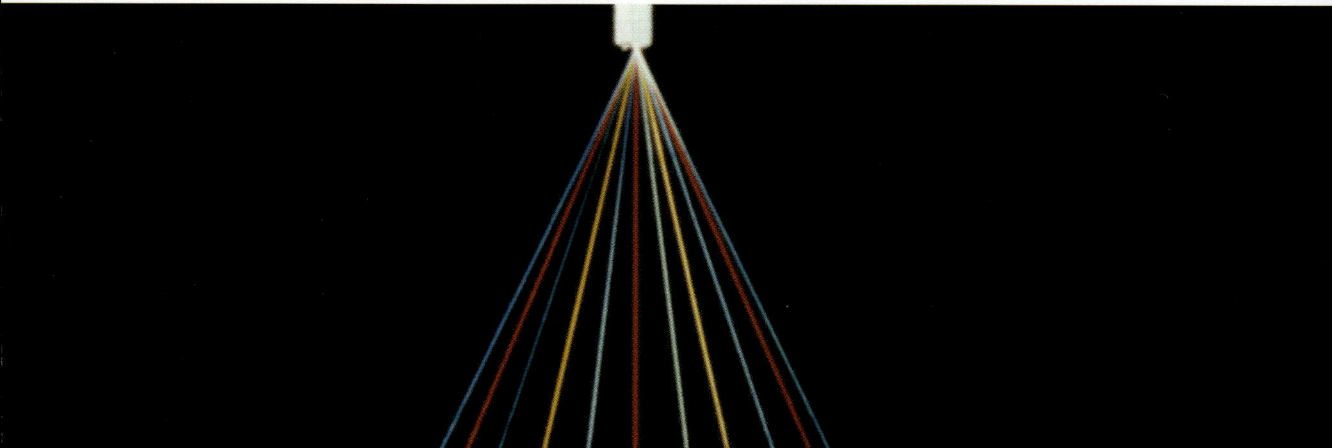

Fig. 4-1 A spring

4.1 Simple harmonic oscillation

Learning objective To understand the simple harmonic oscillations of spring pendulum and simple pendulum by solving the equation of motion as a differential equation.

Elastic force When a solid is deformed, a restoring force acts to cancel the deformation and take it back to its original state. When the magnitude of deformation (for example, stretching and shrinking of the spring) from the natural state where no external force acts is small, the magnitude of the restoring force is proportional to the magnitude of the deformation. This restoring property is called **Hooke's law,** and the restoring force is called the **elastic force**. When the elastic force is F, and the magnitude of deformation is x, Hooke's law is eapressed as

$$F = -kx. \tag{4.1}$$

A positive proportionality constant k is called the **elastic constant** (**spring constant** in the case of spring). The reason for adding a minus sign is to indicate that the direction of restoring force and the direction of deformation are opposite. For example, the restoring force will shrink the stretched spring and extend the shrunk spring. Detailed discussion on the elastic force will be presented in detail in Chapter 10.

Simple harmonic oscillation The oscillation caused by the restoring force which follows Hooke's law is called **simple harmonic oscillation**. Simple harmonic oscillation is the simplest oscillation. An example of simple harmonic oscillation is shown below. As shown in Fig. 4-2 (a), one end of a spring is fixed and the spring is hung vertically. When a weight of mass m is attached to the lower end of the spring, the length of spring stretches by x_0 and the downward gravity mg and the upward elastic force $f = kx_0$ are in balance ($mg = kx_0$). Therefore, the elongation of the spring x_0 is

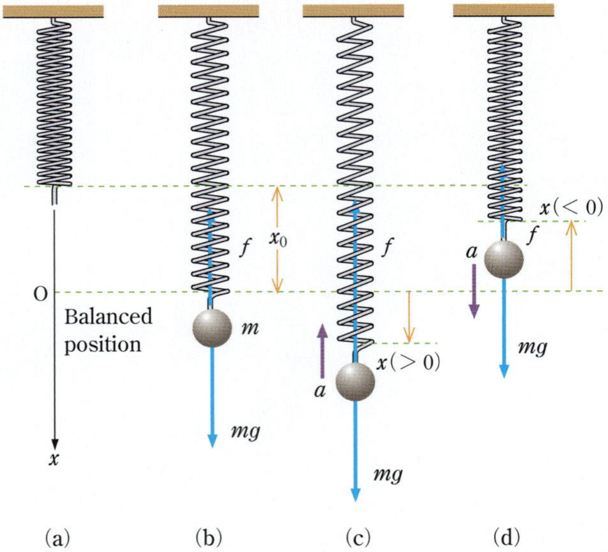

Fig. 4-2 Spring pendulum. The force that acts on the weight is the resultant force of the gravity mg and the elastic force $f = -k(x+x_0)$ of the spring, so it is

$$F = mg-k(x+x_0) = -kx.$$

This is the force (restoring force) that works to return the weight to the balanced position (origin O).

$$x_0 = \frac{mg}{k}.$$

We choose the position of the weight in the balanced state as the origin O, and take the vertically downward direction as $+x$-axis direction [Fig. 4-2 (b)]. Since the force F acting on the weight is the resultant force of gravity mg and elastic force $f = -k(x+x_0)$ of the spring, it follows that

$$F = mg - k(x+x_0) = -kx. \tag{4.2}$$

F is a restoring force proportional to the displacement x from the balanced state. When we pull down the weight downwards and then release the hand, the weight oscillates up and down. This oscillation is an oscillation caused by the restoring force which follows Hooke's law, so it is a simple harmonic oscillation.

Equation of motion of simple harmonic oscillation and its solution

A weight with mass m is oscillating on the x-axis under the action of force $F = -kx$. The equation of motion of the weight is

$$m\frac{\mathrm{d}^2x}{\mathrm{d}t^2} = -kx. \tag{4.3}$$

Setting $\frac{k}{m} = \omega^2$, namely

$$\omega = \sqrt{\frac{k}{m}} \tag{4.4}$$

we obtain from equation (4.3) the following relation :

$$\frac{\mathrm{d}^2x}{\mathrm{d}t^2} = -\omega^2 x. \tag{4.5}$$

This is the standard form of differential equation that the simple harmonic oscillation follows.

To solve equation (4.5) is equivalent to find a function $x(t)$ of the variable t which, when substituted into equation (4.5), gives the same values on both sides of equation (4.5). In other words, equation (4.5) indicates to find a function $x(t)$ which, being differentiated twice with respect to t, leads to $-\omega^2$ time the original function $x(t)$. As a function satisfying such a condition we know

$$x(t) = A\cos(\omega t+\theta_0). \tag{4.6}$$

In this case the velocity $v(t)$ is

$$v(t) = \frac{\mathrm{d}v}{\mathrm{d}t} = -\omega A\sin(\omega t+\theta_0) \tag{4.7}$$

and the acceleration $a(t)$ is $-\omega^2$ times as large as $x(t)$, or

$$a(t) = \frac{\mathrm{d}^2x}{\mathrm{d}t^2} = -\omega^2 A\cos(\omega t+\theta_0) = -\omega^2 x(t). \tag{4.8}$$

Thus the required condition is satisfied, so equation (4.6) is a solution of the differential equation (4.5) of simple harmonic oscillation.

The solution (4.6) representing the position of the oscillating weight is shown in Fig. 4-4 (a). This oscillation is the oscillation that the weight reciprocates between two points $x = A$ and $x = -A$. The

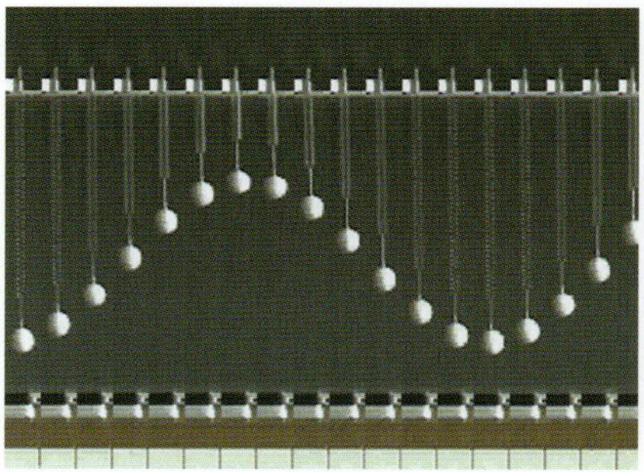
Fig. 4-3 A composite photograph of spring pendulum

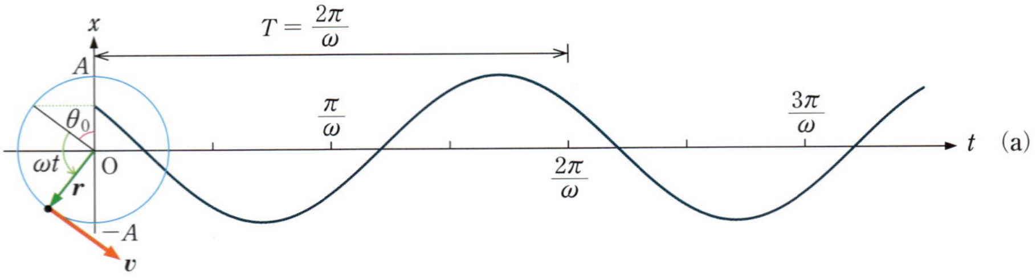

(a) $x = A\cos(\omega t+\theta_0)$

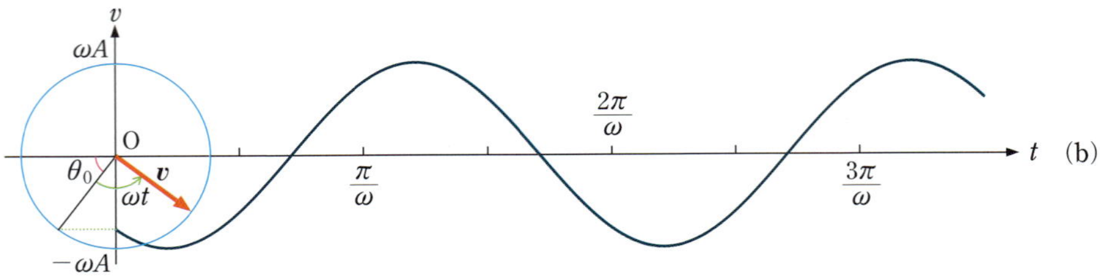

(b) $v = -\omega A \sin(\omega t+\theta_0) = \omega A \cos\left(\omega t+\theta_0+\frac{\pi}{2}\right)$

Fig. 4-4 simple harmonic oscillation

maximum value A of displacement is called the **amplitude**. Looking at Fig. 4-4 (a) we realize that equation (4.6) is equal to the motion of the x component of the object in uniform circular motion with angular velocity ω and radius A. In the case of uniform circular motion, we call ω the angular velocity but in the case of simple harmonic oscillation we call it the **angular frequency**. The term $\omega t+\theta_0$ on the right-hand side of equation (4.6) is called the **phase** of oscillation and θ_0 is referred to as the initial phase. The phase indicates in what state the oscillation is.

Since $\cos(x+2\pi) = \cos x$, the oscillation described by equation (4.6) reciprocates the same motion at every time T satisfying $\omega T = 2\pi$,

$$T = \frac{2\pi}{\omega} = 2\pi\sqrt{\frac{m}{k}} \quad \text{(period).} \tag{4.9}$$

This is a periodic motion with period T. The **frequency** f per unit time is the reciprocal of the period T ($fT = 1$), so it follows

$$f = \frac{1}{T} = \frac{\omega}{2\pi} = \frac{1}{2\pi}\sqrt{\frac{k}{m}} \quad \text{(frequency).} \tag{4.10}$$

Unit of frequency Hz = 1/s

The unit of frequency is 1/s and is called hertz, expressed as Hz. Equation (4.10) shows us that the frequency f is proportional to $\sqrt{k}$ and inversely proportional to $\sqrt{m}$, and therefore, the oscillation of the weight hung by the spring gets faster when the spring is more rigid (k is larger) and the mass is lighter (m is smaller), and it gets slower when the spring is more weak (k is smaller) and the weigh is heavier (m is larger). The period of simple harmonic motion is independent of the amplitude. This is a marked characteristic point of the simple harmonic motion and is called **isochronism** of pendulum.

The solution (4.6) is a general solution of the second order differential equation (4.5) since it contains two arbitrary constants A and θ_0. In order to examine the physical meaning of arbitrary constants A and θ_0, we set $t = 0$ in equations (4.6) and (4.7), then we have

$$x_0 = x(0) = A\cos\theta_0, \qquad v_0 = v(0) = -\omega A\sin\theta_0. \qquad (4.11)$$

Using the additive theorem of trigonometric functions $\cos(\alpha+\beta) = \cos\alpha\cos\beta - \sin\alpha\sin\beta$, we transform equation (4.6) into

$$x(t) = A\cos(\omega t+\theta_0) = A\cos\omega t\cos\theta_0 - A\sin\omega t\sin\theta_0 \qquad (4.12)$$

and then insert equation (4.11) into equation (4.12) to find

$$x = x_0\cos\omega t + \frac{v_0}{\omega}\sin\omega t. \qquad (4.13)$$

It has been shown that the two arbitrary constants A and θ_0 correspond to the position x_0 and velocity v_0 of the weight at $t = 0$. By adjusting A and θ_0, equation (4.6) can correctly represent the motion of the weight irrespective of any values of position x_0 and velocity v_0 at $t = 0$. This demonstrates the meaning that equation (4.6) is the general solution.

Case 1 When we pull the weight down by the distance A and gently release the hand, since $x_0 = A$ and $v_0 = 0$. equation (4.13) becomes

$$x = A\cos\omega t. \qquad (4.14)$$

Example 1 When a weight of 0.10 kg was applied to the spring shown in Fig. 4-2 (a), the spring length stretched by 0.02 m. Find the period and the frequency when the weight oscillates up and down.

Solution Let the stretched length of spring be x_0. From the relation $mg = kx_0$, the spring constant k is

$$k = \frac{mg}{x_0} = \frac{(0.10\ \text{kg})\times(9.8\ \text{m/s}^2)}{0.02\ \text{m}} = 49\ \text{kg/s}^2.$$

Thus from equations (4.9) and (4.10), we obtain the following results:

$$T = 2\pi\sqrt{\frac{m}{k}} = 2\pi\sqrt{\frac{0.1\ \text{kg}}{49\ \text{kg/s}^2}} = 0.28\ \text{s}$$

$$f = \frac{1}{T} = 3.5\ \text{Hz}.$$

Question 1 As illustrated in Fig. 4-5, two springs with spring constants k_1 and k_2 are attached to both sides of a weight with mass m and placed on a smooth horizontal surface and the other ends of the two springs are fixed. In the state at rest, the spring length is assumed to be the natural length. Find the frequency f of the weight when the weight is displaced x with the hand in the direction of the arrow and then released from the hand.

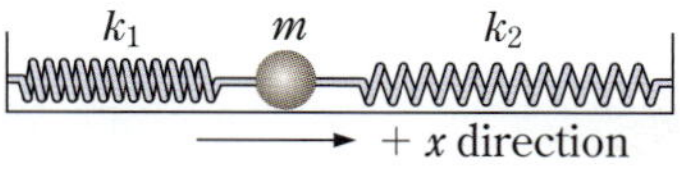

Fig. 4-5

Simple pendulum The second example of single harmonic oscillations is the oscillation of simple pendulum. A device in which one end of a long thread (length L) is fixed while a weight (mass m) is attached to the other end, and which causes the weight to oscillate with a small amplitude in the vertical plane is called a **simple pendulum**. The weight performs the reciprocal motion on an arc of radius L under the action of the thread tension $\boldsymbol{S}$ and gravity $m\boldsymbol{g}$. Since the direction of the thread tension is perpendicular to the direction of motion of the weight, the force to move the weight is the tangential component F of gravity $m\boldsymbol{g}$. When the pendulum is displaced from the vertical line by an angle θ, the relation

$$F = -mg\sin\theta \qquad (g \text{ is the gravitational acceleration}) \qquad (4.15)$$

holds (Fig. 4-6). The negative sign indicates that the direction of force

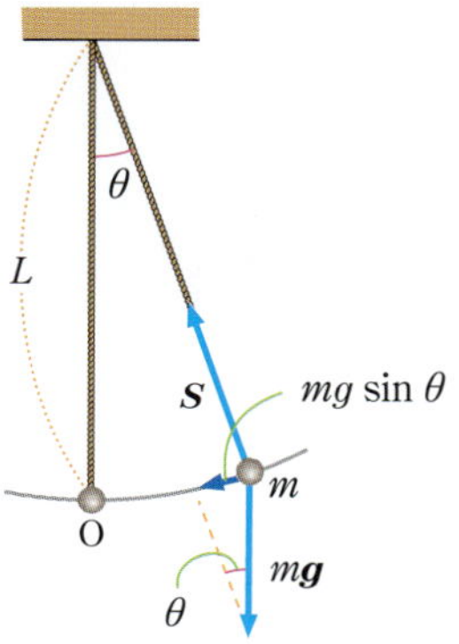

Fig. 4-6 A simple pendulum

is opposite to the direction of displacement of the weight. Due to this force F, the weight performs the reciprocal motion on the arc.

Let the angle formed by the vertical axis and the thread be θ. Then, since the moving distance (the arc length) of the weight from the lowest point O is $L\theta$ [see equation (1.28)], the tangential component of the circle of the acceleration of weight is $\dfrac{d^2(L\theta)}{dt^2} = L\dfrac{d^2\theta}{dt^2}$. Accordingly, the equation of motion of the weight in the tangential direction of the circle is as follows:

$$mL\frac{d^2\theta}{dt^2} = -mg\sin\theta \tag{4.16}$$

$$\therefore\ \frac{d^2\theta}{dt^2} = -\frac{g}{L}\sin\theta. \tag{4.17}$$

Fig. 4-7 A composite photograph of simple pendulum

When the oscillation width of the pendulum is small (when $|\theta|$ is far smaller than 1), since $\sin\theta \approx \theta$ [see equation (1.29)], equation (4.17) becomes

$$\frac{d^2\theta}{dt^2} = -\frac{g}{L}\theta. \tag{4.18}$$

When we set

$$\omega = \sqrt{\frac{g}{L}} \tag{4.19}$$

equation (4.18) takes the same form as equation (4.5), so the general solution of equation (4.18) is the simple harmonic oscillation described as

$$\theta = \theta_{max}\cos(\omega t+\beta). \tag{4.20}$$

The maximum θ_{max} of swing angle and β are arbitrary constants.

The frequency f and the period T are

$$f = \frac{1}{2\pi}\sqrt{\frac{g}{L}}, \qquad T = 2\pi\sqrt{\frac{L}{g}}. \tag{4.21}$$

The shorter the thread length L is, the shorter the period of pendulum is. The fact that the period of oscillation is constant and independent of amplitude θ_{max} is referred to as **isochronism** of the pendulum. When the width of oscillation becomes large, the magnitude of restoring force is $mg|\sin\theta| < mg|\theta|$ and the period T becomes longer than $2\pi\sqrt{\dfrac{L}{g}}$.

Fig. 4-8 The lamp in the Pisa Cathedral

According to the legend, the isochronism of a simple pendulum was discovered in 1583 by Galileo who was watching the lamp of the Pisa Cathedral swinging. Galileo, a 19-year-old student at the University of Pisa, was watching the lamp swing when a large bronze lamp hanging from the ceiling of the cathedral was lighted up by a custodian. Although the amplitude gradually decreased, he noticed that the time for the lamp to complete one cycle of back-and-forth swinging motion was still constant. Galileo is said to have verified the constancy in cycle of oscillation of the pendulum by counting his own pulse.

The period T of a simple pendulum can be measured precisely. From the period measured, the gravitational acceleration g

$$g = \frac{4\pi^2 L}{T^2}$$

is determined with accuracy. This is a good contrast to the case of free fall in which the motion is too fast to carry out precise measurements of g.

Example 2 Find the period of the simple pendulum with thread length $L = 1\,\text{m}$.

Solution From the second equation of equation (4.21),

$$T = 2\pi\sqrt{\frac{L}{g}} = 2\pi\sqrt{\frac{1\,\text{m}}{9.8\,\text{m/s}^2}} = 2.0\,\text{s}.$$

Example 3 Find the length (in meter) of the simple pendulum with period 1 s.

Solution From the second equation of equation (4.21),

$$L = \frac{gT^2}{4\pi^2} = \frac{(9.8\,\text{m/s}^2)\times(1\,\text{s})^2}{4\pi^2} = 0.25\,\text{m}.$$

Question 2 Find the period of the simple pendulum with thread length $L = 2\,\text{m}$.

Question 3 Show that for the simple pendulum the equation of motion in the direction of thread is given

$$mL\left(\frac{d\theta}{dt}\right)^2 = S - mg\cos\theta. \tag{4.22}$$

Question 4 When the weight of simple pendulum moves in a way from a → b → c → d → e, which one of ①, ②, ③, ④, and ⑤ of Fig. 4-9 describes the acceleration of the weight most appropriately? Note that you should consider the tangential acceleration and the centripetal acceleration.

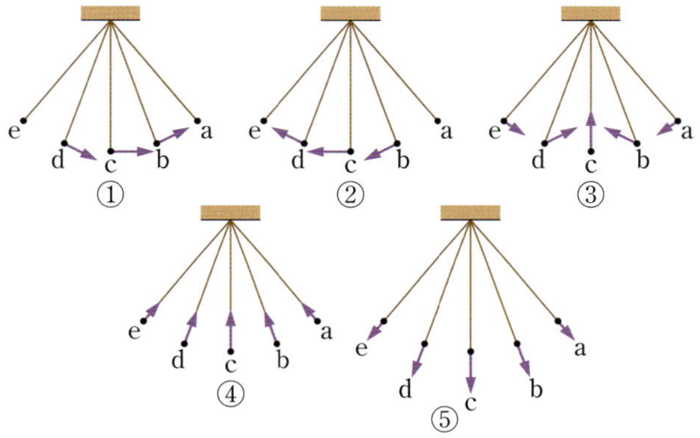

Fig. 4-9

4.2 Damped oscillation and forced vibration

Learning objective To be able to explain what phenomena the damped oscillation, forced vibration and resonance are.

Damped oscillation The simple harmonic oscillation is an endless oscillation with a constant amplitude but in actual oscillations the energy is dissipated due to the air resistance and/or friction. Accordingly, the amplitude damps gradually with time. The oscillation accompanied by the damping amplitude is called **damped oscillation**. The motion of weight when we pull the weight shown in Fig. 4-2 down by a length A and gently release the hand is expressed in Fig. 4-10. If much larger damping due to air resistance is required, it can be attained by attaching a disc beneath the weight and then oscillating it.

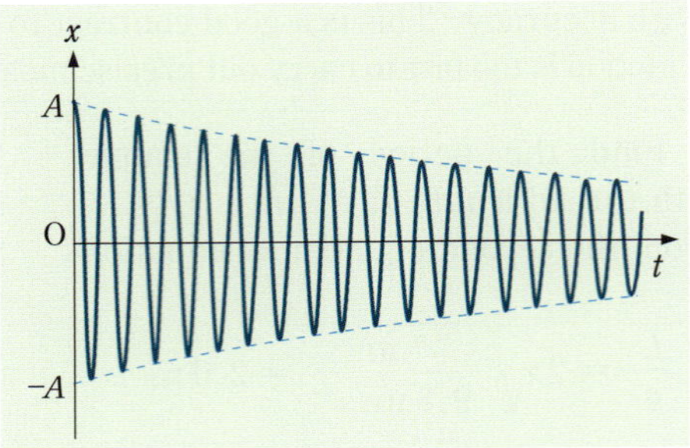

Fig. 4-10 Damped oscillation. Without supplying externally the energy, the oscillation is going to damp.

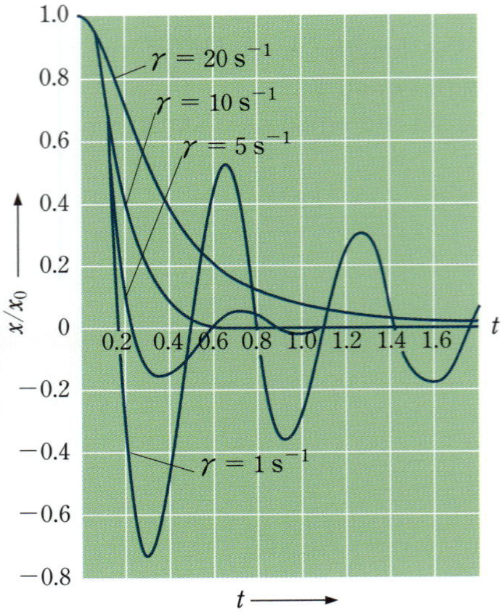

Fig. 4-11 Oscillations of pendulum with $\omega = 10\,\mathrm{s}^{-1}$ under the initial conditions $x = x_0$, and $v = 0$. When $\gamma < 10\,\mathrm{s}^{-1}$, the motion is the damped oscillation and when $\gamma = 10\,\mathrm{s}^{-1}$, the motion is the critical oscillation, and when $\gamma > 10\,\mathrm{s}^{-1}$, the motion is the overdamping.

As an example of the damped oscillation, let us consider a case where resistance (viscous drag) $-2m\gamma v$ proportional to the speed acts on an object of mass m that receives the restoring force $-kx = -m\omega^2 x$ and performs the simple harmonic oscillation. γ is a positive constant.

Fig. 4-11 shows examples of motions when we pull a weight of damping pendulum down by a length x_0 and release it gently at $t = 0$. So long as the viscous drag is small, the oscillation is **damped oscillation** in which the amplitude damps gradually. But when the viscous drag is large, the weight can no longer oscillate. This phenomenon is called **overdamping**. The boundary between damped oscillation and overdamping is referred to as **critical damping**.

Example 4 | Damped oscillation Find the solution of the equation of motion

$$m\frac{\mathrm{d}^2x}{\mathrm{d}t^2} = -kx-2m\gamma\frac{\mathrm{d}x}{\mathrm{d}t} = -m\omega^2x-2m\gamma\frac{\mathrm{d}x}{\mathrm{d}t} \tag{4.23}$$

namely

$$\frac{\mathrm{d}^2x}{\mathrm{d}t^2}+2\gamma\frac{\mathrm{d}x}{\mathrm{d}t}+\omega^2x = 0. \tag{4.24}$$

Solution Setting

$$x = y\mathrm{e}^{-\gamma t} \tag{4.25}$$

we derive the following relations:

$$\frac{\mathrm{d}x}{\mathrm{d}t} = \frac{\mathrm{d}y}{\mathrm{d}t}\mathrm{e}^{-\gamma t}-\gamma y\mathrm{e}^{-\gamma t}$$

$$\frac{\mathrm{d}^2x}{\mathrm{d}t^2} = \frac{\mathrm{d}^2y}{\mathrm{d}t^2}\mathrm{e}^{-\gamma t}-2\gamma\frac{\mathrm{d}y}{\mathrm{d}t}\mathrm{e}^{-\gamma t}+\gamma^2y\mathrm{e}^{-\gamma t}.$$

Substitution of the above equations into equation (4.24) leads to

$$\frac{\mathrm{d}^2y}{\mathrm{d}t^2}+(\omega^2-\gamma^2)y = 0. \tag{4.26}$$

(1) In the case of $\omega > \gamma$ (**damped oscillation**)
When the resistance is small and the relation $\omega > \gamma$ holds, equation (4.26) is the equation of simple harmonic oscillation with angular frequency $\sqrt{\omega^2-\gamma^2}$. Therefore, its general solution is $y = \cos(\sqrt{\omega^2-\gamma^2}\,t+\theta_0)$ and the general solution of (4.23) is

$$x(t) = A\mathrm{e}^{-\gamma t}\cos(\sqrt{\omega^2-\gamma^2}\,t+\theta_0). \tag{4.27}$$

This solution shows the damped oscillation in which the amplitude attenuates following $A\mathrm{e}^{-\gamma t}$. The period T of oscillation is

$$T = \frac{2\pi}{\sqrt{\omega^2-\gamma^2}} \tag{4.28}$$

and is longer than $\dfrac{2\pi}{\omega}$ for the case without resistance.

(2) In the case of $\omega = \gamma$ (**critical damping**)
Since $\omega^2-\gamma^2 = 0$, equation (4.26) becomes $\dfrac{\mathrm{d}^2y}{\mathrm{d}t^2} = 0$. The general solution of this equation is $y = A+Bt$, so the general solution of equation (4.23) is

$$x(t) = (A+Bt)\,\mathrm{e}^{-\gamma t}$$

(A and B are arbitrary constants) (4.29)

(3) In the case of $\omega < \gamma$ (**overdamping**)
When the resistance is large and the relation

$\omega < \gamma$ holds, we set $p=\sqrt{\omega^2-\gamma^2}$. Then equation (4.26) is transformed into $\frac{d^2y}{dt^2} = p^2y$. The general solution of this equation is $y = Ae^{pt}+Be^{-pt}$. Thus

$$x(t) = Ae^{-(\gamma-p)t}+Be^{-(\gamma+p)t}$$

(A and B are arbitrary constants). (4.30)

An application of the damped oscillation is a door closer. This is a damping device that utilizes the restoring force of the air spring and the viscosity of the oil so that it allows the door to be critically damped without making a sound when we gently release our hand from the open door.

Fig. 4-12 Door closer

Forced vibration and resonance If energy is not supplied from the outside, the amplitude of vibration (oscillation) decreases with time. In order to keep vibration of constant amplitude when resistive force or frictional force which damps the oscillation acts, an external force vibrating at a constant period must act to replenish energy. Vibration in the same period as the external force due to the action of external force vibrating at a constant cycle is called the **forced vibration***.

An example of forced vibration is the case where the upper end of the single pendulum yarn is not fixed but it is vibrated in the horizontal direction. If we hold the upper end of a simple pendulum yarn with our hand and vibrate it in the horizontal direction at a frequency much smaller than $f=\frac{1}{2\pi}\sqrt{\frac{g}{L}}$ of the simple pendulum frequency, the weight oscillates with a small amplitude behind the hand movement. Increasing the frequency of reciprocation of the hand increases the amplitude of the weight. When the frequency of reciprocation of the hand is almost equal to the frequency of the single pendulum, the amplitude of the vibration of the weight is maximized. At this time the vibration of the weight is said to **resonate** with the external force. As the frequency of the hand is further increased, the weight moves in a direction opposite to the movement of the hand, and the amplitude of the weight decreases. You are advised to do the experiment by yourself to verify this.

* In addition to the forced vibration, there exists a case in which energy may be supplied by an external force that is not vibrational, as in the case of bowing a string of a violin with a bow, and it may be referred to as the self-excited vibration that continues vibration.

Generally, a vibrating object has a frequency characteristic to the object. When the frequency of an external force coincides with this characteristic frequency, **resonance** occurs in which the amplitude of the forced vibration increases. Resonance is a common phenomenon in daily life. For example, when we carry some water in a shallow container, it is an example of resonance that water moves greatly if walking at the same pace as the characteristic frequency of water in the container. When many people cross over a simple suspension bridge over a valley on a mountain path, they have to walk in such a way that their walking rhythms are non-uniform as a group. It is because when they walk in uniform steps, there is a worry that the bridge will be broken by resonance in case the walking pace of pedestrians and the characteristic vibration of the suspension bridge match. In designing such as buildings and bridges, attention should be paid so that they will not be broken due to self-excited vibration caused by the resonance with earthquakes or strong winds.

Fig. 4-13 Full aeroelastic bridge model test of the super long bridge

Mathematical derivation of forced oscillation A periodically varying force

$$F(t) = F_0 \cos \omega_f t = m f_0 \cos \omega_f t \qquad (F_0 = m f_0) \tag{4.31}$$

acts on an object that performs damped oscillations according to equation (4.23). Let us examine the motion of this object. In this case the equation of motion is

$$\frac{d^2x}{dt^2} + 2\gamma \frac{dx}{dt} + \omega^2 x = f_0 \cos \omega_f t \qquad (\omega > \gamma). \tag{4.32}$$

The general solution of equation (4.32) is

$$x(t) = \frac{f_0}{\sqrt{(\omega_f^2 - \omega^2)^2 + 4\gamma^2 \omega_f^2}} \cos(\omega_f t - \phi) + A e^{-\gamma t} \cos(\sqrt{\omega^2 - \gamma^2}\, t + \theta_0) \tag{4.33}$$

$$\sin \phi = \frac{2\gamma \omega_f}{\sqrt{(\omega_f^2 - \omega^2)^2 + 4\gamma^2 \omega_f^2}}, \qquad \cos \phi = \frac{\omega^2 - \omega_f^2}{\sqrt{(\omega_f^2 - \omega^2)^2 + 4\gamma^2 \omega_f^2}} \tag{4.34}$$

where two arbitrary constants A and θ_0 are included. Equation (4.33) can easily be checked by inserting equation (4.33) into equation (4.32) and using the additive theorem of the trigonometric functions $\cos(\omega_f t - \phi) = \cos \omega_f t \cos \phi + \sin \omega_f t \sin \phi$. The angle ϕ indicates the delay in the phase of vibration with respect to the phase $\omega_f t$ of the externa force. The second term on the right-hand side of equation (4.33) represents the damped oscillation when no external force acts.

The first term on the right-hand side of equation (4.33) refers to the forced vibration caused by the external force and its amplitude

$$x_0 = \frac{f_0}{\sqrt{(\omega_f^2 - \omega^2)^2 + 4\gamma^2 \omega_f^2}} \tag{4.35}$$

varies with the angular frequency ω_f of the external force (Fig. 4-14) ant it takes the maximum at the resonance angular frequency

$$\omega_R = \sqrt{\omega^2 - 2\gamma^2} \tag{4.36}$$

and the maximum value is

$$(x_0)_{max} = \frac{f_0}{2\gamma\sqrt{\omega^2 - \gamma^2}}. \tag{4.37}$$

Resonance is a phenomenon in which the amplitude of forced vibration increases rapidly when the angular frequency ω_f of the external force approaches ω_R.

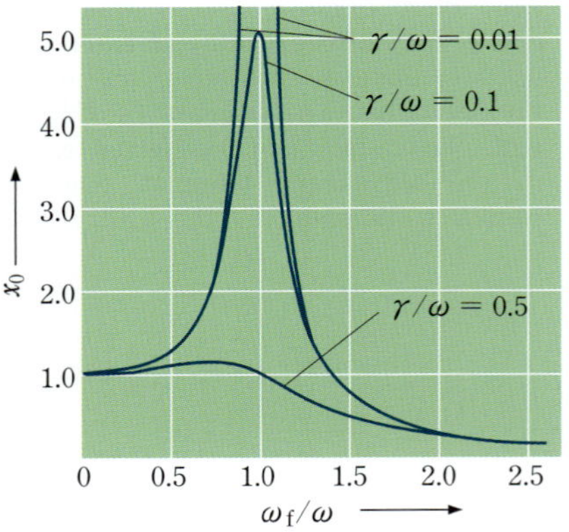

Fig. 4-14 The relationship between the angular frequency ω_f of the external force and the amplitude x_0 of the forced vibration. The unit of vertical axis is $\frac{f_0}{\omega^2}$.

Problems for exercise 4

A

1. A weight of mass m is attached to the end of a spring (spring constant k) and the spring is fixed at one end and is suspended vertically. When the equation of motion

$$ma = -kx$$

holds for the position x of the weight, answer the following questions.

(1) Find the acceleration at $x = 0$.

(2) Find the value of x when the weight is kept at rest.

(3) What kind of oscillation does the solution of this equation show?

2. The period of oscillation in the vertical direction of an object with a mass of 2 kg hung by a spring was 2 s. Find the spring constant.

3. In the experiment of Fig. 4-2, when a weigh of a mass $m = 1.0$ kg was placed on the spring, the elongation x_0 of the spring was 10 cm.

(1) Find the spring constant k.

(2) Find the period of the spring pendulum.

(3) Calculate the maximum value a_{max} of the acceleration of the weight when the weight performs a simple harmonic oscillation of amplitude $A = 5$ cm. How many times is this acceleration as large as the gravitational acceleration g?

4. The gravitational acceleration on the surface of the moon is 0.17 times that on the Earth surface. When we oscillate the same simple pendulum on the surface of the moon how many times is the period of oscillation as long as the period on the Earth surface?

5. Does the period change when the spring pendulum in Fig. 4-2 is oscillated on the surface of the moon?

B

1. A truck of mass 2 t is supported by four springs attached to four wheels (Fig. 1). Let the mass of the truck supported by each spring be 500 kg and the spring constant be 5.0×10^4 N/m. Assume that oscillation occurred because the spring was displaced 1.0 cm from the balanced position. Find the following values of this spring.

(1) Frequency f and period T of oscillation,

(2) Maximum value v_{max} of speed,

(3) Maximum value a_{max} of acceleration.

In reality, the oscillation is reduced rapidly by the action of oscillation-damping equipment.

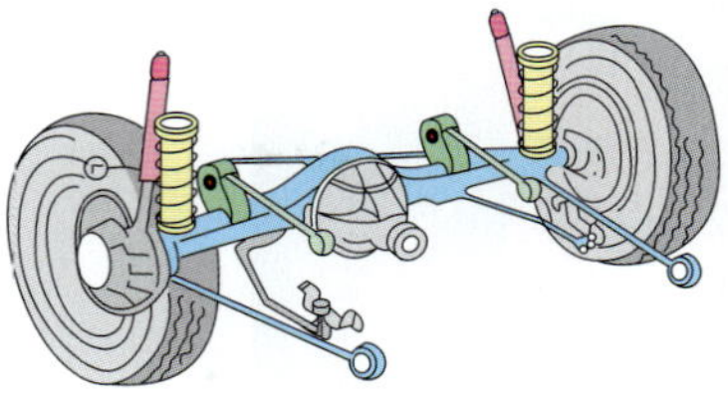

Fig. 1 Springs of rear wheels

2. Newton prepared a simple pendulum with a hollow weight and conducted experiments, putting wood, iron, gold, copper, salt, cloth etc., in the hollow weight. He could find no difference in measured periods of the simple pendulum. What does his finding mean?

3. We prepare the spring pendulums of Fig. 2 (a) and Fig. 2 (b) where the weight of the mass M is suspended with two springs with the same spring constant and the same length in the natural state.

(1) Find the restoring forces due to the two springs in both cases when the displacement of the weight from the balanced position is x.

(2) How many times is the period in the case of Fig. 2 (a) as long as the period in the case of Fig. 2 (b)?

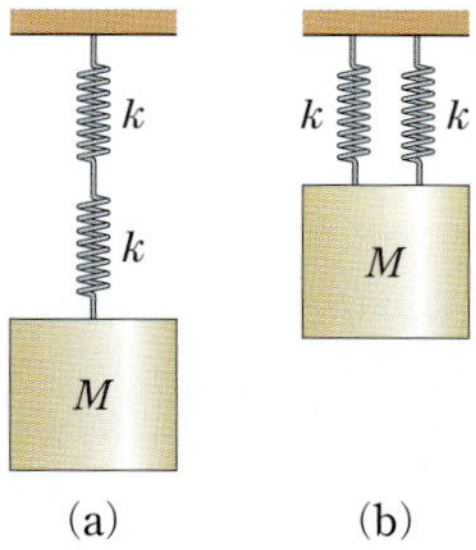

Fig. 2

4. In the experiment of Fig. 4-2, if we select the lower end of the spring having no attached weight as the origin of the x coordinate in the vertically downward direction, the equation of motion is

$$m\frac{d^2x}{dt^2} = mg - kx.$$

Find the solution of this differential equation.

Work and Energy

The range of physics covers various areas such as force and motion, electricity and magnetism, heat, etc. In learning physics, it is customary to learn these various fields after separating them into mechanics, electromagnetism, thermology, etc. However, phenomena discussed in one specific field are not irrelevant to other fields. Physics is the result of human efforts to understand nature in a unified way, based on a few laws. The key to the unified understanding of nature lies in **energy**.

The term energy is used as a daily term, but its etymology is Ergon meaning work in Greek. The meaning of energy as a physical term can be taken as the "ability to do work". There are many different types of energy, but any type of energy is transformed into another type of energy and various forms that can generate or carry energy exist. When the form of energy changes, the work performed by force often acts as an intermediary.

5.1 Work and power

Learning objective To understand that work as a term in physics means "moving-direction component of force" × "moving distance". In other words, to understand that the work performed by force may take a positive value, a negative value, or zero. In addition to understand the definition of power.

Work In everyday life the word "work" is used in many ways, but in physics, when an object under the action of a constant force $\boldsymbol{F}$ moves in a certain direction, the "**work** W that the force $\boldsymbol{F}$ performs" is defined to be the product of "moving-direction component $F_t = F\cos\theta$ of the force $\boldsymbol{F}$" and "moving distance s"

$$W = F_t s = Fs\cos\theta. \quad (5.1)$$

θ is the angle between the force $\boldsymbol{F}$ and the direction of movement (Fig. 5-1). $W = F\times(s\cos\theta)$, so the "work that the force $\boldsymbol{F}$ performs" is "magnitude of force" × "distance moved in the direction of force" (Fig. 5-2).

The unit of work is the product of force unit $\mathrm{N} = \mathrm{kg\cdot m/s^2}$ and length unit m, namely $\mathrm{N\cdot m} = \mathrm{kg\cdot m^2/s^2}$, which is called Joule (symbol J). Joule is also a unit of energy.

$$\mathrm{J} = \mathrm{N\cdot m} = \mathrm{kg\cdot m^2/s^2}. \quad (5.2)$$

Let $\boldsymbol{s}$ be the displacement (a vector with the starting point as the initial point and the arrival point as the end point) of the object. Then the work W performed by the constant force $\boldsymbol{F}$ is expressed as a scalar product (inner product) of $\boldsymbol{F}$ and $\boldsymbol{s}$ described in Appendix.

$$W = \boldsymbol{F}\cdot\boldsymbol{s}. \quad (5.3)$$

This is because when the angle between the vectors $\boldsymbol{A} = (A_x, A_y, A_z)$ and $\boldsymbol{B} = (B_x, B_y, B_z)$ is θ, the scalar product of $\boldsymbol{A}$ and $\boldsymbol{B}$ is

$$\boldsymbol{A}\cdot\boldsymbol{B} = \boldsymbol{B}\cdot\boldsymbol{A} = AB\cos\theta = A_xB_x + A_yB_y + A_zB_z \quad (5.4)$$

(Fig. 5-3).

In the case where the direction of force and the direction of velocity are perpendicular, such as the tension of the thread of single pendulum acting on the weight and the normal force acting on the object on the ground etc., $\cos 90° = 0$, so these forces perform no work ($W = 0$). When a person tries to push a heavy car on a slope with a constant force $\boldsymbol{F}$ to make it climb but the force is insufficient, the car does not move, so the distance s is 0 and hence the work performed is 0. When the car slides by the distance s, the direction of force and the direction of movement are opposite and $\cos 180° = -1$, so the work performed by the force $\boldsymbol{F}$ is a negative quantity $-Fs$. The above-mentioned person performs the positive work on his/her muscles, but the person does not perform any positive work on the car.

Forces acting on an object are not always one kind. In the case of Fig. 5-1, in addition to the force $\boldsymbol{F}$, the normal force of the floor, gravity and friction force act on the object. Equation (5.1) represents the work performed by $\boldsymbol{F}$, one of these forces.

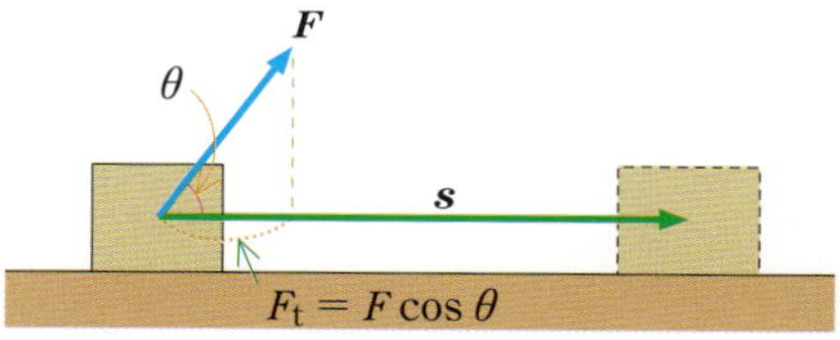

Fig. 5-1 $W = F_t s = Fs\cos\theta = \boldsymbol{F}\cdot\boldsymbol{s}$

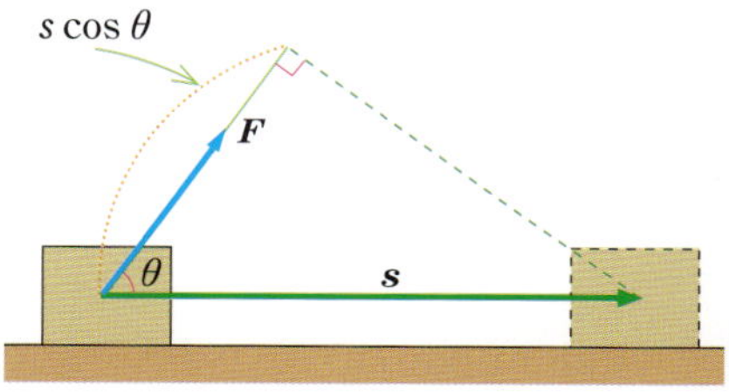

Fig. 5-2 $W = F\times(s\cos\theta)$

Unit of work
$\mathrm{J} = \mathrm{N\cdot m} = \mathrm{kg\cdot m^2/s^2}$
Unit of energy
$\mathrm{J} = \mathrm{N\cdot m} = \mathrm{kg\cdot m^2/s^2}$

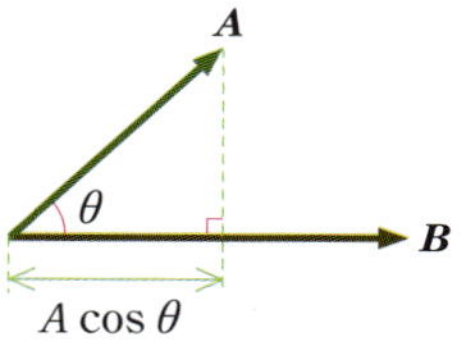

Fig. 5-3 $\boldsymbol{A}\cdot\boldsymbol{B} = AB\cos\theta$

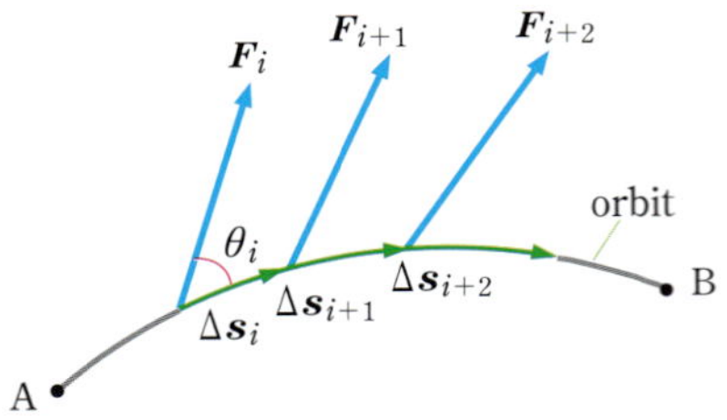

Fig. 5-4 Sum of minute work. When an object moves from point A to point B, the work $W_{A\to B}$ performed by the force $\boldsymbol{F}$ is the sum of minute work performed in each minute section $\Delta W_i = \boldsymbol{F}_i \cdot \Delta \boldsymbol{s}_i = F_i \Delta s_i \cos\theta_i$.

When the object moves from point A to point B, either the magnitude or direction of the force $\boldsymbol{F}$ or both of them may change, or the path may not be straight. In this case, the work $W_{A\to B}$ that the force $\boldsymbol{F}$ performs is obtained by dividing the moving path A → B of the object into N small sections and adding the minute work that the force $\boldsymbol{F}$ performs in each minute section (Fig. 5-4). Let $\Delta \boldsymbol{s}_i$ be a minute displacement vector starting from the initial point of the i-th minute section and ending at the end of that section. When $\Delta \boldsymbol{s}_i$ is small enough, we can consider the force $\boldsymbol{F}$ to be constant in this section and describe it as $\boldsymbol{F}_i$. Accordingly, the minute work ΔW_i which the force $\boldsymbol{F}$ performs in this minute section can be approximated as

$$\Delta W_i = \boldsymbol{F}_i \cdot \Delta \boldsymbol{s}_i = F_i \Delta s_i \cos\theta_i = F_{it} \Delta s_i. \tag{5.5}$$

Therefore, $W_{A\to B}$ is given by the integral defined below :

$$W_{A\to B} = \lim_{N\to\infty} \sum_{i=1}^{N} \boldsymbol{F}_i \cdot \Delta \boldsymbol{s}_i = \int_A^B \boldsymbol{F} \cdot d\boldsymbol{s} = \int_A^B F_t \, ds. \tag{5.6}$$ *1

*1 When the path of integral falls on the x-axis, equation (5.6) is expressed as $W_{A\to B} = \int_{x_A}^{x_B} F_x \, dx$.

F_t is the tangential component of the force $\boldsymbol{F}$ along the path. The integral sign $\int_A^B$ means the integration (line integral) along the path (curve) A → B. This equation of line integral can be understood as the sign to calculate $W_{A\to B}$ by following the procedure shown by the second equality side of equation (5.6).

Unit of power W = J/s

Power The work performed in unit time is called the **work rate** or the **power**. That is, let W be the work performed in time t, then the average power $\bar{P}$ is

$$\bar{P} = \frac{W}{t}. \tag{5.7}$$

Therefore, the unit of power is "unit of work J" divided by "unit of time s" and this is called watt (Symbol W)*2

*2 Pay attention not to confuse the sign W for work with the unit sign W of power.

$$\mathrm{W = J/s = kg{\cdot}m^2/s^2}. \tag{5.8}$$

An object under the action of force $\boldsymbol{F}$ is moving at the velocity $\boldsymbol{v}$. The minute work which the object receives as a result of a minute displacement $\Delta \boldsymbol{s}$ in a minute time Δt is $\Delta W = \boldsymbol{F} \cdot \Delta \boldsymbol{s} = \boldsymbol{F}_i \cdot \Delta \boldsymbol{s}_i$ so the (instantaneous) power P is expressed as

$$P = \lim_{\Delta t\to 0} \frac{\Delta W}{\Delta t} = \lim_{\Delta t\to 0} \frac{\boldsymbol{F} \cdot \Delta \boldsymbol{s}}{\Delta t} = \lim_{\Delta t\to 0} \boldsymbol{F} \cdot \frac{\Delta \boldsymbol{s}}{\Delta t} = \boldsymbol{F} \cdot \boldsymbol{v}. \tag{5.9}$$

In addition, when an object under a constant force $\boldsymbol{F}$ acts at a constant speed $\boldsymbol{v}$ in the direction of force, the power P of this force is expressed as

$$P = Fv. \tag{5.10}$$

Fig. 5-5 A crane

Example 1 (1) A crane lifted a 1000 kg container in 20 s to a height of 25 m. Calculate the average power $\bar{P}$ of this crane.
(2) If a motor with a power of 10 kW is attached to this crane, how many seconds will it take to lift a 1000 kg container to a height of 25 m?

Solution (1) The work W that the crane performed is
$W = mgh = (1000\ \mathrm{kg}) \times (9.8\ \mathrm{m/s^2}) \times (25\ \mathrm{m})$
$= 2.45 \times 10^5\ \mathrm{J}$

$\therefore \quad \bar{P} = \dfrac{W}{t} = \dfrac{2.45\times10^5\ \mathrm{J}}{20\ \mathrm{s}} = 1.2\times10^4\ \mathrm{W}$

$= 12\ \mathrm{kW}.$

(2) $t = \dfrac{W}{P} = \dfrac{245\ \mathrm{kJ}}{10\ \mathrm{kW}} = 24.5\ \mathrm{s}.$

5.2 Work and energy

Learning objective To understand the meaning of the relationship between work and kinetic energy that the work W that the force (resultant force) $\boldsymbol{F}$ performs on the mass point is equal to the increased amount of the kinetic energy $K = \frac{1}{2}mv^2$ of the mass point, namely

$$W_{A\to B} = K_B - K_A = \frac{1}{2}mv_B{}^2 - \frac{1}{2}mv_A{}^2.$$

To understand that the conservative force is a force which gives the same amount of work irrespective of the path it takes between the two points, that the potential energy $U(\boldsymbol{r})$ can be defined for the conservative force and the work $W^{Con}{}_{A\to B}$ performed by the conservative force $\boldsymbol{F}_{Con}$ is equal to the decrease in potential energy $U(\boldsymbol{r})$ of the force $\boldsymbol{F}_{Con}$, and finally that $W^{Con}{}_{A\to B} = U(\boldsymbol{r}_A) - U(\boldsymbol{r}_B)$.

Kinetic energy When the speed of an object of mass is v, the **kinetic energy** of this object is defined as

$$K = \frac{1}{2}mv^2. \tag{5.11}$$

Relationship between work and kinetic energy According to Newton's laws of motion, "force $\boldsymbol{F}$" = "mass m" × "acceleration $\boldsymbol{a}$", so when force acts on an object, acceleration is generated and the speed of the object changes. As shown in Fig. 5-6, when we push a dry ice brock moving on a flat floor in the direction of its movement, the speed of the brock increases, leading to the increase in its kinetic energy. The work W performed by the force is positive ($W > 0$). On the contrary, when we apply the force in the opposite direction of the brock movement, the speed of the brock decreases, causing the decrease in kinetic energy and the work performed by the force is negative ($W < 0$).

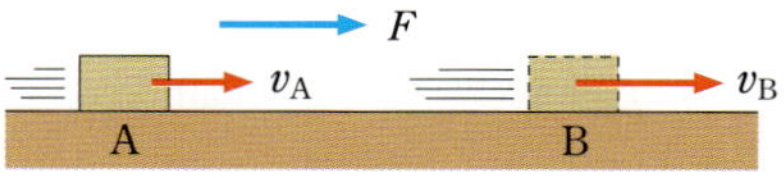

Fig. 5-6 When an object receives work, its kinetic energy increases.
$\frac{1}{2}mv_B{}^2 - \frac{1}{2}mv_A{}^2 = W_{A\to B}$

In general, the kinetic energy of the mass point increases by the amount of work $W_{A\to B}$ which the force $\boldsymbol{F}$ (resultant force of all forces) has performed on it. This is expressed as a formula

$$W_{A\to B} = \int_A^B \boldsymbol{F}\cdot d\boldsymbol{s} = \frac{1}{2}mv_B{}^2 - \frac{1}{2}mv_A{}^2 \tag{5.12}$$

(Fig. 5-6). This relationship is called the **relationship between work and kinetic energy**. Here v_A and v_B are the speed of mass point at points A and B, respectively.

Proof First, we prove the case of linear motion. The equation of motion is

$$m\frac{dv}{dt} = F.$$

We multiply both sides of this equation by $v = \dfrac{dx}{dt}$ and integrate from time t_A to t_B

$$m\int_{t_A}^{t_B} v\frac{dv}{dt}\,dt = \frac{m}{2}\int_{t_A}^{t_B}\frac{dv^2}{dt}\,dt = \frac{mv^2}{2}\bigg|_{t_A}^{t_B} = \frac{1}{2}mv_B{}^2 - \frac{1}{2}mv_A{}^2$$
$$= \int_{t_A}^{t_B} F\frac{dx}{dt}\,dt = \int_{x_A}^{x_B} F\,dx = W_{A\to B}$$

where x_A, x_B, v_A, v_B are the position and velocity at times t_A and t_B, respectively.

In the case of motion in three-dimension, we take a scalar product of both sides of the equation of motion $m\dfrac{d\boldsymbol{v}}{dt} = \boldsymbol{F}$ and $\boldsymbol{v} = \dfrac{d\boldsymbol{r}}{dt}$. Considering

$$v_x\frac{dv_x}{dt} + v_y\frac{dv_y}{dt} + v_z\frac{dv_z}{dt} = \frac{1}{2}\frac{d(v_x{}^2+v_y{}^2+v_z{}^2)}{dt}$$

we can prove equation (5.12) similarly as in the case of linear motion.

Work performed by the conservative force and the potential energy

Consider the case where the force $\boldsymbol{F}$ acting on the mass point located at point $\boldsymbol{r}$ depends only on the position $\boldsymbol{r}$ of mass point and takes the form $\boldsymbol{F}(\boldsymbol{r})$. Gravity $\boldsymbol{F} = m\boldsymbol{g}$, elasticity of spring $F = -kx$, universal gravitation etc. are forces of this type. Kinetic friction force is in the opposite direction of the velocity $\boldsymbol{v}$, so it is not a force in the form of $\boldsymbol{F}(\boldsymbol{r})$. When a mass point moves from point A to point B, the work $W_{A\to B}$ performed by the force $\boldsymbol{F}$ is given by substituting $\boldsymbol{F}(\boldsymbol{r})$ for $\boldsymbol{F}$ in equation (5.6)

$$W_{A\to B} = \int_A^B \boldsymbol{F}(\boldsymbol{r})\cdot d\boldsymbol{s} = \int_A^B F_t(\boldsymbol{r})\,ds. \qquad (5.13)$$

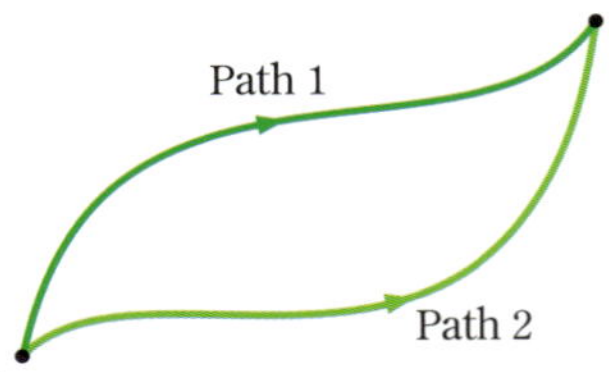

Fig. 5-7 Two paths from point A to point B

There are cases in which the integration (5.13) from an arbitrary point A to an arbitrary point B is determined only by the positions of initial point A and end point B and is independent of the path (Fig. 5-7). Such a force is called the **conservative force**. In other words,

> The conservative force is a force which performs the same constant amount of work regardless of the path which the mass point travels from an arbitrary point A to an arbitrary point B.

In the case of conservative force $\boldsymbol{F}_{Con}$, we first fix the reference point $\boldsymbol{r}_0$ and then define the **potential energy*** of the position vector at $\boldsymbol{r}$ as

$$U(\boldsymbol{r}) = -\int_{\boldsymbol{r}_0}^{\boldsymbol{r}} \boldsymbol{F}_{Con}(\boldsymbol{r})\cdot d\boldsymbol{s} = \int_{\boldsymbol{r}}^{\boldsymbol{r}_0} \boldsymbol{F}_{Con}(\boldsymbol{r})\cdot d\boldsymbol{s}. \qquad (5.14)$$

The right-hand side of equation (5.14) depends only on the position $\boldsymbol{r}$ of the end point and is independent of the path of integration, so we are allowed to define this as the potential energy $U(\boldsymbol{r})$ at point $\boldsymbol{r}$. At the reference point $\boldsymbol{r}_0$, $U(\boldsymbol{r}_0) = 0$. In equation (5.14) the relation $W^{Con}{}_{A\to B} = -W^{Con}{}_{B\to A}$ is used.

Question 1 Explain that $W^{Con}{}_{A\to B} = -W^{Con}{}_{B\to A}$.

The force $\boldsymbol{F}$ is divided into three categories, which are conservative force $\boldsymbol{F}_{Con}$ with the corresponding potential energy, non-conservative force $\boldsymbol{F}_{Non}$ without potential energy, and constraining force $\boldsymbol{F}_{constrain}$.

* Potential energy $U(\boldsymbol{r})$ of the conservative force means its potential ability to perform work on a mass point at point $\boldsymbol{r}$.

Constraining force is the force in which the direction of motion and the direction of force is perpendicular as the case of normal force, so it is a force performing no work.

$$\boldsymbol{F} = \boldsymbol{F}_{\text{Con}} + \boldsymbol{F}_{\text{Non}} + \boldsymbol{F}_{\text{constrain}}. \tag{5.15}$$

Case 1 | Potential energy of gravity When a mass point of mass m starts from an arbitrary point A (position vector $\boldsymbol{r}_\text{A}$) and reaches an arbitrary point B (position vector $\boldsymbol{r}_\text{B}$), the work performed by the gravity $\boldsymbol{F}(\boldsymbol{r}) = m\boldsymbol{g}$

$$W_{\text{A}\to\text{B}}^{\text{gravity}} = \int_{\boldsymbol{r}_\text{A}}^{\boldsymbol{r}_\text{B}} m\boldsymbol{g}\cdot \mathrm{d}\boldsymbol{s} = m\boldsymbol{g}\int_{\boldsymbol{r}_\text{A}}^{\boldsymbol{r}_\text{B}} \cdot \mathrm{d}\boldsymbol{s} = m\boldsymbol{g}\cdot(\boldsymbol{r}_\text{B} - \boldsymbol{r}_\text{A}) \tag{5.16}$$

is determined only by the positions of initial point $\boldsymbol{r}_\text{A}$ and end point $\boldsymbol{r}_\text{B}$ and not depending on the path through which the mass point travels*. Thus, the gravity $m\boldsymbol{g}$ is a conservative force and has the potential energy. Assuming the $+x$ direction as the vertically upward direction, since $m\boldsymbol{g} = (-mg, 0, 0)$, we obtain the following relation :

$$U(\boldsymbol{r}) = -\int_{\boldsymbol{r}_0}^{\boldsymbol{r}} m\boldsymbol{g}\cdot \mathrm{d}\boldsymbol{s} = -m\boldsymbol{g}\cdot(\boldsymbol{r} - \boldsymbol{r}_0) = mg(x - x_0). \tag{5.17}$$

If at the reference point $x_0 = 0$, then it follows that

$$U(x) = mgx \quad \text{(potential energy of gravity)}. \tag{5.18}$$

Here x is the height from the reference point, and readers are advised to memorize equation (5.18) as mgh.

* $\int_{\boldsymbol{r}_\text{A}}^{\boldsymbol{r}_\text{B}} \mathrm{d}\boldsymbol{s} = \boldsymbol{r}_\text{B} - \boldsymbol{r}_\text{A}$ is used.

According to the definition of equation (5.14), the potential energy $U(\boldsymbol{r})$ of the conservative force $\boldsymbol{F}_{\text{Con}}$ is

$$U(\boldsymbol{r}) = W^{\text{Con}}{}_{\boldsymbol{r}\to\boldsymbol{r}_0} \tag{5.19}$$

and $U(\boldsymbol{r})$ is equal to the work that the conservative force $\boldsymbol{F}_{\text{Con}}$ performs when the mass point moves from point $\boldsymbol{r}$ to referencepoint $\boldsymbol{r}_0$.

The path A → B in the integral of equation (5.13) for the conservative force can be chosen arbitrarily, so if we choose the path A → $\boldsymbol{r}_0$ → B, we obtain the following relation :

$$W^{\text{Con}}{}_{\text{A}\to\text{B}} = \int_{\boldsymbol{r}_\text{A}}^{\boldsymbol{r}_\text{B}} \boldsymbol{F}_{\text{Con}}(\boldsymbol{r})\cdot \mathrm{d}\boldsymbol{s} = \int_{\boldsymbol{r}_\text{A}}^{\boldsymbol{r}_0} \boldsymbol{F}_{\text{Con}}(\boldsymbol{r})\cdot \mathrm{d}\boldsymbol{s} + \int_{\boldsymbol{r}_0}^{\boldsymbol{r}_\text{B}} \boldsymbol{F}_{\text{Con}}(\boldsymbol{r})\cdot \mathrm{d}\boldsymbol{s}$$
$$= U(\boldsymbol{r}_\text{A}) - U(\boldsymbol{r}_\text{B}) \tag{5.20}$$

$$\therefore \quad W^{\text{Con}}{}_{\text{A}\to\text{B}} = U(\boldsymbol{r}_\text{A}) - U(\boldsymbol{r}_\text{B}). \tag{5.21}$$

Accordingly, "The work $W^{\text{Con}}{}_{\text{A}\to\text{B}}$ performed by the conservative force $\boldsymbol{F}_{\text{Con}}$ is equal to the decrease in potential energy $U(\boldsymbol{r})$ of the force $\boldsymbol{F}_{\text{Con}}$".

A line without an end like a circle is called a closed curve. In the case where the mass point moves from a point A on a closed curve C along the curve and reaches the initial point A, the work that the conservative force performs is 0, since the right-hand side of equation (5.21) is $U(\boldsymbol{r}_\text{A}) - U(\boldsymbol{r}_\text{B}) = 0$ (Fig. 5-9).

$$\oint_\text{C} \boldsymbol{F}(\boldsymbol{r})\cdot \mathrm{d}\boldsymbol{s} = 0 \quad \text{(In the case where } \boldsymbol{F}(\boldsymbol{r}) \text{ is a conservative force)}. \tag{5.22}$$

$\oint_\text{C}$ is a sign to indicate the integration along the closed curve C. Equation (5.22) is a necessary and sufficient condition for the force $\boldsymbol{F}(\boldsymbol{r})$ to

Fig. 5-8 The track of roller coaster is a closed curve.

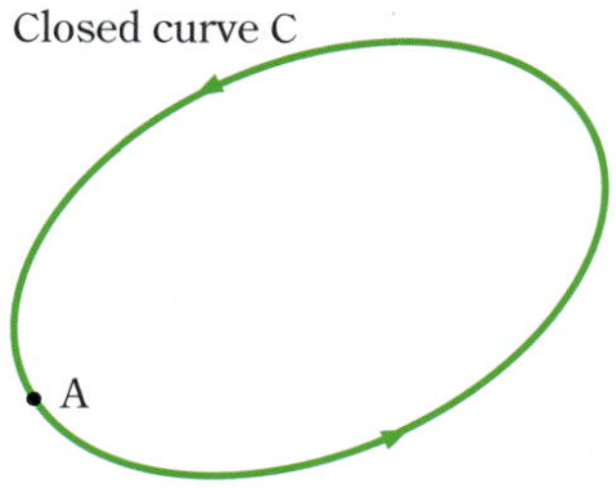

Fig. 5-9 A closed curve C

be conservative. Even a force is expressed in the form of $\boldsymbol{F}(\boldsymbol{r})$, if it does not satisfy equation (5.22), it is a non-conservative force.

Potential energy in one dimensional problems When a mass point moves along the x-axis subject to the force $\boldsymbol{F}(\boldsymbol{r}) = [F(x), 0, 0]$ which is parallel to the x-axis and whose magnitude is determined only by the x coordinate, since $\mathrm{d}\boldsymbol{s} = [\mathrm{d}x, 0, 0]$, equation (5.13) is reduced to the definite integral

$$W_{\mathrm{A}\to\mathrm{B}} = \int_{\boldsymbol{r}_\mathrm{A}}^{\boldsymbol{r}_\mathrm{B}} \boldsymbol{F}(\boldsymbol{r})\cdot \mathrm{d}\boldsymbol{s} = \int_{x_\mathrm{A}}^{x_\mathrm{B}} F(x)\,\mathrm{d}x. \tag{5.23}$$

Since this definite integral is determined only by x_A and x_B of the initial and end points of x coordinate, the force $\boldsymbol{F}(\boldsymbol{r}) = [F(x), 0, 0]$ is a conservative force. In this case the potential energy expressed by equation (5.14) of the force $\boldsymbol{F}(\boldsymbol{r})$ is reduced to

$$U(x) = -\int_0^x F(x)\,\mathrm{d}x. \tag{5.24}$$

Here $x = 0$ is chosen as the reference point. Comparing equation (5.24) with equation (3.11) we can find that $F(x)$ is derived from the potential energy $U(x)$ as

$$F(x) = -\frac{\mathrm{d}U}{\mathrm{d}x}. \tag{5.25}$$

Question 2 Find the magnitude and direction of the conservative force $F(x)$ with the potential energy shown in Fig. 5-10.

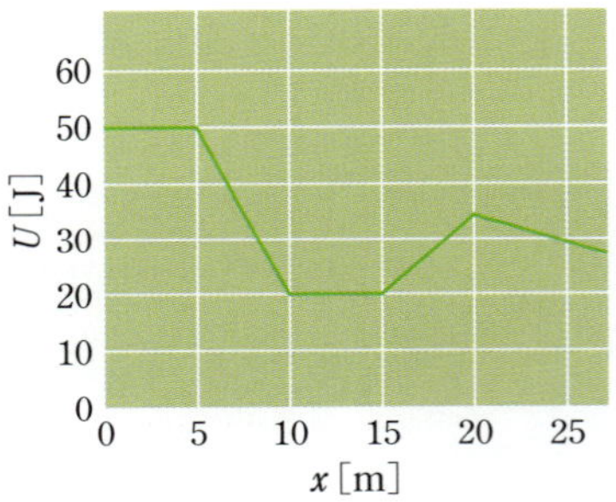

Fig. 5-10

Case 2 | Elastic potential energy of the spring In one dimensional problems, the case with $F(x) = -kx$ is the elastic force of the spring (Fig. 5-11). Therefore, the elastic force of the spring $F = -kx$ is a conservative force. Choosing the natural length ($x = 0$) as the reference point, we obtain for the potential energy $U(x)$

$$U(x) = -\int_0^x (-kx)\,\mathrm{d}x = \frac{1}{2}kx^2. \tag{5.26}$$

Since $U(x)$ is stored in the stretched or compressed spring, it is called the **elastic potential energy**.

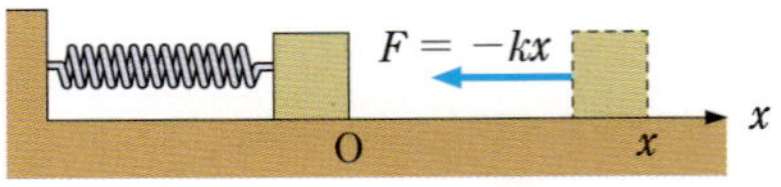

Fig. 5-11 Elastic force of the spring $F = -kx$

Potential energy of universal gravitation Assume that an object with mass m_1 is at the origin O and it exerts the universal gravitation

$$\boldsymbol{F}(\boldsymbol{r}) = -G\frac{m_1 m_2}{r^2}\frac{\boldsymbol{r}}{r} \tag{5.27}$$

on other object with mass m_2 and located at point $\boldsymbol{r}$*. As is shown in Fig. 5-12, since

$$\boldsymbol{r}\cdot\mathrm{d}\boldsymbol{s} = r\,\mathrm{d}s\cos\theta = r\,\mathrm{d}r \tag{5.28}$$

so the work the universal gravitation performs when two objects are slowly separated and the distance is made to be infinite is independent of the path and

$$W_{r\to\infty} = \int_r^\infty \boldsymbol{F}(\boldsymbol{r})\cdot\mathrm{d}\boldsymbol{s} = -\int_r^\infty G\frac{m_1 m_2}{r^2}\frac{\boldsymbol{r}}{r}\cdot\mathrm{d}\boldsymbol{s} = -\int_r^\infty G\frac{m_1 m_2}{r^2}\,\mathrm{d}r.$$

Therefore, if we select the situation in which the distance between two objects with mass m_1 and mass m_2 is infinitely apart as the reference point, the potential energy of universal gravitation $U(r) = W_{r\to\infty}$ is

* $\dfrac{\boldsymbol{r}}{r}$ is a vector of length 1 in the direction of $\boldsymbol{r}$.

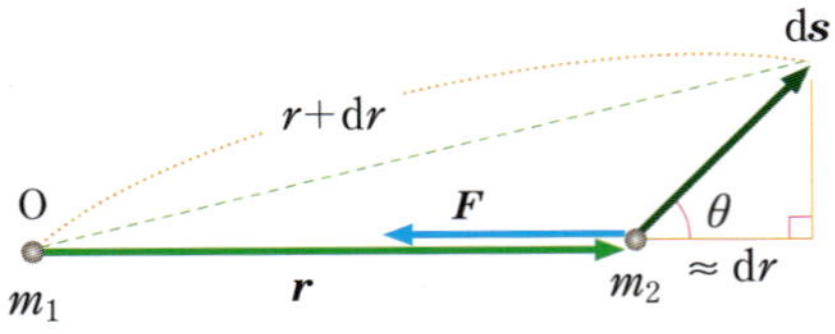

Fig. 5-12 $\boldsymbol{r}\cdot\mathrm{d}\boldsymbol{s} = r\,\mathrm{d}s\cos\theta = r\,\mathrm{d}r$

$$U(r) = -\int_r^{\infty} G\frac{m_1 m_2}{r^2}\,dr = G\frac{m_1 m_2}{r}\bigg|_r^{\infty} = -G\frac{m_1 m_2}{r}. \tag{5.29}$$

When objects spread out in space and the mass distribution of each object is spherically symmetric, the potential energy of universal gravitation corresponds to the situation in which r in equation (5.29) is taken to be the distance between the centers of two objects. The universal gravitation acts between two objects, so the potential energy of universal gravitation is the energy of a system consisting of two objects.

For reference Derivation of conservative force from potential energy

In the case of general conservative force $\boldsymbol{F}$, the relation between conservative force and potential energy corresponding to one dimensional case of equation (5.25) is

$$F_x = -\frac{\partial U}{\partial x}, \qquad F_y = -\frac{\partial U}{\partial y}, \qquad F_z = -\frac{\partial U}{\partial z}. \tag{5.30}$$

$\frac{\partial U}{\partial x}$ is a partial derivative of the function U with respect to x with the other variables y and z kept constant, and is obtained as

$$\frac{\partial U}{\partial x} = \lim_{\Delta x \to 0} \frac{U(x+\Delta x, y, z) - U(x, y, z)}{\Delta x}. \tag{5.31}$$

Equation (5.30) is readily derived from the relation $U(\boldsymbol{r}+\Delta\boldsymbol{r}) - U(\boldsymbol{r}) = W_{\boldsymbol{r}+\Delta\boldsymbol{r}\to\boldsymbol{r}} \approx -\boldsymbol{F}\cdot\Delta\boldsymbol{r}$ and the definition of the partial derivative. For example, when $\Delta\boldsymbol{r} = (\Delta x, 0, 0)$, $-\boldsymbol{F}\cdot\Delta\boldsymbol{r} = -F_x\,\Delta x$ and using equation (5.31) we obtain the first equation of equation (5.30).

Introducing a vector differential operator expressed by the sign ∇ called nabla

$$\nabla = \left(\frac{\partial}{\partial x}, \frac{\partial}{\partial y}, \frac{\partial}{\partial z}\right) \tag{5.32}$$

we can express equation (5.30) as

$$\boldsymbol{F}(\boldsymbol{r}) = -\nabla U(\boldsymbol{r}). \tag{5.33}$$

Using the property of vector product $\nabla \times \nabla = 0$, we can derive from equation (5.33)

$$\nabla \times \boldsymbol{F}(\boldsymbol{r}) = 0 \qquad \text{(In the case where } \boldsymbol{F} \text{ is a conservative force).} \tag{5.34}$$

This equation is a necessary and sufficient condition for $\boldsymbol{F}(\boldsymbol{r})$ to be conservative and is a differential expression of the condition of equation (5.22) in the integral form.

5.3 Energy conservation law

Learning objective To understand that when forces acting on the mass point do not include non-conservative forces, the energy conservation law holds, namely "mechanical energy" = "kinetic energy" + "potential energy" = constant and to be able to apply the law to various cases. To understand that when non-conservative forces act on a mass point, the mechanical energy of the mass point changes by an amount corresponding to the work performed by the non-conservative forces and that the mechanical energy is transformed through an intermediary of non-conservative forces into chemical energy, internal energy, etc.

Fig. 5-13 Yatogawa-Daisan Power Plant, Shimane Prefecture (maximum power 240 kW)

Conservation law of mechanical energy Let us consider the situation in which forces $\boldsymbol{F}$ acting on the mass point do not include non-conservative forces. In this situation, $\boldsymbol{F} = \boldsymbol{F}_{\text{Con}} + \boldsymbol{F}_{\text{constrain}}$. We then combine equation (5.12) of the relationship between work and kinetic energy with equation (5.21) of the relation between work performed by the conservative force and potential energy to derive the following equation

$$\frac{1}{2}mv_{\text{B}}{}^2 - \frac{1}{2}mv_{\text{A}}{}^2 = W^{\text{con}}{}_{\text{A}\to\text{B}} = U(\boldsymbol{r}_{\text{A}}) - U(\boldsymbol{r}_{\text{B}}). \qquad (5.35)$$

From equation (5.35) we can derive the **conservation law of the mechanical energy** which says "mechanical energy" = "kinetic energy" + "potential energy" = constant, namely

$$\frac{1}{2}mv_{\text{B}}{}^2 + U(\boldsymbol{r}_{\text{B}}) = \frac{1}{2}mv_{\text{A}}{}^2 + U(\boldsymbol{r}_{\text{A}}) = \text{constant}. \qquad (5.36)$$

The mechanical energy of the mass point with mass m subject only to gravity and constraining force, namely the sum of potential energy and kinetic energy is constant (Fig. 5-14):

$$\frac{1}{2}mv^2 + mgh = \text{constant}. \qquad (5.37)$$

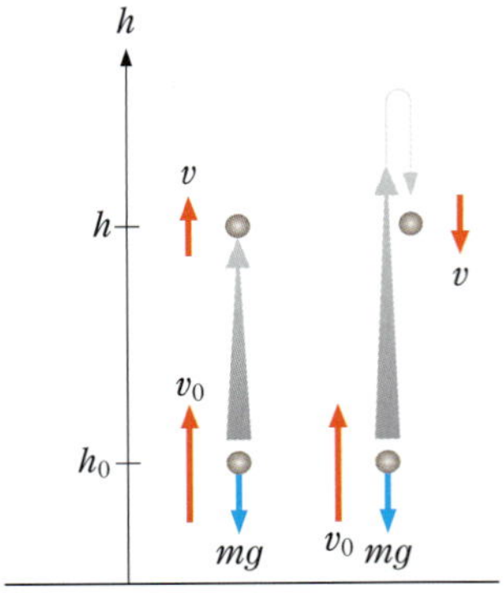

Fig. 5-14 $\frac{1}{2}mv^2 + mgh = \frac{1}{2}mv_0{}^2 + mgh_0$. The speed when objects thrown up at the same speed v_0 pass a fixed height and the speed when they fall passing the same fixed height are the same.

Case 3 When a person riding a bicycle goes down the height $h = 5$ m hill without pedaling at an initial speed zero, the potential energy of gravity mgh at the top of the hill is changed into the kinetic energy $\frac{1}{2}mv^2$, so the speed v at the foot of the hill is

$$v = \sqrt{2gh} = \sqrt{2 \times (9.8\ \text{m/s}^2) \times 5\ \text{m}} = 10\ \text{m/s}$$

(Fig. 5-15).

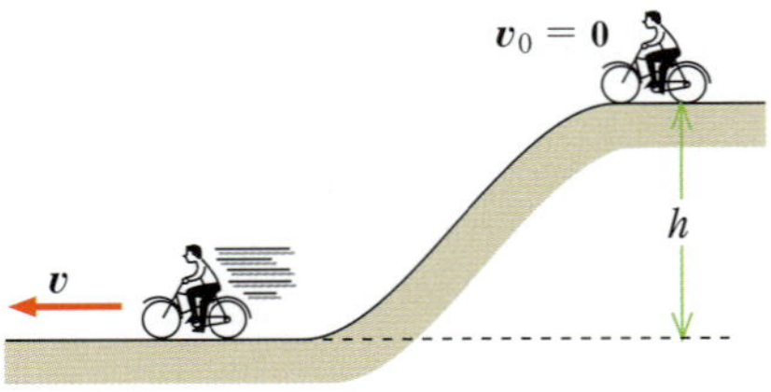

Fig. 5-15 The speed at the foot of the hill

Example 2 ‖ Escape velocity We are wanting to launch a rocket and let it escape from the Earth's gravitational zone and reach a position infinitely far from the Earth. Find the minimum value of the initial speed v of the rocket at launching. The rocket is a one-stage rocket and the effect of rotation of the Earth is assumed to be neglected.

Solution From equation (5.29) and equation (2.12), the potential energy $U(R_{\text{E}})$ of universal gravitation of the object of mass m on the Earth (radius R_E, mass M_E) surface is

$$U(R_E) = -G\frac{M_E m}{R_E} = -mgR_E \qquad (5.38)$$

(Fig. 5-16). Therefore, when the rocket with mass m is launched from the Earth surface at the initial speed v, the mechanical energy E of the rocket is

$$E = \frac{1}{2}mv^2 + U(R_E) = \frac{1}{2}mv^2 - mgR_E. \quad (5.39)$$

While this rocket moves in outer space, the mechanical energy is conserved. When the rocket reaches the infinitely far position with potential energy 0 at speed v_∞, the mechanical energy is wholly occupied only by the kinetic energy $\frac{1}{2}mv_\infty{}^2$. Thus, the condition for the rocket to be able to escape from the Earth's gravitational zone is

$$E = \frac{1}{2}mv^2 - mgR_E = \frac{1}{2}mv_\infty{}^2 \geqq 0. \qquad (5.40)$$

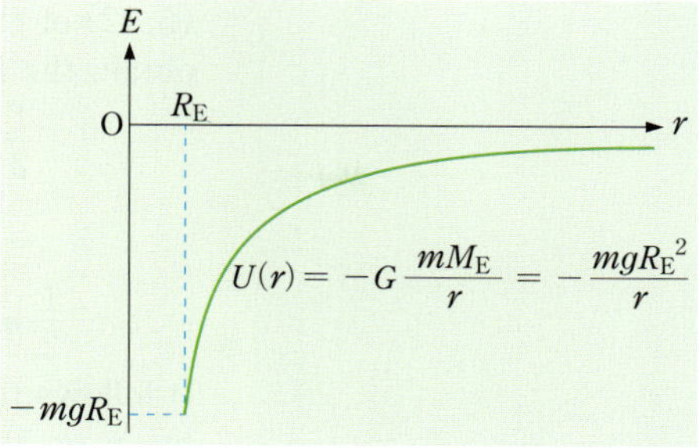

Fig. 5-16 Potential energy of Earth's universal gravitation

$$U(r) = -G\frac{mM_E}{r} = \frac{mgR_E{}^2}{r}$$

Accordingly, the minimum speed (escape velocity) required for the rocket is, from $\frac{1}{2}mv^2 - mgR_E = 0$,

$$v = \sqrt{2gR_E} = \sqrt{2\times(9.8\ \mathrm{m/s^2})\times(6.37\times10^6\ \mathrm{m})}$$
$$= 1.12\times10^4\ \mathrm{m/s} = 11.2\ \mathrm{km/s}. \qquad (5.41)$$

Conservation law of mechanical energy in simple harmonic oscillation due to elastic force Speed of an object which performs the simple harmonic motion with amplitude A

$$x(t) = A\cos(\omega t + \theta_0) \qquad (5.42)$$

due to elastic force $F = -kx$ of a spring is

$$v(t) = -\omega A\sin(\omega t + \theta_0). \qquad (5.43)$$

From equation (5.42) and equation (5.43) it follows that

$$\frac{1}{2}mv^2 + \frac{1}{2}kx^2 = \frac{1}{2}A^2(m\omega^2)\sin^2(\omega t + \theta_0) + \frac{1}{2}A^2k\cos^2(\omega t + \theta_0)$$
$$= \frac{1}{2}kA^2 = \frac{1}{2}m\omega^2A^2 = \text{constant}. \qquad (5.44)$$

Thus, the **conservation law of mechanical energy,**

"kinetic energy"+"elastic potential energy" = constant

is derived ($m\omega^2 = k$ and $\sin^2 x + \cos^2 x = 1$ are used) (Fig. 5-17). In this case, the mechanical energy $\frac{1}{2}m\omega^2A^2$ is proportional to the square of amplitude A as well as to the square of angular frequency ω.

Question 3 For the object performing the simple harmonic oscillation, derive the relation between the maximum velocity v_{max} and the maximum displacement x_{max}

$$v_{max} = \omega x_{max} = 2\pi f x_{max}.$$

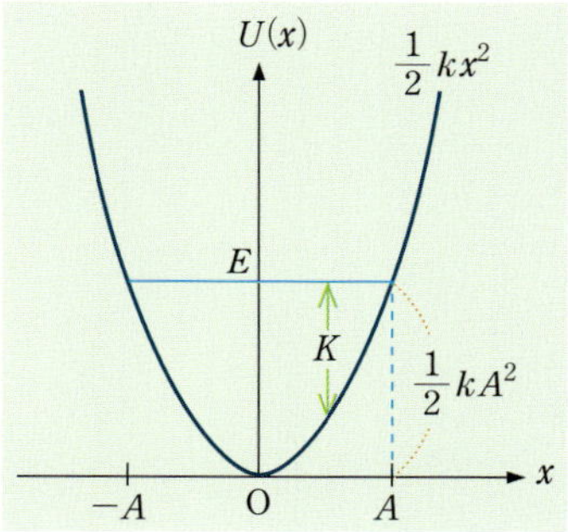

Fig. 5-17 Energy of simple harmonic oscillation.
The work which the weight performs on the spring during stretch and shrink of the spring is changed into the elastic potential energy $U(x)$ of the spring. The work which the elastic force of the spring performs on the weight when the stretched spring shrinks, and the shrunk spring stretches is changed into the kinetic energy K of the weight.

Work performed by non-conservative force and mechanical energy Hand force and frictional force are not the force in the form of $\boldsymbol{F}(\boldsymbol{r})$, so these forces are not conservative. When the non-conservative force $\boldsymbol{F}_{Non}$ acts on a mass point, we substitute equation (5.21) of the relationship between conservative force and potential energy into equation

(5.12) of the relationship between work and kinetic energy, and we obtain the following relation :

$$\frac{1}{2}mv_B^{\ 2}-\frac{1}{2}mv_A^{\ 2} = W^{\text{Con}}{}_{A\to B}+W^{\text{Non}}{}_{A\to B}$$

$$= U(\boldsymbol{r}_A)-U(\boldsymbol{r}_B)+\int_A^B \boldsymbol{F}_{\text{Non}}(\boldsymbol{r})\cdot d\boldsymbol{s}$$

$$\therefore \quad \frac{1}{2}mv_B^{\ 2}+U(\boldsymbol{r}_B) = \frac{1}{2}mv_A^{\ 2}+U(\boldsymbol{r}_A)+\int_A^B \boldsymbol{F}_{\text{Non}}(\boldsymbol{r})\cdot d\boldsymbol{s}. \quad (5.45)$$

It follows therefore that the mechanical energy of mass point increases by an amount equal to $W^{\text{Non}}{}_{A\to B}$ performed by the non-conservative force $\boldsymbol{F}_{\text{Non}}$ (when $W^{\text{Non}}{}_{A\to B}$ is negative, the mechanical energy decreases).

An example in which the work $W^{\text{Non}}{}_{A\to B}$ performed by the non-conservative force $\boldsymbol{F}_{\text{Non}}$ is positive is the case where we hold an object of mass m in our hand and slowly lift it to height h against gravity. In this case the magnitude of force of the hand is roughly mg and the work W performed by the hand is roughly mgh. This work becomes the increased amount mgh in the potential energy of gravity. The source of the work that the hand performs comes from the chemical energy of the arm muscle.

An example in which the work $W^{\text{Non}}{}_{A\to B}$ performed by the non-conservative force $\boldsymbol{F}_{\text{Non}}$ is negative is the raindrop falling at a constant speed (terminal velocity) v_t in the air while receiving the viscous drag. Since raindrops are moving at a constant speed, their kinetic energy is constant. Therefore, when the raindrop of mass m falls by the height h, both the potential energy of gravity and the mechanical energy of the raindrop are reduced by the amount mgh. The cause of this decrease is that the direction of the viscous drag $bv_t(=mg)$ which acts on raindrops is opposite to the direction of motion of the raindrop, so the work of viscous drag acting on the raindrop is a negative amount $-bv_th = -mgh$. In this case, the mechanical energy lost by the negative work caused by the resistance force of the air changes into heat. Strictly speaking, the lost mechanical energy is transformed to the internal energy which is the energy of thermal motion of water molecules in raindrops and gas molecules in the air. In addition, assuming that one calorie (cal), the practical unit of heat, is about 4.2 Joules (J), Joule confirmed in his experiment shown in Fig. 5-18 in 1843 that the sum of

Practical unit of heat
1 cal = 4.2 J

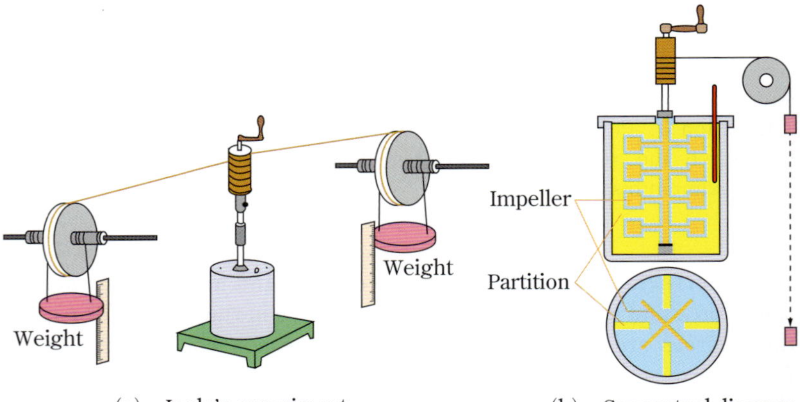

Fig. 5-18 Joule's experiment
The impeller which is rotated by the descent of the weight stirs water and the water temperature rises. Joule confirmed that if 1 cal, which is the amount of heat required to raise the temperature of 1 g of water by 1 °C, is equal to 4.2 J, then the "potential energy of gravity of the weight"+"heat" is constant.

internal energy and mechanical energy is conserved. In this experiment, the potential energy of gravity of the weight is transformed, through the work that the impeller produces, into the kinetic energy of water in the container, eventually causing the rise in water temperature, that is, the internal energy of water (Refer to Section 14.1).

Energy conservation law In addition to heat, energy closely related to our daily life is electric energy and chemical energy. Electric energy is converted into mechanical energy by mechanical work performed by the motor, or it is converted into heat (internal energy) by an electric heater. Also, mechanical energy is converted into electric energy through mechanical work on the generator. Petroleum and coal as energy sources generate heat by combustion, but since combustion is a chemical change, the energy of petroleum and coal is called the **chemical energy**. Human work depends on the chemical energy stored in muscle.

Fig. 5-19 Petrochemical complex at Yokkaichi City

According to the theory of relativity, mass is a form of energy, and when mass m turns into energy E of another form, its amount is $E = mc^2$ (c is the speed of light in vacuum). In nuclear power generation, the mass is reduced by a certain type of nuclear reaction, and the energy is utilized as the kinetic energy of the reaction product.

In this way, when various forms of energy are considered, the form of energy changes and the carrier of energy also changes, but it is confirmed by experiments that the total amount of energy is always constant and does not increase or decrease. This fact is called the **energy conservation law**. The conservation of a certain physical quantity means that the physical quantity does not change with time and is kept constant. The idea of conservation of energy was proposed by Meyer, Jules, Helmholtz etc. by the middle of the 19th century. Later, the law of conservation of energy was confirmed experimentally and is now accepted as one of the most fundamental laws of physics.

Expression of energy conservation law We compare the situation before and after a process. Let the increased amount in kinetic energy of an object system (a collection of one or more objects) be $\Delta K (= K_{\mathrm{after}} - K_{\mathrm{before}})$ and the increased amount in potential energy be ΔU^{Con}. Similarly, we describe the increased amount in internal energy (sum of kinetic energy and potential energy of thermal motion of molecules constituting the object system) as ΔU, increase in chemical energy as $\Delta E_{\mathrm{chemical}}$, increase in work $W^{\mathrm{Non}}_{\mathrm{system}\leftarrow\mathrm{outside}}$ performed on the object system from the outside by the non-conservative force as W and heat $Q_{\mathrm{system}\leftarrow\mathrm{outside}}$ transferred from the outside to the system as Q. Then the energy conservation law ca be expressed as

$$\Delta K+\Delta U^{\mathrm{con}}+\Delta U+\Delta K++\Delta E_{\mathrm{chemical}} = W+Q. \qquad (5.46)$$

If other kinds of energy may change, such factors are added on the left-hand side. W on the right-hand side is the product of displacement distance s of the point of action of the external force (non-conservative force) $\boldsymbol{F}_{\mathrm{non}}$ and the component $F_{\mathrm{non\,t}}$ of $\boldsymbol{F}_{\mathrm{non}}$ in the displacing direction of the point of action, or $W = sF_{\mathrm{non\,t}}$.

When the object system exchanges work and heat, we consider a new system including the part outside the original system. Then in the new system, $W = Q = 0$. So equation (5.46) becomes

$$\Delta K + \Delta U^{\mathrm{con}} + \Delta U + \Delta E_{\mathrm{chemical}} = 0. \tag{5.47}$$

Equation (5.47) is the energy conservation law of the object system isolated from the outside.

Fig. 5-20 On the running car, the resistance force of water and the frictional force from the surface of the road act.

Case 4 | Motion of cars and conservation law of energy Among the external forces acting on cars, non-conservative forces are those of air resistance and friction force with which the road surface acts on the tires. If the tires do not slide at the point of contact between the road surface and the tires, the point of action does not move, so the frictional force acting on the tires on the road surface does not work on the tires. Thus with ignoring the heat transfer Q from the outside to the car and the air resistance, the law of conservation of energy (5.46) of the car becomes

$$\Delta K + \Delta U^{\mathrm{Con}} + \Delta U = -\Delta E_{\mathrm{chemical}}. \tag{5.48}$$

This equation shows that increases in kinetic energy ΔK of the car and in potential energy of gravity ΔU^{Con} and in internal energy ΔU of the car due to temperature increase are due to the decrease $-\Delta E_{\mathrm{chemical}}$ in chemical energy of fuel consumed in the engine. If the Earth and the air are included in the object system, since the external force of the non-conservative force does not act, equation (5.48) holds as is even if the tires slide or the air resistance is not negligible. But in this case ΔU includes the increase in internal energy due to temperature rise of the air and the road surface.

Problems for exercise 5

A

1. (1) When a weightlifting player lifts the barbell of mass $m = 80$ kg slowly to a height of 2.0 m, how much amount of work (in Joule) does the player do on the barbell?
   (2) This player moved 5 m sideways while lifting the barbell. How much amount of work (in Joule) did the player do on the barbell?
   (3) This player gently lowered the barbell to the floor. How much amount of work (in Joule) did the player do on the barbell?
2. A pitcher threw a baseball ball of 0.15 kg at a speed of 144 km/h. How much is the kinetic energy of this ball? How much amount of work (in Joule) did the pitcher do on the ball?
3. When the roller coaster makes one round of the course, how much amount of work does the gravity performs on passengers?
4. Two identical balls were thrown from the roof of the building at the same speed but in different directions When the balls reach the ground, are their speeds different? Ignore the air resistance.
5. A ball was gently released at point A on the frictionless slope, as shown in Fig. 1. When the ball popped out from the point B in Fig. 1, which trajectory is correct, a or b?

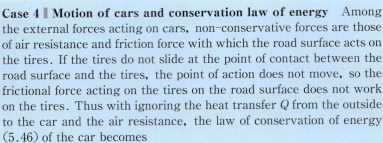

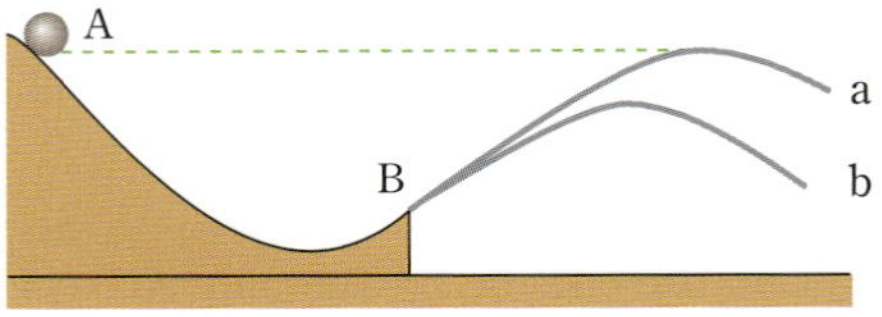

Fig. 1

6. A pendulum with the string length L and the weight of mass m, with its string kept horizontally, was released at an initial velocity of 0. Find the tension S of the string when the string becomes vertical.
7. As shown in Fig. 2, the weight is hung from the ceiling with a thread of 1 m in length. We keep the thread-and-weight at an angle of 30° to the vertical

and then release it gently. After the thread comes into contact with the 50 cm high shelf, the weight reaches the highest point. Find the angle θ between the thread and the vertical line.

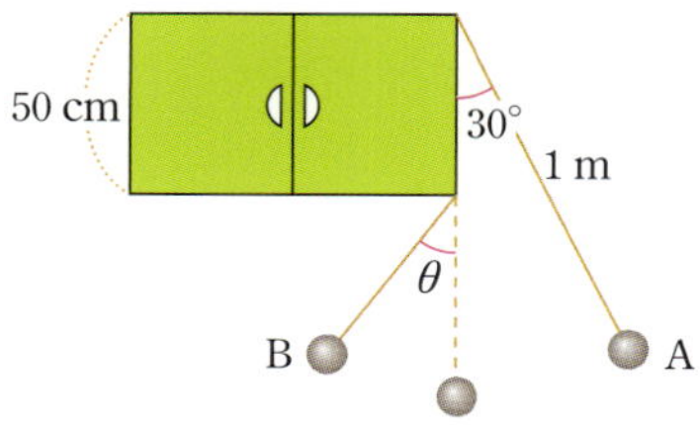

Fig. 2

8. A person with a weight of 50 kg is running up the stairs at a rate of 2 m in height per second. Find the power of the work that the person is doing on himself/herself.
9. We are wanting to raise the steel of mass 1 ton up at the rate of 10 m/min. Assuming that there is no loss due to friction or the like, find the minimum wattage required for the motor of the crane.
10. At the Sudagai Power Plant in Gunma Prefecture, water of 65 m^3/s falls with the effective head of 77 m and rotates the water turbine of the generator to generate electricity of 46000 kW. In this power plant, what percentage of the potential energy of gravity of water is changed into electric energy?
11. A person with a body weight of 40 kg climbs a mountain with a height of 3000 m.
    (1) How much amount of work is performed by that person?
    (2) One kilogram of fat supplies 3.8×10^7 J of energy. If the person converts fat energy to the work with 20 % efficiency, how much amount of fat will be reduced in this mountain climbing?
12. **Joule's experiment** ▌ In the apparatus of Fig. 5-18, the temperature of water rises because the impeller rotated by the descent of the weight stirs the water. Experiments were carried out using 0.5 kg of water and a weight in total of 3.0 kg. When the experiment to let the weight down from the height of 3.0 m is repeated ten times, how much amount will be the rise in water temperature? Assume that the heat capacity of the container is negligible.
13. As shown in Fig. 3, with a toy called "pachinko" using a rubber band, we launch out a small ball. At this time, suppose that the elastic potential energy of the stretched rubber changes to the kinetic energy of the ball. How many times will the initial velocity of the ball be when the rubber is stretched twice? If we launch this ball vertically into the sky, will it reach up to the 4 times higher position? In addition, when we launch the ball horizontally how many times will the reaching distance be longer?

Fig. 3

B

1. A bicycle with a mass of 75 kg including a rider went up a straight road with a tilt angle of 5° at a speed of 10.8 km/h for 2 minutes. Find the height h at which the rider reached and the power necessary to go up to this height. Assume sin 5° = 0.087.
2. When a batter hits the ball thrown by a fast-ball pitcher at the same speed, the kinetic energy does not change. In this case, how much amount of work does the batter do on the ball?
3. A weight is hung by a thread of length L from the ceiling. We keep the thread horizontal as shown in Fig. 4 and release the hand gently from it. When the thread becomes vertical, the thread comes into contact with the rod P, and the weight moves on the arc of the radius r. Show that the condition that the weight, with the thread kept stretched, moves on the circumference with a radius r around the rod P is $d = L - r \geqq \frac{3}{5}L$ in Fig. 4.

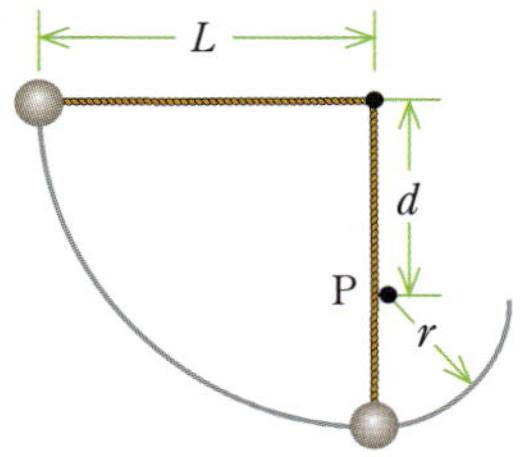

Fig. 4

4. A car running at 80.5 km/h (22.4 m/s) needs 4.85×10^3 W to overcome the air resistance, and 3.1×10^3 W to compensate for the friction with road surface and the mechanical losses.
    (1) Show that the work required to move this car 1 m at 80.5 km/h is 3.6×10^2 J.
    (2) The mass of this car is 1.02×10^3 kg. Show that when the car is running on a horizontal road at 22.4 m/s, the ratio of frictional force and

gravity acting on the road surface is 0.036.

(3) The consumption of gasoline at this time is 17 km/L. The chemical energy of gasoline is 3.3 $\times 10^7$ J/L. What is the percentage of chemical energy of gasoline which changed into the mechanical work?

How much amount of work can a human perform with muscular strength a day?

Since we have learned the law of conservation of energy, so let us consider a human as a machine which performs the work when energy is supplied.

In nutrition science, kcal is commonly used as a unit of energy. For an adult male engaged in ordinary labor, energy of about 2400 kcal seems to be sufficient for ingesting as food as a day. The energy of this food is a chemical energy corresponding to the amount of heat generated when the food burns (food is oxidized). 2400 kcal is 10 million J, or 2.8 kWh-hour (kWh. $1\,\text{kWh} = 3.6\times10^6\,\text{J}$), that is, it is 2800 Wh. So, if all this energy is used for human muscle labor and a human can continue working for 24 hours without intermission, then, he/she will be a working machine whose power is about 120 W (2800 Wh/24 h = 120 W).

However, even while sleeping, humans consume energy for breathing, blood circulation and other activities. The power of a human working muscularly is said to be on average 100 W or less (the average work per second is 100 J or less), and the work that can be performed in one year is only about 100 kWh at best.

A work of 100 W is a work of lifting 10 kg of baggage at a rate of 1 m per second. If a 3600 m tower (nearly the same height as Mt. Fuji) stands on a flat ground, the work to raise 10 kg of baggage up to the tip of this tower is 100 Wh, that is, 0.1 kWh. Assuming electricity cost of 1 kWh is 30 yen, the electric bill is 3 yen. If electric appliances can efficiently perform all 100 kWh of annual muscle work of one human, the electricity bill is 3000 yen.

Up to about 200 years ago, humans had relied mainly on human and cattle muscular forces as power sources. Wind and hydraulic power brought by the nature could only be used in a limited space, like sailing boats and water mills. Even if we use a horse, we can only draw out about 15 human power.

However, new power sources such as steam engines, generators and motors etc., were invented by the development of science and technology, and the situation changed drastically since the industrial revolution occurred. In Japan the amount of generated electric energy in a year per capita is about 10^4 kWh. If we assume that the human muscle labor is 100 kWh per year, through the use of electricity the situation of Japanese people is equivalent to using about 100 slaves per person.

This large amount of energy consumption had a big impact on our social life. The reason why the style of family changed from a large family to a two-generation family is the fact that labor-saving was promoted by the mass consumption of energy, The key to understand the change in Japanese society since the second half of the 20th century is the massive consumption of energy. Large consumption of energy provokes the environmental destruction. Also, resources are finite. Energy problems are one of the major problems faced to humans.

Angular Momentum of Mass Point and Law of Rotational Motion

The cause to change the momentum which is a quantity expressing the intensity of motion of the object is the force. However, when the motion of an object is discussed as a rotation around a certain point, it is appropriate to consider the angular momentum as a quantity representing the intensity of the rotational motion and take the moment of force as the cause of changing the angular momentum.

In this chapter, angular momentum, moment of force and law of rotational motion will be studied. First, we consider the planar motion of a mass point. In other words, the case where the mass point subject to the action of the force $\boldsymbol{F} = (F_x, F_y, 0)$ parallel to the xy plane moves on the xy plane is considered. In rotational motion, the point of action where the force acts on the object and the straight line of action of force which passes through the point of action and is oriented in the direction of force are important.

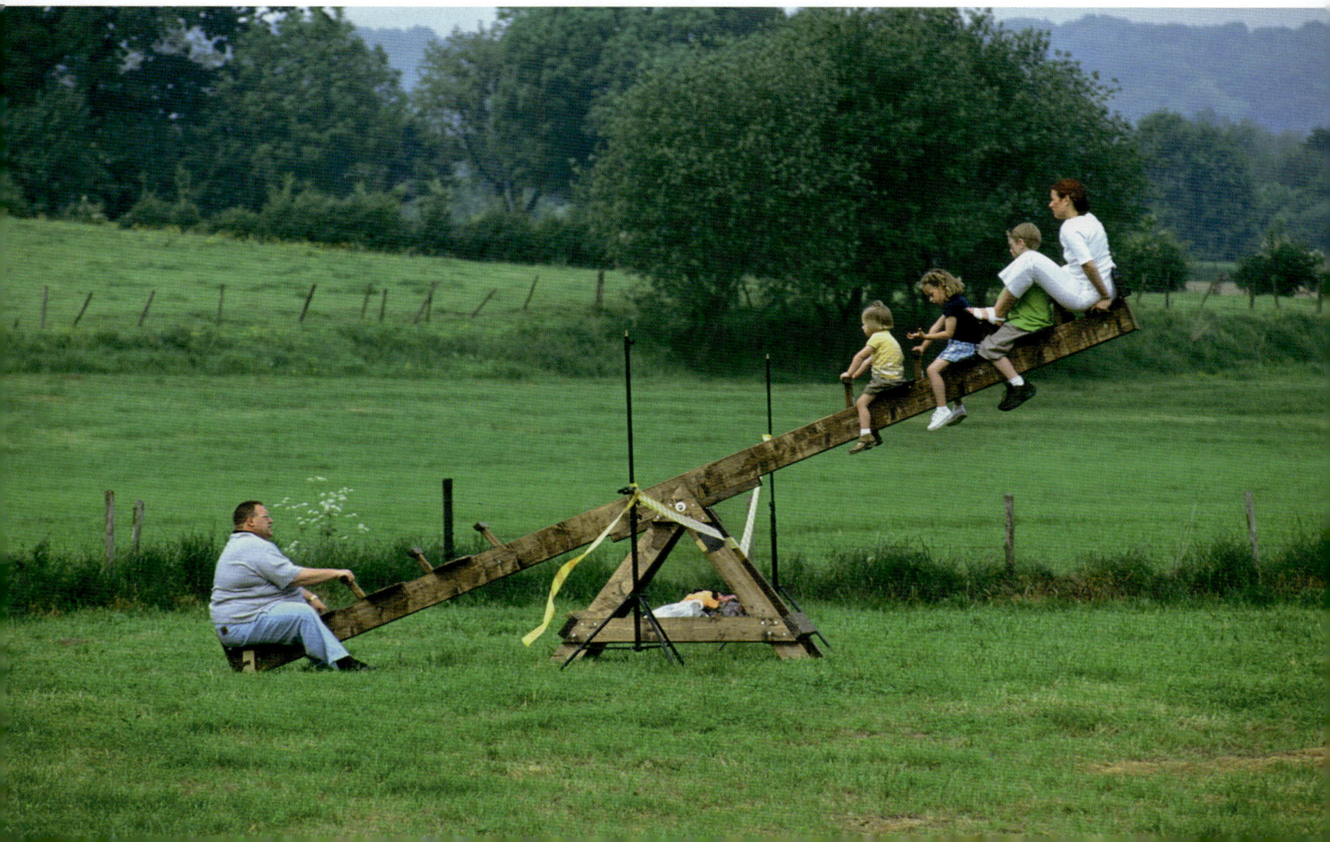

6.1 Rotational motion of mass point–case of planar motion

Learning objective To understand the angular momentum as a quantity representing the intensity of rotational motion and the moment of force as a cause to change the angular momentum, and also the law of rotational motion telling that the time rate of change in angular momentum is equal to the moment of force.

To understand what the central force is and that when a mass point is moving subject solely to the action of the central force, the angular momentum conservation law that the angular momentum around the center of force is constant holds.

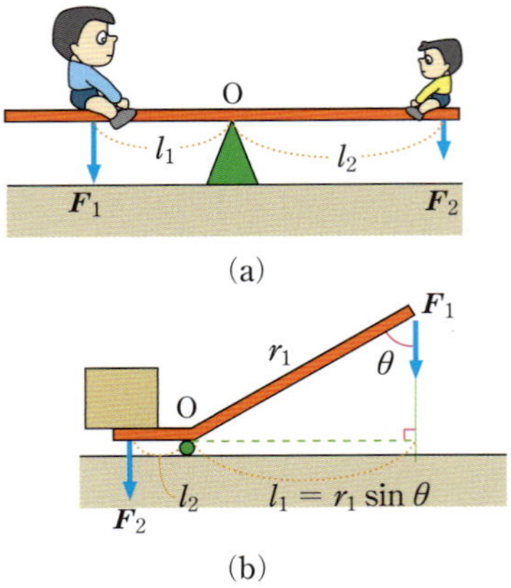

Fig. 6-1 (a) When $F_1 l_1 = F_2 l_2$, the seesaw balances.
(b) When $F_1 l_1 (= F_1 r_1 \sin\theta) > F_2 l_2$, the luggage is lifted.

Unit of moment of force
N·m

Moment of force *F* Through our experience of playing with a seesaw and/or lifting heavy objects with a leverage, we recognize well that the ability of the force acting on the object to rotate it around the fulcrum (rotation axis) is given by

"magnitude F of force"
×"distance l from fulcrum O to the line of action of force"

(Fig. 6-1). We call this quantity

$$N = Fl \tag{6.1}$$

the **moment** of force $\boldsymbol{F}$ or **torque** around the point O. When the angle θ is defined as shown in Fig. 6-2, since the distance from the point O to the line of action of force is $l = r\sin\theta$, equation (6.1) is rewritten as

$$N = Fr\sin\theta. \tag{6.2}$$

The unit of moment of force is N·m.

We distinguish the differences in the direction in which the force $\boldsymbol{F}$ will rotate the object by assigning a plus or minus sign on the moment of force. If the direction of rotation is counterclockwise, we assign a plus sign ($N = Fl$) and in the case of clockwise direction, a minus sign ($N = -Fl$) (Fig. 6-3).

As illustrated in Fig. 6-4, when a force $\boldsymbol{F}$ parallel to the xy plane acts on the point $(x, y, 0)$ on the xy plane, the moment N of the force $\boldsymbol{F}$ around the origin O is

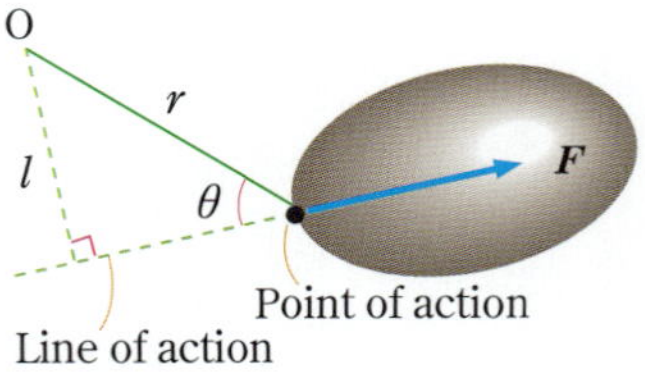

Fig. 6-2 Moment N of force $\boldsymbol{F}$ around the point O
$N = Fl = Fr\sin\theta$

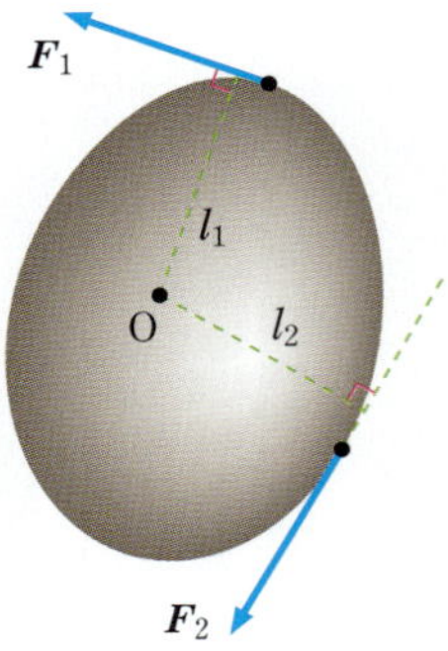

Fig. 6-3 $N = F_1 l_1 - F_2 l_2$

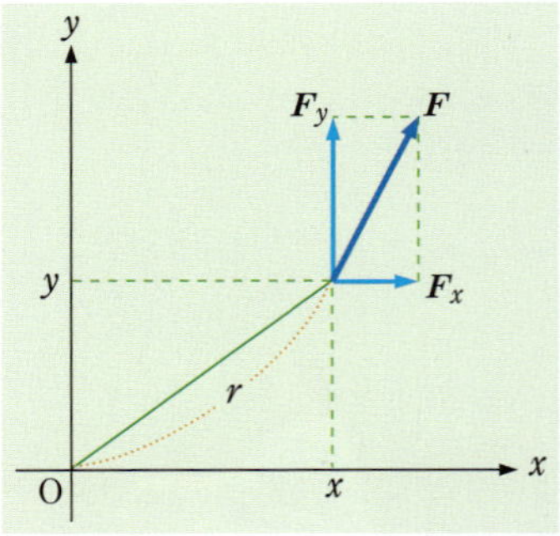

Fig. 6-4 Moment N of the force $\boldsymbol{F}$ around the origin O
$N = xF_y - yF_x$

$$N = xF_y - yF_x. \qquad (6.3)$$

This is because the contribution of the component $\boldsymbol{F}_x$ of the force $\boldsymbol{F}$ on the moment is $-yF_x$, and the contribution of the component $\boldsymbol{F}_y$ is xF_y.

Angular momentum The moment of momentum $\boldsymbol{p} = m\boldsymbol{v}$ around the point O corresponding to the moment of force is called the **angular momentum** around the point O. The magnitude of the angular momentum L around the point O of the mass point at the point P with mass m, velocity $\boldsymbol{v}$, and momentum $\boldsymbol{p} = m\boldsymbol{v}$ shown in Fig. 6-5 is expressed as the product of "magnitude of momentum $p = mv$" and "length of the perpendicular drawn to the velocity vector $\boldsymbol{v}$ passing through the mass point P from the point O, or $d = r \sin \phi$"

$$L = mvd = mvr \sin \phi. \qquad (6.4)$$

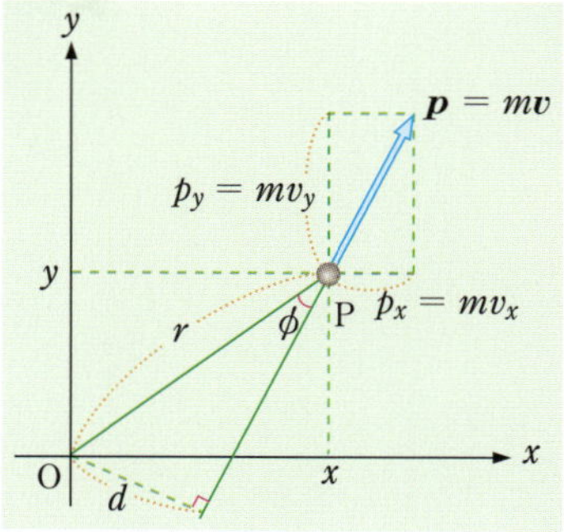

Fig. 6-5 Angular momentum around the point O

$L = pd = mvd = pr \sin \phi$

$L = m(xv_y - yv_x)$

Similarly, as the case of the moment of force, we assign a plus or minus sign to the angular momentum L. The angular momentum is a quantity representing the intensity of rotational motion.

Corresponding to equation (6.3), the angular momentum L of the mass point of mass m moving on the xy plane around the origin O is

$$L = m(xv_y - yv_x). \qquad (6.5)$$

Case 1 The mass point of mass m is moving at a constant speed on the circumference of the circle of radius r centered on point O at angular velocity ω and velocity $v = r\omega$. In this case, the angular momentum L of this mass point around point O is

$$L = mvr = mr^2\omega \qquad (6.6)$$

(Fig. 6-6).

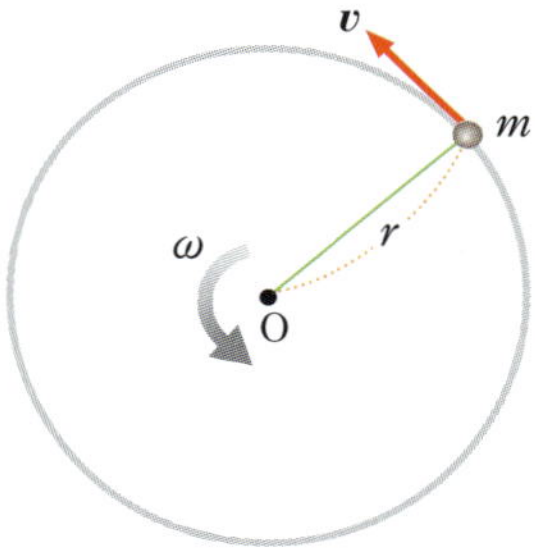

Fig. 6-6 In the case of uniform circular motion.

$L = mvr = mr^2\omega$

Law of rotational motion Differentiating both sides of equation (6.5) with respect to t, we obtain the following relation :

$$\frac{\mathrm{d}L}{\mathrm{d}t} = m\frac{\mathrm{d}}{\mathrm{d}t}(xv_y - yv_x) = m(v_xv_y + xa_y - v_yv_x - ya_x)$$
$$= x(ma_y) - y(ma_x) = xF_y - yF_x = N$$

$$\therefore \quad \frac{\mathrm{d}L}{\mathrm{d}t} = N. \qquad (6.7)$$

Thus, we have derived the **law of rotational motion**. In other words,

The time rate of change in angular momentum of a mass point is equal to the moment of force acting on that mass point.

Here, the angular momentum and the moment of force are not limited to those around the origin. So long as they are those around the same point, this law holds.

Fig. 6-7 Swings moving around the column. Since the resultant force of tension and gravity is the central force, those siting on the swing move in a uniform circular motion.

Central force When the line of action of force acting on a mass point is always on a straight line connecting a fixed point O and the mass point, and its strength is determined only by the distance r between the point O and the mass point, this force is called the **central force**, and the point O is called the center of force. The universal gravity that the Sun exerts on the planet and the electric force that charged particles exert on other charged particles are central forces. Moreover, when we attach a stone to the string and swing the stone to move horizontally at a constant speed in a circular motion, the tension of the string is a central force.

Angular momentum conservation law When a mass point moves under the action of only the central force with the point O as the center of force, the line of action of this force passes through the point O, so the distance between the point O and the line of action of the force is zero. Therefore, the moment of force around the point O is zero. Accordingly, expressing the angular momentum of this mass point around the point O as L we have from equation (6.7) the following relation :

$$\frac{\mathrm{d}L}{\mathrm{d}t} = 0 \qquad \text{(in the case of central force).} \tag{6.8}$$

This shows that the time rate of change in angular momentum L is zero. In this way we can derive the relation

$$L = \text{constant} \qquad \text{(in the case of central force).} \tag{6.9}$$

That is to say

> When a mass point moves under the action of only the central force, the angular momentum around the center of force is constant.

This is called the **angular momentum conservation law**. Moreover, when the mass point moves under the action of only the central force, this mass point moves on the plane including the center of force [see equation (6.15)]. When forces other than the central force act on the mass point, the angular momentum of the mass point varies with time.

The area through which the line segment connecting the point O and the mass point passes in unit time is referred to as the **areal velocity** of this mass point relative to the point O. As it will be confirmed in Question 1 described below, the areal velocity is proportional to the angular momentum. Therefore, the angular momentum conservation law can also be expressed as

> The areal velocity of a mass point moving under only the action of central force is constant.

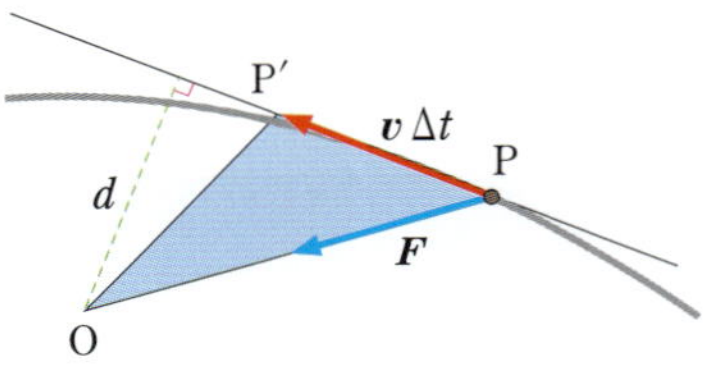

Fig. 6-8

Question 1 Using Fig. 6-8, confirm

$$\text{(angular momentum } L = mvd) = 2(\text{mass } m)\times\left(\text{areal velocity } \frac{dv}{2}\frac{\Delta t}{\Delta t}\right). \tag{6.10}$$

Example 1 We attach a pebble of mass m to the tip of a string which passes through a vertical thin tube and perform a uniform velocity circular motion with a radius r_0 and at a speed v_0 in the horizontal plane (Fig. 6-9). We ignore the gravity acting on the pebble and assume there is no friction between the string and the pipe.

(1) Find the speed v_1 when we pulled this string slowly to reduce the radius of circular motion to r_1. How did the kinetic energy of the pebble change at this time? What caused this change?

(2) How did the angular velocity of the circular motion change at this time?

Solution (1) The tension of the string acting on the pebble is the central force. Therefore, the angular momentum L of the pebble is conserved

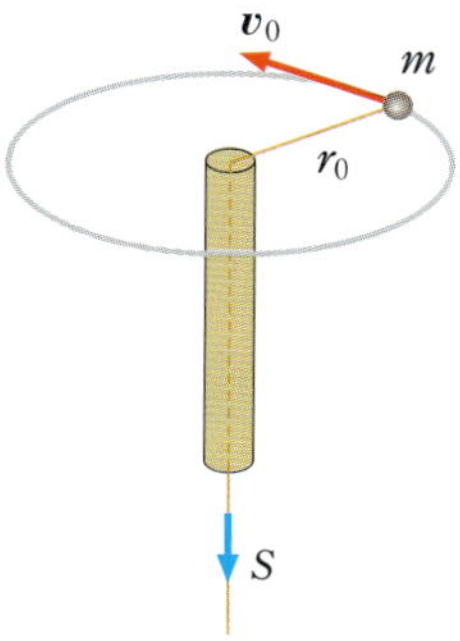

Fig. 6-9

$$L = mr_0v_0 = mr_1v_1.$$

Accordingly

$$v_1 = \frac{r_0}{r_1}v_0.$$

Since $r_1 < r_0$ and $v_1 > v_0$, the kinetic energy of the pebble increases.

$$\frac{1}{2}mv_1^2 - \frac{1}{2}mv_0^2 = \frac{L^2}{2mr_1^2} - \frac{L^2}{2mr_0^2} > 0$$

This increase in kinetic energy is ascribed to the work that the tension S of the string performs :

$$S = \frac{mv^2}{r} = \frac{L^2}{mr^3}$$

$$\text{work} = -\int_{r_0}^{r_1} \frac{L^2}{mr^3}\,\mathrm{d}r.$$

(2) $v_0 = r_0\omega_0$ and $v_1 = r_1\omega_1$. From equation (6.6), we have

$$L = mr_0^2\omega_0 = mr_1^2\omega_1$$

$$\therefore \quad \omega_1 = \frac{r_0^2\omega_0}{r_1^2} > \omega_0$$

Therefore, the angular velocity ω of the circular motion of the object increases.

For reference Law of rotational motion expressed as a vector product*

Since the rotational axis has its direction, the moment of force and the angular momentum are vectors having magnitude and direction and they are expressed in terms of vector product explained in Appendix. When the force $\boldsymbol{F}$ acts on the mass point located at $\boldsymbol{r}$ with mass m and velocity $\boldsymbol{v}$, the moment $\boldsymbol{N}$ of the force $\boldsymbol{F}$ and angular moment $\boldsymbol{L}$ around the origin O are defined respectively as

$$\boldsymbol{N} = \boldsymbol{r} \times \boldsymbol{F} \tag{6.11}$$

$$\boldsymbol{L} = \boldsymbol{r} \times \boldsymbol{p} = \boldsymbol{r} \times m\boldsymbol{v} \tag{6.12}$$

(Fig. 6-10 and Fig. 6-11). The components of $\boldsymbol{N}$ and $\boldsymbol{L}$ are

$$\left.\begin{array}{ll} N_x = yF_z - zF_y, & L_x = m(yv_z - zv_y) \\ N_y = zF_x - xF_z, & L_y = m(zv_x - xv_z) \\ N_z = xF_y - yF_x, & L_z = m(xv_y - yv_x) \end{array}\right\}. \tag{6.13}$$

Therefore, equation (6.13) corresponds to the generalized definition of the rotational motion around the z-axis discussed at the beginning of this section.

The time rate of change in angular momentum $\boldsymbol{L}$ around the origin O is derived by differentiating equation (6.12) with respect to t and using the relations $\boldsymbol{v} \times m\boldsymbol{v} = 0$ and $m\boldsymbol{a} = \boldsymbol{F}$. Thus

$$\frac{\mathrm{d}\boldsymbol{L}}{\mathrm{d}t} = \boldsymbol{N}. \tag{6.14}$$

This equation holds even when the vectors $\boldsymbol{L}$ and $\boldsymbol{N}$ are not parallel.

When only the central force $\boldsymbol{F}$ acts, the moment $\boldsymbol{N}$ of the force $\boldsymbol{F}$ relative to the center of force is $\boldsymbol{0}$ and the angular momentum $\boldsymbol{L}$ relative to the center of force is constant.

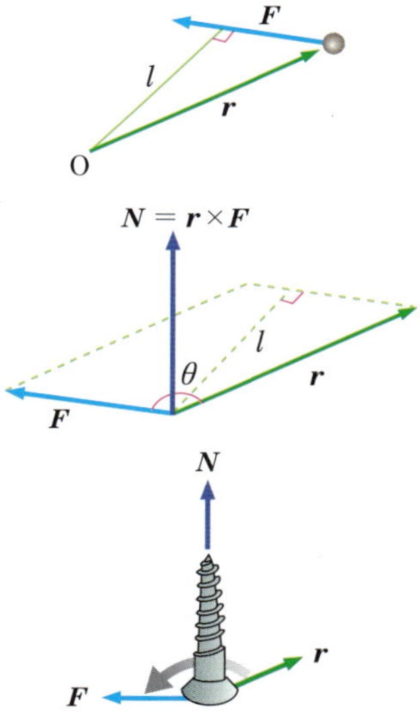

Fig. 6-10 Moment of force $\boldsymbol{N} = \boldsymbol{r} \times \boldsymbol{F}$. The vector product $\boldsymbol{r} \times \boldsymbol{F}$ of two vectors $\boldsymbol{r}$ and $\boldsymbol{F}$ is a vector. Its magnitude is the area of a parallelogram with $\boldsymbol{r}$ and $\boldsymbol{F}$ as two adjacent sides, or $Fr\sin\theta$, and the direction is perpendicular to both $\boldsymbol{r}$ and $\boldsymbol{F}$ and is the traveling direction of the right-hand screw when it is turned from $\boldsymbol{r}$ to $\boldsymbol{F}$ (through an angle less than 180°).

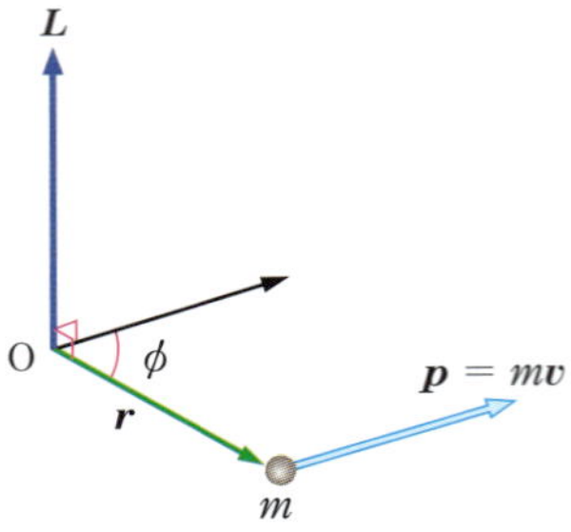

Fig. 6-11 Angular momentum
$\boldsymbol{L} = \boldsymbol{r} \times \boldsymbol{p} = \boldsymbol{r} \times m\boldsymbol{v}$
$L = rp \sin \phi = rmv \sin \phi$

$$\boldsymbol{L} = \boldsymbol{r} \times m\boldsymbol{v} = \text{constant} \quad \text{(in the case of central force)} \tag{6.15}$$

Thus, the position vector $\boldsymbol{r}$ of the mass point is on a plane perpendicular to the constant vector $\boldsymbol{L}$, and the mass point solely subject to the action of the central force moves on the plane including the center of force (see Fig. 6-11).

For reference Couple of forces

A pair of forces $\boldsymbol{F}$, $-\boldsymbol{F}$ which act on two mutually parallel and different lines of action of force, and are equal in magnitude but opposite in direction is referred to as the couple of forces (Fig. 6-12). Let $\boldsymbol{d}$ be a vector whose initial point is the point of action of force $-\boldsymbol{F}$ and the end point is the point of action of force $\boldsymbol{F}$. Then, the moment $\boldsymbol{N}$ of the couple of forces relative to all points is

$$\boldsymbol{N} = \boldsymbol{d} \times \boldsymbol{F}. \tag{6.16}$$

No single force (resultant force) exists that can give the same effect as the couple of forces.

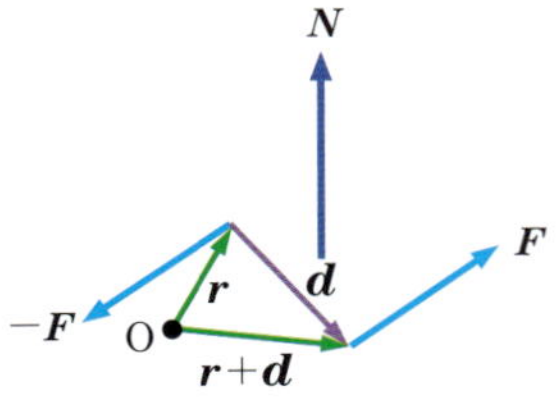

Fig. 6-12 Moment $\boldsymbol{N}$ of couple of forces $\boldsymbol{F}$, and $-\boldsymbol{F}$.
$\boldsymbol{N} = (\boldsymbol{r}+\boldsymbol{d}) \times \boldsymbol{F} + \boldsymbol{r} \times (-\boldsymbol{F})$
$= \boldsymbol{d} \times \boldsymbol{F}$

For reference Motion of planets and satellites, and Kepler's law

In the late 16th century, Tycho Brahe observed the position of the stars, the Sun, the moon, planets etc., for an extended period with unprecedented accuracy. Based on this observed results Kepler who was his assistant discovered the following three laws called **Kepler's law** at the end of the trial and error (Fig. 6.13).

First law The orbit of the planet is an ellipse with the Sun as one of the focal points.
Second law The areas which the line segment connecting the Sun and the planet sweeps out in a fixed period of time are equal (law of constant areal speed).
Third law The ratio of the square of the time (period) T for the planet to make one round of the Sun and the cube of the semi-major axis a of the orbit has the same value for all planets.

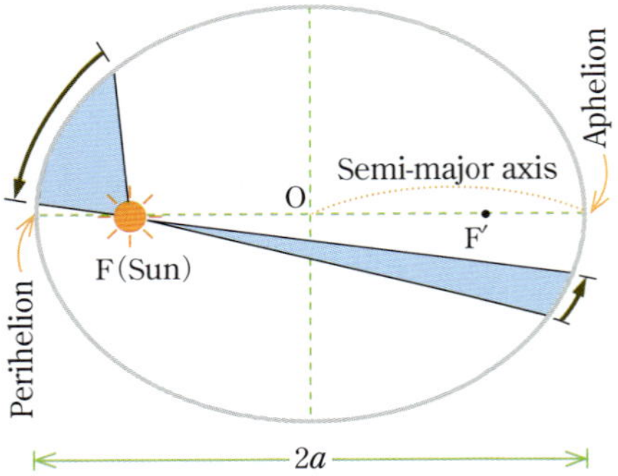

Fig. 6-13 Elliptical orbit of the planet and constant areal velocity. The planet moves on an elliptical orbit with the Sun as one of the focal points (F). The areas through which the line segment connecting the Sun and the planet sweeps out in the same time are constant. As a result, the moving speed of planet is slow near the aphelion far from the Sun, and it is fast near the perihelion near the Sun.

About 100 years after the discovery of Kepler's law, Newton proved Kepler's law using the law of motion on the assumption that the universal gravitation works among all celestial bodies. Conversely, he derived the law of universal gravitation from the law of motion and Kepler's law.

Since the universal gravitation acting between the Sun and the planet is the central force with the Sun as the center of force, the angular momentum conservation law, that is, the law of constant areal velocity holds.

The circle is the case where the two focal points of the ellipse coincide. When the orbit of a planet is a circle, Newton's equation of motion for the planet of mass m is

$$mr\omega^2 = mr\left(\frac{2\pi}{T}\right)^2 = \frac{GmM_S}{r^2} \quad (M_S \text{ is the mass of the sun}).$$

Therefore,

$$\frac{r^3}{T^2} = \frac{GM_S}{4\pi^2} = \text{constant} \tag{6.17}$$

is derived. In this way, Kepler's third law that "The square of the period T is proportional to the cube of the orbit radius r" could be proved.

The proof of Kepler's first law and that of his third law in the case of elliptic motion are omitted.

Problems for exercise 6

A

1. How many times is the semi-major axis of the orbit of the comet with a period of 70 years as long as the semi-major axis of the Earth's orbit?
2. To launch an artificial satellite, we use a multi-stage rocket which accelerates one after another and we modify the orbit to put the satellite on a predetermined orbit. What will happen to the artificial satellite (= one-stage rocket) if we, instead of using a multistage rocket, launch a one-stage rocket (= artificial satellite) and do not change the orbit?
3. For both the years 2010 and 2011, the spring equinox day is March 21 and the autumn equinox day is September 23. What can we say about the date of the perihelion?

B

1. Derive the equation of motion of a single pendulum (4.18) from the law of rotational motion (6.7).
2. Consider the motion of the object subject to the central force $\boldsymbol{F}(\boldsymbol{r}) = -k\boldsymbol{r}$ (k is a positive constant). Show the following.
   (1) The orbit is an ellipse centered on the center of force.
   (2) The angular momentum is constant.
   (3) The period is constant regardless of the orbit.

Mechanics of the System of Particles

Up to this point, we have examined the movement of one small object (mass point). The force, however, acts between two objects. As in the case of the parabolic motion near the Earth surface, when one object (in this case, the Earth) is far larger than the other, we can take that a large object does not move. If the two objects of nearly the same mass collide, we need to consider the motion of both objects. Also, actually existing objects have volumes, deform and rotate. In considering the motion of an object spreading in space, it is appropriate to divide the object into minute portions and treat each as a **mass point**, that is, a point having mass. A collection of mass points is called a **system of particles**.

In this chapter, you will first learn the center of gravity (or center of mass) of the system of particles, and then learn the equation of motion of the center of gravity, the momentum of the system of particles and the collision between two mass points. Problems specific to rigid bodies, a system of particles whose distances between mass points do not change, will be discussed in the next chapter.

7.1 System of particles and center of gravity of rigid body

Learning objective To understand that the center of gravity of the system of particles is defined as the point of action of the resultant force of gravity acting on the mass points composing the system of particles, namely that the center of gravity of the system of particles is defined as the point at which the sum of the moments of gravity acting on the mass points composing the system of particles is **0**.

Two characteristics of the center of gravity of rigid body

Objects are classified into two types : hard objects such as iron and stone, and soft objects such as rubber. A hard object is an object with little deformation when force is applied. A hard object whose deformation can be ignored even if a force is applied from the outside is called a **rigid body**.

When we consider the motion of the system of particles and the balance of the rigid body, the **center of gravity** plays an important role. The center of gravity of the system of particles is defined as the point of action of the resultant force of gravity acting on the mass points composing the system of particles. Therefore, the center of gravity of the system of particles is defined as the point where the sum of the moments of gravity acting on the mass points composing the system of particles is **0**. Using this property, the position of the center of gravity of the rigid body can be obtained as shown in Fig. 7-1.

As in the case of mass which is related to both the strength and the inertia of the gravity received by the object, the center of gravity of the system of particles with the sum M of mass has the property that it performs the same movement as that of the mass point of mass M on which the vector sum of all the external forces acting on the mass points composing the system of particles acts. We will discuss more on this property of center of gravity in the next section.

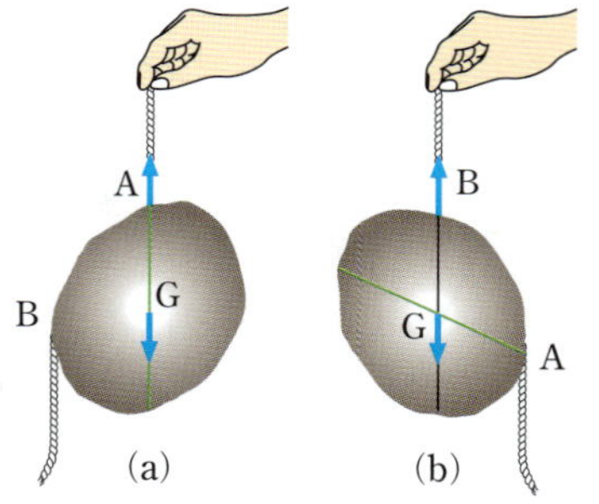

Fig. 7-1 Since the resultant force of gravity acting on each part of the rigid body acts on the center of gravity G, when the rigid body is hung and kept at rest as shown in the figure, the center of gravity G is located directly below the fulcrum of the thread.

Center of gravity of two mass points Consider a rigid body with two heavy small balls (mass m_1 and m_2) attached respectively to both ends P, Q of the light bar shown in Fig. 7-2. On the two balls, the gravities $\boldsymbol{W}_1 = m_1\boldsymbol{g}$ and $\boldsymbol{W}_2 = m_2\boldsymbol{g}$ work vertically downward, respectively. In order to support this bar at one point and keep it at rest, the point G at which the sum of the moments of gravity acting on the two balls becomes **0**, namely the point G which internally divides the line (bar) $\overline{\mathrm{PQ}}$ into $m_2 : m_1$ has only to be supported by a force $\boldsymbol{F}$ of magnitude $(m_1+m_2)g$. Therefore, the effects of gravity $\boldsymbol{W}_1$ and $\boldsymbol{W}_2$ acting on the two spheres attached to both ends of the bar are the same as one force $\boldsymbol{W}$ of magnitude $(m_1+m_2)g$ passing through the point G and working vertically downward. This force $\boldsymbol{W}$ is called the resultant force of the gravitational forces $\boldsymbol{W}_1$ and $\boldsymbol{W}_2$ acting on the rigid body composed of two spheres, and the point G is called the **center of gravity** or center of mass of the two spheres.

Let the position vector of the mass point of mass m_1 be $\boldsymbol{r}_1 = (x_1, y_1, z_1)$ and the position vector of the mass point of mass m_2 be $\boldsymbol{r}_2 = (x_2, y_2, z_2)$. Then, the position vector $\boldsymbol{R} = (X, Y, Z)$ of the

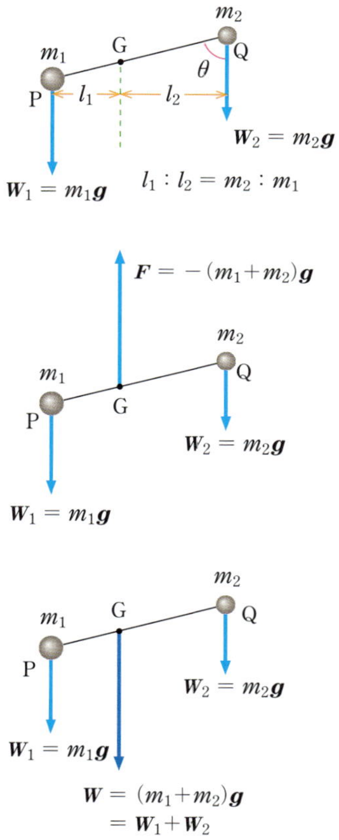

Fig. 7-2 The resultant force of gravities $\boldsymbol{W}_1$ and $\boldsymbol{W}_2$ acting on the two balls fixed to a light bar is the vertical downward force $\boldsymbol{W}_1+\boldsymbol{W}_2$ passing through the center of gravity. Since the sum of moment of gravity around the point G must be **0**, the point G is a point which internally divides $\overline{\mathrm{PQ}}$ into $m_2 : m_1$, accordingly $\overline{\mathrm{PG}} : \overline{\mathrm{GQ}} = m_2 : m_1$.

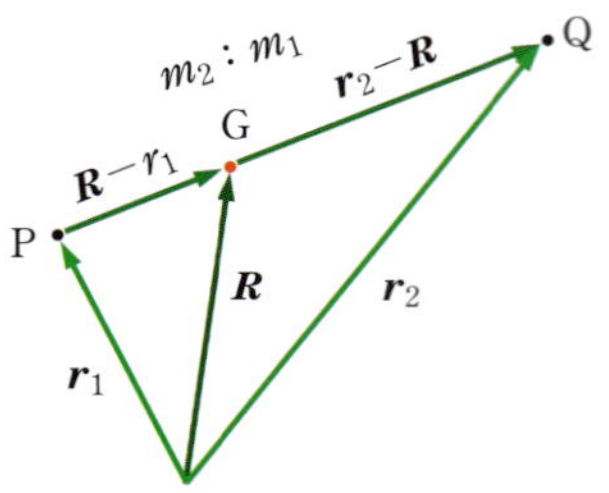

Fig. 7-3 That the position vector $\boldsymbol{R}$ of the center of gravity is

$$\boldsymbol{R} = \frac{m_1\boldsymbol{r}_1+m_2\boldsymbol{r}_2}{m_1+m_2}$$

can be understood if we consider the relation

$$(\overrightarrow{PG} = \boldsymbol{R}-\boldsymbol{r}_1) : (\overrightarrow{GQ} = \boldsymbol{r}_2-\boldsymbol{R}) = m_2 : m_1.$$

It follows that $\boldsymbol{R}$ is obtained by solving $m_1(\boldsymbol{R}-\boldsymbol{r}_1) = m_2(\boldsymbol{r}_2-\boldsymbol{R})$.

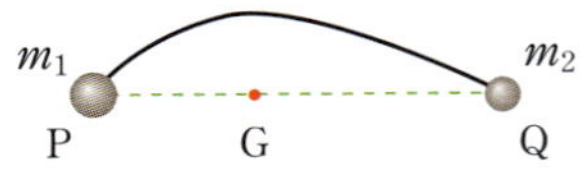

Fig. 7-4 The center of gravity G may sometimes exist outside the object.

center of gravity G of the two mass points is

$$\boldsymbol{R} = \frac{m_1\boldsymbol{r}_1+m_2\boldsymbol{r}_2}{m_1+m_2} \tag{7.1}$$

and the position coordinate of the center of gravity is

$$X = \frac{m_1x_1+m_2x_2}{m_1+m_2}, \quad Y = \frac{m_1y_1+m_2y_2}{m_1+m_2}, \quad Z = \frac{m_1z_1+m_2z_2}{m_1+m_2}. \tag{7.1$'$}$$

The proof of equation (7.1) is illustrated in Fig. 7-3.

Even if the light bar with two attached objects in Fig. 7-2 is bent as shown in Fig. 7-4 and the point G internally dividing the line segment $\overline{PQ}$ into $m_2 : m_1$ is outside the bar, the point G is the center of gravity of the two objects, and the position of the center of gravity is given by equation (7.1).

Question 1 There is a log of 6 m in length. A force of 80 kgf is necessary to lift one end A thereof and a force of 70 kgf is required to lift the other end B. Find the mass and position of the center of gravity of this log.

Center of gravity of more than three mass points When the mass points of masses $m_1, m_2, m_3, \cdots$ are at the points $\boldsymbol{r}_1 = (x_1, y_1, z_1)$, $\boldsymbol{r}_2 = (x_2, y_2, z_2)$, $\boldsymbol{r}_3 = (x_3, y_3, z_3)$, $\cdots$ the position vector $\boldsymbol{R} = (X, Y, Z)$ of the center of gravity G of this rigid body is

$$\boldsymbol{R} = \frac{m_1\boldsymbol{r}_1+m_2\boldsymbol{r}_2+m_3\boldsymbol{r}_3+\cdots}{M} \tag{7.2}$$

and the position coordinates are

$$X = \frac{m_1x_1+m_2x_2+m_3x_3+\cdots}{M},$$
$$Y = \frac{m_1y_1+m_2y_2+m_3y_3+\cdots}{M},$$
$$Z = \frac{m_1z_1+m_2z_2+m_3z_3+\cdots}{M}. \tag{7.2$'$}$$

Here

$$M = m_1+m_2+m_3+\cdots = \sum_i m_i \tag{7.3}$$

is the total mass of the system of particles.

The fact that the resultant force of gravity acting on the system of particles is the vertically downward force $M\boldsymbol{g} = (m_1+m_2+m_3+\cdots)\boldsymbol{g}$ passing through the center of gravity G specified by equation (7.2) is shown by first creating the resultant force of $m_1\boldsymbol{g}$ and $m_2\boldsymbol{g}$, then creating the resultant force of the resultant force of ($m_1\boldsymbol{g}$ and $m_2\boldsymbol{g}$) and $m_3\boldsymbol{g}$, $\cdots$ and repeating this process [regarding the proof that the moment of gravity around the center of gravity G defined by equation (7.2) is **0**, see B4 in Problems for exercise 7].

Case 1 The center of gravity of a thin disc with uniform material property and constant thickness is the center of circle, and the center of gravity of a thin triangular plate with uniform material property and constant thickness is the geometric center of gravity of the triangle, which is the intersection of the three median lines (the line connecting the vertex and the midpoint of the opposite side) (Fig. 7-5).

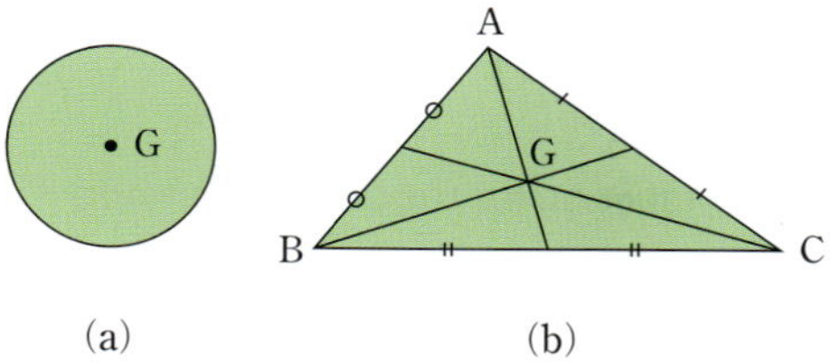

Fig. 7-5 Center of gravity
(a) The center of gravity of thin uniform disc is the center of circle.
(b) The center of gravity of thin uniform triangle plate is the intersection of the three median lines.

7.2 Motion of the system of particles

Learning objective To understand that the equation of motion for the center of gravity of the system of particles is $M\boldsymbol{A} = \boldsymbol{F}$, that is, the center of gravity of the system of particles with total mass M shows the same motion as the mass point of mass M on which the vector sum $\boldsymbol{F}$ of all the external forces acting on the mass points composing the system of particles acts. To understand that when the external force does not act on the system of particles, momentum conservation law holds and the center of gravity of the system of particles performs the linear motion at a constant velocity. Finally, to understand that the momentum conservation law is effective for the collision phenomena.

Equation of motion for the center of gravity of the system of particles There are two types of forces acting on each mass point composing the system of particles : a force (external force) from an object outside the system of particles and a force (internal force) from another mass point in the system of particles (Fig. 7-6). The external force acting on the i-th mass point is expressed as $\boldsymbol{F}_i$, and the force (internal force) acting on the i-th mass point by the j-th mass point is written as $\boldsymbol{F}_{i\leftarrow j}$. According to the law of action and reaction, the internal force satisfies the relation $\boldsymbol{F}_{i\leftarrow j} = -\boldsymbol{F}_{j\leftarrow i}$.

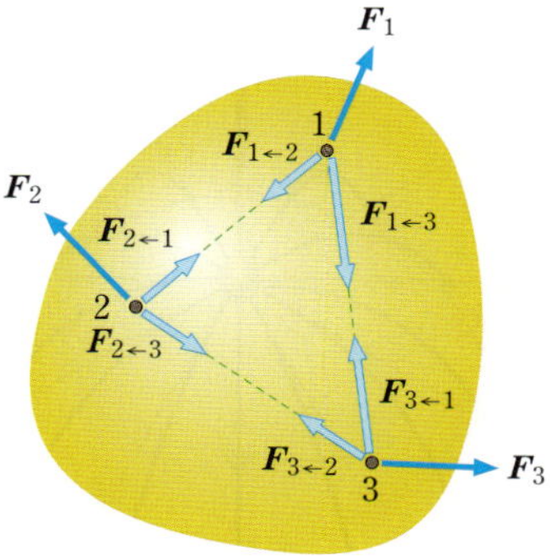

Fig. 7-6 Internal force and external force

For simplicity, we first consider the system of two particles consisting of two mass points 1 and 2. The equations of motion of these mass points 1 and 2 are

$$m_1 \frac{\mathrm{d}^2 \boldsymbol{r}_1}{\mathrm{d}t^2} = \boldsymbol{F}_1 + \boldsymbol{F}_{1\leftarrow 2}, \qquad m_2 \frac{\mathrm{d}^2 \boldsymbol{r}_2}{\mathrm{d}t^2} = \boldsymbol{F}_2 + \boldsymbol{F}_{2\leftarrow 1}. \tag{7.4}$$

Summing up the both sides of the above equations and using the law of action and reaction $\boldsymbol{F}_{1\leftarrow 2} + \boldsymbol{F}_{2\leftarrow 1} = \boldsymbol{0}$, we can derive

$$\frac{\mathrm{d}^2}{\mathrm{d}t^2}(m_1 \boldsymbol{r}_1 + m_2 \boldsymbol{r}_2) = \boldsymbol{F}_1 + \boldsymbol{F}_2 = \boldsymbol{F}. \tag{7.5}$$

$\boldsymbol{F}$ is the vector sum of the external forces acting on the two mass points 1 and 2 and is called the **external force**. Using the relation $(m_1 \boldsymbol{r}_1 + m_2 \boldsymbol{r}_2) = M\boldsymbol{R}$ obtained by multiplying the both sides of equation (7.1) by $M = (m_1 + m_2)$, equation (7.5) is transformed into

$$M\frac{\mathrm{d}^2\boldsymbol{R}}{\mathrm{d}t^2} = \boldsymbol{F} \quad (M\boldsymbol{A} = \boldsymbol{F})$$

(equation of motion for the center of gravity). (7.6)

Here $\frac{\mathrm{d}^2\boldsymbol{R}}{\mathrm{d}\boldsymbol{t}^2} = \boldsymbol{A}$ is the acceleration of center of gravity. Equation (7.6) is the equation of motion for the center of gravity of the system of two particles. Dividing equation (7.6) into its components, we obtain the following relations :

$$M\frac{\mathrm{d}^2X}{\mathrm{d}t^2} = F_x, \quad M\frac{\mathrm{d}^2Y}{\mathrm{d}t^2} = F_y, \quad M\frac{\mathrm{d}^2Z}{\mathrm{d}t^2} = F_z. \tag{7.6$'$}$$

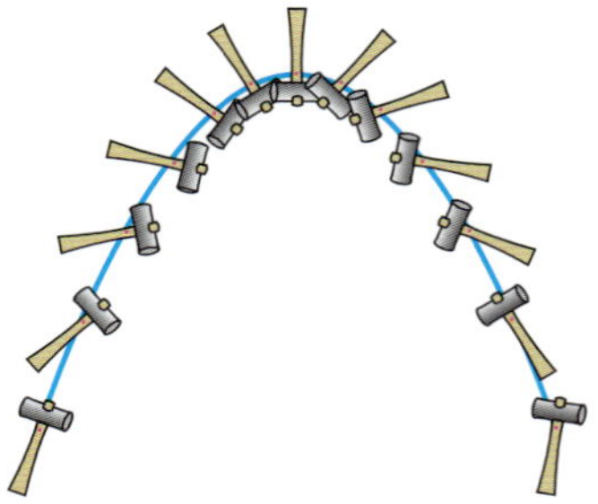

Fig. 7-7 The center of gravity of a hammer performs a parabolic motion.

Also in the case of a system of particles composed of three or more mass points, if we assume that the position vector of the center of gravity of the system of particles is $\boldsymbol{R}$, and the total mass of the system of particles is M, and the vector sum (called the external force) of the external forces $\boldsymbol{F}_1+\boldsymbol{F}_2+\boldsymbol{F}_3+\cdots$ acting on the system of particles is $\boldsymbol{F}$, then equation (7.6) holds. Equation (7.6) is the **equation of motion for the center of gravity of the system of particles**. This equation shows the following description even if each mass point of the system of particles moves in a complex manner (Fig. 7-7) :

The center of gravity of the system of particles performs the same motion as when the mass point of the mass $M = \sum_i m_i$ does under the action of external force $\boldsymbol{F} = \sum_i \boldsymbol{F}_i$.

Question 2 A fireworks ball launched right above the sky exploded at the highest position. Consider the falling motion of the beautiful sphere of fireworks (Fig. 7-8).

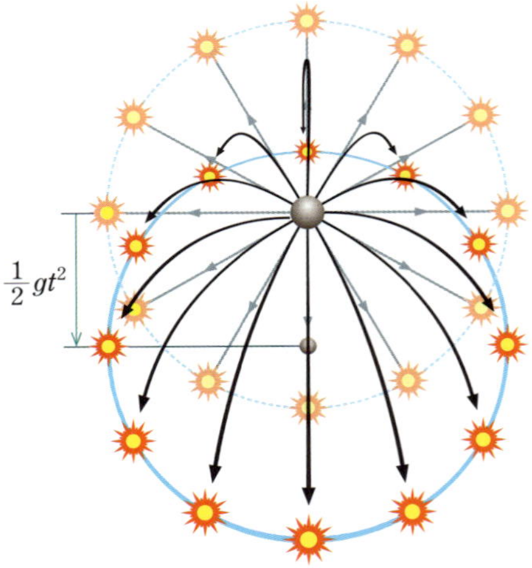

Fig. 7-8

Momentum of the system of particles In Chapter 3, the product $\boldsymbol{p} = m\boldsymbol{v}$ of mass m of a mass point and its velocity $\boldsymbol{v}$ was called the momentum. Sum of the momentum of each mass point composing the system of particles

$$\boldsymbol{P} = m_1\boldsymbol{v}_1+m_2\boldsymbol{v}_2+m_3\boldsymbol{v}_3+\cdots \tag{7.7}$$

is referred to as the total momentum of the system of particles. The following equation

$$M\boldsymbol{R} = m_1\boldsymbol{r}_1+m_2\boldsymbol{r}_2+m_3\boldsymbol{r}_3+\cdots \tag{7.8}$$

is obtained by multiplying both sides of equation (7.2) by M. We differentiate equation (7.8) with respect to t. Since $\frac{\mathrm{d}\boldsymbol{R}}{\mathrm{d}t} = \boldsymbol{V}$ is the velocity of the center of gravity and $\frac{\mathrm{d}\boldsymbol{r}_i}{\mathrm{d}t} = \boldsymbol{v}_i$, it follows that

$$M\boldsymbol{V} = m_1\boldsymbol{v}_1+m_2\boldsymbol{v}_2+m_3\boldsymbol{v}_3+\cdots = \boldsymbol{P}.$$

In short, "total momentum of the system of particles" = "total mass of the system of particles" × "velocity of center of gravity"

$$\boldsymbol{P} = M\boldsymbol{V}. \tag{7.9}$$

Between the mass point and the object spreading in space, the corresponding relations for equation of motion $m\boldsymbol{a} = \boldsymbol{F}$ and $M\boldsymbol{A} = \boldsymbol{F}$, and for momentum $\boldsymbol{p} = m\boldsymbol{v}$ and $\boldsymbol{P} = M\boldsymbol{V}$ hold.

Differentiating equation (7.9) with respect to t and paying attention to the relations $\frac{\mathrm{d}M}{\mathrm{d}t} = 0$ and $\frac{\mathrm{d}\boldsymbol{V}}{\mathrm{d}t} = \boldsymbol{A}$, we obtain

$$\frac{\mathrm{d}\boldsymbol{P}}{\mathrm{d}t} = \frac{\mathrm{d}}{\mathrm{d}t}(M\boldsymbol{V}) = M\frac{\mathrm{d}\boldsymbol{V}}{\mathrm{d}t} = M\boldsymbol{A} = \boldsymbol{F} \tag{7.10}$$

namely,

$$\frac{\mathrm{d}\boldsymbol{P}}{\mathrm{d}t} = \boldsymbol{F}. \tag{7.11}$$

This is an alternative expression of the law of motion of the center of gravity [equation (7.6)].

Momentum conservation law In the case where no external force acts on the system of particles, that is, when $\boldsymbol{F} = \boldsymbol{0}$, from equation (7.11) we have the following relation :

$$\frac{\mathrm{d}\boldsymbol{P}}{\mathrm{d}t} = \boldsymbol{0}, \qquad \text{or} \qquad \boldsymbol{P} = M\boldsymbol{V} = \text{constant} \tag{7.12}$$

which shows that the **momentum conservation law** of the system of particles holds. It follows that

The total momentum $\boldsymbol{P}$ and velocity $\boldsymbol{V}$ of center of gravity of the system of particles on which no external forces act are constant, and the center of gravity performs the linear motion at a constant velocity.

Case 2 | Spacecraft Since the center of gravity of the main body of spacecraft and the fuel is isolated in outer space and is not affected by external forces, it continues to move at a constant velocity. However, when the spacecraft injects fuel backward, its reaction accelerates the body of spacecraft forward. Also, when fuel is injected sideways, the spacecraft can turn the direction.

Fig. 7-9 International Space Station (ISS) and the Earth

Collision In a collision of two objects, a large force acts between the two objects in a very short time, so the effect of external force during that time is negligible compared to the effect of internal force. Hence, the momentum conservation law "the total momentum immediately before and immediately after the collision of two objects is equal"

$$m_{\mathrm{A}}\boldsymbol{v}_{\mathrm{A}} + m_{\mathrm{B}}\boldsymbol{v}_{\mathrm{B}} = m_{\mathrm{A}}\boldsymbol{v}_{\mathrm{A}}' + m_{\mathrm{B}}\boldsymbol{v}_{\mathrm{B}}' \tag{7.13}$$

holds (Fig. 7-10).

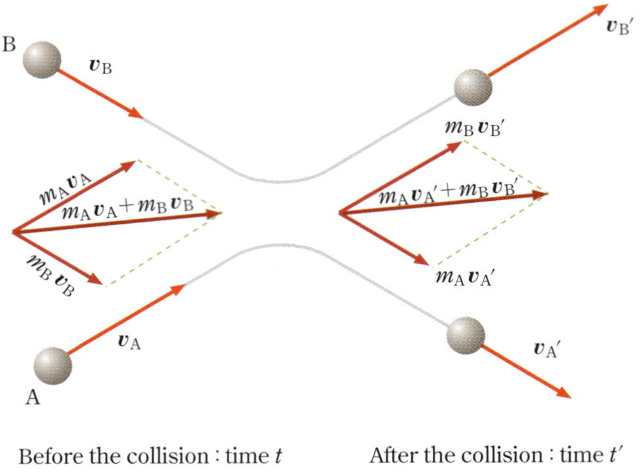

Fig. 7-10

Fig. 7-11 Billiards

Elastic collision In the collision of rigid wooden balls, the balls do not dent and the energy loss due to generation of heat, sound, vibration and the like can be ignored. In this type of collision, since the mechanical energy is conserved, the kinetic energy does not change before and after the collision. Accordingly, in the collision shown in Fig. 7-10, the following relation

$$\frac{1}{2}m_A v_A{}^2 + \frac{1}{2}m_B v_B{}^2 = \frac{1}{2}m_A v_A'^2 + \frac{1}{2}m_B v_B'^2 \text{ (elastic collision)} \quad (7.14)$$

holds. The collision in which kinetic energy is conserved is called **elastic collision**. In the elastic collision, both momentum and kinetic energy are conserved.

Example 1 We have a toy shown in Fig. 7-12. Two metal spheres A and B of the same size and the same mass are independently suspended by thin wires of the same length. We lift the ball A by height h and release it gently. Then, the ball A hits another ball B that was kept at rest. The ball A nearly gets stopped and the ball B starts to move almost up to the same height h. Explain this phenomenon based on the momentum conservation law and the energy conservation law.

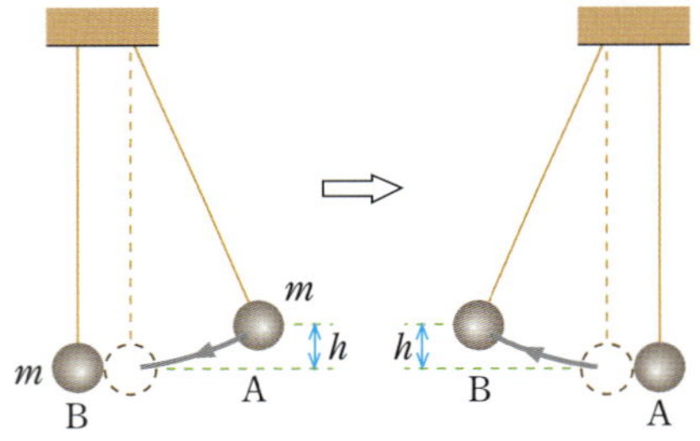

Fig. 7-12

Solution Let the mass of the sphere be m and the velocity of the sphere A immediately before collision be v_A and the velocities of the spheres A and B immediately after the collision be v_A' and v_B', respectively. Then

momentum conservation law

$$mv_A = mv_A' + mv_B' \quad (7.15)$$

and

kinetic energy conservation law

$$\frac{1}{2}mv_A{}^2 = \frac{1}{2}mv_A'^2 + \frac{1}{2}mv_B'^2 \quad (7.16)$$

hold. Substituting into equation (7.16) the relation $v_A' = v_A - v_B'$ obtained from equation (7.15), we derive the following relation :

$$(v_A - v_B')^2 + v_B'^2 - v_A{}^2 = 2v_B'^2 - 2v_A v_B' = 0$$

$$\therefore \ v_B'(v_B' - v_A) = 0.$$

The solution $v_B' = 0$, $v_A' = v_A$ represents the physically improbable situation in which the ball A travels through the ball B, so we can derive

$$v_B' = v_A, \quad v_A' = 0. \quad (7.17)$$

Accordingly, since $v_A' = 0$, it has been shown that the ball A gets stopped after collision. In addition, from $v_B' = v_A$ and the mechanical energy conservation law, the ball B was shown to rise up to the height h.

When an object A having the same mass as an object B at rest hits B in a frontal collision, the object A gets stopped. This result is applied to moderate the neutron in the nuclear reactor. In order to stop the neutron in motion it is only necessary to make the neutron incident on a substance containing many protons (hydrogen nuclei) having almost the same mass as the neutron.

Question 3 Arrange 10-yen coins as shown in Fig. 7-13 and hit a 10-yen coin placed on the most right with your finger in the direction of arrow to collide the other coin (s). What will happen? Conduct an experiment and interpret the result from the physical point of view.

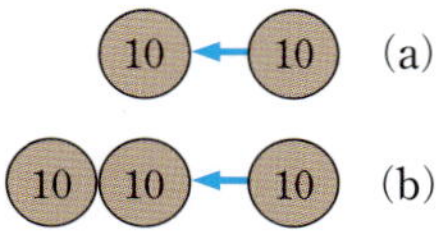

Fig. 7-13

Inelastic collision A case where the kinetic energy of the system decreases due to heat generation or deformation caused by a collision is referred to as inelastic collision. That is, the inelastic collision is a collision in which the total momentum is conserved, while the total kinetic energy is not conserved.

Two-body problem (system of two particles) In the case of a system of two particles free from external forces but interacting through internal forces, the equations of motion of the two mass points are

$$m_1 \frac{d^2 \boldsymbol{r}_1}{dt^2} = \boldsymbol{F}_{1 \leftarrow 2}, \qquad m_2 \frac{d^2 \boldsymbol{r}_2}{dt^2} = \boldsymbol{F}_{2 \leftarrow 1}. \tag{7.18}$$

Subtracting the formula obtained by dividing both sides of the above second expression by m_2, from the formula obtained by dividing both sides of the above first expression by m_1 and using the relation $\boldsymbol{F}_{1 \leftarrow 2} = -\boldsymbol{F}_{2 \leftarrow 1}$, we obtain the following relation :

$$\frac{d^2 \boldsymbol{r}_1}{dt^2} - \frac{d^2 \boldsymbol{r}_2}{dt^2} = \frac{d^2}{dt^2}(\boldsymbol{r}_1 - \boldsymbol{r}_2) = \frac{1}{m_1}\boldsymbol{F}_{1 \leftarrow 2} + \frac{1}{m_2}\boldsymbol{F}_{1 \leftarrow 2} = \frac{m_1 + m_2}{m_1 m_2}\boldsymbol{F}_{1 \leftarrow 2}. \tag{7.19}$$

We introduce here the **relative position vector** (the vector whose initial point is at the mass point 2 and the end point is the mass point 1, Fig. 7-14)

$$\boldsymbol{r} = \boldsymbol{r}_1 - \boldsymbol{r}_2. \tag{7.20}$$

Use of the relative position vector leads to

$$m \frac{d^2 \boldsymbol{r}}{dt^2} = \boldsymbol{F}_{1 \leftarrow 2}. \tag{7.21}$$

That is to say the relative motion only due to the internal force of two mass points is the same as the motion of the mass point 1 of mass m around the stationary mass point 2 under the action of force $\boldsymbol{F}_{1 \leftarrow 2}$. Here

$$m = \frac{m_1 m_2}{m_1 + m_2} \tag{7.22}$$

is referred to as the **reduced mass**. In this case, since the external force does not act, the center of gravity of the system of two particles performs the linear motion at a constant velocity. As the case of the planet (mass m_1) and the Sun (mass m_2), when the mass of one is much larger than the mass of the other $m_1 \ll m_2$, the reduced mass m is approximately equal to the lighter mass ($m \approx m_1$).

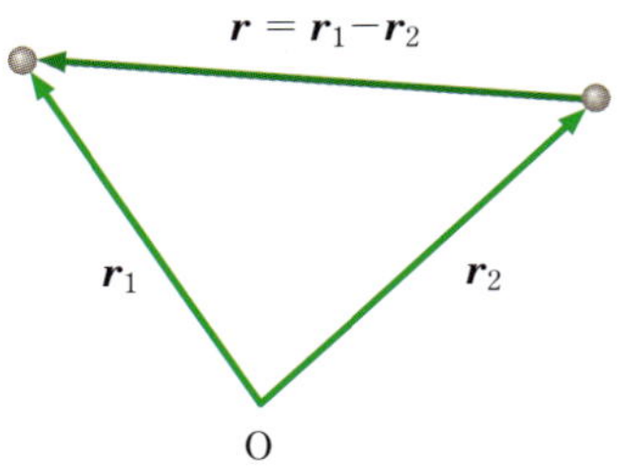

Fig. 7-14 relative position vector $\boldsymbol{r} = \boldsymbol{r}_1 - \boldsymbol{r}_2$

Case 3 | Binary star A binary star is a system in which two stars with mass m_1 and m_2 circle around the center of gravity. In the case of binary star, the equation of motion which is equal to "mass of binary star×centripetal acceleration = universal gravitation" is

$$m\frac{v^2}{r} = G\frac{mM}{r^2} \qquad (M = m_1+m_2). \tag{7.23}$$

Question 4 Show that when the **relative velocity** (velocity of mass point 1 viewed from mass point 2) $\boldsymbol{v}_1-\boldsymbol{v}_2$ of mass point 1 to mass point 2 is represented by $\boldsymbol{v}$ ($\boldsymbol{v} = \boldsymbol{v}_1-\boldsymbol{v}_2$), the kinetic energy of the system of two particles can be expressed as the sum of the kinetic energy of the center of gravity motion and the relative motion :

$$\frac{1}{2}m_1{v_1}^2+\frac{1}{2}m_2{v_2}^2 = \frac{1}{2}MV^2+\frac{1}{2}mv^2. \tag{7.24}$$

Question 5 Show that the kinetic energy of the system of particles in general is the sum of the kinetic energy of the center of gravity and the kinetic energy of the relative motion of each mass point with respect to the center of gravity :

$$\sum_i \frac{1}{2}m_i{v_i}^2 = \frac{1}{2}MV^2+\sum_i \frac{1}{2}m_i|\boldsymbol{v}_i-\boldsymbol{V}|^2. \tag{7.25}$$

7.3 Angular momentum of the system of particles

Learning objective To understand the angular momentum and the law of rotational motion of the system of particles

Let the angular momentum of the mass point i composing the system of particles be $\boldsymbol{L}_i$, and the moment of force acting on the mass point i be $\boldsymbol{N}_i$. Then the law of rotational motion discussed in Section **6.1**

$$\frac{\mathrm{d}\boldsymbol{L}_i}{\mathrm{d}t} = \boldsymbol{N}_i \tag{7.26}$$

holds. Sum of the angular momentum of the mass point composing the system of particles

$$\boldsymbol{L} = \boldsymbol{L}_1+\boldsymbol{L}_2+\boldsymbol{L}_3+\cdots \tag{7.27}$$

is called the **angular momentum of the system of particles,** and the sum of the moment of force acting on the mass point

$$\boldsymbol{N} = \boldsymbol{N}_1+\boldsymbol{N}_2+\boldsymbol{N}_3+\cdots \tag{7.28}$$

is called the **moment of external force***. From the above three equations, we can derive that the time rate of change in angular momentum $\boldsymbol{L}$ of the system of particles is equal to the moment $\boldsymbol{N}$ of external force, or

$$\frac{\mathrm{d}\boldsymbol{L}}{\mathrm{d}t} = \boldsymbol{N}. \tag{7.29}$$

Equation (7.29) is the **equation of motion of the angular momentum of the system of particles**. This law of rotational motion holds for the angular momentum $\boldsymbol{L}$ and the moment $\boldsymbol{N}$ of force around any arbitrary point. This law also holds for a point that is instantaneously kept at rest as in the case of a contact point between a slope and a ball that rolls down the slope without sliding, which will be discussed in the next chapter.

* $\boldsymbol{N}_i$ in equation (7.28) also includes the moments of internal force acting between mass points, but as a result of the law of action-reaction they cancel each other by taking the sum (see B5 of Problems for exercise 7).

Angular momentum conservation law In equation (7.29) if $\boldsymbol{N} = \boldsymbol{0}$, then $\dfrac{\mathrm{d}\boldsymbol{L}}{\mathrm{d}t} = \boldsymbol{0}$ and $\boldsymbol{L} =$ constant. Therefore,

When the moment $\boldsymbol{N}$ of external force around a point is $\boldsymbol{0}$, then the angular momentum $\boldsymbol{L}$ of the system of particles relative to that point is constant regardless of time.

This is called the **angular momentum conservation law**.

Case 4 | A figure skater As a slowly spinning skater, who is spreading both arms widely and is tiptoeing, contracts both arms, the angular velocity ω of rotation increases more and more (Fig. 7-15). Since the moment N_z around the vertical axis (z-axis) of the external force (gravity and normal force) acting on the figure skater spinning with tiptoeing is 0, the angular momentum $L_z = \omega \sum_i m_i l_i^2$ around the vertical axis of the skater is constant. Here l_i is the radius of gyration of body part i of mass m_i. When the skater contracts her arms, the radius of gyration l_i of the arm part decreases, which causes the reduction of $\sum_i m_i l_i^2$, leading to the increase in angular velocity ω. Contraction of arms increases the energy of rotational motion of $\dfrac{1}{2}\omega^2 \sum_i m_i l_i^2$ and this increase is due to the work that the arms performed.

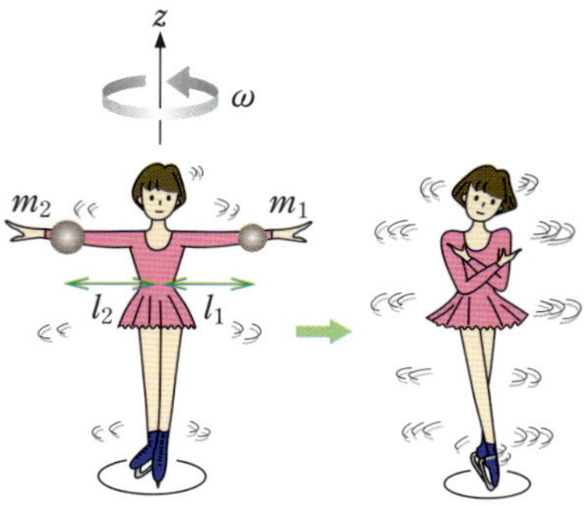

Fig. 7-15 A figure skater

Fig. 7-16 A figure skater

Separation of rotation of the center of gravity around the origin and rotation around the center of gravity The angular momentum $\boldsymbol{L}$ of the system of particles around the origin is the sum of "the angular momentum $\boldsymbol{L}_\mathrm{G}$ of the center of gravity around the origin when the total mass M of the system of particles is taken to be accumulated at the center of gravity" and "the angular momentum $\boldsymbol{L}'$ of the system of particles around its center of gravity", namely

$$\boldsymbol{L} = \boldsymbol{L}_\mathrm{G} + \boldsymbol{L}'$$
$$\sum_i \boldsymbol{r}_i \times m_i \boldsymbol{v}_i = \boldsymbol{R} \times M\boldsymbol{V} + \sum_i (\boldsymbol{r}_i - \boldsymbol{R}) \times m_i(\boldsymbol{v}_i - \boldsymbol{V}). \tag{7.30}$$

In the case of revolution of the Earth around the Sun, $\boldsymbol{L}_\mathrm{G}$ is the angular momentum of revolution of the Earth around the Sun and $\boldsymbol{L}'$ is the angular momentum of rotation of the Earth.

On the other hand, the moment of external force $\boldsymbol{N}$ is the sum of "the moment $\boldsymbol{N}_\mathrm{G}$ around the origin of $\boldsymbol{F}$ when the external force (vector sum of external forces) $\boldsymbol{F}$ is taken to act on the center of gravity" and "the moment $\boldsymbol{N}'$ around the center of gravity", namely

$$\boldsymbol{N} = \boldsymbol{N}_\mathrm{G} + \boldsymbol{N}'$$
$$\sum_i \boldsymbol{r}_i \times \boldsymbol{F}_i = \boldsymbol{R} \times \boldsymbol{F} + \sum_i (\boldsymbol{r}_i - \boldsymbol{R}) \times \boldsymbol{F}_i. \tag{7.31}$$

The equations of motion of the two angular momenta $\boldsymbol{L}_\mathrm{G}$ and $\boldsymbol{L}'$ are

$$\frac{\mathrm{d}\boldsymbol{L}_\mathrm{G}}{\mathrm{d}t} = \boldsymbol{N}_\mathrm{G} \tag{7.32}$$

and

$$\frac{\mathrm{d}\boldsymbol{L}'}{\mathrm{d}t} = \boldsymbol{N}'. \tag{7.33}$$

In other words, even if the center of gravity of the system of particles

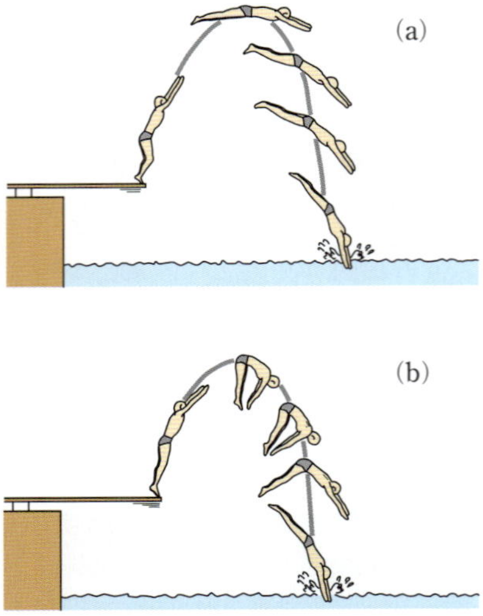

Fig. 7-17 When a dive player jumps in the same direction kicking the board with the same force, the center of gravity of the player travels on the same parabolic curve. Here, the air resistance is taken to be negligible.

performs a complex motion,

The time rate of change in angular momentum $\boldsymbol{L}'$ of the system of particles around the center of gravity is equal to the moment $\boldsymbol{N}'$ around the center of gravity.

Question 6 Using $\boldsymbol{L}_G = \boldsymbol{R} \times \boldsymbol{P} = \boldsymbol{R} \times M\boldsymbol{V}$ and $\boldsymbol{N}_G = \boldsymbol{R} \times \boldsymbol{F}$, derive the law of rotational motion around the origin

$$\frac{d\boldsymbol{L}_G}{dt} = \boldsymbol{N}_G. \tag{7.34}$$

Use the equation of motion $M\boldsymbol{A} = \boldsymbol{F}$.

Case 5 ‖ Dive into the pool In the case where dive players want to dive vertically into water of the pool, it is easier for them to adjust the angle by rounding their body in the halfway of diving into the pool in the style shown in (b) and not (a). This is because as the distance from the center of the body to each part of the body decreases, the angular velocity of rotation increases. The trajectry of the center of gravity in the two cases is the same.

Problems for exercise 7

A

1. Two persons, weight M_A and M_B, are resting on the smooth ice. The person A threw a ball of mass m at a horizontal velocity (against the ice) v and the person B caught this ball. What kind of motion do the two people do?
2. An astronaut (space suit+weight = about 100 kg) is tied to a spacecraft of the weight 900 kg with a 30 m rope. If the astronaut hauls the rope, how long will the spacecraft move during the time (under the gravity-free condition).
3. In a high jump, is it possible to jump over a bar even if the center of gravity of the high jumper does not pass above the bar?
4. **Perfectly inelastic collision ‖** Object A with velocity $\boldsymbol{v}_A$ and mass m_A collided with object B with velocity $\boldsymbol{v}_B$ and mass m_B and adhered each other. Find the velocity $\boldsymbol{v}'$ of the adhered object immediately after the collision. This kind of collision accompanying the adhesion phenomenon is called the perfectly inelastic collision (Fig. 1).

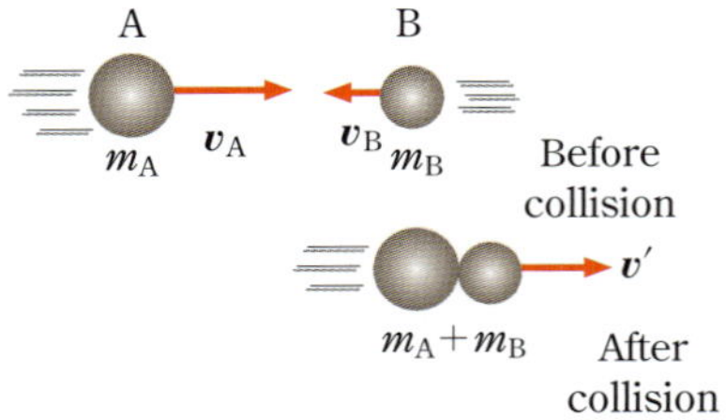

Fig. 1

5. A wood block of mass $M = 1$ kg is suspended from the branch of tree with a light rope. An arrow of mass $m = 30$ g flying horizontally at a speed of $V = 30$ m/s hit the wood block and stuck in (Fig. 2).
   (1) Calculate the speed v of wood block and arrow immediately after the arrow stuck in the wood block.
   (2) A wood block stuck in by the arrow moves on an arc centered on the branch. Find the height h of the highest point.

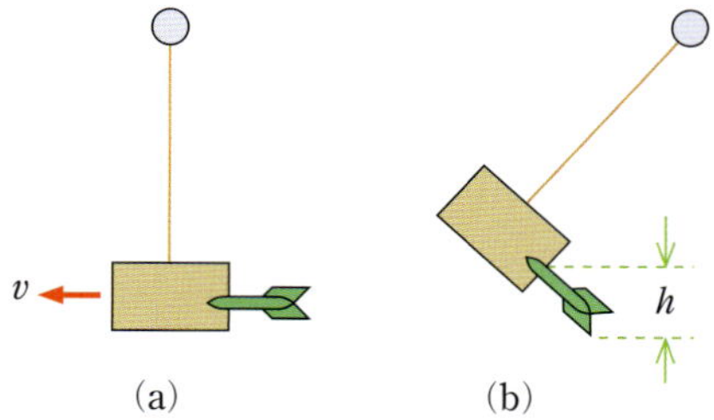

Fig. 2

B

1. When we get on a swing and want to swing widely, we bend the knees near the highest point and stand up near the lowest point (Fig. 3). Explain the reason.

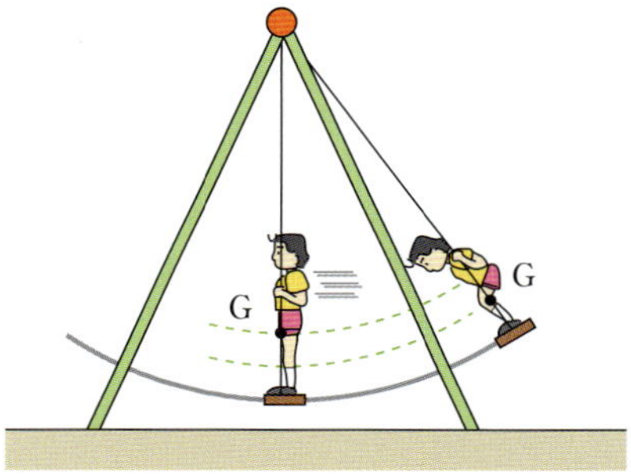

Fig. 3

2. In the case of deformable objects, even if the moment of external force around a certain point is 0, the object can rotate around that point. When a man holds a heavy ball in each hand, sits on a swivel chair, extends both arms to the left and right and twist the body right, the chair turns to the left. Then, if he lowers his hands and returns his body to the left, the chair and the man are rotated to the left. Discuss this relationship with the angular momentum conservation law.
3. When we look at the galaxy consisting of a huge number of stars from afar, the stars are distributed in a disk shape and the stars as a whole rotate around the axis of symmetry perpendicular to the disk. It is thought that the galaxy was initially spherical. Consider the cause why the spherical distribution has changed to discoid distribution from the point based on the angular momentum conservation law.
4. The gravity acting on the mass point of mass m_i at point $\boldsymbol{r}_i$ is $m_i\boldsymbol{g}$, and the moment around the origin O is $\boldsymbol{r}_i \times m_i\boldsymbol{g}$. Therefore, the moment around the origin O of all gravity acting on the system of particles is expressed as

$$\boldsymbol{N} = \sum_i \boldsymbol{r}_i \times m_i \boldsymbol{g} = \left(\sum_i m_i \boldsymbol{r}_i\right) \times \boldsymbol{g}$$
$$= M\boldsymbol{R} \times \boldsymbol{g} = \boldsymbol{R} \times M\boldsymbol{g}. \qquad (1)$$

Therefore, this is equal to the moment around the origin of gravity $M\boldsymbol{g}$ when the total mass $M = \sum_i m_i$ is at the center of gravity. Show that the moment of gravity around the center of gravity is $\boldsymbol{0}$.
5. Show the following : As illustrated in Fig. 7-6, if the internal force $\boldsymbol{F}_{i \leftarrow j}$ acts along a line segment connecting the two mass points i and j that is, if $\boldsymbol{F}_{i \leftarrow j} \,/\!/\, (\boldsymbol{r}_i - \boldsymbol{r}_j)$, the contributions of internal force to the right-hand side of equation (7.28) cancel each other by the law of action and reaction.

Mechanics of Rigid Body

Laws of motion derived in the previous chapter for the system of particles are applied to rigid bodies which are the system of particles with no change in distance between mass points. Mechanics of rigid body is widely applied to various situations in our daily life. An example is the problem where we consider the balance of rigid bodies.

In this chapter, you will study the equation of motion of rigid body, balance of the rigid body, rotational motion of the rigid body around the fixed axis, and planar motion of the rigid body. It is noted that the point of action of the resultant force of gravities acting on each part of rigid body is the center of gravity of the rigid body.

Akita Kanto Festival

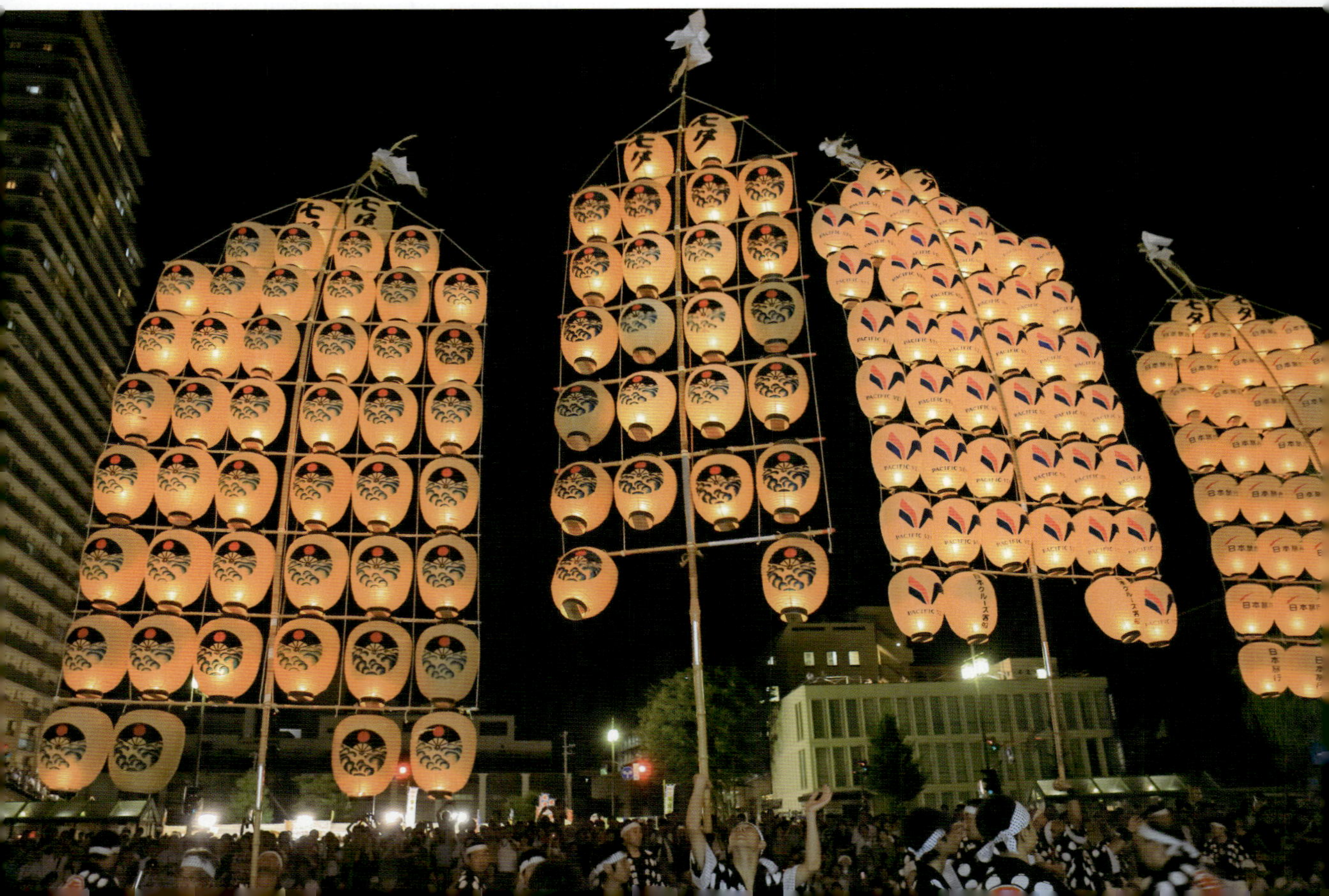

8.1 Equation of motion of rigid body and balance of rigid body

Learning objective To learn the law of motion for the center of gravity of the rigid body and the law of rotational motion. Then to understand the two conditions for balance of rigid body which are derived from the condition for rigid body at rest not to start moving, and to become able to solve problems of balance of the rigid body.

The motion of rigid bodies can be divided into the motion of the center of gravity and the rotational motion around the center of gravity. The kinetic energy of the rigid body is the sum of the kinetic energy of the center of gravity and the rotational energy around the center of gravity [see equation (7.25)].

Law of equation of motion for the center of gravity of a rigid body

The equation of motion for the center of gravity of a rigid body is the equation of motion for the system of particles (7.6)

$$M\frac{\mathrm{d}^2\boldsymbol{R}}{\mathrm{d}t^2} = \boldsymbol{F} \quad (M\boldsymbol{A} = \boldsymbol{F}). \tag{8.1}$$

Here $\boldsymbol{F}$ is the vector sum of all external forces acting on the rigid body.

In corresponding to the relation (5.12) between work and kinetic energy of the mass point which is derived from the equation of motion $m\boldsymbol{a} = \boldsymbol{F}$, from equation (8.1) of the motion of center of gravity, the relation "The work W_{cm} performed by the external force $\boldsymbol{F}$ on the center of gravity of the rigid body is equal to the increase in kinetic energy $K_{\mathrm{cm}} = \frac{1}{2}MV^2$ of the center of gravity", or

$$W_{\mathrm{cm}} = \int_{R_{\mathrm{A}}}^{R_{\mathrm{B}}} \boldsymbol{F}\cdot\mathrm{d}\boldsymbol{R} = \frac{1}{2}MV_{\mathrm{B}}{}^2 - \frac{1}{2}MV_{\mathrm{A}}{}^2 \tag{8.2}$$

is derived. $\mathrm{d}\boldsymbol{R}$ is the displacement of the center of gravity and we must pay attention that W_{cm} is an apparent work which the external force $\boldsymbol{F}$ performs on the center of gravity when the external force is taken to act on the center of gravity of the rigid body. When we consider the energy conservation law, it is necessary to use as the work W_i of each externalforce $\boldsymbol{F}_i$ the work $\boldsymbol{F}_i\cdot\boldsymbol{s}_i$ associated with the displacement $\boldsymbol{s}_i$ of the point of action of the force $\boldsymbol{F}_i$.

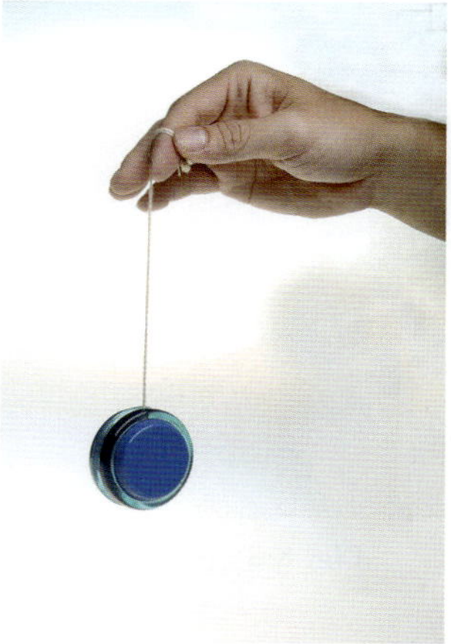

Fig. 8-1 A yo-yo

Case 1 After winding a string around a disk (mass M) we drop the disk like a yo-yo holding the end of the string. The vertically downward gravity Mg and the vertically upward string tension $\boldsymbol{S}$ act on the disk (Fig. 8-2). Therefore, the force to make the disk to drop is the resultant force $F = Mg - S$ in the vertically downward direction. From the equation (8.2), it can be seen that the increment of the energy K_{cm} of the falling motion of the center of gravity of the disk is the work that the resultant force $Mg - S$ performs on the center of gravity. Since the increment of K_{cm} when we drop the disk without holding the string is the work that only gravity Mg does, falling speed of the disk slows when we hold the end of the string.

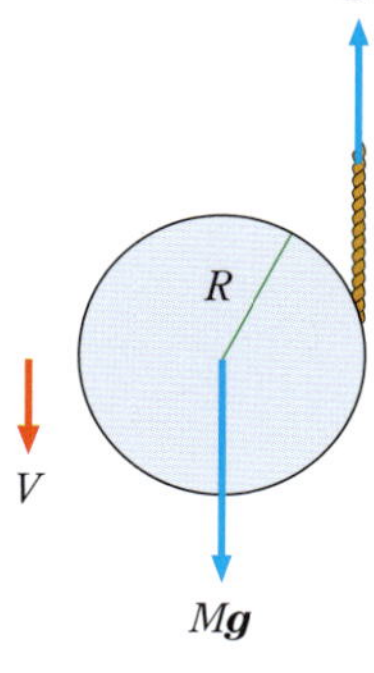

Fig. 8-2

Let us consider the falling movement of height h of this yo-yo from the viewpoint of energy conservation. Since the movement of the disk at each moment is a rotational motion around the contact point of the string and the disk, the string does not perform any work. The decrease Mgh in potential energy of gravitation of the disk is changed, through the work Mgh which gravity performs on the disk into the energy of center of gravity motion of the disk and the energy of the rotational motion.

Case 2 The relationship (8.2) between the work which the external force performs on the center of gravity and the motion of center of gravity also applies to common non-rigid objects spreading in space. When as car moves on a flat road, the external force $\boldsymbol{F}$ appearing on the right side of the equation of motion $M\boldsymbol{A} = \boldsymbol{F}$ is the friction force with which the road surface acts forward. If we assume that the coefficient of friction between the road surface and the tire is μ, the magnitude of the friction force F is less than μMg. Therefore, it can be seen that the magnitude A of the acceleration of the car is less than μg ($A < \mu g$). Therefore, it is derived from equation (8.2) that the increment of kinetic energy when the car moves by the distance d is less than μMgd. If we want to start the car at a large acceleration, some kind of contrivance to increase the friction coefficient of the tire becomes necessary.

When the tire does not slip at the contact point with the road surface, the friction force does not do work on the tire. Therefore, the frictional work Fd on the left side of equation (8.2) is an apparent work. The increase in the kinetic energy of the car is due to the chemical energy of gasoline consumed in the engine.

Fig. 8-3 Auto race

Law of rotational motion of the rigid body Let the angular momentum and moment of external force around an arbitrary point in space be $\boldsymbol{L}$ and $\boldsymbol{N}$, respectively. Then, from equation (7.29)

$$\frac{\mathrm{d}\boldsymbol{L}}{\mathrm{d}t} = \boldsymbol{N} \quad \text{(around an arbitrary point in space).} \tag{8.3}$$

Let the angular momentum and moment of external force around the center of gravity G of the rigid body be $\boldsymbol{L}'$ and $\boldsymbol{N}'$, respectively. Then, from equation (7.33)

$$\frac{\mathrm{d}\boldsymbol{L}'}{\mathrm{d}t} = \boldsymbol{N}' \quad \text{(around the center of gravity).} \tag{8.4}$$

In this way the two rules have been derived.

It should be noted here that in this chapter (excluding precession of tops) only the cases where the direction of rotation axis of the rigid body is always the same and parallel to the z-axis are considered. Accordingly, instead of equations (8.3) and (8.4), the equations of the z direction components of equations (8.3) and (8.4)

$$\frac{\mathrm{d}L}{\mathrm{d}t} = N \quad \text{(around an arbitrary point in space)} \tag{8.3'}$$

and

$$\frac{dL'}{dt} = N' \quad \text{(around the center of gravity)} \tag{8.4'}$$

will be used.

Balance of rigid body When a rigid body under the action of various forces remains at rest, these forces are said to be in balance. Let us find the condition such that the forces $\boldsymbol{F}_1, \boldsymbol{F}_2, \cdots$ are in balance.

If a body remains at rest, since the acceleration $\boldsymbol{A} = \dfrac{\boldsymbol{F}_1+\boldsymbol{F}_2+\cdots}{M} = \boldsymbol{0}$ [equation (8.1)], the vector sum of external forces working on the rigid body must be $\boldsymbol{0}$. Therefore, we obtain the following relation :

$$\text{(condition 1)} \quad \boldsymbol{F}_1+\boldsymbol{F}_2+\cdots = \boldsymbol{0}. \tag{8.5}$$

The second condition is that the sum $\boldsymbol{N}$ of externa forces around an arbitrary point P is $\boldsymbol{0}$, or

$$\text{(condition 2)} \quad \boldsymbol{N}_1+\boldsymbol{N}_2+\cdots = \boldsymbol{0} \quad \text{(around an arbitrary point).} \tag{8.6}$$

This is the condition that the angular momentum around a point P does not change [(equation 8.3)]. So long as the condition (8.6) is satisfied, the rigid body kept at rest never starts rotating around the point P. When the center of gravity remains at rest and does not start rotating around the point P, the rigid body keeps at rest. Therefore, the two conditions (8.5) and (8.6) are conditions for balancing the forces acting on the rigid body. In addition, when the vector sum $\boldsymbol{F}$ of the external forces is $\boldsymbol{0}$, regardless of the selected position of the center of the moment of external forces, the moment $\boldsymbol{N}$ of the external forces is equal.

Because the problem of balance of rigid body discussed in this chapter corresponds to the case where a rigid body on a plane is subjected to forces parallel to the plane, the condition 2 can be derived from equation (8.3′) as

$$\text{(condition 2')} \quad N_1+N_2+\cdots = 0 \quad \text{(around an arbitrary point).} \tag{8.6'}$$

In this case, the moment $\boldsymbol{N}_i$ around the point P of the i-th force $\boldsymbol{F}_i$ acting on the rigid body is the product $F_i l_i$ of the magnitude F_i of force and the distance l_i from the point P to the line of action of the force $\boldsymbol{F}_i$ multiplied by a positive sign or a negative sign depending on the direction of force. The point of action of the resultant force of gravity acting on each part of the rigid body is the center of gravity of the rigid body.

Case 3 A rectangular column is kept at rest on the slope. Normal force $\boldsymbol{N}$, friction force $\boldsymbol{F}$ and gravity $\boldsymbol{W}$ are acting on this rectangular column (Fig. 8-4). The normal force acts on the entire bottom of the rectangular column, but its resultant force is assumed to be $\boldsymbol{N}$. Let A be the intersection of the line of action of gravity and the slope. From the condition that the rectangular column does not rotate around point A, it can be seen that the line of action of the normal force $\boldsymbol{N}$ passes through the point A. Therefore, the condition that this rectan-

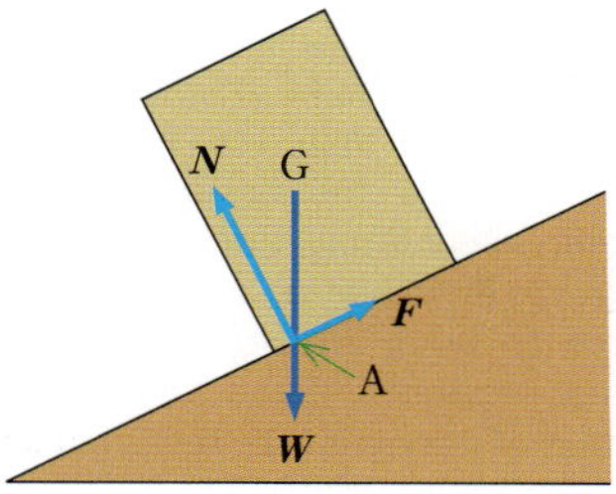

Fig. 8-4

gular column does not fall down is that the point A should lie on the contact surface between the rectangular column and the slope.

Example 1 A ladder of length $L = 4$ m stands leaning against the wall (Fig. 8-5). The friction between the wall and the upper end of the ladder is negligible and the static friction coefficient between the floor and the lower end of the ladder is set to $\mu = 0.40$. The weight of the ladder is 20 kg, and the center of gravity G is in the center of the ladder. A person with a weight of 60 kg started to climb this ladder when the angle between the ladder and the floor was 60°. Can this person reach the upper end of the ladder?

Solution Let $\boldsymbol{W}_1$ and $\boldsymbol{W}_2$ be the gravitational forces that the ladder and the person receive, respectively. Then, $\boldsymbol{W}_2 = 3\boldsymbol{W}_1$. When the person is at a distance x from the lower end, in referring to Fig. 8-5, equation (8.5) showing the condition for the forces to be in balance gives

$$W_1 + W_2 = N_1, \quad \therefore \quad 4W_1 = N_1 \quad \text{(vertical direction)}$$

$$N_2 = F_1 \quad \text{(horizontal direction)}.$$

The condition for balance (8.6) around the lower end of the ladder is

$$W_1 \frac{1}{2} L \cos 60° + W_2 x \cos 60° - N_2 L \sin 60° = 0.$$

Since $\sin 60° = \dfrac{\sqrt{3}}{2}$ and $\cos 60° = \dfrac{1}{2}$, the above equation leads to

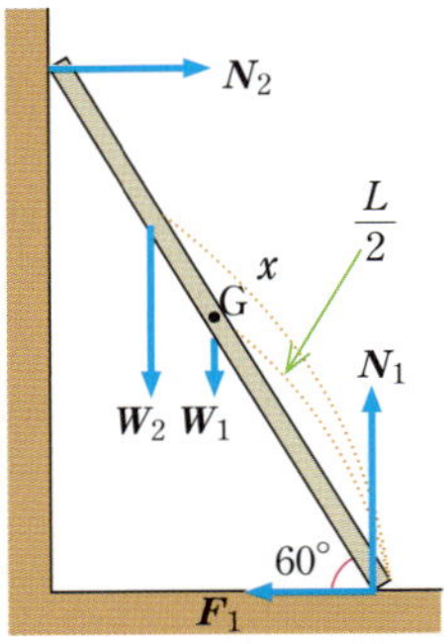

Fig. 8-5

$$\frac{1}{16} N_1 L + \frac{3}{8} N_1 x - \frac{\sqrt{3}}{2} F_1 L = 0.$$

$$\therefore \quad \frac{F_1}{N_1} = \frac{1}{8\sqrt{3}L}(L + 6x).$$

The static frictional coefficient $\mu = 0.40$, so $\dfrac{F_1}{N_1} \leqq \mu = 0.4$. Therefore,

$$\frac{L + 6x}{8\sqrt{3}L} \leqq 0.4, \qquad L + 6x \leqq 5.54L$$

$$\therefore \quad x \leqq 0.76L.$$

The ladder will fall in sliding the wall when the person climbs up by about 3.0 m from the lower end of the ladder.

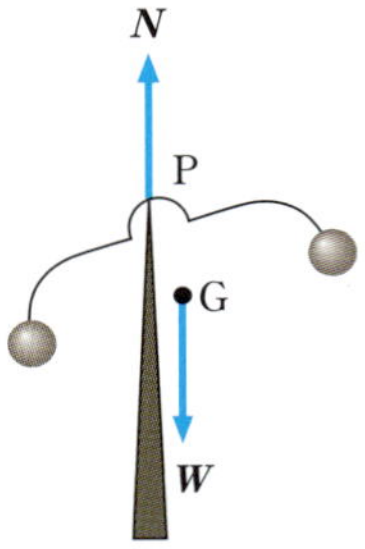

Fig. 8-6 A balancing toy "Yajirobei". The center of gravity G of the toy is lower than the fulcrum P. Therefore, if the balancing toy is tilted, the normal force $\boldsymbol{N}$ and the gravity $\boldsymbol{W}$ will work as the restoring force to bring the toy back to the initial horizontal state. Note that the center of gravity of the toy is outside.

Stable balance and unstable balance When the forces acting on an object are in balance, the two states exist ; one is a stable balance and the other is an unstable balance. When the object is shifted slightly from the state of balance, if the restoring force acts, such a case is called the stable balance, while the other case is called the unstable balance. The balancing toy shown in Fig. 8-6 is a good example of the stable balance.

8.2 Rotational motion of the rigid body around the fixed axis and the moment of inertia

Learning objective To learn that if we consider the correspondences between the moment of inertia I and mass m, the moment N of force and force F, and the angular position θ and position coordinate x, we can understand the rotational motion of the rigid body around the fixed axis by taking the correspondence with the linear motion.

As shown in Fig. 8-7, we put a heavy weight of mass m on one end of a light bar of length l and rotate the weight at angular velocity ω and speed $v = l\omega$ around the rotation axis passing through the other end O of the bar. From equation (6.6), the angular momentum L of the weight around the rotation axis is $L = ml^2\omega$. We express this as

$$L = I\omega \tag{8.7}$$

$$I = ml^2. \tag{8.8}$$

I is called the moment of inertia of the weight around the rotation axis. Since the kinetic energy K of the weight is $K = \frac{1}{2}mv^2 = \frac{1}{2}ml^2\omega^2$, we have

$$K = \frac{1}{2}I\omega^2. \tag{8.9}$$

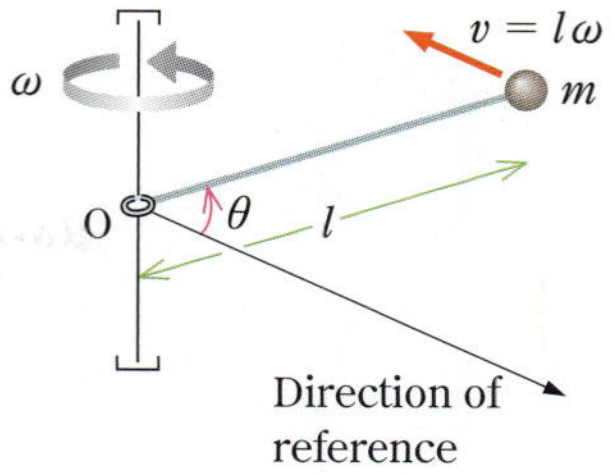

Fig. 8-7

We consider the rotation of a rigid body with a fixed shaft fixed on the z-axis by a bearing, as shown in Fig. 8-8 (a). Since the shaft is fixed by the bearing, each point of the rigid body performs, on the plane perpendicular to the axis (plane parallel to the xy plane), a circular motion centered around the intersection of this plane and the axis.

The position of the rigid body is determined by the angular position θ. The angular velocity $\omega = \frac{d\theta}{dt}$ and the angular acceleration $\alpha = \frac{d\omega}{dt} = \frac{d^2\theta}{dt^2}$ are the same at all points of the rigid body.

We divide the rigid body into minute portions and consider the rigid body to be an assembly of these minute portions. Let l_i be the distance between the i-th minute portion with mass m_i and the rotation axis (z-axis) in Fig. 8-8 (b). Since the speed is $v_i = \omega l_i$, the angular momentum is $m_i l_i^2\omega$. Accordingly, the angular momentum of the rigid body as a whole around the rotation axis is expressed as

$$L = m_1 l_1^2\omega + m_2 l_2^2\omega + \cdots = \omega \sum_i m_i l_i^2. \tag{8.10}$$

Equation (8.10) is rewritten as

$$L = I\omega. \tag{8.11}$$

Here

$$I = m_1 l_1^2 + m_2 l_2^2 + \cdots = \sum_i m_i l_i^2 \tag{8.12}$$

is called the **moment of inertia** of this rigid body around the rotation axis.

If the moment of external force around the rotation axis is denoted as N, from equation (8.3′) we can derive the following law of rotational motion of rigid body around the fixed axis

$$\frac{dL}{dt} = I\frac{d\omega}{dt} = I\frac{d^2\theta}{dt^2} = N. \tag{8.13}$$

Kinetic energy K of the rotational motion of this rigid body is the sum of the kinetic energy of each minute portion

$$K = \frac{1}{2}\sum_i m_i v_i^2 = \frac{1}{2}\sum_i m_i l_i^2\omega^2 = \frac{1}{2}\omega^2\sum_i m_i l_i^2 = \frac{1}{2}I\omega^2. \tag{8.14}$$

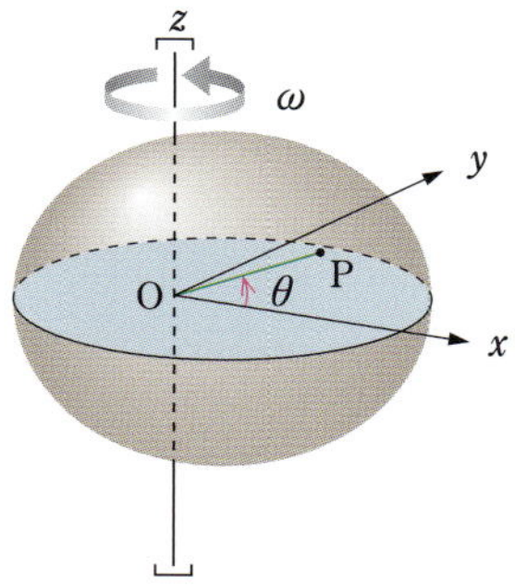

(a) The position of a rigid body is determined by the angular position θ.

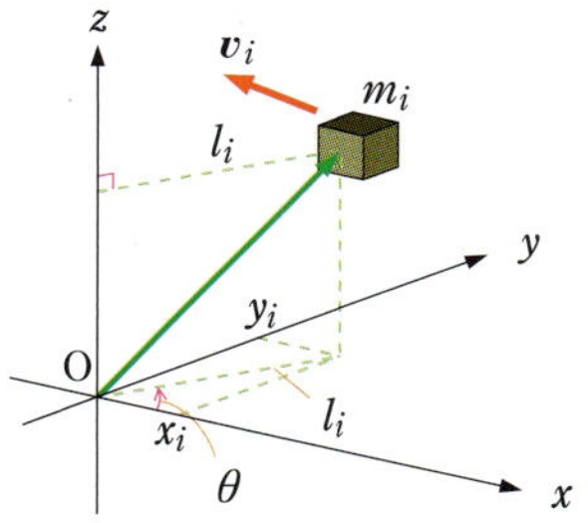

(b)

Fig. 8-8 Motion of rigid body with a fixed axis

Fig. 8-9 An old pendulum clock

Correspondence between the rotational motion of a rigid body around a fixed axis and the linear motion of a mass point

Comparison between the equation of linear motion along the x-axis $ma = m\dfrac{\mathrm{d}v}{\mathrm{d}t} = m\dfrac{\mathrm{d}^2x}{\mathrm{d}t^2} = F$ and the equation of rotational motion around the fixed axis $I\alpha = I\dfrac{\mathrm{d}\omega}{\mathrm{d}t} = I\dfrac{\mathrm{d}^2\theta}{\mathrm{d}t^2} = N$ shows the following corresponding relations :

moment of inertia I	$\Longleftrightarrow$	mass m
angular position θ	$\Longleftrightarrow$	position coordinate x
moment of force (torque) N	$\Longleftrightarrow$	force F
angular velocity $\omega = \dfrac{\mathrm{d}\theta}{\mathrm{d}t}$	$\Longleftrightarrow$	velocity $v = \dfrac{\mathrm{d}x}{\mathrm{d}t}$
angular acceleration $\alpha = \dfrac{\mathrm{d}^2\theta}{\mathrm{d}t^2}$	$\Longleftrightarrow$	acceleration $a = \dfrac{\mathrm{d}^2x}{\mathrm{d}t^2}$
angular momentum $L = I\omega$	$\Longleftrightarrow$	momentum $p = mv$

In addition, the relational expressions of the rotational motion corresponding to those which hold for the linear motion are obtained using the above replacement as follows :

kinetic energy $\dfrac{1}{2}I\omega^2$	$\Longleftrightarrow$	kinetic energy $\dfrac{1}{2}mv^2$
work $W = N\theta$	$\Longleftrightarrow$	work $W = Fx$
power $P = N\omega$	$\Longleftrightarrow$	power $P = Fv$
angular velocity at uniformly accelerated motion $\omega = \omega_0 + \alpha t$	$\Longleftrightarrow$	velocity at uniformly accelerated motion $v = v_0 + at$
angular displacement at uniformly accelerated motion $\theta - \theta_0 = \omega_0 t + \dfrac{1}{2}\alpha t^2$	$\Longleftrightarrow$	displacement at uniformly accelerated motion $x - x_0 = v_0 t + \dfrac{1}{2}at^2$

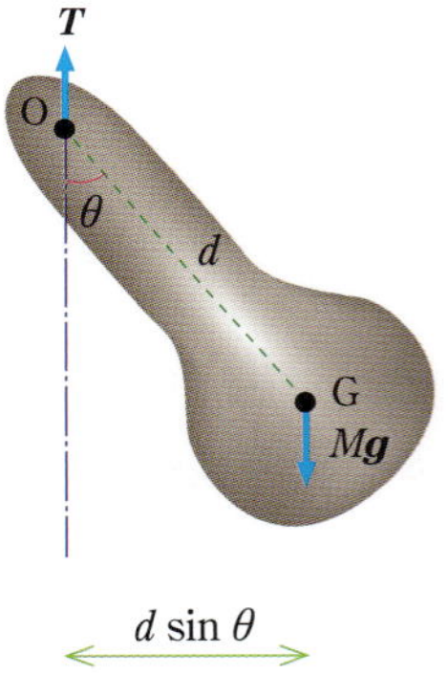

Fig. 8-10 Rigid pendulum

Case 4 ▌Rigid pendulum A rigid body that can freely rotate around a horizontal fixed axis and also oscillate under the action of gravity is called a rigid pendulum (Fig. 8-10). The external force acting on the rigid pendulum is the drag $\boldsymbol{T}$ of the bearing acting on the fixed shaft and the gravity $M\boldsymbol{g}$. Since the line of action of the drag force passes through the fixed axis, the moment of drag around the fixed axis is zero. As shown in Section **7.1**, the effect of gravity acting on a rigid body is the same as when the total gravity $M\boldsymbol{g}$ acting on the rigid body of the mass M acts on the center of gravity G. Here let the distance from the fixed axis O to the center of gravity G be d and let the angle between the line segment $\overline{\mathrm{OG}}$ and the vertical line be θ. It follows that the distance from the fixed axis O to the line of action of gravity passing through the center of gravity G is given by $d \sin\theta$. Accordingly, the moment $N = Fl$ of gravity $M\boldsymbol{g}$ around the fixed axis is, since $F = Mg$ and $l = d\sin\theta$, $N = -Mgd\sin\theta$ (the negative sign means that the gravity works in the direction to restore the swing of the pendulum). Therefore, the equation of motion of a rigid pendulum having a moment of inertia I around the fixed axis is

$$I\frac{\mathrm{d}^2\theta}{\mathrm{d}t^2} = -Mgd\sin\theta \qquad \therefore \quad \frac{\mathrm{d}^2\theta}{\mathrm{d}t^2} = -\frac{Mgd}{I}\sin\theta. \tag{8.15}$$

Since this equation corresponds to the one obtained by replacing $\frac{g}{L}$ in equation (4.18) for the single pendulum with $\frac{Mgd}{I}$, the angular frequency ω of the small oscillation of rigid pendulum is $\sqrt{\frac{Mgd}{I}}$ and the period $T = \frac{2\pi}{\omega}$ is

$$T = 2\pi\sqrt{\frac{I}{Mgd}}. \tag{8.16}$$

Calculated examples of moment of inertia Calculated examples of moment of inertia are given in Fig. 8-11. Even with the same rigid body, the moment of inertia is different when the axis of rotation is different (compare the moments of inertia of the right and left in the top three rows in Fig. 8-11).

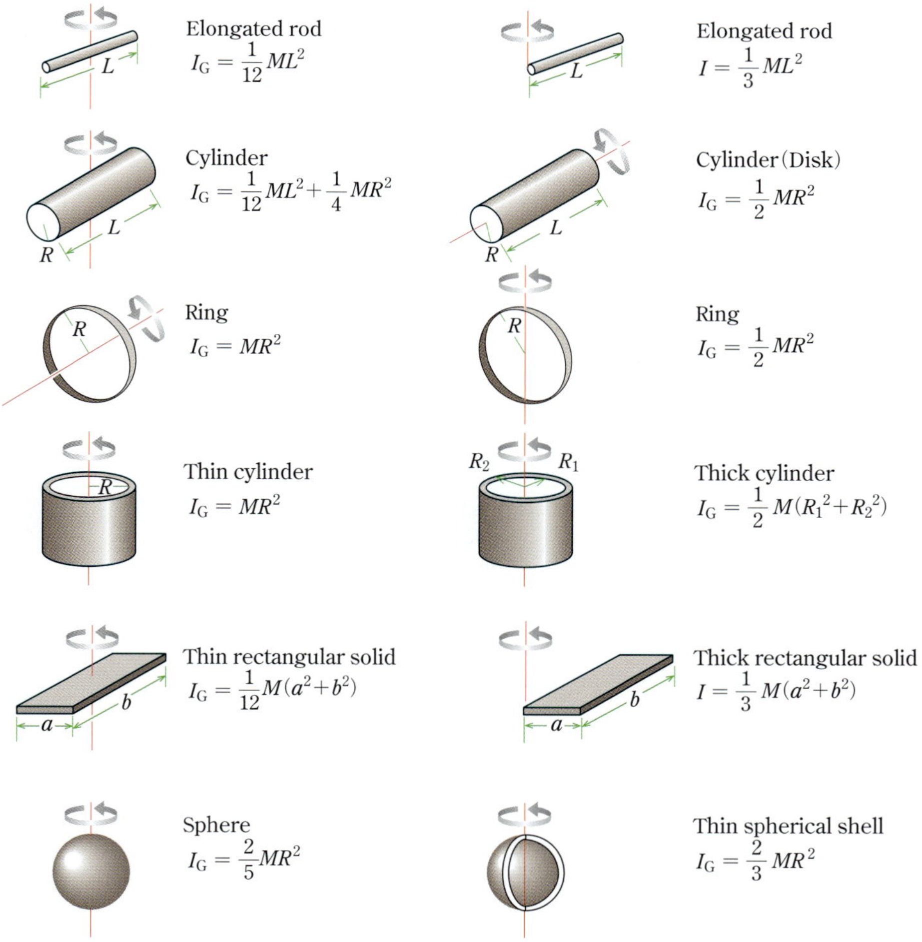

Fig. 8-11 Examples of calculated moment of inertia. The mass of the rigid body is M. I_G is the moment of inertia when the axis of rotation passes through the center of gravity of the rigid body. When the moment of inertia around the certain axis of the rigid body with mass M is set $I = Mk^2$, k is sometimes called the radius of gyration around that axis of the rigid body.

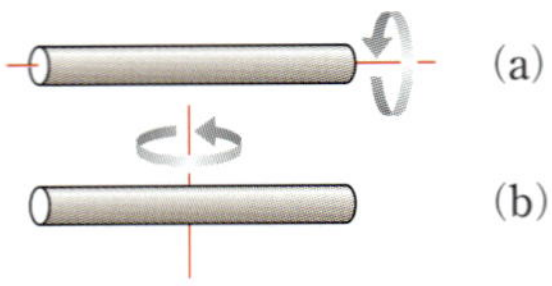

Fig. 8-12

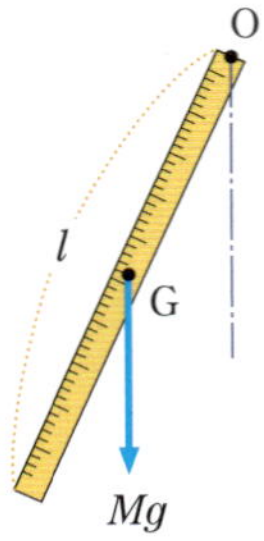

Fig. 8-13

Question 1 In Fig. 8-12 (a) and (b), which one of the rods (a) and (b) has a larger moment of inertia?

Case 5 Let us calculate the period of oscillation in the vertical plane when we hold one end of a ruler of length $l = 30$ cm (Fig. 8-13).

$$I = \frac{1}{3}Ml^2, \qquad d = \frac{1}{2}l. \tag{8.17}$$

So, from equation (8.16)

$$T = 2\pi\sqrt{\frac{I}{Mgd}} = 2\pi\sqrt{\frac{2l}{3g}} = 2\pi\sqrt{\frac{2\times(0.30\,\text{m})}{3\times(9.8\,\text{m/s}^2)}} = 0.90\,\text{s}.$$

Some rigid bodies are easy to rotate and others are hard to rotate. Rigid bodies difficult to rotate mean that a larger work is required to rotate them at the same angular speed ω. Since this work is $\frac{1}{2}I\omega^2$, it is proportional to the moment of inertia I. Moment of inertia I is a quantity that represents the difficulty to rotate a rigid body. As can be seen from the example of the spinning top, the rotating rigid body has the property to keep the same rotational state. It is difficult for rigid bodies with large moment of inertia to change their state of rotation. For this reason, the word inertia is assigned.

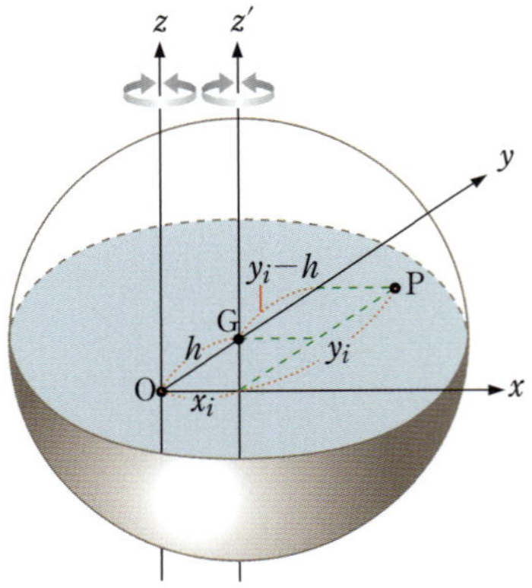

Fig. 8-14 Parallel-axis theorem $I = I_G + Mh^2$

Parallel-axis theorem Let I be the moment of inertia around the axis of rotation (referred to as the z-axis) passing through one point O in a rigid body of mass M, and let the moment of inertia around the axis passing through the center of gravity G and parallel to the z axis be I_G. Then, we have the relation

$$I = I_G + Mh^2. \tag{8.18}$$

h is the distance between the center of gravity G and the z-axis (Fig. 8-14).

Proof We choose the coordinate axes such that the coordinate of the center of gravity is $(0, h, 0)$ (the center of gravity is on the y axis).

$$\begin{aligned} I &= \sum_i m_i l_i^2 = \sum_i m_i(x_i^2 + y_i^2) \\ &= \sum_i m_i\{x_i^2 + (y_i - h)^2 + 2hy_i - h^2\} \\ &= \sum_i m_i\{x_i^2 + (y_i - h)^2\} + 2h\sum_i m_i y_i - h^2\sum_i m_i \\ &= I_G + 2h(Mh) - h^2 M = I_G + Mh^2 \end{aligned}$$

[The distance between the point (x_i, y_i, z_i) and the axis passing through the center of gravity is $\{x_i^2 + (y_i - h)^2\}^{1/2}$, and $h = \frac{1}{M}\sum_i m_i y_i$ and $M = \sum_i m_i$ are used.]

Question 2 Show that for the top case in Fig. 8-11, equation (8.11) holds.

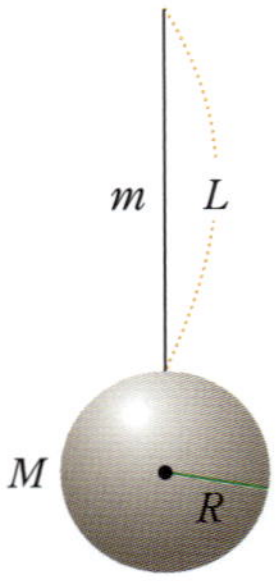

Fig. 8-15 Borda's pendulum

Case 6 | Borda's pendulum A pendulum consisting of a metal ball of radius R and mass M is suspended with a thin wire of length L ($L \gg R$) and mass m. This pendulum is called the Borda's pendulum

(Fig. 8-15). Ignoring the relative motion between the wire and the ball and regarding the whole as a rigid body, we calculate the moment of inertia I of this pendulum. The moment of inertia of the metal ball I_1 is obtained using the parallel-axis theorem and the relations $I_G = \frac{2}{5}MR^2$ and $h = L+R$,

$$I_1 = \frac{2}{5}MR^2+M(L+R)^2.$$

The moment of inertia I_2 of the thin wire is

$$I_2 = \frac{1}{3}mL^2,$$

$$I = I_1+I_2 = \frac{2}{5}MR^2+M(L+R)^2+\frac{1}{3}mL^2. \tag{8.19}$$

8.3 Planar motion of rigid body

Learning objective To become able to write equations describing the movement of rigid bodies rolling down the slope without sliding and the movement of yo-yo, and find the solution using the relationship between the linear motion of the center of gravity and the rotational motion around the center of gravity.

The motion in which all points of a rigid body move on a plane parallel to a certain fixed plane is called the planar motion of rigid body. One example is the motion of a cylinder rolling down a flat slope. When we select this fixed plane as the xy plane and intend to determine the position of the rigid body, we must know in addition to the x, y coordinates X, Y of the center of gravity G the position of another point P of the rigid body in the xy plane including the center of gravity. This position can be determined by the angle θ formed between the directed line segment $\overrightarrow{GP}$ and the $+x$-axis (Fig. 8-16). Therefore, to examine the planar motion of rigid body, the coordinate X, Y of the center of gravity G and the equation of motion describing the rotational angle θ around the center of gravity are required.

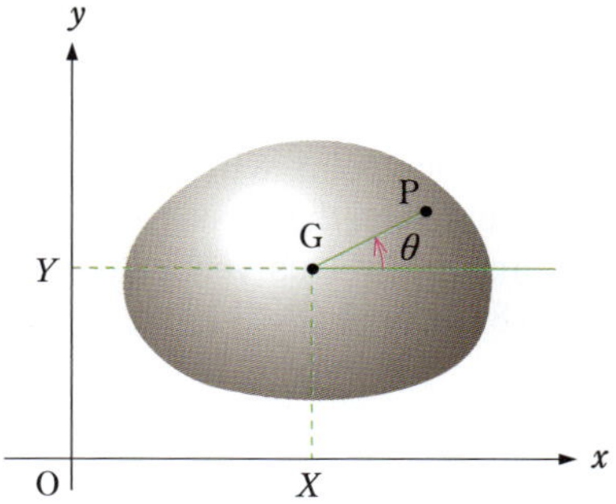

Fig. 8-16 Planar motion of rigid body. The position of rigid body is determined if the coordinate (X, Y) of the center of gravity and the rotational angle θ around the center of gravity are known.

The equation of motion of the center of gravity of a rigid body with mass M subject to the action of external force $\boldsymbol{F}$ is expressed by equation (8.1)

$$M\frac{d^2X}{dt^2} = F_x, \qquad M\frac{d^2Y}{dt^2} = F_y. \tag{8.20}$$

F_x and F_y are the x and y components of the external force $\boldsymbol{F}$, respectively.

Let the moment of inertia around the straight line passing through the center of gravity and parallel to the z-axis be I_G and let the moment of external force relative to the same straight line be N'. Then the equation of the rotational motion around the center of gravity is given by introducing $L' = I_G\omega$ into equation (8.4′). Thus, we obtain the relation :

$$I_G \frac{d\omega}{dt} = I_G \frac{d^2\theta}{dt^2} = N'. \tag{8.21}$$

If non-conservative forces such as friction forces acting on the rigid body perform no work and heat is not generated, the mechanical energy conservation law holds. In other words, if it the height of the center of gravity of the rigid body is h, the sum of the gravitational potential energy Mgh, the energy $\frac{1}{2}MV^2$ of the center of gravity motion of rigid body and the rotational motion energy $\frac{1}{2}I_G\omega^2$ around the center of gravity is constant :

$$\frac{1}{2}MV^2 + \frac{1}{2}I_G\omega^2 + Mgh = \text{constant}. \tag{8.22}$$

Example 2 | Motion of the rigid sphere rolling down the slope without sliding Examine the motion when the sphere with mass M and radius R rolls off the slope without sliding on the slope forming the angle β with the horizontal plane (Fig. 8-17).

Solution Forces acting on the sphere are gravity $\boldsymbol{Mg}$ acting on the center of gravity G, normal force $\boldsymbol{T}$ and friction force $\boldsymbol{F}$ acting at the contact point with the slope. Therefore, when we select the x-axis downward along the slope and the y-axis vertically to the slope, equations (8.20) and (8.21) give

$$M\frac{d^2X}{dt^2} = Mg\sin\beta - F, \tag{8.23}$$

$$0 = T - Mg\cos\beta \qquad \therefore \quad T = Mg\cos\beta,$$

and

$$I_G\frac{d\omega}{dt} = FR. \tag{8.24}$$

When the sphere does not slide at the contact point with the slope, as shown in Fig. 8-18, the following relations hold between the velocity V of center of gravity of the sphere and the angular velocity ω of rotation

$$V = R\omega, \qquad \text{namely} \qquad \frac{dX}{dt} = R\frac{d\theta}{dt}. \tag{8.25}$$

Differentiating both sides of the above equation with respect to t, we obtain the relation

$$\frac{d^2X}{dt^2} = R\frac{d^2\theta}{dt^2}. \tag{8.26}$$

Eliminating F from equations (8.23) and (8.24) and using equation (8.26), we can derive

$$Mg\sin\beta = M\frac{d^2X}{dt^2} + \frac{I_G}{R^2}\frac{d^2X}{dt^2}$$
$$= \left(M + \frac{I_G}{R^2}\right)\frac{d^2X}{dt^2}$$

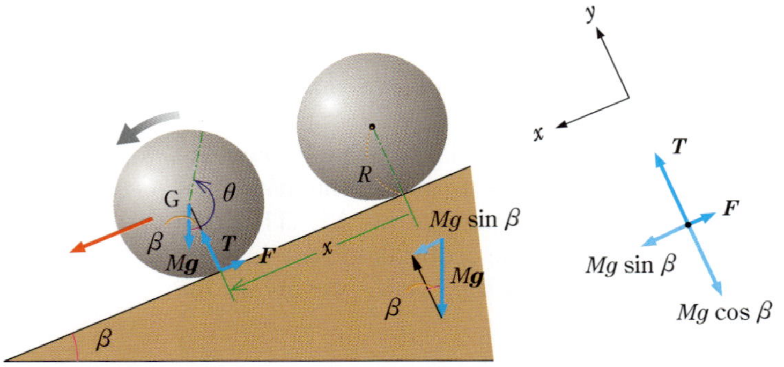

Fig. 8-17 A rigid sphere rolling down the slope.

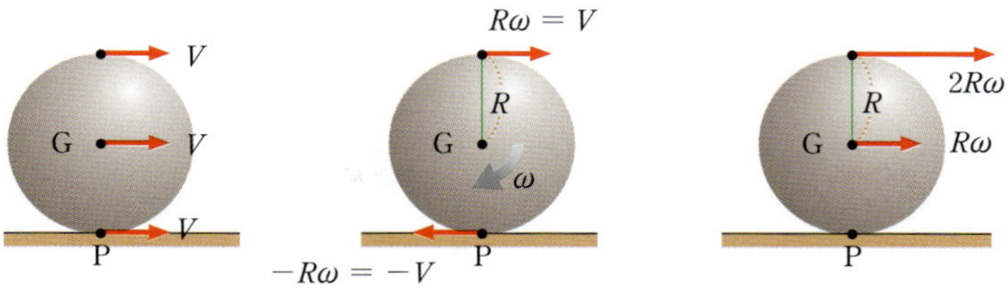

(a) Translational motion of the velocity V

(b) Rotational motion of the angular velocity $\omega = V/R$ around the center of gravity

(c) (a) + (b)

Fig. 8-18 The case (c) where the sphere rolls without sliding is the combination of the cases (a) and (b). Combining the translational motion of the velocity V and the rotational motion of the angular velocity ω around the center of gravity, the velocity $V - R\omega$ at the contact point P is 0, so the velocity of the center of gravity is $V = R\omega$. The motion of the sphere at each moment is the rotational motion of the angular velocity $\omega = V/R$ centered on the contact point P between the sphere and the slope.

$$\therefore \quad \frac{d^2X}{dt^2} = \frac{g \sin\beta}{1 + \dfrac{I_G}{MR^2}}. \tag{8.27}$$

In short, the center of gravity of the sphere moves with the acceleration which is $\dfrac{1}{1+(I_G/MR^2)}$ times the acceleration $g \sin\beta$ of the case where there is no friction between sphere and slope and the sphere slides down without rolling. In general, we can understand when the rigid objects have smaller $\dfrac{I_G}{MR^2}$, they fall fast, while those having large $\dfrac{I_G}{MR^2}$ fall slowly. According to Fig. 8-11, the I_G of the sphere is $\dfrac{2}{5}MR^2$, so the acceleration is $\dfrac{5}{7}$-times $g \sin\beta$.

Question 3 Show that in Example 2, the ratio of kinetic energy of center of gravity and rotational energy is $1 : \dfrac{I_G}{MR^2}$.

Question 4 Show that in Example 2, the speed of the center of gravity at the same falling distance is $\dfrac{1}{\sqrt{1+\dfrac{I_G}{MR^2}}}$ times as large as the occasion where the ball slides down the slope without rolling.

The moments of inertia I_G of thin cylinder, thin spherical shell and cylinder around the central axis are MR^2, $\dfrac{2}{3}MR^2$, and $\dfrac{1}{2}MR^2$, respectively. Therefore, when a thin cylinder, a thin spherical shell and a cylinder roll down without sliding the slope of an angle β relative to the horizontal line (Fig. 8-19), the $\dfrac{g \sin\beta}{1+(I_G/MR^2)}$-times the fall acceleration of the center of gravity are

thin cylinder : $\dfrac{1}{2} g \sin\beta$, thin spherical shell : $\dfrac{3}{5} g \sin\beta$,

cylinder : $\dfrac{2}{3} g \sin\beta$.

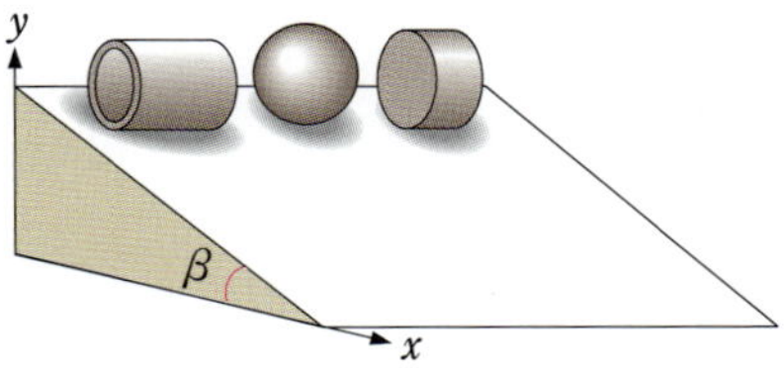

Fig. 8-19 A rigid body rolling down the slope

Accordingly, the sliding down speed becomes faster in this order. Since the fall acceleration of the sphere is $\frac{5}{7}g\sin\beta$, the sphere rolls down much faster than these rigid bodies.

When raw eggs and boiled eggs are compared, since the egg white and yolk of raw egg do not rotate at the same angular velocity as that of the shell even when the shell rotates, so raw eggs are substantially smaller in moment of inertia than boiled eggs. Thus, raw eggs roll off the slope faster than the boiled ones. With this method, we can distinguish raw eggs and boiled eggs without breaking eggs.

Example 3 Examine the motion of the situation shown in Fig. 8-20. We wrap the thread around a uniform disk (radius R, mass M) and fix the other end of the thread, and then bring vertically the thread part that is not in contact with the disk and release the disk to fall. Find the tension S of the thread.

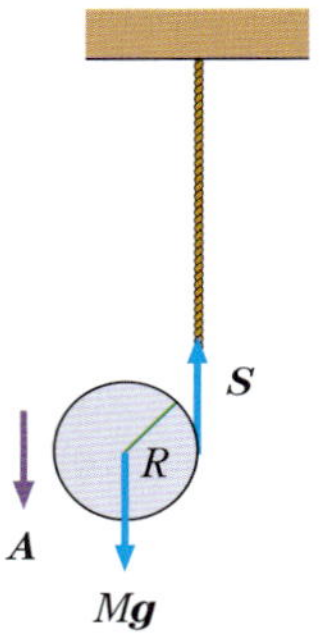

Fig. 8-20

Solution Let the vertical downward direction be the $+x$ direction. Then, equations (8.20) and (8.21) give

$$M\frac{d^2X}{dt^2} = Mg - S \tag{8.28}$$

and

$$I_G\frac{d^2\theta}{dt^2} = SR. \tag{8.29}$$

We calculate the acceleration $A = \frac{d^2X}{dt^2}$ and the tension S using equation (8.26) $\frac{d^2X}{dt^2} = R\frac{d^2\theta}{dt^2}$ ($A = R\alpha$) of the relation between acceleration and angular velocity :

$$A = \frac{d^2X}{dt^2} = \frac{g}{1+\frac{I_G}{MR^2}} \tag{8.30}$$

$$S = \frac{I_G}{MR^2+I_G}Mg. \tag{8.31}$$

Since $I_G = \frac{1}{2}MR^2$ for the case of disk,

$$A = \frac{d^2X}{dt^2} = \frac{2}{3}g, \quad S = \frac{1}{3}Mg. \tag{8.32}$$

Example 4 Find the acceleration A of the center of gravity in the falling motion of the yo-yo (mass M, moment of inertia I_G, radius of axis R_0) shown in Fig. 8-21.

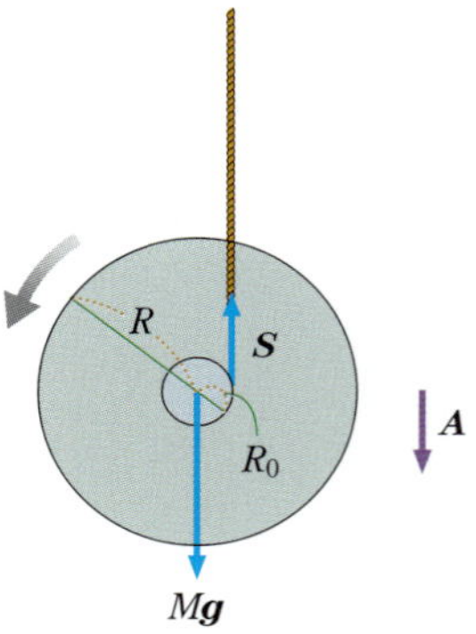

Fig. 8-21 A yo-yo

Solution The vertically downward gravity Mg and the vertically **upward tension S of the thread** act on the yo-yo. The equation of motion in the vertical direction of the center of gravity and the equation of the rotational motion around the axis of the yo-yo are

$$MA = Mg - S \tag{8.33}$$

and

$$I_G\alpha = SR_0. \tag{8.34}$$

Using the relation $A = R_0\alpha$ and eliminating S from equations (8.33) and (8.34), we are led to the following formula :

$$A = \frac{g}{1+\frac{I_G}{MR_0^{\,2}}}. \tag{8.35}$$

If the axis is narrowed (to make R_0 small) to increase $\frac{I_G}{MR_0^{\,2}}$, the fall acceleration A becomes much smaller than the gravitational acceleration g, and the yo-yo falls slowly. When the mass of the axis part of yo-yo in Fig. 8-21 is negligible, since I_G of yo-yo is $\frac{1}{2}MR^2$, we obtain the relation

$$A = \frac{2R_0^{\,2}}{2R_0^{\,2}+R^2}g. \tag{8.36}$$

For reference Precession of the top

Let us find the period T of the precession of the top shown in Fig. 8-22 (a). Let θ be the angle between the vertical direction and the axis of the top rotating at a large angular velocity ω_0 and let I be the moment of inertia around the axis. The gravity $\boldsymbol{Mg}$ acts on the center of gravity G of the top and the drag $\boldsymbol{T}$ acts on the contact point O of the top with the ground. If d is the distance between the point G and the point O, the magnitude of moment $\boldsymbol{N}$ of the external force around the point O is $N = Mgd \sin\theta = N_0 \sin\theta$ and $\boldsymbol{N}$ is on the horizontal plane and is directing perpendicular to the angular momentum $\boldsymbol{L}$ [Fig. 8-22 (b)]. Therefore, the magnitude $L \approx I\omega_0$ of the vector $\boldsymbol{L}$ does not change and the tip of the vector $\boldsymbol{L}$ moves drawing a circle with a radius of $L \sin\theta$ in the horizontal plane. If its angular velocity is Ω, since

$$\Delta L = L \sin\theta\, \Omega\, \Delta t = I\omega_0 \Omega \sin\theta\, \Delta t \tag{8.37}$$

[Fig. 8-22 (c)], from equation (8.3)

$$\Delta L = I\omega_0 \Omega \sin\theta\, \Delta t = N\, \Delta t = N_0 \sin\theta\, \Delta t \tag{8.38}$$

$$\therefore \quad \Omega = \frac{N_0}{I\omega_0} = \frac{Mgd}{I\omega_0},$$

$$T = \frac{2\pi}{\Omega} = \frac{2\pi I\omega_0}{Mgd} \quad \text{(period)}. \tag{8.39}$$

If the rotation of top is counterclockwise viewed from above, the upper end of the top will perform a uniform (constant velocity) circular motion of angular velocity $\dfrac{Mgd}{I\omega_0}$ of counterclockwise rotation in the horizontal plane when viewed from above.

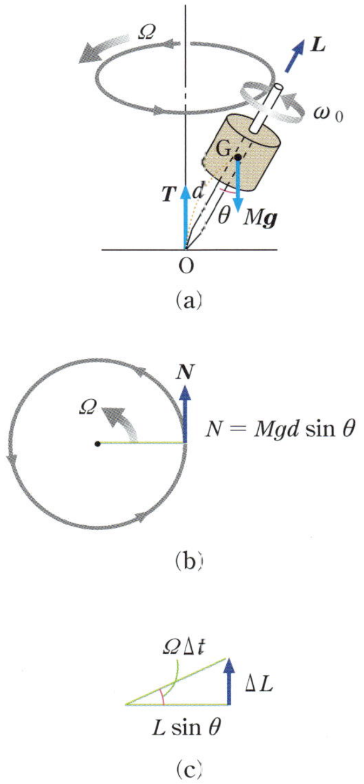

Fig. 8-22 Precession of the top

Problems for exercise 8

A

1. To break a kind of tree nut it is necessary to apply more than 3 kgf from both sides. Find the force needed to break the nut with the tool shown in Fig. 1.

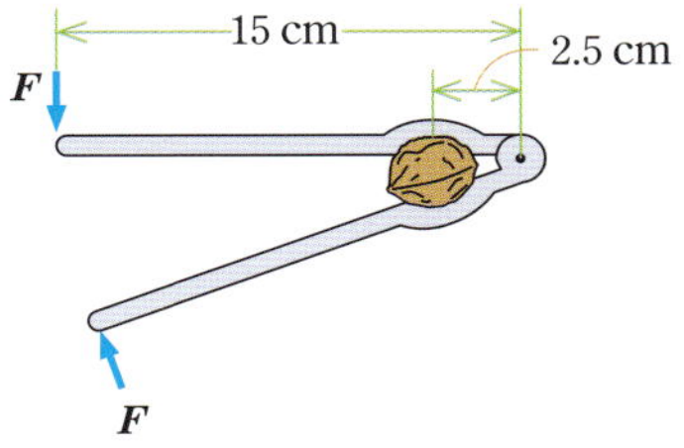

Fig. 1

2. A uniform rectangular plate of 2.0 m in length, 2.4 m in width, and 40 kg in weight is attached to a horizontal bar of length $l = 3.0$ m as shown in Fig. 2. The bar is fixed with a hinge and a rope fixed to the wall.

(1) Find the tension S of the rope.

(2) Find the force with which the hinge acts on the bar.

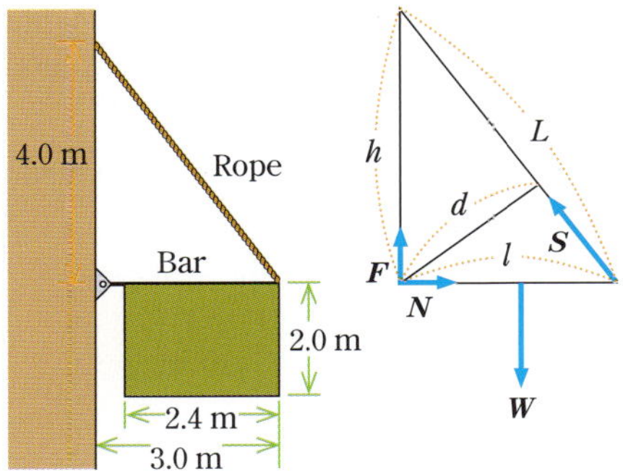

Fig. 2

3. Fig. 3 shows a conceptual diagram of the forces acting on the spinal column when a person bends down to lift the baggage of mass M. Assume that the body weight is W, the trunk weight W_1 is about

0.4 W, and the weight of the head and arms W_2 is about 0.2 W. $\boldsymbol{R}$ is the force which the sacrum exerts on the spinal column, and $\boldsymbol{T}$ is the force which the spinal muscle exerts on the spinal column. Express T in terms of W, M, and θ. How many kgf is T when $W = 60$ kg, $M = 20$ kg, and $\theta = 30°$? Use $\sin 12° = 0.208$.

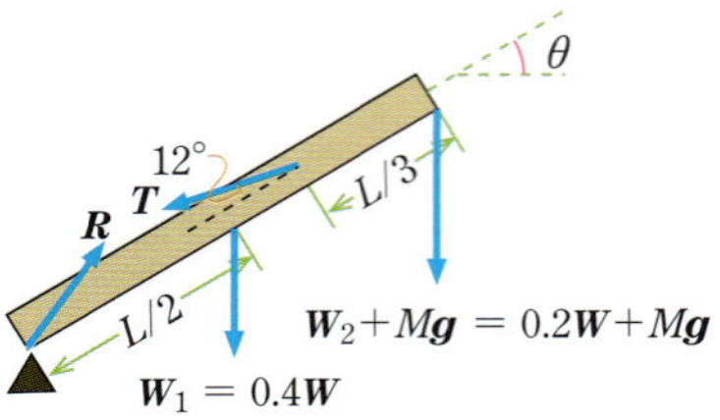

Fig. 3

4. An iron column with a radius of 1 m and a height of 1 m (specific gravity is 8 g/cm^3) rotates around its central axis at 600 revolutions per minute. Find the energy of this rotational motion.
5. Each of the three rotor blades of a helicopter has a length L of 5.0 m and a mass M of 200 kg (Fig. 4). Find the energy of rotational motion when the rotor blades rotate at 300 revolutions per minute.

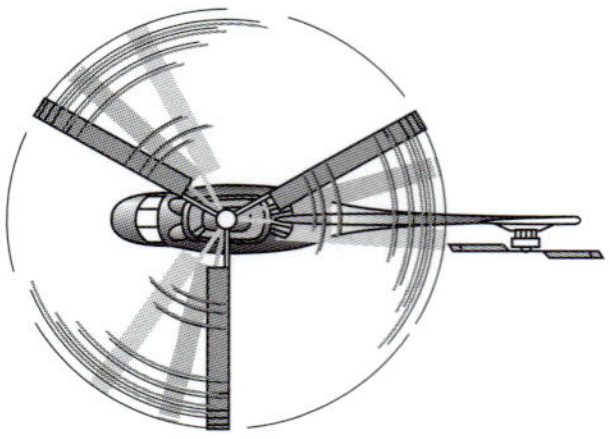

Fig. 4

6. An iron bar and an aluminum bar of the same length and the same thickness were bonded as shown in Fig. 5. Consider two cases, (a) to rotate around the point O, and (b) to rotate around the point O′ Which case is easier to rotate?

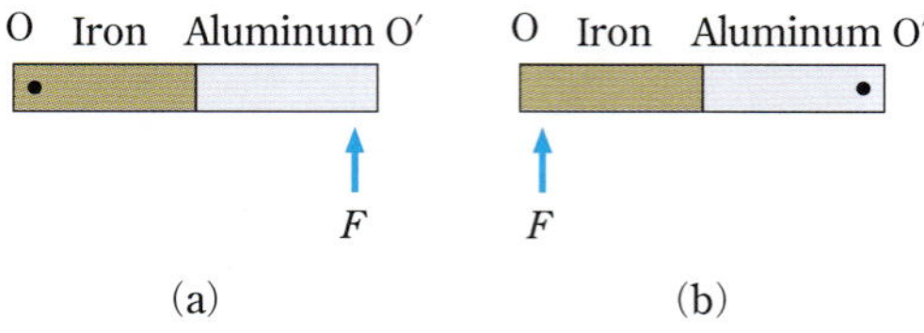

Fig. 5

7. When walking on a high fence, why do we stretch both our arms to the right and left?
8. An acrobat holds a long rod and is walking on a tightrope.
   (1) When the body begins to tilt to the right, to which side should the rod be inclined?
   (2) What role do the weights at the both ends of the rod play?
9. Compare the speed of the ball descending the hill without friction and reaching the bottom of the hill under two different conditions ; the case where the ball slides down the hill without rolling, and the case where the ball rolls off the same inclined hill from the same height without sliding.
10. When we gently roll down three beer-cans, namely can containing beer, can in which beer is frozen, and empty can, from the top end of the slope, which beer-can will roll on the slope most quickly?

B

1. The center of gravity of a hemisphere of radius R and mass M is at distance $\frac{3}{8}R$ from the center of the circle. To this hemisphere a light weight stick of negligible mass was attached, and also a weight of mass m was attached to the stick (Fig. 6). When the position of this weight is raised, the standing hemisphere becomes unstable. Examine this.

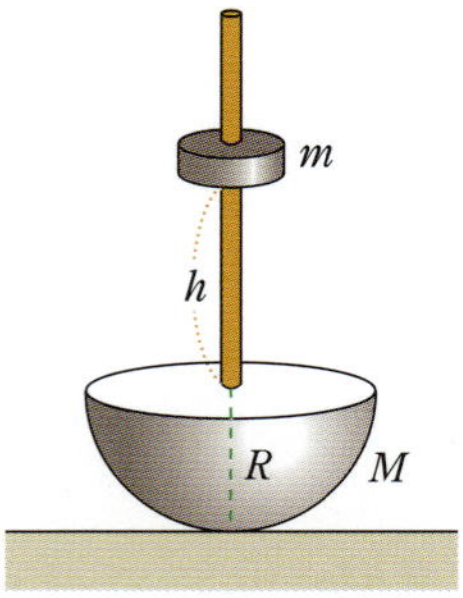

Fig. 6

2. Fig. 7 is a conceptual diagram of a racing car. Show that the maximum acceleration A_{max} that this car can attain is μg and to attain this maximum acceleration, the relation $l_2 = \mu h$ is required. μ is the static friction coefficient between the tires and the road.

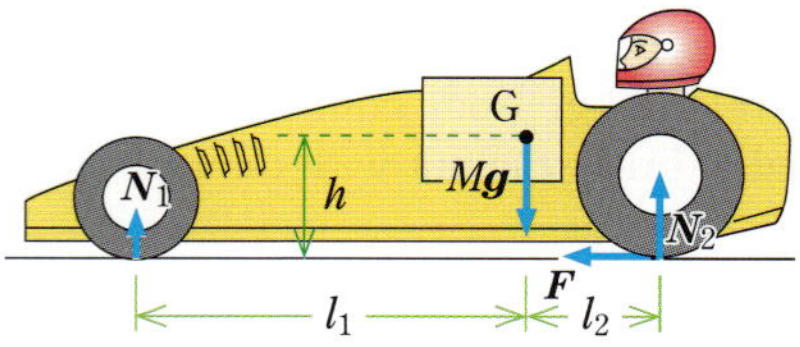

Fig. 7 A conceptual diagram of a racing car

3. The moment of inertia I around the side of the length c of the rectangular parallelepiped with the lengths of three sides a, b, and c and mass M shown in Fig. 8 is

$$I = \frac{1}{3}M(a^2+b^2).$$

Find the period T when this rectangular parallelepiped is oscillated around the axis shown in the figure as a rigid pendulum.

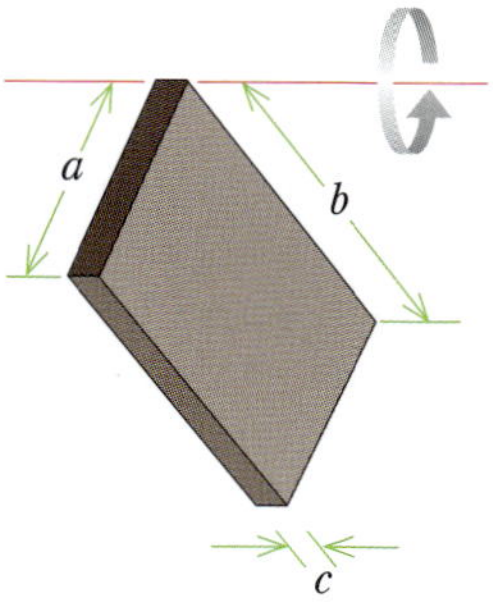

Fig. 8

4. It is said that when we hit horizontally the point $\frac{2}{5}R$ above the center of a sphere of radius R in billiards, the ball rolls without slipping. Explain this fact.

5. As shown in Fig. 9, when we pull the end of a thread wound on the reel, the direction of movement of the thread winding reel varies depending on the pulling direction. What will happen in the case of $\boldsymbol{F}_1$, $\boldsymbol{F}_2$, and $\boldsymbol{F}_3$ in Fig. 9? Consider the moment of external forces around the point of contact with the floor

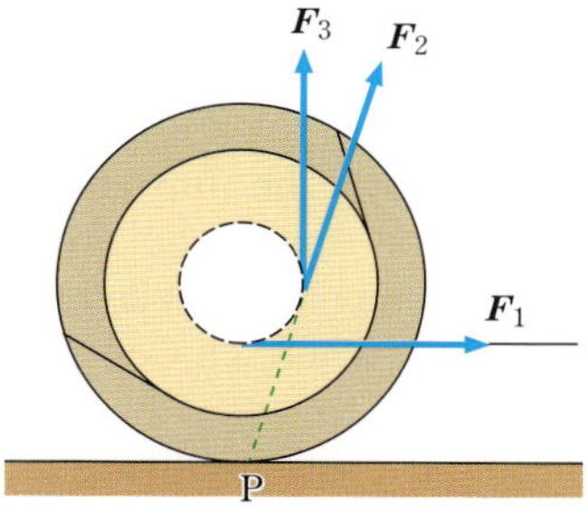

Fig. 9

Equation of motion that cars follow

The center of gravity G of a car (mass M) follows the same equation of motion as a mass point of mass M on which the vector sum of all external forces acting on the car acts [equation (8.1)]. When the car runs on a flat straight road, the external force $\boldsymbol{F}$ (if the resistance of the air is neglected) oriented in the horizontal direction is only the frictional force that the road surface exerts on the tires of the driving wheels at the contact point. Therefore, the equation of motion is

$$M\frac{\mathrm{d}\boldsymbol{V}}{\mathrm{d}t} = \boldsymbol{F} \quad (\boldsymbol{F} \text{ is a friction force}). \tag{1}$$

Thus, it follows that the driving force of the engine does not appear.

Therefore, we consider an electric car in which wheels (moment of inertia I_G, radius R) are rotated with torque $2Kr$ due to the couple $(\boldsymbol{K}, -\boldsymbol{K})$ which the motor exerts on the axles (radius r) (Fig. 8-A). Equation of motion of the wheel is

$$I_G\frac{\mathrm{d}\omega}{\mathrm{d}t} = 2Kr - FR. \tag{2}$$

The second term $-FR$ of the right-hand side is the torque of the friction force hindering the rotation.

Using the condition that the tire does not slip at the contact point with the road surface

$$V = R\omega \tag{3}$$

from equations (1) and (2) we can derive the **equation of motion of the car** which has the apparent weight $M+\frac{I_G}{R^2}$ and runs with the driving force $2\frac{r}{R}K$ of the motor,

$$\left(M+\frac{I_G}{R^2}\right)\frac{\mathrm{d}V}{\mathrm{d}t} = 2\frac{r}{R}K.$$

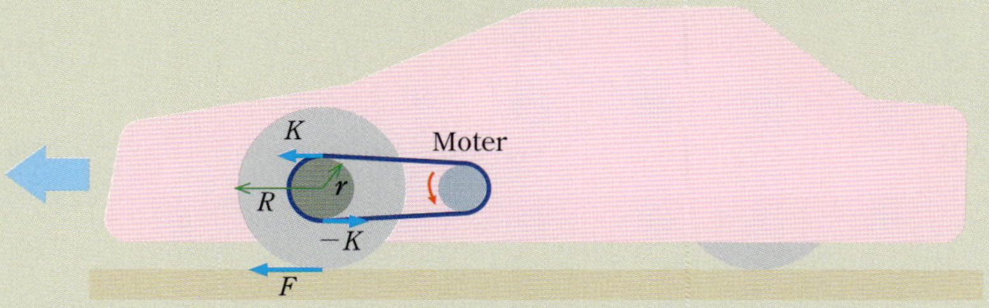

Fig. 8-A

Why is a running bicycle not falling down?

Even if we try to stand a non-spinning top on a floor, the top will fall down, but the well-spinning top does not fall down so readily. Similarly like this, a bicycle at rest falls unless it is supported but the running bicycle is rather hard to fall down. This is because in both cases, rotating objects have the inertia to keep rotating. Even if the spinning top is about to be toppled, the top does not topple immediately and the upper end of the top shaft performs a circular movement called the precession (see Reference of Section **8.3**). Consider the case of bicycle.

Before considering this problem, let us review the law of rotational motion in vector notation :

$$\frac{d\boldsymbol{L}}{dt} = \boldsymbol{N}. \tag{1}$$

In the above equation the angular momentum $\boldsymbol{L}$ and moment $\boldsymbol{N}$ (torque) of external force appear.

The angular momentum $\boldsymbol{L}$ is a quantity representing the intensity and direction of rotational motion and its magnitude L is "moment of inertia I" × "angular velocity ω". Its direction is the travelling direction of a screw when the screw is turned in the direction of the rotation of object (Fig. 8-B).

Torque $\boldsymbol{N}$ is the quantity to represent the ability of force which acts on an object to rotate. As shown in Fig. 8-C, the torque of the couple $\boldsymbol{F}$ and $-\boldsymbol{F}$ with distance d to the line of action has the magnitude of Fd and is directed in the traveling direction of the screw turned by the couple.

Equation (1) of the law of rotational motion can be transformed into $\Delta\boldsymbol{L} = \boldsymbol{N}\,\Delta t$ and we understand this equation shows "When the force of torque $\boldsymbol{N}$ is applied for a time Δt to an object rotating with angular momentum $\boldsymbol{L}$, the rotational state of the object changes to the one with angular momentum $\boldsymbol{L}+\Delta\boldsymbol{L} = \boldsymbol{L}+\boldsymbol{N}\,\Delta t$". Here, $\boldsymbol{N}\,\Delta t$ is in the direction of the torque $\boldsymbol{N}$, and its magnitude is "magnitude of torque N" × "time Δt" represented by the arrow symbols.

The angular momentum $\boldsymbol{L}$ of the rotating bicycle wheels is facing (as can be seen by rotating the screw in the direction of the wheel) (Fig. 8-D) from the right to the left. When the rider tilts the body to the right, the torque $\boldsymbol{N}$ of the gravity acting on the rider and the bicycle and the normal force of the ground is directed toward the front from the back (in the traveling direction of the screw by this force). As a result, the angular momentum $\boldsymbol{L}+\boldsymbol{N}\,\Delta t$ of the wheel is directed to the forward left and the bicycle turns right.

This theoretical conclusion can be proved by an experience when a person is riding a bicycle and lifts his/her hands from the handlebar and tilts the body slightly to the right, the bicycle turns to the right without being toppled.

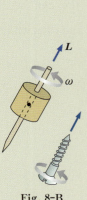

Fig. 8-B

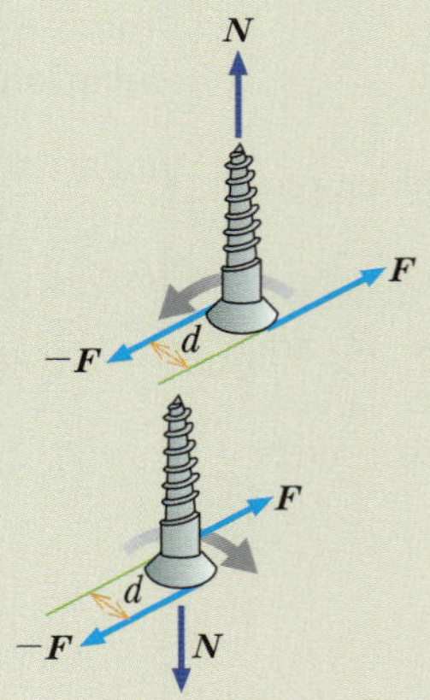

Fig. 8-C

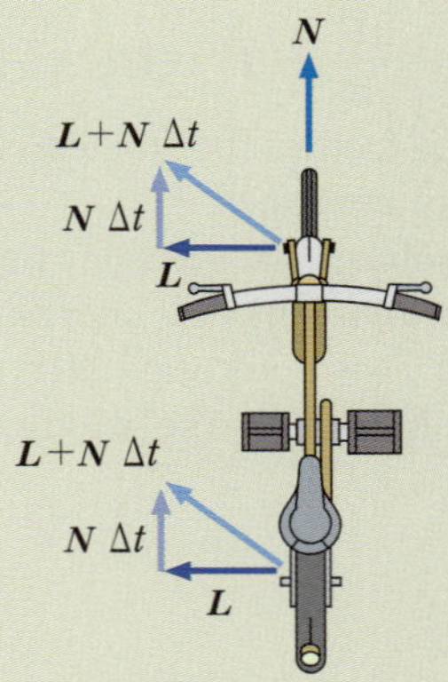

Fig. 8-D

Force of Inertia

To measure the position and velocity of an object, we must select the reference coordinate axis (coordinate system). When train passengers want to describe the phenomenon in the train, coordinate axes fixed to the train floor or wall are convenient. The first law of Newton's equation of motion does not hold in an arbitrary coordinate system. The first law asserts that there exists a coordinate system in which an object not affected by a force is kept at rest or in a linear motion at a constant velocity. The coordinate system in which the law of inertia (the first law) holds is called the **inertial frame**, and the coordinate system in which the law of inertia does not hold is called the **non-inertial frame**. The second law of motion (the law of motion) holds only in the inertial frame. If we want to make the law of motion hold in the non-inertial frame, we must introduce an apparent force called the **force of inertia**.

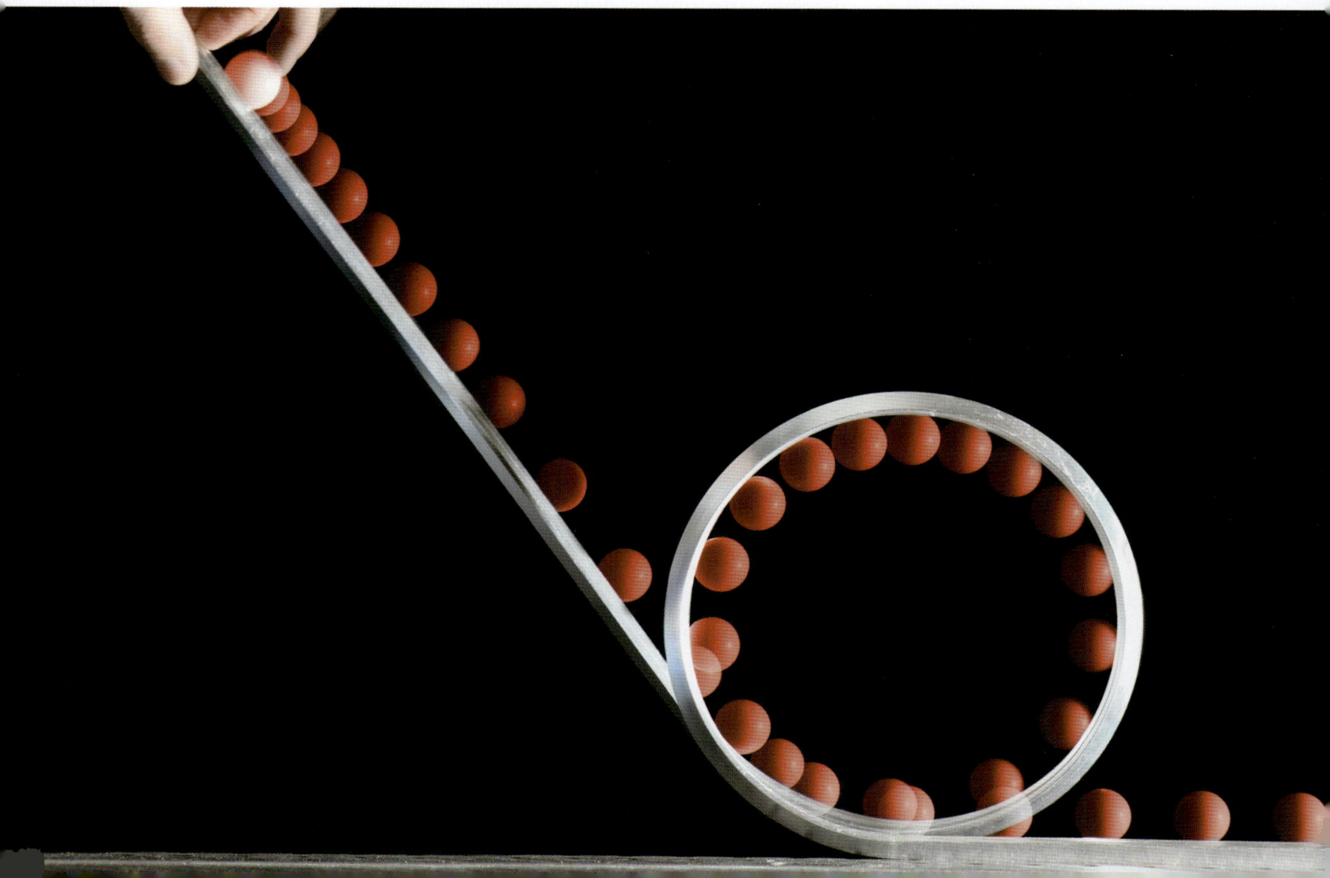

9.1 Non-inertial frame and force of inertia (apparent force)

Learning objective To understand that the coordinate system in which the law of inertia holds is called the inertial frame and that the coordinate system which performs an accelerated motion at constant acceleration $\boldsymbol{a}_0$ with respect to the inertial frame is a non-inertial frame and to make the law of motion apparently hold in the non-inertial frame, we must introduce an apparent force called the force of inertia $-m\boldsymbol{a}_0$.

(a) A train at rest

(b) A suitcase does not move even the train begins to move.

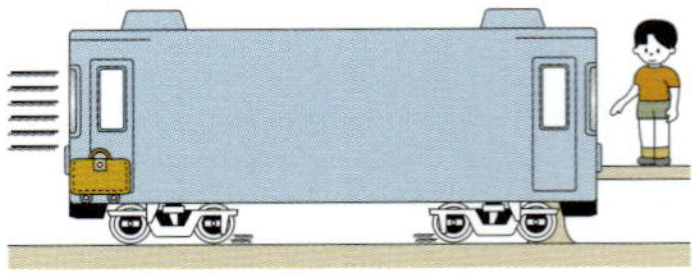

(c) After the suitcase hits the train wall, it moves together with the train.

Fig. 9-1 A suitcase viewed from the platform

Non-inertial frame and force of inertia A suitcase with casters is placed parallel to the traveling direction on the train floor (Fig. 9-1). When the train begins to move, the suitcase moves on the floor in the direction opposite to the traveling direction of the train. When this suitcase is viewed from the platform, although the train begins to move, the suitcase does not seem to move (until it hits the rear wall) relative to the platform (in the case of negligible friction). Observers on the platform understand this phenomenon as follows : Because the force does not act on the suitcase, the suitcase which was at rest with respect to the platform while the train was stopped remains at rest relative to the platform, even if the train begins to move. In other words, the equation of motion for this suitcase is

$$m\boldsymbol{a} = \boldsymbol{0}. \tag{9.1}$$

However, if the acceleration of train is set to be $\boldsymbol{a}_0$, the suitcase moves with acceleration $\boldsymbol{a}' = -\boldsymbol{a}_0$ in the direction opposite to the train passengers*. Therefore, the coordinate system based on the train floor or wall (the coordinate system fixed to the train) is a non-inertial frame because the object with no force applied performs the accelerated motion. However, since it is convenient for most people to put themselves in the center and to understand various phenomena around them, the passengers are apt to think that Newton's law of motion also holds in the coordinate system (non-inertial frame) fixed to the train. Thus, train passengers feel that "The suitcase is subjected to the "force" acting backwards, so the suitcase starts to move backwards". This "force" is called the force of inertia. In other words, the **force of inertia** refers to a force that must be introduced so that the law of motion holds formally in the non-inertial frame performing the accelerated motion, although no object causing the force exists.

In the coordinate system fixed to the train immediately after departure, the equation of motion of the suitcase is

$$m\boldsymbol{a}' = \text{force of inertia} \tag{9.2}$$

and since $\boldsymbol{a}' = -\boldsymbol{a}_0$, in the coordinate system performing the constant acceleration $\boldsymbol{a}_0$, the following relation holds :

$$\text{force of inertia} = -m\boldsymbol{a}_0. \tag{9.3}$$

In many cases, the coordinate system fixed to the Earth can be taken as the inertial frame. However, since the Earth rotates and revolves, the Earth is not exactly the inertial frame. In treating large motions on the global scale and/or high precision experiments, we must consider the force inertia accompanying the rotation of the Earth

* The relationship between the position vector $\boldsymbol{r}'$ in the coordinate frame fixed to the train and the position vector $\boldsymbol{r}$ in the coordinate system fixed to the ground is

$$\boldsymbol{r}' = \boldsymbol{r} - \frac{1}{2}\boldsymbol{a}_0 t^2.$$

Therefore, the velocity $\boldsymbol{v}'$ and acceleration $\boldsymbol{a}'$ in the coordinate system fixed to the train are

$$\boldsymbol{v}' = \boldsymbol{v} - \boldsymbol{a}_0 t \quad \text{and} \quad \boldsymbol{a}' = \boldsymbol{a} - \boldsymbol{a}_0.$$

Since $\boldsymbol{a} = \boldsymbol{0}$, we have $\boldsymbol{a}' = -\boldsymbol{a}_0$.

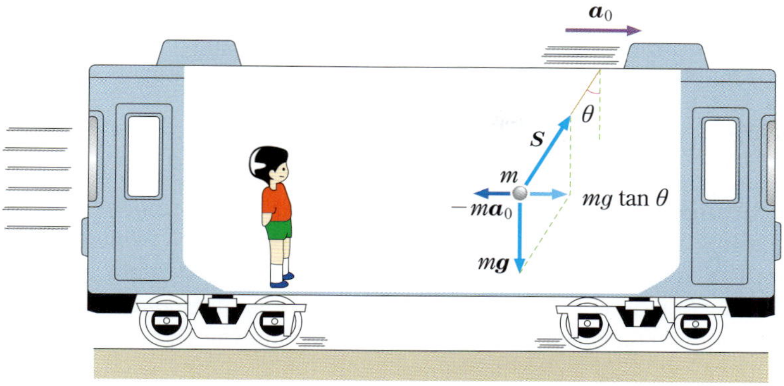

Fig. 9-2 A train moving at acceleration $\boldsymbol{a}_0$

Question 1 As shown in Fig. 9-2, a train is moving at a uniform acceleration of $\boldsymbol{a}_0$.

(1) When a string with a weight attached is hung from the ceiling of the train, the string does not point in the vertical direction. How do observers on the ground and observers in the train explain this phenomenon?

(2) When the string breaks, the weight falls. Find the angle θ between the fall direction observed in the train and the vertical direction. Observers in the train may think that the weight will fall freely due to the apparent gravity facing the fall direction.

(3) When we fix the end of the string attached to the balloon to the train floor, in which direction does the string lean? Consider this issue using the apparent buoyancy acting on the balloon.

Question 2 When we descend from the top floor of a skyscraper using an elevator, we feel as if our body got lighter right after starting to descend (Fig. 9-3). Find the normal force in the units of N and kgf that a person with a weight of 50 kg in a stationary state receives from the elevator floor when the downward acceleration is 1 m/s^2. The sense that humans feel for their own weight comes from the force to support their weight. If the elevator rope breaks and the elevator starts to fall freely, it is said that passengers in the elevator will be in weightless condition. Explain what this means.

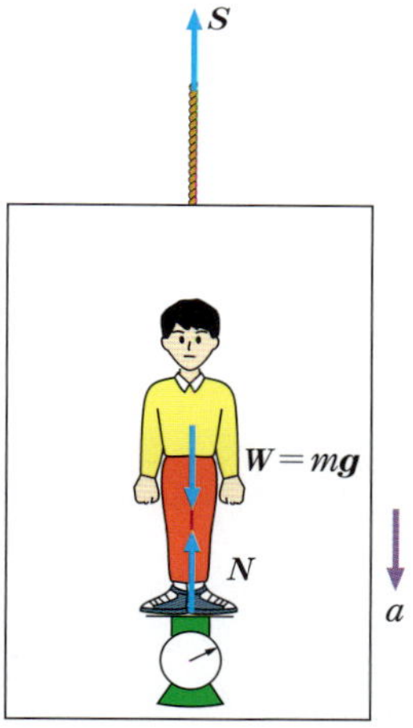

Fig. 9-3 When the elevator accelerates, the needle of weight scale swings and the measured weight will be lighter or heavier. At this time, the weight of the treadle of the weight scale also changes, so the needle of the weight scale that nobody is on can swing. Thus, correction for this amount is necessary.

Galilean principle of relativity Situation is different if a train is in linear motion at a constant velocity. When the train is running in the long tunnel, passengers sometimes do not know in which direction the train is running. The acceleration of the train in linear motion at a constant velocity is $\boldsymbol{a}_0 = \boldsymbol{0}$, and the force of inertia does not act on the passengers, because the coordinate system fixed to the train is an inertial frame.

In mechanics, it is thought that "The coordinate system that performs the linear motion at a constant velocity with respect to an inertial frame is the inertial frame, and in all inertial frames, acceleration, mass and force are the same, and the same law of motion $m\boldsymbol{a} = \boldsymbol{F}$ holds". This is called the **Galilean principal of relativity**. This terminology comes from the fact that in order to persuade opponents to the heliocentric theory who say "If the Earth is moving,

Fig. 9-4 A person riding the elevator feels like he/she has lost or got some weight.

the object cannot fall down vertically in the air", Galileo is said to have shown that the stone falling from the mast of the moving ship dropped onto the deck directly underneath.

9.2 Centrifugal force and the Coriolis force

Learning objective To understand that a coordinate system rotating at a constant angular velocity with respect to the inertial frame is a non-inertial frame, and when we want to formally establish the law of motion in the rotating frame, we must introduce two types force of inertia called the centrifugal force and the Coriolis force.

Fig. 9-5 A merry-go-round

Centrifugal force The mass point of mass m moving in a uniform circular motion on a circle of radius r at velocity v and angular velocity $\omega = \dfrac{v}{r}$ is subjected to the centripetal acceleration with its magnitude $\dfrac{v^2}{r} = r\omega^2$ and this acceleration is caused by the centripetal force of magnitude $\dfrac{mv^2}{r} = mr\omega^2$.

Consider an object P of mass m which is connected by a string of length r with a center pillar of a merry-go-round rotating at an angular velocity ω and is rotating with the floor at the angular velocity ω in a uniform circular motion. When a person standing on the merry-go-round observes the object P, that person feels, "This object is at rest with respect to me, because

the outward force (centrifugal force) with the magnitude of $mr\omega^2$ (9.4)

balanced with the centripetal force (tension of the string) acts" (Fig. 9-6). As in this case of merry-go-round, the coordinate system (**rotational frame**) fixed to the object in rotational motion relative to the inertial frame (Earth surface) is a non-inertial frame, and if we want to realize Newton's law of motion in the rotational frame, introduction of the **force of inertia** facing outward is indispensable. Since this force of inertia acts in the direction to move an object doing the circular motion away from the center of circle, it is called the **centrifugal force**.

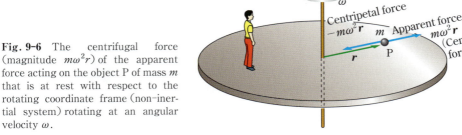

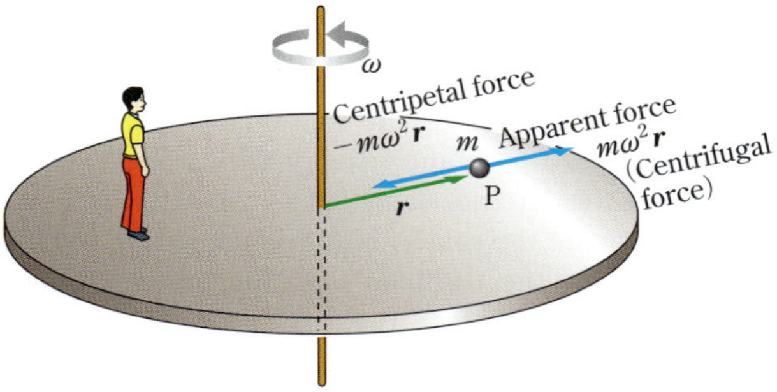

Fig. 9-6 The centrifugal force (magnitude $m\omega^2 r$) of the apparent force acting on the object P of mass m that is at rest with respect to the rotating coordinate frame (non-inertial system) rotating at an angular velocity ω.

Since the Earth rotates around its axis, we, who are rotating with the Earth, feel that the centrifugal force is acting on objects that are at rest on the Earth surface. However, since both the centrifugal force and gravitational attraction acting on the object are proportional to the mass of the object, we cannot distinguish them. Thus, strictly speaking the gravity acting on the object is the resultant force of the universal gravitation of the Earth and the centrifugal force (Fig. 9-7).

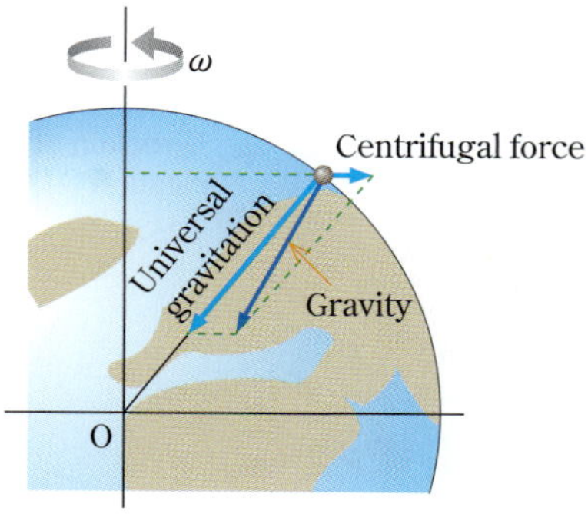

Fig. 9-7 Gravity on the Earth surface is the resultant force of universal gravitation and centrifugal force. The magnitude of the centrifugal force differs depending on the latitude, but it is less than or equal to 0.4 % of the universal gravitation.

Coriolis force When viewed in the coordinate system (rotating coordinate frame) rotating at a constant angular velocity ω with respect to the inertial system, the centrifugal force of the force of inertia acts on the object at rest in this frame. When the object is in motion with respect to the rotating coordinate system, in addition to the centrifugal force, another force of inertia, **Coriolis force,** appears.

In Fig. 9-8 when we throw a ball from the center of the turntable towards the point P at the edge of the table, the ball does not move on the line segment OP drawn on the table, instead it deviates to the right from this line. Although the ball viewed on the ground moves straight [Fig. 9.8 (a)], since the line segment OP rotates during that time, the ball deviates to the right of the segment OP when seen on the table [Fig. 9.8 (b)]. The inertial force that causes this deviation is called the Coriolis force. That is to say, when an object moves relative to a rotating coordinate system, the centrifugal force and the Coriolis force appear as the force of inertia.

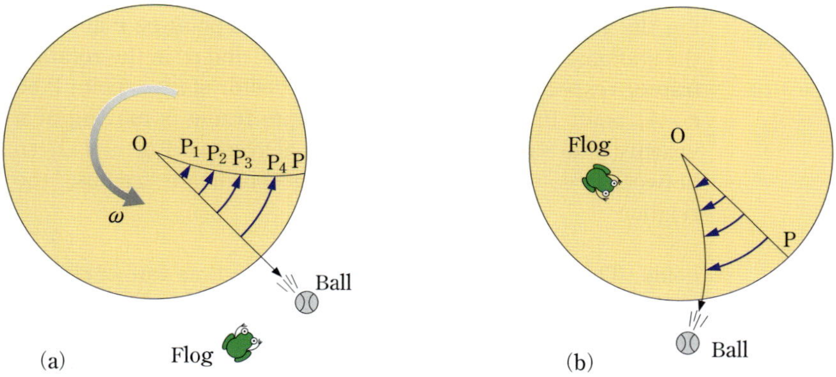

Fig. 9-8 A ball is thrown from the center O of the turntable to the point P at the edge of the table.
(a) When viewed on the ground, the ball moves straight but the line segment OP is rotating.
(b) When viewed on the turntable, the ball deviates gradually to the right from the line segment OP.

The direction of the Coriolis force $\boldsymbol{F}$ acting on the object of mass m moving at velocity $\boldsymbol{v}'$ with respect to the coordinate system rotating at the angular velocity ω is perpendicular to both the rotation axis and the velocity $\boldsymbol{v}'$ and its magnitude F is

$$F = 2m\omega v' \sin\theta. \tag{9.5}$$

θ is the angle between the axis of rotation and the velocity $\boldsymbol{v}'$ (Fig. 9-9). In terms of the vector product notation, the Coriolis force $\boldsymbol{F}$ is expressed as

$$\boldsymbol{F} = 2m\boldsymbol{v}' \times \boldsymbol{\omega}. \tag{9.6}$$

$\boldsymbol{\omega}$ is a vector with its magnitude of angular velocity ω, parallel to the rotating axis and having the traveling direction of the screw when the right screw is turned.

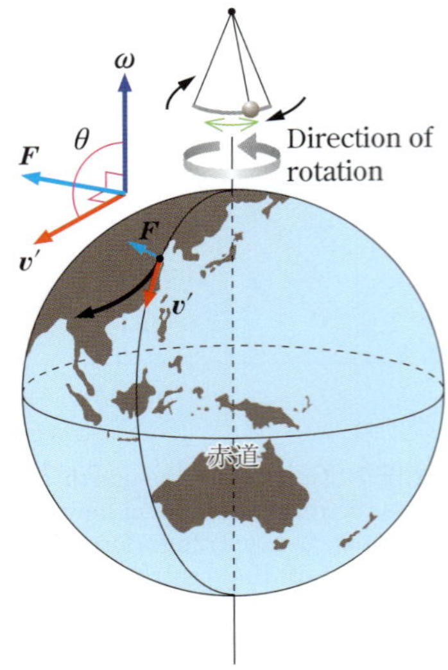

Fig. 9-9　The Coriolis force. When an object is launched southward, the object deviates to the right. The reason is that the speed of rotation is larger in the southern than in the northern areas.

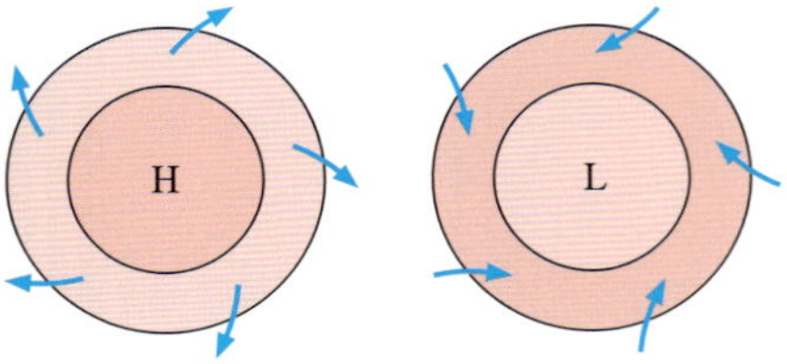

Fig. 9-10　The direction of winds in the northern hemisphere

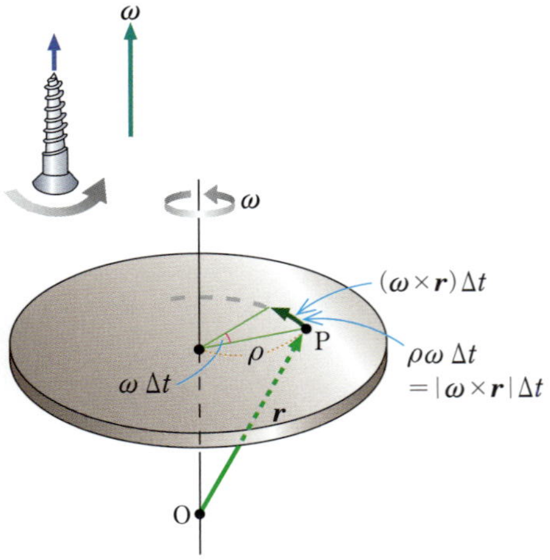

Fig. 9-11　At point $\boldsymbol{r}$, the rotational frame is moving at velocity $\boldsymbol{\omega}\times\boldsymbol{r}$ and at speed $\rho\omega = |\boldsymbol{\omega}\times\boldsymbol{r}|$.

Since the Earth rotates on its own axis, forces of inertia must be considered when we deal with large scale movements in a coordinate system fixed to the Earth. The trade winds and the air currents near high pressure/low pressure are examples where the remarked influence of the Coriolis force is observed. The area in the vicinity of the equator can receive more heat from the Sun than the other areas. The warm air rises and then the wind from the temperate zone blows. In the northern hemisphere, the wind blowing southward is deviated to the west by the influence of the Coriolis force. This is the wind called the trade winds blowing southwestward.

If we observe from the meteorological satellite the direction of the wind blown from the high pressure system (H) and the wind blown into the low pressure system (L), both directions of the wind are not perpendicular to the isobaric line. In the northern hemisphere, as shown in Fig. 9-10, they deviate to the right side of the initial direction and in the southern hemisphere they deviate to the left side. This phenomenon is also due to the Coriolis force. In the vicinity of the eye of typhoon, the force due to difference in pressure is nearly balanced with the resultant force of the Coriolis force and the centrifugal force, and the wind blows almost parallel to the isobaric line.

For reference　Proof of equation (9.6)

We consider a rotating coordinate frame whose origin is fixed to the origin of the inertial frame (stationary coordinate system) and which is rotating uniformly at angular velocity $\boldsymbol{\omega}$. Let the velocity and acceleration of a mass point be $\boldsymbol{v}$ and $\boldsymbol{a}$ in the stationary coordinate system, and $\boldsymbol{v}'$ and $\boldsymbol{a}'$ in the rotating frame, respectively. Let the position vector of mass point P be $\boldsymbol{r}$. At point P, the rotating coordinate frame is moving at a velocity of $\boldsymbol{\omega}\times\boldsymbol{r}$ relative to the inertial system (Fig. 9-11), so between two velocities of the two coordinate frames, the relation

$$\boldsymbol{v} = \boldsymbol{v}'+\boldsymbol{\omega}\times\boldsymbol{r} \tag{9.7}$$

holds. The acceleration $\boldsymbol{a}$ is the sum of the time rate of change in the rotating coordinate system with velocity $\boldsymbol{v} = \boldsymbol{v}'+\boldsymbol{\omega}\times\boldsymbol{r}$ and the rate of change in velocity $\boldsymbol{v} = \boldsymbol{v}'+\boldsymbol{\omega}\times\boldsymbol{r}$ due to the rotation of the rotating frame. Since the angular velocity vector $\boldsymbol{\omega}$ is constant, it follows that

$$\begin{aligned}\boldsymbol{a} &= (\boldsymbol{a}'+\boldsymbol{\omega}\times\boldsymbol{v}')+\boldsymbol{\omega}\times(\boldsymbol{v}'+\boldsymbol{\omega}\times\boldsymbol{r})\\ &= \boldsymbol{a}'+2\boldsymbol{\omega}\times\boldsymbol{v}'+\boldsymbol{\omega}\times(\boldsymbol{\omega}\times\boldsymbol{r}).\end{aligned} \tag{9.8}$$

The equation of motion $m\boldsymbol{a} = \boldsymbol{F}$ in the internal frame is expressed in the rotational frame as

$$\begin{aligned}m\boldsymbol{a}' &= m\boldsymbol{a}-2m\boldsymbol{\omega}\times\boldsymbol{v}'-m\boldsymbol{\omega}\times(\boldsymbol{\omega}\times\boldsymbol{r})\\ &= \boldsymbol{F}+2m\boldsymbol{v}'\times\boldsymbol{\omega}-m\boldsymbol{\omega}\times(\boldsymbol{\omega}\times\boldsymbol{r}).\end{aligned} \tag{9.9}$$

The second term on the right-hand side is the Coriolis force and the third term is the centrifugal force*.

* $-m\boldsymbol{\omega}\times(\boldsymbol{\omega}\times\boldsymbol{r}) = m\omega^2\boldsymbol{r}-m\boldsymbol{\omega}(\boldsymbol{\omega}\cdot\boldsymbol{r}) = m\omega^2\boldsymbol{\rho}$

Problems for exercise 9

A

1. In an elevator a weight of 100 g in mass was measured using a spring balance and it was 120 gf. Find the acceleration of this elevator.
2. Answer whether the first law of motion holds or not for the next observer.
   (1) A person falling with a parachute at a constant speed.
   (2) A person falling with a parachute just after jumping out of the plane.
   (3) A pilot of the jet aircraft that is revers-jetting after landing on the runway.
3. A weight is hung in the train. When this train runs a rail curve of radius 800 m at a speed of 30 m/s, how many degrees will the thread hanging the weight lean from the vertical line?
4. A lead ball was dropped without initial velocity from the tip of the mast of the ship anchored at the port located on the equator. If the ship is completely at rest in a windless condition, will the ball fall right under the tip of the mast?

B

1. On the floor of a merry-go-round, an iron ball is tied to a central pillar with a light rope and moves with the floor. The rope broke at a certain moment. How can a person on the merry-go-round and the one on the ground interpret the motion of the iron ball after breaking of the rope? Can a person riding on the merry-go-round explain the motion of the iron ball only with the centrifugal force?
2. Explain the reasons why when a rocket is launched from the North Pole to the south, it deviates to the west and when the rocket is launched to the north from the Equator, it deviates to the east (see Fig. 9-9). [**Hint** : Use the following fact that the Earth rotates from the west to the east and by this rotation, the point of latitude θ moves from the west to the east at the speed of $\omega R \cos\theta$ (the speed is different depending on the latitude θ). R is the radius of the Earth
3. Explain that when the pendulum is swung in the vertical plane at the Earth's north pole, the oscillating plane of the pendulum rotates, in a period of 24 hours, opposite to the rotation of the Earth (see Fig. 9-9) due to the rotation of the Earth. This phenomenon of rotation of the oscillation plane of the pendulum is that at the latitude θ, (the vertical component of the angular velocity ω of the Earth is $\omega \sin\theta$) this rotation occurs at a period of $T = \dfrac{24}{\sin\theta}$ hours. In 1851, Foucault created a pendulum (Foucault's pendulum) hanging a 28 kg weight on a thin wire with a length of 67 m, and conducted this experiment as an evidence of the Earth rotation.

Mechanics of Elastic Body

When a force is applied to solids from the outside, they deform, but at this time the restoring force to restore the deformation acts. When the deformation is small, if the applied external force is removed, solids return to their original shapes by the restoring force. This property is called the **elasticity**. This type of deformation of solids is called the **elastic deformation** and solids which can elastically deform are referred to as **elastic bodies**. Elastic deformation of solids includes elongation, shrinkage, shear, etc.

It is because of the elasticity that solids around us keep their shapes.

In this chapter, elastic deformation of elastic bodies will be studied.

In general, when the force from the outside is small and the elastic deformation is small,

the extent of deformation is proportional to the applied external force.

We call this **Hooke's law**.

If the deformation becomes too large, solids do not restore their original forms even if the external force is removed. This property is called the **plasticity**. The limit below which only elasticity is observed is called the **elastic limit**. The upper limit of Hooke's law is called the proportional limit. With further increase in external force after appearing of plastic behavior, solids are destroyed (Fig. 10-1).

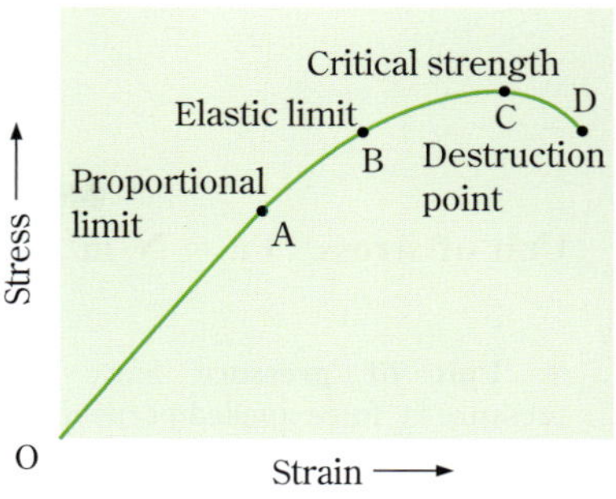

Fig. 10-1 Relationship between stress and strain of easily extensible metals (ductile metal)

10.1 Stress

Learning objective To understand that, when an external force acts on an object, the force acting between adjacent parts inside the object is expressed as stress, and that the stress consists of two kinds, one the normal stress (tensile stress and compressive stress) and the other, the tangential stress (shear stress).

When a force is applied to an elastic body, adjacent parts inside the elastic body exert forces. As an example, let us consider the case to pull both ends A and B of the bar with force F of the same magnitude. If the bar is divided into two parts with an arbitrary cross section C, the parts on both sides of the surface C pull each opposite side with force F. This fact is shown by the balancing condition obtained from the situation that the AC and CB parts of the bar are kept at rest (see Fig. 10-2).

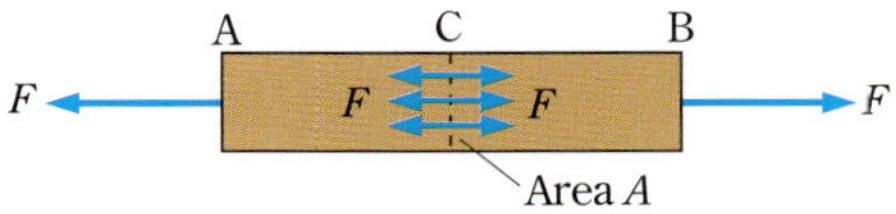

Fig. 10-2 Stress F/A at surface C. Because the left part AC of the bar is kept at rest, from the balancing condition of the force acting on the AC part, the right part CB exerts the force with magnitude F on the left part AC through the surface C with area A to the right direction. Likewise, the left part AC exerts the force with magnitude F on the left part of the right part CB through the surface C to the left direction.

In general, when a force acts on an elastic body from the outside, both sides of an arbitrary surface passing through an arbitrary point P inside the elastic body exerts through this surface the forces equal in magnitude but opposite in direction to each other. The force acting through the unit area including the point P located on this surface is referred to as the **stress** relative to this surface at point P. The component perpendicular to the stress plane is called the **normal stress**, and the component parallel to the plane is called the **tangential stress**. The tangential stress is also called the shear stress. Normal stress is divided into **tension** (tensile stress) and **pressure** (compressive stress). The unit of stress is "unit of force"/"unit of area", thus

Unit of stress Pa = N/m^2

Table 10-1 (Unit : kgf/cm^2 $\approx$ 10^5 Pa)

	Wood	Stone	Iron	Human femur
Against pressure	500	1 000	4 000	1 600
Against tensile force	400	60	4 000	1 200

* **Unit of pressure** Since the pressure is "force applied perpendicular to the surface"/"area", the unit of pressure in the SI units is N/m^2 = Pa, but other units as follows are also used.

Standard atmospheric pressure (atm) = 1 atm = 101325 Pa
= 760 mmHg = 760 Torr

Historically, 1 atm was defined as the pressure that the height 760 mm column of mercury of density 13.5951 g/cm^3 generates.

In the weather forecast, as a unit of atmospheric pressure 1 hPa (hectopascal) = 10^2 Pa is used.

N/m^2, and is called Pascal, written as Pa*.

As examples of the magnitude of stress, Table 10-1 shows the limiting magnitudes of stress up to which various materials can endure. Units are all kgf/cm^2 $\approx$ 100 000 Pa. Although the piano wire is made of iron, it can endure the tension up to 15 000 kgf/cm^2.

Question 1 Show in the case of Fig. 10-3, at the cross section D $\left(\text{area } \dfrac{A}{\cos\theta}\right)$ shown by the dotted line the normal stress is $\dfrac{F\cos^2\theta}{A}$ and the tangential stress is $\dfrac{F\sin\theta\cos\theta}{A}$. Keep in mind that in specifying the stress at the same point, specification of the orientation of the plane is required.

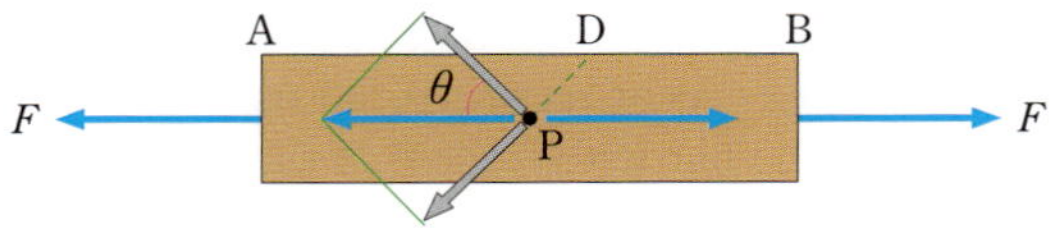

Fig. 10-3 Normal stress $\dfrac{F\cos^2\theta}{A}$, and tangential stress $\dfrac{F\sin\theta\cos\theta}{A}$

10.2 Elastic constants

Learning objective To understand that the proportional constant of stress and strain when the external force acts on an object is the elastic constant. To understand the definition of Young's modulus and the shear modulus.

Strain When the force acts on an elastic body from the outside, stress acts inside the elastic body, and the elastic body deforms according to the stress, but the magnitude of deformation is proportional to the size of the elastic body. For example, when a rubber cord is pulled, it will stretch, but in the case where the rubber cord is pulled with the same force, the elongation length of 20 cm rubber cord is twice that of 10 cm one. Therefore, it is appropriate to consider the amount of "deformation of the elastic body" divided by the "size of the elastic body" as a quantity unrelated to the size of object, and this quantity is called the **strain** of elastic body. Hook's law is expressed as

When the strain is small, the stress and strain are proportional.

The proportional constant is called the **elastic constant**. Explanation of the important elastic constants is given below.

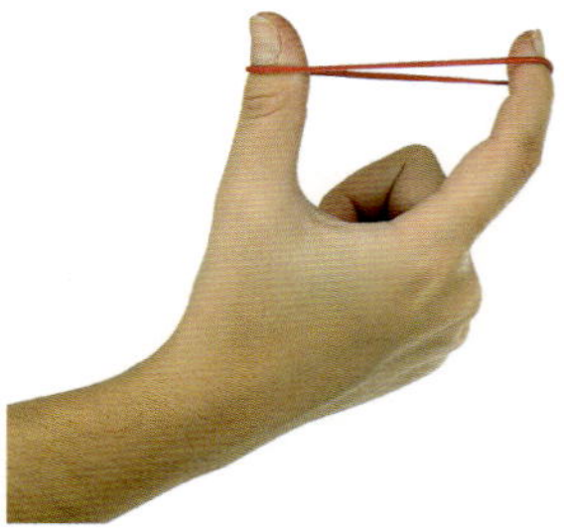

Fig. 10-4 The rubber cord will stretch when pulled.

Young's modulus When we apply a force F to both ends of a uniform bar of length L and cross-sectional area A to stretch the bar length by ΔL, the strain is $\frac{\Delta L}{L}$ and the stress is $\frac{F}{A}$, so Hooke's law is

$$\frac{F}{A} = E\frac{\Delta L}{L} \tag{10.1}$$

(Fig. 10-5). The proportional constant E is a constant depending on the substance and is referred to as **Young's modulus** or the **modulus of elasticity**. Its unit is Pa = N/m^2. By transforming equation (10.1) into

$$\Delta L = \frac{1}{E}\frac{FL}{A} \tag{10.2}$$

it is understood that elongation ΔL of the bar is proportional to the magnitude F of the pulling force and the length L of the bar, and is inversely proportional to the cross-sectional area A of the bar.

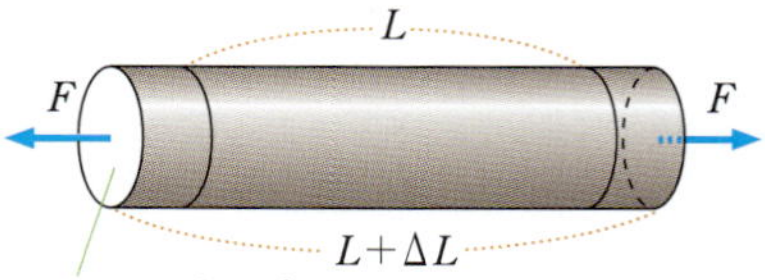

Fig. 10-5 $\Delta L = \frac{1}{E}\cdot\frac{FL}{A}$

Unit of Young's modulus (modulus of elasticity) Pa = N/m^2

Poisson ratio If a uniform bar is pulled and stretched ($L \to L+\Delta L$) in the longitudinal direction, it shrinks in the lateral direction ($w \to w+\Delta w$, $h \to h+\Delta h$. If ΔL is positive, then Δw and Δh are negative quantities. Fig. 10-6) The ratio σ of the proportion $\varepsilon' = -\frac{\Delta w}{w} = -\frac{\Delta h}{h}$ of shrinkage in the lateral direction to the proportion $\varepsilon = -\frac{\Delta L}{L}$ of elongation in the longitudinal direction, namely the quantity σ defined as

$$\varepsilon' = -\frac{\Delta w}{w} = -\frac{\Delta h}{h} = \sigma\varepsilon = \sigma\frac{\Delta L}{L} \tag{10.3}$$

is called the **Poisson ratio**. σ is a constant depending on the substance and is a dimensionless quantity[*1].

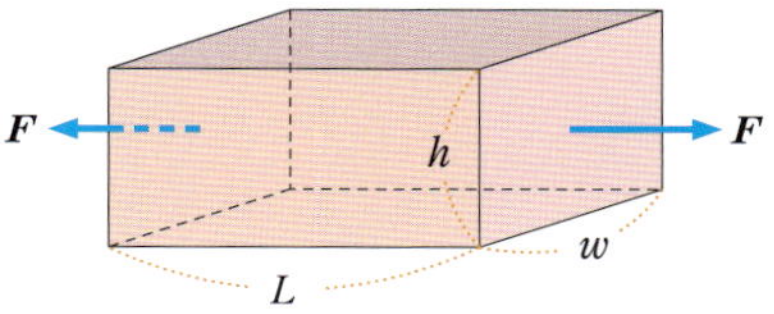

Fig. 10-6

Bulk modulus Compressibility When a uniform pressure p is applied to the surface of the elastic body, the volume of the elastic body changes from V to $V+\Delta V$. If pressure is positive, then the volume decreases and ΔV is negative. Accordingly, we express Hooke's law as

$$p = -k\frac{\Delta V}{V}. \tag{10.4}$$ [*2]

The proportional constant k is a constant determined by the substance and called the **bulk modulus** and has the unit of Pa. Moreover, $\frac{1}{k}$ is called the **compressibility**.

Shear modulus When the tangential stress of magnitude τ is applied to the four side faces of the square pillar as shown in Fig. 10-7, the square on the bottom deforms into a diamond shape. Such a deformation without volume change is called the **shear strain**[*3]. If the apex angle of the diamond is given as $\frac{\pi}{2}\pm\theta$, Hooke's law is expressed as

*1 If volume does not change, $\sigma = 0.5$. Most substances have $\sigma > 0$ and many of them have $\sigma \approx 0.3$. But the case of $\sigma < 0$ is known for some polymers such as those having a foam structure.

*2 Strictly speaking, the pressure p should be described as Δp meaning the change in pressure (from such as the atmospheric pressure).

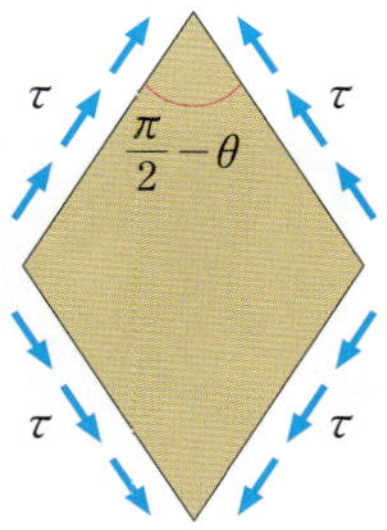

Fig. 10-7

*3 In Fig. 2 in Problems for exercise 10, B2. $\overline{\mathrm{AH}}\times\overline{\mathrm{AE}} \approx \overline{\mathrm{A'H'}}\times\overline{\mathrm{A'E'}}$

$$\tau = G\theta. \tag{10.5}$$

The proportional constant G is called the **shear modulus** or **rigidity**. Its unit is Pa.

Relationship among elastic constants The four elastic constants $E, \sigma, k,$ and G are connected by the following two relationships

$$k = \frac{E}{3(1-2\sigma)}, \qquad G = \frac{E}{2(1+\sigma)}. \tag{10.6}$$

Accordingly, two of the four are independent (for proof, refer to B3, 4 of Problems for exercise 10). In Table 10-2 are listed the elastic constants of several representative materials. Most substances have $\sigma > 0$, so $E > G$.

Table 10-2 Elastic constants of several representative materials

Materials	E [Pa]	G [Pa]	σ	k [Pa]
	$\times 10^{10}$	$\times 10^{10}$		$\times 10^{10}$
Aluminum	7.03	2.61	0.345	7.55
Copper	12.98	4.83	0.343	13.78
Iron (steel)	20.1～21.6	7.8～8.4	0.28～0.30	16.5～17.0
Silver	8.27	3.03	0.367	10.36
Polyethylene	0.04～0.13	0.026	0.458	—
Rubber (elastic)	$(1.5\sim5.0)\times10^{-4}$	$(5\sim15)\times10^{-5}$	0.46～0.49	—

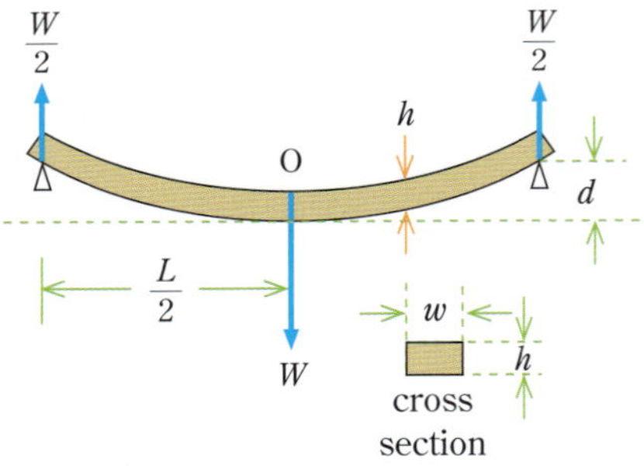

Fig. 10-8 Young's modulus $E = \dfrac{WL^3}{4h^3wd}$

Deflection of the bar As shown in Fig. 10-8, the rectangular parallelepiped bar of length L, height h, and width w are supported at both ends and the weight is hung at the center of the bar. If the gravity acting on the weight is W, and the deflection of the bar is d, Young's modulus E of this object is given by the calculation as

$$E = \frac{WL^3}{4h^3wd} \quad \left(d = \frac{WL^3}{4Eh^3w}\right). \tag{10.7}$$

This deflection method is used to measure Young's modulus E. When deflection occurs, the intermediate surface does not stretch nor shrink, but the portion above the intermediate surface shrinks, while the portion below the intermediate surface stretches, and the amount of stretch or shrinkage is proportional to the distance from the intermediate surface.

If this object is weaker to the pressure than the tensile force, with the increase in mass of the weight, when the elongation of the portion below the intermediate surface exceeds a certain limit, a crack is formed in it. In the case of breaking, since the deformation amount exceeds the elastic limit, equation (10.7) is not applicable. However, equation (10.7) suggests that the strain due to the elongation of the lower portion and the shrinkage of the upper portion is inversely proportional to $(\text{height})^2 \times (\text{width}) = h^2w$. The reason why the strain is inversely proportional to h^2w instead of h^3w is that the amount of stretch and shrinkage of the top and bottom surfaces is proportional to h even with the same deflection d. Therefore, in the case of bars with the same cross section (when hw is the same), the strength against such

Fig. 10-9 The bar and the plate will deflect if the force is applied.

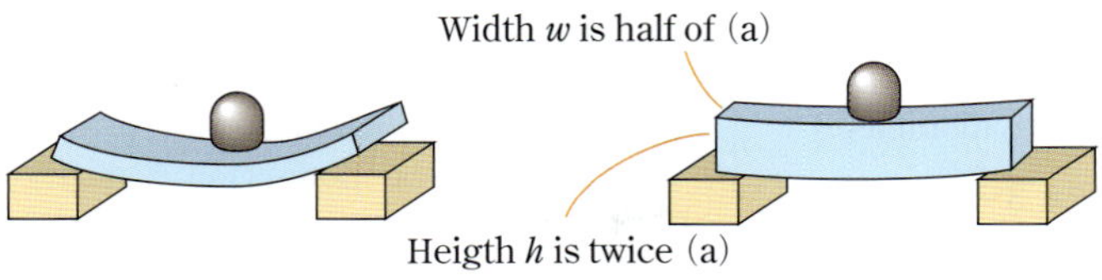

(a) $h = 1,\ w = 2,\ h^2w = 2$ (b) $h = 2,\ w = 1,\ h^2w = 4$

Fig. 10-10 (b) is twice as strong as (a).

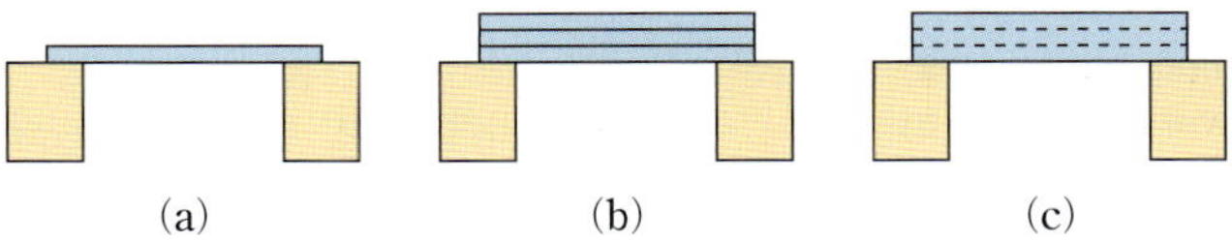

Fig. 10-11 (a) One sheet of plate. (b) When three plates are stacked, the strength is tripled. (c) When three plates are laminated into plywood, the strength becomes nine times.

deflection is proportional to h. So, even in the case of the same lumber, the strength against deflection changes (Fig. 10-10), and when laminated into plywood, the object gets stronger in proportion to the square of the number of lamination (Fig. 10-11).

Torsion of cylinder The upper end of a thin wire with radius R and length L is fixed and vertically suspended and the weight is attached at the lower end of the wire. A moment of force is applied to the weight by the couple of N (Fig. 10-12) to twist the wire. In this state the equation for the twisted angle θ is

$$\theta = \frac{2LN}{\pi G R^4} \tag{10.8}$$

and this can be derived by calculation. This fact is applied for the measurement of shear modulus G.

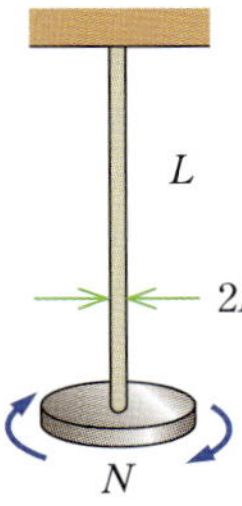

Fig. 10-12

Problems for exercise 10

A

1. When two persons able to exert the same magnitude of force pulled a single rope, the rope was torn. When the same rope as this one is fixed at one end to the wall and pulled till it is torn, how many persons with the same ability as the above mentioned one will be needed?
2. How much will the wire be stretched when 1 kg weight is hung on a steel wire with a length of 1 m and a diameter of 0.2 mm? Young's modulus is 2×10^{11} Pa.
3. When a force of 0.1 kgf ($0.1\times 9.8\ \mathrm{N} = 0.98\ \mathrm{N}$) was applied to each face of a gelatin cube with a side length of 30 cm as shown in Fig. 1, the surfaces of the cube were displaced 1.0 cm in parallel.
   (1) How much is the shear angle θ?
   (2) How much is the shear stress?
   (3) Find the shear modulus of this gelatin.

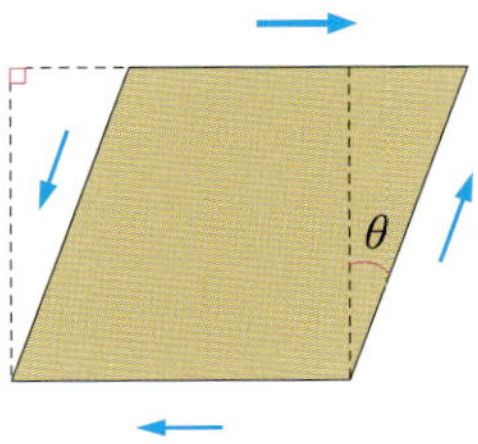

Fig. 1

4. Human femur fractures at a pressure above 1.6×10^8 Pa or a tensile force above 1.2×10^8 Pa. Young's modulus is 1.7×10^{10} for tensile force and 0.9×10^{10} Pa for pressure.
   (1) Assuming the cross-sectional area of the thinnest part of the adult femur is 6 cm^2, calculate the magnitude of tensile force to cause the femur fracture and express it in kgf.
   (2) How long will it stretch before it fractures?

5. When a 20 cm diameter steel ball is sunk to the bottom of 10 000 m in depth, how much does the diameter D of the ball change? The bulk modulus is 1.7×10^{11} Pa. Use the relation $\frac{\Delta V}{V} \approx 3\frac{\Delta D}{D}$ which holds when the diameter increases by ΔD and the increase in volume is ΔV.

B

1. When a force acts on a uniform bar of length L, cross-sectional area A, Young's modulus E, and the bar is stretched by ΔL, how much amount of energy is stored in the bar (consider the work W done by the force)?
2. When an elastic body is stretched in one direction by the applied tension, it shrinks in the lateral direction. If the elastic body is pulled in one direction such that the length in the lateral direction does not change, then the modulus of elasticity is given as

$$k+\frac{4}{3}G = \frac{(1-\sigma)E}{(1+\sigma)(1-2\sigma)}$$

Prove the above equation. Use equation (10.6).
3. Consider the case to apply a uniform pressure p to all faces of a cube with a side length L and derive the relationship of the bulk modulus

$$k = \frac{E}{3(1-2\sigma)}.$$

4. Referring to Fig. 2, derive the equation

$$G = \frac{E}{2(1+\sigma)}.$$

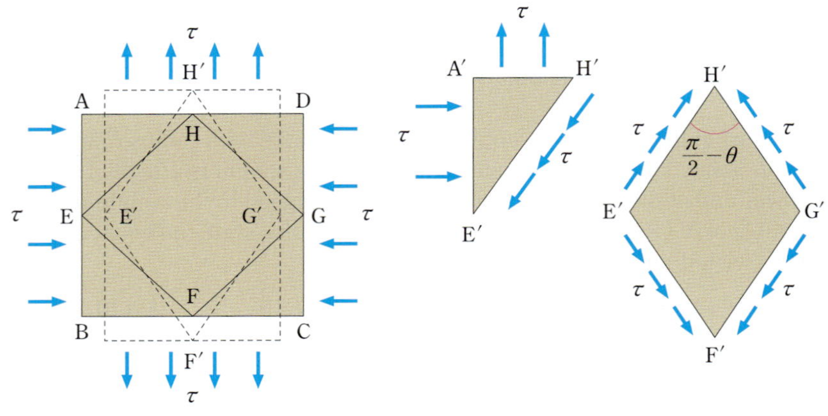

(c) Tangential stress

Fig. 2

Mechanics of Fluid

Gases and liquids such as air and water do not have a fixed shape but freely deform according to the shape of their container, so they are collectively referred to as **fluids**. Since most of the motions we experience in our daily life occur in the air or in the water, fluids are familiar to us, but there are facts that many people are not aware of. For example, the word "pressure" is commonly used in the weather forecast. The atmospheric pressure is linked to the vector quantity of pressure which is due to the action of air, but in fact it is a scalar quantity.

In this chapter, first the pressure in static fluid (fiuid at rest) and then Bernoulli's law and the pressure in the moving fluid will be studied. In the next step, topics of lift and resistance force acting on moving objects in fluid will be discussed. Change in temperature during the motion of fluid may occur, but in this chapter the temperature of fluid is assumed to be constant.

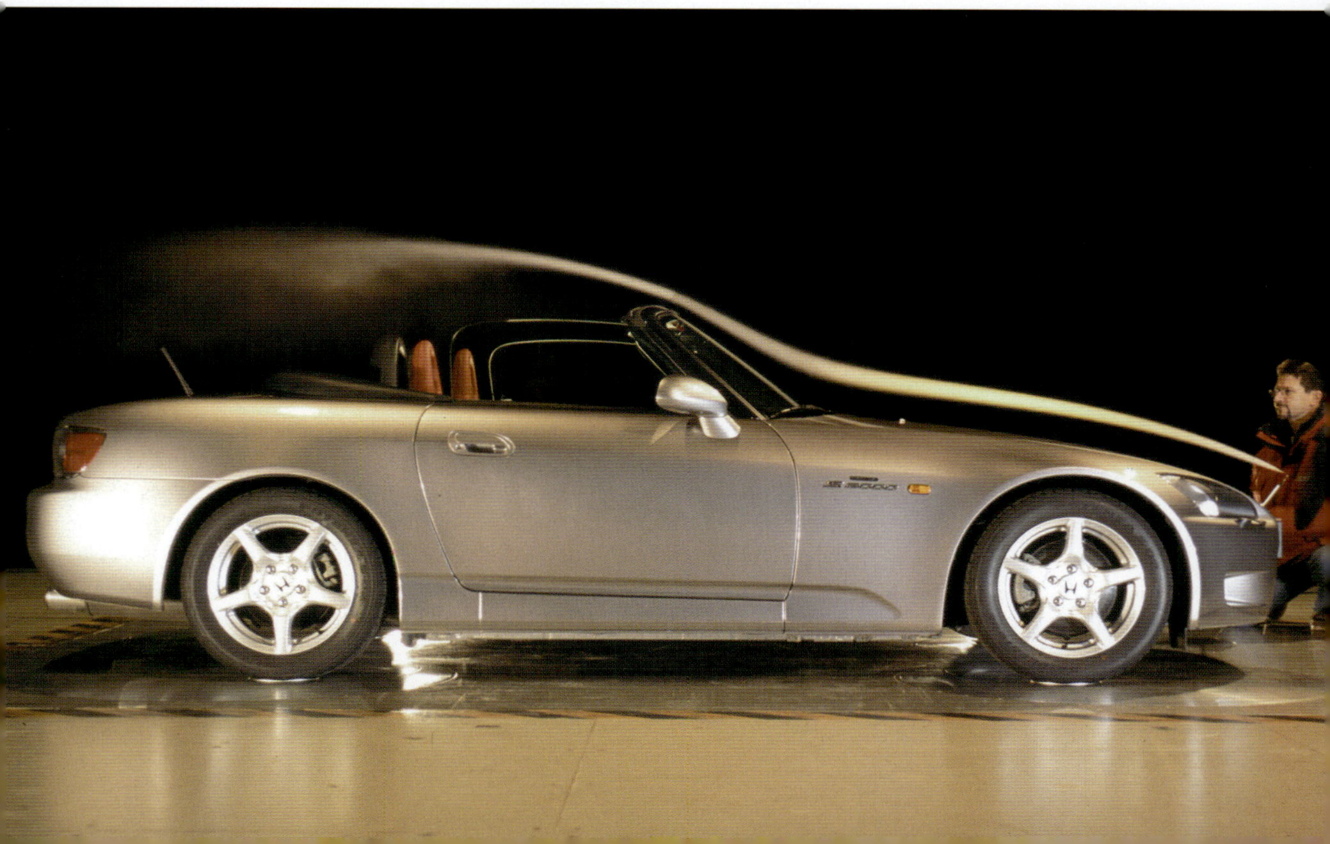

Examining the air flow around the car in the wind tunnel room.

Applying the term "stress" that we have learned in the previous chapter to the freely deformable nature of fluid, we can say that

In a static fluid, the stress is a pressure perpendicular to the plane considered.

In a case where the tangential stress parallel to a certain surface in a static fluid appears, a flow along this surface would occur, which is contradictory to the state at rest.

In a moving fluid, stress can have a component parallel to the plane under consideration. For example, when tea in a tea cup is stirred with a spoon, the tea starts spinning around in the tea cup. When the spoon is removed out of the tea, the spinning speed of the tea gradually decreases and after a while spinning stops. This is because in real fluids, when adjacent parts have different speeds, tangential stress acts to eliminate the difference in speed. The property to generate a tangential stress in the case where differences in speed exist is called the **viscosity** of fluid. Due to the forces acting between the molecules, fluids are commonly viscous. A hypothetical fluid that is non-viscous is called a **perfect fluid**. Since viscosities of water and air are low, we can take them as the perfect fluid in many cases.

Fig. 11-1 A teacup containing tea and poured milk

11.1 Pressure in static fluids

Learning objective To understand that the pressure in a static fluid is described as a scalar quantity called the hydrostatic pressure determined by the height, and to understand Archimedes' principles of buoyancy.

When a large stone is placed on the ground, the stone exerts a large force on the ground, but it does not exert a large force on the air nearby. However, when we want to store water on the ground, we need to set up a tank and put water in it. The water in the tank exerts a force on its bottom, but it also exerts a force on the sides of the tank. When we are in water, we feel the pressure of water through our skin. At the same point in the water the pressure is the same in magnitude no matter which direction the skin is facing. That is, static fluids have the following property.

When a surface passing through an arbitrary point in the static fluid is considered, the stress on this surface is always the pressure perpendicular to the surface, and at the same point it has an equal value regardless of the direction of the surface considered.

This stress is called the **hydrostatic pressure** at that point. The hydrostatic pressure is a scalar quantity.

The property that at the same point in the fluid, regardless of the orientation of the face the pressure is constant can be derived from the balance of pressure acting on the three rectangles of the small isosceles triangular prism in Fig. 11-2. The pressure acting on the plane of any orientation is shown to be equal to the pressure acting on the horizontal plane (which is equal to p_C in the figure). Although gravity also acts on

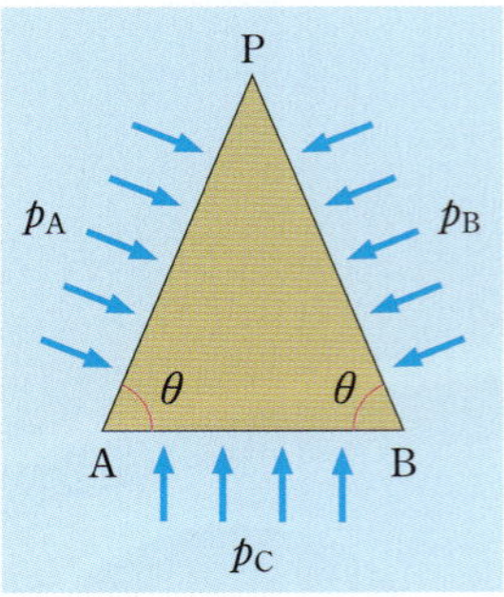

Fig. 11-2 Balance in isosceles triangular prism PAB

$\overline{PA}\, p_A \sin\theta = \overline{PB}\, p_B \sin\theta$

$\therefore\ p_A = p_B$

$2\,\overline{PA}\, p_A \cos\theta = \overline{AB}\, p_C$

$\therefore\ p_A = p_C$

this triangular prism, gravity is proportional to volume. For small objects, gravity proportional to volume is negligible compared to pressure proportional to surface area.

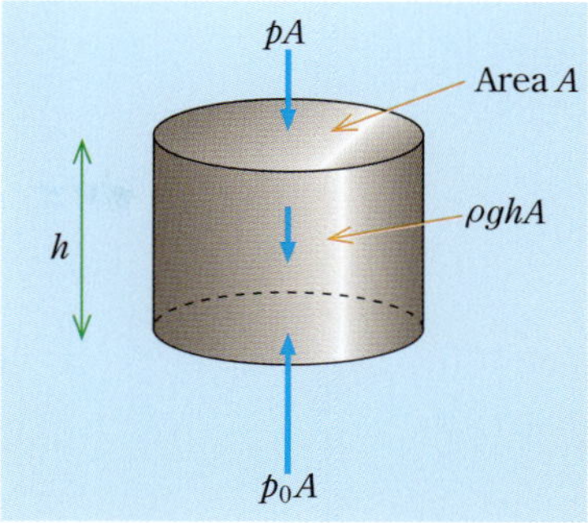

Fig. 11-3 Pressure p decreases with height h.

$p = p_0 - \rho gh$

* "force acting on the surface" = "hydrostatic pressure" × "surface area"

Relationship between hydrostatic pressure and height As shown in Fig. 11-3, we consider the portion of the cylinder with height h and bottom area A in a static fluid. Since the fluid in this cylinder is at rest, the forces acting on the fluid in this cylinder are balanced. From the equilibrium of forces acting on the side, it is seen that the hydrostatic pressure of the fluid at the same height is equal (it is easier to understand to use a prism rather than a cylinder). Assume that the hydrostatic pressure on the upper surface of the cylinder is p and the hydrostatic pressure on the lower surface is p_0. Then the magnitude of the force acting downward on the upper surface A of the cylinder is pA, and the one acting upward on the lower surface is p_0A*. When the density ρ of the fluid is assumed to be constant, since the gravity acting on the fluid with volume hA and mass ρhA is ρghA, from the condition of balance $pA + \rho ghA - p_0A = 0$, we can derive

$$p = p_0 - \rho gh \quad (p + \rho gh = \text{constant, when density } \rho \text{ is constant}). \tag{11.1}$$

Equation (11.1) is a relationship between the hydrostatic pressure and the height for fluids called non-compressible fluids for which effects of compression can be fully ignored. Therefore, it is seen

The hydrostatic pressure in static fluids decreases with height but is equal at the same height.

Pascal's principle Equation (11.1) indicates that if the pressure (hydrostatic pressure) at a point in a static incompressible fluid sealed in a vessel is increased by a certain amount, the pressure at all points in the fluid will increase by the same amount. This fact was discovered by Pascal in 1653, so it is called **Pascal's principle**.

Pascal's principle is applied to various machines. For example, in the hydraulic jack of Fig. 11-4, when we press the small surface A of area A_A with a force F, the pressure of the fluid increases by $\frac{F}{A_A}$, so that the force

$$\frac{F}{A_A} \times A_B = F\frac{A_B}{A_A}$$

acts on the large surface B of area A_B. If A_B is made much larger than A_A, then $F\frac{A_B}{A_A}$ gets much larger than F, and heavy objects lying on surface B can be lifted without applying too much force on surface A. When we step on the brake pedal of an automobile, the pressure due to the foot makes the shoes (or pads) tightly contact to the drums (or discs) rotating with the wheels, but at that time the hydraulic pressure is also used to transmit the pedaling force to the braking system.

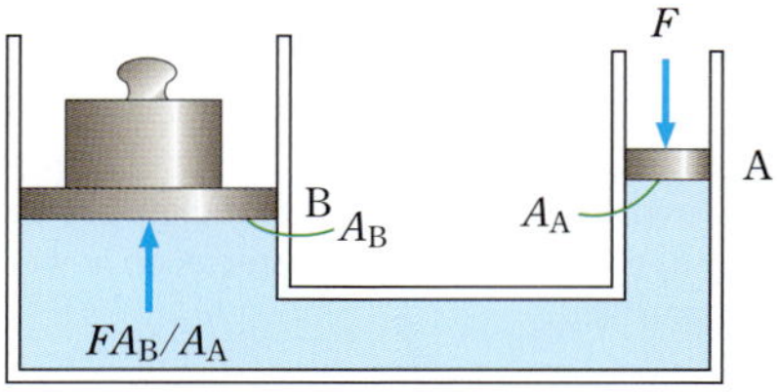

Fig. 11-4 A hydraulic jack

Question 1 Explain that when a diver is on work in the sea, the pressure of the air in the lungs will be the same as the pressure of ocean water

Fig. 11-5 The Dead Sea. The Dead Sea, the lake located at the border between Jordan and Israel contains a lot of salt and has a density $\rho = 1.17\ \mathrm{g/cm^3}$. Accordingly, the buoyancy force is large and humans and eggs do not sink.

pressing the body.

Buoyancy and Archimedes' Principle When entering the pool, we have an impression that our body gets lighter. A wood piece put in the water floats. A stone put into the water sinks to the bottom, but when we lift the stone in the water it is lighter than lifting it up in the air. The pressure from the fluid acts on the surface of the object placed in the fluid. The pressure increases with the increase in depth, so the resultant force of pressure acting on the object in the fluid works upward (see Fig. 11-3). This upward resultant force is called the **buoyancy**. Since the buoyancy is balanced with gravity acting on the fluid at the location of the object,

> the magnitude of buoyant force acting on an object in the fluid is equal to the magnitude of gravity acting on the fluid displaced by the object.

This is referred to as **Archimedes' principle**. This is because Archimedes living at Syracuse, Sicily Island found this principle around 220 B.C.

11.2 Bernoulli's law

Learning objective To understand the equation of continuity and Bernoulli's law and to acquire the ability to solve simple application problems

Fig. 11-6 A technique called "sumi-nagashi" used in art. By gently swirling the fluid, the Indian ink or dye moves to form patterns on a wet surface.

Steady flow Since the perfect fluid is not viscous and no tangential stress acts, even when the fluid is moving, the pressure at an arbitrary point takes the same value regardless of the direction of surface. In general, the velocity and density of a fluid changes with time, but a flow in which the velocity and density of the fluid at each point does not change with time is called the **steady flow**. If we drop a small amount of ink in this flow, the ink flows and many lines are formed. A line representing such a flow is called the streamline. A **streamline** is a curve with direction in which the velocity vector of each point on it is tangentially oriented.

Consider first a closed curve in a steady flow, and then consider a group of all streamlines passing through this closed curve. In this way, one tube is formed. This is called the **stream-tube**.

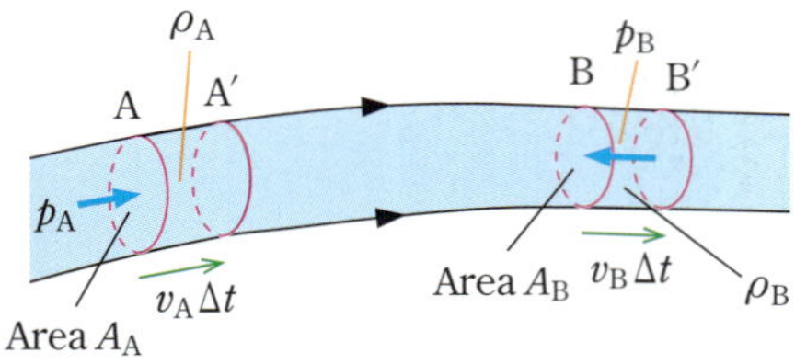

Fig. 11-7 Equation of continuity $\rho_A A_A v_A = \rho_B A_B v_B$

Equation of continuity We consider the two vertical cross sections A and B of a thin stream-tube and let the respective cross sectional areas be A_A and A_B (Fig. 11-7). Let the density and velocity of the fluid at A be ρ_A and v_A, respectively, and the corresponding values at B be ρ_B and v_B, respectively. The volume of fluid passing through cross section A in the minute time Δt is $A_A v_A \Delta t$ and the mass is $\rho_A A_A v_A \Delta t$, while the volume of fluid passing through cross section B in the minute time Δt is $A_B v_B \Delta t$ and the mass is $\rho_B A_B v_B \Delta t$. If in the part formed by the two cross sections A and B the fluid does not spring out or is not sucked, the inflow mass and the outflow mass remain the same namely the following relation holds :

$$\rho_A A_A v_A \Delta t = \rho_B A_B v_B \Delta t, \text{ or } \rho_A A_A v = \rho_B A_B v_B.$$

Along one stream-tube, for all cross sections the relation

$$\rho A v = \text{constant} \tag{11.2}$$

holds. This relation is called the **equation of continuity** of fluid.

For incompressible fluids the density ρ is constant, and along one stream-tube, for all hols cross sections the relation

$$Av = \text{constant} \quad \text{(incompressible fluid)} \tag{11.3}$$

holds (Fig. 11-8). This relation is also called the equation of continuity of fluid.

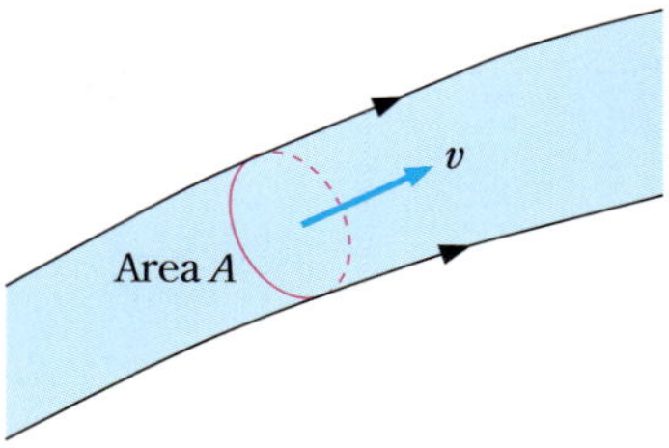

Fig. 11-8 Equation of continuity $Av =$ constant

Bernoulli's law We consider the steady flow of an incompressible perfect fluid. In other words,

(1) The density of fluid is constant and does not change.

(2) In the fluid no viscous force works. That is, there is no mechanical energy loss due to friction within the fluid

(3) At any point the velocity of the fluid is constant over time and the streamline does not change with time (the turbulence which will be discussed in Section **11.4** does not occur).

Under the above-mentioned conditions, applying the principle that the work done when the fluid moves is equal to an increase in mechanical energy [equation (5.45)], we have

$$p + \frac{1}{2}\rho v^2 + \rho g h = \text{constant} \tag{11.4}$$

for every point on the stream line of steady flow of incompressible complete fluids (Fig. 11-9). This is referred to as **Bernoulli's law**. The pressure p on an arbitrary stream line is smaller as the height h is higher and as the flow velocity v is higher.

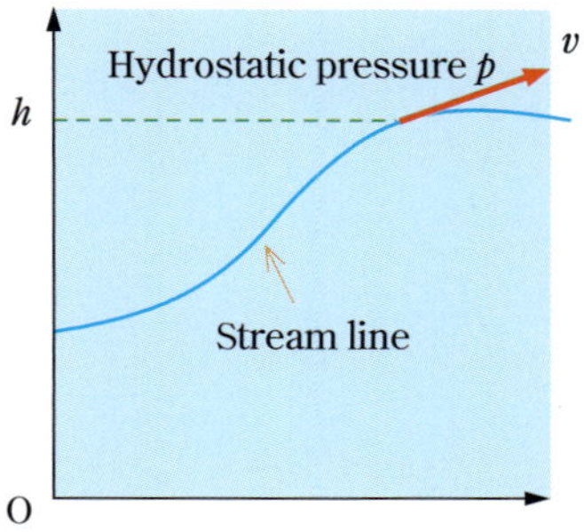

Fig. 11-9 Bernoulli's law $p + \frac{1}{2}\rho v^2 + \rho g h =$ constant

To apply Bernoulli's law to static fluids, it suffices to set $v = 0$ in equation (11.4), which leads to equation (11.1).

Examples of an application of Bernoulli's law are given below.

Case 1 | Torricelli's law When the outflow from a small hole near the bottom of a water tank of depth h is small and the descent speed of the water surface is small, the flow can be regarded as a steady flow, so Bernoulli's law is applicable (Fig. 11-10). Let the atmospheric pressure be p_0 and the outflow velocity of water be v. Then Bernoulli's law leads to the following relationship between the water surface and the hole :

$$p_0 + \rho g h = p_0 + \frac{1}{2}\rho v^2 \tag{11.5}$$

$$\therefore \quad v = \sqrt{2gh}\,. \tag{11.6}$$

As the water level drops and h becomes smaller, the outflow velocity v of water decreases

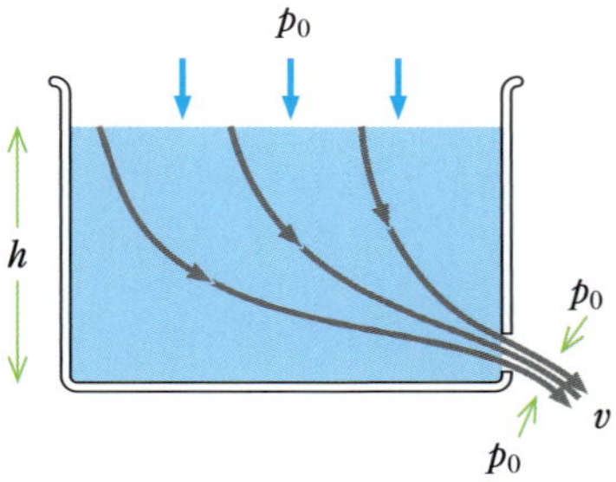

Fig. 11-10 Torricelli's law $v = \sqrt{2gh}$

Case 2 | Venturi tube In the Venturi tube shown in Fig. 11-11 through which the incompressible fluid of density ρ flows, since the

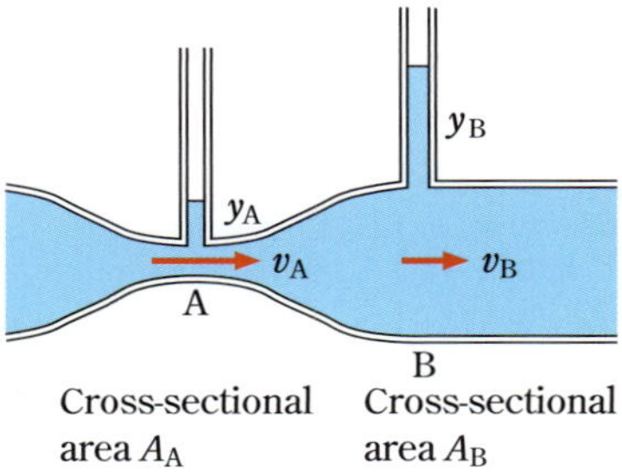

Fig. 11-11 A venturi tube

height h of flow is constant, Bernoulli's law becomes

$$p_A + \frac{\rho v_A^2}{2} = p_B + \frac{\rho v_B^2}{2} \tag{11.7}$$

and the continuity equation is

$$A_A v_A = A_B v_B. \tag{11.8}$$

The pressure p_A is low at A where the cross-sectional area is small and the flow is fast, while the pressure p_B is high at B where the cross-sectional area is large and the flow is slow. Substituting equation (11.8) into equation (11.7) and eliminating v_B or v_A, we obtain

$$p_B - p_A = \rho v_A^2 \frac{A_B^2 - A_A^2}{2A_B^2} = \rho v_B^2 \frac{A_B^2 - A_A^2}{2A_A^2}. \tag{11.9}$$

Using equation (11.1), the pressure difference between the two points A and B is determined from the difference $y_B - y_A$ in height of the liquid column as

$$p_B - p_A = \rho g (y_B - y_A). \tag{11.10}$$

Using equation (11.9), we can find the velocity of fluid.

For reference Proof of Bernoulli's Law (11.4)

The work performed by the surrounding fluid when the fluid in the stream-tube in Fig. 11-7 moves from AB part to A′B′ part are $p_A A_A v_A \Delta t$ for A, and $-p_B A_B v_B \Delta t$ for B. Since the shape of the stream-tube does not change with time, the side wall of the tube does not move, and the pressure acting on the side wall does not perform any work. The increase in kinetic energy of the fluid is the difference between the kinetic energy of the BB′ part of the mass $\rho A_B v_B \Delta t$ and the AA′ part of the mass $\rho A_B v_B \Delta t$, namely $\frac{1}{2}\rho v_B^2 A_B v_B \Delta t - \frac{1}{2}\rho v_A^2 A_A v_A \Delta t$. If the height of the surface A is h_A and the height of the surface B is h_B, the increase in gravitational potential energy is due to the increase at BB′ and the decrease at AA′, that is $\rho g h_B A_B v_B \Delta t - \rho g h_A A_A v_A \Delta t$. Thus, from the relation (5.45) between mechanical energy and work, we obtain the following relation :

$$\frac{1}{2}\rho v_B^2 A_B v_B \Delta t - \frac{1}{2}\rho v_A^2 A_A v_A \Delta t + \rho g h_B A_B v_B \Delta t - \rho g h_A A_A v_A \Delta t = p_A A_A v_A \Delta t - p_B A_B v_B \Delta t. \tag{11.11}$$

Transforming the above equation using the equation of continuity $v_A A_A = v_B A_B$ we can derive

$$p_A + \frac{1}{2}\rho v_A^2 + \rho g h_A = p_B + \frac{1}{2}\rho v_B^2 + \rho g h_B. \tag{11.12}$$

11.3 Lift

Learning objective To understand the outline of generation mechanism of lift.

When a plane flies horizontally at a constant speed and the pilot looks out of the plane, the flow of air coming from the front is a constant flow in time. The air flow coming slightly upward from the front descends slightly after passing the main wing (Fig. 11-12). This air flow is a superposed flow of a uniform horizontal flow coming from the front and a flow circulating clockwise around the main wing. Therefore, the air flow above the wing is faster than the air flow below the wing. So, according to Bernoulli's law, the air pressure p_{lower} pushing the lower wing surface upwards is larger than the air pressure p_{upper} pushing the upper wing surface downwards. This difference in pressure is the **lift** that acts on the wing.

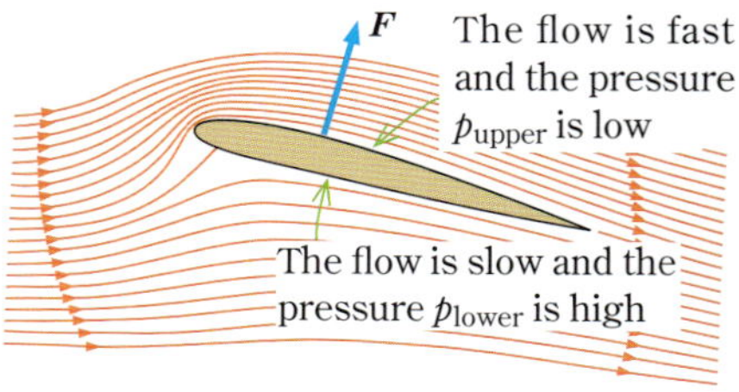

Fig. 11-12 The air flow seen from the pilot is a superposed flow of a uniform horizontal flow coming from the front and an air flow circulating clockwise around the wing. The vertically upward component of the force F that the fluid acts on the wing is the lift F_{L}.

Why does the air flow circulating around the wing that is the cause of lift to flow? The air flow around the wing of an airplane that started to move on the runway is a non-circulating flow with a stagnation point (a point with zero velocity) at point S on the back side near the rear end T of the wing [Fig. 11-13 (a)]. When the plane is accelerating on the runway to take off, the flow cannot go around the rear end T [Fig. 11-13 (b)] and separates from the wing and forms a counterclockwise vortex, while the stagnation point S retreats and when it reaches the rear end, the air flow becomes steady. Therefore, the circulating air flow flowing clockwise around the wing, which is the cause of the lift, generates in pairs with the vortex generated at the rear end of the wing when the airplane takes off [Fig. 11-13 (c)]. The counter-rotating vortices (starting vortices) remain near the runway but disappear due to the air viscosity.

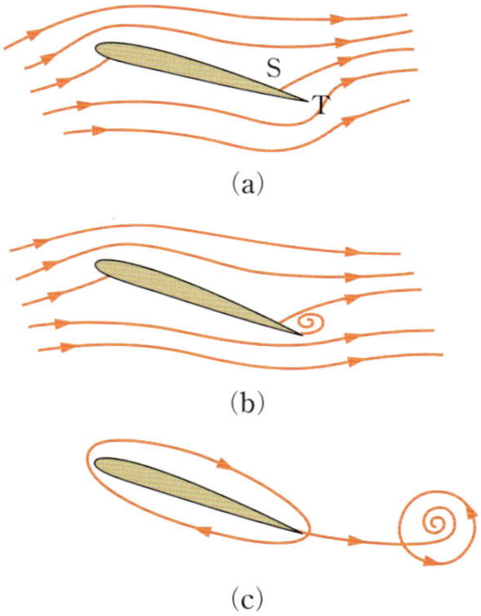

Fig. 11-13 Circulating air flow is generated around the wing in pairs with the vortex.

Since the lift F_{L} acting on the wing is generated by the pressure difference $p_{\text{lower}} - p_{\text{upper}}$ on the upper and lower sides of the wing, assuming that the area of the upper surface and the lower surface of the wing is A, we can derive

$$F_{\text{L}} = (p_{\text{lower}} - p_{\text{upper}})A = \frac{1}{2}\rho({v_{\text{upper}}}^2 - {v_{\text{lower}}}^2) \tag{11.13}$$

using Bernoulli's law for the steady flow shown in Fig. 11-12. Both the airflow speeds v_{upper} above the wing and v_{lower} below the wing are considered to be proportional to the speed (wing speed) v of the air stream far from the plane, so the magnitude of the lift is expressed as

$$F_{\text{L}} = \frac{1}{2}C\rho A v^2. \tag{11.14}$$

The proportional constant C is called the **lift constant** and is related to the shape of the wing and the attack angle (angle of inclination of the wing relative to the forward direction) α. As long as the attack angle α is small, the lift constant is approximately proportional to α, but when the attack angle is large, a vortex is created behind the wing, and the lift decreases. In addition, the resistance suddenly increases, and the airplane stalls.

Question 2 The area to mass ratio of the wing is larger for low speed propeller planes than for high speed jet planes. Why?

Fig. 11-14 A propeller plane

11.4 Viscous drag and inertial resistance

Learning objective First, understand the viscous force. Then, to understand that two types of force, viscous drag and inertial drag, exist as resistance acting on objects moving in the fluid and to understand the causes of these resistances.

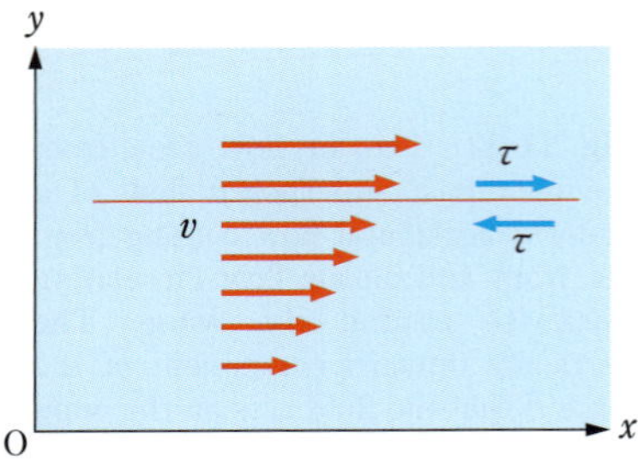

Fig. 11-15 Viscous force $\tau = \eta \dfrac{dv}{dy}$.
The upper fast-flow part exerts a rightward force on the lower slow-flow part, and the lower slow-flow part exerts a leftward force on the upper fast-flow part.

Unit of viscosity
N·s/m² = Pa·s

Table 11-1 Viscosity values of some substances [Pa·s]

Air	1.82×10^{-5}
Carbon dioxide	1.47×10^{-5}
Water	1.002×10^{-3}
Mercury	1.56×10^{-3}
Ethanol	1.197×10^{-3}
Glycerin	1.495×10^{3}

(20 °C, 1 atm)

Viscous force The speed of flow generally depends on the location. As shown in Fig. 11-15, it is assumed that the speed v of the flow parallel to the x-axis changes with height y. $\dfrac{dv}{dy}$ is the velocity gradient. If a plane parallel to the flow through any one point in this flow is considered, between the upper and lower portions of the surface the tangential stress parallel to the surface acts to reduce the velocity gradient and make the velocity uniform. This tangential stress is called the **viscous force**. Let the viscous force per unit area be τ, then for fluids composed of low molecular weight molecule such as water, air, and oil, experiments have shown that the following proportional relationship

$$\tau = \eta \frac{dv}{dy} \tag{11.15}$$

holds. The proportionality factor η is a constant determined by the substance and is called the **viscosity** (or **viscosity coefficient**) of the substance. The unit of viscosity is N·s/m² = Pa·s. Viscosity values of some substances are listed in Table 11-1.

Bernoulli's law cannot be applied if the work performed by the viscous force is non-negligible in comparison with the change in the mechanical energy of the fluid.

One characteristic of viscosity is the **nonslip condition**. This is an experimental fact that at the interface between fluid and solid the fluid moves at the same velocity as the solid, that is, the relative velocity of the solid surface and the fluid in contact with it is zero. For this reason, no matter what the smooth wall of the tube is made of, no difference is observed in the viscous behavior of the fluid flowing through it.

Case 3 We calculate the magnitude of viscous force F acting on an air track with a base area A of 500 cm² that moves linearly at a constant speed of 1 m/s on an air cushion with a thickness Δy of 1 mm on a horizontal floor. The viscosity η of air is 1.8×10^{-5} Pa·s, so

$$F = \tau A = \eta \frac{\Delta v}{\Delta y} A$$
$$= (1.8 \times 10^{-5}\,\text{Pa·s}) \times \frac{1\,\text{m/s}}{10^{-3}\,\text{m}} \times (0.05\,\text{m}^2)$$
$$= 9 \times 10^{-4}\,\text{Pa·m}^2 = 9 \times 10^{-4}\,\text{N}$$

Hagen-Poiseuille's law When pressures p_A and p_B ($p_A > p_B$) are applied to both ends A and B of a horizontal circular tube of inside diameter R and length L through which a viscous fluid flows, the amount Q of the fluid flowing in the direction from A to B in a unit time is

$$Q = \frac{\pi R^4}{8\eta L}(p_A - p_B) \tag{11.16}$$

(Fig. 11.16). This is called **Hagen-Poiseuille's law**. The viscosity η can be determined experimentally using this law. Equation (11.16) holds for laminar flows in water pipes and blood vessels.

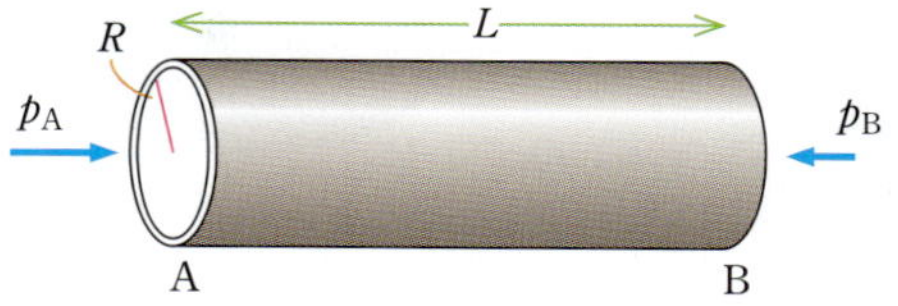

Fig. 11-16 Hagen-Poiseuille's law $Q = \frac{\pi R^4}{8\eta L}(p_A - p_B)$

Viscous drag We consider an object of length L moving at a small speed v in a viscous fluid (a length here means the representative length of an object such as radius, diagonal, etc.). As a result of the nonslip condition, the relative velocity of the object surface and the fluid in contact with it is zero. Therefore, when an object moves in the fluid, the object drags the fluid near its surface, and the velocity gradient is generated in the fluid, and the viscous force acts to stop the movement of the object. This is called the **viscous drag**.

Viscous resistance F is proportional to "fluid viscosity η"×"velocity gradient"×"surface area of the object". The surface area of objects of similar shape is proportional to L^2 and the velocity gradient is considered to be proportional to $\frac{v}{L}$, so the viscosity resistance is

$$F \propto \eta \frac{v}{L} L^2 = \eta v L. \tag{11.17}$$

Detailed calculation has shown that when a spherical object of radius r moves at a speed v in the fluid of viscosity η the viscosity resistance F is

$$F = 6\pi\eta r v. \tag{11.18}$$

This is called **Stokes' law**. This law is applicable when the object moves very slowly in the fluid.

Inertial resistance The situation changes as the speed v of the object increases. Due to the nonslip condition, the fluid near the object surface has a part with a very large velocity gradient. This part is called the **boundary layer** because it is a thin layer. Since the velocity gradient is small outside the boundary layer, the fluid can be taken to behave as the perfect fluid. We consider at this point the flow of the surrounding perfect fluid viewed in the coordinate system which moves with the object (Fig. 11-17). The fluid which was decelerated to the speed zero at the front top point A of the object is accelerated while passing through the shoulder part of the object, then the speed becomes maximum at point B, and it is decelerated again to the zero at the rear end C of the object. But in real fluid, with the increase in speed v of the fluid, the flow is decelerated due to viscosity, and the flow speed becomes 0 (stall) before reaching the rear end C, and the boundary layer peels off the object surface to generate vortexes behind the object (Fig. 11-18).

The fluid pressure at the position behind the object at which vortexes are generated is roughly equal to the pressure p_∞ far away from the object. The fluid far away from the object flows at speed v, while the flow speed at the front top point A of the object is 0, so according to Bernoulli's law, the pressure p at point A is $p = p_\infty + \frac{1}{2}\rho v^2$.

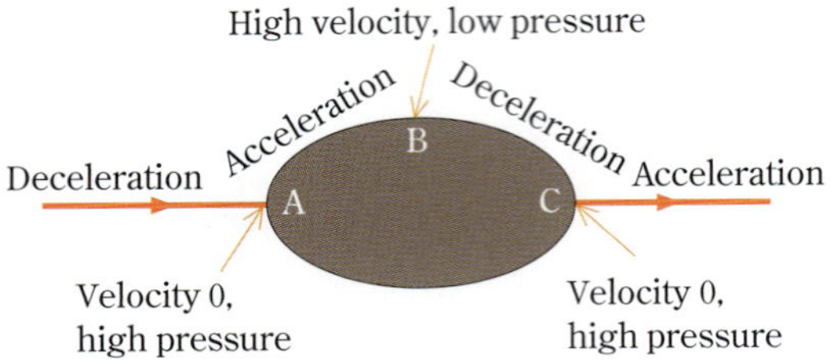

Fig. 11-17 Flow of the perfect fluid around the object

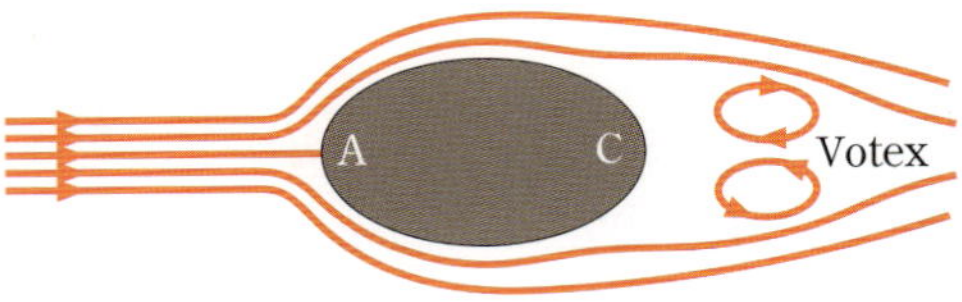

Fig. 11-18 Generation of vortex. Since the fluid inside the boundary layer of the extremely large velocity gradient is rotating, the vortex is generated as a result of conservation of angular momentum when the boundary layer peels off the object surface.

Therefore, the resistance of the fluid acting on the object with the cross-sectional area A moving at speed v in the fluid of density ρ is expressed as

$$F_L = \frac{1}{2} C\rho A v^2. \tag{11.19}$$

The resistance coefficient C is approximately 0.5 for the sphere. This resistance is called the **inertial resistance** or the pressure resistance.

The shape which weakens the rate of pressure rise from the shoulder to the rear end part to reduce the possibility of peeling-off of the boundary layer so that the inertial resistance is significantly reduced is called the **streamline shape**.

Fig. 11-19 Kármán's vortex

Laminar and turbulent flows There are two types of flow, namely smooth, regular flow called the **laminar flow**, and disturbed, irregular flow called the **turbulent flow**. In laminar flow, fluid particles move downstream in a smooth and regular trajectory, with less mixing between different fluid layers. In turbulent flow, the irregular and random motion overlaps the average downstream motion of fluid, and the momentum exchange is carried out between different mean streamlines.

Mechanical similarity rule and Reynolds number Let us consider the resistance that a similar body receives from a fluid. When the flow is slow, the viscous drag whose magnitude is proportional to ηvL works, and the flow is laminar. As the flow becomes faster, the inertial resistance proportional to $\rho v^2 L^2$ works and the flow becomes turbulent. Therefore, the ratio of inertial resistance to viscous resistance

$$Re = \frac{\rho v^2 L^2}{\eta v L} = \frac{\rho v L}{\eta} \tag{11.20}$$

is a number that characterizes the flow, and it is appropriate to think that the flow pattern will be the same if the above ratio is the same. This is called the **mechanical similarity rule**. This ratio Re is a dimensionless number called the **Reynolds number**. The transition from laminar flow to turbulent flow occurs around $Re \approx 3\,000$. Model experiments of ships and airplanes are performed with the same Reynolds number as in the real world.

Problems for exercise 11

A

1. Find the contact area between each tire and the ground when the pressure of four tires of a 1-t car is 2 atm (in this case, the air pressure of the tire is actually 3 atm).
2. Densities of air and helium at 1 atm and 0 °C are 1.29 kg/m^3 and 0.178 kg/m^3, respectively. Calculate the mass that a 1 m^3 balloon with a mass of 200 g filled with helium gas can lift up. The temperature is assumed to be 0 °C.
3. We filled a vinyl tube with water, closed both ends, placed one end 0.20 m below the surface of the water tank, and placed the other end 0.30 m below the water outside the tank. Find the speed of water flowing out of the tube when both ends are open (this tube is sometimes called a siphon tube).
4. We have a Venturi tube with a radius of 1 cm for the thin part and a radius of 3 cm for the thick part. When the water speed at the wide part is 0.2 m/s, calculate (1) the speed at the narrow part, and (2) the pressure depression.
5. In 1654, Goericke, the mayor of Magdeburg, conducted the following experiment at Magdeburg : the rims of two hollow copper hemispheres with a 40 cm diameter were brought into contact and firmly sealed and the air was removed from the interior of the hollow spheres of the overlapped hemisphere. Then he pulled both sides simultaneously with eight horses and showed that the two hemispheres could not be pulled apart. Show that the two hemispheres must be pulled with a force of 1300 kgf or more to separate them.
6. Is there any advantage in view of hydrodynamics when we ride and run a bicycle right behind a truck?
7. Explain, using Fig. 1 as seen from the observer moving at the same speed as the ball, that the ball thrown to the left with topspin receives the downward force from the air. A force perpendicular to the flow direction acts on the moving object which is rotating in the fluid. This is called the **Magnus effect**.

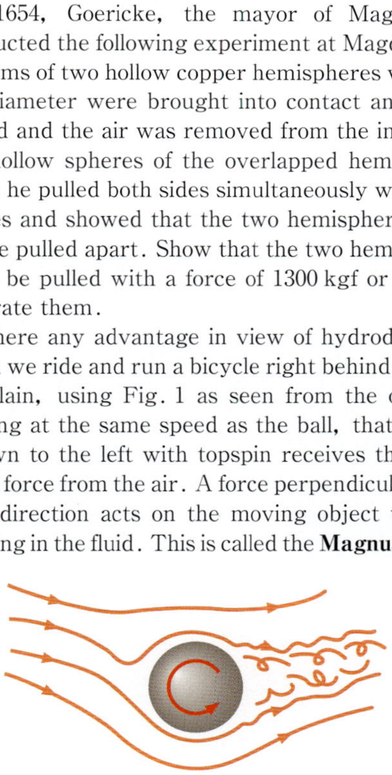

Fig. 1

B

1. When a submarine lands on soil or sand of the seabed, there may be the case in which it cannot rise up by itself. What causes this incident?
2. Fig. 2 shows the pitot tube used to measure the speed of aircraft. If the flow speed just outside the small hole B is assumed to be the aircraft speed u, show that $u = \sqrt{\dfrac{2\rho_0 gh}{\rho_{air}}}$. ρ_0 is the density of the liquid in the U-shaped tube.

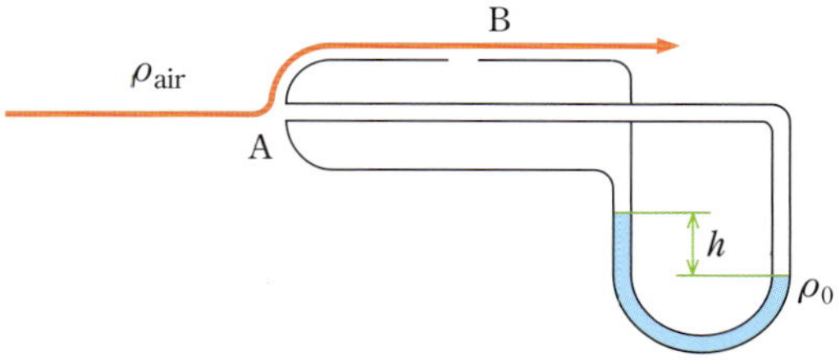

Fig. 2

3. There is a small hole of area S at the bottom of a cylindrical tank of bottom area A. When this tank contains water to depth H, show that the time until the water is drained is

$$T = \sqrt{\frac{2H}{g}}\frac{A}{S}.$$

4. The tank described in the previous problem was made empty and then the small hole in the bottom was plugged. We poured water to the tank at a fixed rate from the water supply tap and it took 20 minutes to fill it. When we remove the plug of the hole at the bottom of the filed tank, the tank becomes empty in 10 minutes. How many minutes will it take for the filled tank to become empty if water is kept being supplied and the plug is removed?
5. When a large iron ball and a small iron ball fall free simultaneously from the same height in the air, the two iron balls arrive at the ground almost simultaneously. If these two balls fall free from the sea surface at the same time, will the two balls reach the seeled at the same time?
6. Even if we plunge one of our hands into the water, we do not feel much resistance. However, when an airplane falls into the sea, the seawater is said to be hard. Why does the nature of the response of water depend on the differences in speed, even with the same water?

Wave

When we throw a stone into a still water surface, the water surface starts to vibrate. The vibration of the water surface spreads in the form of concentric ripples centered on the point where the stone fell. In this way, a vibration that occurs in a certain place (**wave source**) of a continuum causes the vibration in the surrounding part, and a phenomenon in which the vibration is transmitted to the next part one after another is called the **wave**.

Like water in the case of water waves, those that have the property to transmit waves are called the **medium** (*pl*. media). When the wave travels through the medium, each part of the medium vibrates near its original position, but the medium does not move with the wave. However, energy is transmitted along with the vibration of the medium.

In this chapter, you will learn waves as a transmission of mechanical vibration of the medium.

12.1 Properties of waves

Learning objectives To understand the difference between longitudinal and transverse waves. To become able to explain the quantities used to represent waves, such as frequency, wavelength, period, and speed, and their relationships, and to understand how to represent waves using trigonometric functions.

Longitudinal waves and transverse waves When a long string is kept horizontal, its one end being fixed and the other end is reciprocated in the direction perpendicular to the string, the vibration generated at the end of the string is transmitted to the next part successively at a speed of v (Fig. 12-1). As like this, when the direction of vibration of the medium (string) is perpendicular to the traveling direction of the wave, this wave is called the **transverse wave**.

When one end of the long spiral spring is fixed and the other end is reciprocated in the direction of the spring, the vibration is transmitted at a speed of v in the spring (Fig. 12-2). As like this, when the traveling direction of the wave coincides with the vibrating direction of the medium (spring), this wave is called the **longitudinal wave**. Longitudinal wave is also referred to as the **compression wave**, because the sparse region of the medium and the dense region of the medium appear and the sparse and dense states of the medium are transmitted.

Waves traveling on strings of violins and S waves (secondary waves of vibration called the principal motion following the preliminary tremor) of earthquakes are transverse waves, and sound waves and P waves (preliminary tremor) of earthquakes are longitudinal waves.

A longitudinal wave is a propagation of changes in compression and

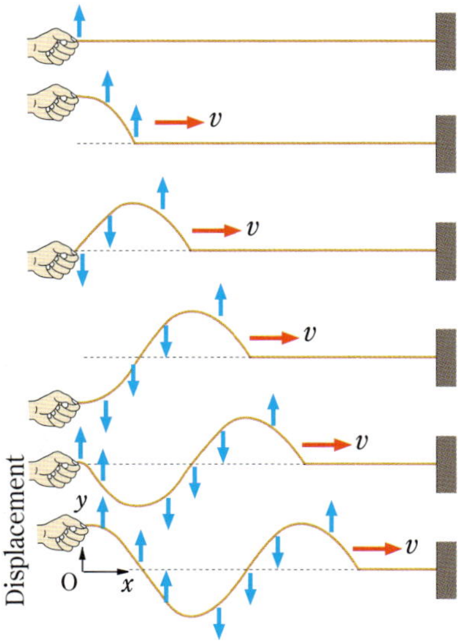

Fig. 12-1 A wave (transverse wave) traveling the string

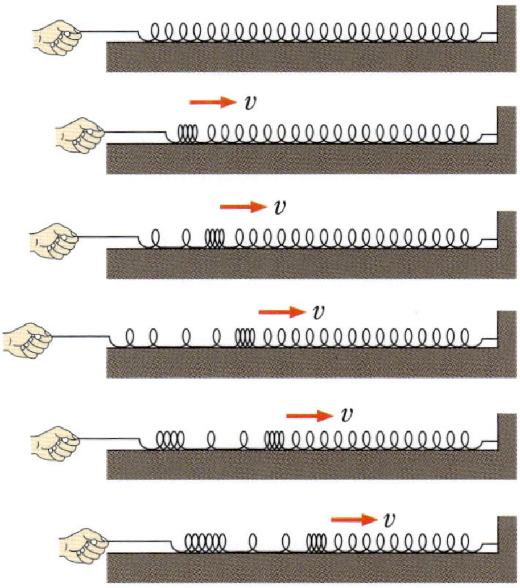

Fig. 12-2 A wave (longitudinal wave) traveling the spiral spring

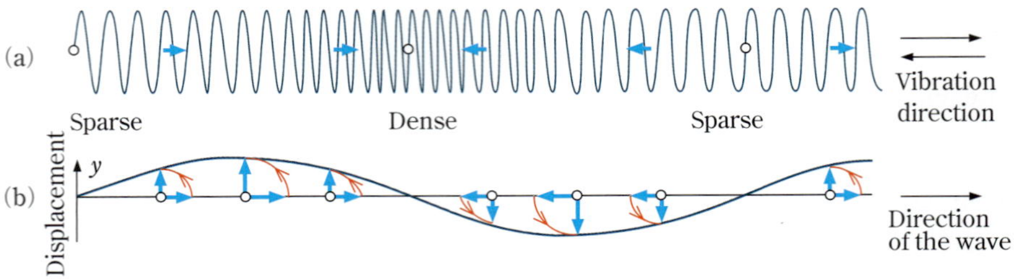

Fig. 12-3 Representation of the longitudinal wave traveling the spiral spring. (a) displacement (the arrow shows the displacement) of the medium at a certain time. (b) wave form of the longitudinal wave (note that the displacement is 0 in both dense and sparse locations).

expansion of the medium, so it travels in all solids, liquids, and gases. Propagation of the transverse wave requires the tangential stress between adjacent parts. The transverse wave travels through solid, but because there exists no tangential stress in the fluid (gas, liquid), the transverse wave does not travel through the fluid.

Speed of waves The speed of wave is determined by the restoring force to restore the deformation of the medium and the inertia to prevent the deformation from changing, that is, the density of the medium. The speed of wave is higher as the restoring force is larger, and higher as the density is smaller.

Representation of waves To represent waves, we choose the original position of the medium on the horizontal axis and the displacement of the medium on the vertical axis. In order to represent the longitudinal wave, as shown in Fig. 12-3, the direction of displacement of the medium is rotated 90° so that the displacement is perpendicular to the traveling direction of the wave (in this case, the displacement to the right is taken as a positive displacement).

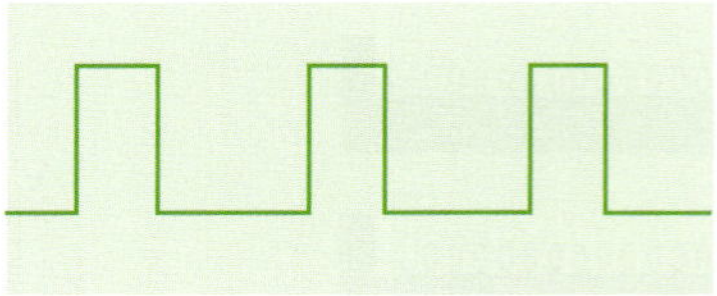

Fig. 12-4 A train of pulse

Wave form As shown in Fig. 12-3 (b), a **wave form** is a curve connecting the displacement of each point of the medium at a certain time t. The highest point of the wave form is called the **crest** and the lowest point is called the **trough**. When the wave form is a sine curve, this wave is called a **sine wave**. The maximum value of the displacement of the medium is called the wave **amplitude**. As in the case of only one crest in the wave form, an isolated wave is called a **pulse** (Fig. 12-4).

Quantities representing the nature of waves In order to form a train of wave, the wave source must vibrate continuously. When the wave source vibrates f times per second, all points of the medium also vibrate f times per second one after another. The number f of vibrations per unit time is called the wave frequency or **frequency**. The unit of frequency is 1/s and this is called **hertz** (symbol Hz).

Unit of frequency
1/s = Hz

The time T required for each point of the medium to vibrate once

$$T = \frac{1}{f} \tag{12.1}$$

is called the **period** of the wave.

The distance λ between the crests and the troughs created in pair by one time of vibration of the wave source is called the **wavelength**.

When the wave source vibrates f times per second, f pairs of crests and troughs of length λ are generated. Therefore, the length of wave generated in one second is λf. This is the **speed of wave,** as it is the distance that a crest or trough travels in a second. Therefore, the following relationship exists among the wave speed v, wavelength λ, frequency f and period T.

$$v = \lambda f = \frac{\lambda}{T}. \tag{12.2}$$

Equation of sine wave In Fig. 12-1, let the center of vibration at the left end of the string be the origin O and let the right direction be the positive direction of the x-axis. The left end (wave source) of the string is performed a simple harmonic oscillation along the y-axis with amplitude A, frequency $f = \dfrac{\omega}{2\pi}$, and period T, namely

$$y = A \sin \omega t = A \sin 2\pi f t = A \sin \frac{2\pi t}{T} \tag{12.3}$$

(ω is called the angular frequency).

As the wave source O starts to oscillate, each part of the string also starts to oscillate at the same frequency f one after another. We assume that the wave speed is v. Then the time for the wave to travel from the wave source to point P at distance x is $\dfrac{x}{v}$, so the displacement y of point P at time t is equal to the displacement of the wave source at time $t - \dfrac{x}{v}$ which is earlier by time $\dfrac{x}{v}$. Accordingly, the relationship between the displacement y of each point of the string and the time t is

$$y = A \sin \omega\left(t - \frac{x}{v}\right) = A \sin 2\pi f\left(t - \frac{x}{v}\right). \tag{12.4}$$

Using equations (12.1) and (12.2), equation (12.4) is expressed as

$$y = A \sin 2\pi\left(\frac{t}{T} - \frac{x}{\lambda}\right). \tag{12.5}$$

$2\pi\left(\dfrac{t}{T} - \dfrac{\lambda}{v}\right)$ is referred to as the **phase** of wave at time t and position x*.

The wave forms (sinusoidal curves) at time $t = 0$, and $\dfrac{1}{4}T$, $\dfrac{1}{2}T$, $\dfrac{3}{4}T$, T are shown in Fig. 12-5. If we consider equation (12.5) as a function of t under the condition of $x =$ constant, this equation shows that point P performs a simple harmonic oscillation delayed by time $\dfrac{x}{v}$ compared to the wave source. Also, if we consider equation (12.5) as a function of x under the condition of $t =$ constant, this equation represents the wave form at time t. That is, when the wave source performs a simple harmonic oscillation, a sine wave is generated.

* The phase of the wave indicates where the state of periodically changing wave is in the wave period. Two waves are said to be in phase when they move in the same way at the same moment. Two waves are said to be in opposite phase when they behave exactly opposite at the same moment.

The sine function sin$x$ is a periodic function [$\sin(x+2\pi) = \sin x$] of period 2π. Therefore, the phase is in phase when the phase differs by integer multiple of 2π (namely when n is an integer, difference of $2\pi n$), while the phase is in opposite phase when the phase differs by $(2n+1)\pi$.

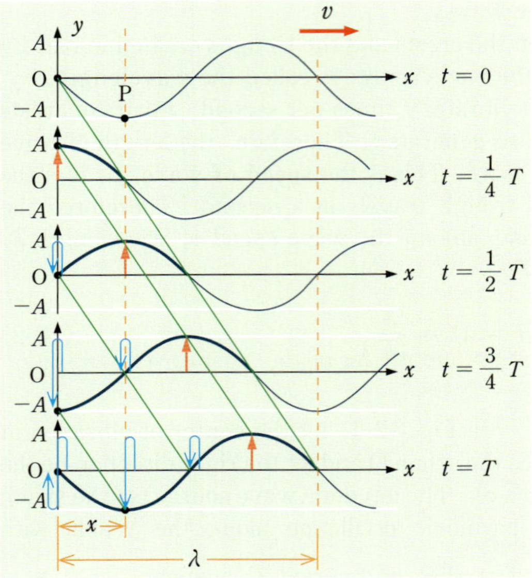

Fig. 12-5

Unit of the power in wave
W/m²

*1 In the case of the wave shown in Fig. 12-3, the speed and displacement gradient of the medium are the largest at dense and sparse locations, so the kinetic energy and potential energy are the largest and the power in wave is the largest. On the other hand, at the maximum displacement, the speed of medium and the displacement gradient of the medium are zero, so the kinetic energy, potential energy and power in wave are also zero.

The periodic change in power in wave is caused by the periodic change in power Fv that the external force F performs at the wave source with respect to medium (v is the speed of wave source).

*2 Substitute ρv for mass m of the weight in the equation of mechanical energy of the spring pendulum :

$$\frac{1}{2}m\omega^2A^2. \qquad (5.44)$$

Energy of waves When the wave is propagated, the medium at rest starts to vibrate, and the kinetic energy and deformation potential energy of the medium are generated. A wave is not only a transmission of vibration of a medium, but also a transfer of energy. The energy I that passes through a unit area perpendicular to the traveling direction of the wave in unit time is called the **power in wave**. The unit is $\mathrm{W/m^2}$.

As the wave propagates, the kinetic energy of each point of the medium and the potential energy of the force that restores the displacement gradient of the medium synchronously change periodically. For this reason the power in wave at each point is not constant*1. The power in sine wave with amplitude A, frequency f, and speed v which travels in the medium of density ρ is averaged over time and is given by

$$I = 2\pi^2 f^2 A^2 \rho v = \frac{1}{2}\omega^2 A^2 \rho v \qquad (\omega = 2\pi f). \qquad (12.6)^{*2}$$

The power in wave is proportional to the product of the square of amplitude and the square of frequency.

If wave energy is absorbed by the medium, the wave amplitude decreases.

12.2 Wave equation and speed of waves

Learning objective To understand that the propagation of waves is described by a function like $y(x, t)$ as equation (12.5) which represents the displacement of the medium, that the equation of motion which this function follows is the wave equation, and that the derivation of wave equation leads to the speed of waves.

Wave equation and speed of transverse wave traveling through a string A string of length L and linear density (mass per unit length)

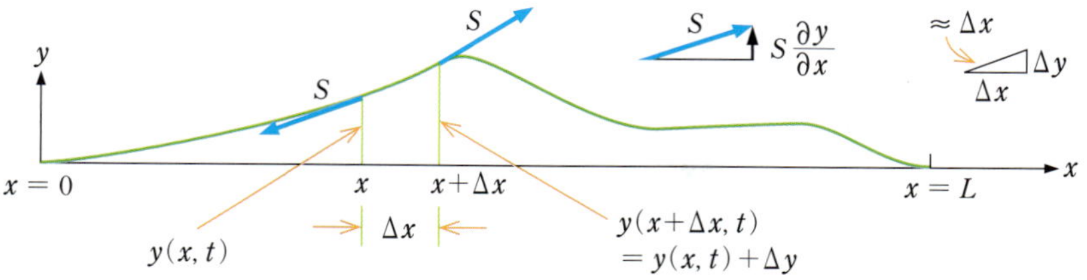

Fig. 12-6

μ is pulled with tension S along the x-axis (Fig. 12-6). Let us derive Newton's equation of motion for a minute part (mass $\mu\,\Delta x$) of the length Δx between the two points x and $x+\Delta x$ of this string. The vibration direction of each point of the string is perpendicular to the string. Taking the perpendicular direction as the y direction, we express the displacement of the string as $y(x, t)$. Therefore, the acceleration of the string is obtained by differentiating $y(x, t)$ twice with respect to t while keeping x constant (that is, partial differentiation), and is denoted as $\frac{\partial^2 y}{\partial t^2}$. The "mass"×"acceleration" of the minute part of the string is

$$\mu\,\Delta x\frac{\partial^2 y}{\partial t^2} \tag{12.7}$$

(read the symbol ∂ der or rounded d).

Since the tangent slope at point x of the curve $y(x, t)$ is $\frac{dy}{dx}$, the slope of the string at point x is obtained by differentiating (partially differentiating) $y(x, t)$ with respect to x while t is kept constant, or $\frac{\partial y(x,t)}{\partial x}$. In the case where the gradient is small and $\sin\theta \approx \tan\theta$, the y component of the tension $\mathbf{S}$ of the string is $S\frac{\partial y}{\partial x}$. Therefore, the y component of the resultant force acting on both sides of the minute part of the string is

$$S\frac{\partial y}{\partial x}(x+\Delta x, t)-S\frac{\partial y}{\partial x}(x, t). \tag{12.8}$$

Using the relation

$$\frac{\partial y}{\partial x}(x+\Delta x, t) = \frac{\partial y}{\partial x}(x, t)+\Delta x\frac{\partial^2 y(x,t)}{\partial x^2}+O((\Delta x)^2)^*$$

we obtain for the resultant force in the y-axis direction acting on the minute part the relation

$$S\,\Delta x\frac{\partial^2 y(x,t)}{\partial x^2}. \tag{12.9}$$

Therefore, from equations (12.7) and (12.9) the equation of motion of the minute part is

$$\frac{\partial^2 y}{\partial x^2} = \frac{1}{v^2}\frac{\partial^2 y}{\partial t^2} \qquad \left(v = \sqrt{\frac{S}{\mu}}\right). \tag{12.10}$$

This is referred to as the **wave equation** of the string.

The general solution of the wave equation (12.10) can be expressed as

$$y(x, t) = f(x-vt)+g(x+vt) \tag{12.11}$$

where f and g are two arbitrary functions. The validity of this solution

Fig. 12-7 Ripples on the water surface

* $O((\Delta x)^2)$ means a quantity which decreases in proportion to $(\Delta x)^2$ in the limit $\Delta x \to 0$.

is confirmed using the property that

$$\frac{\partial f(x-vt)}{\partial x} = f'(x-vt) \text{ and } \frac{\partial f(x-vt)}{\partial t} = -vf'(x-vt). \quad (12.12)$$

Here $f'(u) = \dfrac{\mathrm{d}f(u)}{\mathrm{d}u}$. The solution (12.11) represents the fact that the waves in wave forms $f(x)$ and $g(x)$ at time $t=0$ are moving at speed

$$v = \sqrt{\frac{S}{\mu}} \quad \text{(speed of transverse wave traveling through a string)} \quad (12.13)$$

in the positive and negative directions of the x-axis, respectively.

Question 1 As shown in Fig. 12-8, a tension of 1 kg is applied to a rubber string of 4 m in length and 0.3 kg in mass. Calculate the speed of the transverse wave that travels through this rubber string.

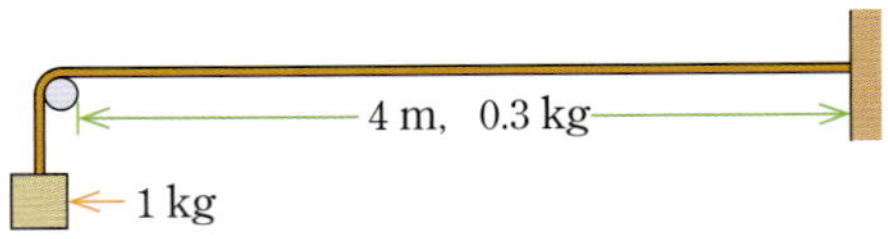

Fig. 12-8

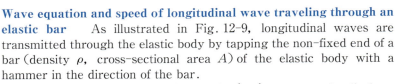

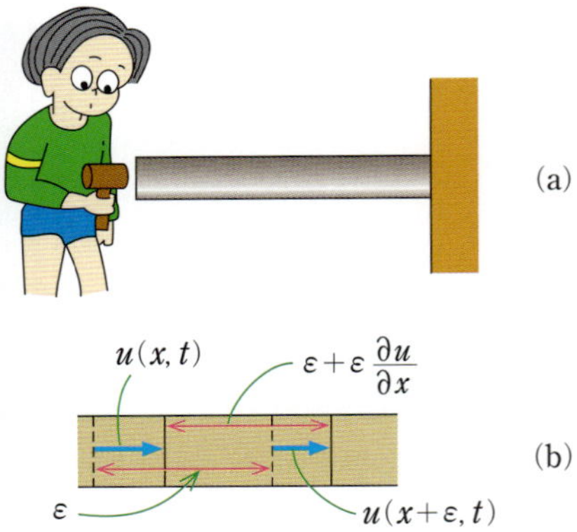

Fig. 12-9 To generate longitudinal waves in the bar.

Wave equation and speed of longitudinal wave traveling through an elastic bar As illustrated in Fig. 12-9, longitudinal waves are transmitted through the elastic body by tapping the non-fixed end of a bar (density ρ, cross-sectional area A) of the elastic body with a hammer in the direction of the bar.

Let x be the direction of the bar, and $u(x, t)$ represent the displacement of the part at point x of the bar when it is kept at rest. In this state, the elongation of the part whose both ends are x and $x+\varepsilon$ (length ε) is

$$u(x+\varepsilon, t) - u(x, t) \approx \varepsilon \frac{\partial u}{\partial x}(x, t) \quad (12.14)$$

[Fig. 12-9 (b)]. Thus, using Young's modulus E of the elastic body, we can find from equation (10.1) the elastic force $F(x, t)$ that both sides exert each other across through the cross section including the point x:

$$F(x, t) = AE\frac{\Delta L}{L} = AE\frac{1}{\varepsilon}\left[\varepsilon\frac{\partial u}{\partial x}(x, t)\right] = AE\frac{\partial u}{\partial x}(x, t). \quad (12.15)$$

Accordingly, from the equation of motion of the part whose both ends at rest are at x and $x+\Delta x$ and whose length is Δx and mass is $\rho\,\Delta x\,A$, the wave equation of the longitudinal wave traveling through the elastic bar with density ρ and the wave speed v are derived as follows:

$$\rho\,\Delta x\,A\frac{\partial^2 u}{\partial x^2} = F(x+\Delta x, t) - F(x, t)$$

$$= AE\left[\frac{\partial u}{\partial x}(x+\Delta x, t) - \frac{\partial u}{\partial x}(x, t)\right] = AE\,\Delta x\frac{\partial^2 u}{\partial x^2}$$

$$\therefore \quad \frac{\partial^2 u}{\partial x^2} = \frac{1}{v^2}\frac{\partial^2 u}{\partial t^2} \quad (12.16)$$

$$v = \sqrt{\frac{E}{\rho}}. \qquad (12.17)$$

Wave equation and speed of transverse wave traveling through an elastic bar In this case it is enough to replace Young's modulus E with rigidity G.

$$v = \sqrt{\frac{G}{\rho}} \quad \text{(transverse wave).} \qquad (12.18)$$

Since equation (10.6) shows $E > G$, the longitudinal waves travel faster than the transverse waves.

Wave speed traveling through an infinitely wide elastic body In an elastic body, the state of deformation and stress is transmitted in a form of wave. This is called the **elastic wave**. Among the elastic waves, the simple case is the one in which states of deformation and stress in a plane perpendicular to the traveling direction of wave remain in the same at all locations. This is called the plane wave. The speed of the longitudinal wave is found in the same way as the longitudinal wave traveling through the bar. In the case of bar, along with the longitudinal compression and expansion, the lateral expansion and compression corresponding to Poisson's ratio occur, but in the case of an infinitely wide elastic body, lateral expansion and compression do not occur. For this reason, it is sufficient to use in equation (12.17) instead of Young's modulus the elongational elastic modulus $\frac{(1-\sigma)}{(1+\sigma)(1-2\sigma)}E = k+\frac{4}{3}G$ which holds for the case without lateral deformation (Refer to Problems for exercise 10, B2). In this way, the speed of the longitudinal wave traveling through an infinitely large elastic body of density ρ is

$$v_{\text{longtudinal}} = \sqrt{\frac{(1-\sigma)}{(1+\sigma)(1-2\sigma)}}\sqrt{\frac{E}{\rho}}. \qquad (12.19)$$

The speed of the transverse wave is the same as in the case of the transverse wave traveling through the same elastic bar :

$$v_{\text{transverse}} = \sqrt{\frac{G}{\rho}} = \frac{1}{\sqrt{2(1+\sigma)}}\sqrt{\frac{E}{\rho}}$$

$$(\therefore \quad v_{\text{longtudinal}} > v_{\text{transverse}}). \qquad (12.20)$$

12.3 Principle of superposition of waves and interference

Learning objective To understand the principle of superposition of waves and the interference

When two stones are thrown at the same time into the still water of a pond, concentric waves appear from the two points A and B where the stones fell as shown in Fig. 12-11. The two waves encounter to form the striped pattern shown in Fig. 12-11. The striped pattern is formed because the displacement of medium when two waves come simultaneously is a superposed result of the displacement of medium when those

Fig. 12-10 An interference of two waves

waves come alone. In other words. If two waves $y_1(\boldsymbol{r}, t)$ and $y_2(\boldsymbol{r}, t)$ are solutions of the wave equation (12.10), then the wave obtained by combining the two waves

$$y(\boldsymbol{r}, t) = y_1(\boldsymbol{r}, t) + y_2(\boldsymbol{r}, t) \tag{12.21}$$

is also the solution of the wave equation. This is called the **principle of superposition of waves**. So long as the amplitude of wave is small, the principle of superposition holds satisfactorily*. The small wave amplitude means, for example in the case of a wave traveling through an elastic body, that the deformation of the elastic body is less than the proportional limit.

* Waves that do not conform to the principle of superposition are called nonlinear waves. Examples of nonlinear waves are stable pulse-like solitary waves called the soliton and shock waves.

When two or more waves encounter, a resultant wave is the one obtained by superposing these waves, and a phenomenon that strengthens or weakens each other is called the **interference of wave**.

Case 1 Let l_1 and l_2 be the distances from points A and B to a certain point in Fig. 12-11 and let λ be the wavelength. At locations like point P where the differences between distances l_1 and l_2 are an even multiple of a half wavelength (integer multiple of wavelength)

$$|l_1 - l_2| = \frac{\lambda}{2} \cdot 2n = n\lambda \qquad (n = 0, 1, 2, \cdots) \tag{12.22}$$

the phases of the two waves are the same, such as crest and crest and trough and trough, so the amplitude oscillates twice as large as the amplitude of one wave.

At locations like point Q where the difference between distances l_1 and l_2 is an odd multiple of a half wavelength

$$|l_1 - l_2| = \frac{\lambda}{2}(2n+1) \qquad (n = 0, 1, 2, \cdots) \tag{12.23}$$

the phases of the two waves cancel each other all time, so the oscillation does not occur. In this case the two phases are said to be opposite.

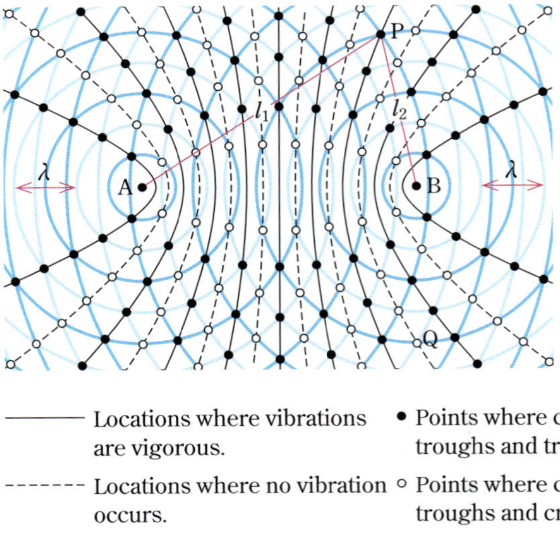
P
l_1
l_2
A
B
λ
λ
Q
Locations where vibrations are vigorous.
Locations where no vibration occurs.
Points where crests and crests, and troughs and troughs overlap.
Points where crests and troughs, and troughs and crests overlap.

Fig. 12-11

12.4 Reflection and refraction of waves

Learning objective To understand the law of reflection and the law of refraction when a wave comes to the boundary of a medium and is reflected at the boundary or refracted through the boundary.

Wave front When a wave travels through space, the surface formed by connecting the points with the same phase is called a wave front. The wave with a flat wave front is called the **plane wave,** and the wave with a spherical face is called the **spherical wave**. The wave front travels at speed v of the wave. The traveling direction of the wave at each point of the wave front is perpendicular to the wave front.

Fig. 12-12 Coast waves

Law of reflection Waves traveling on the water surface of the pool are reflected when they hit the edge of the pool. The frequency, speed and wavelength of the wave (incident wave) incident on the boundary surface (edge of the pool) and the wave reflected (reflected wave) are equal. The angle θ_i between the traveling direction of the incident wave and the straight line (normal line) perpendicular to the boundary surface is called the **angle of incidence (incident angle),** and the angle θ_r between the traveling direction of the reflected wave and the normal to the boundary surface is the **angle of reflection (reflected angle)** (Fig. 12-13). For the wave reflection the **law of reflection** holds :

$$\text{angle of incidence } \theta_i = \text{angle of reflection } \theta_r. \tag{12.24}$$

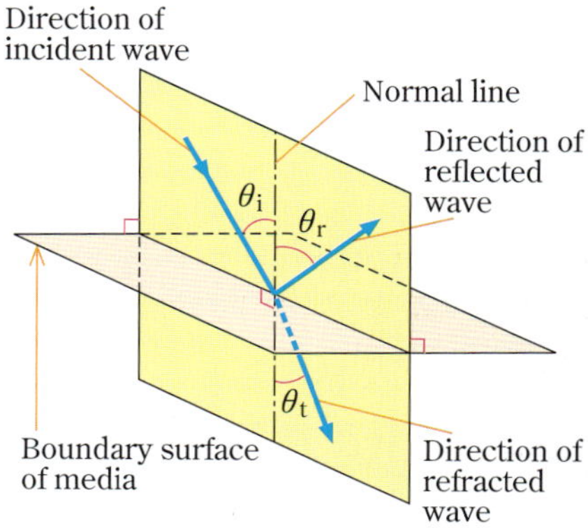

Fig. 12-13 Reflection and refraction

Law of refraction When waves come diagonally from the distance towards the sandy beach of the coast, their wave front becomes parallel to the coastline as they approach the shore. This is because the wave traveling on the sea surface travels more slowly as the depth (h) of water becomes shallow (in the case of wavelength $\lambda > h$, $v \approx \sqrt{gh}$. g is the gravitational acceleration). Thus, when a wave travels to a place where the speed is different, the phenomenon that the traveling direction of the wave changes, called the **refraction of wave,** is observed.

In general, when a wave is incident on the boundary between two types of media, part of the wave is reflected at the interface but the rest passes through the interface. Waves are refracted when they pass the interface. The frequency does not change during refraction. Therefore, the wavelength is proportional to the speed of the wave. The angle θ_i between the traveling direction of the refracted wave and the normal to the interface is called the **refracted angle**. When a wave travels from medium 1 (wave speed v_1, wavelength λ_1) to medium 2 (wave speed v_2, wavelength λ_2), it is refracted according to the following **law of refraction** (Fig. 12-14) :

$$\frac{\sin\theta_i}{\sin\theta_t} = \frac{v_1}{v_2} = \frac{\lambda_1}{\lambda_2} = n_{1\to2} \quad \text{(a constant).} \tag{12.25}$$

A constant $n_{1\to2}$ is called the **refractive index of medium 2 relative to medium 1 (relative refractive index)**. The relation

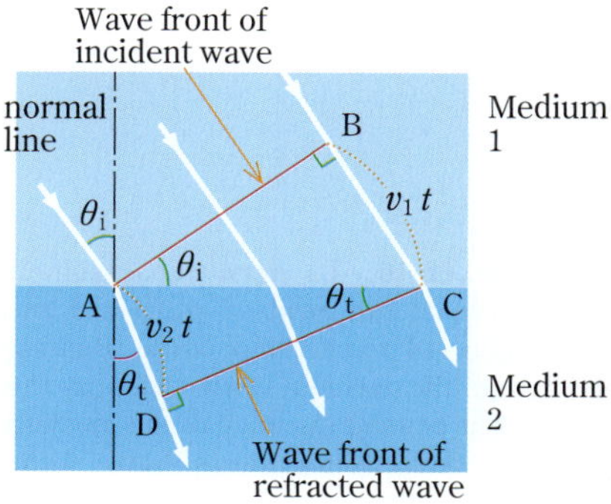

Fig. 12-14 Law of refraction

$$v_1 t = \overline{BC} = \overline{AC}\sin\theta_i$$
$$v_2 t = \overline{AD} = \overline{AC}\sin\theta_t$$
$$\therefore \quad \frac{\sin\theta_i}{\sin\theta_t} = \frac{v_1}{v_2}$$

$$n_{1\to2} = \frac{1}{n_{2\to1}} \tag{12.26}$$

holds.

Question 2 A wave was incident from medium 1 to medium 2 ($n_{1\to2} = 1.41$) at an incident angle $\theta_i = 45°$. Find is the refraction angle θ.

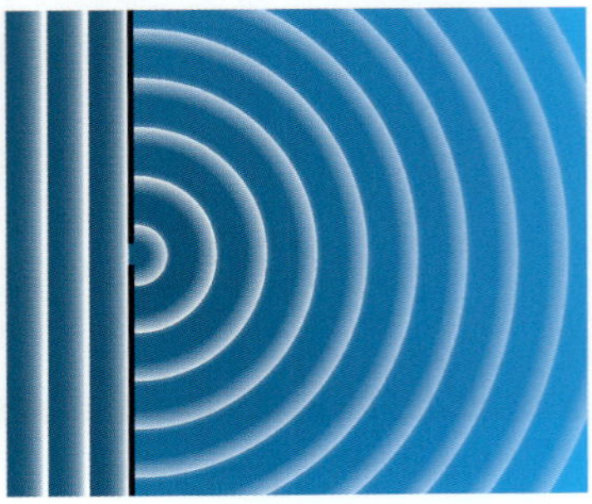

Fig. 12-15 Diffraction of wave

Diffraction Sound waves are also transmitted to the shadow area of objects. Waves on the sea surface fall behind the breakwater. When there is an obstacle in the path of waves, the waves do not go straight, but they go around into the shadow area of the obstacle (Fig. 12-15). This phenomenon is called the **diffraction**. Diffraction phenomena are significant when the wavelength is about the same as or longer than the size of the obstacle or gap.

12.5 Standing wave

Learning objective To understand the generation mechanism of standing wave.

Phase of the reflected wave When the wave comes to the boundary of the medium, part of the energy of the incident wave is reflected and changes to energy of the reflected wave, and the rest is absorbed by the boundary or passes through the boundary. When the medium is fixed at the boundary as shown in the right end (**fixed end**) of the string in Fig. 12-1 and when the medium can vibrate freely in the direction of vibration at the boundary as shown in the left end (**free end**) of the bar in Fig. 12-9, the energy of the incident wave is completely reflected. Therefore, in these cases, the amplitudes of the incident wave and the reflected wave are equal.

(1) **Reflection at fixed end** The displacement of the medium at the fixed end is zero. As shown in Fig. 12-16 (a), when the incident wave of speed v arrives at the fixed end of the medium and the fixed end receives a force from the medium, the medium receives the force of the same magnitude but in the opposite direction from the fixed end (reaction). Therefore, the reflected wave of speed v in the opposite direction to the incident wave is generated. The wave traveling toward the end of the medium (incident wave) and the reflected wave follow the principle of superposition. Expressing the incident wave as

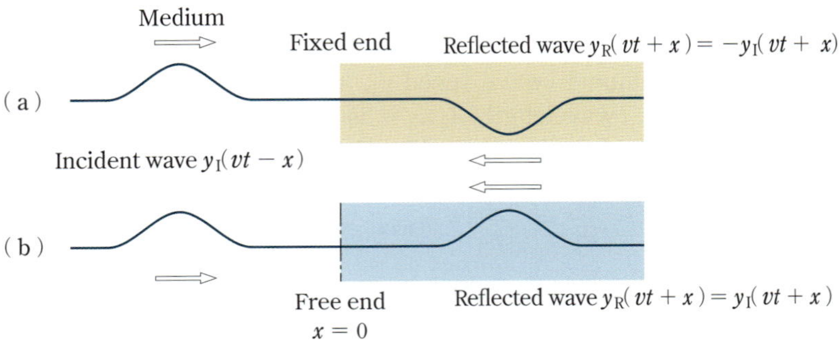

Fig. 12-16 To draw the resultant wave of the incident wave and the reflected wave, think of the situation that the incident wave goes to the right beyond the edge of the medium and the reflected waves shown in the figure appears in the region without medium and goes to the left, and superpose these two waves. The figure illustrates the case in which the incident wave has not reached the boundary and the reflected wave has not been generated yet on the medium.

$y_I(vt-x)$ and the reflected wave as $y_R(vt+x)$, we have for the resultant wave $y(x, t) = y_I(vt-x)+y_R(vt+x)$. If the fixed end is $x = 0$, the displacement $y(0, t) = 0$ at the fixed end, so $y_I(vt) = -y_R(vt)$ is satisfied. Thus, the reflected wave is expressed as follows :

$$y_R(vt+x) = -y_I(vt+x). \tag{12.27}$$

The reflected wave from the fixed end is a wave obtained by shifting point-symmetrically with respect to the fixed end the hypothetical incident wave that is taken to travel beyond the fixed end, as shown in Fig. 12-16 (a). The wave combining the incident wave and the reflected wave becomes the wave that actually travels through the medium.

Case 2 When the incident wave is a sine wave of wavelength λ, $-A\sin\left\{\frac{2\pi}{\lambda}(vt-x)+\alpha\right\}$, the reflected wave at the fixed end of $x = 0$ is $A\sin\left\{\frac{2\pi}{\lambda}(vt+x)+\alpha\right\}$, so the resultant wave is

$$\begin{aligned} y(x, t) &= -A\left[\sin\left\{\frac{2\pi}{\lambda}(vt-x)+\alpha\right\}-\sin\left\{\frac{2\pi}{\lambda}(vt+x)+\alpha\right\}\right] \\ &= -2A\cos\left(\frac{2\pi vt}{\lambda}+\alpha\right)\sin\frac{2\pi x}{\lambda}. \end{aligned} \tag{12.28}$$

Here the relation $\sin(A\pm B) = \sin A\cos B\pm\cos A\sin B$ (double sign in same order) is used. Each point of the string performs a simple harmonic oscillation of the same phase with amplitude $2A\sin\frac{2\pi x}{\lambda}$.

(2) Reflection at free end As shown in Fig. 12-16 (b), when the wave travels towards the free end at speed v and reaches the free end, the free end receives no force from the right side of the free end. Therefore, a reflected wave of speed v is generated so that the displacement of the medium near the free end is kept constant, and the medium is also deformed so that no force acts on the free end. As a result, **the gradient of wave form at free end is zero**.

This fact is easy to understand in considering the case in Fig. 12-17. Let us consider the two adjacent minor portions AB and BC near the free end C of the spring. If the length $\overline{BC}$ is minute, its mass is negligible. The force does not act on the minor portion BC from the boundary (right side). If a force acts on BC from the minor portion AB on left side, the difficulty that the acceleration of BC would become extremely large a rises. Therefore, the minor portion AB does not act on BC. In case where a difference as shown in Fig. 12-17 (a) occurs in the displacement of the two points A and B, the minor portion AB is compressed, so it would expand and exert a rightward force F on BC. On the other hand, as shown in Fig. 12-17 (b), if the displacements of the two points A and B are equal, BC receives no force from AB. Therefore, the displacements of the two points A and B are equal, and the gradient of wave form at free end is 0*.

A reflected wave that satisfies such conditions is a wave obtained by moving symmetrically with respect to the boundary surface the hypothetical incident wave that is taken to travel beyond the free end

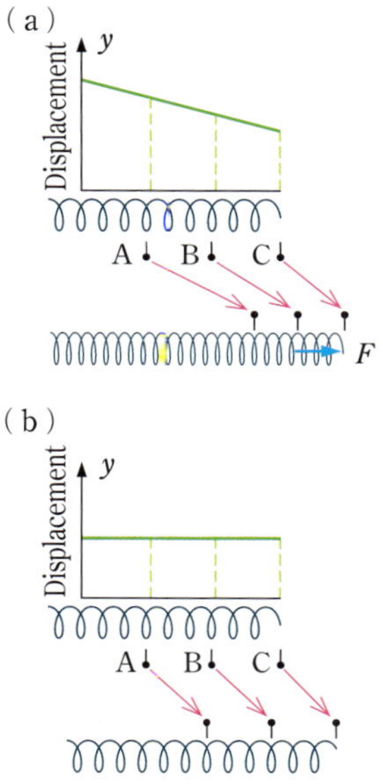

Fig. 12-17

* Since stress = (elastic modulus) $\times\frac{y(0, t)-y(-\Delta x, t)}{\Delta x}$, the boundary condition of the free end $x = 0$ is $\left.\frac{\partial y(x, t)}{\partial x}\right|_{x=0} = 0$.

[Fig. 12-16 (b)]. If the incident wave is $y_{\mathrm{I}}(vt-x)$ and the reflected wave is $y_{\mathrm{R}}(vt+x)$, the reflection wave is expressed as follows :

$$y_{\mathrm{R}}(vt+x) = y_{\mathrm{I}}(vt+x). \tag{12.29}$$

Case 3 When the incident wave is a sine wave of wavelength λ, $A\sin\left\{\frac{2\pi}{\lambda}(vt-x)+\alpha\right\}$, the reflected wave at the free end of $x=0$ is $A\sin\left\{\frac{2\pi}{\lambda}(vt+x)+\alpha\right\}$, so the resultant wave is

$$y(x, t) = 2A\sin\left(\frac{2\pi vt}{\lambda}+\alpha\right)\cos\frac{2\pi x}{\lambda}. \tag{12.30}$$

In this case each point of the medium performs a vibration of the same phase with amplitude $2A\cos\frac{2\pi x}{\lambda}$.

Standing wave Consider the cases where a sine wave (wavelength λ, period T) transmitted through a string is reflected at the fixed end and the free end. Fig. 12-18 shows the waveforms at time 0, $\frac{1}{8}T$, $\frac{2}{8}T$, $\frac{3}{8}T$, and $\frac{4}{8}T$ using the method shown in Fig. 12-16 (——

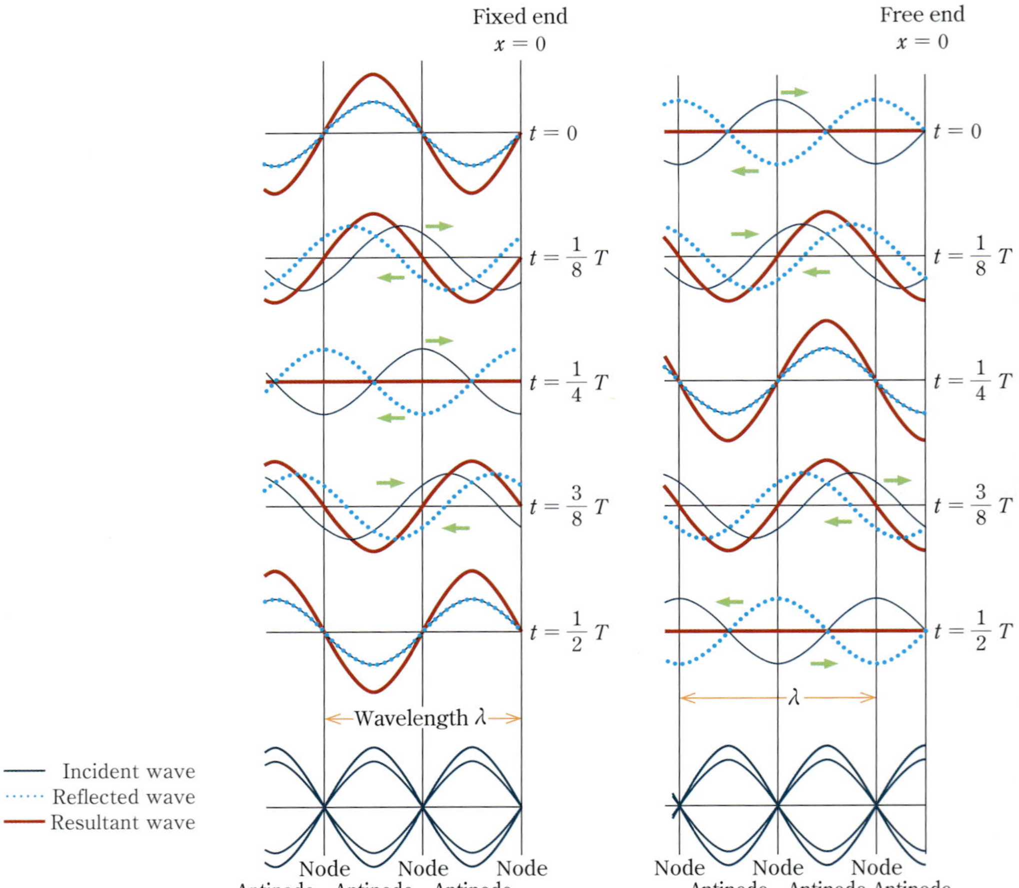

Fig. 12-18 Standing wave

incident wave, ······ reflected wave, and —— resultant wave). From these wave forms it can be understood that as shown in the bottom of the figure the **vibration is done at a fixed amplitude that depends on the position**. Like these resultant waves, waves which are formed by a superposition of two sine waves traveling in opposite directions at the same wavelength, frequency and amplitude, and which vibrate at the same location and do not travel are called **standing waves** (or stationary waves). The locations where the amplitude of standing waves is maximum are called **antinodes**, and the locations where no vibration occurs are called **nodes**. As is apparent from Fig. 12-18, the fixed end is a node and the free end is an antinode. In contrast to standing waves, waves that travel are called **traveling waves**. The distance between adjacent nodes and between antinodes of the standing waves, is half the wavelength of the incident wave and the reflected wave.

Characteristic vibration of the string As shown in Fig. 12-19, when we flip a string of length L of a violin, generally the superposed vibration of the standing wave of wavelength λ with nodes at both ends and n antinodes is obtained. The wavelength λ satisfies the relation $n\lambda = 2L (n = 1, 2, \cdots)$, or

$$\lambda_n = \frac{2L}{n} \qquad (n = 1, 2, \cdots). \tag{12.31}$$

The vibration of standing waves is called the **characteristic vibration** and its frequency is called the **characteristic frequency**. Since $f_n \lambda_n = v$, using equation (12.13) the frequency f_n of the standing wave with n antinodes is

$$f_n = \frac{n}{2L}\sqrt{\frac{S}{\mu}} \qquad (n = 1, 2, \cdots). \tag{12.32}$$

The vibration of $\lambda_1 = 2L$ at $n = 1$ is referred to as the **fundamental vibration** and vibrations of $\lambda_2 = L$ at $n = 2$, $\lambda_3 = \frac{2}{3}L$ at $n = 3$, $\cdots$ are called the **second, third, harmonic vibrations**, respectively. Vibrations other than the fundamental vibration are called harmonic vibrations or overtone vibrations.

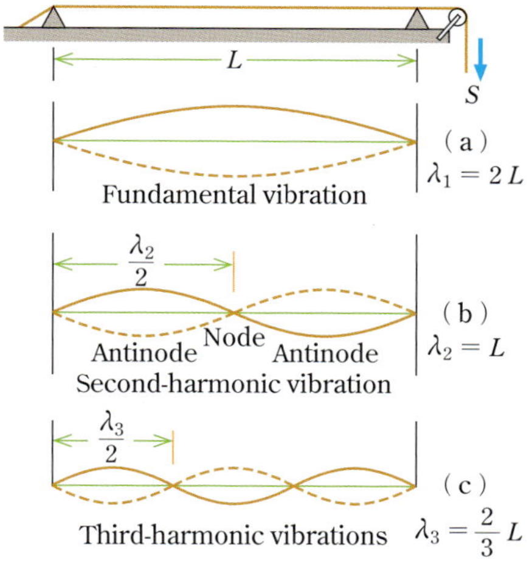

Fig. 12-19 Characteristic vibration of the string

Example 1 A piano wire with a length of 50 cm and a mass of 5 g is stretched at a tension of 400 N. Calculate the fundamental frequency.

Solution $\mu = \frac{5\times10^{-3}\,\text{kg}}{0.5\,\text{m}} = 10^{-2}\,\text{kg/m}$. Accordingly

$$f_1 = \frac{1}{2L}\sqrt{\frac{S}{\mu}} = \frac{1}{2\times(0.5\,\text{m})}\sqrt{\frac{400\,\text{N}}{10^{-2}\,\text{kg/m}}}$$
$$= \frac{1}{1\,\text{m}}\sqrt{\frac{(200)^2\,\text{kg}\cdot\text{m/s}^2}{\text{kg/m}}} = 200\,\text{s}^{-1}$$
$$= 200\,\text{Hz}.$$

For reference Mathematical derivation of the wavelength of the characteristic frequency of the string (12.31)

Assume that both ends of a string placed along the x-axis are at $x = 0$ and $x = L$. These points are the fixed ends, so the relation $y(0, t) = y(L, t) = 0$ is satisfied. Adding the condition $y(L, t) = 0$ to the equation (12.28) satisfying the relation $y(0, t) = 0$, we can derive the condition as

$$\sin\frac{2\pi L}{\lambda}\cos\left(\frac{2\pi vt}{\lambda}+\alpha\right)=0. \qquad (12.33)$$

In order that the above condition is satisfied at an arbitrary time, the following condition is required :

$$\sin\frac{2\pi L}{\lambda}=0 \qquad \therefore \quad \frac{2\pi L}{\lambda}=n\pi \qquad (n \text{ is an integer}). \qquad (12.34)$$

For this reason, in the case of a string of length L fixed at both ends, only the waves with the wavelength

$$\lambda_n=\frac{2L}{n} \qquad (n=1,2,\cdots) \qquad (12.35)$$

are transmitted.

Therefore, from equations (12.28) and (12.35), the vibrations allowed for this string are

$$y(x,t)=A_n\cos\left(\frac{n\pi vt}{L}+\alpha_n\right)\sin\frac{n\pi x}{L} \qquad (n=1,2,\cdots) \qquad (12.36)$$

and those obtained by superposing these vibrations. That is

$$y(x,t)=\sum_{n=1}^{\infty}A_n\sin\frac{n\pi x}{L}\cos\left(\frac{n\pi vt}{L}+\alpha_n\right) \qquad (12.37)$$ [*1]

where A_n and α_n are arbitrary constants.

Equation (12.37) is, under the boundary condition

$$y(0,t)=y(L,t)=0 \qquad (x=0 \text{ and } L \text{ are the fixed ends}) \qquad (12.38)$$

the general solution of the wave equation :

$$\frac{\partial^2 y}{\partial x^2}=\frac{1}{v^2}\frac{\partial^2 y}{\partial t^2}. \qquad (12.39)$$

*1 As in the case of the wave form generated for a string of length L, an arbitrary continuous function $y(x)$ which is defined in the region $0 \leqq x \leqq L$ and also satisfies the boundary condition at $x=0$ and $x=L$

$$y(0)=y(L)=0$$

can be expressed in terms of superposition of the sine function

$$\sin\left(\frac{n\pi x}{L}\right) \qquad (n=1,2,3,\cdots)$$

multiplied by a constant A_n, namely in the form of infinite series

$$y(x)=\sum_{n=1}^{\infty}A_n\sin\left(\frac{n\pi x}{L}\right).$$

This series is referred to as the Fourier series and the expansion into Fourier series is called the Fourier expansion.

12.6 Sound waves

Learning objective To deepen the understanding of the nature of waves by taking the example of sound waves which are the longitudinal waves traveling through the air and are familiar to our daily life. To understand the Doppler effect applied to measure the speed of objects in motion.

When sound sources (sounders) such as drum membranes and piano strings vibrate, the surrounding air also repeats compression and expansion, so compression waves (longitudinal waves) are transmitted through the air. They are sound waves. When sound waves enter the human ear they vibrate the tympanic membrane and are heard as sounds by the auditory organ[*2]. Sounds are heard in water and through thin walls, so we can recognize that sound waves travel through liquids and solids. However, in vacuum where no substance exists, sound waves are not transmitted.

*2 The frequency of sounds that can be heard by humans (audible sounds) is in the range of approximately 20 to 20 000 Hz. Sounds whose frequency is higher than audible sounds are called **ultrasonic waves**. The propagating speed of ultrasonic waves is the same as the speed of ordinary sound waves.

Speed of sound waves The speed of sound waves in the air is independent of the atmospheric pressure and frequency but varies with temperature. According to the experimental results at around 0 °C, the speed of sound waves in the air at a temperature of t [°C] and at 1 atmospheric pressure is represented as[*3]

*3 In this book the speed of sound waves in the air is expressed by the symbol V.

$$V = 331.45 + 0.607\,t\ [\mathrm{m/s}]. \tag{12.40}$$

When the temperature is 14 °C, the speed of the sound waves is about 340 m/s.

According to the theoretical calculation, the speed of sound waves passing through an ideal gas, with molecular weight M, ratio of the molar heat capacity at constant pressure and the molar heat capacity at constant volume[*1] $\dfrac{C_p}{C_v} = \gamma$ and absolute temperature T, is

$$V = \sqrt{\frac{\gamma RT}{M}} \quad \text{(ideal gas)}. \tag{12.41}$$

(R is the gas constant[*2]) (see Problem for exercise 15, B5). Assume that the air is an ideal gas. Since $M = 28.8$ g/mol, $\gamma = 1.40$, and $R = 8.31$ J/(mol·K), at 300 K, ie, at 26.85 °C, the speed of sound wave is theoretically

$$V = 348\ \mathrm{m/s} \qquad (T = 300\ \mathrm{K} = 26.85\ ^\circ\mathrm{C}).$$

As seen from equation. (12.41), the speed of sound wave in the gas increases in proportion to the square root of absolute temperature T, regardless of pressure and frequency.

The speed of sound wave in liquids is, setting aside a few exceptions, 1000–1500 m/s.

The equation for the speed of the longitudinal wave traveling in solids has been obtained in Section **12.2**.

Sound speeds in some substances are shown in Table 12-1.

Sound waves are propagated through the medium at a constant speed, and they do not depend on the moving speed of the sound source. Therefore, when the wind is blowing, the sound is propagated quickly downwind.

*1 As for the molar heat capacity C_p at constant pressure and the molar heat capacity C_v at constant volume, refer to Section **15.2**.

*2 As for the gas constant R, refer to Section **14.3**.

Table 12-1 Sound Speeds

Substance	Density (0 °C) [kg/m^3]	Speed of sound (0 °C) [m/s]
Air (dry) (1 atm)	1.2929	331.45
Hydrogen (1 atm)	0.08988	1269.5
Distilled water (25 °C)	1000	1500
Sea water (20 °C)	1021	1513
Mercury (25 °C)	1.37×10^4	1450
Aluminum[1)]	2.67×10^3	6420
Iron[1)]	7.86×10^3	5950

1) speed of the longitudinal wave in free solid

Vibration of the air column When we put out lip on the edge of the test tube and blow it strongly, the tube makes a sound characteristic to it. This is because standing waves of sound waves (longitudinal waves) are generated in the air column in the tube (or air-filled tube). Sound waves are reflected in both the closed end of the tube (closed end) and the open end of the tube (open end). At the closed end, the gas cannot vibrate in the direction of the pipe, so the closed end becomes a fixed end and works as a node of the standing wave. At the open end, the gas can move freely, so the open end becomes a free end and works as an

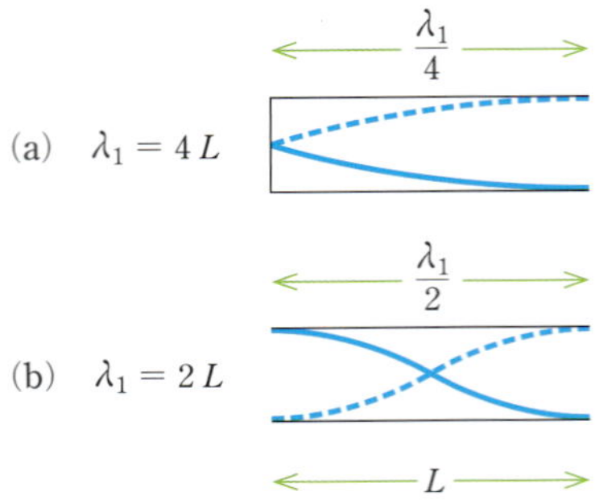

Fig. 12-20 Characteristic vibration of the air column

(a) Closed tube : The frequency of the harmonic vibration is an odd multiple of the fundamental frequency.

$$\lambda_n = \frac{4L}{2n-1}, \qquad f_n = \frac{(2n-1)V}{4L}$$

(b) Open tube : The frequency of the harmonic vibration is an integer multiple of the fundamental frequency.

$$\lambda_n = \frac{2L}{n}, \qquad f_n = \frac{nV}{2L}$$

$$n = 1, 2, 3, \cdots$$

antinode of the standing wave. In reality, since the sound waves are emitted from the open end to the outside of the tube, the position of antinodes of the standing waves deviates slightly from the open end and the aninodes locate slightly outside the open end. This deviation (**open end correction**) ΔL is proportional to the radius r of the tube, and $\Delta L \approx 0.6r$ for a thin tube.

A tube whose one end is closed is called a **closed tube**, and a tube whose both ends are open is called an **open tube**. Fig. 12-20 shows the standing waves of the fundamental vibration generated in the closed and open tubes.

Question 3 Show that the wavelength λ_n and frequency f_n of the characteristic vibration with n nodes generated in the air column are given as below.

In the case of closed tube $$\lambda_n = \frac{4L}{2n-1}, \quad f_n = \frac{(2n-1)V}{4L} \qquad (12.42)$$

In the case of open tube $$\lambda_n = \frac{2L}{n}, \quad f_n = \frac{nV}{2L}. \qquad (12.43)$$

Fig. 12-21 Noise

Unit of sound pressure level dB

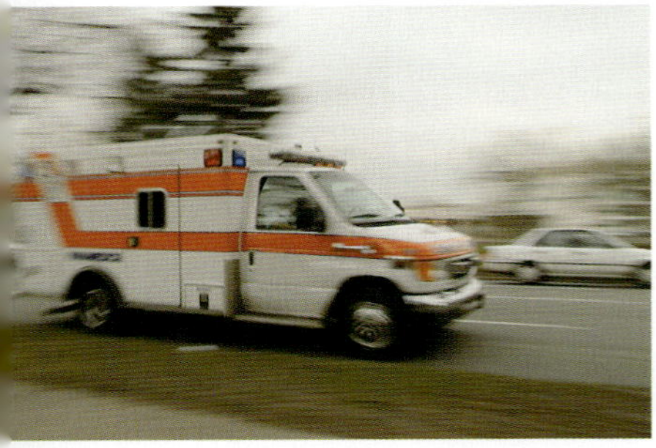

Fig. 12-22 An ambulance sounding its siren

Sound pressure and sound pressure level As a value representing quantitatively the magnitude of sound wave, the effective value of P_e is used. P_e is obtained by dividing by $\sqrt{2}$ the maximum value of Δp $(= p - p_0)$, which is the pressure difference between. air pressure p and static pressure p_0.

The effective value P_{e0} of the minimum sound pressure at which a young person with good ears can hear a 1 kHz sound is roughly

$$P_{e0} = 2 \times 10^{-6}\ \mathrm{N/m^2}. \qquad (12.44)$$

The sound volume that we feel through our auditory organ is considered to be close to the logarithmic scale of sound pressure. Based on this fact the effective value of sound pressure as a quantity representing the sound volume is defined from the ratio of the effective values of P_e and P_{e0} as

$$L_p = 20 \log_{10} \frac{P_e}{P_{e0}} \qquad (12.45)$$

The quantity L_p is called the **sound pressure level** and its unit is decibel (symbol dB).

The sound pressure of sounds that we commonly hear is often about $0.1\ \mathrm{N/m^2}$ and this corresponds to $20 \log_{10} 5000 = 74$ dB. Increase in the sound pressure level of 20 dB means that the sound pressure increases tenfold. Although the sound volume perceived by the human sense is mainly related to the magnitude of the sound pressure of the sound wave, it is also related to the sound frequency and the discussion on this issue is not so straightforward.

Question 4 Show that the sound pressure of the sound with 74 dB is approximately 10^{-6} atm.

Doppler effect When a police car running in the lane on the opposite side of the expressway with sounding the siren passes by, the pitch of the siren sound suddenly drops. The pitch (frequency) of the sound heard when one or both of the sound source and the person hearing the sound are moving is generally different from the frequency of the sound source. This phenomenon is called the **Doppler effect**, because Doppler pointed out in 1842 the possibility of this phenomenon for light and sound waves.

For simplicity, we consider the case where the velocity $\boldsymbol{v}_S$ of the sound source S and the velocity $\boldsymbol{v}_L$ of the person (observer) L hearing the sound are on a straight line as shown in Fig. 12-23. The velocity of the sound source and the observer is expressed by the symbols v_S and v_L which can take positive or negative values, and the positive sign is assigned when the sound source and the observer are facing in the direction of approaching each other (in the case of Fig. 12-23). It is assumed that the air of sound medium is motionless in a windless condition.

In Fig. 12-23 the sound source which is at point A at time $t=0$ moves the distance $v_S t$ in time t and comes to point B. The sound travels in the medium (air) at a constant speed V (the sign is always positive) relative to the medium regardless of the motion of the sound source and the observer, so the wave front of the sound which the sound source emits from point A at $t=0$ becomes a spherical surface of radius Vt centered at point A when the time t passes. Let the frequency of the sound source be f_S. Then $f_S t$ waves that the sound source emits between the two points A and B enter the section BD of the length

$$\overline{\mathrm{BD}} = (V - v_S)t.$$

Therefore, the wavelength λ' of the sound wave is

$$\lambda' = \frac{(V-v_S)t}{f_S t} = \frac{V-v_S}{f_S}. \tag{12.46}$$

This is the wavelength of the sound wave that travels through the medium when the sound source moves. As is clear from Fig. 12-23, in the section BD, the waves are dense, so the wavelength is short. Because the observer is moving at a speed v_L relative to the medium, the relative speed of the sound wave to the observer is $V+v_L$. If this is divided by the wavelength λ' of equation (12.46), the frequency f_L of the sound wave observed by the observer can be obtained as follows :

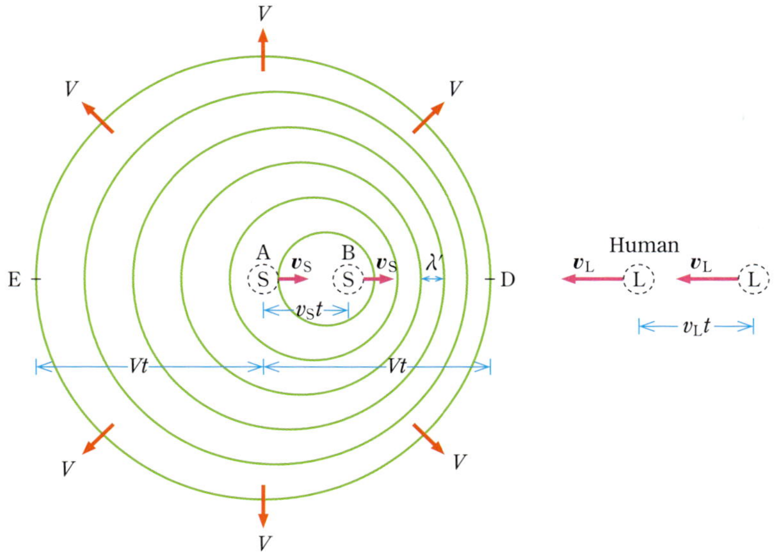

Fig. 12-23 Doppler effect

$$f_L = \frac{V+v_L}{\lambda'} = \frac{V+v_L}{V-v_S} f_S. \tag{12.47}$$

When the direction of movement of one or both of the sound source and the observer is opposite (in the direction of leaving away) to that in Fig. 12-23, the f_L corresponds to the situation in which the sign of one or both of v_S and v_L is negative in equation (12.47). Thus, it can be seen when the observer approaches the sound source ($v_L > 0$) and also when the sound source approaches the observer ($v_S > 0$), the frequency f_L observed by the observer is higher than the frequency f_S of the sound source.

When the air (medium) moves at a speed v in the direction from the sound source to the observer (when the wind is blowing), it is sufficient to replace V in (12.47) with $V+v$, namely

$$f_L = \frac{V+v+v_L}{V+v-v_S} f_S. \tag{12.48}$$

When the wind is blowing from the observer in the direction of the sound source, replace v with $-v$.

Example 2 | Doppler effect of the reflected sound from the object in motion As shown in Fig. 12-24, directing at a car moving at a constant velocity $\boldsymbol{v}_r$ on a straight road, the ultrasonic wave of frequency f_S was emitted from an ultrasonic source S installed on the ground near the road. The ultrasonic wave reflected by the car is detected by a receiver installed on the ground near the sound source. Find the frequency f_L of the reflected wave detected. The position of the car when it reflected the detected reflected wave is R. Assume that $\overrightarrow{RS}$ and $\boldsymbol{v}_r$ are almost parallel and it is windless*.

Solution The frequency f_R of the ultrasonic wave measured with a detector installed on a car running at v_r is given by equation (12.47) with $v_S = 0$ and $v_L = v_r$:

$$f_R = \frac{V+v_r}{V} f_S.$$

The frequency of the reflected sound that the car emits is also f_R.

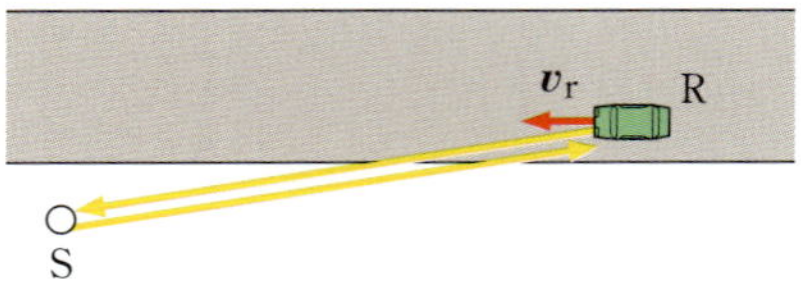

Fig. 12-24

Because the sound-reflecting source approaches at a speed v_r for a detector at rest, substituting the relations $v_S = v_r$ and $v_L = 0$ into equation (12.47) gives

$$\therefore \quad f_L = \frac{V}{V-v_r} f_R = \frac{V+v_r}{V-v_r} f_S. \tag{12.49}$$

If the angle between $\overrightarrow{RS}$ and $\boldsymbol{v}_r$ is θ, it is sufficient to replace v_r in equation (12.49) with $v_r \cos\theta$.

From the measured value of the difference in frequency $\Delta f = f_L - f_S$

$$\Delta f = \frac{2v_r}{V-v_r} f_S \approx \frac{2v_r}{V} f_S$$

between the reflected sound and the emitted sound, the speed v_r of the approaching object can be obtained.

In an ultrasonic blood flow meter, ultrasonic waves are reflected by red blood cells in blood vessels, and the Doppler effect is used to measure the flow speed of blood. The sound speed in the blood vessel is $V = 1570$ m/s.

* Commercially available speed guns use the Doppler effect of electromagnetic waves (microwaves) of 2.4×10^{10} Hz and from the relation

$$\Delta f = f_L - f_S \approx \frac{2v_r}{c} f_S$$

the speed of the approaching object is obtained. c is the speed of light. If the ultrasonic waves are used in a speed gun, the sound speed V varies with temperature and the measured results is also affected by the wind.

12.7 Group velocity and beat

Learning objective To become able to calculate the frequency of beat.

A sine wave is an infinitely long wave. The actual length of wave is finite, and waves with finite length are called **wave packets**. A wave packet is a resultant wave combining an infinite number of sine waves with different frequencies.

For simplicity let us consider two waves of equal amplitude traveling in the positive direction of the x-axis :

$$\left.\begin{aligned} y_1 &= A \sin(\omega_1 t - k_1 x) \\ y_2 &= A \sin(\omega_2 t - k_2 x). \end{aligned}\right\} \tag{12.50}$$

ω_1 and ω_2 ($\omega = 2\pi f$) are angular frequencies, and k_1 and k_2 $\left(k = \dfrac{2\pi}{\lambda}\right)$ are wave numbers*. Using the relation of trigonometric function

$$\sin a + \sin b = 2 \sin \frac{a+b}{2} \cos \frac{a-b}{2},$$

the resultant wave of the two waves y_1 and y_2 is, with the help of superposition principle, as follows :

$$\begin{aligned} y(x, t) &= A\{\sin(\omega_1 t - k_1 x) + A \sin(\omega_2 t - k_2 x)\} \\ &= \left[2A \cos\left\{\left(\frac{\omega_1-\omega_2}{2}\right)t - \left(\frac{k_1-k_2}{2}\right)x\right\}\right] \\ &\quad \times \sin\left\{\left(\frac{\omega_1+\omega_2}{2}\right)t - \left(\frac{k_1+k_2}{2}\right)x\right\}. \end{aligned} \tag{12.51}$$

In Fig. 12-25 are drawn the two waves y_1 and y_2 and the resultant wave $y = y_1 + y_2$. The vibration shown in Fig. 12-25 (b) and in equation (12.51) lasts for an infinitely long time, so the length of wave produced by this vibration is not finite. However, consider one aggregated wave enclosed by the dashed lines in Fig. 12-25 (b) as a wave packet, which is expressed in bracket [ ] in equation (12.51). What is in the bracket [ ] shows that the aggregated wave moves in space at a speed

$$v = \frac{\omega_1 - \omega_2}{k_1 - k_2}. \tag{12.52}$$

If the difference between ω_1 and ω_2 is small, equation (12.52) is reduced to

$$v = \frac{d\omega}{dk}. \tag{12.53}$$

* The reciprocal $\dfrac{1}{\lambda}$ of the wavelength λ is the number of waves between unit lengths, so it is called the **wave number**, but here 2π times $\dfrac{1}{\lambda}$, namely $k = \dfrac{2\pi}{\lambda}$ is called the wave number.

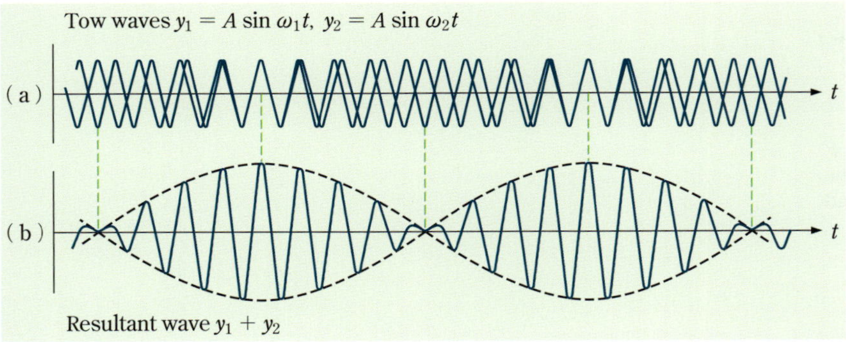

Fig. 12-25

This is the propagating speed of the wave packet and is referred to as the **group velocity**. In contrast, the traveling speed $v = \dfrac{\omega}{k} = f\lambda$ of each sine wave is called the **phase velocity**.

Up to this point, the wave speed has been considered to be constant regardless of the wavelength. However, in general, the wave speed often varies depending on the wavelength. The case where a wave whose wavelength is shorter in comparison with the depth of water travels on the water surface is such an example. In such cases, it can be shown that when a wave packet of finite length is formed by superposing an innumerable number of sine waves, the propagating speed of the wave packet is different from the phase velocity, and the length of wave packet gradually increases with time and the amplitude decreases gradually. It is at the group velocity that the waves carry energy.

Media showing differences in wave speed depending on the wavelength are called dispersive media, and the phenomena that occur due to this property are generally called dispersion. The origin of the term dispersion is the dispersion of light by prisms, which is an issue of study in the next chapter.

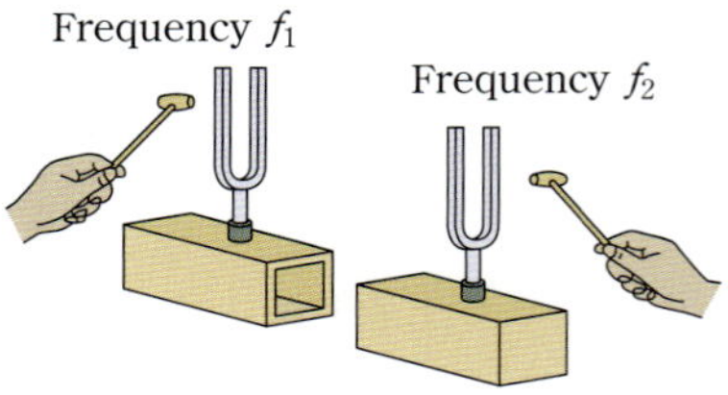

Fig. 12-26

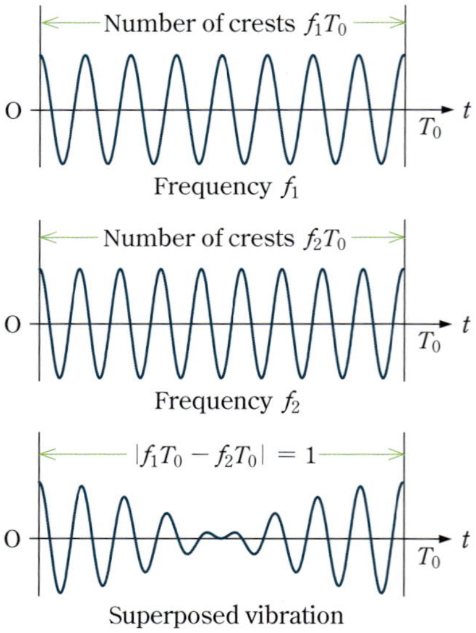

Fig. 12-27 The beat frequency F resulting from the superposition of the two vibrations of frequencies f_1 and f_2 is $F = |f_1 - f_2|$.

Beat If we strike two tuning forks of approximately the same frequency simultaneously (Fig. 12-26), we will hear a sound like roar at a frequency that is neither of them. This **beat** (roar) is a phenomenon that occurs because the strength of the superposed sound of two sounds increases or decrease.

Let the frequencies of the two tuning forks be f_1 and f_2. When the two vibrations superpose, the resultant vibration of the air repeats the periodical strong-weak change as shown in Fig. 12-27. This superposed vibration is heard as a beat sound. If the period of the beat is T_0, the number of crests f_1T_0 of frequency f_1 vibration and the number of crests f_2T_0 of frequency f_2 vibration in one cycle T_0 differ by just one, so the relation $|f_1T_0 - f_2T_0| = 1$ holds. Since the number F of beats per second is $F = \dfrac{1}{T_0}$, it follows that

$$F = |f_1 - f_2|. \tag{12.54}$$

The number of beats per unit time is equal to the difference between the frequencies of the two sounds.

Problems for exercise 12

A

1. A wave motion is expressed as

$$y = (3\text{ cm})\sin 2\pi\left(\frac{t}{0.2\text{ s}} - \frac{x}{20\text{ cm}}\right)$$

Find the followings :
(1) amplitude, (2) frequency,
(3) wavelength, (4) wave speed.
The units of x, y, and t are cm, cm, and s, respectively.

2. **Shock wave** When the wave source moves in the medium at a speed v greater than the speed V of the wave, the wave front becomes a conical surface with the wave source at the top. This wave front travels with great energy, so when it hits an obstacle it gives a large impact to the obstacle and is called a shock wave. The phenomenon that is caused by the instantaneous rise in pressure when the shock front that the supersonic aircraft creates in supersonic flight reaches the ground is known as the sonic boom ($M = \frac{v}{V}$ is called the **Mach number**). Assume that the vertex angle is 2θ (Fig. 1).
(1) Find $\sin\theta$.
(2) Find the speed v of the sound source when $V = 340$ m/s and $\theta = 30°$.

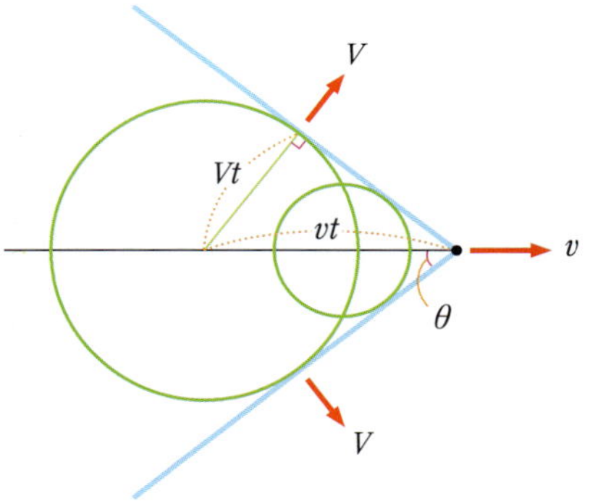

Fig. 1

3. In Fig. 12-18 (right), at $t = \frac{1}{2}T$ the wave vanishes. What happened to the wave energy?

4. When the violin string and the 440 Hz tuning fork were operated to sound at the same time, a 6 Hz beat was heard. The beat frequency decreased with a slight decrease in the string tension. Find the frequency of the strings.

5. Thunder was heard 3.0 seconds after seeing the lightning. Find the distance to thunderclouds. Assume the sound speed to be 340 m/s.

6. Two trains that ran at 72 km/h passed each other. One of the trains was emitting a beep sound of frequency 500 Hz. What frequency (in Hz) of sound did the passengers on the other train hear? Set the sound speed to 340 m/s.

B

1. Illustrate the vibration speed of the medium at the state give in Fig. 12-3.

2. Explain that when the displacement of the medium is expressed by equation (12.5), the vibration speed v of each point of the medium is expressed as

$$v = \frac{2\pi A}{T}\cos 2\pi\left(\frac{t}{T} - \frac{x}{\lambda}\right).$$

3. If the weight attached at the bottom of the cylinder in Fig. 2 is twisted, is the twist instantaneously transmitted to the top of the cylinder?

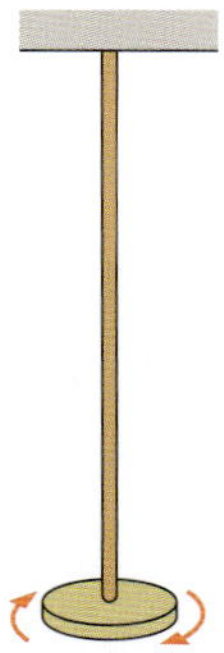

Fig. 2

4. The maximum value P of the sound pressure that human ears can tolerate is about 28 Pa.
(1) How much is this sound pressure in dB?
(2) How much is this sound pressure in atm?

Light

Current information-communication technology is based on the application of electromagnetic waves including light. The waves studied in the previous chapter are phenomena in which mechanical vibrations are transmitted as waves in elastic bodies, strings, air, etc. On the other hand, electromagnetic waves are phenomena in which the vibrations of electric and magnetic fields are transmitted in the form of waves.

Human visual organs perceive electromagnetic waves within a wavelength of $(3.8\text{–}7.7)\times10^{-7}$ m at a frequency of $(3.9\text{–}7.9)\times10^{14}$ Hz as light. Because the wavelength of light is so short compared to the size of common objects, the light appears to travel almost straight, so the occurrence of reflection, refraction, and diffraction is easily to realize. For this reason, in this chapter we will study the wave nature of light, including also an intention to deepen the understanding of reflection, refraction, and diffraction of waves that we learned in the previous chapter.

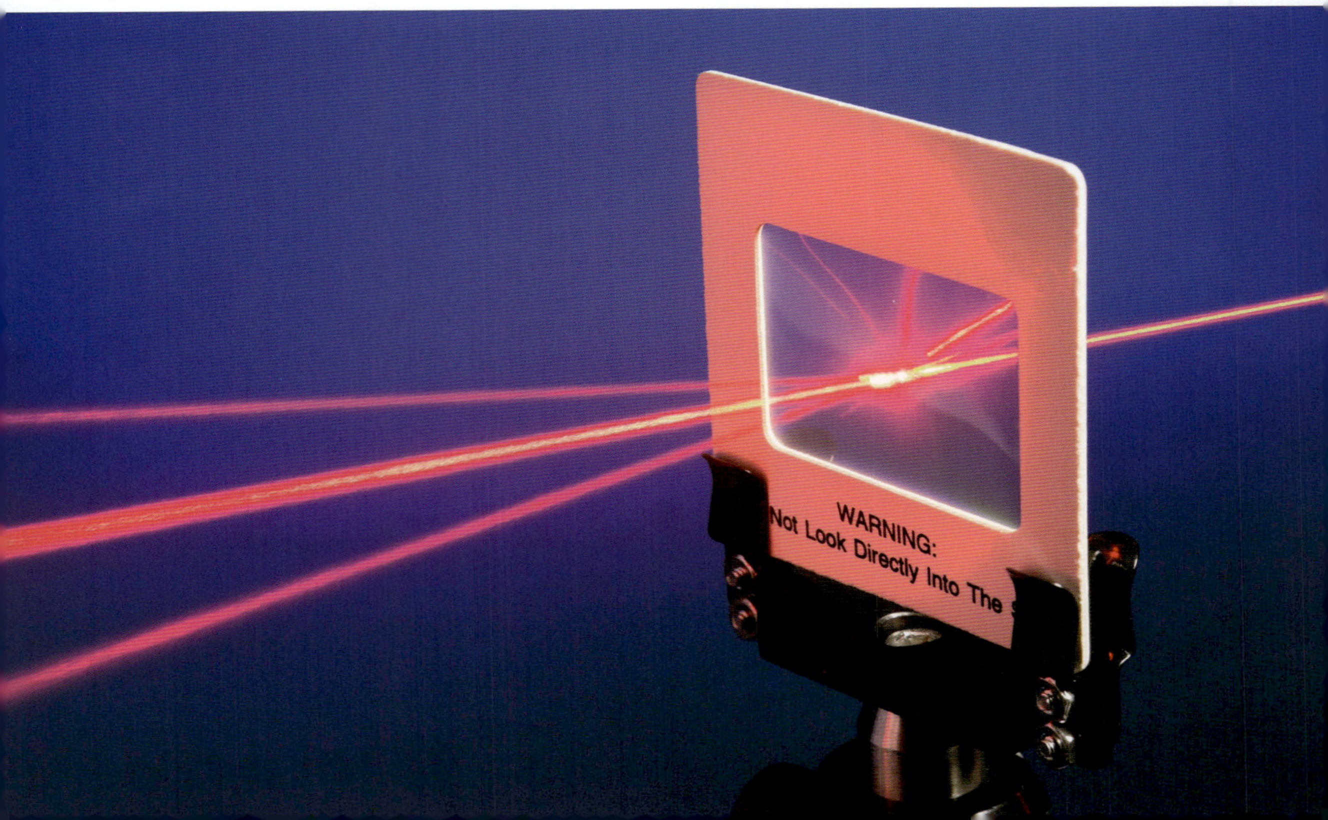

Laser light diffracted by a diffraction grating

Electric fields, magnetic fields, electromagnetic waves, etc. will be learned in the electromagnetism discussed in the latter half of this book, but electromagnetic waves can be understood intuitively as the phenomena of transmission in space of interacting vibration between electric and magnetic fields (refer to Figs. 22-4 and 22-5). The source of electromagnetic waves is a current that vibrates in the antenna and in atoms.

The major difference between waves of mechanical vibration which travel through a substance as a medium and electromagnetic waves is that mechanical waves are not transmitted in vacuum where no substance exists, but electromagnetic waves are transmitted in vacuum. In contrast to mechanical waves travelling at a constant speed relative to the medium, electromagnetic waves travel at a constant speed in vacuum with respect to any observer*. This issue will be studied in Chapter 23.

Another important difference is that electromagnetic waves travel through space as waves, but they behave not only as waves, but also exhibit particle-like properties when absorbed or emitted by substances. We will learn about this issue in Chapter 24 (see Fig. 24-8).

* Therefore, equation (12.47) of the Doppler effect for sound waves does not hold for light waves. The equation of the Doppler effect for a light wave in vacuum is an equation in which only the speed c of light and the relative velocity v between the light source and the observer appear :

$$f_{\mathrm{L}} = f_{\mathrm{S}}\sqrt{\frac{c+v}{c-v}} \quad \text{(in approaching)}$$

$$f_{\mathrm{L}} = f_{\mathrm{S}}\sqrt{\frac{c-v}{c+v}} \quad \text{(in leaving away).}$$

13.1 Reflection and refraction of light

Learning objective To understand the definition of 1 meter. To understand the law of refraction of light. To become able to explain under what condition the total reflection occurs and why the light is dispersed by the prism.

Speed of light in vacuum Because light travels about 3.0×10^5 km per second, in everyday life the light is felt to travel instantaneously and the measurement of the speed of light was difficult. However, the speed of light in air or in vacuum was measured by the method shown in B2 and B3 of Problems for exercise 22 and by other methods, and it has been confirmed that regardless of the wavelength, the motional states of light source and observer, the speed of light (symbol c) in vacuum takes the following value,

$$c = 2.997\,924\,58\times10^8\ \mathrm{m/s} \quad \text{(definition)} \tag{13.1}$$

(refer to Section **23.1**).

Speed of light in vacuum
$c = 2.997\,924\,58\times10^8$ **m/s**

Therefore, since 1983 this figure has been used as a definition of the speed of light. In addition, using the time that can be measured precisely using an atomic clock and the speed of light in vacuum that can be measured accurately, 1 m, the unit of length, is defined as the distance that light in vacuum travels in 1/200 792 458 second.

Reflection and refraction of light When a thin laser beam or a thin bundle of light passing through a narrow hole is incident on water or a transparent plastic plate, we can see the light beam is reflected and refracted at the boundary.

Fig. 13-1 Reflection and refraction of light

Light waves are reflected according to the law of reflection (12.24) :

$$\theta_{\mathrm{i}} = \theta_{\mathrm{r}} \quad \text{(angle of incidence = angle of reflection).} \tag{13.2}$$

Table 13-1 Refractive indices [for the yellow light of sodium (wavelength 5.893×10^{-7} m)]

Gas (0 °C, 1 atm)		Liquid (20 °C)		Solid (20 °C)	
Air	1.000292	Water	1.333	Diamond	2.42
Carbon dioxide	1.000450	Ethanol	1.362	Ice (0 °C)	1.31
Helium	1.000035	Paraffin oil	1.48	Glass	about 1.5

Remark : Refractive index varies slightly depending on the wavelength.

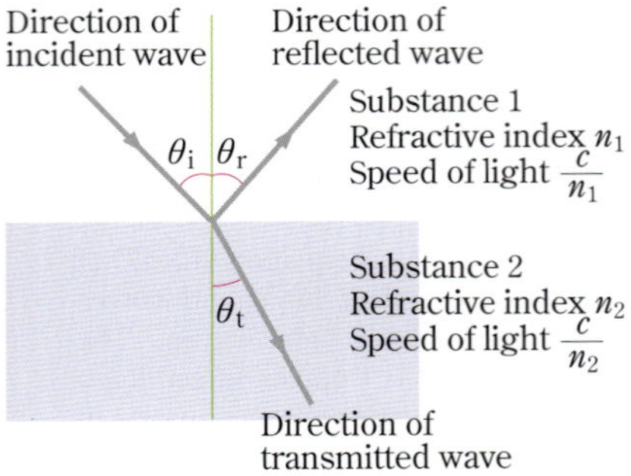

Fig. 13-2 Reflection and refraction of light

* A vacuum is not a substance but for convenience this style of expression is adopted.

When a light wave travels from a substance 1 having a light velocity c_1 to a substance 2 having a light velocity c_2 and is refracted at the boundary, the light follows the law of refraction (12.25) :

$$\frac{\sin\theta_i}{\sin\theta_t} = \frac{c_1}{c_2} = n_{1\to2} \qquad \text{(constant)} \tag{13.3}$$

(Fig. 13-2). The constant $n_{1\to2}$ is the refractive index of substance 2 with respect to substance 1. In the case of light, the refractive index $n_{\text{vacuum}\to2}$ of substance 2 with respect to vacuum is called the refractive index of the substance 2 and is denoted as n_2. The refractive index of vacuum is 1. Table 13-1 shows the refractive indices of several substances.

If substance 1 is a vacuum*, and the speed of light in vacuum is c, then from equation (13.3) the relation $\frac{c}{c_2} = n_2$ is derived. Expressing the speed of light in a substance with refractive index n as c_n, we have

$$c_n = \frac{c}{n} \qquad \text{(speed of light in a substance with refractive index } n\text{).} \tag{13.4}$$

According to the theory of relativity, the speed of information exchange cannot exceed the speed of light in vacuum, so the speed of light c_n in a substance is slower than the speed of light in vacuum. Accordingly, for the refractive index of any substance the relationship $n > 1$ holds.

The law of refraction tells us that when light is incident from substance 1 of refractive index n_1 to substance 2 of refractive index n_2, since the ratio of the speed of light is $\frac{c_1}{c_2} = \frac{c/n_1}{c/n_2} = \frac{n_2}{n_1}$, equation (13.3) is changed to

$$\frac{\sin\theta_i}{\sin\theta_t} = \frac{n_2}{n_1}. \tag{13.5}$$

When light from a substance of refractive index n_1 is incident on the boundary of another substance of refractive index n_2, the **reflection rate** (or **reflectivity**) R is

$$R = \left(\frac{n_2 - n_1}{n_2 + n_1}\right)^2. \tag{13.6}$$

This equation holds when the relative permeability μ_r of both media is equal to 1.

Question 1 Find the reflectivity when light is perpendicularly incident on a glass plate of the refractive index about 1.5 from the air.

Total reflection As in the case of light entering the air from water or glass, when the light travels from a substance with a high refractive index to a substance with a low refractive index, the refraction angle θ_t is greater than the incident angle θ_i. As the incident angle increases, it becomes larger than the **critical angle** (incident angle when the angle of refraction reaches 90°) θ_c given by the following formula :

$$\sin\theta_c = n_{1\to 2} = \frac{n_2}{n_1} < 1 \quad (\text{when } n_1 > n_2), \tag{13.7}$$

In this situation the refracted light disappears and the incident light is completely reflected (Fig. 13-4). This phenomenon is called the **total reflection**. Total reflection is a phenomenon found in all types of waves.

Fig. 13-3 A soap bubble

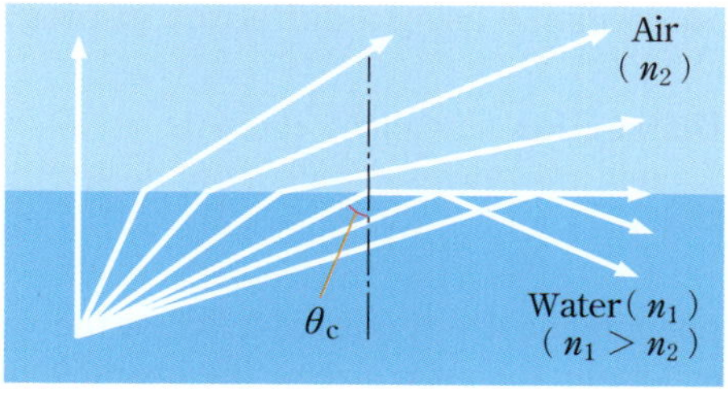

Fig. 13-4 Total reflection

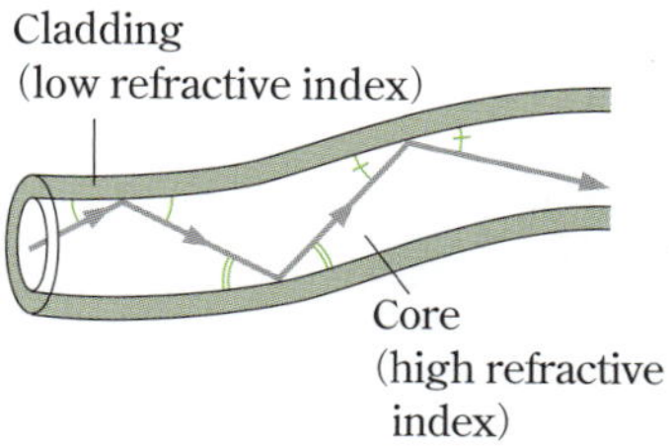

Fig. 13-5 Conceptual illustration of optical fiber

Optical fibers that transmit light to the distance are based on the total reflection of light. The thickness of an optical fiber made of an elongated fine glass wire is about the thickness of a human hair (about 100 μm). The refractive index of the central part (core) is higher than that of the outer (cladding). Therefore, light entering from one end of the optical fiber is transmitted to the other end without going out of the core (Fig. 13-5). Optical fibers are used for optical communications, and also used for endoscopes such as gastroscope.

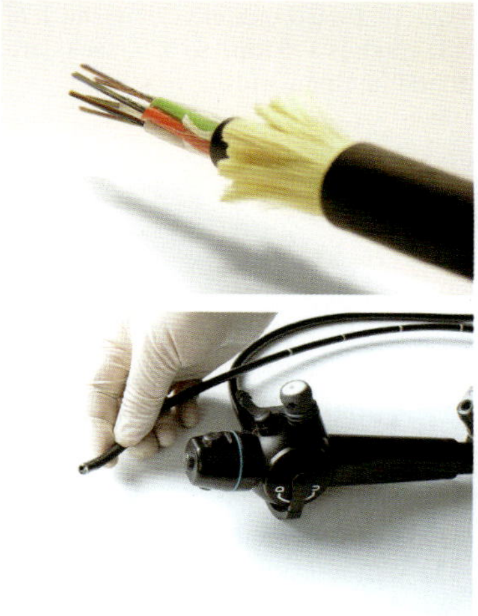

Fig. 13-6 Optical fibers (top) and a gastroscope (bottom)

Dispersion of light The refractive index of glass and water slightly depends on the wavelength of light, and the shorter the wavelength, the higher the refractive index. As shown in Fig. 13-7, when the sunlight passing through a slit is incident on a prism and is refracted and the emitted light is projected on a screen, a series of color patterns, in as cending order of refraction, from red, orange, yellow, green, blue, purple colors, that is, **spectra** are generated. This experiment demonstrates that the difference in color of light reflects the difference

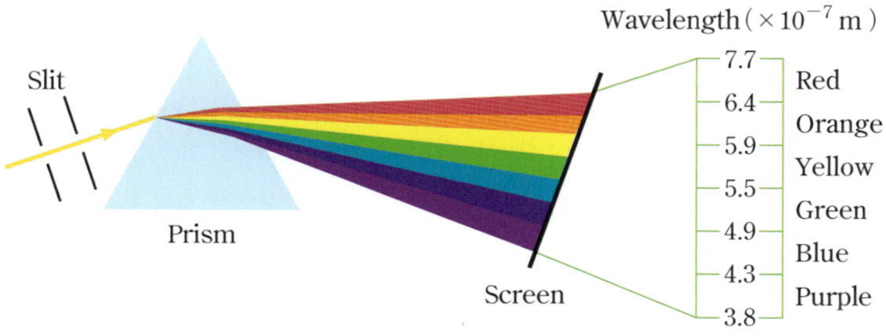

Fig. 13-7 Light dispersion and wavelength and color of spectrum

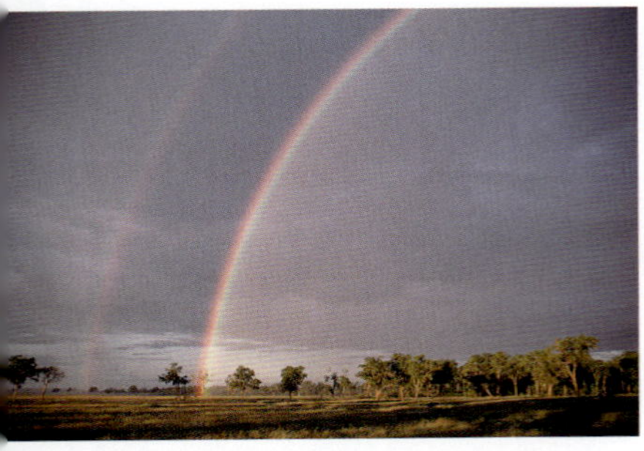

Fig. 13-8 The rainbow arises from the dispersion of sunlight by water droplets in the air

Fig. 13-9 Reflection and interference that a CD exhibits. The width of CD track is 1.6 μm, so 625 tracks are found in 1 mm width.

in wavelength. As just described, phenomenon of light splitting into components with different wavelength by refraction due to the difference in speed (refractive index) in substances depending on the wavelength is called the **dispersion** of light.

13.2 Diffraction and interference of light waves

Learning objective To be able to explain the basis of the wave theory of light that although the light travels straight, it travels through space as a wave. To understand the relationship between the size of holes and slits in light obstacle and the diffraction angle. To understand the principle of diffraction gratings to become able to explain the fact that the wavelength of a light wave can be determined using a diffraction grating.

Diffraction When there is an obstacle in the travelling path of waves, the waves can no more go straight, but they can go around into the area which would have been a shadowed area of the travelling waves. This phenomenon is called the **diffraction**. Diffraction phenomena are remarkable when the wavelength is about the same as or longer than the size of the obstacle or the hole in the obstacle.

The wavelength of light is $(3.8\text{–}7.7)\times 10^{-7}$ m, which is very short compared to the size of common objects. Accordingly, the diffraction is usually hard to notice, so the light appears to travel straight. However, when light from a point light source passes through a narrow slit with a width of about 0.01 mm or less, the light is diffracted to right and left, resulting in a series of bright-and-dark streaks. Thus, the light is diffracted, and we can recognize that the light travels as a wave.

Diffraction due to slit When the incident light beam is perpendicularly incident on the slit of width D, if the light travels only straight, the light will not be diffracted in any direction other than $\theta = 0$. However, as shown in Fig. 13-10 the intensily of light is $I(\theta) \neq 0$ in the direction $\theta \neq 0$. In other words, diffraction of light actually occurs. When the incident light is monochromatic light of wavelength λ, the light intensity on the screen far from the slit is given as*

$$I(\theta) = (\text{constant})\frac{\sin^2\left(\dfrac{\pi D}{\lambda}\sin\theta\right)}{\left(\dfrac{\pi}{\lambda}\sin\theta\right)^2}. \tag{13.8}$$

* The diffraction in this case is called the **Fraunhofer diffraction**. The diffraction when the distance between the slit and the screen is not so large is called the **Fresnel diffraction**.

Equation (13.8) and Fig. 13-10 show that as $\dfrac{\lambda}{D}$ increases, that is, as the slit width D decreases, the diffraction increases.

At an angle θ at which the difference in distance to both ends of the slit is an integer multiple of the light wavelength λ, namely

$$D\sin\theta = m\lambda \qquad (m = \pm 1, \pm 2, \cdots), \tag{13.9}$$

$I(\theta) = 0$ becomes dark. This is because the phases of the waves coming from each part of the slit take all phases from 0 to 2π and they overlap to cancel each other.

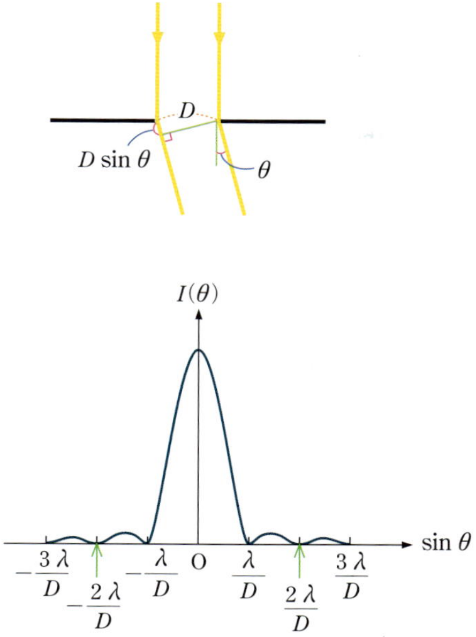

Fig. 13-10 Diffraction of light by a slit of width D

Diffraction due to a round hole Diffraction also occurs when monochromatic light of wavelength λ incident vertically to a round hole of radius R open on the obstacle plate. The diffraction angle θ is

$$\theta \lesssim 0.61\frac{\lambda}{R}. \tag{13.10}$$

Diffraction grating A diffraction grating is a device that separates light composed of waves of various wavelengths into monochromatic lights. The **diffraction grating** is obtained by cutting a large number of parallel grooves (gratings) at equal intervals on one side of a glass plate at a rate of 500 to 10000 per cm*. Diffuse reflection occurs in the groove part and it becomes opaque, so the transparent part (width D) between the grooves acts as a slit.

A parallel beam (wavelength λ) is vertically incident on the glass surface of the diffraction grating (spacing of grating d, number of gratings N) (Fig. 13-11). If the angle θ between the traveling direction of the transmitted light and the normal to the grating plane satisfies the following condition

$$d\sin\theta = m\lambda \qquad (m = 0, \pm 1, \pm 2, \cdots), \tag{13.11}$$

since the difference $d\sin\theta$ in the distance from the point P on the distant screen to the adjacent slit is an integer multiple of the wavelength λ, the phases of the light waves arriving from all the slits to the point P coincide, and the amplitude of the light wave at the point P is N times that of a single slit. Therefore, the intensity of the light wave at point P is N^2 times the intensity of a single slit and the point P becomes extremely bright.

Since the intensity of light passing through the entire diffraction

* A glass plate on which surface metals such as Al are vapor-deposited and grooves are formed at equal intervals is also used.

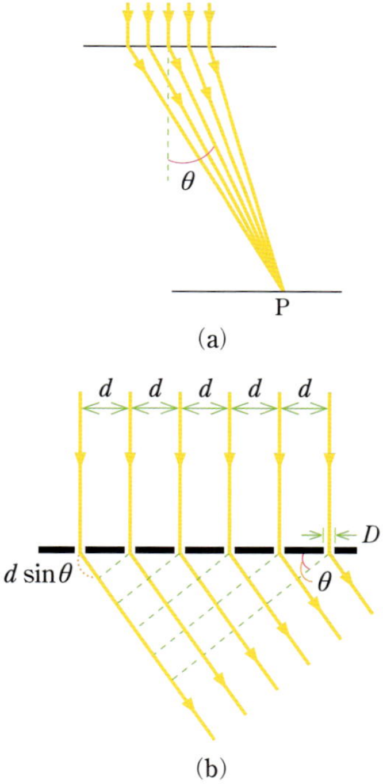

Fig. 13-11 Diffraction of light by a diffraction grating. If the distance from the diffraction grating to the screen is larger than Nd, the light collected at P can be considered parallel.

grating with N gratings is proportional to N, the width of the bright line narrows in inverse proportion to N, $\left(\frac{N}{N^2} = \frac{1}{N}\right)$. When the angle θ deviates slightly from the angle that satisfies equation (13.11), the light waves from many slits cancel each other out quickly, and the width of the bright line becomes extremely narrow. Therefore, the wavelength of light can be accurately determined by measuring the diffraction angle θ using a diffraction grating. If the wavelengths are different, the diffraction angles at which the diffracted lights enforce each other are different, so when the light of mixed waves with different wavelengths like sunlight is shed to the diffraction grating, it is resolved by diffraction to generate a spectrum.

Problems for exercise 13

A

1. In Fig. 1, in which direction does the person A feel the object B on the opposite side of the rectangular parallelepiped glass?

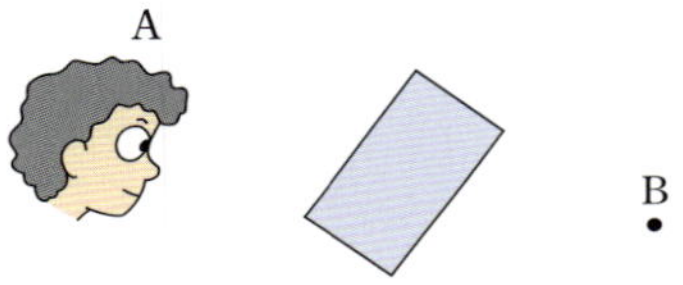

Fig. 1

2. Find the critical angle, when sound waves are incident on water ($V_2 = 1500$ m/s) from air ($V_1 =$ 340 m/s).
3. Find the critical angle of total reflection of diamond ($n = 2.42$) in the air.
4. One slit of width 10^{-5} m was illuminated with the light of wavelength 5×10^{-7} m. How wide is the width of the central diffracted light on the 1 m-apart screen?
5. When a monochromatic light with a wavelength of 0.5 μm was irradiated perpendicularly to the diffraction grating, the first bright line appeared in the direction making an angle of 30° with the normal. How many slits for diffraction grating are drawn in 1 cm?

B

1. How do fish in water observe the movement of the sun?
2. The light with a wavelength of 5.00×10^{-7} m was vertically incident on the circular hole of radius 10^{-4} m. How large is the size of the spots formed on the 1 m-apart screen?
3. The interference effect of the sound from the two speakers of stereo device is not usually detected. Why?

Heat

We know many facts about heat and temperature through our daily experiences. Temperature is one of the physical quantities of an object in thermal equilibrium. When a high temperature object is brought into contact with a low temperature object, the temperature of the high temperature object decreases, while the temperature of the low temperature object rises, and eventually the temperatures of the two objects become the same. This is a state of thermal equilibrium. At this time, we say that heat is transferred from the high temperature object to the low temperature object. What is then the entity of heat?

In this chapter, first, you will learn various facts related to heat and temperature, such as thermal equilibrium, phase and phase transition, heat capacity and specific heat capacity, thermal expansion, heat transfer (heat conduction, convection, thermal radiation), and then understand physical properties of gases from the standpoint of molecular theory.

Fig. 14-1 Melted pig iron will reach 1500–1600 °C.

14.1 Heat and temperature

Learning objective To have a deep understanding of physical quantities such as thermal equilibrium, temperature, heat, heat of transition, internal energy, heat capacity, specific heat capacity, molar heat capacity etc.

Thermal equilibrium and heat As a physical quantity that represents the warmth and coldness of an object, **temperature** is used. When a hot object is brought into contact with a low temperature object, such as when hot water is put into a cold tea cup, the high temperature object falls in temperature, and the low temperature object rises in temperature, so eventually the temperatures of the two objects become the same. At this time, it is said that heat has moved from the high temperature object to the low temperature object. Heat transfer stops when the temperatures of two objects in contact become the same. This state is called **thermal equilibrium,** and the two objects are said to be in thermal equilibrium with each other.

The following rule of thumb holds for thermal equilibrium.

When there are three objects A, B, and C, and if A and B are in thermal equilibrium and B and C are in thermal equilibrium, then, when A and C are in direct contact, they are at every case in thermal equilibrium.

This rule of thumb is referred to as the **zeroth law of thermodynamics**. At this time the temperatures of A, B, and C are the same. In addition, without bringing directly the objects A and C in contact, we can examine whether the temperatures of A and C are the same or not using the object B as a thermometer.

If the temperature changes, the internal state of substance also changes, causing changes in volume, pressure, etc. Also, at specific temperatures, phase transitions, namely solid (solid phase) $\rightleftarrows$ liquid (liquid phase), and liquid $\rightleftarrows$ gas (gas phase), such as melting, evaporation, solidification, and condensation occur. Based on these phenomena, a temperature scale is determined that can represents the temperature numerically. The **Celsius temperature scale** is a scale established by Celsius in which at 1 atmospheric pressure the freezing point and the boiling point of water are selected as to be 0 and 100 °C, respectively. The temperature scale using the volume change of a dilute gas (ideal gas) which follows Boyle-Charles' law is the absolute temperature scale (the unit of temperature is kelvin and the symbol is K)* (see pages 181 and 205).

The phase transition between gas and liquid below the critical temperature (see the next page) is accompanied by volume change and inflow heat and outflow heat (heat of evaporation, heat of condensation). These heats are called **transition heats**. A phase transition accompanied with inflow and outflow of transition heat is called a **first-order phase transition**. Solid-liquid and solid-gas phase transitions are always first order phase transitions. The heat of transition (heat of fusion, heat of vaporization) absorbed by a substance in the first-order

* The Celsius scale of temperature T_C and the absolute temperature T are related as

$$T_C = T - 273.15.$$

For example, 30 °C is 303.15 K. In this footnote, for simplicity T_C and T stand for only the numerical part.

Table 14-1 Various temperatures (°C) (Boiling points and melting points are at 1 atm.)

Core temperature of the Sun	about 1.55×10^7
Surface temperature of the Sun	about 5 800
Melting point of gold	1 064.18
Melting point of silver	961.78
Melting point of zinc	419.527
Melting point of tin	231.928
Boiling point of water[1)]	99.974
Triple point of water[2)]	0.01
Boiling point of nitrogen	−195.8
Boiling point of hydrogen	−252.87
Boiling point of helium	−268.934

1) In the International System of Units, the boiling point of water at 1 atm is no longer 100 °C (see p. 205).

2) Triple point is the state at which gas, liquid and solid phases coexist.

phase transition is the energy required for the molecules in the substance to overcome the intermolecular attraction to become free. In the reverse process, the substance releases the same amount of heat of transition (heat of solidification, heat of condensation) to the outside.

Phase and phase transition A diagram that shows in which phase, gas (gas phase), liquid (liquid phase) or solid (solid phase), a substance is at various temperatures and pressures is called a **phase diagram**. An example is Fig. 14-3. Regions expressed as Gas, Liquid, and Solid indicate the regions in which the corresponding phases are stable. The SG curve forming the boundary between the solid and gas regions corresponds to the coexistence of solid and gas. The LG curve forming the boundary between liquid and gas regions corresponds to the coexistence of liquid and gas. The SG curve is called the vapor pressure curve of solid, and the LG curve is called the vapor pressure curve of liquid. The SL curve corresponds to the coexisting state of solid and liquid. The point of intersection of the SL curve showing the boundary between the solid and liquid regions and the straight line of pressure p = constant indicates the melting point at pressure p. In the case of water, at 0.01 °C and 0.0060 atm, three phases of water, ice and water vapor are in equilibrium. Such points are called the **triple point***.

The LG curve ends at the rightmost point marked as the **critical point**. The temperature and pressure at critical point are called the critical temperature and the critical pressure, respectively. At temperatures above the critical temperature, even if the pressure is changed at a constant temperature, or at pressures above the critical pressure, even if the temperature is changed at a constant pressure, properties of the substance show the change between (low density and high compressibility) gas-like properties and (high density, low compressibility) liquid-like properties and discontinuous changes cannot be observed.

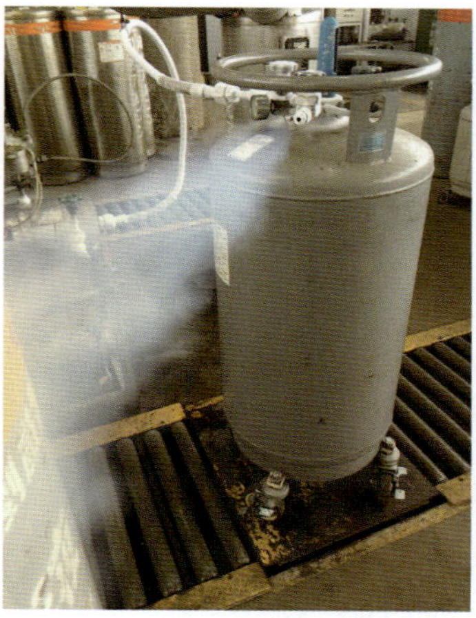

Fig. 14-2 A scene filling a cryogenic container with liquid nitrogen. The cold nitrogen gas at −196 °C makes the water vapor in a white and visible form in the air.

* The triple point of water is 273.16 K.

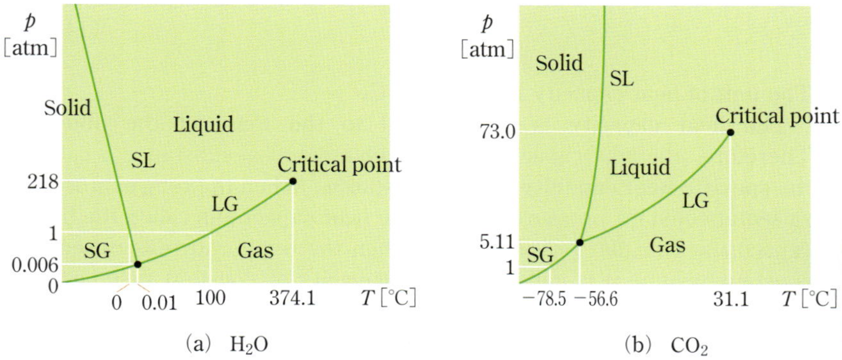

Fig. 14-3 Phase diagram

Heat and molecular motion, internal energy Using a gas or electric kettle we can heat and boil water. Also, when two pieces of hard wooden board are rubbed together the contact surface is heated. These facts indicate that chemical energy, electrical energy, and mechanical work have been converted into heat, and thus heat is a form of energy.

Fig. 14-4 A long slide. When you slide down the slide, your buttocks are heated.

All objects are composed of molecules, and in them molecules perform random motions called the thermal motion. The temperature of an object is a physical quantity that represents the magnitude of the thermal motion energy per molecule composing the object. Therefore, when the object is heated, the temperature rises because the energy applied from the outside becomes the energy of the thermal motion of molecules. If the thermal motion of molecules gets vigorous, the substance expands, and eventually the solid melts into a liquid, and the liquid evaporates into a gas.

In thermology, the sum of the kinetic energy of the thermal motion of molecules that make up the object and their potential energy is called the **internal energy** of the object*. And when energy is transferred from a high temperature object (or high temperature part) to a low temperature object (or low temperature part), this form of transferred energy is called **heat**.

* Macroscopic energies such as the energy of the center of gravity motion and the rotational motion around the center of gravity of the object as well as the gravitational potential energy of the object are not included in the internal energy.

Unit of amount of heat J
Practical unit of amount of heat cal
Unit of heat capacity J/K = J/°C

Heat capacity and specific heat capacity Because heat is a transferring form of energy, the unit of the amount of heat transferred (heat) is J. Historically, the amount of heat required to raise the temperature of 1 g water by 1 °C was taken as a unit of heat, which was called one calorie (cal). According to the current Measurement Act of Japan, use of calorie is admitted only to the amount of nutritional heat and the heat consumed in the basal metabolism and in these cases it is understood as

$$1\ \text{cal} = 4.184\ \text{J}. \tag{14.1}$$

Table 14-2 Specific heat capacities of substances [J/(g·K)]

Substance	Specific heat capacity
Iron (0 °C)	0.437
Cupper (0 °C)	0.380
Silver (25 °C)	0.236
Silicon (25 °C)	0.712
Sulfur (rhombic, 25 °C)	0.705
Water (15 °C)	4.19
Sea water (17 °C)	3.93
Methanol (12 °C)	2.5
Water vapor (100 °C) (constant pressure)	2.051
Air (20 °C) (constant pressure)	1.006
Hydrogen (0 °C) (constant pressure)	14.191

How much the temperature of an object rises when a certain amount of heat is given depends on the type and mass of the object. The amount of heat required to raise the temperature of an object by 1 °C is called the **heat capacity** of the object. If an amount ΔQ of heat is given and the rise in temperature is ΔT, the heat capacity C is given by

$$C = \frac{\Delta Q}{\Delta T}. \tag{14.2}$$

The unit of heat capacity is J/K = J/°C.

The heat capacity is proportional to the mass of the object. Therefore, the heat capacity of a certain amount of substance is called the **specific heat capacity** of that substance. When the temperature is raised at constant pressure the specific heat capacity is called the heat capacity at constant pressure, and when the temperature is raised at constant volume it is called the heat capacity at constant volume. In general, the specific heat capacity is defined as the heat capacity of 1 g of substance. In this case, "heat capacity" = "specific heat capacity"× "mass (gram number)". The heat capacity of 1 mol of substance containing about 6×10^{23} particles of molecule is called the **molar heat capacity**. Even for the same substance, the heat capacity varies with temperature. For example, the molar heat capacity of solid is about 25 J/(K·mol) at room temperature regardless of the substance but it takes a much smaller value at low temperatures (see Fig. 14-5).

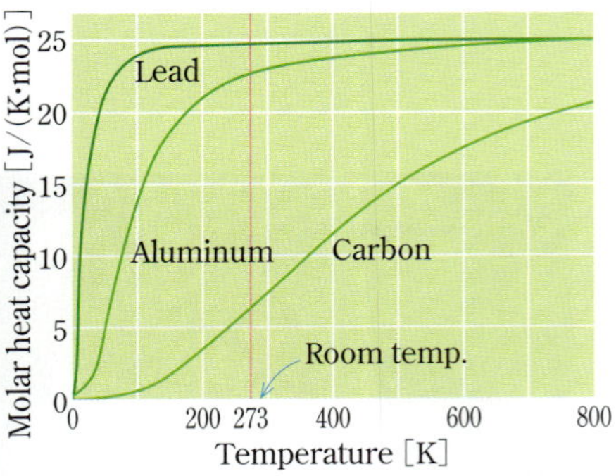

Fig. 14-5 Molar heat capacity of solid (Pb, Al, C). With the rise in temperature, the molar heat capacities of these three elements take a value of $3R \approx 25$ J/(K·mol) (see Section **15.2**).

Thermal expansion When substances are heated, their volumes commonly expand. In general, between the change in temperature $T \to T+\Delta T$ and the change in length $L \to L+\Delta L$ the relation

$$\Delta L = \alpha L \, \Delta T \tag{14.3}$$

holds. The constant α is called the **coefficient of linear expansion**. Between the change in temperature $T \to T+\Delta T$ and the change in volume $V \to V+\Delta V$ the relation

$$\Delta V = \beta V \, \Delta T \tag{14.4}$$

holds. The constant β is called the **coefficient of cubical expansion**. The coefficients α and β are related as

$$\beta = 3\alpha \tag{14.5}$$

(see Problems for exercise 14, B2).

The thermal expansion of water is special. Below 3.98 °C, the coefficient of cubical expansion of water is negative and water shrinks when the water temperature increases from 0 °C. At 3.98 °C the density becomes the highest value (1.000 g/cm^3) and with further increase in temperature, water expands. The cause of this behavior is that in ice, adjacent water molecules (H_2O) are bonded to each other by hydrogen bonds, so they are arranged in space leaving some room around them, so the density is small, but if the temperature rises and the ice changes into a liquid, hydrogen bonds between adjacent water molecules are broken and more molecules are packed in a narrow space*.

Table 14-3 Coefficients of linear expansion (20 °C)

Substance	α [K^{-1}]
Aluminum	2.31×10^{-5}
Copper	1.65×10^{-5}
Iron	1.18×10^{-5}
Platinum	8.8×10^{-6}
Glass (average)	$(8\sim10) \times 10^{-6}$
Glass (Pyrex)	2.8×10^{-6}

* The density 0.917 g/cm^3 of ice at 0 °C is smaller than the density 0.9984 g/cm^3 of water at 0 °C. The density of water is the highest at 3.98 °C, which is 0.999973 g/cm^3 more precisely.

14.2 Transfer of heat

Learning objectives To understand the mechanism of heat conduction, convection, and thermal radiation, which are the three types of heat transfer when there is a temperature difference between two objects or within one object. To have a deep understanding of Planck's law, Wien's displacement law, and Stefan-Boltzmann's law.

Heat conduction **Heat conduction** is the transfer of heat between contacting objects or within objects and the energy of thermal motion of the atoms in the high temperature part is transferred by the action of interatomic force to the next atom successively and ultimately reaches the low temperature part. In this case, atoms and ions do not move, but in metals, electrons can move, so metals generally have high thermal conductivity.

When two objects at temperatures T_1 and T_2 ($T_1 < T_2$) are connected by a rod of length L and cross-sectional area A, the heat flow H transferred through this rod is expressed as

$$H = \frac{\Delta Q}{\Delta t} = kA \, \frac{T_2 - T_1}{\Delta t}. \tag{14.6}$$

Here k is a proportionality constant called the thermal conductivity of

Table 14-4 Thermal conductivities

Substance	κ [W/(m·K)]
Aluminum	236
Copper	403
Stainless steel	15
Water (80 °C)	0.673
Wood (dry) (room temp.)	0.14-0.18
Paper (room temp.)	0.06
Soda glass	0.55-0.75
Air	0.0241

Otherwise mentioned, values are those at 0 °C.
Thermal conductivity of gases is nearly independent of the pressure.

this rod.

Air is a poor conductor of heat. Clothes capture air in the fabric, and the air acts as a heat insulator.

Convection In liquids and gases heat is transferred to some extent by heat conduction but most of the heat is transferred through the motion of the fluids themselves. The movement of fluid caused by the difference in density between the hot and cold parts is called **convection**.

Fig. 14-6 A flame of gas burner

Thermal radiation Electromagnetic waves such as light, infrared rays, and ultraviolet rays (see Chapter 22) are radiated from high-temperature objects, and these electromagnetic waves travel through a space and are absorbed by the low-temperature objects to transfer energy. Electromagnetic waves travel in vacuum at the speed of light $c = 3\times10^8$ m/s, so in the case of thermal radiation energy travels at the speed of light.

When iron is heated by an acetylene burner, it first turns red when the temperature rises and with further increase in temperature it starts to glow pale. Thus, although high-temperature objects radiate light, the radiated light color changes with temperature, and the radiated electromagnetic waves have shorter wavelengths as the temperature of the objects becomes higher (the wavelengths are shorter in the order, infrared → red light → purple light → ultraviolet light). Since gases heated to a high temperature radiate colored lights specific to the type of gas, we only consider here solids and liquids.

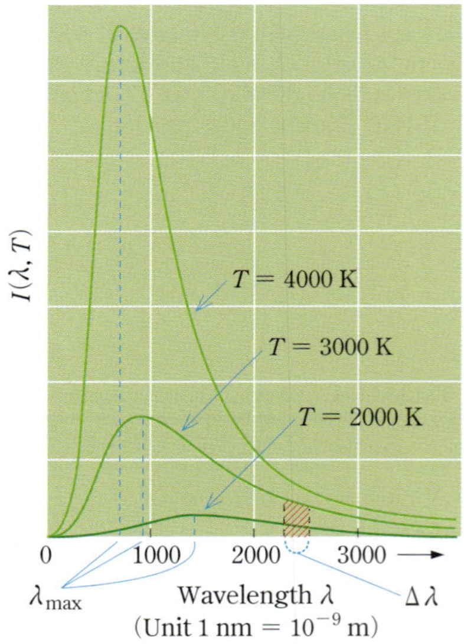

Fig. 14-7 Planck's law. The relationship between the wavelength λ and the amount of radiated energy $I(\lambda, T)$

In 1900, Planck discovered a formula that successfully describes the experimental results (Fig. 14-7) of energy measured at each wavelength of visible light, infrared light, ultraviolet light, and other electromagnetic waves radiated from furnaces at various temperatures. The area of the hatched portion between the wavelength λ and $\lambda+\Delta\lambda$ for the curve of the absolute temperature T in Fig. 14-7 represents the amount of energy per second of the electromagnetic waves of wave length between λ and $\lambda+\Delta\lambda$ radiated from the surface area of 1 m^2 of the object at the absolute temperature T. This quantity is expressed as

$$I(\lambda, T)\Delta\lambda = \frac{2\pi hc^2}{\lambda^5}\frac{1}{e^{hc/\lambda kT}-1}\Delta\lambda. \tag{14.7}$$

This equation is referred to as **Planck's law**. Strictly speaking, this law holds for radiation from the object which can absorb all incident electromagnetic waves, so it is also called the law of **black body radiation**. The parameter k is the Boltzmann constant (see the next section) and h is the constant called the Planck constant of the following value,

$$h = 6.626\times10^{-34}\ \text{J·s}. \tag{14.8}$$

Planck constant
$h = 6.626\times10^{-34}$ J·s

In the case of an object which cannot completely absorb the electromagnetic waves and has an absorbance a smaller than 1, the radiated energy is a times the amount given by equation (14.7), so it becomes smaller.

Planck's law leads to two important conclusions. The first conclusion is that the wavelength of the electromagnetic wave most strongly radiated at each temperature, that is, the wavelength λ_{max} corresponding to the peak of the curve in Fig. 14-7, is inversely proportional to the absolute temperature T, and the relationship between these two quantities is expressed as

$$\lambda_{\max} T = 2.9 \times 10^{-3}\ \mathrm{m \cdot K}. \tag{14.9}$$

In short, equation (14.9) telling that hotter objects radiate electromagnetic waves of shorter wavelength is consistent with our experience. This relationship had already been found by Wien prior to the discovery of Planck's law, so it is called **Wien's displacement law**.

The temperature of very hot objects, such as the Sun and distant stars, can be determined by measuring the energy of the radiated electromagnetic waves at each wavelength and comparing it with Planck's law. In the case of the Sun, $\lambda_{\max}$ is 500 nm, or 5×10^{-7} m, which corresponds to green, so it can be understood that the surface temperature of the Sun is 5800 K. Since the temperature of the tungsten filament of lamp is about 2000 K, the lamp radiates more infrared ray than it radiates light, so the efficiency as a light source is poor.

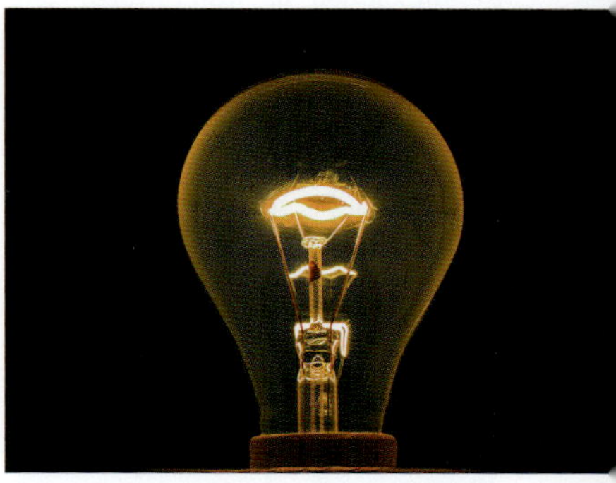

Fig. 14-8 An incandescent bulb which uses the light radiated from the filament.

The second important conclusion is that the area under the curve in Fig. 14-7 represents the total energy W of the electromagnetic waves radiated per second from the surface 1 $\mathrm{m^2}$ of the object at the absolute temperature T:

$$W(T) = \int_0^\infty I(\lambda, T)\, d\lambda = \sigma T^4 \tag{14.10}$$

$$\sigma = 5.67 \times 10^{-8}\ \mathrm{W/(m^2 \cdot K^4)}. \tag{14.11}$$

The relationship that W is proportional to the fourth power of the absolute temperature T is referred to as **Stefan-Boltzmann's law**, since it had been found by Stephan and Boltzmann before Planck's law was discovered.

Case 1 ▌ The surface temperature of the Sun and the surface temperature of the Earth If we know the surface temperature of the Sun, we can find from equation (14.10) the total amount of energy radiated by the Sun per second and calculate the amount of energy reaching the Earth. Assuming that the surface temperature of the Sun is 5800 K, we can estimate the amount of energy radiated per second from 1 $\mathrm{m^2}$ of the surface of the Sun:

$$W = 5.67 \times 10^{-8} \times 5800^4\ \mathrm{W/m^2} = 6.4 \times 10^7\ \mathrm{W/m^2}.$$

When this energy comes from the Sun with a radius of 70×10^4 km to the Earth at a distance of 150×10^6 km, the energy density decreases in inverse proportion to the square of the distance, so the amount of energy from the Sun that an area of 1 $\mathrm{m^2}$ facing the Sun on the Earth receives per second is 6.4×10^7 J times $(70 \times 10^4 / 15000 \times 10^4)^2 = 6.4 \times 10^7$ J times 1/46000, which is equal to 1400 J.

According to actual measurements, the amount of energy that a surface of 1 $\mathrm{m^2}$ facing the Sun outside the atmosphere of the Earth receives per second is 1.37 kJ (the total amount of solar radiation that 1 $\mathrm{cm^2}$ receives per minute is about 2 cals). This is called the **solar constant**.

If the radius of the Earth is R_E, the area of the Earth viewed from the Sun is $\pi R_E{}^2$ of the area of the circle of radius R_E, but the surface area of the Earth is the surface area $4\pi R_E{}^2$ of the sphere with radius R_E. The radiation of the sun that 1 $\mathrm{m^2}$ of the Earth surface receives

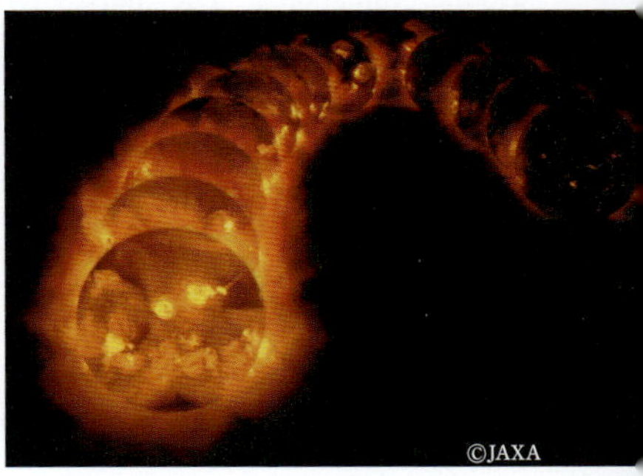

Fig. 14-9 Photos of the Sun observed by the soft X-ray telescope of the Orbiting Solar Observatory SOLAR-A (「ようこう」, Yohkoh). Photos taken on every 81 days are arranged from immediately after the launching (left: on November 1991) to the end of 1995 (right).

from the Sun is 1/4 of the above value on average.

Since the heat balance holds between the Earth and the outer space, 1 m^2 of the surface of the Earth radiates 1/4 of 1.37 kJ, or 343 J, per second on average. If this value is put in the left side of equation (14.10), the values of T on the right side is 279 K, that is, about 6 °C. A considerable part of the radiation from the Sun is reflected by the surface of the atmosphere and others. If the reflectivity is 30 %, then $T \approx 255$ K, that is, −18 °C. The average temperature of the Earth surface is 15 °C, and the average temperature of the atmosphere is −18 °C*.

* The reason why the average temperature on the Earth surface is high is the greenhouse effect caused by the action of water vapor and carbon dioxide in the atmosphere which are easy for the sunlight to pass but hard for the infrared ray to pass. The small temperature difference between the high-sunshine tropical zone and the low-sunshine cold zone is due to heat transfer by the atmospheric and seawater circulations on the global scale.

Fig. 14-10 Microwave antennas

Case 2 | Temperature of the universe The temperature of the universe is −270 °C, or 3 K. This temperature has been obtained from the following fact : In spite of much effort made to reduce the transceiver noise for intercontinental communication, the noise is not reduced below a certain level, and the same level of noise is received no matter in which direction in the universe the antenna is pointed. This is to say, uniform and homogeneous microwaves coming from all directions of the universe are received. As the observation results of the scientific satellite COBE given in Fig. 14-11 and others show, the wavelength distribution of this microwave follows Planck's law with an absolute temperature of 2.73 K (λ_{max} is 1.1 mm). Everything in the universe, except for stars, is filled with microwaves called cosmic background radiation.

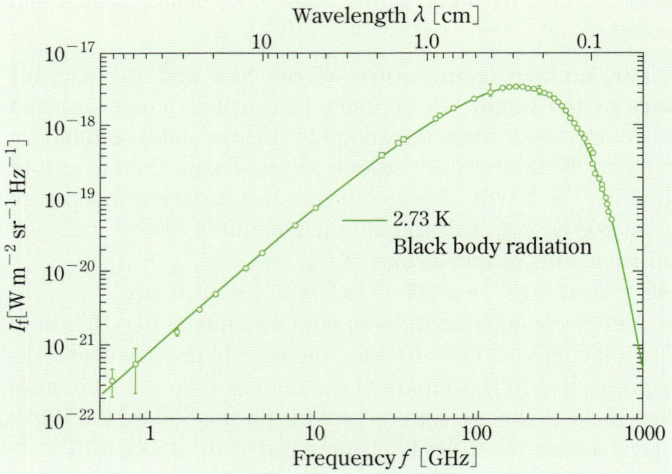

Fig. 14-11 The temperature of the universe is 2.73 K (cosmic background radiation). The horizontal axis is the frequency (lower side) and the wavelength (upper side). The curve is the theoretical value of 2.73 K black body radiation (Planck's law). I_f of the vertical axis is the quantity which can gives the amount of energy of light radiated between the frequency f and $f+\Delta f$ as $I_f \Delta f$.

14.3 Kinetic theory of gases

Learning objective To understand the equation of state of the ideal gas describing the relationships of pressure, volume, temperature and amount of substance. To understand that the internal energy of the ideal gas can be derived by applying the molecular theory to gases.

Equation of state of the ideal gas We know several facts concerning the gas properties.

At a constant temperature, volume V and pressure p are in inverse proportion (Fig. 14-12).

$$pV = \text{constant} \qquad \text{(constant temperature)}. \tag{14.12}$$

Since this relationship was found by Boyle, it is call **Boyle's law**.

When gases are heated under constant pressure condition, the volume increases by expansion. When the temperature of the gas is raised by 1 °C, the volume increases by 1/273.15 of the volume V_0 at 0 °C. When the temperature decreases by 1 °C, the volume decreases by 1/273.15 of V_0 (Fig. 14-13). Therefore, the volume V at t °C is expressed as

$$V = \left(1+\frac{t}{273.15}\right) V_0 \qquad \text{(constant pressure)}. \tag{14.13}$$

This relationship was found by Charles and is called **Charles' law**.

A gas is an assembly of molecules. As the temperature rises, the molecular motion becomes active, so the pressure at which the gas pushes the piston increases, and the volume of the gas increases. Conversely, as the temperature decreases, the molecular motion weakens, so the pressure at which the gas pushes the piston decreases, and the volume of gas decreases.

At this point we introduce the **absolute temperature** (the unit is kelvin, symbol K) obtained by adding 273.15 to the temperature T_C in Celsius temperature*1

$$T = T_C + 273.15. \tag{14.14}$$

Using the absolute temperature, Charles' law is expressed as

$$V = \frac{T}{273.15} V_0 \qquad \text{(constant pressure)}. \tag{14.15}$$

When the pressure p is kept constant, the volume V of gas is proportional to the absolute temperature T.

For dilute gases at the same temperature, pressure, and volume, any gas always contains the same number of molecules. This property is called the **Avogadro law**.

A gas for which Boyle's, Charles', and Avogadro's laws hold is called an ideal gas. In the standard state (1 atm = 1.01325×10^5 N/m², 0 °C = 273.15 K), one mole of molecules, that is,

$$N_A = 6.022\times10^{23}/\text{mol} \tag{14.16}$$

molecules are contained in 22.414 L of ideal gas. The parameter N_A is called the **Avogadro constant***2.

The numerical part of the mass expressed in gram of 1 mol of substance is called the molecular weight of the substance. Therefore, the mass of a 22.414 L gas at 273.15 K and at 1 atm is numerically equal to the molecular weight of the gas.

The volume V of the ideal gas in the container (cylinder) is inversely proportional to the pressure p according to Boyle's law when the absolute temperature T and the amount of substance n are constant,

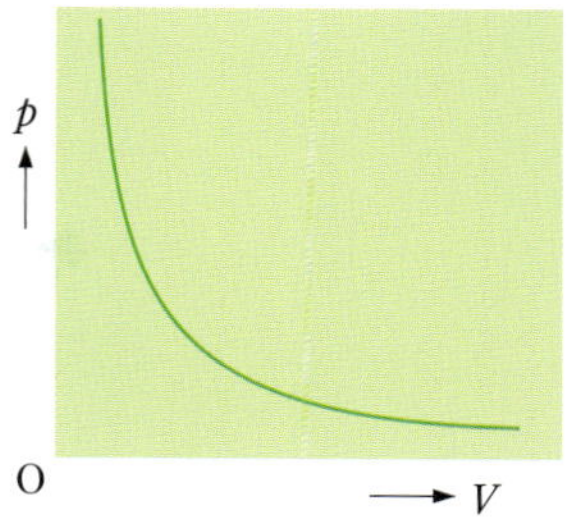

Fig. 14-12 Boyle's law
pV = constant.

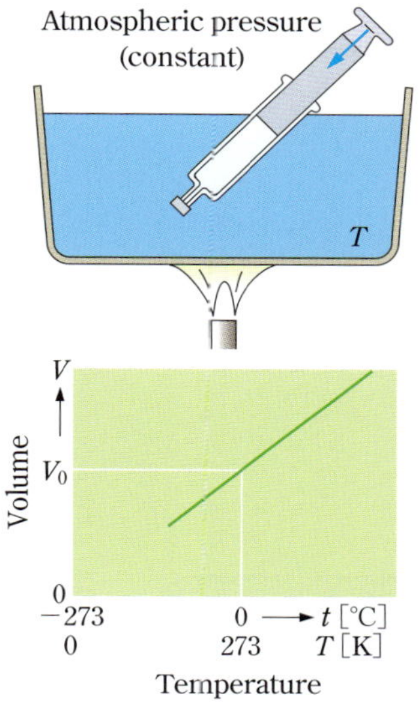

Fig. 14-13 Charles' law
$\frac{V}{T}$ = constant.

Unit of absolute temperature K

*1 Strictly speaking, the absolute temperature is defined as the thermodynamic temperature (refer to Section **15.4**).

Avogadro constant $N_A = 6.022\times10^{23}$/mol

*2 Historically, Avogadro's constant N_A is defined as the number of carbon atoms contained in 0.012 kg (12 g) of ^{12}C, and when a substance composed of one kind of constituent such as molecule, atom, or ion contains N_A constituent particles, the amount of substance was defined as one mole (1 mol).

Since 2019, N_A has a predefined value:

$N_A = 6.022\,140\,76\times10^{23}$/mol.

and is proportional to the absolute temperature T when the pressure p and the amount of substance n are constant, and when pressure p and absolute temperature T are constant, it is proportional to the amount of substance n. Therefore, by putting these three relations together, we have a relationship between the pressure p, volume V, and absolute temperature T of n moles of gas :

$$V = \text{constant} \times \frac{nT}{p}. \tag{14.17}$$

Setting the constant as R transforms equation (14.17) into

$$pV = nRT. \tag{14.18}$$ *

* From this formula on, T contains the unit K and n contains the unit mol.

This is referred to as **Boyle–Charles' law**. R is a constant called the **gas constant** and irrespective of the type of gas, its value is

Gas constant
$R = 8.31\ \mathrm{J/(K \cdot mol)}$

$$R = 8.31\ \mathrm{J/(K \cdot mol)} \tag{14.19}$$

$$\left[\frac{(1.013 \times 10^5\ \mathrm{N/m^2}) \times (2.24 \times 10^{-2}\ \mathrm{m^3})}{(273\ \mathrm{K}) \times (1\ \mathrm{mol})} = 8.31\ \mathrm{N \cdot m/(K \cdot mol)}\right].$$

Although real gases deviate from equation (14.18) at low temperatures and high densities, equation (14.18) expresses the gas state satisfactorily for dilute gases at high temperatures and low densities. Based on the assumption of the existence of a gas that always satisfies equation (14.18), this gas is called the **ideal gas**. Equation (14.18) is called the **equation of state of the ideal gas** because it is the relationship between physical quantities expessing ideal gas states such as pressure, volume, temperature, amount of substance, etc.

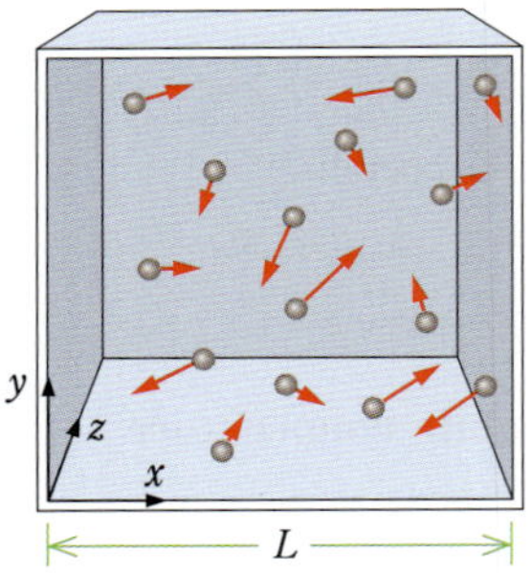

(a) Gas confined in a container

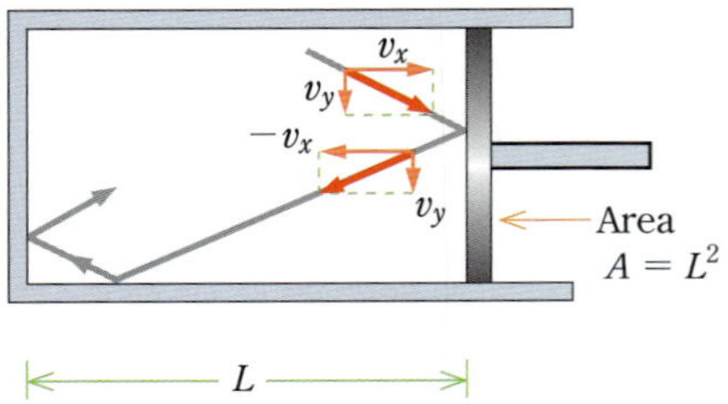

(b) Collision of gas molecules with wall

Fig. 14-14 Motion of gas molecules

Kinetic theory of gases Boyle–Charles's law can be explained from the standpoint of molecular theory. Let us understand from the microscopic standpoint the pressure exerted on the wall by the gas in the vessel as the action of the gas molecules colliding with the wall. We assume that n mol gas, that is, nN_A gas molecules, are contained in a cubic container of length L of one side in Fig. 14-14. These molecules change their state of motion when they collide with walls and/or with other molecules, but for the sake of simplicity, the gas is so dilute that collisions between molecules are assumed to be negligible. We focus at this point on one molecule that collides elastically with the right wall of Fig. 14-14 at velocity $\boldsymbol{v} = (v_x, v_y, v_z)$. In this elastic collision, the v_y and v_z components of the velocity $\boldsymbol{v}$ parallel to the wall do not change, but the x component perpendicular to the wall changes from v_x to $-v_x$. Accordingly, the component perpendicular to the wall of momentum $m\boldsymbol{v}$ of the molecule of mass m changes by $(-mv_x)-(mv_x) = -2mv_x$. This is equal to the leftward impulse ("force" × "action time") that the molecule receives during the collision, due to the relationship between the momentum change and the impulse. On the other hand, according to the law of action and reaction, the impulse that this molecule exerts on the wall is $2mv_x$ pointing to the right. This molecule collides with the other wall and returns to the right side wall again. The time for one reciprocation is $\dfrac{2L}{v_x}$, so the collision frequency of this molecule with the same wall during the time t is $\dfrac{t}{2L/v_x} = \dfrac{v_x t}{2L}$, and during this time, the impulse that one molecule exerts on the wall is

$$2mv_x \times \frac{v_x t}{2L} = \frac{m{v_x}^2}{L} t. \quad (14.20)$$

If this contribution is summed over all molecules, the impulse that the gas exerts on the wall can be obtained. Denoting the mean value of ${v_x}^2$ of nN_A molecules as $\langle {v_x}^2 \rangle$ and adding ${v_x}^2$ of all molecules gives $nN_A\langle {v_x}^2 \rangle$. Thus, the impulse that the entire molecules exert on the wall during time t is $\frac{nN_A m\langle {v_x}^2 \rangle}{L} t$. On the other hand, if the mean force that the entire molecules exert on the wall is F, then the impulse during time t is Ft. The mean force F is

$$F = \frac{nN_A m\langle {v_x}^2 \rangle}{L} \quad (14.21)$$

(Fig. 14-15). Since the area of the wall is L^2, the gas pressure p is

$$p = \frac{F}{L^2} = \frac{nN_A m\langle {v_x}^2 \rangle}{V}. \quad (14.22)$$

Here, the relation $V = L^3$ is used.

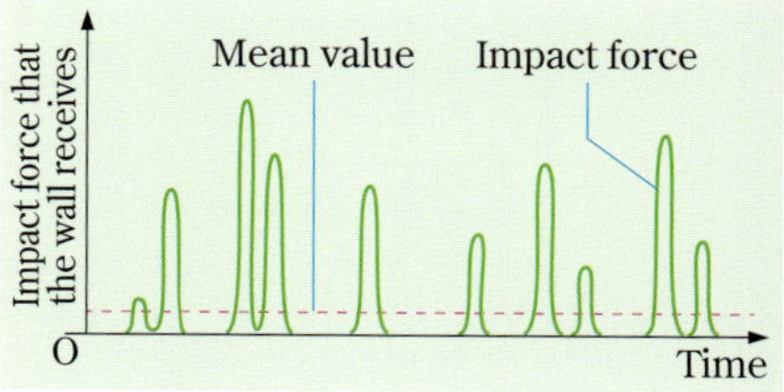

Fig. 14-15 Impact forces and the mean value (mean force) that the wall receives from gas molecules

The motion of gas molecules as a whole is isotropic, and it seems reasonable to assume that $\langle {v_x}^2 \rangle = \langle {v_y}^2 \rangle = \langle {v_z}^2 \rangle$. Using the Pythagorean theorem $v^2 = {v_x}^2 + {v_y}^2 + {v_z}^2$ and considering the mean, we have the relation

$$\langle v^2 \rangle = \langle {v_x}^2 \rangle + \langle {v_y}^2 \rangle + \langle {v_z}^2 \rangle = 3\langle {v_x}^2 \rangle. \quad (14.23)$$

Equation (14.22) combined with equation (14.23) gives

$$pV = \frac{1}{3} nN_A m\langle v^2 \rangle = \frac{2}{3} E. \quad (14.24)$$

Here

$$E = \frac{1}{2} nN_A m\langle v^2 \rangle \quad (14.25)$$

is the total kinetic energy of gas molecules.

Comparison of equation (14.24) with Boyle-Charles' law (14.18) leads to the relation

$$\frac{1}{2} nN_A m\langle v^2 \rangle = E = \frac{3}{2} nRT. \quad (14.26)$$

Therefore, it can be seen that the total kinetic energy E of the gas molecules is proportional to the absolute temperature T. Equation (14.26) also indicates that the mean kinetic energy of one gas molecule is

$$\frac{1}{2} m\langle v^2 \rangle = \frac{3}{2} \frac{R}{N_A} T. \quad (14.27)$$

The term $\frac{R}{N_A}$ on the right-hand side is a constant we encounter frequently in the molecular theory and is called the **Boltzmann constant** and denoted as k (or k_B).

Boltzmann constant
$k\ (= k_B) = 1.38 \times 10^{-23}$ **J/K**

$$k = 1.38 \times 10^{-23}\ \mathrm{J/K}. \quad (14.28)$$

Using Boltzmann constant k, equation (14.27) is expressed as

$$\frac{1}{2} m\langle v^2 \rangle = \frac{3}{2} kT. \quad (14.29)$$

In this way, it has been shown that the kinetic energy E of gas molecules is proportional to the absolute temperature T. Conversely, if the absolute temperature T is defined by equation (14.29), Boyle-Charles' law can be derived from the kinetic theory of gases.

Case 3 | Mean speed of gas molecules The mean speed (strictly speaking, root mean square speed) of hydrogen molecule H_2 [mass $m(H_2) = 3.35\times10^{-27}$ kg] and mercury molecule Hg [mass $m(\mathrm{Hg}) = 3.33\times10^{-25}$ kg] at 300 K (27 °C) is calculated from equation (14.29) as follows :

$$(\mathrm{H_2})\quad \sqrt{\langle v^2\rangle} = \sqrt{\frac{3kT}{m}} = \sqrt{\frac{3\times(1.38\times10^{-23}\,\mathrm{J/K})\times(300\,\mathrm{K})}{3.35\times10^{-27}\,\mathrm{kg}}} = 1.93\times10^3\,\mathrm{m/s}$$

$$(\mathrm{Hg})\quad \sqrt{\langle v^2\rangle} = \sqrt{\frac{3kT}{m}} = \sqrt{\frac{3\times(1.38\times10^{-23}\,\mathrm{J/K})\times(300\,\mathrm{K})}{3.33\times10^{-25}\,\mathrm{kg}}} = 1.93\times10^2\,\mathrm{m/s}.$$

What will happen when oxygen gases at different temperatures are mixed? Because the temperatures are different, at first, the high-speed oxygen molecules and the low-speed oxygen molecules are mixed, but as a result of collisions between molecules speeds of the oxygen molecules are gradually averaged and eventually settle to an intermediate speed. This is the actual process of heat transfer and the state of the thermal equilibrium when gases at different temperatures are mixed. In this way, the properties of the ideal gas can be understood by considering the gas as an assembly of molecules.

The speed of gas molecules distributes around the mean value. This speed distribution of gas molecules can also be calculated theoretically. When the speed of gas molecules is measured, the probability that the speed is between v and $v+\Delta v$ is

$$Nv^2 \exp\left(-\frac{mv^2}{2kT}\right)\Delta v, \qquad N = \sqrt{\frac{2m^3}{\pi k^3 T^3}} \tag{14.30}$$

(Fig. 14-16). Here, $\exp(x) = \mathrm{e}^x$. Since this speed distribution was theoretically derived by Maxwell, it is called the **Maxwell distribution**.

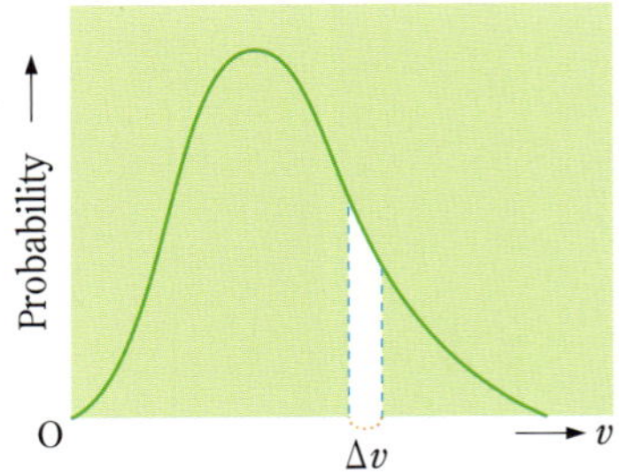

Fig. 14-16 Maxwell's speed distribution

For reference Boltzmann distribution

Equation (14.30) can also be expressed as

$$\frac{N}{4\pi}\exp\left(-\frac{m(v_x{}^2+v_y{}^2+v_z{}^2)}{2kT}\right)\Delta v_x \Delta v_y \Delta v_z. \tag{14.31}$$

In this distribution function, the energy $E = \frac{1}{2}mv^2$ of molecule appears in the form of

$$\mathrm{e}^{-E/kT}. \tag{14.32}$$

In general, that the probability that a molecule in a substance in thermal equilibrium at temperature T has energy E is proportional to $\mathrm{e}^{-E/kT}$ can be derived from dynamics and probability theory. This distribution is called the **Boltzmann distribution**. A scientific field to study the properties of matter based on the assumption of this probability distribution is called **statistical mechanics**.

In classical statistical mechanics based on the Boltzmann distribution, the kinetic energy of $\frac{1}{2}kT$ per degree of freedom is assigned on average to each molecule. This is called the **law of equipartition of energy**. When the temperature becomes low and the quantum effect becomes significant, quantum statistical mechanics must be used and the law of equipartition of energy no longer holds (see the column given at the end of this chapter).

Mean free path In liquids, molecules are considered to be in contact with each other. When a liquid changes into a gas, the volume increases significantly, so the average spacing of molecules in the gas state gets much larger than the diameter of the molecules. In a gas at ordinary temperature and pressure, molecules are scattered at intervals of about 10 times their diameter. They change the direction of motion when they collide with other molecules or collide with the walls of container, but between collisions they travel straight at a speed of several hundred meters per second. The average distance L that one gas molecule travels in the interval between the first collision with other molecule and the second collision with another molecule is called the **mean free path**.

When N gas molecules of diameter d are contained in a unit volume, the mean free path is

$$L = \frac{1}{\sqrt{2}\pi N d^2}. \tag{14.33}$$

A collision occurs if there is a center of another molecule in the range of distance d from the center of the gas molecule, so the condition that, on average, one other molecule is contained in a cylinder with a base area πd^2 and a length L, namely the condition $N(\pi d^2 L) = 1$ holds. From this we get $L = \dfrac{1}{\pi N d^2}$ (Fig. 14-17). In deriving this formula, the relative motion of the gas molecules has been ignored. The correct formula is equation (14.33).

In the standard state, since $N = N_A/0.0224\ \mathrm{m^3} = 2.7\times10^{25}$ particles/m^3, if the diameter of the air molecule is set to be about 4×10^{-10} m, we find the mean free path L in the air of the standard state is

$$L = \{\sqrt{2}\pi\times2.7\times10^{25}\times(4\times10^{-10})^2\}^{-1}\ \mathrm{m} = 5\times10^{-8}\ \mathrm{m}.$$

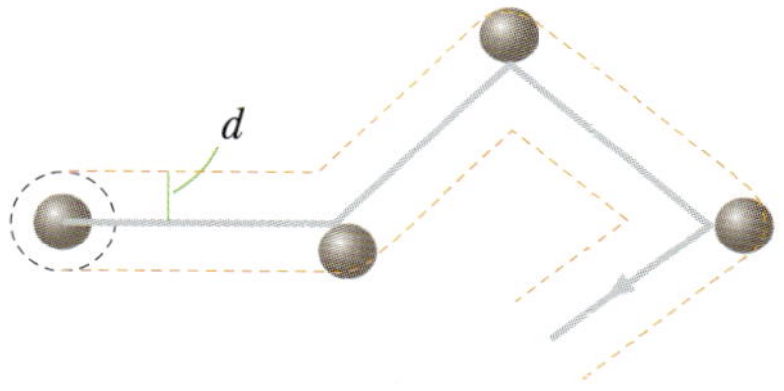

Fig. 14-17 While a gas molecule of diameter d moves by distance L, it collides with, on average, $\pi N d^2 L$ gas molecules inside the bent cylinder of volume $\pi d^2 L$, so the mean travel distance L until the molecule collides once is

$$L = \frac{1}{\pi N d^2}.$$

Question 1 The density of liquid oxygen at -182.5 °C is 1.12 g/cm^3. (1) Show that the diameter d of the oxygen molecule is about 4×10^{-10} m. (2) The ideal gas of 22.4 L (more exactly 22.41402 L) in the standard state (1 atm $= 1.01325\times10^5$ Pa, 0 °C $=$ 273.15 K) contains 1 mol of gas molecules. Calculate the density ρ of oxygen gas in the standard state, and show that the ratio of the average spacing of the oxygen molecules to the diameter of oxygen molecule is about 10 : 1. The molecular weight of oxygen is 32.0.

Internal energy of the ideal gas The sum of the kinetic energy and potential energy of the thermal motion of molecules in a substance is called the **internal energy** of the substance. Since in the kinetic theory of gases discussed in this chapter the intermolecular forces between molecules are ignored, gas molecules do not have the potential energy of thermal motion. The kinetic energy considered so far is the kinetic energy of the center of gravity motion of gas molecules.

The motion of a monoatomic molecule composed of one constituent atom is only the center of gravity motion, so E in equation (14.26) represents the total energy of the thermal motion of n mol of gas molecules composed of monoatomic molecules such as helium He, neon Ne, and argon Ar, that is, it means the internal energy (symbol U). In other words,

$$U = \frac{1}{2} m\langle v^2 \rangle n N_{\mathrm{A}} = \frac{3}{2} nRT \qquad \text{(monoatomic gas)}. \tag{14.34}$$

In the case of diatomic molecules composed of two atoms and triatomic molecules composed of three atoms, etc., the rotation and vibration of the molecule can be considered. Thus, gases composed of such polyatomic molecules are expected to have the internal energy greater than that given by equation (14.34). We, therefore, instead of equation (14.34) express as

$$U = \frac{f}{2} nRT. \tag{14.35}$$

As will be shown in Section **15.2**, experimental results have demonstrated that, except for the cases of low temperatures, $f = 3$ for monoatomic molecules (He, Ar etc.) as predicted by the theory, $f \approx 5$ for diatomic molecules (O_2, N_2, CO, etc.), and $f \gtrsim 6$ for triatomic molecules (CO_2, SO_2, etc.). This fact is interpreted as due to the energy of the rotational motion of the molecule. In the case of diatomic molecules, the energy of rotational motion around the axis connecting two atoms is negligible, so the energy of rotational motion is less than in the case of triatomic molecules. In molecules, atomic vibrations do not occur unless the temperature is very high. Conversely, as the temperature rises, various new types of thermal motion occur, so the value of f increases with temperature (see Fig. 15-8).

14.4 van der Waals' equation of state

Learning objective To learn van der Waals' equation of state which is a model of equation of state applicable for high density gases and to understand the basic concepts of phase transitions between gas and liquid.

When the equation of state $pV = nRT$ was derived, the volume of gas molecules was neglected. However, with the increase in density of gase, the volume of gas molecules becomes no longer negligible. Clausius pointed out that in place of the volume V of container the quantity $V - nb$ should be used. b can be taken as the volume of 1 mol of gas molecules (ie. volume of 1 mol of liquid).

In addition, in deriving the equation of state $pV = nRT$, effects of intermolecular forces were neglected. When the molecule is not in the immediate vicinity of the container wall, the molecule is uniformly surrounded by other molecules and the forces from these molecules are considered to cancel out and the resultant force will be zero. However, if the molecule is highly close to the wall of the container, other molecules are only inside the container, so the resultant force of the forces from those molecules will not be zero, and it becomes the force oriented towards the center of container. For this reason, the force that changes the momentum of the molecule when it collides with the wall is the resultant force of the force exerted by the wall and the intermolecular force, so the force that the wall receives during the collision with the molecule becomes smaller than the case of ignoring the intermolecular forces. Accordingly, with the consideration of the intermolecular forces, the pressure the wall receives from the gas also decreases. This decrease in pressure is proportional to the number of molecules that are within a certain distance from the wall and subject to the force directed towards the center of container, and the number of molecules that are within a certain distance from this molecule and exert a force on this molecule. Since the number of these two kinds of molecules is in proportion to the density $\frac{nN_A}{V}$, van der Waals considered that the decrease in pressure p is expressible as $-\frac{an^2}{V^2}$.

In considering these two effects, it follows that $p = \frac{nRT}{V}$ is transformed into

$$p = \frac{nRT}{V-nb} - \frac{an^2}{V^2} \tag{14.36}$$

namely,

$$\left(p + \frac{an^2}{V^2}\right)(V-nb) = nRT. \tag{14.37}$$

This is referred to as **van der Waals' equation of state**. This equation of state holds satisfactorily for real gases.

In Fig. 14-18 are shown the **isotherm** showing the relationship between pressure p and volume V of equation (14.36) under the condition of $T =$ constant. When the temperature is high, the isotherm has neither a minimum nor a maximum. In addition, when the pressure p is changed while the temperature T is kept constant, the continuous change occurs between the low-density gas-like state and the high-density liquid-like state.

When the temperature is reduced and reaches

$$T_C = \frac{8a}{27bR} \tag{14.38}$$

the isotherm takes the stationary value p_C at volume V_C. This temperature T_C is called the **critical temperature**, and the pressure p_C the **critical pressure**. At the eritical point, the following relationship holds :

$$V_C = 3nb, \qquad p_C = \frac{a}{27b^2}. \tag{14.39}$$

[For the proof of equations (14.38) and (14.39), see Problems for exercise 14, B4].

At temperatures below the critical temperature T_C the isotherm has one maximum and one minimum, so the isotherm intersects the horizon

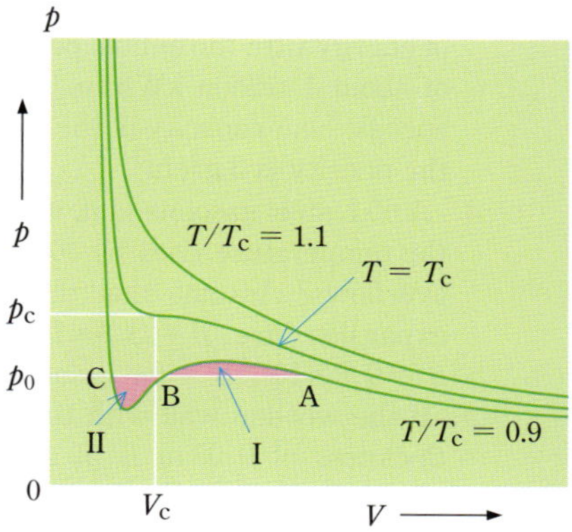

Fig. 14-18 Isothermal curves given by van der Waals' equation of state

at three points A, B, and C for a certain range of p. If the area of the region I surrounded by the isothermal curve and the horizontal line is larger than the area of the region II, A is stable, and if it is smaller, C is stable, and B is always in an unstable state. A pressure p_0 at which the areas of I and II are equal gives a saturated vapor pressure at temperature T. This is called **Maxwell's rule** (see Problems for exercise 15, B8). When the gas is compressed while the temperature T is kept constant, liquefaction starts at point A at a pressure of p_0 and liquefaction is complete at point C. The horizontal part between A and C is a state in which the states of gas and liquid coexist at a certain ratio.

The high-density side to the left of point C of the isotherm represents the liquid state, but this equation of state of the liquid state does not serve as a good approximation for real liquids. The phase transition between the gas phase and the liquid phase below the critical temperature T_C is a first-order phase transition.

Problems for exercise 14

A

1. Explain why the lake water starts to freeze from the lake surface when it cools.
2. There are red stars and blue stars. What information about these stars can we obtain from this fact?
3. There is a high-temperature cubic rock body with each side 4.5 km in length under the ground. If the energy corresponding to the temperature difference of 100 °C is taken out from it, how much amount of energy is obtained? Compare this extracted amount of energy with the annual power generation capacity of about 1 trillion kWh in Japan. Assume that the specific heat capacity of the rock is 0.8 J/(g·K) and the density is 3 g/cm^3.
4. A 60 L steel gasoline tank was filled at 10 °C. When the temperature rises by 30 °C, how much gasoline overflows? Assume that the coefficients of cubical expansion are $\beta = 3.8\times10^{-5}\,\mathrm{K}^{-1}$ for steel and $\beta = 9.5\times10^{-4}\,\mathrm{K}^{-1}$ for gasoline.
5. If the wooden wall with the area of 20 m^2 and the thickness of 0.05 m is at 20 °C inside and -5 °C outside, how much amount of heat is transberred through the wall per second? Let the thermal conductivity of the wood be $k = 0.15$ J/(m·s·K)
6. Show that the mean free path L of the molecules of the ideal gas at temperature T and pressure p is

$$L = \frac{kT}{\sqrt{2}\,\pi p d^2}$$

$$\left(\text{Use } N = \frac{nN_A}{V} = \frac{pN_A}{RT} = \frac{p}{kT}\right).$$

B

1. Using the phase diagram of Fig. 14-3 (b), explain the fact that solid carbon dioxide CO_2 (dry ice) does not become a liquid even when it melts at room temperature.
2. Prove the relationship $\beta = 3\alpha$. between the linear expansion coefficient α and the cubical expansion coefficient β.
3. A naked person (surface area A and body temperature T_1) is standing in a room at temperature T_2. The human body radiates $a\sigma A T_1^4$ heat and absorbs $a\sigma A T_2^4$ heat per second. Assuming $A = 1.2\,\mathrm{m}^2$, $T_1 = 36$ °C, $T_2 = 20$ °C, and $a = 0.7$, calculate the heat that the person loses per second.
4. Derive equations (14.38) and (14.39).

The discovery of Planck's law and the birth of modern physics

The laws that appear up to Chapter 22 of this book were discovered by the end of the 19th century. Physics composed of mechanics, electromagnetism, thermodynamics, etc. established by the end of the 19th century based on these laws are called classical physics. Planck's law that the readers have learned in this chapter was discovered in 1900, the last year of the 19th century. However, this law which satisfactorily explains the radiation of light from high temperature objects cannot be derived from classical physics. If this law can be theoretically derived, it can explain why low temperature stars glow red and high temperature stars glow blue. Red light is an electromagnetic wave that has a longer wavelength and a lower frequency than blue light, But does some relation exist between the difference in wavelength and the temperature?

Planck noticed, in order to theoretically derive the law he had found himself, the requirement of the assumption that "The magnitude E of the energy that light with frequency f can have is limited to an integer multiple of "Planck's constant h" × "frequency f", namely

$$E = nhf \qquad (n = 0, 1, 2, \cdots).$$

This means that the light of frequency f is a group of energy particles with magnitude hf.

In classical physics, the amplitude of the wave can be made as small as possible, so the energy of the light wave should be able to have any small value. In mathematical terms, light energy values need to be continuous. However, just as there are atoms as the smallest unit in a substance that seems continuous in our eyes, there exists actually a minimum unit in the energy of light that looks continuous. This minimum unit of energy is called the energy quantum. The discovery of energy quantum was the first step of the modern physics based on the quantum theory and of the development of electronics underlying the modern society.

The energy hf of the energy quantum of light differs depending on the frequency f, and the larger the frequency f, the larger the energy hf. The higher the temperature, the larger the internal energy of the object, so the higher the temperature, the larger energy of the particles can be extracted. Therefore, the higher the temperature, the higher the frequency f and the shorter the wavelength λ.

In the kinetic theory of gases, we learned that the average kinetic energy of a monoatomic molecule of gas at an absolute temperature T is $\frac{3}{2}kT$. The measure of the amount of energy absorbed and released by an object at an absolute temperature T is kT. Atoms cannot absorb or release particles with energy greater than about kT. Dr. Shin-ichiro Tomonaga explained this fact using an illustration shown in Fig. 14-A. When the horizontal axis of the diagram of the light emission intensity is not the wavelength λ as in Fig. 14-7, but the frequency f is selected as shown in Fig. 14-11 and the frequency corresponding to the peak of the curve is described as $f_{\max}$, then Wien's displacement law (14.9) can be rewritten as

$$hf_{\max} = 2.82\,kT,$$

corresponding to this fact.

One more comment to add is to pay attention to that in Chapter 24 the symbol ν is used in place of f to express the frequency of light.

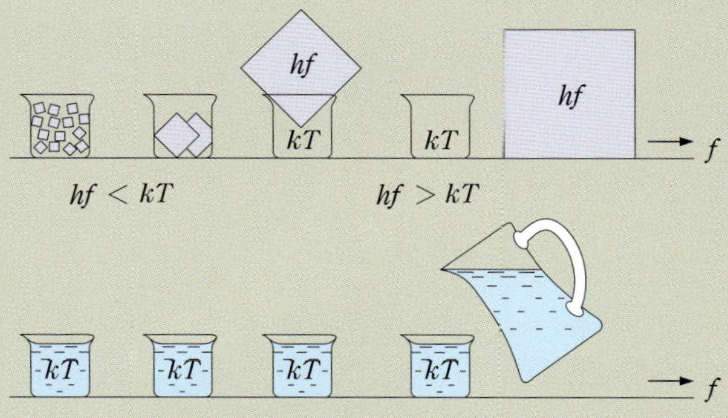

Fig. 14-A Absorption and release of energy by atoms of the substance at an absolute temperature T. The upper part is a representation in quantum theory and the lower part is the one in classical theory.

Thermodynamics

Thermodynamics is a field of science that deals with concrete problems by putting general properties of heat into several rules and using them as a starting point regardless of the molecular structure of matter. In thermodynamics, the energy conservation law in the form of state change of an object (system) exchanging heat and/or work with the outside (environment) is called the first law of thermodynamics.

Heat naturally transfers from hot objects to cold objects, but heat does not transfer from cold objects to hot objects. As like this, a state change in which the change in the opposite direction does not proceed spontaneously is called the **irreversible change**. The law for the direction of state change is the second law of thermodynamics, which is formulated quantitatively in the form of "the entropy always increases". The reason why only part of the chemical energy of fuel can be converted to work in a heat engine such as a steam engine is that the change of heat into work is irreversible.

15.1 The first law of thermodynamics

Learning objective To be able to derive the first law of thermodynamics, which is an energy conservation law when heat is involved, from the principle of energy conservation. To understand how this law is expressed in changes such as isobaric change, isochoric change, isothermal change, and adiabatic change.

Quantity of state and equation of state Physical quantities that represent objects (systems) in thermal equilibrium include temperature, pressure, volume, internal energy, etc. Physical quantities that represent the state of an object are called **quantities of state** or **state variables**. The relationship between state variables such as an equation $pV = nRT$ which holds for the ideal gas is called the **equation of state**. The relation between internal energy and temperature such as equation (14.35) is also an equation of state. Due to the equation of state it is not possible to change arbitrarily all of the state variables such as volume V, pressure p, temperature T and internal energy U of the object.

First law of thermodynamics The law of conservation of energy in the case where an object (system) exchanges heat with the outside (the environment) or performs work to the outside or work is performed to the system from the outside is called the **first law of thermodynamics**. Since the entity of heat is the energy moving from the high temperature object to the low temperature object, when the heat $Q_{\text{system}\leftarrow\text{outside}}$ enters the object from the outside and the outside performs work $W_{\text{system}\leftarrow\text{outside}}$ to the object, the internal energy U which is the energy of the atoms and molecules that constitute the object increases.

First law of thermodynamics When heat $Q_{\text{system}\leftarrow\text{outside}}$ enters the object from the outside and the outside performs work $W_{\text{system}\leftarrow\text{outside}}$ to the object, the change in internal energy U of the object before and after the change is

$$U_{\text{after}} - U_{\text{before}} = Q_{\text{system}\leftarrow\text{outside}} + W_{\text{system}\leftarrow\text{outside}}. \tag{15.1}$$

If heat $Q_{\text{outside}\leftarrow\text{system}}$ goes out of the object, $Q_{\text{system}\leftarrow\text{outside}} = -Q_{\text{outside}\leftarrow\text{system}} < 0$, and if the object performs work $W_{\text{outside}\leftarrow\text{system}}$ to the outside, then according to the law of action and reaction, $W_{\text{system}\leftarrow\text{outside}} = -W_{\text{outside}\leftarrow\text{system}} < 0$. In the form of minute quantities, equation (15.1) is expressed as

$$\Delta U = \Delta Q_{\text{system}\leftarrow\text{outside}} + \Delta W_{\text{system}\leftarrow\text{outside}}. \tag{15.2}$$

In addition, it is noted that heat $Q_{\text{system}\leftarrow\text{outside}}$ and work $W_{\text{system}\leftarrow\text{outside}}$ are not quantities of state whose values are determined for each state. Because heat and work coming in and going out when the object changes from state A to state B depend on the path of the state change between state A and state B, heat $Q_{\text{system}\leftarrow\text{outside}}$, $\Delta Q_{\text{system}\leftarrow\text{outside}}$, and work $W_{\text{system}\leftarrow\text{outside}}$, $\Delta W_{\text{system}\leftarrow\text{outside}}$ are not determined only by the initial and final states alone*.

The first law of thermodynamics was independently found in the

* To emphasize that the heat $\Delta Q_{\text{system}\leftarrow\text{outside}}$ and the work $\Delta W_{\text{system}\leftarrow\text{outside}}$ are not changes in quantity of state, some textbooks adopt the notation $\Delta' Q_{\text{system}\leftarrow\text{outside}}$ and $\Delta' W_{\text{system}\leftarrow\text{outside}}$.

1840s by Joule of the United Kingdom, as well as Meyer and Helmholtz of Germany in the form of invention of the conserved physical quantity "energy" including heat (see Section **5.3**).

Various types of interactions exist between objects and the outside. In the following, we will learn how the first law of thermodynamics is expressed in processes of isobaric change, isochoric change, isothermal change, adiabatic change, etc.

Isobaric change Changes in temperature and volume that occur when the pressure applied to an object is constant are called isobaric changes.

Example 1 As shown in Fig. 15-1, the gas is put into the cylinder attached with a piston and the gas volume is increased slowly by only ΔV. Show that the work performed by the environment to the gas is

$$\Delta W_{\text{system}\leftarrow\text{outside}} = -p\,\Delta V. \tag{15.3}$$

Here, p is the pressure of the gas.

Solution Assume that the area of the piston is A. The force with which the piston pushes the gas is pA. When the piston moves slowly to the right by Δx, the force acting on the gas by the piston and the direction of movement of the piston are opposite, so "work performed by the outside (piston) to the gas $\Delta W_{\text{system}\leftarrow\text{outside}}$" = "force pA"×"distance moved in the direction of force $-\Delta x$" = $-pA\,\Delta x$. Since $A\,\Delta x$ is the increase in gas volume ΔV, it follows that $\Delta W_{\text{system}\leftarrow\text{outside}} = -p\,\Delta V$.

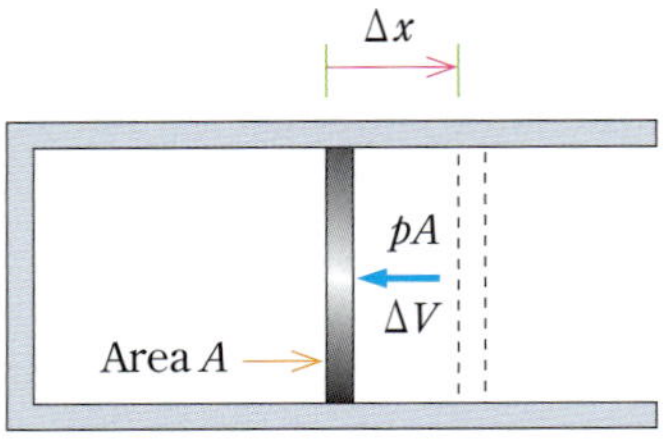

Fig. 15-1 $\Delta W_{\text{system}\leftarrow\text{outside}} = -p\,\Delta V$

From Example 1, equation (15.2) is transformed into

$$\Delta U = \Delta Q_{\text{system}\leftarrow\text{outside}} - p\,\Delta V. \tag{15.4}$$

When the volume of gas changes from V_1 to V_2 in the isobaric change, $\Delta V = V_2 - V_1$, so the work that the outside performs to the gas is

$$W_{\text{system}\leftarrow\text{outside}} = -p(V_2 - V_1). \tag{15.5}$$

In the case of volume expansion ($V_2 > V_1$), $W_{\text{system}\leftarrow\text{outside}} < 0$, and in the case of compression ($V_2 < V_1$), $W_{\text{system}\leftarrow\text{outside}} > 0$, and the work $W_{\text{outside}\leftarrow\text{system}}$ that the gas performs to the outside is $p(V_2 - V_1)$.

In the cases other than isobaric change, since the pressure p changes with the change in volume V, the process of volume change must be considered by dividing into minute portions. When the gas volume increases gradually from V_1 to V_2, the work $W_{\text{system}\leftarrow\text{outside}}$ performed by the outside to the gas is the limiting value in the limit of $\Delta V_i \to 0$ of the sum of the work at each minute change

$$W_{\text{system}\leftarrow\text{outside}} = -\sum_i p_i\,\Delta V_i \tag{15.6}$$

namely (−1) times "the area of the colored section ■" of Fig. 15-2 :

$$W_{\text{system}\leftarrow\text{outside}} = -\int_{V_1}^{V_2} p\,\mathrm{d}V. \tag{15.7}$$

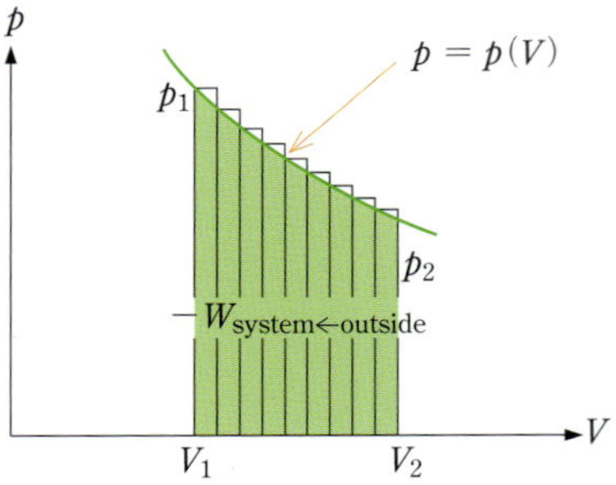

Fig. 15-2 When the gas volume expands from V_1 to V_2, the work $W_{\text{system}\leftarrow\text{outside}}$ performed by the outside to the gas $W_{\text{system}\leftarrow\text{outside}} = -\int_{V_1}^{V_2} p\,\mathrm{d}V$ is (−1) times "the area of the colored section ■" and is a negative quantity. When the gas is compressed, $V_2 < V_1$ and $\Delta V < 0$, so $p\,\Delta V < 0$ and $\int_{V_1}^{V_2} p\,\mathrm{d}V < 0$, thus $W_{\text{system}\leftarrow\text{outside}} = -\int_{V_1}^{V_2} p\,\mathrm{d}V > 0$.

Isochoric change Changes in temperature and pressure that occur when the volume of an object is constant are called isochoric changes. This change is easily attained for gases but for liquids and solids it is difficult to change the temperature without changing the volume. In the isochoric change, the outside performs no work ($W_{\text{system}\leftarrow\text{outside}} = 0$) to the object. Therefore, equations (15.2) and (15.1) become

$$\Delta U = \Delta Q_{\text{system}\leftarrow\text{outside}},\ U_{\text{after}} = U_{\text{before}} + Q_{\text{system}\leftarrow\text{outside}} \quad \text{(isochoric change)}. \tag{15.8}$$

Isothermal change Changes in which an object is placed in a large constant-temperature bath (container kept at a constant temperature), and the volume and pressure change slowly while care is taken to keep the object temperature constant are called isothermal changes.

Example 2 Calculate the work $W = W_{\text{outside}\leftarrow\text{system}}$ to the outside performed by 1 mol of ideal gas during the isothermal change.

Solution Assume that contacting with the heat reservoir at T_0 the ideal gas changed isothermally from state (p_1, V_1, T_0) to state (p_2, V_2, T_0). In this case, the equation of state of ideal gas is $pV = RT_0 =$ constant. Accordingly, the work that the ideal gas performs to the outside is

$$W_{\text{outside}\leftarrow\text{system}} = \int_{V_1}^{V_2} p\,dV = \int_{V_1}^{V_2} \frac{RT_0}{V}\,dV$$
$$= RT_0 \int_{V_1}^{V_2} \frac{1}{V}\,dV$$
$$= RT_0(\log V_2 - \log V_1)$$
$$= RT_0 \log \frac{V_2}{V_1} = RT_0 \log \frac{p_1}{p_2}. \tag{15.9}$$

The change is isothermal, so the internal energy does not change (as discussed later, the internal energy of the ideal gas is a function of temperature alone). Therefore, the amount of heat that the ideal gas absorbed from the heat reservoir is $Q_{\text{system}\leftarrow\text{outside}} = -W_{\text{system}\leftarrow\text{outside}} = W_{\text{outside}\leftarrow\text{system}}$. When the ideal gas absorbs heat, $W_{\text{outside}\leftarrow\text{system}} = Q_{\text{system}\leftarrow\text{outside}} > 0$ and the volume increases ($V_2 > V_1$), and when the ideal gas releases heat, $W_{\text{outside}\leftarrow\text{system}} = Q_{\text{system}\leftarrow\text{outside}} < 0$ and the volume decreases.

Adiabatic change Changes in the state of an object in a situation where the heat transfer from and to the outside can be neglected, such as when the object is surrounded by thermal insulators, are called adiabatic changes. Since $\Delta Q_{\text{system}\leftarrow\text{outside}} = 0$, equations (15.2) and (15.1) become

$$\Delta U = \Delta W_{\text{system}\leftarrow\text{outside}},\ U_{\text{after}} = U_{\text{before}} + \Delta W_{\text{system}\leftarrow\text{outside}} \quad \text{(adiabatic change)}. \tag{15.10}$$

When the gas is compressed adiabatically, the outside performs work to the gas ($\Delta W_{\text{system}\leftarrow\text{outside}} > 0$), so the internal energy of the gas increases and the temperature of the gas rises. When the gas is adiabatically expanded, the gas performs work to the outside ($\Delta W_{\text{system}\leftarrow\text{outside}} < 0$), so the internal energy decreases and the temperature of the gas decreases. When the gas changes its volume adiabatically, the change in pressure is more intense than when the temperature is constant because the temperature also changes (Fig. 15-3).

In summer, when the moist air on the ground is heated, it expands and becomes less dense, resulting in an updraft. Since the pressure in

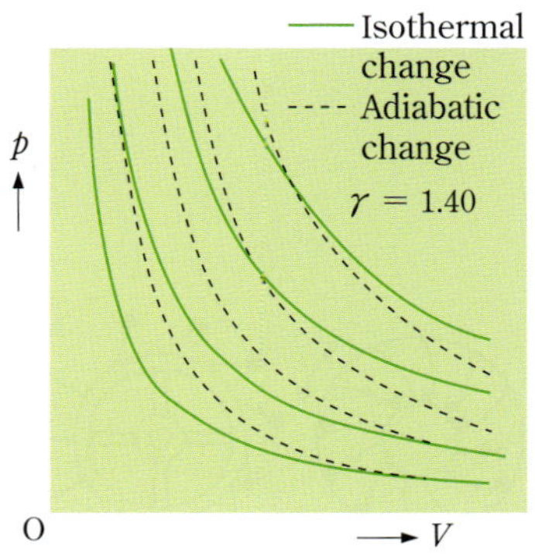

Fig. 15-3 Isothermal change (pV = constant) and adiabatic change (pV^{γ} = constant) of the ideal gas. $\gamma = \dfrac{C_p}{C_V}$ and $\gamma = 1.40$ for air. In the adiabatic change, the relationship between T and V is $TV^{\gamma-1} =$ constant.

Fig. 15-4 Cumulonimbi

the upper air is low, the air undergoes adiabatic expansion and the temperature drops. At this time, water vapor in the air condenses to form ice particles. This is a cumulonimbus that is produced in summer.

In the adiabatic change of ideal gas the following relationship holds (for proof, see the end of the next section).

$$pV^{\gamma} = \text{constant}, \quad TV^{\gamma-1} = \text{constant}, \quad \frac{T^{\gamma}}{p^{\gamma-1}} = \text{constant}. \qquad (15.11)$$

γ is the ratio of the molar heat capacity C_p at constant pressure to the molar heat capacity C_V at constant volume, $\gamma = \frac{C_p}{C_V}$.

Question 1 To bring the air at 10 °C to the temperature 100 °C by adiabatically compressing it in an insulated container, to how many times as small as the original volume should the volume be compressed? The γ of the air is 1.40.

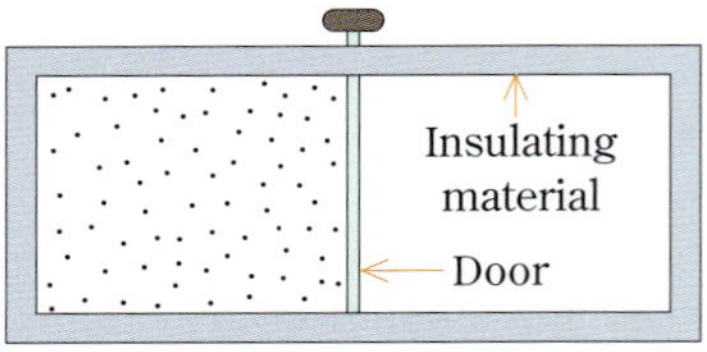

Fig. 15-5 Adiabatic free expansion of gas

Adiabatic free expansion of gas As shown in Fig. 15.5, we put gas in one of the two compartments of the container which has a partition door in the center. The container as a whole is covered with a heat insulating material. We turn the central door of the container to expand the gas to the other compartment of vacuum. This phenomenon is called **adiabatic free expansion of gas**.

The adiabatic free expansion of gas is an adiabatic change ($Q_{\text{outside}\leftarrow\text{system}} = Q_{\text{system}\leftarrow\text{outside}} = 0$) and also the gas does not perform work to the outside ($W_{\text{outside}\leftarrow\text{system}} = W_{\text{system}\leftarrow\text{outside}} = 0$). Therefore, equation. (15.1) becomes $U_{\text{before}} = U_{\text{after}}$, and the internal energy of the gas does not change. If the internal energy of the gas is assumed to be only a function of temperature, and not depending on pressure nor volume, that is, if $U = U(T)$ is assumed, then from the relation $U_{\text{before}} = U(T_{\text{before}}) = U_{\text{after}} = U(T_{\text{after}})$ we can conclude that the temperature remains constant ($T_{\text{before}} = T_{\text{after}}$) in the adiabatic free expansion.

In the adiabatic free expansion of real gases, the temperature changes slightly but the change is only a negligible extent. The ideal gas is defined as a virtual gas which satisfies the equation of state $pV = nRT$ and whose internal energy is a function of temperature only.

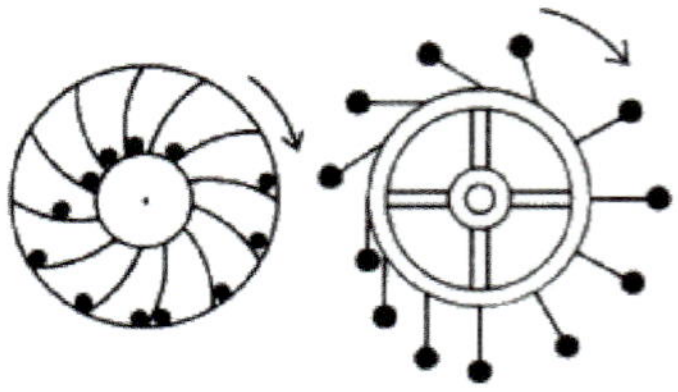

Fig. 15-6 Perpetual motion machine? The above tool was invented in the 13th century in Europe. It is based on the idea that it can rotate perpetually as a result of the displacement of the center of gravity synchronized with the rotation of wheel. This is a perpetual motion machine called the overbalanced wheel type.

For reference Perpetual motion machine

It would be convenient if we could have a machine that would keep working perpetually without energy being supplied from the outside. Many people tried to invent such machines for a long time, but no one succeeded. According to the first law of thermodynamics, when the heat engine repeating the same process operates for one cycle $U_{\text{after}} = U_{\text{before}}$, so "net heat quantity $Q_{\text{heat engine}\leftarrow\text{outside}}$ supplied from the outside to the heat engine" = "work $W_{\text{outside}\leftarrow\text{heat engine}}$ performed by the heat engine to the outside". Therefore, there exists no heat engine known as the **perpetual motion machine of the first kind** which performs no action other than performing work to the outside.

It would be useful if we could have a machine that could take heat from one heat reservoir and change all the heat into work without causing other changes. For example, if it would be possible to take heat from seawater and change it into work to turn a screw, it would be unnecessary to load ships with fuel. This heat engine called the **perpetual motion machine of the second kind** does not violate the first law of thermodynamics (energy conservation law) but it is contradictory to our daily experiences. The perpetual motion machine of the second kind is prohibiled by the second law of thermodynamics which the readers will learn in Section **15.3**.

15.2 Molar heat capacity of ideal gas

Learning objective To understand the derivation of the molar heat capacity at constant volume C_V and molar heat capacity at constant pressure C_p of the ideal gas and to understand the relationship between these two quantities C_p and C_V.

Molar heat capacity at constant volume When the temperature of 1 mol gas is raised by ΔT by addition of heat ΔQ at constant volume, the outside performs no work to the gas ($\Delta W = 0$), so the internal energy increases $\Delta U = \Delta Q$*. Dividing this equation by ΔT, since "heat capacity" = "amount of heat added ΔQ" divided by "rise in temperature ΔT", we have for the heat capacity of one mole of gas or molar heat capacity C_V the following relation :

$$C_V = \left(\frac{\Delta Q}{\Delta T}\right)_{\text{constant volume}} = \frac{\Delta U}{\Delta T} \qquad (15.12)$$

(here, U is the internal energy of 1 mol gas).

* In Section **15.2**
$\Delta Q = \Delta Q_{\text{gas}\leftarrow\text{outside}}$
$\Delta W = \Delta W_{\text{gas}\leftarrow\text{outside}}$

Molar heat capacity at constant pressure When ΔQ is added to the gas, the temperature rises by ΔT and the volume increases by ΔV, so the outside performs work $-p\,\Delta V$ to the gas. Equation (15.2) becomes

$$\Delta U = \Delta Q - p\,\Delta V \qquad (15.13)$$

[see equation (15.4)]. Since the equation of state of 1 mol of gas is $pV = RT$, addition of heat ΔQ at constant pressure p gives the relation

$$p(V+\Delta V) = R(T+\Delta T).$$

Combining the above equation with the equation $pV = RT$ leads to

$$p\,\Delta V = R\,\Delta T \quad \text{(isobaric change of ideal gas)}. \qquad (15.14)$$

Thus, equation (15.13) is changed into

$$\Delta U = \Delta Q - R\,\Delta T \quad \text{(isobaric change of ideal gas)}. \qquad (15.15)$$

Dividing the above equation by ΔT, we have the molar heat capacity C_p, that is, heat capacity of 1 mol of gas, expressed as

$$C_p = \left(\frac{\Delta Q}{\Delta T}\right)_{\text{constant pressure}} = \frac{\Delta U}{\Delta T} + R = C_V + R. \qquad (15.16)$$

It follows that between molar heat capacity C_V at constant volume and molar heat capacity C_p at constant pressure, Mayer's relation

$$C_p - C_V = R = 8.31\ \text{J/(K·mol)} \qquad (15.17)$$

is obtained (Fig. 15-7). The molar heat capacities of some gases are shown in Table 15-1, in which all gases satisfy equation (15.17).

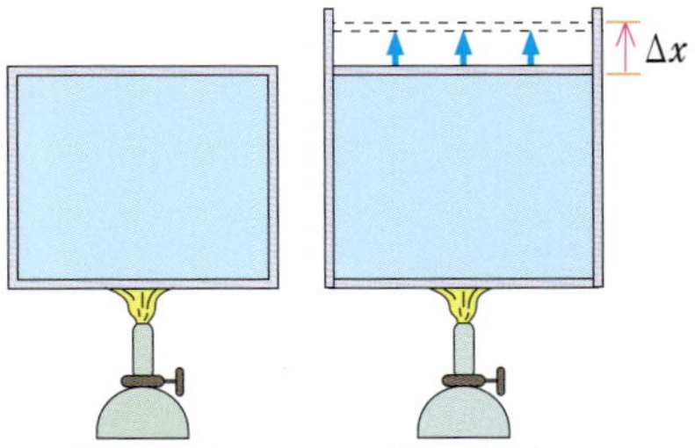

Fig. 15-7 Molar heat capacity at constant volume C_V and molar heat capacity at constant pressure C_p.
$C_p - C_V = R$

Question 2 Explain the reason why the molar heat capacity at constant pressure C_p is larger than the molar heat capacity at constant volume C_V.

In the kinetic theory of gases, as shown in equation (14.35), the internal energy of 1 mol gas is $U = \frac{f}{2}RT$, so it follows that $C_V = \frac{\Delta U}{\Delta T} = \frac{f}{2}R$. For monoatomic gases, since $f = 3$, we have

$$C_V = \frac{\Delta U}{\Delta T} = \frac{3}{2}R = 12.5\ \mathrm{J/(K\cdot mol)} \qquad \text{(monoatomic gas)}. \tag{15.18}$$

Table 15-1 Molar heat capacities of gases
[Values at 1 atm and 15 °C. Unit in J/(K·mol)]

Gas	C_p	$\gamma = \frac{C_p}{C_V}$	$C_p - C_V$
He	20.94	1.66	8.3
Ar	20.9	1.67	8.4
O_2	29.50	1.396	8.37
N_2	28.97	1.405	8.35
CO_2	36.8	1.302	8.5
SO_2	40.7	1.26	8.4

Table 15-1 demonstrates that equation (15.18) holds exactly for monoatomic molecular gases. Equation (15.18) does not hold for diatomic gases and triatomic gases. The reason is that, as described in Section **14.3**, the internal energy is increased by the rotational motion of gas molecules when the temperature rises (Fig. 15-8). As for the diatomic molecules in the table, $C_V \approx \frac{5}{2}R$, so $f \approx 5$, and for triatomic molecules $C_V \gtrsim 3R$, so $f \gtrsim 6$.

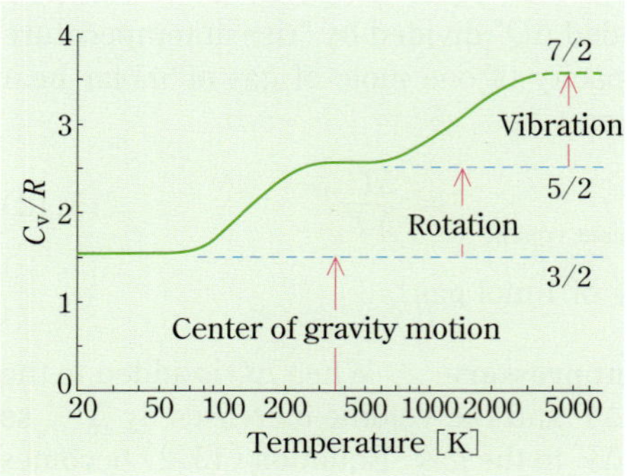

Fig. 15-8 Temperature variation of $\frac{C_V}{R}$ of hydrogen gas H_2. With the increase in temperature, the molar heat capacity increases as a result of rotation and vibration of the molecule.

Dulong-Petit's Law In 1819, Dulog and Petit found that the molar heat capacity of solid elements is $3R \approx 25\ \mathrm{J/(K\cdot mol)}$ regardless of the type of constituent atoms. This is called **Dulog-Petit's law**. This phenomenon is because the mean value of potential energy and kinetic energy due to thermal vibration of each atom is $\frac{3}{2}kT$. At low temperatures, the molar heat capacity becomes less than $3R$ due to quantum effects. Carbon, boron, etc. show the deviation from this law even at room temperature (see Fig. 14-5).

For reference Proof of equation (15.11) for the adiabatic change of ideal gas

We take 1 mole of ideal gas. In the adiabatic change, since $dQ_{system\leftarrow outside} = 0$, the first law of thermodynamics becomes $dU = dW_{system\leftarrow outside}$. The work that the outside performs to the gas is $dW_{system\leftarrow outside} = -p\,dV$, and from $C_V = \frac{dU}{dT}$ the relation $dU = C_V\,dT$ holds, so the relation

$$C_V\,dT = -p\,dV \qquad (15.19)$$

is obtained. When p, V, T changes to $p+\Delta p$, $V+\Delta V$, $T+\Delta T$, respectively, the equation of state $pV = RT$ changes as follows:

$$(p+dp)(V+dV) = R(T+dT)$$

$$\therefore \quad pV+p\,dV+V\,dp+dp\cdot dV = RT+R\,dT.$$

The quantity $dp\cdot dV$ is only a minute amount and can be neglected. Using the relation $pV = RT$, we have

$$p\,dV+V\,dp = R\,dT. \qquad (15.20)$$

From equation (15.19) we have $dT = -\frac{p}{C_V}dV$ and using this and equation (15.17), $R\,dT$ is expressed as

$$R\,dT = -\frac{R}{C_V}p\,dV = \frac{C_V-C_P}{C_V}p\,dV = \left(1-\frac{C_P}{C_V}\right)p\,dV. \qquad (15.21)$$

The ratio of C_P to C_V is now expressed as γ:

$$\gamma = \frac{C_P}{C_V} \qquad (15.22)$$

[If the equation $C_V = \frac{f}{2}R$ is used, since $\gamma = \frac{C_V+R}{C_V} = \frac{f+2}{f}$, we obtain for monatomic molecules ($f = 3$), $\gamma = 1.67$, for diatomic molecules ($f \approx 5$), $\gamma \approx 1.40$, and for triatomic molecules ($f \gtrsim 6$), $\gamma \lesssim 1.33$]. Using equations (15.21) and (15.22), equation (15.20) is changed into $\gamma p\,dV+V\,dp = 0$. This formula is modified into

$$\frac{dp}{p}+\gamma\frac{dV}{V} = 0. \qquad (15.23)$$

Integration (since the primitive function of $\frac{1}{x}$ is $\log|x|$)* of equation (15.23) leads to

$$\log p+\gamma\log V = \log p+\log V^{\gamma} = \log pV^{\gamma} = \text{constant}.$$

Namely, in an adiabatic change, we have the relation

$$pV^{\gamma} = \text{constant}. \qquad (15.24)$$

Applying $pV = RT$ to equation (15.24) gives the relations

$$TV^{\gamma-1} = \text{constant, and } \frac{T^{\gamma}}{p^{\gamma-1}} = \text{constant}. \qquad (15.25)$$

* $\log x$ is a logarithm whose base is e. In electric calculators and in many physical textbooks published in Europe and the US, the natural logarithm $\log_e x$ is expressed as $\ln x$.

15.3 The second law of thermodynamics

Learning objective To become able to list some examples of irreversible changes. To understand the two statements of the second law of thermodynamics which is related to the direction of irreversible changes.

Fig. 15-9 Spinning tops lying on the floor never stand up to start spinning naturally.

Reversible changes and irreversible changes As in the case of oscillation of a pendulum where air resistance and friction can be ignored, if a motion of some phenomenon is video-filmed and the reproduced images in the reverse direction correspond to the actually attainable motion, the phenomenon of that motion is said to be reversible. An object sliding on a floor with friction is decelerated and stopped, but when this motion is video-filmed and played back in the reverse rotation, the object that was at rest appears to start moving spontaneously and accelerate. Similarly like this, if the images reproduced in the reverse rotation correspond to the motion that is not realized in practice, the phenomenon is called irreversible.

Strictly reversible changes occur only in idealized situations, such as motion without friction or air resistance.

When a hot object is brought into contact with a cold object, the heat transfer from the hot object to the cold object always occurs. Heat transfer from a low temperature object to a high temperature object is not prohibited by the energy conservation law but it never happens naturally. To transfer heat from a low temperature object to a high temperature object to further cool the low temperature object and to further heat the high temperature object, work from the outside is indispensable, similarly as in the case of a refrigerator. Thus, heat conduction from hot objects to cold objects is an irreversible process*.

The reverse process of heat generation by friction is the process of taking heat from one heat reservoir and transforming it all into work, but this also never happens spontaneously.

* In thermodynamics, similarly as in such cases as heat conduction between objects with infinitesimally small temperature difference and expansion due to infinitesimally small pressure difference, if an infinitesimal change in temperature difference or pressure difference occurs in the opposite direction, the process is said to be reversible. In other words, changes that proceed slowly and can be considered to be all the time in thermal equilibrium on the way, namely the changes called quasi-static changes are considered as reversible changes.

Second law of thermodynamics The law that indicates the direction of irreversible changes in which heat is involved is the **second law of thermodynamics**, and there are following two expressions based on the two irreversible changes mentioned just above.

Clausiu statement Heat never transfers from a cold object to a hot object without causing changes elsewhere.
Thomson statement The heat extracted from one heat reservoir is never wholly converted to work without changes elsewhere.

Since we can derive one statement from the other, the two statement are equivalent (see Problems for exercise 15, B6).

15.4 Heat engine and its efficiency

Learning objective To understand that a heat engine is composed of three components : a high-temperature heat reservoir, a low-temperature heat reservoir, and a working substance, and that it is the device in which the working substance carries out a circulating process to convert part of the heat supplied by the high-temperature heat reservoir into work. To memorize the theoretical upper limit value of the heat engine.

Heat engine Power units are called engines in English and "kikan" in Japanese. A **heat engine** is a device that receives heat from the outside and performs work. Historically, steam engines were first developed to burn coal, convert chemical energy into heat, and turn it into mechanical work. Currently thermal power plants burn oil and/or coal to change the chemical energy into heat, and nuclear power plants change the nuclear energy released when nuclear fission of uranium nuclei takes place into heat, and these heats are then changed into the work of turbines. In ships and cars, diesel engines and gasoline engines are used. Heat engines constitute an integral part of industry and everyday life.

Fig. 15-10 A steam locomotive

A heat engine which changes heat Q into work W as much as possible is desirable. The ratio $\frac{W}{Q}$ at which heat Q becomes work W is called the **efficiency of heat engine**. How high can we improve the efficiency of the heat engine? Is it possible to raise it to 1? Carnot was the first to study this problem. The industrial revolution which began in the 18th century was driven by the use of steam engines as power sources instead of human power and horse power. Carnot considered that in order to enhance technological methods of generating heat to the stage of science, it is required to carry out the study of heat phenomena as a whole from a general point of view, regardless of individual engines, machines, etc.

Consider steam engines. The cross section of a steam engine is shown in Fig. 15-11. The explanation of the operation of this steam engine is given below the figure. The steam engine has a boiler that heats water into high-temperature, high-pressure steam and a condenser that cools steam into water. In general, heat engines have two heat reservoirs : (1) a **high-temperature heat reservoir** that releases heat like a boiler, and (2) a **low-temperature heat reservoir** that absorbs heat like a condenser that cools steam. Furthermore, (3) like a steam there is a **working substance** that expands and contracts to performs work to the outside. Thus in total, heat engines have three elements, high-temperature heat reservoir, low-temperature heat reservoir, and working substance.

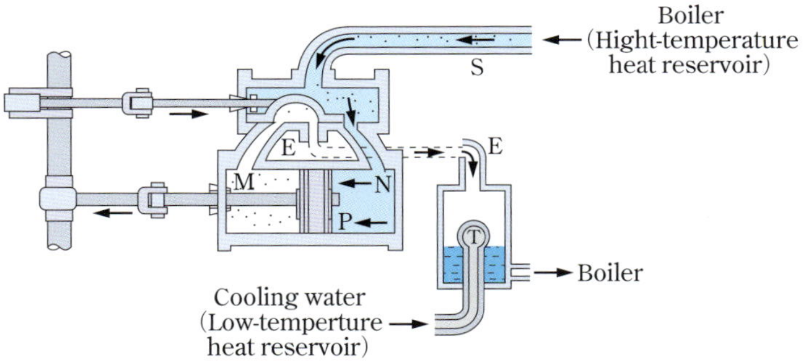

Fig. 15-11 A steam engine. The high-temperature high-pressure steam coming from the boiler through the pipe S moves the piston P through the pipe N (or M). The vapor on the opposite side is discharged to the outside through M (or N) and E. T is a condenser using cooling water, which cools to condense the vapor to be discharged.

All heat engines other than steam engines have these three elements. In gasoline engines and diesel engines, air is used as the working substance and the working substance is directly heated by burning the fuel in the engine and the working substance is released to the atmosphere without being cooled, but here the atmosphere works as a low-temperature heat reservoir. Therefore, they also have three elements*.

* The Thomson statement of the second law of thermodynamics indicates the necessity of a low temperature heat reservoir.

In short, a heat engine has three elements : a high-temperature heat reservoir (temperature T_H), a low-temperature heat reservoir (temperature T_L) and a working substance. The working substance undergoes a **cyclic process (cycle)** starting from one state and returning to the original state again. During each cycle, the working substance receives heat Q_H from the high-temperature heat reservoir, converts part of it into work W, and releases the remaining $Q_L = Q_H - W$ as heat to the low-temperature heat reservoir (Fig. 15-12). Therefore, the efficiency η of this heat engine is

$$\eta = \frac{W}{Q_H} = \frac{Q_H - Q_L}{Q_H}. \qquad (15.26)$$

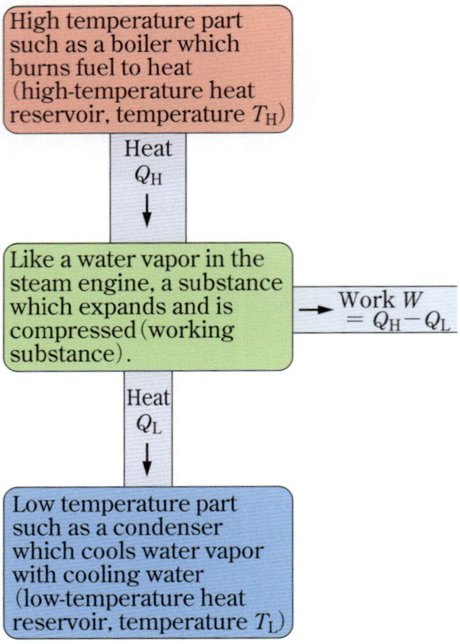

Fig. 15-12 The three elements of engine. W is the work that the working substance performs to the outsides.

Carnot's cycle At this point the thought experiment that Carnot studied to find the limits of the efficiency of heat engines will be described. This is the study written by Carnot in 1819 and published in 1824 as a book with the title, "Reflections on the motive power of fire and on a suitable engine for generating this power".

Carnot selected the ideal gas as the working substance of the heat engine, put the gas in a cylinder attached a frictionless piston, used the large high-temperature heat reservoir of constant temperature T_H and the large low temperature heat reservoir of constant temperature T_L ($T_H > T_L$), and considered the reversible cyclic process which combines isothermal expansion, adiabatic expansion, isothermal compression, and adiabatic compression stages (Fig. 15-13). For simplicity, the amount of ideal gas is taken to be 1 mol.

(1) While the cylinder is in contact with the high-temperature heat reservoir at temperature T_H and the working substance expands slowly, the working substance receives heat Q_H and isothermally

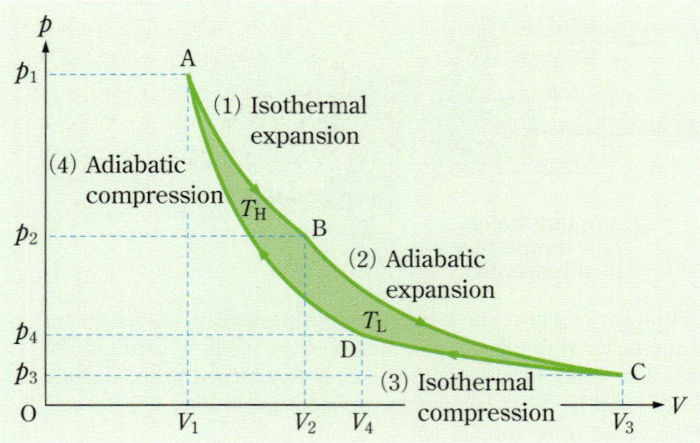

Fig. 15-13 Carnot's cycle. The surface area of the ■ part is equal to the net work done by Carnot's heat engine in one cycle [The area of the lower part of the curve A → B → C represents the work that the heat engine performs to the outside in the isothermal expansion of (1) and the adiabatic expansion of (2), and the area of the lower part of the curve C → D → A represents the work done by the outside in the isothermal compression of (3) and the adiabatic compression of (4).

expands from the state (p_1, V_1, T_H) to the state $(p_2, V_2, T_H)(V_2 > V_1)$. In this process, the equation of state $pV = RT_H$ holds, and the pressure decreases $(p_2 < p_1)$. Since the isothermal change does not change the internal energy of ideal gas, the work W_1 performed by the working substance is equal to Q_H. From Example 2 in Section **15.1**, we have

$$Q_H = W_1 = RT_H \log \frac{V_2}{V_1}. \tag{15.27}$$

(2) The cylinder is separated from the high-temperature heat reservoir, and the working substance is adiabatically expanded slowly to change from the state (p_2, V_2, T_H) to the state (p_3, V_3, T_L). The working substance performs work W_2 to the outside so the temperature drops $(T_H > T_L)$. Since there is neither entering nor going-out heat,

$$W_2 = U(T_H) - U(T_L). \tag{15.28}$$

(3) When the working substance is compressed slowly while the cylinder is in contact with a low temperature heat reservoir of temperature T_L, the working substance releases heat Q_L, and the gas is isothermally compressed from the state (p_3, V_3, T_L) to the state (p_4, V_4, T_L). Because the volume decreases $(V_4 < V_3)$, the work W_3 that the working substance performs to the outside is negative $(W_3 = -Q_L < 0)$ and

$$-Q_L = W_3 = RT_L \log \frac{V_4}{V_3}. \tag{15.29}$$

(4) The cylinder is separated from the low-temperature heat reservoir and the working substance is adiabatically compressed slowly to return from the state (p_4, V_4, T_L) to the initial state (p_1, V_1, T_H). There is no entering heat nor going-out heat, and the temperature rises from T_L to T_H, because the outside performs a positive work to the working substance. Therefore, in this change the work W_4 that the working substance performs to the outside is minus $(W_4 < 0)$:

$$W_4 = U(T_L) - U(T_H). \tag{15.30}$$

This completes one cycle. This cycle is called **Carnot's cycle**. In the adiabatic change, since equation (15.11) holds, the relations

$$T_H V_2^{\gamma-1} = T_L V_3^{\gamma-1} \quad \text{and} \quad T_H V_1^{\gamma-1} = T_L V_4^{\gamma-1}$$

are derived. The relation $\dfrac{V_2^{\gamma-1}}{V_1^{\gamma-1}} = \dfrac{V_3^{\gamma-1}}{V_4^{\gamma-1}}$ obtained by eliminating T_H and T_L from the above equations leads to

$$\frac{V_2}{V_1} = \frac{V_3}{V_4}. \tag{15.31}$$

Thus, from equations (15.29) and (15.31)

$$Q_L = -W_3 = -RT_L \log \frac{V_4}{V_3} = RT_L \log \frac{V_2}{V_1} \tag{15.32}$$

is derived.

Therefore, when the reversible engine considered by Carnot performs a reversible circulation process once, the total work W performed to the outside is

$$W = W_1 + W_2 + W_3 + W_4 = Q_H - Q_L = R(T_H - T_L) \log \frac{V_2}{V_1}. \tag{15.33}$$

The work W that Carnot's heat engine performs in one cycle is the area of the colored section ■ in Fig. 15-13.

Since the heat Q_H that Carnot's heat engine receives from the high-temperature heat reservoir is given by equation (15.27), the efficiency $\eta = \dfrac{W}{Q_H}$ of Carnot's reversible engine is

$$\eta = \frac{W}{Q_H} = \frac{T_H - T_L}{T_H} = 1 - \frac{T_L}{T_H}. \tag{15.34}$$

T_H is the temperature of the high-temperature heat reservoir and T_L is the temperature of the low temperature heat reservoir.

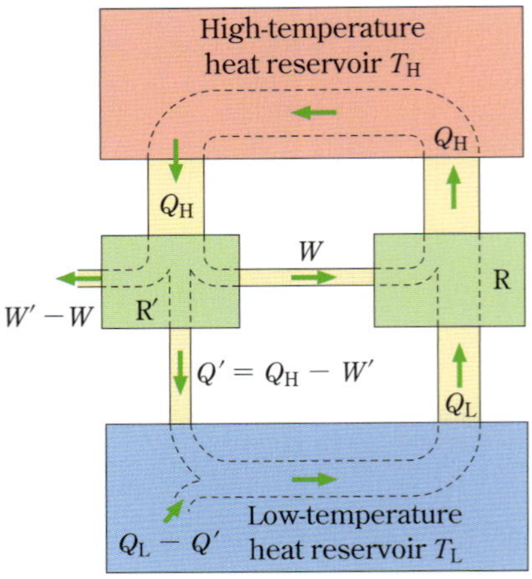

Fig. 15-14 If another working substance R′ were used to create a heat engine with higher efficiency than Carnot's reversible engine, then ⋯.

Efficiency of heat engines and Carnot's principle If a substance other than the ideal gas R is used as a working substance, can we create a more efficient heat engine? Let us consider the following situation : A heat engine (which can be an irreversible engine) using another working substance R′ is operated between the high-temperature heat reservoir T_H and the low-temperature heat reservoir T_L. The heat Q_H is received from the high-temperature heat reservoir and the work W' is performed to the outside, with the remaining $Q' = Q_H - W'$ being released to the low temperature heat reservoir as heat. Since Carnot's heat engine is a reversible engine, when it is operated reversely, it gets work W from the outside and receives heat Q_L from the low-temperature heat reservoir and releases heat Q_H to the high-temperature heat reservoir. Thus, if we create a combined heat engine by combining these two heat engines as shown in Fig. 15-14, what this combined heat engine does is only to receive heat $Q_L - Q'$ from the low temperature heat reservoir and convert it all into work $W' - W = Q_L - Q'$ to the outside. Therefore, if $W' - W > 0$, this is contradictory to the second law of thermodynamics (Thomson statement). Thus, it should follow that $W' \leqq W$. In other words, it has been proved that there exists no heat engine that is more efficient than Carnot's reversible engine.

If this combined heat engine having an efficiency of $\dfrac{W'}{Q_H}$ is assumed to be a reversible engine and is operated in reverse, then it is possible to prove $W \leqq W'$. Accordingly, the relation $W = W'$ is obtained, so it has been proved that the efficiency is the same, irrespective of the working substance used in constructing the reversible engine. In practice, due to generation of the heat loss in heat conduction from the high-temperature heat reservoir to the low-temperature heat reservoir and the frictional heat, etc., any actual heat engine is irreversible and its efficiency is less than that of the reversible engine. These results are summarized as follows :

Carnot's principle Among heat engines that receive heat Q_H from a constant temperature heat reservoir (high-temperature heat reservoir, temperature T_H) and release heat Q_L to a constant temperature heat acceptor (low-temperature heat reservoir, temperature T_L) and perform work W, the most efficient is the reversible engine, and its efficiency is

$$\eta = \frac{W}{Q_H} = \frac{Q_H - Q_L}{Q_H} = \frac{T_H - T_L}{T_H}. \quad (15.35)$$

Mechanical energy and electrical energy can be converted into work with 100 % efficiency, but the efficiency of heat engines cannot be 100 %. Currently, steam turbines that rotate the blades of the turbine with high-temperature and high-pressure steam are often used rather than steam engines shown in Fig. 15-11. However, the upper limit of the steam turbine efficiency is also given by equation (15.35).

In order to operate a large heat engine to perform large-scale work, a huge amount of heat must be generated by a great amount of petroleum, natural gas, coal, nuclear fuel, and so on. However, since only part of the heat is converted into work, a large amount of heat is released to the atmosphere (a low temperature heat reservoir) such as the atmosphere, rivers, and seas. In addition, in using fossil fuels, it is desirable to improve efficiency to reduce the amount of carbon dioxide emitted into the surrounding.

To increase the efficiency of the heat engine, it is necessary to reduce $\frac{T_L}{T_H}$. That is to say it is necessary to lower the temperature T_L of the low-temperature heat reservoir and to increase the temperature T_H of the high-temperature heat reservoir. But since the low-temperature heat reservoir is cooling water or air that cools the working substance, its temperature T_L cannot be set to 270 to 300 K or less. Therefore, the only way left to improve efficiency is to raise the temperature of the high-temperature heat reservoir T_H. When the temperature of the high-temperature heat reservoir is raised, the pressure p_1 of the working substance at the high-temperature heat reservoir will also increase, so the heat engine must be constructed with materials that can withstand high-temperature and high-pressure conditions.

The steam of a high-performance thermal power plant is about 600 °C and the efficiency is about 43 %. As a system with higher efficiency, the combined cycle power generation is available. In this system a generator is rotated with both a gas turbine rotating with the gas power arising from burning city gas and a steam turbine rotating with the steam power generated by the high-temperature exhaust gas. When the gas temperature at the inlet of the gas turbine is about 1600 °C, the power generation efficiency is about 60 %.

©2019 Toshiba Energy Systems & Solutions Corporation

Fig. 15-15 A turbine used in the power plant

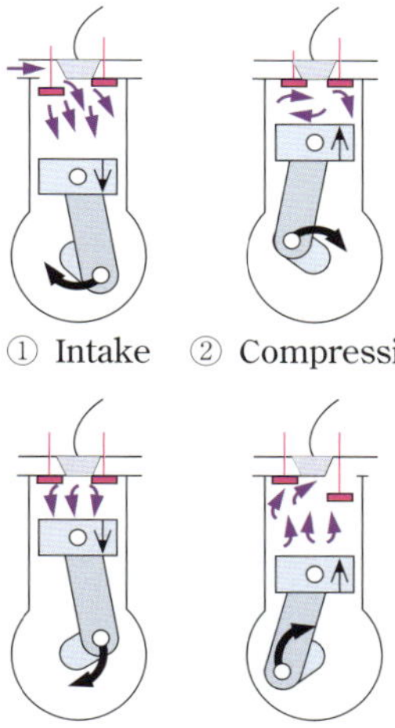

① Intake ② Compression

③ Combustion ④ Exhaust

Fig. 15-16 Otto cycle

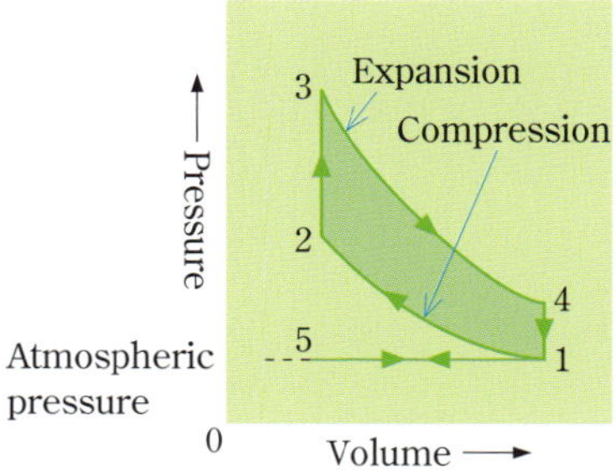

Fig. 15-17 Diagram of volume (V)-pressure (p) relation of the Otto cycle

Case 1 In Fig. 15-16 is illustrated the operation of gasoline engine called the **Otto cycle**. In this engine, in stage ① the fuel gas is drawn into the cylinder during the outward movement of the piston, and in stage ② the drawn gas is compressed during the inward movement of the piston, then in stage ③ the gas is ignited at the moment when it is most compressed, and the gas expands during the second outward movement, and finally in stage ④ the burnt gas is expelled during the second inward movement. The engine repeats these processes.

The change in volume and pressure of air that is the working substance of the engine is shown in Fig. 15-17. In Fig. 15-17, 5 → 1 is the intake process of ①, and 1 → 2 is the compression process of ② and at point 2 the gas is ignited. 2 → 3 → 4 is the explosion process of

③, and 4 → 1 → 5 is the exhaust process of ④.

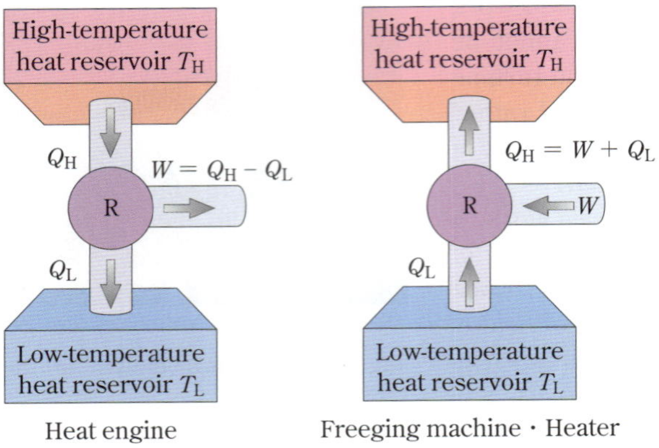

Fig. 15-18 A heat engine and a cooler · heater

Refrigerator and heater Carnot's heat engine is a machine in which a working substance receives heat Q_H from a high-temperature heat reservoir, converts W part of Q_H into mechanical work, and releases the remaining energy $Q_L = Q_H - W$ as heat to a low-temperature heat reservoir. When Carnot's heat engine is operated reversely and work W is applied to the working substance from the outside, the working substance receives heat Q_L from the low-temperature heat reservoir and releases heat $Q_H = Q_L + W$ to the high-temperature heat reservoir (Fig. 15-18). If we pay attention to the low-temperature heat reservoir, this machine is a refrigerator (freeging machine) or an air-conditioner that draws heat from a low-temperature heat reservoir to further reduce the temperature, and it is a (heat pump type) heater if attention is paid to the high-temperature heat reservoir. In the case of a cooler, indoor air is the low-temperature heat reservoir, and outdoor air is the high-temperature heat reservoir. In the case of a heat pump type heater, indoor air is the high-temperature heat reservoir, and outdoor air is the low-temperature heat reservoir.

Defining the performance of the refrigerator (freezer, cooler) as $\frac{Q_L}{W}$, we have

$$\text{performance of the refrigerator} = \frac{Q_L}{W} = \frac{Q_L}{Q_H - Q_L} \leqq \frac{T_L}{T_H - T_L}. \tag{15.36}$$

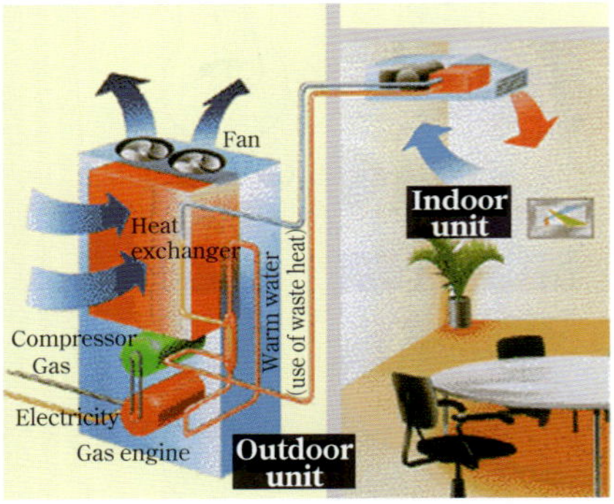

Fig. 15-19 Mechanism of heat pump

Defining the performance of the heat pump type heater as $\frac{Q_H}{W}$, we have

$$\text{performance of the heat pump type heater} = \frac{Q_H}{W} = \frac{Q_H}{Q_H - Q_L} \leqq \frac{T_H}{T_H - T_L}. \tag{15.37}$$

An electric heater that generates Joule heat from the current supplied to the nichrome wire can generate only the same amount of heat as the consumed electric power, but in the case of a heat pump heater, heat is transferred from the low-temperature heat reservoir to the high-temperature heat reservoir and the amount of heat Q_H obtained is larger than the electric power consumption W.

In most air conditioners, the working material repeats adiabatic compression and adiabatic expansion. The driving force for this process comes from the compressor, and the work performed by the compressor is the work from the outside.

Question 3 When the air temperature is −5 °C and the room temperature is 25 °C, what amount of work (in J) is required to supply 1 J of heat to the room with a heat pump type heater?

Thermodynamic temperature According to Carnot's principle no matter what working substances are used to construct a reversible engine between the heat Q_H received by the high-temperature heat reservoir at the temperature T_H and the heat Q_L released by the low-temperature heat reservoir at the temperature T_L the relationship

$$\frac{Q_H}{Q_L} = \frac{T_H}{T_L} \tag{15.38}$$

holds. Therefore, if we operate a reversible engine with two objects consisting of an object at reference temperature T_0 and another object at unknown temperature T as heat reservoirs and let Q_0 and Q be the magnitudes of the heat quantity received by the objects at temperatures T_0 and T, respectively, the unknown temperature T can be determined as $T = \frac{Q}{Q_0} T_0$. The temperature defined by Kelvin in this way based on a reversible engine is called the **thermodynamic temperature**. This temperature does not depend on the type of thermometer (i.e. working substance of the reversible engine). In the International System of Units, the temperature of the triple point of water was chosen to be 273.16 K as the standard temperature T_0*. The thermodynamic temperature is the same as the absolute temperature T that appears in the equation of state $pV = nRT$ of ideal gas.

* Currently, the unit kelvin K of the thermodynamic temperature is set by defining the Boltzmann constant k appearing in the equation (14.32) exactly as $1.380\,649 \times 10^{-23}$ J/K.

15.5 Principle of increase of entropy

Learning objective To understand that the second law of thermodynamics is quantitatively formulated in the form of “Entropy always increases” by introducing a quantity of state called entropy.

Entropy In a reversible engine that absorbs heat Q_1 from a high-temperature heat reservoir at temperature T_1 and releases heat Q_2 to a low-temperature heat reservoir at temperature T_2, the following relationship holds :

$$\frac{Q_1}{T_1} = \frac{Q_2}{T_2}. \tag{15.39}$$

This fact indicates that the quantity $\frac{Q}{T}$ plays an important role in reversible changes. Therefore, Clausius introduced the physical quantity called **entropy** (symbol S) which can satisfy the following three properties.

(1) When an amount of heat Q is reversibly released from a system of temperature T, the entropy of the system decreases by $\frac{Q}{T}$.

(2) When a system of temperature T absorbs heat Q reversibly, the entropy of the system increases by $\frac{Q}{T}$.

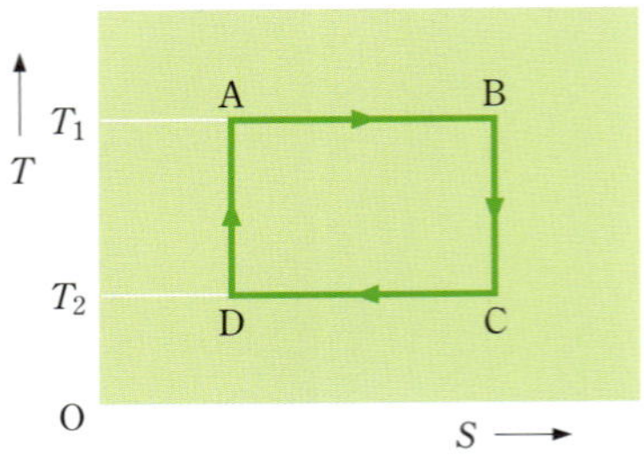

Fig. 15-20 Entropy (S)-temperature (T) diagram

* $Q_i = Q_{\text{system}\leftarrow i}$

(3) The entropy of the system remains unchanged even if the energy of the system is transferred reversibly to the outside of the system as work.

Then, the entropy change in Carnot's cycle in Fig. 15-13 is as shown in Fig. 15-20. When the system changes from state A to state C, the change in entropy is the same $\left(\dfrac{Q_1}{T_1} = \dfrac{Q_2}{T_2}\right)$ irrespective of the path A → B → C or A → D → C that the system takes. Thus, if the entropy S_A of state A is selected as a reference, the entropy S_C of state C can be decided uniquely as $S_C = S_A + \dfrac{Q_1}{T_1} = S_A + \dfrac{Q_2}{T_2}$.

In general, when a system receives heat Q_1, Q_2, ⋯ from a number of heat reservoirs (T_1, T_2, ⋯) in a reversible change (from now on Q_i is defined as a negative quantity when it is released)* from state A to state B, the difference $S_B - S_A$ in entropy of the two states A and B is given as

$$S_B - S_A = \sum_i \frac{Q_i}{T_i} \quad \text{(reversible change)}. \tag{15.40}$$

If the system changes reversibly from state A to state B, then equation (15.40) holds regardless of how A changes to B. In other words, for any A → B reversible change it can be proved, using Carnot's principle and Carnot's cycle, that the right-hand side of equation (15.40) is constant regardless of the path. Therefore, entropy can be defined as a function of state, namely a quantity of state.

If the system changes irreversibly from state A to state B, equation (15.40) does not hold. Therefore, as in the case of adiabatic free expansion of the ideal gas to vacuum, which is an irreversible change of the system, since there is no heat exchange ($Q = 0$) with the outside, the entropy may appear unchanged, but equation (15.40) does not hold for irreversible change, and in reality $S_A \neq S_B$.

Unit of entropy J/K

In this way, we could define a new quantity of state, **entropy**, by considering a reversible change connecting two states. The origin of the word entropy is Greek, meaning change. The unit of entropy is J/K.

In atomic theory, **entropy** is a quantity that represents the randomness of the molecular assembly that composes a system. The expressions (1) and (2) given above reflect the fact that if heat is absorbed the randomness of the molecular assembly increases, and if heat is released the randomness of the molecular assembly is reduced, and the change in randomness is more remarkable as the temperature is lower. The expression (3) reflects the fact that work is due to the ordered motion of the molecules.

Example 3 Calculate the change in entropy when 1 kg of water at 0 °C is heated to 100 °C.

Solution The amount of heat required to raise the temperature of an object with a specific heat capacity C and mass m by dT is $dQ = mC\,dT$. When the temperature of this object is raised from T_A to T_B, the change in entropy of this object is

$$S_B - S_A = \int_A^B \frac{dQ}{T} = mC\int_{T_A}^{T_B} \frac{dT}{T}$$

$$= mC \log T \Big|_{T_A}^{T_B} = mC \log \frac{T_B}{T_A} \tag{15.41}.$$

Accordingly, when $T_A = 273$ K, $T_B = 373$ K, $m = 1000$ g, and $C = 4.2$ J/(g·K), we have

$$S_B - S_A = (10^3\ \text{g}) \times \{4.2\ \text{J/(g·K)}\} \times \log \frac{373\ \text{K}}{273\ \text{K}} = 1.3 \times 10^3\ \text{J/K}.$$

Principle of increase of entropy So far, we have considered changes in entropy of the systems when they change reversibly. These systems reversibly exchange heat with an external heat reservoir. When a system receives heat Q from a heat reservoir at temperature T in an isothermal reversible change, the entropy of the system increases by $\frac{Q}{T}$, but the entropy of the heat reservoir that released the heat Q decreases by $\frac{Q}{T}$ $\left(\text{increase by } -\frac{Q}{T}\right)$. Thus, the overall entropy of the system and heat reservoir does not change.

At this place, we consider the case where an isolated system which does not exchange heat nor work with the outside changes irreversibly from state B to state A. Since this change is irreversible, no reverse change from state A to state B occurs so long as the system is isolated. Let us show that in the irreversible change, the entropy S_A of the final state is greater than the entropy S_B of the initial state B, that is,

$$S_A > S_B \qquad \text{(irreversible change B} \to \text{A).} \qquad (15.42)$$

Here we change the subscripts of Carnot's principle [equation (15.35)] H to 1 and L to 2 and also change the sign of Q_2 :

$$\frac{Q_1 + Q_2}{Q_1} < \frac{T_1 - T_2}{T_1} \qquad \text{(irreversible engine)} \qquad (15.43)$$

[Since the low-temperature heat reservoir is assumed to absorb heat Q_2 ($Q_2 < 0$), the sign of Q_2 is different from that of equation (15.35)]. Equation (15.43) is modified into

$$\frac{Q_1}{T_1} + \frac{Q_2}{T_2} < 0 \qquad \text{(irreversible engine).} \qquad (15.44)$$

For a heat engine exchanging heat with three or more heat reservoirs,

$$\sum_i \frac{Q_i}{T_i} < 0 \qquad \text{(irreversible engine)} \qquad (15.45)$$

holds for one cycle of operation (Q_i is the amount of heat that the heat engine absorbs from the heat reservoir at temperature T_i). This is called **Clausius' inequality**. In the case of reversible change, the temperature T_i of the heat reservoir is also the temperature of the system.

Let us assume that a system isolated from the outside changes irreversibly from state B to state A, then exits the isolated state, and changes from state A to state B in a reversible process while exchanging heat with the heat reservoir outside the system (Fig. 15-21). In this case, the inequality (15.45) becomes

$$\sum_{i(\text{B}\to\text{A})}^{\text{irreversible}} \frac{Q_i}{T_i} + \sum_{i(\text{A}\to\text{B})}^{\text{reversible}} \frac{Q_i}{T_i} < 0 \qquad (15.46)$$

(Q_i is the amount of heat that the engine absorbs from the heat reservoir at T_i).

Since the change, state A $\to$ state B, is a reversible change, the second term on the left side is $S_B - S_A$ from equation (15.40). In the

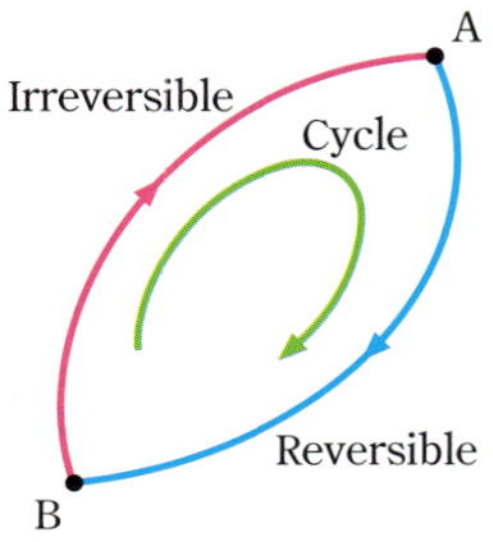

Fig. 15-21

change, state B → state A, the system is isolated and there is no exchange of heat with the outside, so $Q_i = 0$. Thus, the first term of the left side is 0. Consequently, equation (15.46) becomes $0+(S_B-S_A) < 0$,

$$\therefore \quad S_A > S_B \quad \text{(in the reversible change, the equality is used).} \tag{15.47}$$

In other words, when the system isolated from the outside changes irreversibly from state B to state A, the entropy of the system increases. When the isolated system changes reversibly, the entropy of the system does not change ($S_A = S_B$), but the reversible change is the idealized case of the limit where the friction, and the heat conduction from high temperature part to low temperature part can be ignored, and in reality such changes do not exist. In this way, the **principle of increase of entropy** has been derived.

The entropy of an isolated system increases in any case.

The principle of increase of entropy indicates the direction of irreversible change and is a quantitative expression of the second law of thermodynamics. The reason why equation (15.44) becomes an inequality is because in the irreversible engine, the heat released to the low-temperature heat reservoir is too much compared to that in the reversible engine. Therefore, the increase in entropy means waste of heat.

Example 4 Mixing 1 kg of water at 20 °C with 1 kg of water at 80 °C gives 2 kg of water at 50 °C. Calculate the change in entropy of this mixing.

Solution Using equation (15.41), the entropy of m kg of water at temperature T is

$$S = (m \times 4.2 \times 10^3 \text{ J/K}) \log \frac{T}{T_0}.$$

T_0 is the reference temperature for measuring the entropy, which can be chosen arbitrarily. If we choose $T_0 = 298$ K, then the entropy S_i of the initial state is

$$S_i = (4.2 \times 10^3 \text{ J/K}) \log \frac{353 \text{ K}}{298 \text{ K}} = 782 \text{ J/K}$$

and the entropy S_f of the final state is

$$S_f = (8.4 \times 10^3 \text{ J/K}) \log \frac{323 \text{ K}}{298 \text{ K}} = 819 \text{ J/K}.$$

$$\therefore \quad S_f - S_i = 37 \text{ J/K} > 0.$$

Example 5 Calculate the change in entropy when the ideal gas adiabatically expands freely into vacuum.

Solution As mentioned earlier, $Q = 0$ in this change but $S_f \neq S_i$ because it is an irreversible change. Since the temperatures of the initial and final states are the same, the initial state can be expressed as (T, V_i) and the final state as (T, V_f) $(V_f > V_i)$. To calculate $S_f - S_i$, it is necessary to connect the initial and final states with a reversible change. As this reversible change, a path is selected in which the ideal gas comes into contact with a heat reservoir at temperature T and expands isothermally and reversibly (Fig. 15-22). Since the internal energy of the ideal gas is a function of temperature only, $dU = 0$ in the isothermal process. Therefore, from the first law of thermodynamics,

$$0 = dU = dQ + dW = dQ - p\,dV,$$

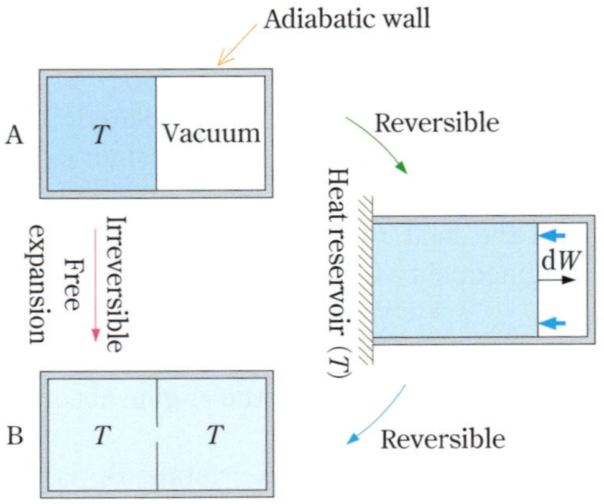

Fig. 15-22

thus, $dQ = p\,dV$. Using the equation of state $pV = nRT$ of ideal gas, we get

$$S_f - S_i = \int_{i \to f} \frac{dQ}{T} = \int_{V_i}^{V_f} \frac{p\,dV}{T} = nR \int_{V_i}^{V_f} \frac{dV}{V} = nR \log \frac{V_f}{V_i}. \quad (15.48)$$

Since $V_f > V_i$, it follows that $S_f > S_i$.

Examples 4 and 5 dealing respectively with the mixing of two substances kept at different temperatures and the free expansion of gas show that the entropy of an isolated system without exchange of heat nor work with the outside increases with irreversible changes.

In isolated systems, entropy increases with irreversible changes. When a materially and energetically isolated system is in maximum entropy, the system is in the most random and chaotic thermal equilibrium state with no heat transfer or material transfer within the system.

However, around us, various kinds of orders are locally formed by life activities of living things, human production activities and social activities. Furthermore, although the fall of water is an irreversible change, the fact that hydropower is continuously possible means that there are order-forming processes in the realm of nature. The process of order formation is a change in which entropy decreases.

If we consider the Earth as one system, we can take it with regard to the material as an isolated system having no exchange with the outside. However, as for energy, energy is received from the Sun with a surface temperature of 5800 K in the form of electromagnetic waves, and electromagnetic waves (infrared rays) are emitted from the surface of the Earth of about 270 K into space, so it is not an isolated system but an open system. In the Earth system, the flow of energy from the solar radiation to the heat dissipation to space is the driving force that brings about the circulation of air and water, and supports the life of living things. The energy flowing in from the Sun and the energy emitted by the Earth are equal, but the entropy associated with the heat from the hot Sun is much smaller than the entropy associated with the heat emitted by the cold Earth. In this way, the local order formation process on the Earth becomes possible.

Fig. 15-23 A photo taken from the International Space Station. It shows us how the Sun illuminates the Earth. Unlike seen on the Earth ground, the Sun looks pure white because the sunlight does not pass through the atmospheric layer.

For reference Statistical mechanics and entropy

Entropy is a quantity of state introduced in thermodynamics and is independent of the molecular structure of matter. It is statistical mechanics that applies the theory of probability to the motion of molecules making up a substance and derives various properties of the substance as statistical rules that the molecular assembly represents. Boltzmann showed that when the macroscopic quantities of state of the system such as energy, volume, temperature, and particle number are given, in statistical mechanics the entropy S is expressed as the number W of possible microscopic states. Namely,

$$S = k \log W. \quad (15.49)$$

The greater the number of microscopic states W, the greater the degree of randomness. According to statistical mechanics, a system in a macroscopic state with a larger number of microscopic states W has a higher probability of existence. The state of great W is the state

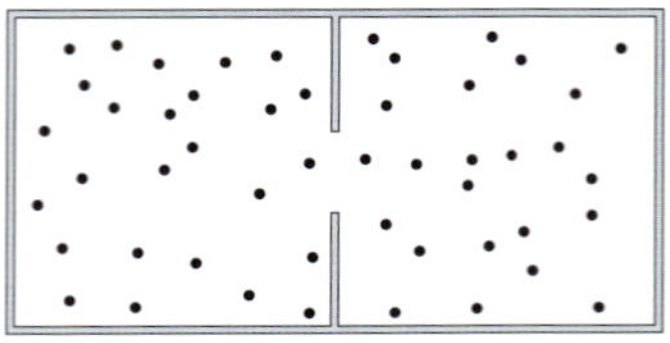

(a)

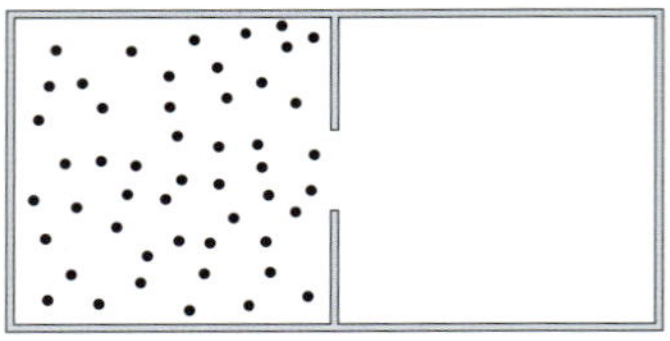

(b)

Fig. 15-24

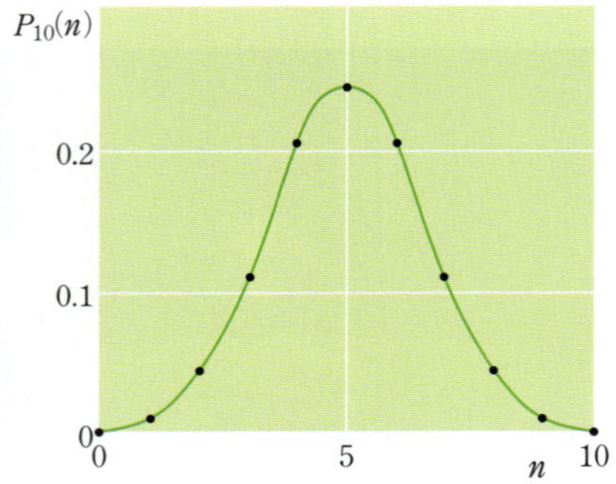

(a) In the case of $N = 10$

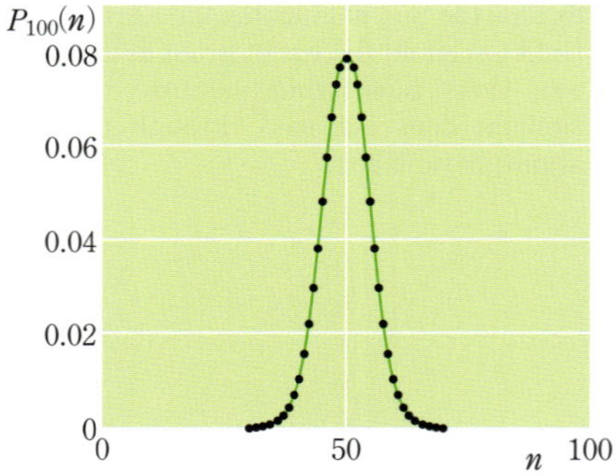

(b) In the case of $N = 100$

Fig. 15-25 The probability $P_N(n)$ of n gas molecules existing in the left side region when the total number of gas molecules is N.

* If the total entropy of the system and the environment (external) increases, there occurs the case in which the entropy of the system decreases.

of large entropy S according to equation (15.49). This is the stochastic meaning of entropy.

As an example we consider a gas containing N molecules as shown in Fig. 15-24. With time averaging, the probability that each molecule is in the right region is $\frac{1}{2}$ and the probability that it is in the left region is $\frac{1}{2}$. With the assumption that there is no correlation in molecular motion, the probability that n molecules exist in the left region $P_N(n)$ is the probability of the binomial distribution :

$$P_N(n) = \frac{N!}{n!\,(N-n)!}\left(\frac{1}{2}\right)^N. \qquad (15.50)$$

The expected value of the random variable n of this distribution is $\frac{N}{2}$, and the standard deviation (fluctuation) is $\frac{\sqrt{N}}{2}$ (Fig. 15-25). Therefore, in the case of 1 mol of gas containing 6×10^{23} molecules, the relative mass difference between the left and right regions is about $\frac{1}{\sqrt{N}} \approx 10^{-12}$, so it can be ignored. The probability that the region of the right side is vacuum as shown in Fig. 15-24 (b) is $\left(\frac{1}{2}\right)^N \approx 0$.

15.6 Direction of thermodynamic phenomena —isothermal process and free energy

Learning objective To understand that in the changes at constant volume and temperature, the Helmholtz free energy always decreases, and the Gibbs free energy always decreases in the changes at constant pressure and temperature.

In the previous section, it was shown that when an isolated system without heat and work exchange with the outside performs irreversible changes, the entropy of the system increases. Heat conduction from the high-temperature part to the low-temperature part of the system, heat generation due to friction inside the system, free expansion of gas from the high-pressure part to the low-pressure part when there is a pressure difference in the system, etc., these are irreversible changes. In actual thermodynamic phenomena, such phenomena are inevitable, so the entropy of isolated systems increases. That is, the direction of the thermodynamic phenomenon that occurs in an isolated system is the direction in which the entropy of the system increases. This is the principle of increase of entropy.

However, in practice, changes that involve the exchange of heat and work with the outside are often matters of interest and to consider both system entropy and environment (outside) entropy is inconvenient*. In this section, we will examine the direction of the thermodynamic phenomena that occur in the non-isolated system when the system changes isothermally while exchanging heat with a large heat reservoir of temperature T.

For irreversible change B → A from state B to state A, the following inequality

$$\sum_{i(B\to A)} \frac{Q_i}{T_i} < S_A - S_B \tag{15.51}$$

is derived from Clausius' inequality (15.46). In the case of minute change ($T_A \approx T_B = T$)

$$S_A - S_B \approx dS, \qquad \sum_{i(B\to A)} \frac{Q_i}{T_i} \approx \frac{dQ}{T}, \tag{15.52}$$

the inequality (15.51) becomes

$$dQ < T\,dS. \tag{15.53}$$

dQ is the heat that the system absorbs from the (outside) heat reservoir. Substitution of the first law of thermodynamics $dU = dQ + dW_{\text{system}\leftarrow\text{outside}}$ into equation (15.53) gives

$$dU - T\,dS < dW_{\text{system}\leftarrow\text{outside}}. \tag{15.54}$$

For isothermal change ($dT = 0$), $d(TS) = (T+dT)(S+dS) - TS \approx T\,dS + S\,dT = T\,dS$. Accordingly, equation (15.54) becomes

$$d(U-TS) < dW_{\text{system}\leftarrow\text{outside}} = -dW_{\text{outside}\leftarrow\text{system}} \quad \text{(isothermal chage).} \tag{15.55}$$

Isothermal isochoric change and Helmholtz free energy In equation (15.55)

$$F = U - TS \tag{15.56}$$

is called the **Helmholtz free energy**. If F is used, equation (15.55) is transformed as

$$-dF > dW_{\text{outside}\leftarrow\text{system}} \quad \text{(isothermal change).} \tag{15.57}$$

In other words, **the work $dW_{\text{outside}\leftarrow\text{system}}$ that the system performs to the outside in the isothermal change is always smaller than the reduced amount $-dF$ of Helmholtz free energy**. Thus, it is not the internal energy but the Helmholtz free energy that can be used as work in isothermal changes. It is called free energy, because it means the energy which is freely usable as work.

If the force acting on the system is only uniform pressure, then $dW_{\text{outside}\leftarrow\text{system}} = p\,dV$. If the system performs the isothermal isochoric change, $dV = 0$, so $dW_{\text{outside}\leftarrow\text{system}} = p\,dV = 0$, and equation (15.57) becomes

$$dF < 0 \quad \text{(isothermal isochoric change).} \tag{15.58}$$

Thus, the Helmholtz free energy of the system necessarily decreases, and the state in which the Helmholtz free energy of the system is minimized is the equilibrium state.

Isothermal isobaric change and Gibbs free energy We then consider the change in isothermal and isobaric (constant pressure) condition. In this case the work $dW_{\text{outside}\leftarrow\text{system}}$ that the system performs to the outside is the sum of the work $p\,dV$ which the uniform pressure p performs with the volume change of the system and the substantial work $dW^{\text{substantial}}_{\text{outside}\leftarrow\text{system}}$ that other forces do. Thus equation (15.55) becomes

$$d(U-TS) < -dW^{\text{substantial}}_{\text{outside}\leftarrow\text{system}} - p\,dV \quad \text{(isothermal change).} \tag{15.59}$$

In the case of isobaric change, since $dp = 0$, using

$$d(pV) = (p+dp)(V+dV) - pV \approx p\,dV + V\,dV = p\,dV,$$

equation (15.59) can be expressed as

$$d(U-TS+pV) < -dW^{\text{substantial}}_{\text{outside}\leftarrow\text{system}} \quad \text{(isothermal isobaric change)}. \tag{15.60}$$

$$G \equiv U-TS+pV = F+pV \tag{15.61}$$

is referred to as the **Gibbs free energy**. In this way, equation (15.60) can be expressed as

$$-dG > dW^{\text{substantial}}_{\text{outside}\leftarrow\text{system}} \quad \text{(isothermal isobaric change)}. \tag{15.62}$$

Therefore, the substantial work $dW^{\text{substantial}}_{\text{outside}\leftarrow\text{system}}$ that the system performs to the outside in the isothermal isobaric change is always smaller than the reduced amount $-dG$ of Gibbs free energy. Thus, in isothermal isobaric changes, it is not the internal energy but the Gibbs free energy that can be used as substantial work.

Problems for exercise 15

A

1. Niagara Falls is about 50 m high, and the average amount of water flow is $4\times10^5\ \text{m}^3/\text{min}$.
   (1) How different is the water temperature at the top and bottom of the waterfall?
   (2) Assume that about 20 % of water is used for hydropower generation and determine the output power of the power plant.
2. Fig. 1 shows the p-V curve of a gas (1 mol). The gas was changed in the order of ABCDA.
   (1) Where was the isothermal change?
   (2) Where was the isobaric change?
   (3) Where was the work performed to this gas?
   (4) Among these changes, where did the internal energy change?
   (5) In the change from A to B, how much amount of work was done by this gas? Also, how much amount of heat was given to the gas in this change. Let the molar heat capacity at constant pressure be C_p.

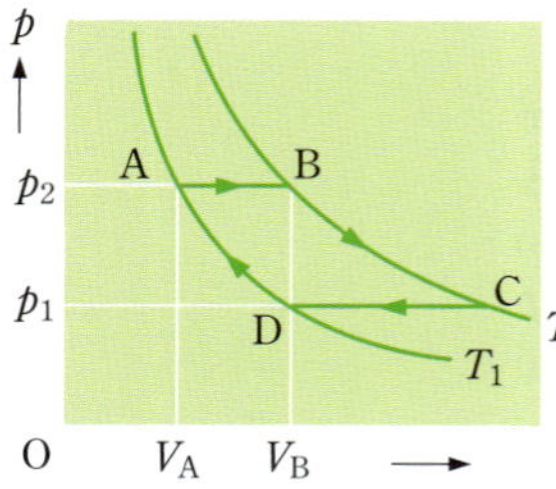

Fig. 1

3. Find the temperature in °C when the air at 27 °C is adiabatically compressed to 1/20 of its original volume.
4. A mass of dry air at 20 °C was blown up to several thousand meters above the flat ground by the updraft, and the volume adiabatically expanded till twice as large as at the initial state. Calculate the temperature of the air.
5. What is the maximum efficiency of the heat engine working between the high-temperature heat reservoir at 400 °C and the low-temperature heat reservoir at 50 °C?

B

1. 1 kg of ice at 0 °C melted into water at 0 °C. Calculate the change in entropy. Let the heat of fusion of 1 g of ice be 80 cal.
2. Heat Q was generated in the object of absolute temperature T by friction. Find the change in entropy. Assume that the heat capacity of the object is large, and the change in temperature is negligible.
3. Answer the following questions concerning the entropy of ideal gas.
   (1) Show that the change in entropy when the temperature changes from T_A to T_B by heating the ideal gas (n moles) in a container of constant volume is

$$S_B - S_A = nC_V \log \frac{T_B}{T_A}.$$

   (2) The state (p_A, V_A, T_A) and the state (p, V, T) of the ideal gas are linked by two steps of reversible changes, namely the isochoric change (p_A, V_A, T_A) → (p_B, V_A, T) at volume V_A and the iso-

thermal change $(p_B, V_A, T) \to (p, V, T)$ at temperature T. Use this fact to show that the **entropy of ideal gas** of n mole is given by the following equation：

$$S = nC_V \log T + nR \log V + \text{constant}.$$

4. If the enthalpy of a substance (1 mol) is defined as $H = U + pV$, show that the molar heat capacity at constant pressure of this substance can be expressed as $C_p = \left(\frac{\Delta H}{\Delta T}\right)_{\text{isobaric}}$.

5. Let the bulk modulus of the gas be $B = -V\frac{\Delta p}{\Delta V}$ and the density of the still gas be ρ_0, then the speed of the sound wave is $c = \sqrt{\frac{B}{\rho_0}}$. In the case of audible sound, the compression and the expansion are fast, so the change is not an isothermal change but an adiabatic change. Using the relations $pV^\gamma =$ constant, $pV = nRT$, and $\rho_0 = \frac{nM}{V}$, derive

$$B = -V\left(\frac{dp}{dV}\right)_{\text{adiabatic}} = \gamma p.$$

$$\therefore \quad c = \sqrt{\frac{\gamma RT}{M}}.$$

Furthermore, show that for the isothermal change, $c = \sqrt{\frac{RT}{M}}$.

6. Look at Figs. 2 (a) and 2 (b) and show that if one of the two statements of the second law of thermodynamics does not hold, the other statement does not hold, that is, the two statements are equivalent.

7. A heat engine is performing the cyclic process shown in the entropy-temperature diagram given in Fig. 3. What is the efficiency of this heat engine?

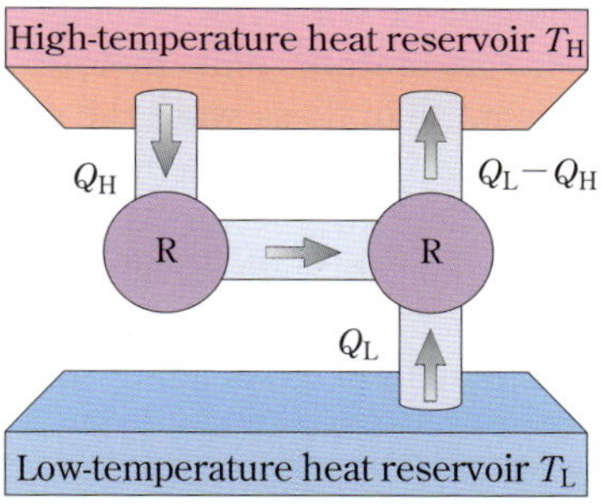

Freezing machine・heater

(a) If the Thomson statement does not hold (left), then using a freezing machine・heater (right), we can show that the Clausius statement does not hold.

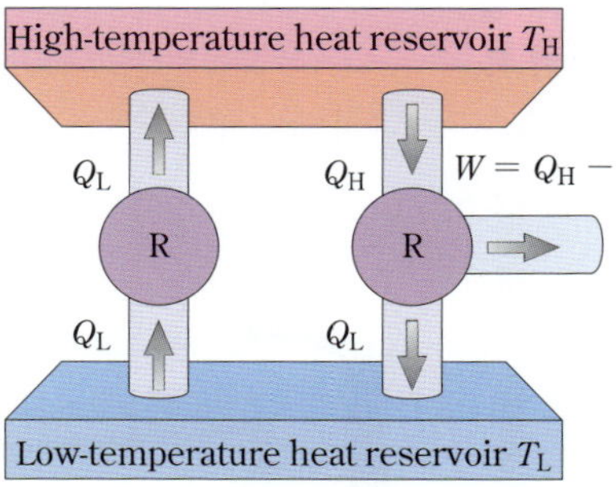

Heat engine

(b) If the Clausius statement does not hold (left), then using a heat engine (right), we can show that the Thomson statement does not hold.

Fig. 2

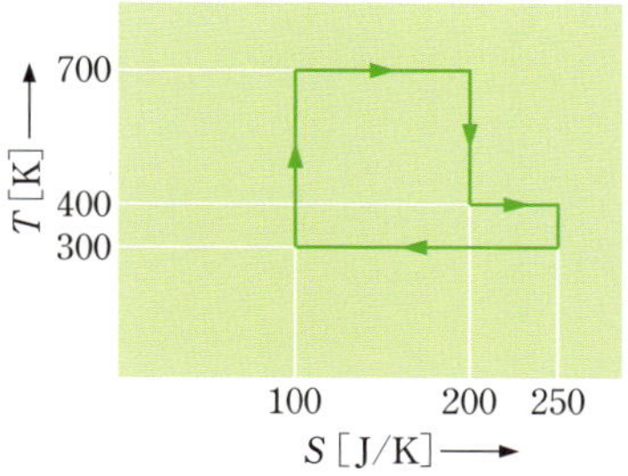

Fig. 3

8. Show that when one system performs an isothermal reversible cyclic process, the work done by the system is zero. Using this result, show that to find the pressure p_0 at which the vapor and liquid phases of the substance described by the van der Waals equation of state coexist, it is appropriate to draw a horizontal line so that the areas of regions I and II in Fig. 4 are equal (Maxwell's rule).

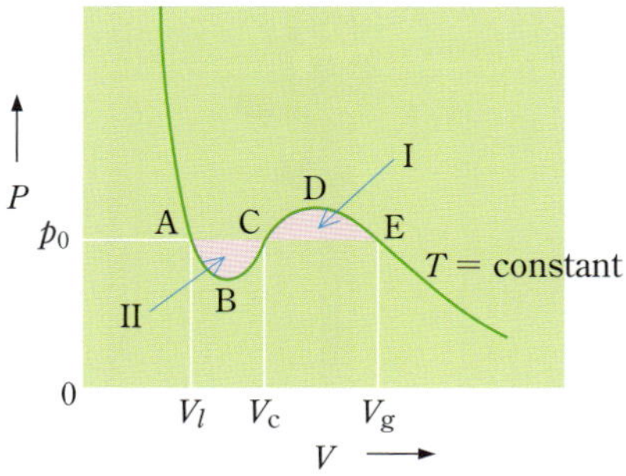

Fig. 4

Electrostatic Field in Vacuum

The need for electricity in everyday life will be obvious when we imagine what would happen if no electrical energy were supplied by a generator or battery. In the latter half of the 19th century, generators were put to practical use, and the power supplied from the generators was used first to turn on the light bulbs and to drive motors. Hertz successfully generated and detected electromagnetic waves in 1888 and it became possible to use electromagnetic waves as a means to transmit information. The invention of generators and motors, and the application of electromagnetic wave, all of these were the results of the study of electromagnetism, and it was indispensable to understand the concepts of electric field and magnetic field.

In seven chapters from this chapter through Chapter 22, we will discuss the fundamentals of electromagnetism in which charge and current, electric field and magnetic field play a leading role.

A charged sphere. It is close to one of the two electrodes.

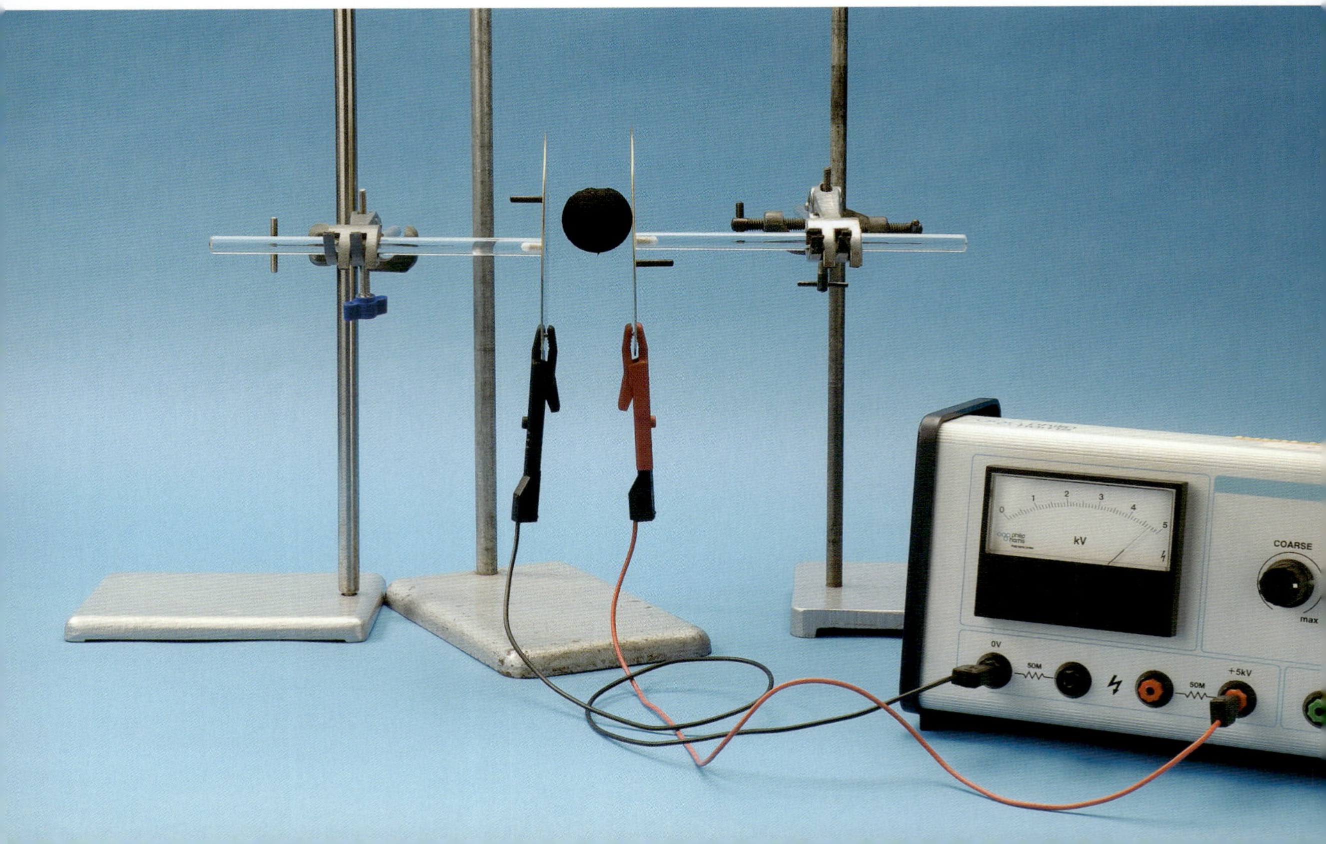

The key physical quantities to understand electromagnetic phenomena are the **charge** of objects, and the **electric field** and the **magnetic field** that are the properties of the space. The fundamental laws of electromagnetism are the laws that indicate what kind of electric field and magnetic field the charge creates around it. They are the **Maxwell's equations** consisting of four laws which are also the motion laws of electric field and magnetic field, and the law of the force (**Lorentz force**) that electric and magnetic fields act on charged particles.

When we learned mechanics, the fundamental law of Newton's law of motion played a role of the starting point. In the study of electromagnetism in this book, while systematically learning the laws of the electric and magnetic fields that the electromagnetic phenomena follow, we will describe all the electromagnetic phenomena in a unified way, and aim to reach Maxwell's laws (Maxwell's equations) which are the fundamental laws of electromagnetism and indicate the presence of electromagnetic waves.

In this chapter, we learn the electric force acting between the charges resting in vacuum, then learn the nature of the electric field created by the charges kept at rest in vacuum, and finally learn the potential which is the electric potential energy per unit positive charge.

Fig. 16-1 Static electricity

16.1 Charge and charge conservation law

Learning objective To understand, by connecting with the fact that the matter is composed of protons, neutrons and electrons, that there are positive charges and negative charges in charge and charges are conserved.

The physical quantity held by an object as the source of electromagnetic phenomena is called the **electric charge**. What is the charge? The author hopes that the readers understand the charge through learning how the charge behaves in various electromagnetic phenomena.

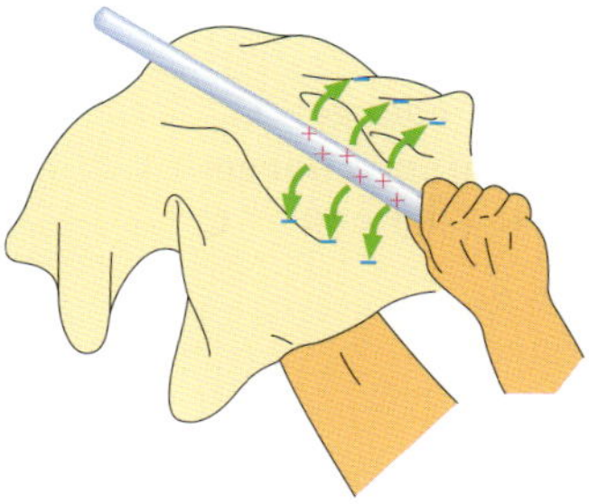

Fig. 16-2 When a glass stick is rubbed with a silk cloth, the glass stick is positively charge and the silk cloth is negatively charged. When a rubber stick is rubbed with fur, the rubber stick is negatively charged and the fur is positively charged.

The first electrical phenomenon that humans encountered was frictional electricity. For more than 2000 years, it has been known that amber (a pine resin fossil) sticks rubbed with fur attract light objects such as nearby dust and hair. The origin of the word "electric" in English is the Greek "elektron" meaning amber. The word "electricity" in English meant "those which cause rubbed things to attract light objects".

From the study of electric force acting between charged objects, following facts were found (Fig. 16-2, Fig. 16-3).

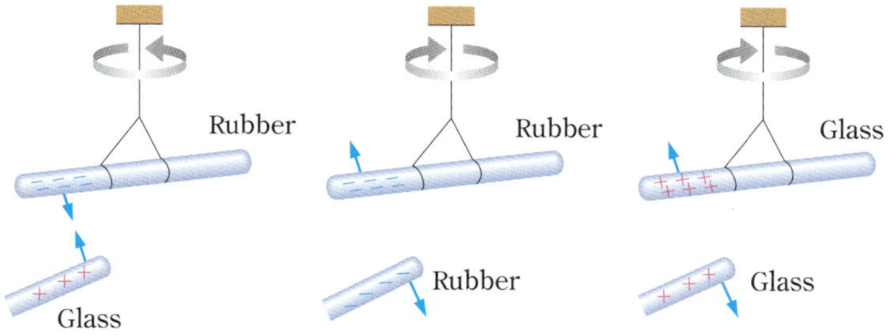

Fig. 16-3

(1) There are two types of charge : positive charge and negative charge. Repulsion acts between charges of the same type, and attraction works between charges of different type.
(2) When two uncharged objects are rubbed together, one object is positively charged, and the other object is negatively charged by the same amount. When a positively charged object is brought into contact with an object which is negatively charged by the same amount, positive and negative charges are neutralized.

Therefore, if the sum of charges considering positive and negative signs is called the total charge,

The total charge of a closed system is constant.

This is called the **charge conservation law**, which is one of the fundamental laws of nature.

The charge conservation law can be understood from the structure of substances. Substances are composed of atoms, and an atom of atomic number Z is composed of the nucleus with a positive charge Ze at the center and Z electrons having $-e$ charge each around it, and the nucleus consists of uncharged neutrons and positively charged protons having $+e$ charge each. Therefore,

"total charge of substance" = "total number of protons" $\times e$
+ "total number of electrons" $\times (-e)$

In friction and chemical reactions, no annihilation nor generation of protons and electrons occur. Therefore, the charge conservation law can be explained by the fact that the total number of protons and the total number of electrons are unchanged in these phenomena. Charges are preserved even when protons and electrons are generated or annihilated as in the case of β (beta) decay of nuclei.

Unit of charge C
Elementary charge
$e \approx 1.60 \times 10^{-19}$ C

The unit of charge, **coulomb** (symbol C), is set by accurately determining the **elementary charge** e, which is the magnitude of the charge of electrons and protons, as $1.602\,176\,634 \times 10^{-19}$ C.

Electrons are much lighter than nuclei, and electrons leak like thin clouds on the surface of objects. Since the force to bind electrons to an object differs depending on the substance, when objects are rubbed, movement of electrons occurs on the object surface, and one is positively charged, while the other is negatively charged.

16.2 Coulomb's law

Learning objective To understand Coulomb's law which is the law of electric force acting between charges and to understand the principle of superposition.

In 1785 Coulomb, using a sensitive torsion balance to measure the electric force acting between the charged small balls, found "The magnitude of the electric force F acting between two small charged objects is proportional to the product of the charges q_1 and q_2 of the two charged objects and inversely proportional to the square of distance r (Fig. 16-4)".

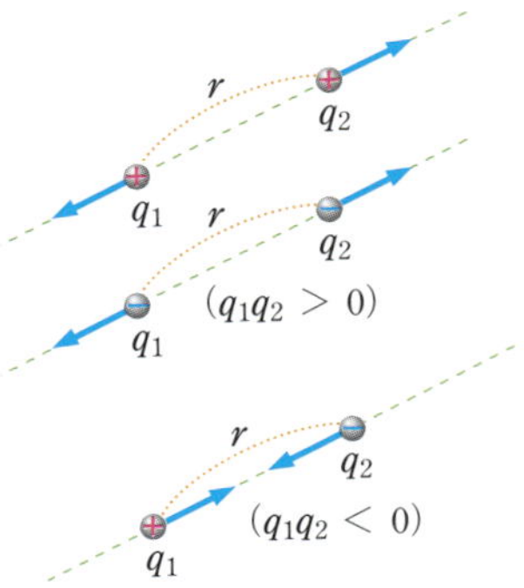

Fig. 16-4 Coulomb's law

$$F = \frac{q_1 q_2}{4\pi\varepsilon_0 r^2}$$

This satisfies the law of action and reaction.

$$F = \frac{q_1 q_2}{4\pi\varepsilon_0 r^2} \quad \text{(in vacuum)}. \tag{16.1}$$

This is called **Coulomb's law**. The electric force following Coulomb's law is called the **Coulomb force**. The term $\frac{1}{4\pi\varepsilon_0}$ on the right-hand side is the proportional constant in the International System of Units, in which the unit of charge is coulomb (C), the unit of length is meter (m), and the unit of force is Newton (N). The following value is given to the proportional constant :

$$\frac{1}{4\pi\varepsilon_0} \approx 8.988 \times 10^9\ \mathrm{N \cdot m^2/C^2}. \tag{16.2}$$

This is the value in vacuum but it is nearly the same in air. The constant ε_0 (read as epsilon zero)

$$\varepsilon_0 \approx 8.854 \times 10^{-12}\ \mathrm{C^2/(N \cdot m^2)} \tag{16.3}$$

is referred to as the **electrical constant**[*1]. The numerical part of $\frac{1}{4\pi\varepsilon_0}$, 8.988×10^9 is, to be exact, expressed as $c^2/10^7$ using the numerical part of the speed of light in vacuum c, which is $2.997\,924\,58 \times 10^8$ m/s (see Section **22.2**).

Electrical constant (dielectric constant of vacuum)
$\varepsilon_0 \approx 8.854 \times 10^{-12}\ \mathrm{C^2/(N \cdot m^2)}$

*1 Although the electrical constant ε_0 has historically been called the dielectric constant of vacuum, it is not a constant that represents the nature of vacuum as a dielectric substance. ε_0 is a conversion factor for matching the independently defined mechanical units [m], [kg], and [s] with the numerical values measured in the electromagnetic unit [A].

The unit of charge, coulomb, is the quantity of electricity that flows in one second across the cross section of wire in which a current of one ampere (A) flows.

Small charged objects are objects whose size is small compared to the distance to other charged objects, so the effect of electrostatic induction that will be discussed in the next chapter can be neglected. The charge in such a case is idealized to be called the **point charge**[*2]. Coulomb's law is the law of electric force acting between point charges. The law of action and reaction holds also for the Coulomb force.

*2 In this book symbols Q and q are used for charge. In principle, the symbol q is used for a point charge.

Case 1 Two small glass beads, each 1 μC (= 10^{-6} C) positively charged, are placed 10 cm apart. The magnitude of the electric force acting between them is

$$F = (9.0 \times 10^9\ \mathrm{N \cdot m^2/C^2}) \times \frac{(10^{-6}\ \mathrm{C})^2}{(0.1\ \mathrm{m})^2} = 0.9\ \mathrm{N} \quad \text{(repulsive force)}.$$

The magnitude of this electric force is equal to the magnitude of gravity acting on an object of about 90 g.

The above calculation shows that 1 C is a very large amount of charge. An object with a large charge attracts an object with a charge of the opposite sign and/or it discharges, so it is difficult to maintain a large charge. The electric charge of 1 mol of monovalent ion called the Faraday constant is 96485 C.

Faraday constant
$F = eN_A = 96485$ C/mol

Coulomb's law expressed in vectors Equation (16.1) gives no information of the direction of force. Let us express the direction of force in terms of vectors. The direction of the electric force is the direction of the line segment connecting the two charges. Assume that the position vector of point charge q_1 is $\boldsymbol{r}_1$ and the position vector of

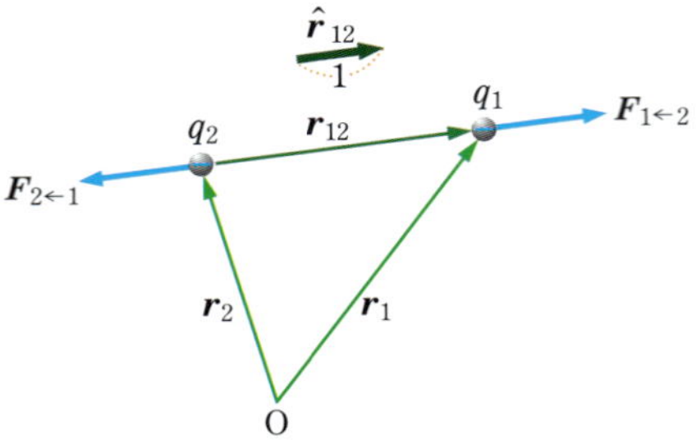

Fig. 16-5 Coulomb's law (in the case of $q_1q_2 > 0$)

point charge q_2 is $\boldsymbol{r}_2$. Then, the vector starting at charge q_2 and ending at charge q_1 is $\boldsymbol{r}_{12} = \boldsymbol{r}_1 - \boldsymbol{r}_2$ (Fig. 16-5). Since the distance between charge q_1 and charge q_2 is $r_{12} = |\boldsymbol{r}_1 - \boldsymbol{r}_2|$, the unit vector (vector of length 1) directed from charge q_2 to charge q_1 is $\hat{\boldsymbol{r}}_{12} = \dfrac{\boldsymbol{r}_{12}}{r_{12}} = \dfrac{\boldsymbol{r}_1 - \boldsymbol{r}_2}{|\boldsymbol{r}_1 - \boldsymbol{r}_2|}$.

Therefore, the electric force $\boldsymbol{F}_{1\leftarrow 2}$ which the point charge q_2 exerts on the point charge q_1 is, including the direction,

$$\boldsymbol{F}_{1\leftarrow 2} = F\hat{\boldsymbol{r}}_{12} = \frac{1}{4\pi\varepsilon_0}\frac{q_1q_2}{r_{12}{}^2}\hat{\boldsymbol{r}}_{12} = \frac{1}{4\pi\varepsilon_0}\frac{q_1q_2}{|\boldsymbol{r}_1 - \boldsymbol{r}_2|^2}\frac{\boldsymbol{r}_1 - \boldsymbol{r}_2}{|\boldsymbol{r}_1 - \boldsymbol{r}_2|} \tag{16.4}$$

(Fig. 16-5). This is Coulomb's law expressed in terms of vectors. Since $\boldsymbol{F}_{2\leftarrow 1} = -\boldsymbol{F}_{1\leftarrow 2}$, Coulomb's force satisfies the law of action and reaction.

Since the x component, the y component, and the z component of the vector $\boldsymbol{r}_1 - \boldsymbol{r}_2$ are $x_1 - x_2$, $y_1 - y_2$, and $z_1 - z_2$ respectively, each component of the electric force $\boldsymbol{F}_{1\leftarrow 2}$ is given by the equation obtained by replacing the numerator $\boldsymbol{r}_1 - \boldsymbol{r}_2$ of equation (16.4) with each component.

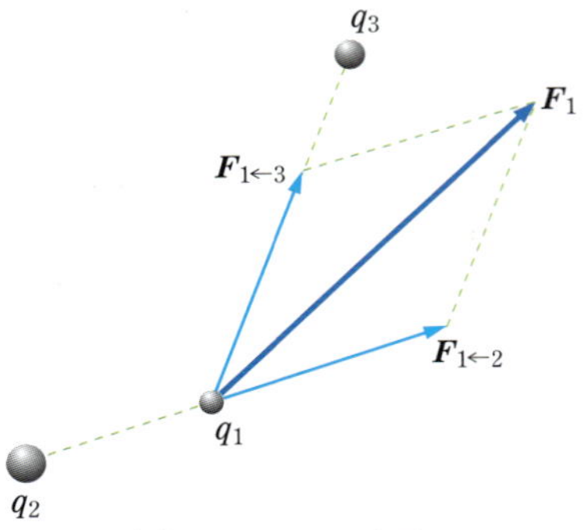

(a) $\boldsymbol{F}_1 = \boldsymbol{F}_{1\leftarrow 2} + \boldsymbol{F}_{1\leftarrow 3}$

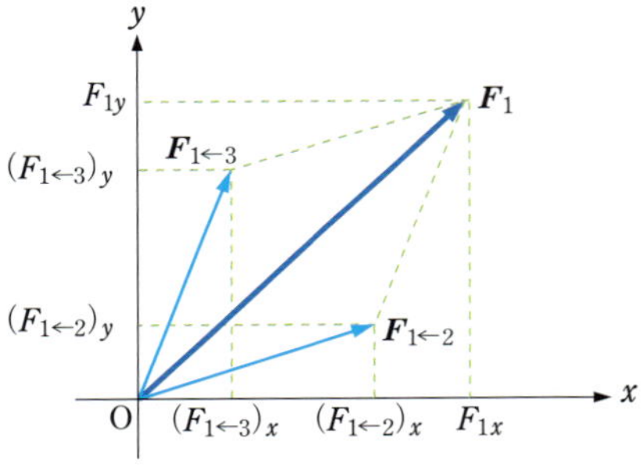

(b) $F_{1x} = (F_{1\leftarrow 2})_x + (F_{1\leftarrow 3})_x$,
$F_{1y} = (F_{1\leftarrow 2})_y + (F_{1\leftarrow 3})_y$

Fig. 16-6 Principle of superposition of electric force (in the case where all point charges q_1, q_2, and q_3 are on the xy plane, and $q_1q_2 > 0$ as well as $q_1q_3 < 0$).

Electric force in the case of more than or equal to three charges

When there are three charges q_1, q_2, and q_3, the electric force $\boldsymbol{F}_1$ acting on the charge q_1 is found experimentally to be the vector sum of the electric force $\boldsymbol{F}_{1\leftarrow 2}$ when only the charges q_1 and q_2 exist and the electric force $\boldsymbol{F}_{1\leftarrow 3}$ when only the charges q_1 and q_3 exist. That is to say

$$\boldsymbol{F}_1 = \boldsymbol{F}_{1\leftarrow 2} + \boldsymbol{F}_{1\leftarrow 3} \tag{16.5}$$

[see Fig. 16-6 (a)]. This is called the **principle of superposition of electric force**. Expressing equation (16.5) in terms of components we have

$$\begin{aligned} F_{1x} &= (F_{1\leftarrow 2})_x + (F_{1\leftarrow 3})_x, \\ F_{1y} &= (F_{1\leftarrow 2})_y + (F_{1\leftarrow 3})_y, \\ F_{1z} &= (F_{1\leftarrow 2})_z + (F_{1\leftarrow 3})_z \end{aligned} \tag{16.6}$$

[see Fig. 16-6 (b)]. Even when there are more than or equal to four charges, the principle of superposition of electric force still holds.

Question 1 As shown in Fig. 16-7, the charges q and $-q$ are placed at each vertex of a square whose one side length is L. Calculate the force $\boldsymbol{F}$ acting on the charge q placed at the upper left vertex.

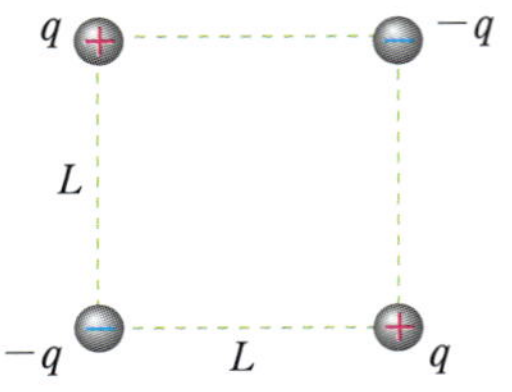

Fig. 16-7

16.3 Electric field

Learning objective To understand the concept that the electric force does not act directly between charges, but the charge creates around it a state with an electrical property called an electric field, and the electric field acts on the other charges there. To understand the nature of lines of electric force representing the electric field.

In physics, a space in which "physical quantity" is specified for each point is called a field of that physical quantity. For example, at each

低 − Low
高 − Hight
快晴 − Clear
晴 − Sunny
曇 − Cloudy
雨 − Rainy
雪 − Snowy
霧 − Foggy
風向風力 − Wind direction and wind power
28 日 18 時 − 18 : 00 on the 28 th

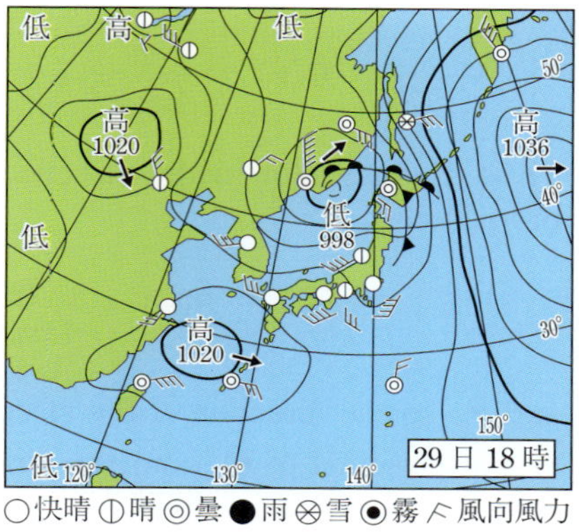

Fig. 16-8 Weather map
The isobar lines represent the atmospheric pressure (scalar quantity) field (scalar field) near the ground surface, and the arrows indicating the wind direction and the wind speed represent the wind velocity field (vector field) near the ground surface.

point in the atmosphere, temperature, pressure, wind velocity, etc. are determined at each time, so the atmosphere can be regarded as a temperature field, a pressure field, and a wind velocity field (Fig. 16-8).

In physics, it is thought that the electric force does not act directly between charges :

The first charge creates around it a state with an electrical property called an **electric field**, and an electric field around the second charge exerts an electric force on the second charge ; a change of an electric field travels through space at the speed of light.

In this chapter, we consider an electric field called the **electrostatic field** that the charge at rest creates around it and an electric force that the electric field acts on the electric charge at rest there. Electric fields exist in vacuum and in matter, but this chapter describes the electrostatic field in vacuum.

Due to the nature of Coulomb's force, when a point charge q is brought in at a point $\boldsymbol{r}$, the electric force acting on this charge is proportional to the brought-in charge q, so we can express

$$\boldsymbol{F} = q\boldsymbol{E}(\boldsymbol{r}). \tag{16.7}$$

The vector field $\boldsymbol{E}(\boldsymbol{r})$ which is unrelated to q

$$\boldsymbol{E}(\boldsymbol{r}) = \frac{\boldsymbol{F}}{q} \tag{16.8}$$

is called the electric field at the point $\boldsymbol{r}$*. In other words, the electric force acting on charge +1 C is the electric field at that point. The unit of electric field is N/C.

Unit of electric field N/C

* We are considering the case where the surrounding charge distribution is not modified by the brought-in point charge q.

From equation (16.7), it can be seen that positive charges receive electric force in the same direction as the electric field, while negative charges receive electric force in the opposite direction to the electric field (Fig. 16-9).

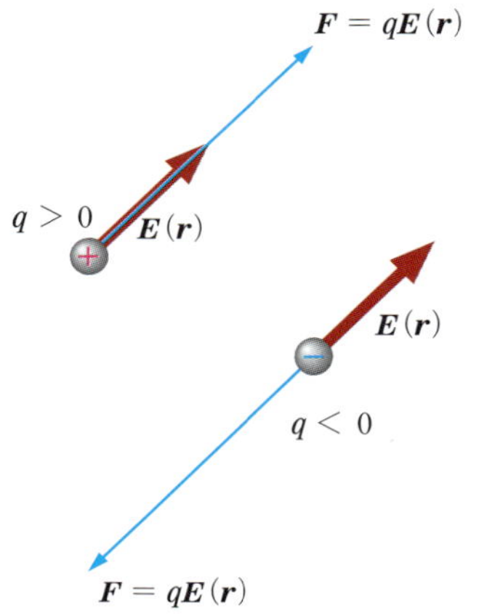

Fig. 16-9 The relationship between the electric field $\boldsymbol{E}(\boldsymbol{r})$ at point $\boldsymbol{r}$ and the electric force $\boldsymbol{F}$ acting on the charge q at point $\boldsymbol{r}$, $\boldsymbol{F} = q\boldsymbol{E}(\boldsymbol{r})$.

Case 2 From equations (16.4) and (16.8), when the point charge q_0 is at the origin, the electric field $\boldsymbol{E}(\boldsymbol{r})$ of point P of the position vector $\boldsymbol{r}$ is

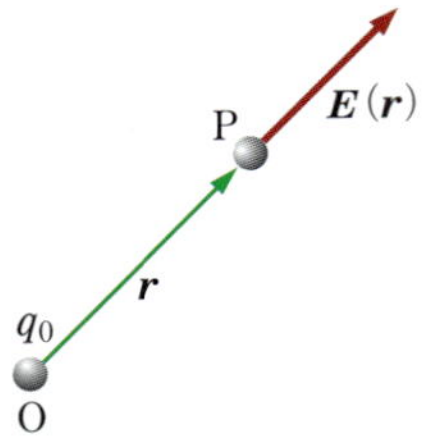

Fig. 16-10 The electric field that the point charge q_0 at the origin creates at point $\boldsymbol{r}$ (in the case $q_0 > 0$. In the case $q_0 < 0$, the direction of $\boldsymbol{E}$ is opposite).

$$\boldsymbol{E}(\boldsymbol{r}) = \frac{q_0}{4\pi\varepsilon_0 r^2}\hat{\boldsymbol{r}}$$

$$\boldsymbol{E}(\boldsymbol{r}) = \frac{q_0}{4\pi\varepsilon_0 r^2}\hat{\boldsymbol{r}} = \frac{q_0}{4\pi\varepsilon_0 r^2}\frac{\boldsymbol{r}}{r} \tag{16.9}$$

($\hat{\boldsymbol{r}}$ is a unit vector in the direction of $\boldsymbol{r}$). The electric field is the product of

$$E(r) = \frac{q_0}{4\pi\varepsilon_0 r^2} \tag{16.10}$$

and $\hat{\boldsymbol{r}}$, so the direction of the electric field $\boldsymbol{E}(\boldsymbol{r})$ is radial from the origin (direction of $\boldsymbol{r}$), and the direction is outward if $q_0 > 0$, and inward if $q_0 < 0$ (Fig. 16-10).

Question 2 Show that the electric field $\boldsymbol{E}(\boldsymbol{r})$ that the point charge q_P whose position vector is at the point of $\boldsymbol{r}_P$ creates at the point of $\boldsymbol{r}$ is

$$\boldsymbol{E}(\boldsymbol{r}) = \frac{1}{4\pi\varepsilon_0}\frac{q_P}{|\boldsymbol{r}-\boldsymbol{r}_P|^2}\frac{\boldsymbol{r}-\boldsymbol{r}_P}{|\boldsymbol{r}-\boldsymbol{r}_P|}.$$

Case 3 The electric field strength at a point 1 m away from the charge of 10^{-6} C is

$$E = (9.0\times10^9\ \mathrm{N\cdot m^2/C^2})\times\frac{10^{-6}\ \mathrm{C}}{(1\ \mathrm{m})^2} = 9.0\times10^3\ \mathrm{N/C}.$$

Assume that the electric field that the point charge q_1 creates is $\boldsymbol{E}_1(\boldsymbol{r})$ when only the point charge q_1 is present, and that the point charge q_2 creates is $\boldsymbol{E}_2(\boldsymbol{r})$ when only the point charge q_2 is present[*1]. Then, the electric field $\boldsymbol{E}(\boldsymbol{r})$ in the presence of two point charges q_1 and q_2 is

$$\boldsymbol{E}(\boldsymbol{r}) = \boldsymbol{E}_1(\boldsymbol{r})+\boldsymbol{E}_2(\boldsymbol{r}) \tag{16.11}$$

according to the principle of superposition of electric forces (Fig. 16-11). This is called the **principle of superposition of electric field**.

*1 Although described as "the electric field that the charge q creates", note the existence of the electric field $\boldsymbol{E}(\boldsymbol{r}) = \boldsymbol{0}$ even when no charge exists.

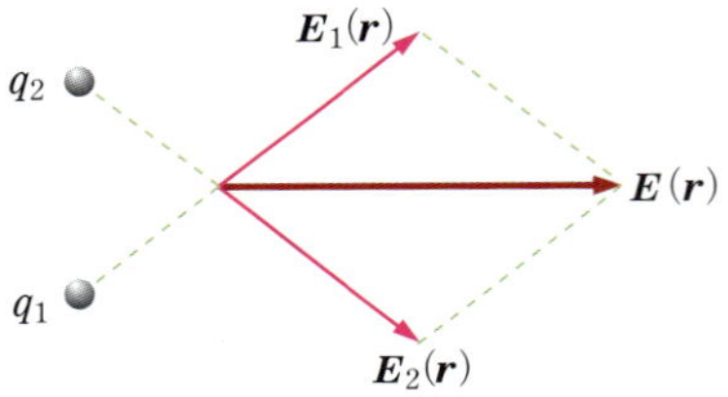

Fig. 16-11 $\boldsymbol{E}(\boldsymbol{r}) = \boldsymbol{E}_1(\boldsymbol{r})+\boldsymbol{E}_2(\boldsymbol{r})$

Lines of electric force We draw at each point in space an arrow representing the electric field at that point [Fig. 16-12 (a)] and then draw a curve with an orientation that makes the arrow of the vector representing the electric field at that point on the line be tangent. These curves are called the **lines of electric force** (in some books it is expressed as electric field lines) [Fig. 16-12 (b)]. The positive charge is the starting point of the lines of electric force, and the negative charge is the end point of the lines of electric force[*2]. In drawing lines of electric force, the density of lines of electric force should be drawn to be proportional to the electric field strength. Using lines of electric force, we can know the direction of the electric field from the direction of these lines, and compare the electric field strength by comparing the density of lines of electric force. In short, the electric field can be illustrated by the lines of electric force. The lines of electric force in some cases are shown in Fig. 16-13.

*2 Therefore, we can say that the electric charge on the object represents the ability to create an electric field and is a quantity representing the ability to be affected by the electric field.

When two lines of electric force intersect, there will be two directions of the electric field at the intersection point, so lines of electric force

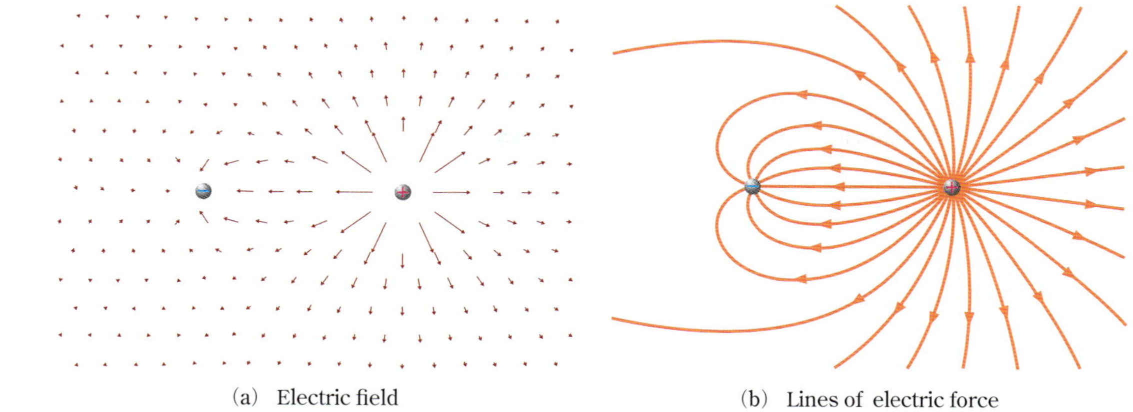

Fig. 16-12 Electric field and lines of electric force created by the positive point charge +3 C and the negative point charge −1 C.

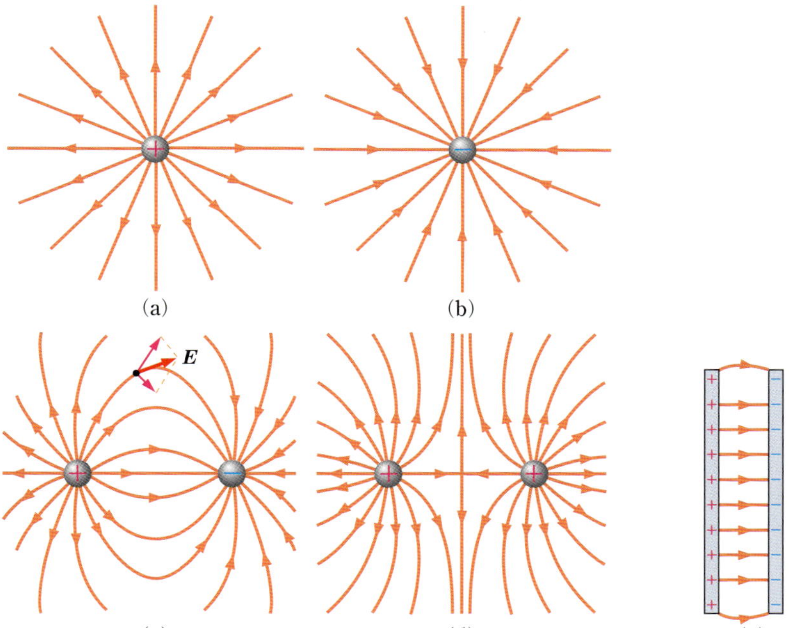

Fig. 16-13 Examples of lines of electric force. In (d), lines of electric lines are drawn to intersect, but at the point of intersection the electric field is **0**. If two lines of electric force intersect, there will be two directions of the electric field at the point of intersection, so the lines of electric force never intersect nor branch except at the places where the electric charge exists and the electric field is **0**.

never intersect nor branch except at the point where the electric charge exists or the electric field is **0**. In other words, lines of electric force are generated at the positive charge and vanish at the negative charge, but they are not interrupted nor newly generated. In the case where the sum of charges is not 0, the electric lines of force extend without limit [Fig. 16-13 (a), (b), (d)].

An electric field with a constant direction and a constant strength regardless of location is called a **uniform electric field**. The lines of electric force of uniform electric field are parallel and the spacing is constant [see Fig. 16-13 (e)].

* The "flux of electric line of force" is a term created by the present author and in some books it is expressed as flux of the electric field, electric field flux, or electric flux.

16.4 Gauss' law of electric field and its application

Learning objective To understand the definition of the flux of electric line of force*. To understand Gauss' law of electric field that the flux of electric line of force arising from the closed surface is $\frac{1}{\varepsilon_0}$ times the total amount of electric charge inside of the closed surface, and to become able to derive the formulae of the electric field when the charge distribution is spherically symmetric and also when the charge is uniformly distributed on the infinitely extending plane.

In this section, we first define the flux of electric line of force corresponding to the number of lines of electric force penetrating the plane S when we draw the lines of electric force at a density such that E lines of electric force pass through a unit area of a plane perpendicular to the electric field $\boldsymbol{E}$. The starting point and the end point of the lines of electric force are the positive charge and the negative charge, respectively. Therefore, when there is a balloon-like closed surface in the electric field, the flux of electric line of force corresponding to the net number ("number of going out" − "number of entering") of lines of electric force coming outside from the inside of this closed surface is proportional the net amount ("positive charge" − "negative charge") of electric charge inside the closed surface. This proportional relationship is Gauss' law of electric field.

Flux of electric line of force In a uniform electric field $\boldsymbol{E}$, the plane S with the defined front and back of area A is placed perpendicular to the electric field. When the direction of the normal vector $\boldsymbol{n}$ of the plane S and the direction of the electric field $\boldsymbol{E}$ are the same, the quantity

$$\Phi_E = EA \tag{16.12}$$

is called the **flux of electric line of force** penetrating the plane S [Fig. 16-14 (a)]. A normal vector is a vector of length 1 perpendicular to the plane and directing from the back to the front.

When the normal vector $\boldsymbol{n}$ of the plane S and the electric field $\boldsymbol{E}$ do not have the same direction and form an angle θ, the flux of electric line of

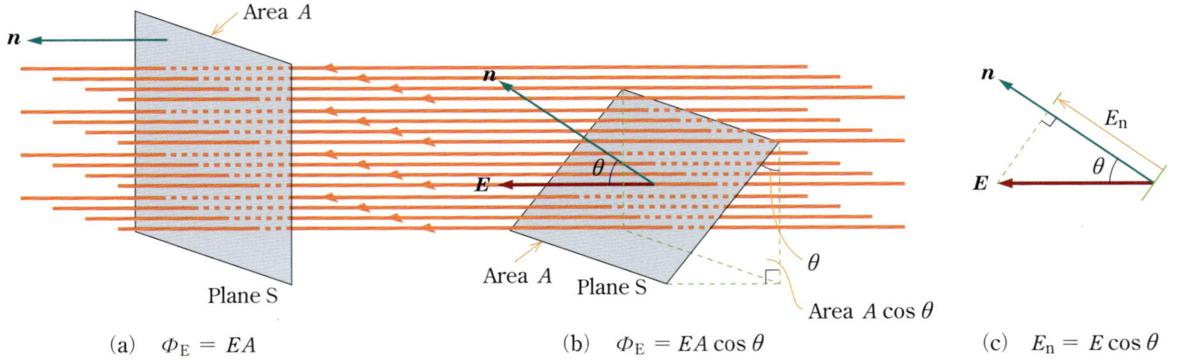

(a) $\Phi_E = EA$ (b) $\Phi_E = EA\cos\theta$ (c) $E_n = E\cos\theta$

Fig. 16-14 Flux of electric line of force Φ_E

force Φ_E penetrating the plane S from the back to the front is defined as

$$\Phi_E = EA\cos\theta, \tag{16.13}$$

[Fig. 16-14 (b)]. Since $E_n = E\cos\theta$ is the normal force component of the electric field $\boldsymbol{E}$ [Fig. 16-14 (c)], equation (16.13) is expressed as

$$\Phi_E = E_n A. \tag{16.14}$$

When $\Phi_E < 0$, it means $E_n < 0$, and the lines of electric force pass through the plane S in the direction of the front → the back.

When the surface S is not a plane and the electric field $\boldsymbol{E}$ is not uniform, the flux of electric line of force passing through the surface S is defined as the sum of the flux of electric line of force passing through each minute plane obtained by dividing the curved surface S into lots of minute part (minute plane). By numbering the minute planes as 1, 2, ⋯ and assuming that the area of the i-th minute plane S_i is ΔA_i, the normal vector $\boldsymbol{n}_i$, the electric field $\boldsymbol{E}_i$, and the normal direction component E_{in} (Fig. 16-15) of the electric field, the flux of electric line of force $\Delta\Phi_{E_i}$ passing through this minute plane S_i is, corresponding to equation (16.14), expressed as

$$\Delta\Phi_{E_i} = E_{in}\,\Delta A_i. \tag{16.15}$$

The flux of electric line of force Φ_E passing through the entire surface S is given by first taking the sum of each flux of electric line of force $\Delta\Phi_{E_1}$, $\Delta\Phi_{E_2}$, ⋯ passing through each minute plane, namely calculating $\Delta\Phi_{E_1}+\Delta\Phi_{E_2}+\cdots$ and then considering the limit when each minute plane is made infinitesimally small and the number N of minute planes is made infinite. We express this value using the surface integral symbol as

$$\Phi_E = \lim_{\Delta A_i\to 0,\ N\to\infty} \sum_{i=1}^{N} E_{in}\,\Delta A_i = \iint_S E_n\,dA. \tag{16.16}$$

In this integration, instead of the minute length dx, the area dA of the minute plane appears and the integration is called the surface integral. As points on the xy plane are specified by two variables x and y, points on the curved surface are also specified by two variables, so the double integration symbol is used. The S in the lower right of the integration symbol indicates that the integration region is the surface S.

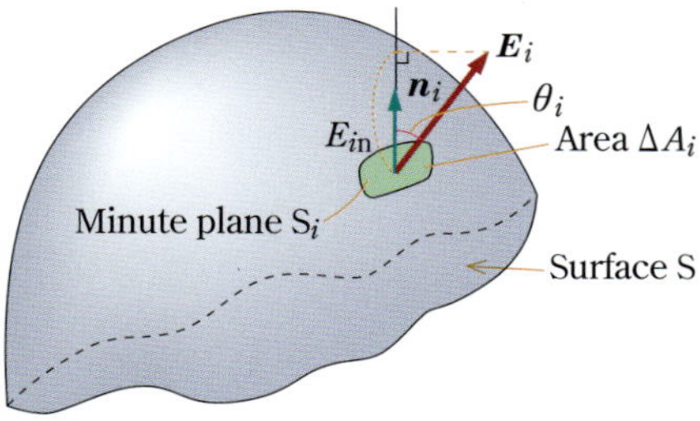

Fig. 16-15 Flux of electric line of force passing through the i-th minute plane when the surface S is divided into minute curved surfaces (approximately minute planes)
$\Delta\Phi_{E_i} = E_{in}\,\Delta A_i\ (E_{in} = E_i\cos\theta_i)$.

The flux of electric line of force passing through the spherical surface centered on the point charge q is $\Phi_E = \dfrac{q}{\varepsilon_0}$. In the electric field created by the point charge q, the electric field strength at a point r away from the point charge q is $E = \dfrac{|q|}{4\pi\varepsilon_0 r^2}$ (Fig. 16-16). The direction of the normal $\boldsymbol{n}$ of a spherical surface of radius r centered on a point charge is decided to be from the inside to the outside of the sphere. Since the electric field $\boldsymbol{E}$ on this sphere is perpendicular to the sphere, the normal component E_n of the electric field including the sign is

$$E_n = \frac{q}{4\pi\varepsilon_0 r^2}. \tag{16.17}$$

Therefore, the flux of electric line of force Φ_E passing through the spherical surface of area $A = 4\pi r^2$ from the inside to the outside is

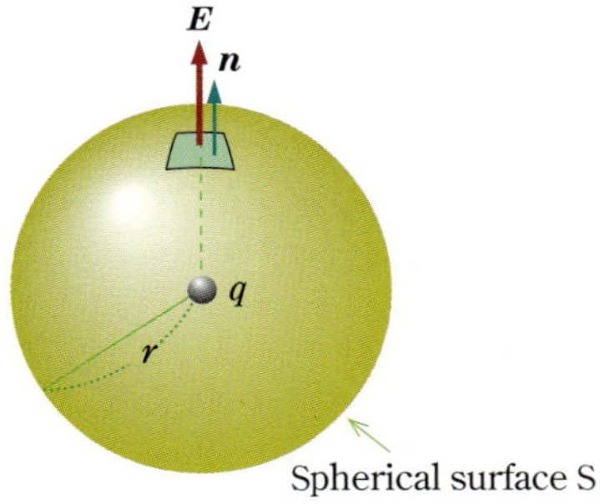

Fig. 16-16 Electric field $\boldsymbol{E}$ on a spherical surface of radius r centered on the point charge q (in the case $q > 0$).

$$\Phi_E = \iint_{\text{spherical surface}} E_n \, dA = E_n \iint_{\text{spherical surface}} dA$$
$$= E_n \times (\text{spherical surface area}) = \frac{q}{4\pi\varepsilon_0 r^2}(4\pi r^2) = \frac{q}{\varepsilon_0} \quad (16.18)$$

and is constant regardless of the radius of sphere. This fact indicates that the positive charge q is the starting point of $\frac{q}{\varepsilon_0}$ lines of electric force and the negative charge q is the end point of $\frac{|q|}{\varepsilon_0}$ lines of electric force and that at any location without charge, the lines of electric force are never generated nor annihilated.

Gauss' law of electric field One of the fundamental laws of electromagnetism is **Gauss' law of electric field**, which relates the total electric charge inside the closed surface with the flux of electric line of force through the closed surface, saying that

"The net flux of electric line of force Φ_E coming outside from the inside of closed surface S"

$$= \frac{\text{"total charge } Q_{in} \text{ inside the closed surface S"}}{\varepsilon_0}$$

$$\iint_S E_n \, dA = \frac{q}{\varepsilon_0}. \quad (16.19)$$

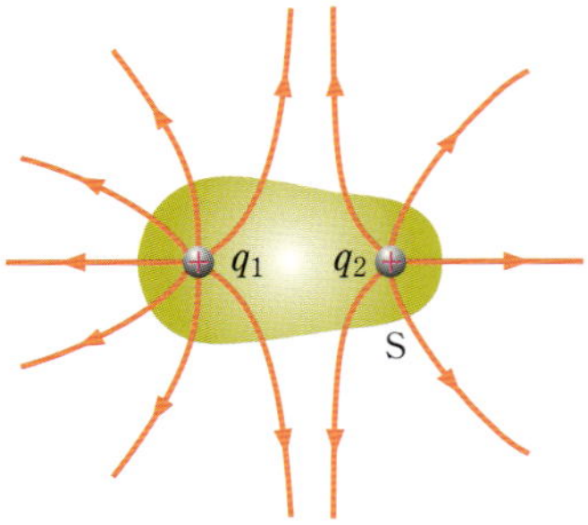

Fig. 16-17 Total flux of electric line of force coming outside from the inside of the closed surface $= \frac{q_1+q_2}{\varepsilon_0}$.

The case where $\Phi_E < 0$ means that the lines of electric force as a whole enter from the outside to the inside of the closed surface S. A closed surface is an interface which, like a spherical surface or a floating bag, clearly separates the space into its inner and outer regions, so that for something to move from one region to the other it must pass through that interface (In the figures in this section, the closed surface is shown by ■).

Equation (16.18) is an example of Gauss' law. When two charges exist inside the closed surface S, from equation (16.18) and the principle of superposition of electric field the net number of lines of electric force (flux of electric line of force) is $\frac{1}{\varepsilon_0}$ times the total charge Q_{in} (Fig. 16-17).

Question 3 Apply Gauss' law to the electric field created by the positive charge Q and the negative charge $-Q$ shown in Fig. 16-18 to show that in the case of closed surface S_1, $\Phi_E = \frac{Q}{\varepsilon_0}$, while in the case of closed surface S_2, $\Phi_E = -\frac{Q}{\varepsilon_0}$, and in the case of closed surface S_3 and S_4, $\Phi_E = 0$.

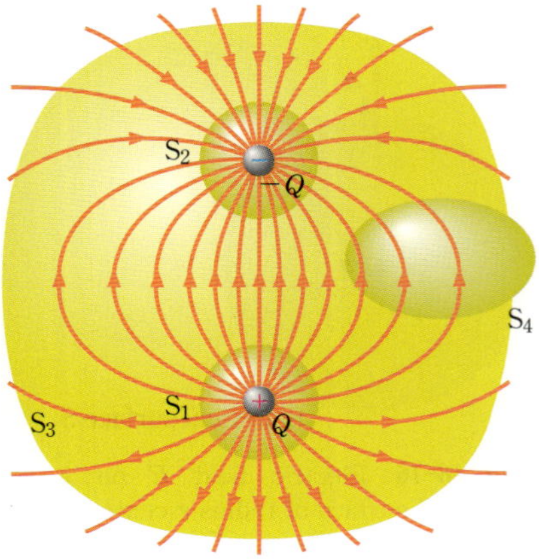

Fig. 16-18

Gauss' law of electric field can be derived exactly from Coulomb's law, but Gauss' law of electric field is one of the four fundamental laws of electromagnetism called Maxwell's equations, and it is regarded as a fundamental law which holds in any situation. Conversely, Coulomb's law of electric force acting between two charges at rest is, theoretically, a law in a position derived from Maxwell's equations.

Application of Gauss' law of electric field In physics, there are cases in which problems may be easily solved by using the symmetry of the object. Even in the field of electromagnetism, if the charge distribution is spherically symmetric or if the charge is uniformly distributed on an infinitely wide plane, the electric field can be readily calculated using Gauss' law of electric field.

In the case where the charge distribution is spherically symmetric In the case where the charge distribution is spherically symmetric, from the fact that the starting point and the end point of lines of electric force are charges and the rotational symmetry, the lines of electric force are distributed radially as shown in Fig. 16-19. Thus, we choose a spherical surface of radius r from the center of symmetry as a closed surface of Gauss' law. Since the electric field strength is constant on the spherical surface and the electric field is perpendicular to the spherical surface, the component of the outward normal direction of the electric field on the spherical surface with radius r can be expressed as $E_n = E(r)$. The flux of electric line of force passing through the spherical surface with an area of $A = 4\pi r^2$ is $\Phi_E = E_n A = E(r)(4\pi r^2)$. Since the total electric charge inside the sphere of radius r is $Q(r)$, Gauss' law of electric field becomes

$$\Phi_E = E_n A = E(r)(4\pi r^2) = \frac{Q(r)}{\varepsilon_0}. \tag{16.20}$$

Accordingly,

$$E(r) = \frac{Q(r)}{4\pi\varepsilon_0 r^2} \tag{16.21}$$

(Fig. 16-20). If $Q(r) > 0$, the electric field is in the outward direction and if $Q(r) < 0$, the electric field is in the inward direction. From these facts

"The electric field at the point $\boldsymbol{r}$ created by the spherically symmetric charge distribution around the origin is equal to the electric field in the case where the total charge within the sphere of radius r is at the origin."

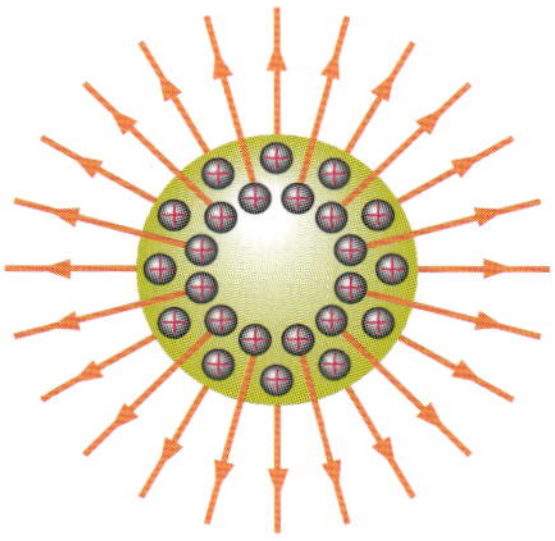

Fig. 16-19 Electric field created by the spherically symmetric charge distribution

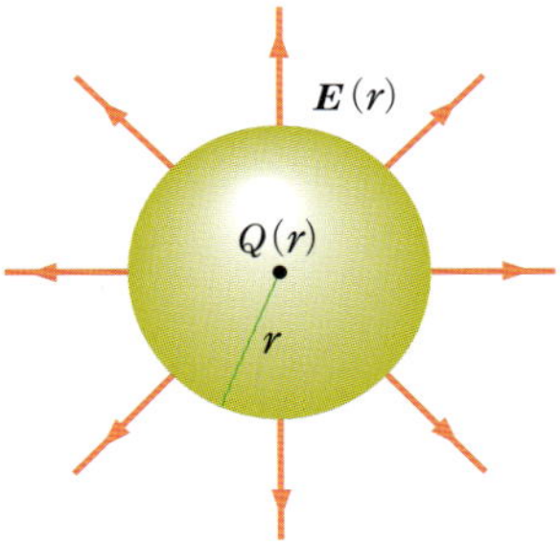

Fig. 16-20 In the case where charge distribution is spherically symmetric.

$$E(r) = \frac{Q(r)}{4\pi\varepsilon_0 r^2}$$

If $Q(r) > 0$, $\boldsymbol{E}(\boldsymbol{r})$ is in the direction of the arrow ; if $Q(r) < 0$, $\boldsymbol{E}(\boldsymbol{r})$ is in the opposite direction of the arrow.

Case 4 If the charge Q is uniformly distributed on a spherical surface of radius R,

$$Q(r) = \begin{cases} 0 & r < R \\ Q & r > R \end{cases}. \tag{16.22}$$

Therefore, substituting equation (16.22) into equation (16.21), we have the following results :

(1) inside the sphere, $Q(r) = 0$, so the electric field is 0 [Fig. 16-21 (a)],

(2) outside the sphere, $Q(r) = Q$, so the electric field is the same as the electric field when the total charge Q on the spherical surface is in the center of the sphere [Fig. 16-21 (b)],

$$E(r) = \begin{cases} 0 & \text{if } r < R \\ \dfrac{Q}{4\pi\varepsilon_0 r^2} & \text{if } r > R \end{cases}. \tag{16.23}$$

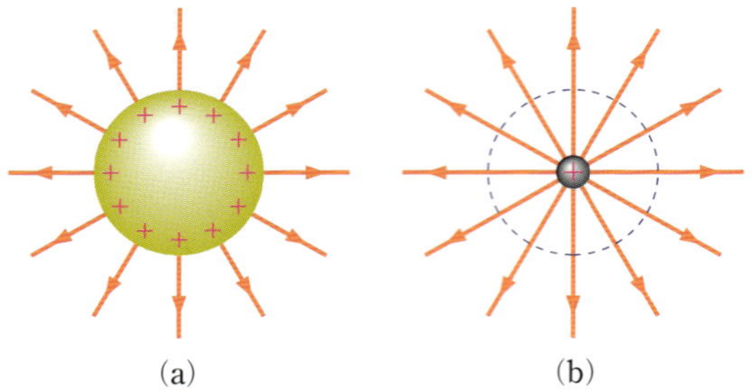

Fig. 16-21 (a) Electric field when the charge Q ($Q > 0$) is uniformly distributed on the spherical surface. The electric field inside the spherical surface is 0. The electric field outside the spherical surface is the same as when the total charge Q is at the center of sphere. (b) Electric field when the charge Q ($Q > 0$) is in the center of the sphere.

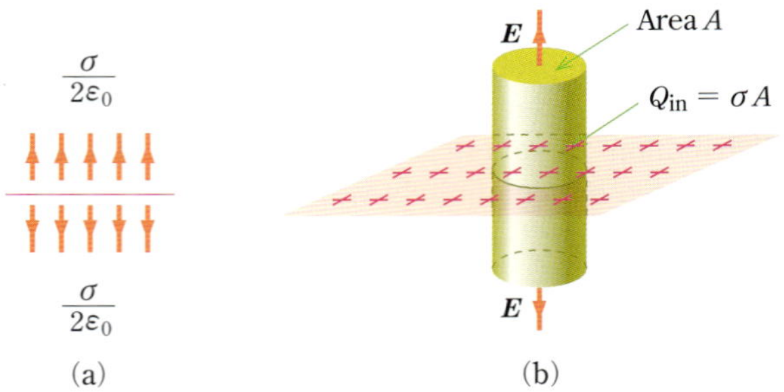

Fig. 16-22 Electric field created by the charge uniformly distributed at the surface density σ on the infinitely wide plane (the case of $\sigma > 0$)

In the case where the charge is uniformly distributed on the infinitely wide and thin insulator plate Let the charge per unit area, namely, "surface density of charge" = "total charge" divided by "area of the charged surface" be σ. As shown in Fig. 16-22 (a), the lines of electric force extend vertically from the charged surface up and down with uniform density (in the case of positive charge). As a closed surface, we consider here a cylinder with base area A as shown in Fig. 16-22 (b). Half of $\frac{Q_{in}}{\varepsilon_0}$ lines of electric force from the charge $Q_{in} = \sigma A$ inside the cylinder, or $\frac{Q_{in}}{2\varepsilon_0}$ lines, vertically penetrate the upper base of the area A, and the remaining $\frac{Q_{in}}{2\varepsilon_0}$ lines vertically penetrate the lower base of area A. From equation (16.19), the electric field strength E is given as

$$E = \frac{Q_{in}}{2\varepsilon_0 A} = \frac{\sigma}{2\varepsilon_0}. \tag{16.24}$$

In the case of a negative charge ($Q < 0$), the direction of the electric field is opposite to that in Fig. 16-22 (a) and is directed to the charged surface.

Example 1 (1) Two infinitely wide flat thin plates are uniformly charged with surface densities σ and $-\sigma$, respectively. Find the electric field when these two plates are arranged in parallel.

(2) In this case, find the electric force that the charge on the unit area of one plate receives from the charge on the other plate.

Solution (1) In Fig. 16-23, since $\boldsymbol{E} = \boldsymbol{E}_1 + \boldsymbol{E}_2$

$$E(r) = \begin{cases} \dfrac{\sigma}{\varepsilon_0} & \text{downward} \quad \text{between the two plates} \\ 0 & \qquad\qquad \text{outside the two plate} \end{cases}. \tag{16.25}$$

(2) The electric force of the electric field $\boldsymbol{E}_2$ created by the charge of the lower plate acting on the charge σ on the unit area of the upper plate is downward and the strength is

$$\sigma E_2 = \sigma \frac{\sigma}{2\varepsilon_0} = \frac{\sigma^2}{2\varepsilon_0}. \tag{16.26}$$

On the charge $-\sigma$ on the unit area of the lower plate the electric force of strength $\frac{\sigma^2}{2\varepsilon_0}$ acts upward.

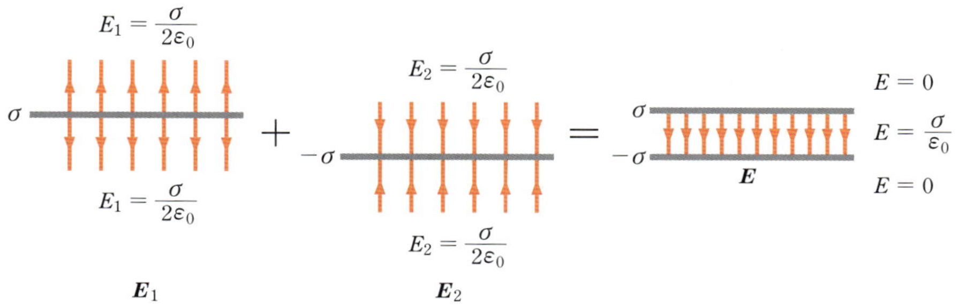

Fig. 16-23 $\boldsymbol{E}_1 + \boldsymbol{E}_2 = \boldsymbol{E}$

In the case where the charge is uniformly distributed over the infinitely long column Let λ be the linear density of charges (charge per unit length) of a column of radius R. Due to the cylindrical symmetry of the charge distribution, the lines of electric force extend radially and perpendicular to the central axis (Fig. 16-24). As a closed surface we consider a cylinder of radius r and length L coaxial with the column ($r \geqq R$). The flux of electric line of force $\frac{Q_{\text{in}}}{\varepsilon_0}$ from the charge $Q_{\text{in}} = \lambda L$ in the cylinder penetrates the side of the cylinder with area $2\pi rL$, and the electric field strength $E(r) = E_{\text{n}}$ on the cylinder is from Gauss' law, $\frac{Q_{\text{in}}}{\varepsilon_0} = E(r) \times (2\pi rL)$. Thus,

$$E(r) = \frac{\frac{Q_{\text{in}}}{\varepsilon_0}}{2\pi rL} = \frac{\lambda}{2\pi\varepsilon_0 r}. \tag{16.27}$$

If $r < R$, then use the Q_{in} appropriate to that case.

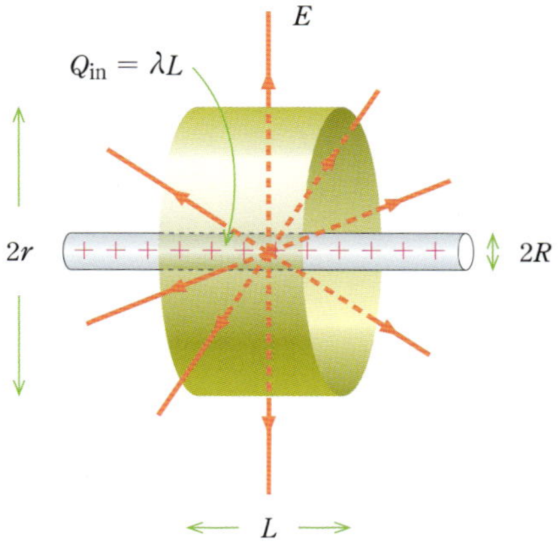

Fig. 16-24 Electric field created by the axisymmetric charge distribution

16.5 Electric potential

Learning objective To understand the definition of the electric potential, which is the electric potential energy per unit charge, and its characteristics, in particular, that the electric field (lines of electric force) is orthogonal to the equipotential surface and the equipotential line and the electric field strength is equal to the electric potential gradient (slope).

Like the universal gravitation $-G\frac{m_1 m_2}{r^2}$ having the potential energy $-G\frac{m_1 m_2}{r}$, Coulomb's force $\frac{q_1 q_2}{4\pi\varepsilon_0 r^2}$ working between point charges is the central force inversely proportional to the square of the distance r. Thus it is a conservative force and it has a potential energy called the **Coulomb potential** or the **Coulomb energy**.

$$U(r) = \frac{q_1 q_2}{4\pi\varepsilon_0 r}. \tag{16.28}$$

This potential energy is equal to the work performed by the electric force in the process where two point charges at a distance r are separated infinitely [$W_{r\to\infty} = U(r)$; see equation (5.29)].

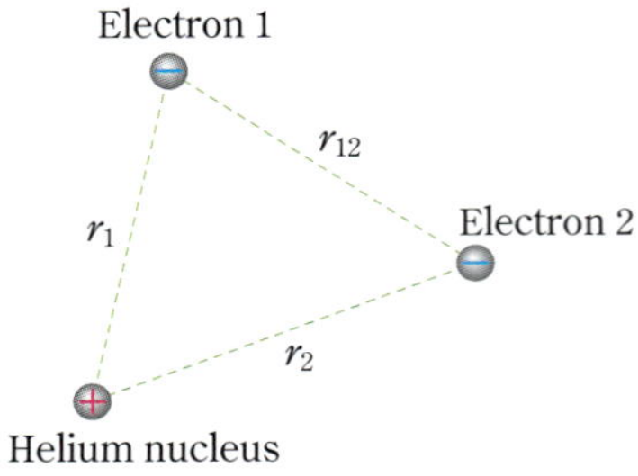

Fig. 16-25

Case 5 A helium atom consists of a helium nucleus with a charge of $2e$ and two electrons with a charge of $-e$. The Coulomb force acting between three charged particles (particles having charges) follows the principle of superposition, so the potential energy is the sum of three Coulomb potentials

$$U = -\frac{2e^2}{4\pi\varepsilon_0 r_1} - \frac{2e^2}{4\pi\varepsilon_0 r_2} + \frac{e^2}{4\pi\varepsilon_0 r_{12}}. \tag{16.29}$$

Here, r_1 and r_2 are the distances between the nucleus and electrons 1 and 2, respectively, and r_{12} is the distance between electrons 1 and 2 (Fig. 16-25).

The electric force $q\boldsymbol{E}$ acting on the charge q in the electric field $\boldsymbol{E}$ is created by the superposition of Coulomb's forces that the surrounding charges exert. Therefore, the electric potential energy U exists and when the charge q moves from point P to point A, the work $W_{P\to A}$ that the electric force $\boldsymbol{F} = q\boldsymbol{E}$ performs is equal to the difference between the electrical potential energy U_P at point P and U_A at point A.

$$W_{P\to A} = \int_P^A qE_t \, ds = U_P - U_A. \tag{16.30}$$

E_t is the tangential component of the path of the electric field $\boldsymbol{E}$ [see equation (5.21)].

By the way, in electromagnetics, what is more important than the electric potential energy itself is the electric potential energy per unit positive charge (1 C) :

$$V = \frac{U}{q} \tag{16.31}$$

Unit of electric potential
V = J/C

and this is called the **electric potential** or simply potential. Since the unit of energy and work is joule (J) and the unit of charge is coulomb (C), the unit of potential is J/C, which is called volt (symbol V). That is to say

$$\mathrm{V} = \mathrm{J/C}. \tag{16.32}$$

From equations (16.30) and (16.31), the potentials V_P and V_A at points P and A satisfy the relation

$$\int_P^A E_t \, ds = V_P - V_A \tag{16.33}$$

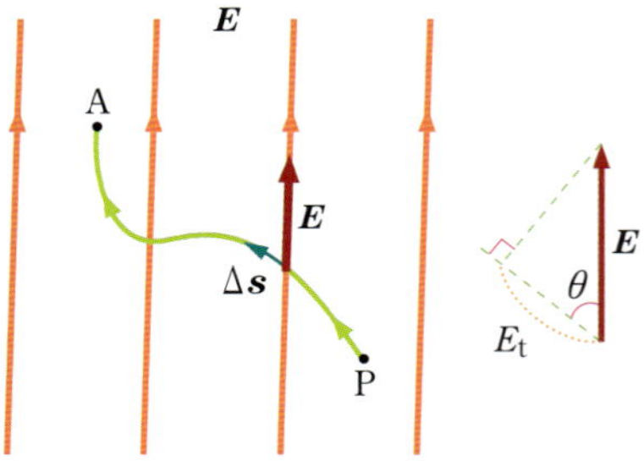

Fig. 16-26 $\int_P^A E_t \, ds = V_P - V_A$

(Fig. 16-26). $V_P - V_A$ on the right-hand side is called the **potential difference** between points P and A. If $V_P > V_A$, then the point P is said to have a higher potential than the point A and if $V_P < V_A$, then the point P is said to have a lower potential than the point A.

From equations (16.30) and (16.31), the work $W_{P \to A}$ performed by the electric force $q\boldsymbol{E}$ when the charge q moves from point P at which the potential difference is $V = V_P - V_A$ to point A is

$$W_{P \to A} = qV = q(V_P - V_A). \tag{16.34}$$

For example, when the charge q moves in the conducting wire from the positive electrode P to the negative electrode A of the battery having the electromotive force $V = V_P - V_A$, the work $W_{P \to A}$ performed by the electric field is $W_{P \to A} = qV$. Conversely, when the charge q moves from the negative electrode to the positive electrode in the battery, the work performed by the electromotive force of battery is also qV.

When an object of mass m with charge q moves from point A to point B of the potential difference $V_A - V_B$ under the action of electric force $q\boldsymbol{E}$ only, the work $W_{A \to B}$ performed by the electric force is, according to equation (5.12) describing the relationship between work and kinetic energy, equal to the amount of increase in kinetic energy :

$$\frac{1}{2}mv_B{}^2 - \frac{1}{2}mv_A{}^2 = W_{A \to B} = q(V_A - V_B). \tag{16.35}$$

This relation means that the sum of kinetic energy and electric potential energy is constant

$$\frac{1}{2}mv_A{}^2 + qV_A = \frac{1}{2}mv_B{}^2 + qV_B. \tag{16.36}$$

That is to say it indicates the energy conservation law.

Electron volt In atomic physics, electric fields are used to accelerate charged particles. Therefore, the amount of increase in kinetic energy when a charged particle with elementary charge e passes through the potential difference of 1 V is selected as a practical unit of energy in atomic physics and written as 1 electron volt or 1 eV. Since $e = 1.602 \times 10^{-19}$ C, it follows that

$$1\ \text{eV} = 1.602 \times 10^{-19}\ \text{J} \quad (\text{C V} = \text{J}). \tag{16.37}$$

One kiloelectron volts $1\ \text{keV} = 10^3\ \text{eV}$, and one megaelectron volts $1\ \text{MeV} = 10^6\ \text{eV}$ are also used.

Case 6 In an X-ray generator, a high voltage is applied between the negative and positive electrodes of the X-ray tube to accelerate electrons emitted from the negative electrode to collide with the positive electrode to generate X-rays (see p.348). When the potential difference between the positive and negative electrodes is 10 000 V, the kinetic energy of the electrons immediately before colliding with the positive electrode is $10\,000\ \text{eV} = 1.602 \times 10^{-15}$ J.

Case 7 | Potential of the uniform electric field In the case of Fig. 16-27 in which the point charge q moves from point P to point A in the uniform electric field $\boldsymbol{E}$, the work $W_{P \to A}$ performed by the electric force $q\boldsymbol{E}$ is, since the magnitude of force is qE and the moving distance of point charge in the direction of force is d, given by the relation

$$W_{P \to A} = qEd = q(V_P - V_A).$$

$$\therefore \quad V_P - V_A = Ed. \tag{16.38}$$

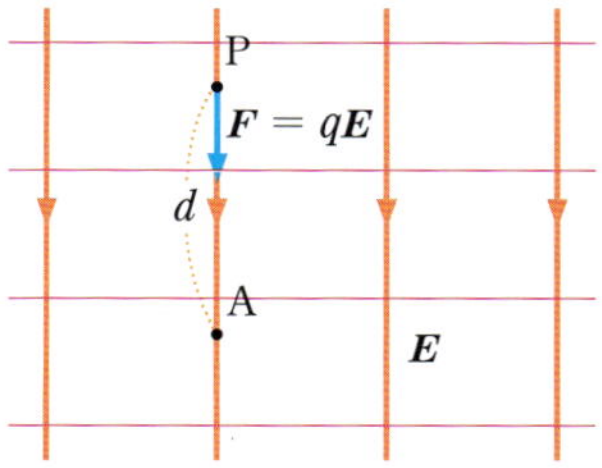

Fig. 16-27 Potential and electric field

$V_P - V_A = Ed, \; E = \dfrac{V_P - V_A}{d}$

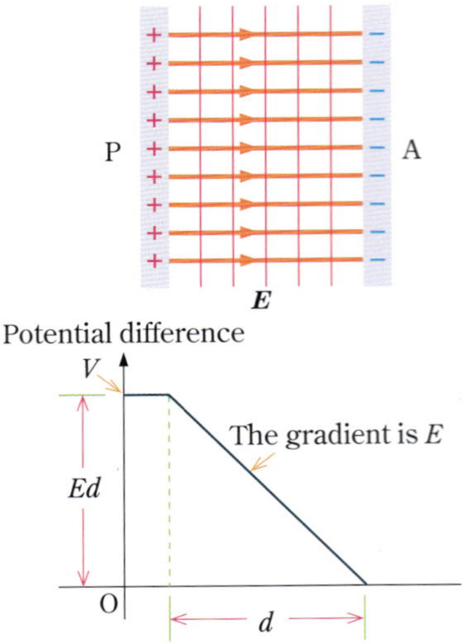

Fig. 16-28 Electric field and potential

$V = Ed, \ d = \dfrac{V}{E}$

Unit of electric field (re-description) N/C = V/m

Equation (16.38) is transformed into

$$E = \frac{V_{\mathrm{P}} - V_{\mathrm{A}}}{d}. \tag{16.39}$$

This shows that the electric field strength E at a certain point is equal to the potential gradient at that point (Fig. 16-28). The direction of the electric field points from the high potential side to the low potential side.

Equation (16.39) tells that the unit of electric field is alternatively expressed by dividing volt (V) of the unit of potential by meter (m) of the unit of length. In other words

$$\text{unit of electric field} = \mathrm{N/C} = \mathrm{V/m}. \tag{16.40}$$

So far, we have been discussing exclusively the potential differences in two points. Let the position vector of point P be $\boldsymbol{r}$, and the position vector of point A be $\boldsymbol{r}_0$. Then, equation (16.33) is changed into

$$V(\boldsymbol{r}) = -\int_{\boldsymbol{r}_0}^{\boldsymbol{r}} E_{\mathrm{t}}\, \mathrm{d}s + V(\boldsymbol{r}_0). \tag{16.41}$$

To measure the potential $V(\boldsymbol{r})$ at point P, it is necessary to determine a reference point (a point at which the potential is 0). If point A is chosen to be a reference point for $V(\boldsymbol{r}_0) = 0$, the potential $V(\boldsymbol{r})$ at point P is expressed as

$$V(\boldsymbol{r}) = -\int_{\boldsymbol{r}_0}^{\boldsymbol{r}} E_{\mathrm{t}}\, \mathrm{d}s = -\int_{\boldsymbol{r}_0}^{\boldsymbol{r}} \boldsymbol{E} \cdot \mathrm{d}\boldsymbol{s}. \tag{16.42}$$

In theoretical calculations it is convenient to choose a point that is infinitely far from the charge as the reference point. In the following examples, the reference points for measuring the potential are infinitely far away. In addition, since the potential of the Earth is almost constant, in practice, the potential of a conductor grounded to the Earth is often chosen as 0. If the reference point changes, the potential at each point changes, but the potential difference between two points does not change

Case 8 | Potential created by a point charge q The potential at point $\boldsymbol{r}$ when there is a point charge q at the origin is obtained using equations (5.29) and (16.9) as follows:

$$V(r) = -\int_{\infty}^{r} \frac{q}{4\pi\varepsilon_0 r^2} \frac{\boldsymbol{r}}{r} \cdot \mathrm{d}\boldsymbol{r} = -\int_{\infty}^{r} \frac{q}{4\pi\varepsilon_0 r^2}\, \mathrm{d}r = \frac{q}{4\pi\varepsilon_0 r}. \tag{16.43}$$

The potential $V(\boldsymbol{r})$ at point $\boldsymbol{r}$ with the point charge q at point $\boldsymbol{r}_0$ is obtained by setting $r = |\boldsymbol{r} - \boldsymbol{r}_0|$ in equation (16.43):

$$V(\boldsymbol{r}) = \frac{q}{4\pi\varepsilon_0 |\boldsymbol{r} - \boldsymbol{r}_0|}. \tag{16.44}$$

Case 9 | Potential created by point charges $q_1, q_2, \cdots, q_N$
The electric field that many charges create at point $\boldsymbol{r}$ is the sum of the electric fields that each charge creates at point $\boldsymbol{r}$. Therefore, the

potential in this case is the sum of the potentials of the electric fields created by each charge. When point charges $q_1, q_2, \cdots, q_N$ are at $\boldsymbol{r}_1, \boldsymbol{r}_2, \cdots, \boldsymbol{r}_N$, respectively, the potential at point $\boldsymbol{r}$ is

$$V(\boldsymbol{r}) = \frac{1}{4\pi\varepsilon_0} \sum_{i=1}^{N} \frac{q_i}{|\boldsymbol{r}-\boldsymbol{r}_i|}. \tag{16.45}$$

Equipotential surface A surface formed by connecting points of equal potential is called an **equipotential surface**, and any curve on the equipotential surface is called an **equipotential line**. Because all points on the equipotential surface are equal in potential, the electric force does not work when the charge moves on the equipotential surface. Therefore, the electric force has no component in the direction of the equipotential surface. Accordingly,

The electric field and the equipotential surface are orthogonal. The electric field is also orthogonal to the equipotential lines.

The potential difference $\Delta V = V_Q - V_P$ between point P (position vector $\boldsymbol{r}$) and point Q (position vector $\boldsymbol{r}+\Delta\boldsymbol{s}$) in Fig. 16-29 is the work $-\boldsymbol{E}\cdot\Delta\boldsymbol{s} = -E_t\,\Delta s$ that the electric field $\boldsymbol{E}$ performs when the unit positive charge is displaced from point Q to point P by $-\Delta\boldsymbol{s}$. That is to say,

$$\Delta V = V_Q - V_P = V(\boldsymbol{r}+\Delta\boldsymbol{s}) - V(\boldsymbol{r}) = -E_t\,\Delta s. \tag{16.46}$$

Therefore, E_t which is the $\Delta\boldsymbol{s}$ component of the electric field $\boldsymbol{E}$ is

$$E_t = -\frac{\Delta V}{\Delta s} \quad \text{(in the general case of } \Delta\boldsymbol{s}\text{)}. \tag{16.47}$$

If the displacement is in the direction of the electric field $\boldsymbol{E}$, $E_t = E$ and the electric field strength is expressed as

$$E = -\frac{\Delta V}{\Delta s} \quad \text{(in the case of } \boldsymbol{E} /\!/ \Delta\boldsymbol{s}\text{)}. \tag{16.48}$$

Since the electric field is perpendicular to the equipotential lines and is directed from high to low potentials, the electric field points in the steepest direction of the downward gradient and the magnitude of the gradient is the electric field strength at that point.

Question 4 Show the direction of the electric field at points a and b in Fig. 16-30. At which point is the electric field stronger?

Question 5 In Fig. 16-31, which of the electric fields at point P or point Q is stronger? In addition, draw on the figure the direction of the electric field at each point.

In Section **5.2**, it was shown that the conservative force is derived by partial differentiation of potential energy [equation (5.30)]. Similarly as this, the electric field is derived by partial differentiation of potential:

$$E_x = -\frac{\partial V}{\partial x}, \quad E_y = -\frac{\partial V}{\partial y}, \quad E_z = -\frac{\partial V}{\partial z}. \tag{16.49}$$

The three relations of equation (16.49) correspond to the case in which Δs in equation (16.47) is Δx, Δy, and Δz, respectively.

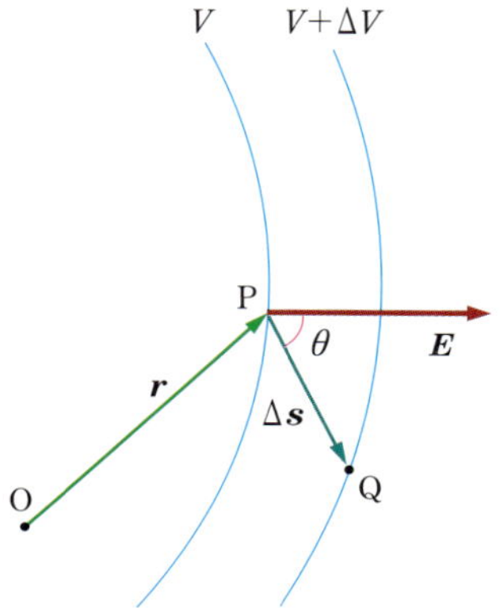

Fig. 16-29 E_t which is the $\Delta\boldsymbol{s}$ component of the electric field $\boldsymbol{E}$ is $E_t = -\frac{\Delta V}{\Delta s} = E\cos\theta$ (In this figure ΔV is negative).

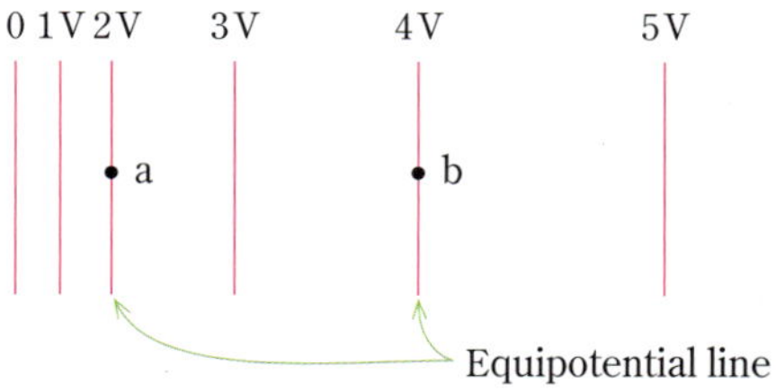

Fig. 16-30

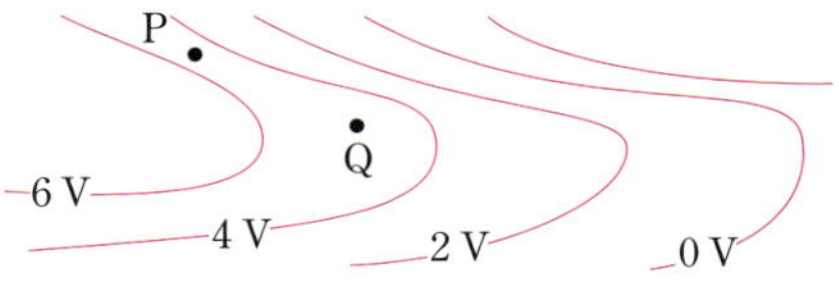

Fig. 16-31

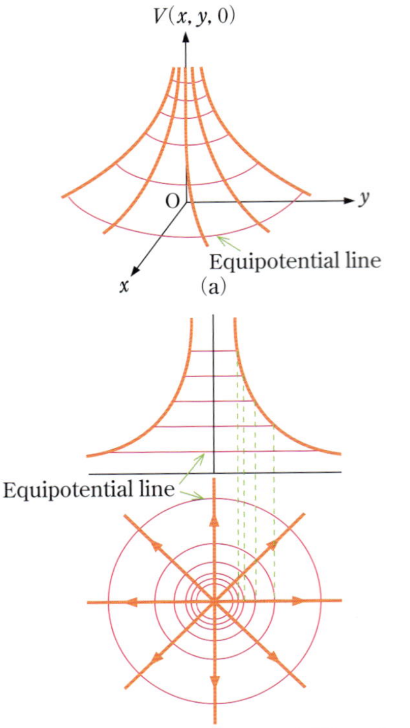

(b) Density of equipotential lines is proportional to the electric field strength. Lines of electric force are orthogonal to the equipotential lines.

Fig. 16-32 Potential $V(x, y, 0)$ on the xy plane when there is a positive point charge at the origin.

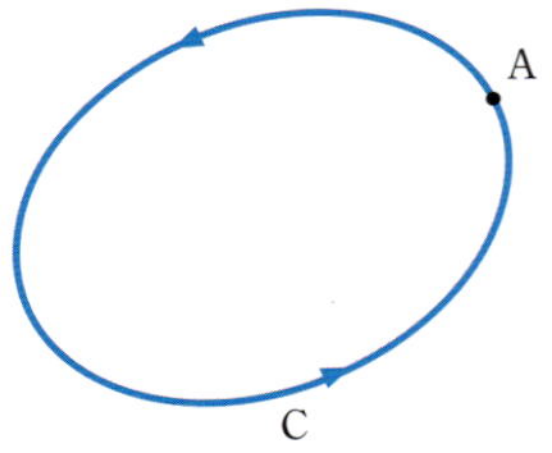

Fig. 16-33 $\oint_C \boldsymbol{E}\cdot d\boldsymbol{s} = \oint_C E_t\, ds = 0$

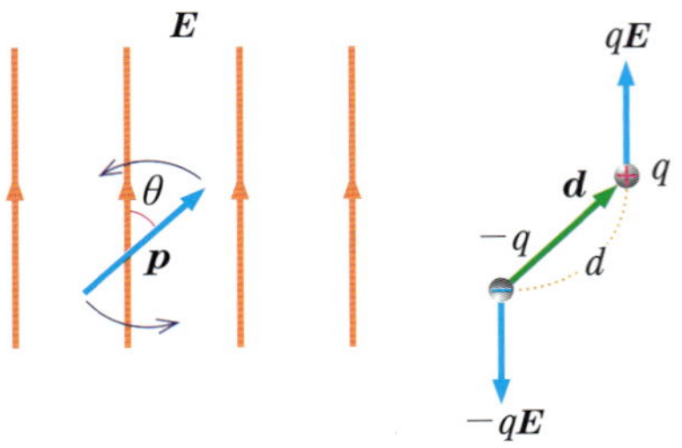

Fig. 16-34 Electric force acting on the electric dipole
$\boldsymbol{N} = \boldsymbol{p}\times\boldsymbol{E} \quad (N = pE\sin\theta)$

If we draw the equipotential surface every time the potential difference ΔV takes a certain fixed value, the electric field is strong where the equipotential surfaces are close (small Δs), and the electric field is weak where the surface intervals are large (large Δs). Equation (16.47) is the generalized form of equation (16.39). In Fig. 16-32 is shown the potential $V(x, y, 0)$ on the xy plane when there is a positive point charge at the origin. The equipotential lines in this figure correspond to the contour lines of the map.

As shown in Fig. 16-33, when one travel along the closed curve C is completed the potentials of the starting point A and the end point A are equal and according to equation (16.33), the sum of work performed by the electrostatic field when the unit positive charge (1 C) travels around the closed curve C is 0, namely

$$\oint_C \boldsymbol{E}\cdot d\boldsymbol{s} = \oint_C E_t\, ds = 0 \tag{16.50}$$

is derived. Equation (16.50) is a necessary and sufficient condition that the potential can be defined, and is a condition meaning that there is no closed curve with neither a starting point nor an end point in the electric field created by the charge at rest. The electric field induced when the magnetic field fluctuates with time does not satisfy equation. (16.50). Therefore, the potential cannot be defined in the case where an induced electric field is created (see Chapter 21).

Electric dipole A pair of positive and negative charges q and $-q$ which are in close proximity is called the **electric dipole**, and $\boldsymbol{p} = q\boldsymbol{d}$ is called the **electric dipole moment**. $\boldsymbol{d}$ is a vector that starts at negative charge and ends at positive charge.

Case 10 | Electric force that a uniform electric field acts on an electric dipole When an electric dipole with electric dipole moment $\boldsymbol{p} = q\boldsymbol{d}$ is placed in a uniform electric field $\boldsymbol{E}$, the electric force $q\boldsymbol{E}$ acts on the positive charge q and the electric force $-q\boldsymbol{E}$ acts on the negative charge $-q$. Therefore, the electric force acting on this electric dipole is a couple with $\boldsymbol{0}$ vector sum of forces. By referring to equation (6.16), it can be seen that the moment $\boldsymbol{N}$ of the couple is expressed as

$$\boldsymbol{N} = \boldsymbol{p}\times\boldsymbol{E} \qquad (N = pE\sin\theta) \tag{16.51}$$

(Fig. 16-34). This couple acts to turn the direction of the electric dipole moment $\boldsymbol{p}$ towards the direction of the electric field $\boldsymbol{E}$.

The electric force potential energy V of the electric dipole in this case can be expressed as

$$V = -\boldsymbol{p}\cdot\boldsymbol{E} \qquad (V = -pE\cos\theta) \tag{16.52}$$

and is minimum when $\boldsymbol{p}$ and $\boldsymbol{E}$ have the same direction.

Problems for exercise 16

A

1. Describe the similarities and the differences between universal gravitation and Coulomb's force. Although the magnitude of universal gravitation acting between protons and electrons is negligibly weak compared to the electric force, the universal gravitation is important between celestial bodies, and the electric power can be ignored. Describe the reason.
2. Three charges are placed on a straight line as in Fig. 1. The resultant electric force acting on the middle charge is zero. Find the distance x.

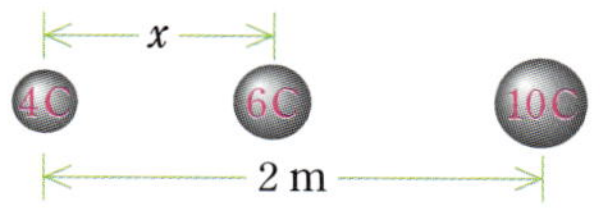

Fig. 1

3. When a point charge of 3×10^{-6} C was placed at a point in space, a force of 6.0×10^{-4} N worked.
   (1) Find the electric field strength at this point?
   (2) What kind of force acts when the charge of -6.0×10^{-6} C is put on the same point?
4. A charge of 1.0 μC is at point $x = 9.0$ cm on the x-axis and a charge of 4.0 μC is at the origin (Fig. 2).

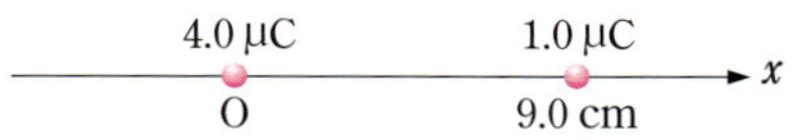

Fig. 2

   (1) Where is the point of electric field $E = 0$?
   (2) Find the electric field at point $x = 15$ cm on the x-axis.
5. The charge of the electron is -1.6×10^{-19} C and its mass is 9.1×10^{-31} kg.
   (1) Find the electric field strength that gives 9.8 m/s^2 acceleration to the electron.
   (2) Find the acceleration due to electric force when the electron is in a uniform electric field of 10 000 N/C.
6. If the Earth were a hollow sphere, could a person stand on the inner ground surface of the spherical shell?
7. Show that the electric field inside the infinitely long hollow column charged uniformly is zero.
8. Two infinitely wide flat plates are uniformly charged with surface density σ respectively. Find the electric field when these two plates are arranged in parallel. Moreover, show that in this case, the force the charge on the unit area of one plate receives from the charge on the other plate is $\dfrac{\sigma^2}{2\varepsilon_0}$ (repulsive force).
9. **Tube of electric force** ▌ Lines of electric force passing through a closed curve in the electric field create a tube in the electric field [Fig. 3 (a)]. This tube is called the tube of electric force. As shown in Fig. 3 (b), if no charge exists between the two cross sections of this tube of electric force, "The flux of electric line of force ($-E_{1n}A_1$) entering through cross section 1 is equal to the flux of electric line of force ($E_{2n}A_2$) exiting through cross section 2". Prove this description.

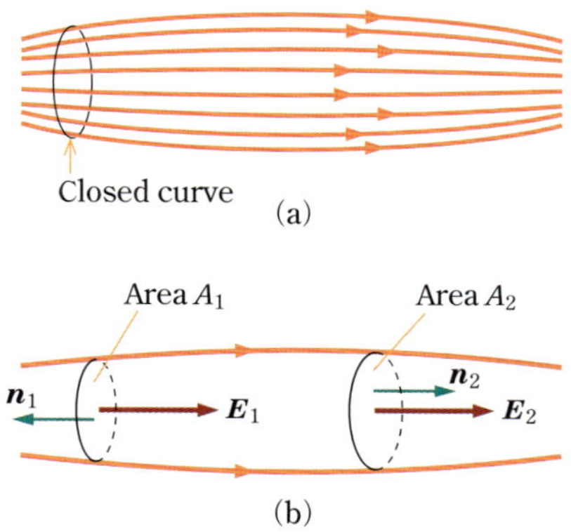

Fig. 3

10. (1) In each of Fig. 4 (a), (b) and (c), show the direction of the electric field at points A and B and the direction of the electric force acting when a charge of -1 μC is brought to each.
    (2) Compare the electric field strength at point B in (a), (b) and (c).

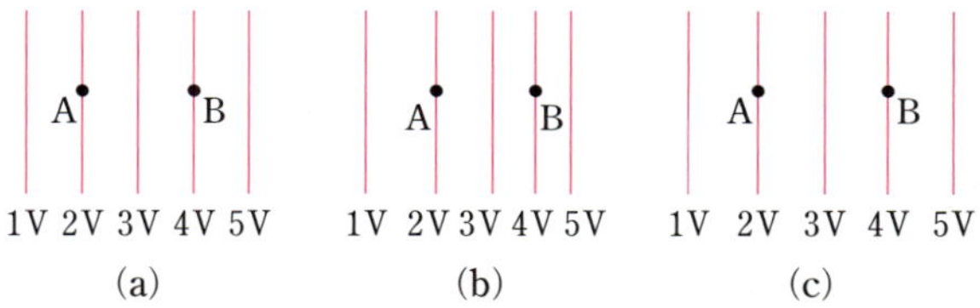

Fig. 4

11. Find the potential difference $V_A - V_B$ between two points A and B in Fig. 5.

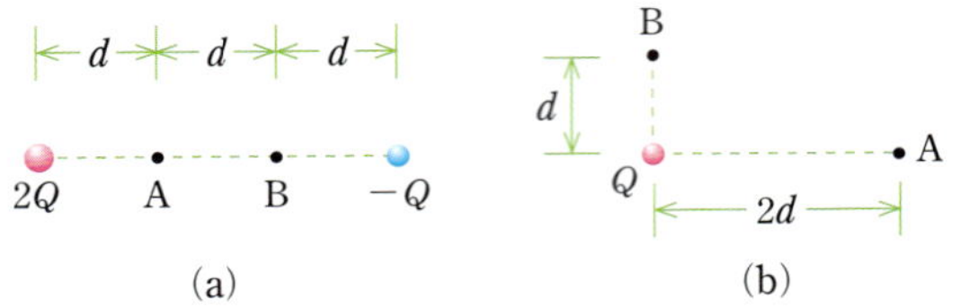

Fig. 5

12. There are two parallel metal plates as shown in Fig. 6. The spacing between the plates is 8 cm. A voltage of 24 V was applied to these plates.

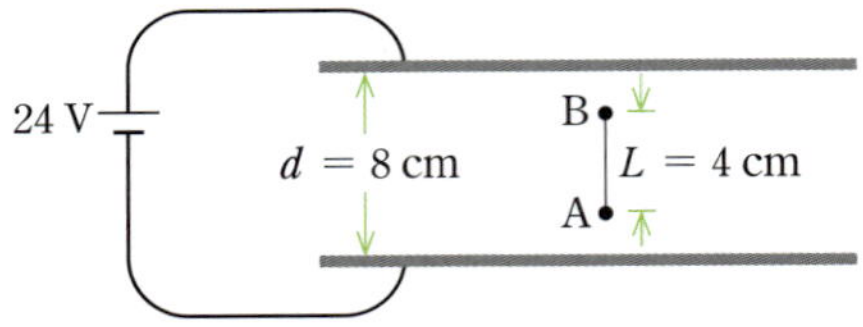

Fig. 6

(1) Find the electric field strength between these parallel metal plates.
(2) A charge of 3×10^{-6} C was placed at point A. Find the electric force acting on this charge.
(3) Calculate the work required to carry this charge from point A to point B.
(4) Find the potential difference $V_B - V_A$ between point B and point A.

B

1. An electric charge of 10 μC is uniformly distributed on a semicircular thin rod with radius $r = 10$ cm. Find the electric field at the center O of this semicircle (Fig. 7).

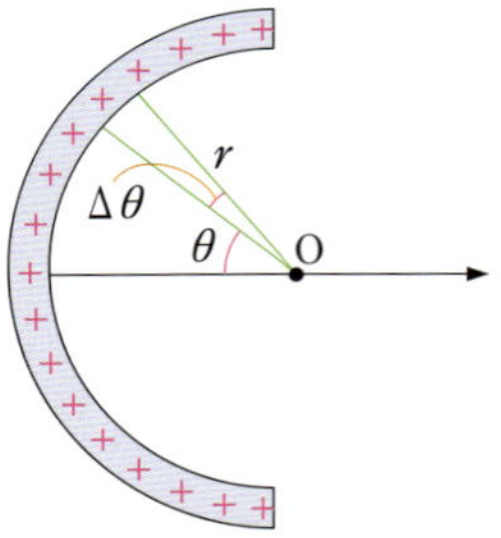

Fig. 7

2. On a sphere of radius R centered at the origin, the positive charge $Q = \frac{4\pi}{3}\rho R^3$ is uniformly distributed with a density of ρ.
(1) Show that the electric field $\boldsymbol{E}(\boldsymbol{r})$ at point $\boldsymbol{r}$ in this sphere is given by

$$\boldsymbol{E}(\boldsymbol{r}) = \frac{\rho r}{3\varepsilon_0}\frac{\boldsymbol{r}}{r} = \frac{\rho \boldsymbol{r}}{3\varepsilon_0} \qquad (r \leqq R).$$

(2) Find the balanced position of the mass point having the negative charge $-q$ placed in this sphere. It is assumed that no force other than electric force acts on the mass point.
(3) If this mass point is slightly moved from the balanced position and is released gently, how will the mass point move? Let the mass of the mass point be m.

3. (1) Assume that the same amounts of positive and negative charges are uniformly distributed with the charge densities ρ and $-\rho$ and overlap on a sphere of radius R. When this distribution of positive charges is displaced in parallel by δ in the x direction, show using the result of the previous problem that the internal electric field becomes uniform.
(2) Assuming that the displacement δ is sufficiently small, find the charge appearing on the surface (Fig. 8).

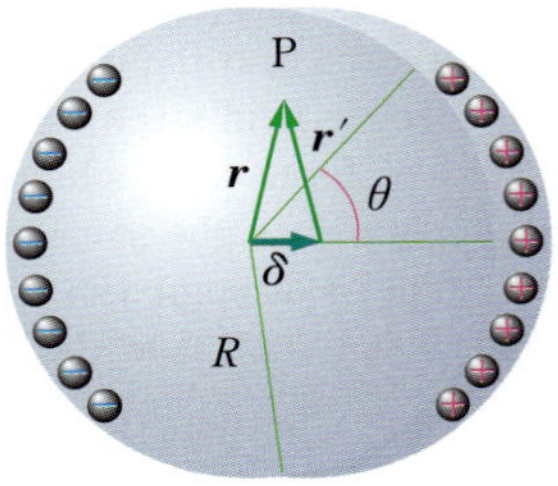

Fig. 8

4. Two thin infinitely spreading insulating plates are placed in parallel. The charge is distributed uniformly on one plate with a surface density of 2σ and on the other plate with a surface density of $-\sigma$. Find the electric field.

5. Prove that the strength F of the universal gravitation acting between two objects A and B with mass m_A and m_B and having a spherically symmetric mass distribution, when the central distance between the two objects is r, is given by $F = G\frac{m_A m_B}{r^2}$.

6. When there is a point charge q at the origin, the potential is

$$V(r) = \frac{q}{4\pi\varepsilon_0 r}$$

[equation (16.43)]. Using equation (16.49), find the electric field.

7. **Potential around the electric dipole** ▌ Charges q and $-q$ are placed at points $(a, 0, 0)$ and $(-a, 0, 0)$, respectively [Fig. 9 (a)]. Show that in this case the potential at point (x, y, z) is

$$V(x, y, z) = \frac{px}{4\pi\varepsilon_0(x^2+y^2+z^2)^{3/2}} = \frac{p\cos\theta}{4\pi\varepsilon_0 r^2}$$

$$(p = 2aq). \qquad (1)$$

Here, $r^2 = x^2+y^2+z^2$, $x = r\cos\theta$ (θ is the angle between the $+x$ direction and the position vector $\boldsymbol{r}$), and it is assumed that $|a| \ll x$ and $(y^2+z^2)^{1/2}$.

Using this result and equation (16.49), derive the electric field that the electric dipole creates on the xy plane. This electric field is shown in Fig. 9 (b).

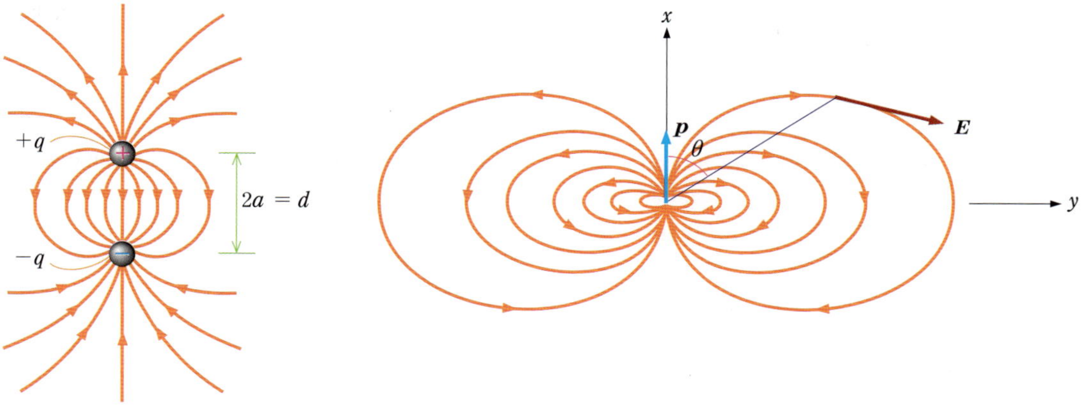

(a) Electric dipole

(b) Electric field created by the electric dipole
(Note that the directions of x-axis and y-axis are different from the usual cases)

Fig. 9 Electric field created by the electric dipole

Franklin and Faraday who laid the foundation of electromagnetism

Franklin and Faraday, who laid the foundations of research of electricity and magnetism, have a common point that they both engaged in work since their childhood and did not complete the elementary school programs.

Benjamin Franklin (1706-1790)

Franklin was born in the United States in the British colonial era. He engaged in labor from an early age and succeeded in printing business. He retired from business at the age of 42 to concentrate on scientific research, but he was deeply involved in the independence movement since the late forties and actively played a role as a politician as well as a diplomat.

In 1746, he heard about electricity experiments, and got intrigued by them, and started the experiments. He immediately understood correctly the function of the Leiden bottle (capacitor) which is a device for storing the electric charge. The bottle had the metal foil on the inside and the outside of the glass bottle. He came to know that the inside and the outside of the glass bottle were oppositely charged and the phenomenon that these two cancel each other was a discharge. In this way, he named the glass electricity and the resin electricity used until then as positive electricity and negative electricity, respectively.

He focused on the similarities between discharge and lightning and his experiment (1752) using a kite that confirms that lightning is an electrical phenomenon and a thundercloud is a charged cloud is well-known. He found that when a grounded metal needle is brought close to a charged metal sphere, the needle absorbs electricity from the non-touching metal sphere, and invented a lightning rod based on this phenomenon. He is said to be the first physicist of the United States.

He is also a distinguished writer and the following his words are famous:
"God helps them that help themselves."
"Keep your eyes wide open before marriage, half shut afterwards."

It was Franklin at the age of 82 who invented the bifocal lens. He is said to have advertised the bifocal lens by saying "Wearing glasses with this lens at a dinner table, you can see well both the facial expression of other people and the dishes".

Michael Faraday (1791-1867)

He was born as a child of a poor blacksmith and the primary education he received was only basic reading and writing skills and calculation lessons. At the age of 12 he was cut off from school and became an apprentice of a bookmaker, but he spent his free time reading books of science enlightenment and trying simple experiments. He happened to receive a ticket for a public lecture held at the Royal Institution, was fascinated by the world of science, wrote a letter to Davy of the Royal Institution who was a speaker, and Faraday was recruited as an assistant at the Royal Institution in 1813. He continued his research for a lifetime there.

Faraday is a physicist who studied electromagnetics, including the discovery of electromagnetic induction and the invention of lines of electric and magnetic forces, and also is one of the founders of organic chemistry, analytical chemistry, electrochemistry, and magnetic chemistry. Because Faraday did not know mathematics at all, his 450 articles have no differential equations.

The Royal Institution is an institution for scientific enlightenment, founded in 1799. In 1826, to disseminate science and improve the public's scientific knowledge Faraday launched a Christmas lecture for small children and a Friday lecture for members of the Royal Institution and its invited guests, and it was a great success. Faraday gave about six Friday lectures each year, but the wide variety of his topics is amazing. He also appeared 19 times in Christmas lectures, among which the most famous series "The Chemical History of a Candle" (translated into Japanese『ローソクの科学』) is available as a book published by Iwanami Shoten.

Conductor and Electrostatic Field

Substances are classified into conductors that conduct electricity and insulators (or non-conductors) that do not conduct electricity. A **conductor** has a **free electric charge**, which is the charge that can move freely in it.

When a conductor is charged or brought into the electrostatic field, the charge distribution of the conductor and the electric field around the conductor exhibit characteristic properties due to free charge.

In this chapter, we will learn the electrostatic field inside and near the surface of charged conductors in vacuum.

Faraday cage
Lines of electric force between two electrodes. Lines of electric force cannot penetrate into the ring-form electrode on the right.

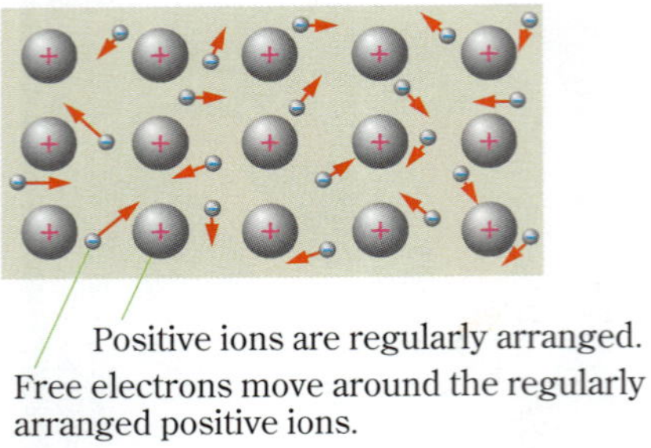
Positive ions are regularly arranged.
Free electrons move around the regularly arranged positive ions.

Fig. 17-1 Structure of metals

Metals and electrolytic solutions are typical examples of conductors. In metals, some electrons are bound to atoms, but some electrons leave atoms and move around in an ordered array of positive ions (Fig. 17-1). These electrons are called **free electrons** or **conduction electrons**. The electrolytic solutions contain positive and negative ions that can move freely in solutions.

Since substances are composed of positive ions and electrons, or positive ions and negative ions, the electric field changes drastically inside the substances. In electromagnetism that targets the macroscopic world, the microscopic electric field (micro-electric field) that changes rapidly on the atomic scale is too small, so we consider the electric field which is still invisible to the human eye but is averaged over a range long enough on the atomic scale. The resulting macroscopically averaged **macroscopic electric field** (macro-electric field) is called the electric field of the substances.

17.1 Conductor and electric field

Learning objective To understand several characteristic properties that the electric field inside the conductor and the electric field on the conductor surface exhibit.

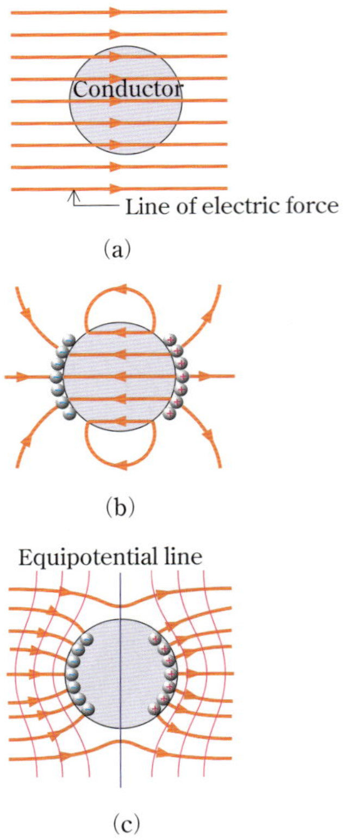
Conductor
Line of electric force
(a)
(b)
Equipotential line
(c)

Fig. 17-2 Electrostatic induction. The electric field inside the conductor in equilibrium is **0**. (a) Electric field applied from the outside, (b) Electric field that charges induced on the conductor surface creates, (c) Electric field when the conductor is placed in a uniform electric field.

When a conductor is placed in the electrostatic field, the positive free charge of the conductor moves in the direction of the electric field, while the negative free charge moves in the opposite direction of the electric field, and positive and negative free electric charges appear on the surface of the conductor. The movement of free charges continues until the electric field created by the surface charge cancels the electric field created by the charge outside the conductor and eventually the electrostatic field inside the conductor becomes **0** (Fig. 17-2). This phenomenon is called **electrostatic induction**. Therefore, when there is an electric field inside the conductor, free charge movement occurs in the conductor, so

In equilibrium, the electric field in the conductor is $\mathbf{0} : \boldsymbol{E} = \mathbf{0}$.

As a result of this, the potential difference between any two points inside the conductor is 0. Therefore,

In equilibrium, all points of one conductor are equipotential.

From this fact, it is understood that if we choose the ground as a reference for measuring the electric potential, the potential of the conductor connected to the ground, that is, the grounded conductor, is equal to the potential of the Earth and is always 0 (although the weak ground current flows through the Earth. The Earth can be considered as an approximately equipotential conductor).

Attention What has been shown above holds when there is no temperature gradient or when there are different parts of the components inside the conductor. If there is a temperature gradient, the thermoelectromotive force is generated between high temperature part and low temperature part. Also, if there exist parts different in component such as inside the battery, the electromotive force due to chemical origin will occur between them. In the equilibrium state, the force due to these effects balances the electric force and $\boldsymbol{E} \neq \boldsymbol{0}$.

Question 1 How can we negatively charge a conductor using only a positively charged conductor (Fig. 17-3)?

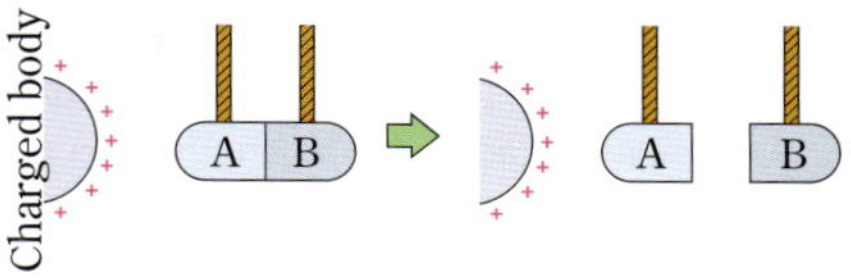

Fig. 17-3

Let us consider an arbitrary closed surface S inside a conductor, and apply Gauss' law (16.19) of electric field. Then, since the electric field $\boldsymbol{E}$ is $\boldsymbol{0}$ at all points on the closed surface S, we have

"Total electric charge inside the closed surface S"$= \varepsilon_0 \iint_S E_n \, dA = 0$

(E_n is the normal component of the electric field). In this way, the total charge inside the closed surface S is derived to be 0. Because a closed surface of any size at any location can be selected as a closed surface S of Gauss' law,

Inside the conductor, positive and negative charges cancel each other, and the charge density is zero.

Electric field on the conductor surface The conductor surface is one equipotential surface. As shown in Section **16.5**, the electric field is perpendicular to the equipotential surface, so in equilibrium,

The electric field on the conductor surface is perpendicular to the conductor surface.

When a conductor is placed in an electrostatic field, movement of the free charge occurs inside the conductor, and charges are distributed on the conductor surface so that the electrostatic field inside the conductor becomes $\boldsymbol{0}$. If the surface density of charge at point P on the conductor surface is σ, then the electric field strength at point P is

$$E = \frac{\sigma}{\varepsilon_0}. \tag{17.1}$$

The electric field on the conductor surface refers to the electric field just outside the conductor surface.

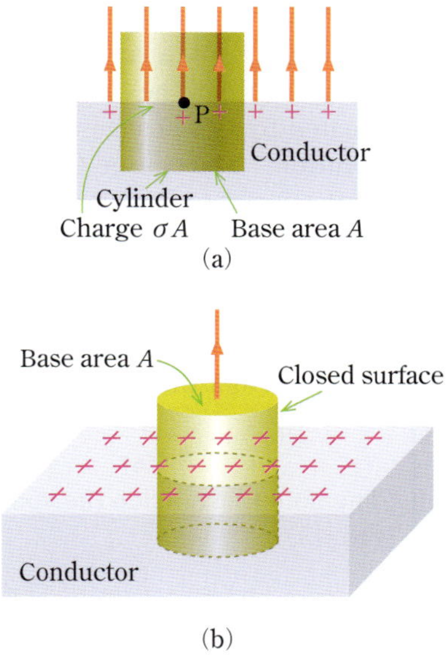

Fig. 17-4 Electric field on the conductor surface $E = \dfrac{\sigma}{\varepsilon_0}$ (in the case $\sigma > 0$).

Proof of equation (17.1) We apply Gauss' law to the cylinder shown in Fig. 17-4. Inside the cylinder, the relation $\boldsymbol{E} = \boldsymbol{0}$ holds. The side of the cylinder is parallel to the electric field ($E_n = 0$) and the flux of electric line of force passing through the upper base of area A is EA, and the amount of electricity in the cylinder Q_{in} is $Q_{in} = \sigma A$, so Gauss' law of electric field is $EA = \dfrac{\sigma A}{\varepsilon_0}$, that is, $E = \dfrac{\sigma}{\varepsilon_0}$ is obtained.

Case 1 The electric current flows through the Earth and the Earth can be taken to be a conductor. Assume that an electric field of the intensity $E = 130$ N/C with a vertical downward direction exists near the Earth surface. Since $E_n = -130$ N/C, the surface density σ

* If the electric field exists in the cavity inside the conductor, the starting point A of lines of electric force and their end point B are on the surface of the cavity and there occurs a contradiction that the potential difference ($V_A > V_B$) is observed on the cavity surface that is to be equipotential.

of charge on the Earth surface in this state is

$$\sigma = \varepsilon_0 E_n = \{8.9\times10^{-12}\,C^2/(N\cdot m^2)\}\times(-130\,N/C)$$
$$= -1.2\times10^{-9}\,C/m^2.$$

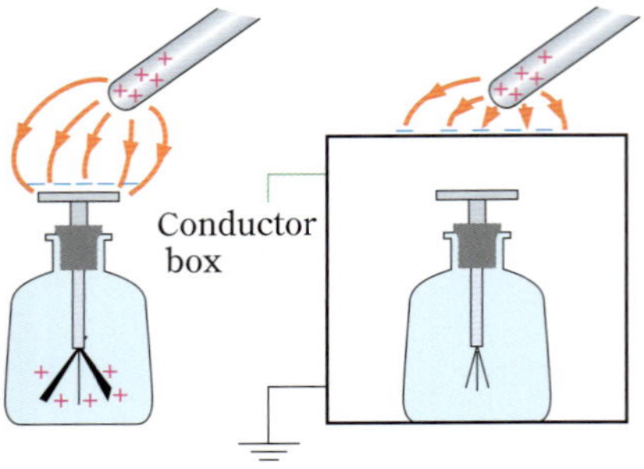

Fig. 17-5 Electric shielding (shield)

Even if there is an electric field outside the conductor, the electric field inside the conductor is **0**. The situation is also the same when there is a cavity inside the conductor. If no charge exists in the cavity, no charge appears on the wall of the cavity and the electric field is **0** inside the cavity, and the cavity and the conductor are equipotential*. This property indicates that the space surrounded by the conductor is not affected by the electric field outside the conductor. This is called the **electrostatic shielding** (**shield**) (Fig. 17-5). Devices for precise electrostatic measurement are wrapped in a grounded metal plate to avoid external electrical effects. Instead of a metal plate a wire mesh is also usable to avoid the influence of the external electric field. Our daily experience that the radio sound is hard to hear inside a reinforced concrete building is an example of electrostatic shielding. The shielding wire for blocking electrostatic induction of electronic devices is also an application of electrostatic shielding.

The inner space of a deep can is approximately considered to be the inner part of the cavity in the conductor. Franklin suspended a charged metal ball with a thread, placed it in a deep empty can as shown in Fig. 17-6, brought it into contact with the bottom and then pulled it up, and showed that no electric force acts between the metal ball and the empty can. This result proves that the metal ball is not charged. The charge of the metal ball moved to the outer surface of the can.

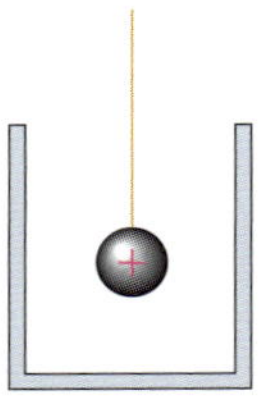

Fig. 17-6

Example 1 A metal sphere of radius R has a charge Q. Find the electric field and the potential created by this charge distribution.

Solution The charge Q is uniformly distributed on the surface of a sphere of radius R. Using the results of Case 4 in Section **16.4**, the electric field outside the metal sphere is the same as the electric field with the charge Q in the center of the sphere, so the strength at the point whose distance from the center is r is

$$E(r) = \frac{Q}{4\pi\varepsilon_0 r^2} \quad (r > R). \qquad (17.2)$$

The potential outside the sphere is from equation (16.43)

$$V(r) = \frac{Q}{4\pi\varepsilon_0 r} \quad (r \geqq R) \qquad (17.3)$$

(Fig. 17-7). Inside the sphere, the electric field is 0 and all points are equipotential :

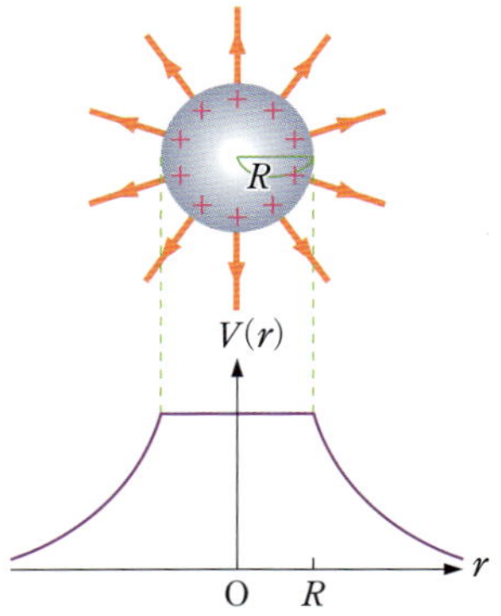

Fig. 17-7 Potential created by a positively charged metal sphere.

$$V(r) = V(R) = \frac{Q}{4\pi\varepsilon_0 R} \quad (r \leqq R). \qquad (17.4)$$

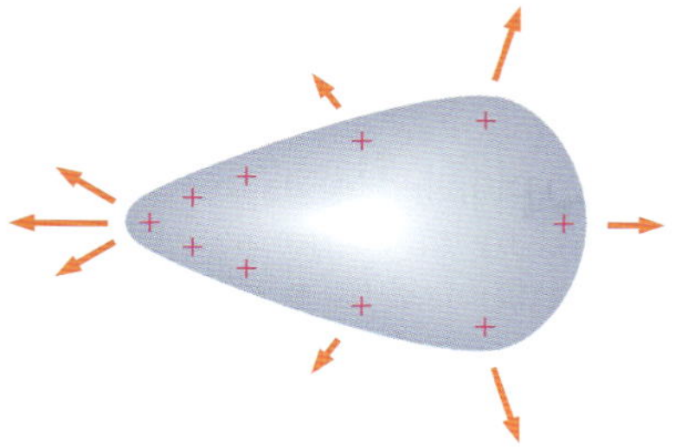

Fig. 17-8 The electric field is the strongest at the sharp part.

The electric field strength on the surface of a conducting sphere of radius R having a charge Q is $E = \dfrac{Q}{4\pi\varepsilon_0 R^2}$ and the potential is $V = \dfrac{Q}{4\pi\varepsilon_0 R}$. Since $E = \dfrac{V}{R}$, it is understood that for conductor spheres with the same potential V, the smaller the radius R, the stronger the electric field on the surface. Moreover, when one conductor is isolated in space, the electric field on the surface of pointed part with a small radius of curvature is strong. In addition, it can be seen that since $\sigma = \varepsilon_0 E$, the surface charge density at the sharp part is the largest (Fig. 17-8). Therefore, when the conductor is brought to be at a high potential, discharge is likely to occur at the sharpest point.

Franklin discovered that the charge escapes from the sharp point of the conductor to the outside, and invented the lightning rod. If a lightning rod is installed on the roof and grounded with a thick conducting wire that can withstand a large current, the charge induced by the thundercloud escapes from the lightning rod, with the result of reducing the possibility of lightning stroke. In case of lighting stroke, charges from thunderclouds flow from the lightning rod to the ground through the thick conducting wire. In this way, it is possible to prevent a large current from flowing through the building to cause damage.

Fig. 17-9 A lightning falling to the Eiffel tower

17.2 Capacitor

Learning objective To understand that capacitors, frequently called condensers, are devices that store positive and negative charges and are also devices that store energy. To understand that the electric capacity of a capacitor is proportional to the area of the electrode plates and inversely proportional to the spacing between the electrode plates.

When a conductor is charged, the charges repel each other and it is difficult to store a large amount of electric charge in one conductor. However, if two conductors A and B are brought close to each other and positive and negative charges are given to the conductors A and B, respectively, the electric charges are attracted, making it easy to accumulate a large amount of electric charge. Devices that can store charges in this way are called **capacitors** or **condensers** and conductors A and B are called electrode plates.

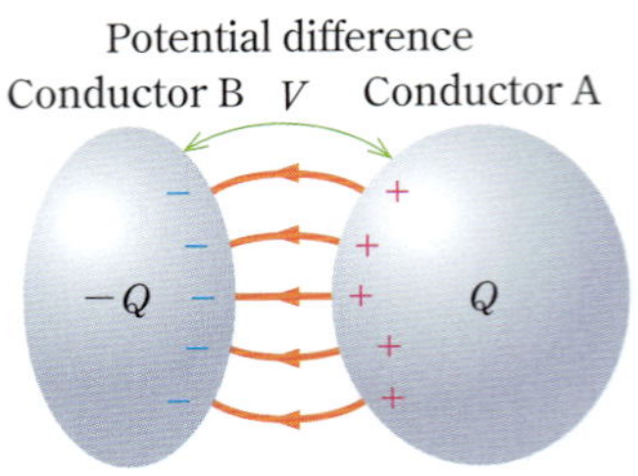

Fig. 17-10 A capacitor consisting two conductors A and B. $Q = CV$

When the positive charge Q is given to electrode plate A and the negative charge $-Q$ is given to electrode plate B, all lines of electric force start at plate A and end at plate B (Fig. 17-10). If the charges Q and $-Q$ at A and B are multiplied by n, the electric fields at all points in space are also multiplied by n, so the potential difference V between plates A and B is also multiplied by n. Therefore, the charges $\pm Q$ of the plates A and B are proportional to the potential difference V and their relation is expressed as

$$Q = CV. \tag{17.5}$$

The proportionality constant C is called the **electric capacity (capacitance)** of a capacitor. The larger the electric capacity C of the capacitor, the larger the amount of charge that can be accumulated with the same potential difference. To increase the electric capacity, in most capacitors, dielectric substances (insulators) such as plastic films and ceramics are inserted between the electrode plates. If the potential difference becomes greater than a certain level, a discharge occurs through the surrounding air or a dielectric substance inserted between the electrode plates and the charge escapes. Thus, it is necessary to increase the electric capacity to accumulate a large amount of charge. The electric capacity C of a capacitor in which the space between the plates is filled with a dielectric substance is the product $C = \varepsilon_r C_0$ of the electric capacity C_0 when the space between the plates is vacuum and the relative dielectric constant ε_r of the dielectric substance ($\varepsilon_r > 1$) (see the next chapter).

Unit of electric capacity
$F = C/V = C^2/J = C^2/(N \cdot m)$

$\varepsilon_0 = 8.854 \times 10^{-12}\ C^2/(N \cdot m^2)$
$= 8.854 \times 10^{-12}\ F/m$

The unit of electric capacity is electric capacity when the electric charge of 1 coulomb (C) is stored by a potential difference of 1 volt (V), which is called 1 farad (symbol F).

$$F = C/V = C/(J/C) = C^2/(N \cdot m). \tag{17.6}$$

Since the unit of 1 F is too large to use, in practice 1 μF (microfarad) $= 10^{-6}$ F and 1 pF (picofarad) $= 10^{-12}$ F are often used.

The electric capacity of a capacitor is determined by the geometrical conditions such as the shape, size and distance of the two conductors, and the type of dielectric substances inserted between the plates.

Example 2 | Parallel-plate capacitor A capacitor in which two metal plates (electrode plates) face each other in parallel is called a parallel-plate capacitor. Find the electric capacity of a parallel-plate capacitor with an area of the plate A and a spacing d (Fig. 17-11). Assume that the size of the plates is much larger than the spacing, so that the electric field between the plates can be considered uniform.

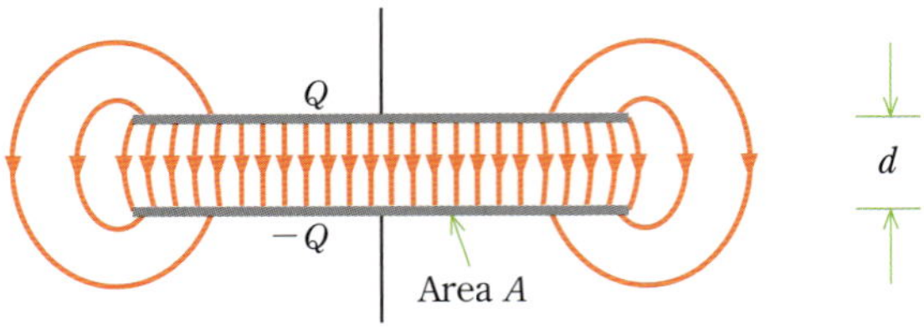

Fig. 17-11 Electric field of the parallel-plate capacitor

Solution The charge $\pm Q$ on the plate is distributed on the inner side of each electrode plate of area A with a uniform surface density $\pm\sigma = \pm Q/A$. Therefore, according to equation (17.1), the electric field strength E between the electrode plates is

$$E = \frac{\sigma}{\varepsilon_0} = \frac{Q}{\varepsilon_0 A}.$$

The potential difference between the two electrode plates with spacing d is

$$V = Ed = \frac{Qd}{\varepsilon_0 A}.$$

Therefore, the electric capacity of the parallel-plate capacitor $C = Q/V$ is

$$C = \frac{\varepsilon_0 A}{d}. \qquad (17.7)$$

The larger the area A of the plate and the smaller the distance d, the larger the electric capacity C.

Case 2 The electric capacity of a capacitor consisting of two square metal plates with a side length of 10 cm and separated by 1 mm is

$$C = \frac{\varepsilon_0 A}{d} = \frac{(8.85\times10^{-12}\,\mathrm{F/m})\times(0.1\,\mathrm{m})^2}{10^{-3}\,\mathrm{m}} = 8.85\times10^{-11}\,\mathrm{F}.$$

Case 3 | Spherical capacitor A capacitor consisting of a metal spheres (radius a) and a concentric metal spherical shell (radius b) as shown in Fig. 17-12 is called a spherical capacitor. Let the charge of the metal sphere be Q and that of the metal spherical shell be $-Q$. Using the result of example 1, since the potential between the electrode plates can be set as $V(r) = \frac{Q}{4\pi\varepsilon_0 R}$ +constant, the potential difference V between the two electrode plates is

$$V = V(a) - V(b) = \frac{Q}{4\pi\varepsilon_0 a} - \frac{Q}{4\pi\varepsilon_0 b} = \frac{Q(b-a)}{4\pi\varepsilon_0 ab}.$$

Accordingly, $C = \frac{Q}{V}$ is

$$C = \frac{4\pi\varepsilon_0 ab}{b-a}. \qquad (17.8)$$

This result corresponds to the case where the spherical shell is grounded (see Problems for exercise 17, B3).

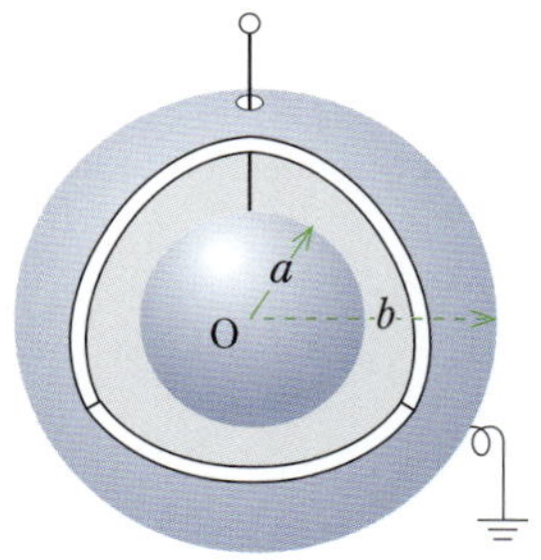

Fig. 17-12 Spherical capacitor

Case 4 | Isolated spherical conductor If a spherical conductor (radius R) isolated in space is considered as the limit of the radius $b\to\infty$ of the spherical shell of the spherical capacitor, it is also a capacitor. Taking the limit $b\to\infty$ in equation (17.8) of the spherical capacitor and using that $\frac{ab}{b-a}\to\infty$ is a in this limit, and setting $a = R$, we obtain the following relation :

$$C = 4\pi\varepsilon_0 R. \qquad (17.9)$$

If we take the Earth (radius $R_E = 6.4\times10^6$ m) as a capacitor, its electric capacity is

$$C = 4\pi\varepsilon_0 R_E = \frac{6.4\times10^6\,\mathrm{m}}{9.0\times10^9\,\mathrm{N\cdot m^2/C^2}} = 7.1\times10^{-4}\,\mathrm{C^2/(N\cdot m)}$$
$$= 7.1\times10^{-4}\,\mathrm{F}.$$

The surface area of the electrode plate of a parallel-plate capacitor having the same electric capacity as the Earth is $8.4\times10^4\,\mathrm{m^2}$ on the assumption that the distance d between the plates is 1 mm. This result shows that it is much more difficult to store positive or negative charges in an isolated conductor than to store positive and negative charges on two closely spaced conductors.

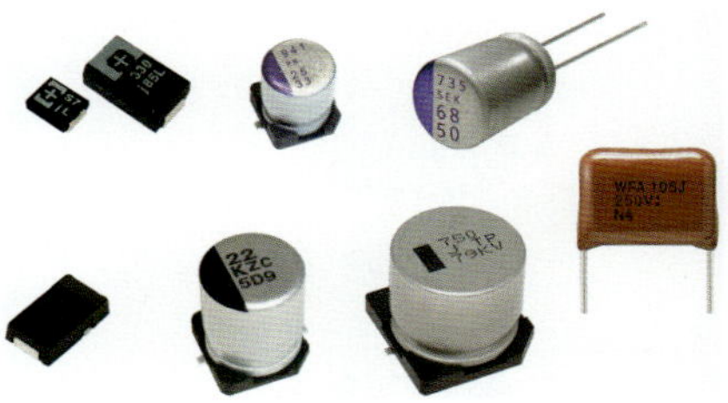

Fig. 17-13 Capacitors

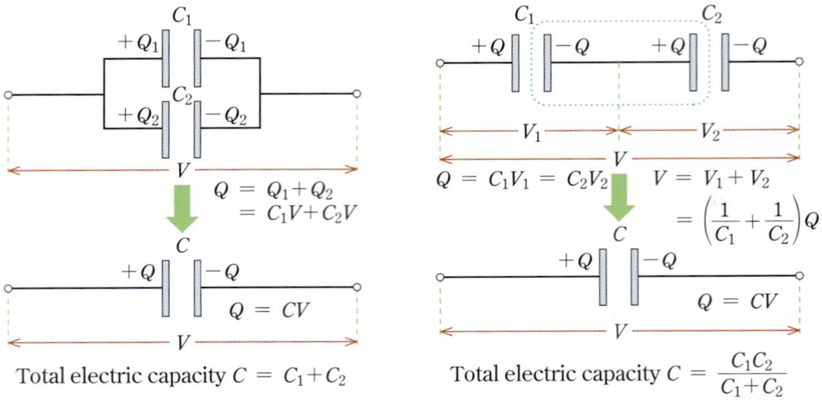

Fig. 17-14 Connection of capacitors

(a) Parallel connection (b) Series connection

Connection of capacitors In the connection of two capacitors, there are parallel and series connections. When two capacitors (electric capacity C_1 and C_2) are connected in parallel as shown in Fig. 17-14 (a), the electric capacity (total electric capacity) C is

$$C = C_1+C_2 \tag{17.10}$$

and when connected in series as in Fig 17-14 (b), the total electric capacity C is

$$\frac{1}{C} = \frac{1}{C_1}+\frac{1}{C_2}, \qquad C = \frac{C_1C_2}{C_1+C_2}. \tag{17.11}$$

Energy stored in a capacitor In order that the charges Q and $-Q$ are stored respectively on plates A and B of a capacitor of electric capacity C, the charge Q must be moved from plate B to plate A against the electric field. The work W necessary for this movement is stored in the capacitor as electric potential energy U.

When the charges stored on plates A and B are q and $-q$, respectively, the potential difference between plates A and B is $v = \frac{q}{C}$. At this time the work ΔW necessary to move the minute charge Δq from the electrode plate B to the electrode plate A to change the amount of charges into $q+\Delta q$ and $-(q+\Delta q)$, respectively, is

$$\Delta W = v\,\Delta q = \frac{q\,\Delta q}{C} \tag{17.12}$$

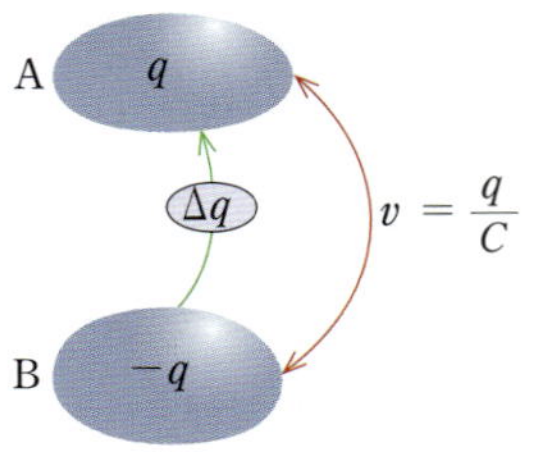

Fig. 17-15 Work necessary to move the minute charge Δq

$\Delta W = v\,\Delta q = \frac{1}{C}\,q\,\Delta q$

(Fig. 17-15). The work W necessary to move the charge to change the state from $q = 0$ to $q = Q$ is given by integrating equation (17.12)

$$W = \int_0^Q \frac{1}{C}\,q\,dq. = \frac{Q^2}{2C} = \frac{1}{2}VQ = \frac{1}{2}CV^2. \tag{17.13}$$

This work is stored in the capacitor as electric potential energy U. In other words a capacitor with an electric capacity C which has a potential difference V between the electrode plates and stores the charge $\pm Q$ on the electrode plates has the following energy :

$$U = \frac{Q^2}{2C} = \frac{1}{2}VQ = \frac{1}{2}CV^2 \tag{17.14}$$

$(Q = CV)$.

Capacitors are devices that store charge, but they are also devices that store energy. When we touch an ungrounded washing machine, we may feel a kind of prick shock because the energy stored between the charged washing machine and the Earth is discharged through the human body (conductor). When a conductor is insulated from the Earth, the electric capacity between the conductor and the Earth is called the stray capacitance.

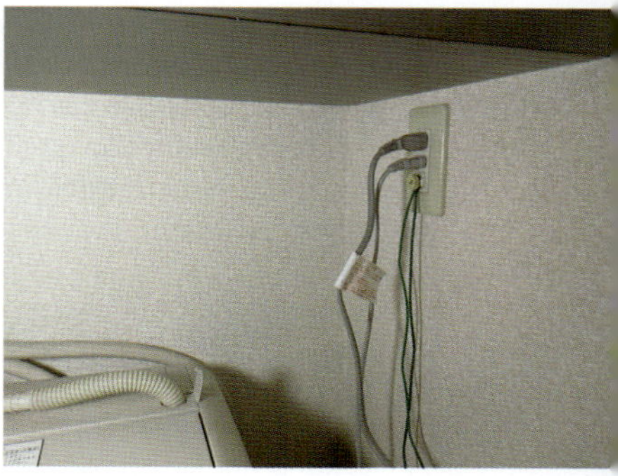

Fig. 17-16 A ground wire of the washing machine.

Energy of the electric field The electric capacity of a parallel-plate capacitor with an electrode area of A and a spacing of d has been derived in Example 2 as to be

$$C = \frac{\varepsilon_0 A}{d}. \qquad (17.7)$$

Since $V = Ed$, the energy U stored in the parallel-plate capacitor is expressed as

$$U = \frac{1}{2}CV^2 = \frac{1}{2}\frac{\varepsilon_0 A}{d}(Ed)^2 = \frac{1}{2}\varepsilon_0 E^2(Ad). \qquad (17.15)$$

The inner volume of the parallel-plate capacitor is Ad, so equation (17.15) indicates that the unit volume of capacitor stores the following energy of electric field :

$$u_E = \frac{1}{2}\varepsilon_0 E^2 \qquad \text{(in vacuum)} \qquad (17.16)$$

Not only the inside of the charged parallel plate capacitor, but also in general, the electric field in vacuum stores the **energy of electric field** of energy density given by equation (17.16).

Problems for exercise 17

A

1. For a conductor with a cavity inside, if a charge is placed inside the cavity, the charge of the same amount but opposite sign appears on the wall of the cavity, and the same amount of charge of the same sign appears on the outer surface of the conductor. Prove this.
2. Around 1775, Franklin, using a cork ball suspended by the string and a charged metal can, conducted experiments as shown in Fig. 1 (a) and (b). Explain why tilts of the strings are different in the two experiments.
3. Fig. 2 is a conceptual diagram of the Van de Graaff generator invented by Van de Graaff in 1930. As the insulator belt rubs the glass cylinder (or contacts the charged metal contact-point), the negative charge

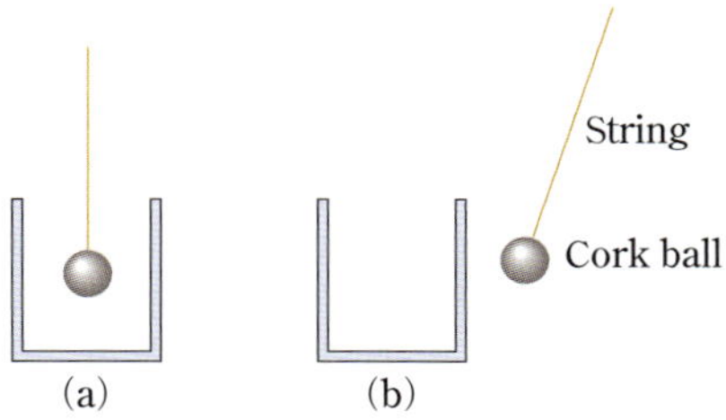

Fig. 1

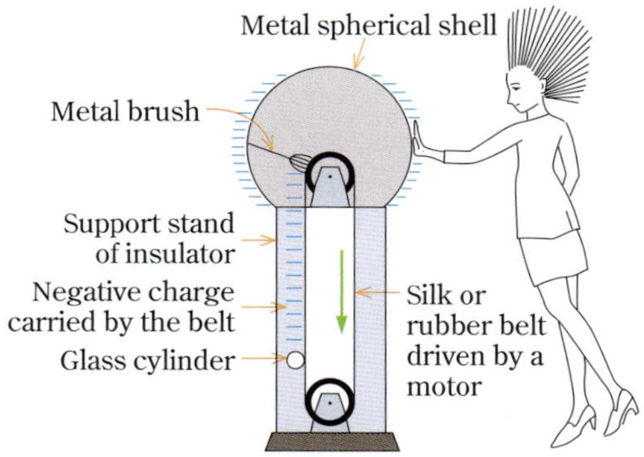

Fig. 2

collected at the bottom of the support is carried to the top of the metal spherical shell. The negative charge on the belt moves to the outer surface of the metal spherical shell. The metal spherical shell has a large amount of negative charge, so the potential difference relative to the ground reaches about 10 million V.

(1) Into what kind of energy was the work of the motor that turns the belt transformed?

(2) Why does the charge of the belt move to the outer surface of the metal spherical shell?

(3) Why is the hair of the person who is touching the metal spherical shell standing upside down?

4. A property is known that even if one equipotential surface of the electric field is replaced with a conductor surface, the electric field outside the conductor does not change (at this time, a charge with a surface density of $\sigma = \varepsilon_0 E$ appears on the conductor surface). Show that using this property and comparing Figs. 3 (a) and 3 (b), when a charged body of charge Q approaches an infinitely large conductor plate, the strength of attraction between the charged body and the conductor plate (surface charge of the opposite sign due to electrostatic induction) is

$$F = \frac{Q^2}{4\pi\varepsilon_0(2d)^2}.$$

d is the distance between the charged body and the conductor plate.

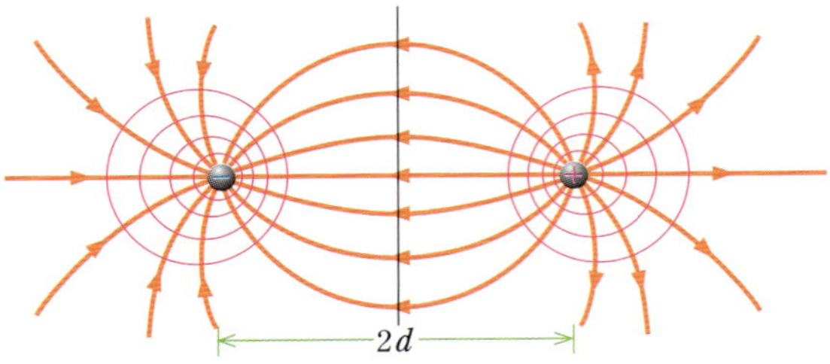

(a)

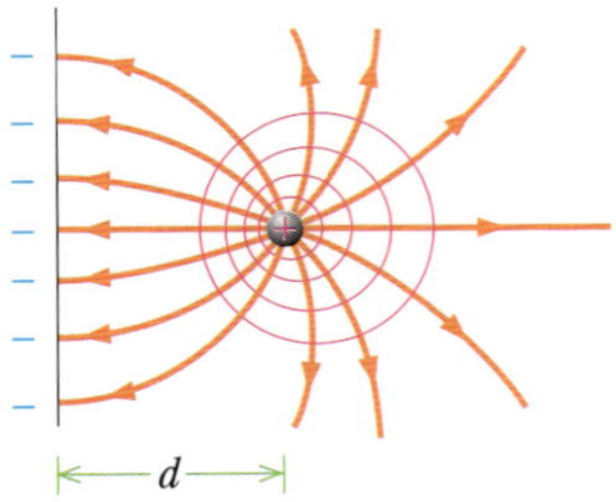

(b)

Fig. 3

5. Find the electric capacity of a capacitor consisting of two square metal plates with one side length 5 cm and separated by 1 mm.

6. Find the electric capacity of a sandwich-form capacitor shown in Fig. 4.

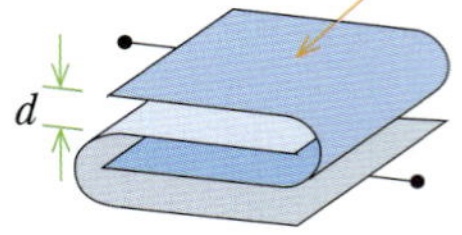

Fig. 4

7. Capacitors A, B, and C with electric capacities 40, 20, and 20 μF, respectively, are connected as shown in Fig. 5. Calculate the total capacity. Find the potential difference between the two plates of C when a potential difference of 10 V is applied to both ends.

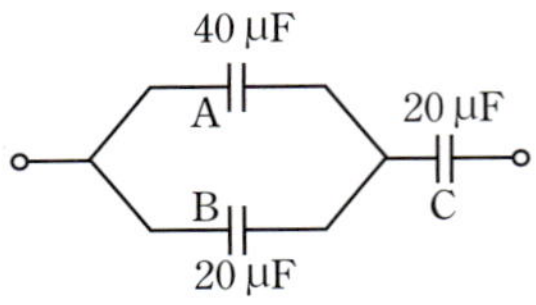

Fig. 5

8. A 20 μF capacitor was charged to 200 V and discharged through a large resistance wire. How much amount of heat was generated within this wire?

B

1. There are many 100 μF capacitors. Connect some of these to prepare a 550 μF capacitor.

2. Find the total electric capacity between the terminals 1 and 2 of the circuit of Fig. 6.

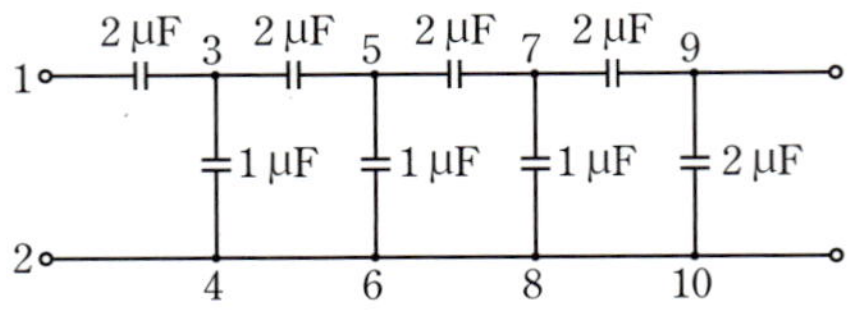

Fig. 6

3. Find the electric capacity C when the conductor sphere of radius a located inside the spherical capacitor is grounded as shown in Fig. 7. Assume that the spherical shell is sufficiently far from the Earth surface.

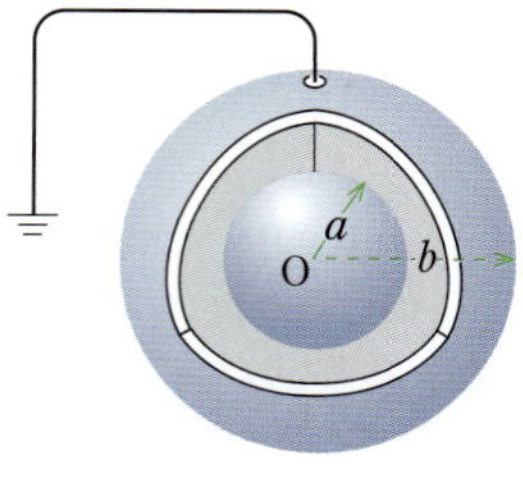

Fig. 7

4. How much amount of energy of the electric field can be stored in an air capacitor with a volume of 1 m^3? Assume that the electric field strength is attainable up to 10^6 V/m.

5. A capacitor A with an electric capacity of 100 μF, which had been charged to 100 V, was connected in parallel with a non-charged capacitor B of the same electric capacity with a wire of large electrical resistance. Answer the following questions.
   (1) Find the voltage of A and B.
   (2) How many joules is the sum of the electric potential energy possessed by A and B?
   (3) Compare the energy obtained above with the electric potential energy that A had at the initial state, and explain what happened to the energy corresponding to this difference.

6. **Electrostatic tension** Let us investigate the force acting on the charge distributed at the surface density σ on the surface of the conductor. The charge σA inside a small circle (area A) centered on point P on the conductor surface creates an electric field of strength $E_1 = \dfrac{\sigma}{2\varepsilon_0}$ on both sides of the surface near the point P. The electric field $\boldsymbol{E}_2$ created by the charge outside the small circle is continuous at point P, so the electric field strength is $E_2 = \dfrac{\sigma}{2\varepsilon_0}$. For this reason, the electric force acting on the charge σA inside the small circle is $F = (\sigma A)E_2 = \dfrac{1}{2}\varepsilon_0 E^2 A$, and the strength of electric force acting on the unit area of the conductor surface is $f = \dfrac{1}{2}\varepsilon_0 E^2$. Explain the contents of the above description. The direction of this electric force is perpendicular to the conductor surface and acts in the direction to push the charge out of the conductor, so it is called the electrostatic tension (see Fig. 8, which is the case $\sigma > 0$).

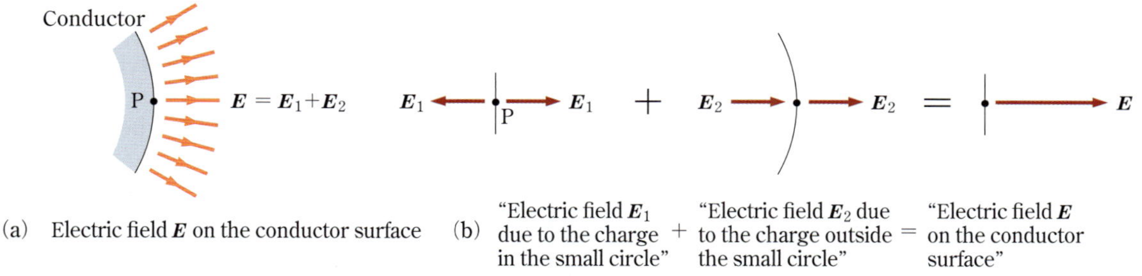

Fig. 8

Dielectric Substance and Electrostatic Field

Among substances there exist groups of substances called **insulators** or **non-conductors** which do not conduct electricity, such as glass and acrylic resin. Unlike a conductor with a free charge that can move around in a substance, in an insulator no electron can move around in the substance. Therefore, even if a charged object is brought close to an insulator, charge transfer over the entire insulator does not occur. However, in each microscopic stractural unit such as molecule, atom, or unit cell of crystal, electrons receive the electric force from the charged object and the electric distribution is biased to one side. As a result a charge with the opposite sign to the charged object appears on the side which is closer to the charged object and a charge of the same sign as that of the charged object appears on the far side. This phenomenon is called the **dielectric polarization**. Because dielectric polarization occurs in insulators, insulators are also called **dielectric substances** when we discuss the electric properties.

In this chapter you will learn the electric field in the presence of dielectric substances.

A scene of assembling minute transistors in a clean room. Semiconductors and insulators (dielectric substances) are used for circuit elements, and various substances and fabrication techniques are being studied.

18.1 Dielectric substances and polarization

Learning objective To understand the following : When a dielectric substance is placed in an electric field, the electric field applied from the outside is weakened inside the dielectric substance on account of the electric field created by the charges produced by the dielectric polarization, so insertion of the dielectric substance between the electrode plates of capacitor causes the increase in electric capacity.

Dielectric substance If a dielectric substance is inserted between the electrode plates of capacitor, electric capacity of the capacitor increases. Let us consider the cause. When a battery of electromotive force V is connected to a parallel-plate capacitor with distance d between the electrode plates, the two plates are charged and the potential difference becomes V [Fig. 18-2 (a)]. Then, the switch is opened to disconnect the battery from the capacitor. If the charge on the electrode plate of area A is Q and $-Q$, the charge density of the plate of area A is $\sigma = \pm\dfrac{Q}{A}$, and the electric field strength E between the plates is $E = \dfrac{\sigma}{\varepsilon_0}$ [see equation (17.1)].

Next, an uncharged dielectric substance such as glass or paraffin is inserted between the two electrode plates [Fig. 18-2 (b)]. The dielectric substance almost completely fill the space between the plates but it does not touch the plates. At this time the charge on the plate remains $\pm Q$, the same as before insertion of the dielectric substance, but the potential difference measured is reduced. When glass fills most of the space between the plates, the potential difference is reduced to less than half. The rate of decrease $\dfrac{1}{\varepsilon_r}$ is a constant determined only by the type of substance and temperature, not by the initial potential difference between the plates or the shape of the capacitor. ε_r is referred to as the **relative dielectric constant**, which is always larger than 1 ($\varepsilon_r > 1$). Values of the relative dielectric constant for several substances are given in Table 18-1.

If a dielectric substance of relative dielectric constant ε_r is inserted between the electrode plates of a capacitor whose electric capacity without any inserted dielectric substance is C_0, the charge $\pm Q$ on the

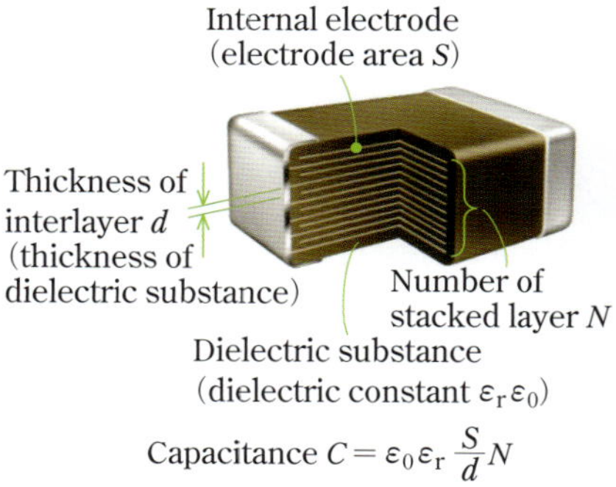

Fig. 18-1 A multi-layer ceramic chip capacitor using dielectric substances. Thinning and miniaturization of capacitors are progressing by making full use of nanotechnology.

Table 18-1 Relative dielectric constants ε_r

Substance	Relative dielectric constant
air (20 °C, 1 atm)	1.000536
water	~80
soda glass	7.5
paraffin	2.2
kraft paper	2.9
Rochelle salt	~4000
barium titanate	~5000

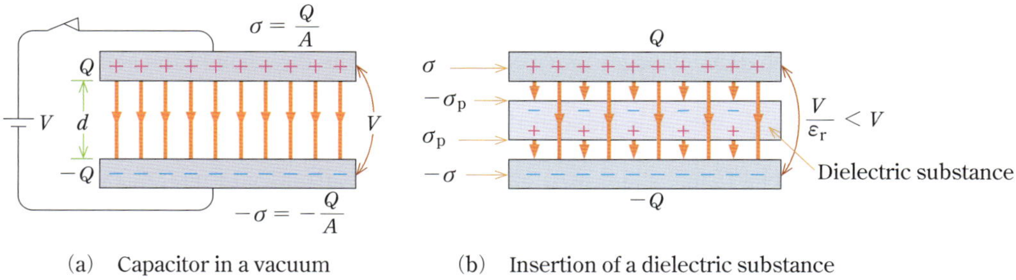

Fig. 18-2

plate does not change but the potential difference between the plate V is multiplied by $\frac{1}{\varepsilon_r}$ and becomes $\frac{V}{\varepsilon_r}$, so the electric capacity C becomes C_0 multiplied by ε_r :

$$C = \varepsilon_r C_0. \tag{18.1}$$

The fact described above that the electric charge on the electrode plates does not change but the potential difference $V = Ed$ between the plates decreases means that the electric field strength E between the plates decreases. This phenomenon can be explained as the dielectric polarization of the dielectric substance which results in the appearance of polarization charges on the surface, as shown below. In Fig. 18-2 (b) are shown the lines of electric force between the electrode plates. Since charges with the surface density σ_p, $-\sigma_p$ appear on the surfaces of the dielectric substance, and the lines of electric force are annihilated and generated on the surface of the dielectric substance, the electric field inside the dielectric substance becomes $\frac{\sigma - \sigma_p}{\sigma}$ times and is weakened.

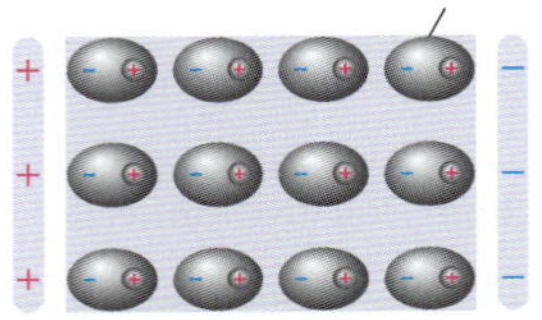

Fig. 18-3 Dielectric polarization (polarization of dielectric substance)

Polarization When dielectric substances are placed in the electric field $\boldsymbol{E}$, positively charged particles in the microscopic structural units such as molecule, atom, or unit cell of crystal move in the direction of the electric field, while negatively charged particles move in the opposite direction of the electric field. However, due to attraction between particles, positively charged particles and negatively charged particles cannot be separated so much. Moving charged particles are positive and negative ions in unit cell or molecule, and electrons in atom (Fig. 18-3). Accordingly, if the separated charges are q and $-q$, and the average central distance is d, each unit becomes an electric dipole and the electric dipole moment is given by

$$p = qd. \tag{18.2}$$

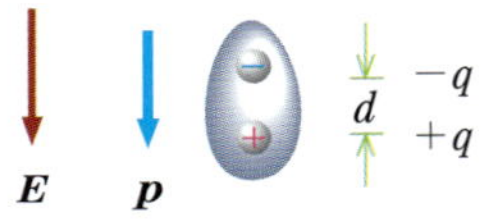

Fig. 18-4

As shown in Fig. 18-4, the electric dipole moment $\boldsymbol{p}$ is a vector of magnitude $p = qd$ and pointed from the negative charge to the positive charge.

Because the size of the microscopic structural unit is so small, when we look at an object macroscopically, the discontinuities arising from the microscopic structure of the object seem to be uniformly smeared out. If we assume the number of microscopic structural units per unit volume is N, positive and negative charges seem to be uniformly distributed on the object with density $\rho = qN$ and $-qN$, respectively. When the electric field $\boldsymbol{E}$ is applied to this object, the positive and negative charges move a distance d in the direction of the electric field $\boldsymbol{E}$. As shown in Fig. 18-5, on the surface of the dielectric substance with area A is induced the charge

$$\rho Ad = \pm qNAd = \pm pNA. \tag{18.3}$$

The induced charge is called **polarized charge** and the phenomenon is referred to as dielectric polarization.

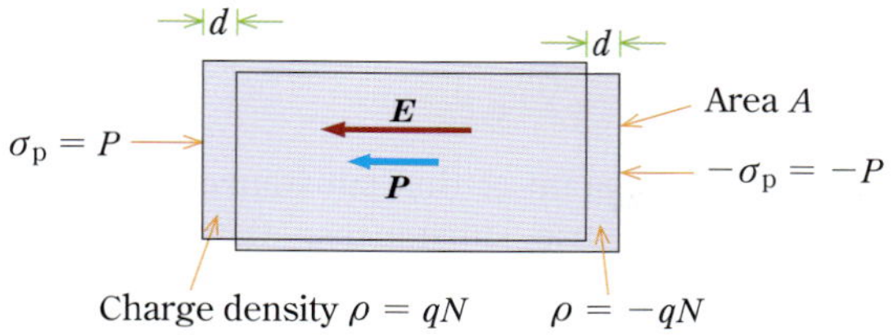

Fig. 18-5 "magnitude of polarization P" = "density of polarized charge σ_p"

Question 1 Why do charged objects attract small, uncharged pieces of paper (see Fig. 18-6)?

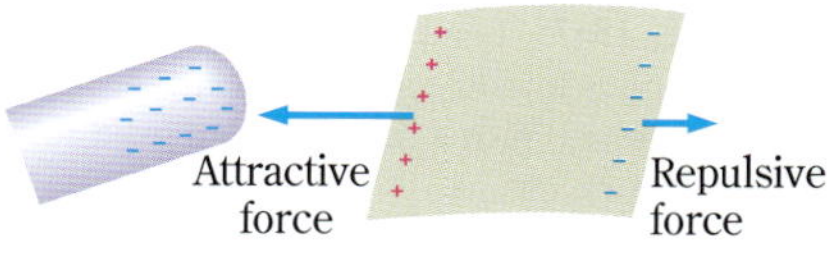

Fig. 18-6

If the surface density of the polarized charge is $\pm\sigma_p$, since the polarized charge $\pm pNA$ on the surface of area A is expressed as $\pm\sigma_p A$, it follows that

$$\sigma_p = pN \equiv P. \tag{18.4}$$

Thus, the surface density of the polarized charge σ_p has the magnitude equal to the sum of minute electric dipole moments $\boldsymbol{p}_j$ $(j = 1, 2, \cdots, N)$ in the unit volume of the dielectric substance :

$$\boldsymbol{P} \equiv \boldsymbol{p}N = \sum_{j\,(\text{unitvolume})} \boldsymbol{p}_j. \tag{18.5}$$

The macroscopic vector field $\boldsymbol{P}$ in the dielectric substance defined by equation (18.5) is called **polarization**. The unit of electric dipole moment is "unit of charge C" × "unit of length m" or C·m, so the unit of polarization $\boldsymbol{P}$ which is the electric dipole moment per unit volume (1 m^3) is C/m^2.

Unit of polarization C/m^2

When the space between the electrode plates of a parallel-plate capacitor is filled with a dielectric substance of relative dielectric constant ε_r, the electric field strength inside it is $E = \dfrac{\sigma - \sigma_p}{\varepsilon_0}$, and is $\dfrac{1}{\varepsilon_r}$ times the electric field strength $E = \dfrac{\sigma}{\varepsilon_0}$ created only by the free charge on the electrode plates (see Fig. 18-2). Therefore,

$$E = \frac{\sigma - \sigma_p}{\varepsilon_0} = \frac{\sigma}{\varepsilon_r \varepsilon_0}. \tag{18.6}$$

Changing this equation as $\sigma - \sigma_p = \varepsilon_0 E$ and $\sigma = \varepsilon_r \varepsilon_0 E$, and combining these, we derive

$$P = \sigma_p = \sigma - \varepsilon_0 E = (\varepsilon_r - 1)\varepsilon_0 E. \tag{18.7}$$

Thus, the magnitude P of polarization is proportional to of the electric field strength E.

In isotropic dielectric substances, $\boldsymbol{P}$ is pointed in the same direction as $\boldsymbol{E}$, so

$$\boldsymbol{P} = (\varepsilon_r - 1)\varepsilon_0 \boldsymbol{E} = \chi_e \varepsilon_0 \boldsymbol{E}. \tag{18.8}$$

Here

$$\chi_e = \varepsilon_r - 1 \tag{18.9}$$

is referred to as the **electric susceptibility**. The product of relative

Fig. 18-7 Lead zirconate titanate which is ferroelectric. This material is processed into a thin film and used for memory etc.

dielectric constant and electric constant, that is

$$\varepsilon = \varepsilon_r \varepsilon_0 \tag{18.10}$$

is called the dielectric constant of that substance.

Note that among dielectric substances, there are groups of substances called **ferroelectrics** that are polarized even in the absence of applied electric field. In such materials, the proportional relationship (18.8) between electric field and polarization does not hold.

Electric flux density The electric field is created by charges, namely free and polarized charges. Therefore, given that in the closed surface S the sum of free charges is Q_0 and the sum of polarization charges is Q_p, Gauss' law of electric field (16.19) becomes

"flux of electric line of force coming outside from the inside of the closed surface S"$\times \varepsilon_0 = Q_0 + Q_p$

$$\varepsilon_0 \iint_S E_n \, dA = Q_0 + Q_p. \tag{18.11}$$

At this point, in addition to the electric field $\boldsymbol{E}$ created by both free and polarized charges, it is convenient to introduce a field related only to free charge. Since the electric field $\boldsymbol{E}$ of the dielectric substance in the capacitor is $\frac{1}{\varepsilon_r}$ times the electric field $\boldsymbol{E}_0$ without the dielectric substance, $\boldsymbol{E}_0 = \varepsilon_r \boldsymbol{E}$ is a candidate. Here, the ε_0 times $\boldsymbol{E}_0$, that is, $\varepsilon_r \varepsilon_0 \boldsymbol{E}$ is adopted and the new field $\boldsymbol{D}$ defined as

$$\boldsymbol{D} = \varepsilon_0 \boldsymbol{E} + \boldsymbol{P} \tag{18.12}$$

is called the **electric flux density** [see equation (18.8)]. The unit of electric flux density is, C/m^2, the same as the unit of polarization.

Just as the state of the electric field $\boldsymbol{E}$ can be expressed in terms of lines of electric force, so the state of the electric flux density $\boldsymbol{D}$ can be expressed in terms of **lines of electric flux**. Because the starting point of the lines of electric force is a positive charge and the end point is a negative charge, some lines of electric force may start from a positive polarized charge and end at a negative polarized charge [in Fig. 18-8 (b), they are cancelled by the lines of electric force of free charge], but they are in the opposite direction of the force line of polarization $\boldsymbol{P}$ [Fig. 18-8 (c)] which is directed from a negative polarized charge to a positive polarized charges. Therefore, lines of electric flux start at the positive free charge and end at the negative free charge [Fig. 18-8 (a)]. Since Q_0 (because $\boldsymbol{E}$ is involved in $\boldsymbol{D}$ in the form of $\varepsilon_0 \boldsymbol{E}$ and not $\frac{Q_0}{\varepsilon_0}$) lines of electric flux comes out of the free charge Q_0,

"Electric flux ϕ_E coming out of the closed surface S"
= "Total free charge Q_0, inside the closed surface S"

$$\iint_S D_n \, dA = Q_0. \tag{18.13}$$

This is called **Gauss' law of electric flux density**. D_n is the component of the electric flux density $\boldsymbol{D}$ normal to the surface S. Instead of using the term "flux of lines of electric flux", we use the term electric flux.

In most substances, polarization $\boldsymbol{P}$ is proportional to $\boldsymbol{E}$ and since the relation $\boldsymbol{P} = (\varepsilon_r - 1)\varepsilon_0 \boldsymbol{E}$ holds, in these substances, equation (18.12)

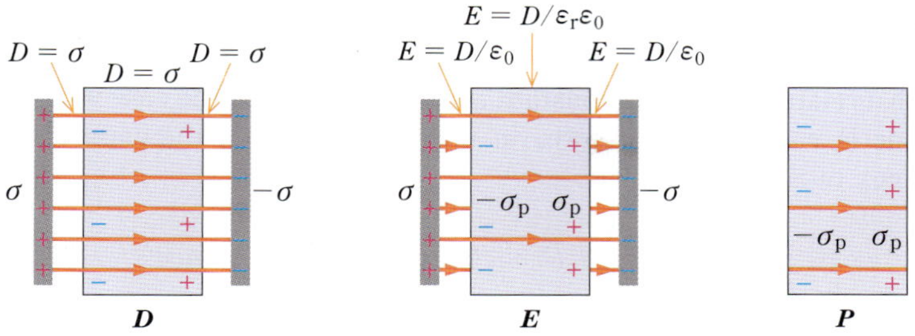

(a) (b) (c)

Fig. 18-8 Electric flux density $\boldsymbol{D}=\varepsilon_0\boldsymbol{E}+\boldsymbol{P}$

(a) In a dielectric substance $D=\varepsilon_r\varepsilon_0E=\varepsilon_r\varepsilon_0\dfrac{\sigma}{\varepsilon_r\varepsilon_0}=\sigma$. In a void space $D=\varepsilon_0E=\varepsilon_0\dfrac{\sigma}{\varepsilon_0}=\sigma$. Thus anywhere in the capacitor, $D=\sigma$. Lines of electric flux are created at positive free charges and annihilated at negative free charges.

(b) In a dielectric substance $E=\dfrac{\sigma-\sigma_p}{\varepsilon_0}$, and in a void space $E=\dfrac{\sigma}{\varepsilon_0}$. Lines of electric force are created at positive charges and annihilated at negative charges.

(c) In a dielectric substance $P=\sigma_p$, and in a void space $P=0$. Lines showing $\boldsymbol{P}$ are created at negative polarized charges and annihilated at positive polarized charges.

can be also expressed as

$$\boldsymbol{D}=\varepsilon_r\varepsilon_0\boldsymbol{E}=\varepsilon\boldsymbol{E}. \tag{18.14}$$

$\varepsilon=\varepsilon_r\varepsilon_0$ is the dielectric constant of the dielectric substance.

Example 1 Find the electric force acting between two point charges q_1 and q_2 in a liquid (relative dielectric constant ε_r).

Solution Let the point charge q_1 be at the origin. The electric flux emerging from this point charge is q_1 according to equation (18.13), so the strength D of the electric flux density at the point of distance r from the origin is "electric flux" divided by "surface area of sphere", which is $\dfrac{q_1}{4\pi r^2}$. Thus

$$D=\frac{q_1}{4\pi r^2}=\varepsilon_r\varepsilon_0E. \tag{18.15}$$

The strength of electric force acting on the point charge q_2 at distance r from the origin is $F=q_2E=\dfrac{q_2D}{\varepsilon_r\varepsilon_0}$. Therefore, the strength of electric force acting between two point charges q_1 and q_2 in a liquid or gas with a relative dielectric constant ε_r is $\dfrac{1}{\varepsilon_r}$ times the value in vacuum, namely

$$F=\frac{q_1q_2}{4\pi\varepsilon_r\varepsilon_0r^2}. \tag{18.16}$$

Since $\varepsilon_r>1$, the electric field inside the dielectric substance is weaker than in vacuum. The cause of the weakening is the polarized charges of the opposite sign induced around the charge (Fig. 18-9). The relative dielectric constant of water is very large, about 80. For this reason, in water, the bonding strength of ionic bonds gets extremely weak, and separation into positive ions and negative ions is easy to occur.

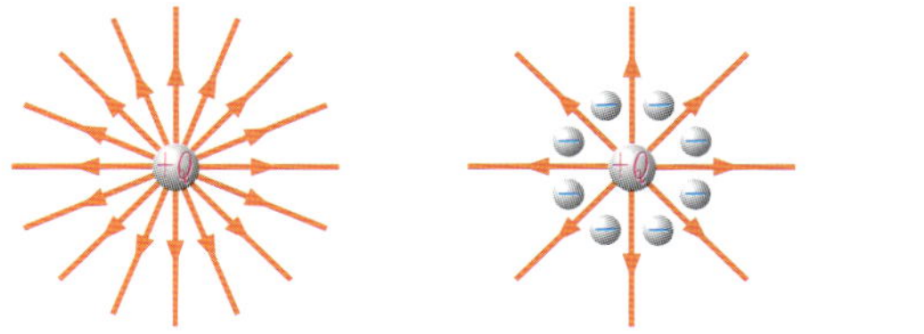

(a) Electric field in vacuum (b) Electric field in a dielectric substance

Fig. 18-9

Energy of the electric field in dielectric substances The electric capacity of a capacitor whose inside is filled with a dielectric substance with a relative dielectric constant ε_r is ε_r times that of the case where the inside is vacuum. Therefore, in the unit volume of the capacitor the energy of the electric field

$$u_E = \frac{1}{2}\varepsilon_r\varepsilon_0 E^2 = \frac{1}{2}ED \quad \text{(in dielectric substance)} \tag{18.17}$$

is stored, which is equal to ε_r times equation (17.16).

Problems for exercise 18

A

1. A voltage of 100 V was applied to a capacitor with an electrical capacity of 1 μF. How much amount of charge can be stored on the electrode plate?
2. Assume that C_1 and C_2 are capacitors of the same shape and same size but a dielectric plate is inserted in C_1. C_1 is charged and its potential difference V_1 is measured. Then, after the battery is removed, C_1 and C_2 are connected in parallel to measure the common potential difference V_2. Find the relative dielectric constant ε_r of the dielectric substance.

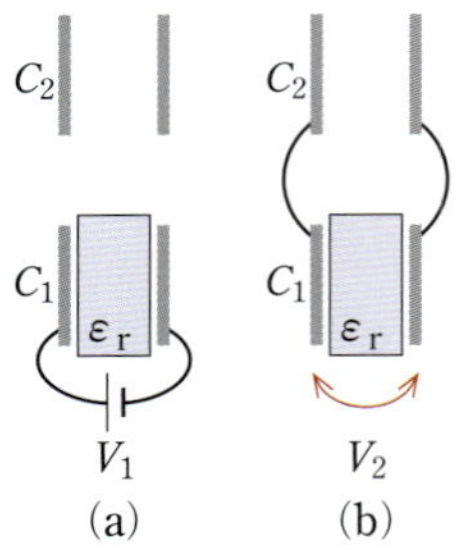

Fig. 1

3. Find the electric capacity of a capacitor made of two sheets of metal foil sandwiching a paper with a surface area of 1 m^2 and a thickness of 0.1 mm. Let the relative dielectric constant of the paper be 3.5.
4. Ions inside and outside the cell are separated by a flat cell membrane with a thickness of 10^{-8} m (relative dielectric constant 8) (Fig. 2).
   (1) Calculate the electric capacity per 1 cm^2 of cell membrane.
   (2) If the potential difference between both sides of the cell membrane is 0.1 V, how much amount of energy can be stored in a 1 cm^2 cell membrane?
   (3) Find the electric field strength E in the cell membrane and the charge Q per 1 cm^2 in layer of both sides of the membrane.

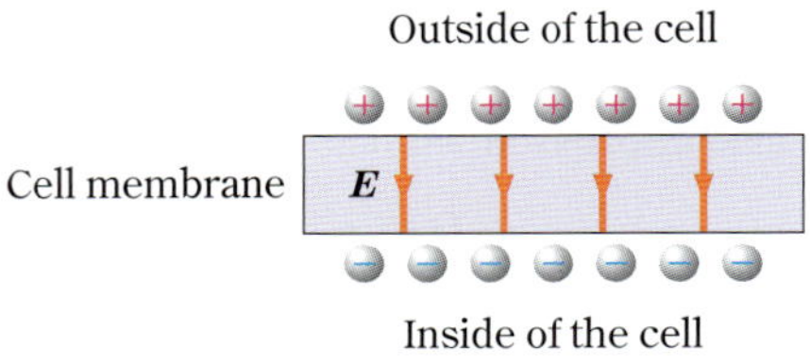

Fig. 2

B

1. As shown in Fig. 3, of the distance between the plates of the parallel-plate capacitor, d_1 is filled with a dielectric substance of relative dielectric constant ε_1, and the remaining d_2 is filled with a dielectric substance of relative dielectric ε_2. Find the electric capacity C of this capacitor. Let the area of the plate be A.

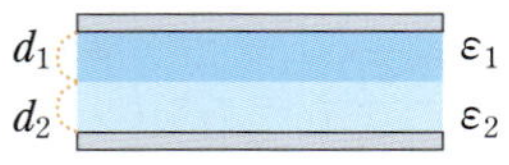

Fig. 3

2. How does equation (17.1) of the electric field strength on the conductor surface change when the conductor is in contact with the dielectric substance?

Electric Current

Because electric power is a power source and an energy source, a power failure paralyzes social activities and causes various inconvenience in our daily life. Electric energy converted from other forms of energy at the power plant travels along the wires to houses and factories where it puts on lights, sounds speakers, operates motors and generates heat with heaters and is transformed into other forms of energy.

In these processes electric current plays an important role and electrons carry the current flowing through the wire. In order for the current to flow steadily, a power supply (electromotive force) is required.

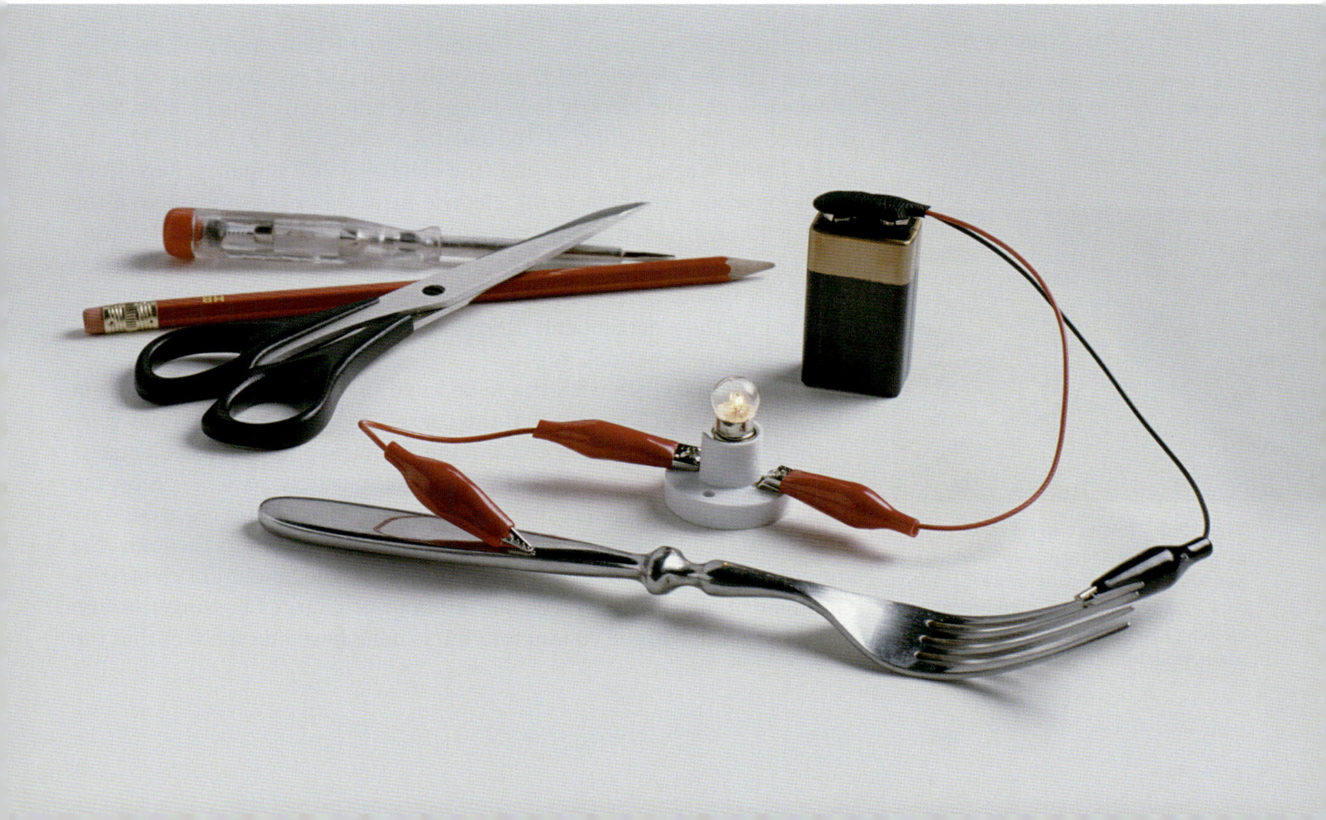

19.1 Electric current and electromotive force

Learning objective To understand that the current is a flow of electric charge accompanying the movement of charged particles, and an electromotive force (power source) is required for the current to flow without intermission.

Current The current is a flow of electric charge accompanying the movement of charged particles. In metal wires, negatively charged free electrons (conduction electrons) move. In electrolyte solutions, positive ions and negative ions move, and in the discharge tube, current flows by electrons moving in vacuum.

To represent the current flowing through a cross section S of a wire, the first thing to do is to determine the positive direction of the wire at that location. If the amount of electricity passing through surface S during time Δt is ΔQ, the net current I passing through surface S in the positive direction is defined as

$$I = \frac{\Delta Q}{\Delta t}. \qquad (19.1)^*$$

* If the positive direction of the conductor is chosen reversely, the positive and negative signs of the current are reversed.

The unit of current is ampere (symbol A), which is the current intensity C/s when a charge of 1 coulomb (C) moves for 1 second (s):

$$\mathrm{A} = \mathrm{C/s}. \qquad (19.2)$$

Unit of current A = C/s

Equation (19.1) is transformed to show that the amount of electricity ΔQ passing through the cross section of the wire when the current I flows for time Δt is expressed as

$$\Delta Q = I\,\Delta t. \qquad (19.3)$$

Accordingly, C = A·s.

The direction of the current is the direction of movement of positive charge, which is opposite to the direction of movement of negative charge. In the electric field, positively charged positive ions move in the direction of the electric field under the influence of the electric force directed in the direction of the electric field [Fig. 19-1 (a)]. On the other hand, negatively charged free electrons and negative ions move in the opposite direction of the electric field under the influence of the electric force directed in the opposite direction of the electric field [Fig. 19-1 (b)]. Thus, in both cases the current and the electric field are in the same direction. Since the electric potential drops in moving in the direction of the electric field, the current flows from the high potential side to the low potential side.

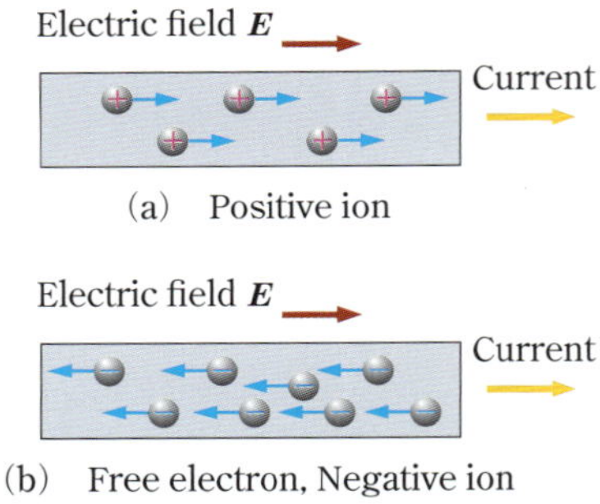

Fig. 19-1 The directions of electric field and current are the same.

We cannot observe with the naked eye that the current is a flow of charged particles. The fact that the current is flowing can be known by the heat generation phenomena and the chemical phenomena (electrolysis) caused by the current, but it can be accurately recognised by the magnetic action of the magnetic field that the current creates around it. Therefore, the measurement of the current flowing through the conductor is not based on the measurement of the amount of electricity (number of charged particles and number of charge) passing through the cross section of the conductor, but is done based on the fact that the

Fig. 19-2 Inspection of the underground power transmission lines (Takamatsu City, Kanagawa Prefecture). They have a high durability against storms and snow, etc., but the installation cost is high.

strength of magnetic action of the current is proportional to the current strength. For this reason, the exact definition of the unit of current, ampere, is also made by the magnetic action of current (see Section **20.5**). Regardless of the positive or negative nature of the charge of particles carrying the current, the generated magnetic field is identical if the current, including the direction, is the same. The positive or negative nature of the charge of the charged particle that carries the current can be known through the Hall effect (see p. 280).

Current flowing through the conducting wire We prepare a uniform conducting wire of cross-sectional area A with a metal containing n free electrons per unit volume. When free electrons of charge $-e$ are moving at an average velocity v in this wire, $nA(v\,\Delta t)$ free electrons pass through the cross section of the wire during time Δt, so the intensity of this current I is

$$I = envA \tag{19.4}$$

(Fig. 19-3).

When an electric current of 1 A flows through a copper wire with about 10^{29} free electrons in 1 m^3 of cross-sectional area A of 2 mm^2, the average velocity v of free electrons (charge $-e = -1.6\times10^{-19}$ C) is 1/300 cm per second.

$$v = \frac{I}{neA} = \frac{1\ \text{C/s}}{(10^{29}\ \text{m}^{-3})\times(1.6\times10^{-19}\ \text{C})\times(2\times10^{-6}\ \text{m}^2)}$$
$$\approx 3\times10^{-5}\ \text{m/s} \qquad \left(\frac{1}{300}\ \text{cm/s}\right).$$

Hence, the flow of free electrons in the wire is very slow.

Free electrons in metal are moving straight in various directions with a high speed of about 10^6 m/s. When an electric field is created in the wire, free electrons are accelerated by the electric force from the electric field, but they are immediately scattered by colliding with thermally oscillating positive ions and/or impurities. Free electrons repeat acceleration, collision, and scattering, and move, on average, at a constant speed proportional to the electric field strength (Fig. 19-4). As a result, a current of constant magnitude flows in the wire. The effect of collision and scattering plays the role of resistance that balances the electric force. This averaged moving velocity $\boldsymbol{v}$ is called the **drift velocity**. The term "drift" means being adrift.

Electromotive force A current that does not change with time is called a **stationary electric current**. To allow a stationary current to flow through a wire, it is necessary to connect a direct current (DC) power supply, such as a battery, to both ends of the wire and keep the potential difference at both ends constant to create an electric field of constant magnitude inside the wire. The function of a power source allowing the current to flow through the circuit is called the **electromotive force**. The unit of electromotive force is the unit of potential difference, volt (symbol V). Power sources include batteries (chemical cells, solar cells, fuel cells), generators, thermocouples, etc. The symbol of battery is shown in Fig. 19-5. It is quite often that the magnitude of electromotive force of the power supply, or the potential

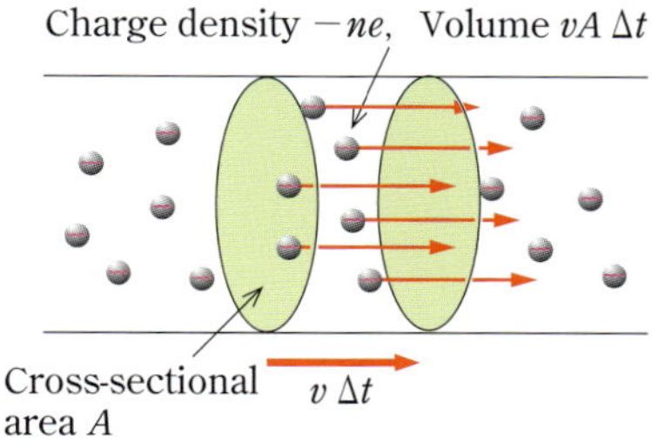

Fig. 19-3 $I = \dfrac{enA(v\,\Delta t)}{\Delta t} = envA$

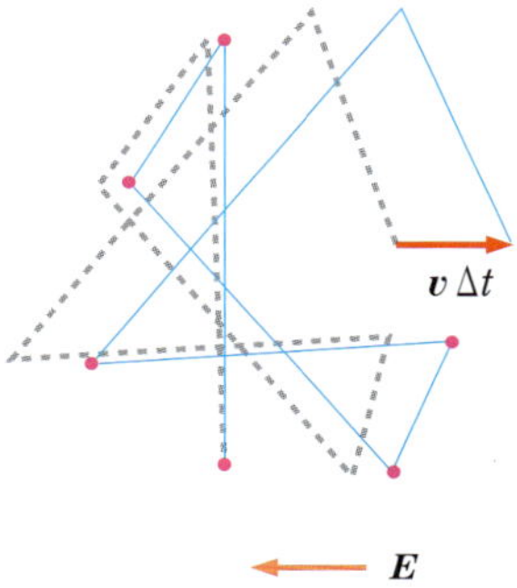

Fig. 19-4 Drift velocity. In the wire, free electrons are moving between positive ions, repeating processes such as acceleration, collision with thermally oscillating positive ions, scattering, acceleration, ⋯, and they move, on average at a constant drift velocity $\boldsymbol{v}$, in the opposite direction of the electric field $\boldsymbol{E}$. The difference in the free electron displacement during time Δt in the absence (dotted line) and in the presence (solid line) of an electric field is $\boldsymbol{v}\,\Delta t$.

Fig. 19-5 Symbol of battery. The long bar represents the cathode (positive electrode) and the short bar represents the anode (negative electrode).

difference between two points of the circuit, is called the **voltage**.

19.2 Ohm's law

Learning objective To understand Ohm's law, which is the proportional relationship between voltage and current, and electric resistivity.

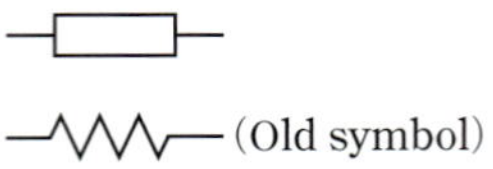

Fig. 19-6 Symbol of resistor

Resistance The action that hinders the flow of current is called the electric resistance or simply resistance. Although any wire has some resistance, elements of electric circuit that play the role of resistance are also used. These elements are called **resistors** which are quite often just called "resistances" in place of the correct term "resistors". Resistors are made of ceramics, oxide, carbon, and alloy coils. The symbol shown in Fig. 19-6 is used as a symbol of resistor (resistance).

Ohm's law As shown in Fig. 19-7, if a direct current (DC) power supply is connected to both ends of the resistor and the voltage V of the power supply is changed while the temperature of the resistor is kept constant, the current I flowing through the resistor will be proportional to the voltage V. This proportional relationship is called **Ohm's law** because Ohm discovered it in 1827. This law is expressed as

Unit of resistance $\Omega = \text{V/A}$

$$V = RI \tag{19.5}$$

and the proportionality constant R is referred to as **electric resistance** or **resistance**. Since the unit of resistance is "unit of voltage" divided by "unit of current", so it is V/A, which is called ohm (symbol Ω).

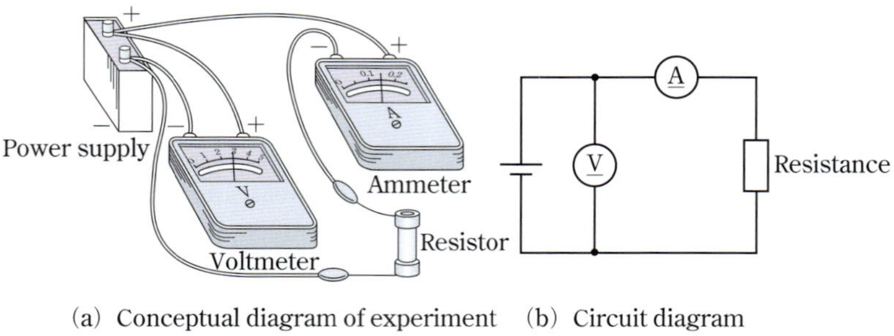

Fig. 19-7 Ohm's law

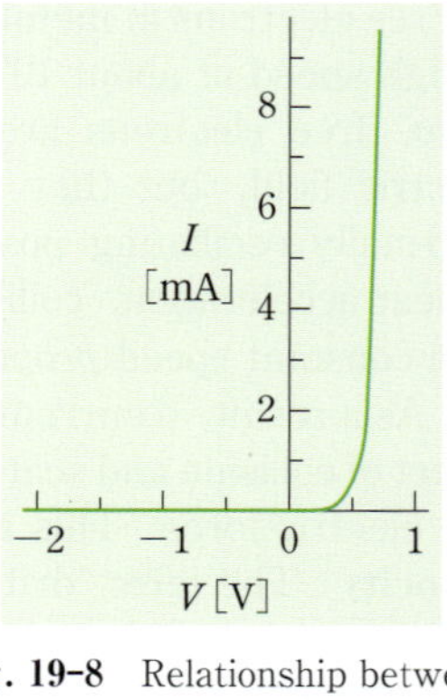

Fig. 19-8 Relationship between voltage and current for diodes

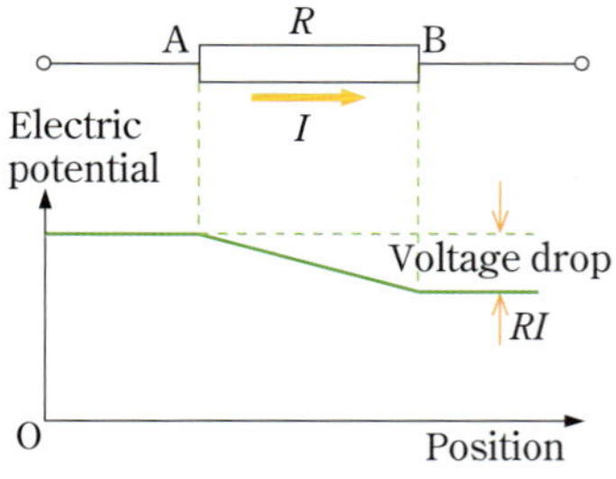

Fig. 19-9 Voltage drop, $V = RI$

Ohms' law holds satisfactorily for metals and alloys, but not for electrolyte solutions, diodes, discharge tubes, etc. For example, in a diode, the relationship between current and voltage is not proportional, and the magnitude of the current that flows at the same voltage differs with the direction in which the voltage is applied (Fig. 19-8).

When the current I flows inside an object with electric resistance, the potential drops in the direction of the current. This is called the **voltage drop**. The voltage drop in the section where the electric resistance is R is, of course, RI (Fig. 19-9).

Electric resistivity Metals conduct electricity well but their resistance is not zero. Since the electric resistance R of a uniform conducting wire at a constant temperature is proportional to its length L and inversely proportional to the cross-sectional area A, the electric resistance R of the conducting wire can be written as

$$R = \rho \frac{L}{A}. \tag{19.6}$$

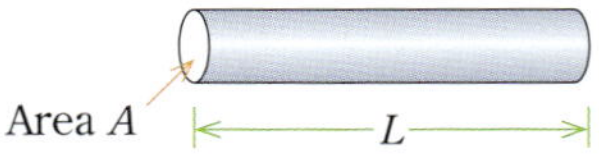

Fig. 19-10 $R = \rho \frac{L}{A}$

The proportional constant ρ is a constant determined only by the wire material and wire temperature (Fig. 19-10). ρ is called the **electric resistivity** of the material at that temperature. The unit of electric resistivity is Ω·m.

Unit of electric resistivity
Ω·m

Electric resistivity changes with temperature. If the electric resistivity at 0 °C is ρ_0, the electric resistivity ρ at t °C around normal temperature is approximately expressed as

$$\rho = \rho_0(1+\alpha t). \tag{19.7}$$

Here, α is called the **temperature coefficient of electric resistivity**. The reciprocal $\sigma = \frac{1}{\rho}$ of the electric resistivity ρ is called the **electric conductivity** (unit $\Omega^{-1}\cdot\text{m}^{-1}$).

Unit of electric conductivity
$\Omega^{-1}\cdot\text{m}^{-1}$

Substituting equation (19.6) of the electric resistivity into equation (19.5) of Ohm's law, we have

$$\frac{V}{L} = \rho \frac{I}{A}. \tag{19.8}$$

$j = \frac{I}{A}$ is the current per unit area of the cross section of the wire, which is called the **electric current density**. When the potential difference between both ends of the wire of length L is V, $\frac{V}{L}$ is the electric field strength E in the conductor, so from equation (19.8), a vector equation representing the relationship between electric field $\boldsymbol{E}$ and current density $\boldsymbol{j}$

$$\boldsymbol{E} = \rho \boldsymbol{j}, \qquad \boldsymbol{j} = \frac{1}{\rho}\boldsymbol{E} = \sigma \boldsymbol{E} \tag{19.9}$$

is derived.

Question 1 From the fact that the electric resistance of the uniform copper wire is inversely proportional to the cross-sectional area of the wire, infer that the current flows uniformly throughout the wire, not just on the surface of the wire. Hint ; Use the relation that "current" = "constant"× "cross section"×"electric field strength".

Fig. 19-11 Copper wires

In metals and alloys, free electrons (charge $-e$, mass m) are scattered by colliding with thermally oscillating positive ions and/or impurities, but between the first collision and the next collision, they perform the uniformly accelerated motion with the acceleration of $-\frac{e}{m}\boldsymbol{E}$ under the action of electric force $-e\boldsymbol{E}$ in the electric field (Fig. 19-4). Immediately after the free electrons collide with thermally oscillating positive ions and/or impurities and are scattered randomly, they have no preference to a specific direction of velocity, and the average

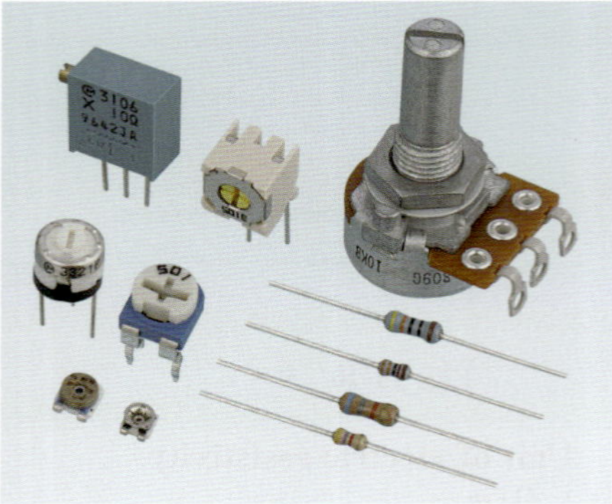

Fig. 19-12 Resistors

velocity as a vector is $\mathbf{0}$. Thus, if the average time from one collision to the next collision (collision time interval) is assumed to be τ, the average velocity $-\frac{e\tau}{m}\boldsymbol{E}$ for this time interval is the drift velocity $\boldsymbol{v}$ (see Problems for exercise 19, B2).

$$\boldsymbol{v} = -\frac{e\tau}{m}E. \tag{19.10}$$

Substituting equation (19.10) of the drift velocity into the current density $\boldsymbol{j} = -en\boldsymbol{v}$ derived from equation (19.4), we obtain

$$\boldsymbol{j} = -en\boldsymbol{v} = -en\left(-\frac{e\tau}{m}\boldsymbol{E}\right) = \frac{ne^2\tau}{m}\boldsymbol{E}. \tag{19.11}$$

Comparing equations (19.9) and (19.11), we can see that the electric resistivity of metals and alloys is expressed as

$$\rho = \frac{m}{ne^2\tau}. \tag{19.12}$$

Since the drift velocity, which is the average velocity of free electron, is much slower than the individual free electron speed of about 10^6 m/s, the collision time interval τ is determined by the speed of about 10^6 m/s of free electrons and does not change with the electric field strength. Consequently, according to equation (19.12), the electric resistivity ρ of metals and alloys is independent of the electric field strength E. This is the reason why Ohm's law holds for metals and alloys.

The electric resistivity of pure metals at room temperature is

$$\rho \approx 10^{-8}\ \Omega\cdot\text{m}.$$

The reason for the low electric resistivity of metals and alloys is that the mass m of free electron is small and the density n of free electron is high. Table 19-1 shows the electric resistivities and their temperature coefficients of some metals. In the case of metals and alloys, the thermal oscillations of positive ions become intense with the rise in temperature, and the collisions between free electrons and positive ions increase, so the collision time interval τ in equation (19.12) decreases. Hence, the electric resistivity of metals and alloys increases as the temperature rises.

Table 19-1 Electric resistivities (20 °C) and their temperature coefficients of metals

Metal	Electric resistivity ρ [Ω·m]	Temperature coefficient*) α
Silver	1.59×10^{-8}	4.1×10^{-3}
Copper	1.68×10^{-8}	4.3×10^{-3}
Gold	2.21×10^{-8}	4.0×10^{-3}
Aluminum	2.71×10^{-8}	4.2×10^{-3}
Tungsten	5.3×10^{-8}	5.3×10^{-3}
Platinum	10.6×10^{-8}	3.9×10^{-3}

*) Expressing the electric resistivity at 0 °C and 100 °C as ρ_0 and ρ_{100}, respectively, we define the temperature coefficient of the electric resistivity as $\alpha = \frac{\rho_{100}-\rho_0}{100\,\rho_0}$.

The electric resistivity of semiconductors at room temperature is

$$\rho = 10^{-4}\text{–}10^{7}\ \Omega\cdot\text{m}.$$

Semiconductors have free electrons, but their density is much lower than that of metals. In the case of semiconductors, current is transmitted not only by free electrons but also by holes. In semiconductors, as the temperature increases, the free electron density n increases, so the electric resistivity decreases with the rise in temperature.

Insulators can carry to some extent electricity. The electric resistivity at room temperature is

$$\rho = 10^{7}\text{–}10^{17}\ \Omega\cdot\text{m}.$$

The electric resistivity of metals and insulators differs by more than 14 digits. The reason for the high electric resistivity of insulators is that $n = 0$ because there are no free electrons. The electric resistivity of insulators decreases with the rise in temperature.

For reference Superconductivity

According to quantum mechanics, which is the mechanics of the atomic world, when positive ions are at rest and regularly aligned on a crystal lattice, (waves of) electrons do not collide with them and the change in their traveling direction does not occur. Therefore, it can be expected that at zero temperature (0 K), the thermal oscillations of positive ions aligned regularly on the crystal lattice disappear, so the electric resistance of the metal becomes zero. In reality, however, for many metals, alloys, and even ceramics, it has been found that the electric resistance becomes zero at low temperatures. This phenomenon is referred to as **superconductivity**. Superconductivity was discovered in 1911 by Kamerlingh Onnes. When he cooled mercury with liquid helium and measured the electric resistance, he discovered that the electric resistance suddenly disappeared at about 4.2 K.

If we prepare a ring made of a substance which can become a superconductor and flow the current through the ring in the superconducting state (superconductor), the electric resistance is zero, so the current continues to flow without attenuation. This current is called the permanent current.

The temperature of change from the normal electric conduction state following Ohm's law to the superconductive state occurs is called the critical temperature [Fig. 19-13 (a)]. The mechanism by which metals and alloys take the superconductive state can be explained by the BCS theory proposed by Bardeen, Cooper, and Schrieffer. The theory says, "At extremely low temperatures, two electrons are mediated by crystal oscillations and move in pairs, and can pass through the material without resistance."

Since 1986, a series of copper oxides (ceramics) have been discovered to become superconductive. The critical temperature of some materials exceeds 100 K. Substances having the critical temperature above 77.3 K, the boiling point of liquid nitrogen, become superconductive when cooled with inexpensive liquid nitrogen. The superconductive behavior of ceramic materials cannot be explained by the BCS

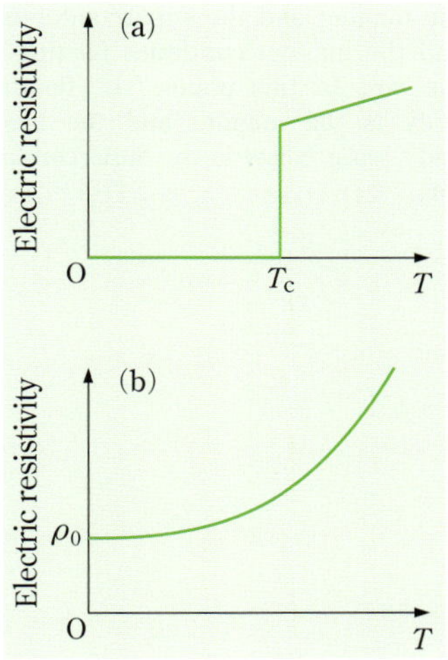

Fig. 19-13 Electric resistivity at extremely low temperatures.
(a) Conceptual diagram of temperature change of electric resistivity of superconductor at extremely low temperatures (T_c is the critical temperature.)
(b) Conceptual diagram of temperature change of electric resistivity of non-superconductor at extremely low temperatures. The electric resistance of metals that do not become superconductive at extremely low temperatures, such as Cu and Ag, should theoretically disappear at absolute zero (0 K), but due to impurities and disorder in the arrangement of ions, the electric resistance becomes nearly constant at temperatures below about 10 K.

Fig. 19-14 Magnetic levitation. When a magnet is placed above a superconductor, a permanent current flows on the surface of the superconductor due to the electromagnetic induction phenomenon which will be discussed in Chapter 21 and the superconductor becomes a magnet. As a result the repulsive force is generated between the magnet and the superconductor, and the magnet continues floating in the air. In this photo, the floating body is the magnet and the black body lying below is the superconductor.

theory. A molecular crystal called fullerene with a cage-like structure formed by linkage of 60 carbon atoms has also been discovered to exhibit the superconductivity.

19.3 Direct current (DC) circuit

Learning objective To understand that the circuit through which the current flows is a device that transfers the electric energy supplied by the power source to the circuit elements and converts it into other forms of energy. To understand how to find the combined resistance in series connection and in parallel connection of two resistors. To understand Kirchhoff's laws and to become able to calculate the current flowing through a simple direct current (DC) circuit.

Circuit A path through which electric current flows is called a **circuit**. The circuit includes, in addition to power source that provides energy, light bulbs, electric heaters, speakers, electrolyte solutions, motors, and so on. These convert electric energy into light, heat, sound, chemical energy, work, etc. The electric circuit through which current flows is not only a device that converts various forms of energy to another form of energy, but also a device that transfers energy to another location. In electromagnetics the circuit is considered to be a system connecting resistors, capacitors, coils (inductors), diodes, transistors, power sources, etc., and resistors, capacitors, coils, diodes, transistors, etc., are called **circuit elements**. A device that converts the electric energy supplied by a power source into other types of energy is called a load.

In this section, we consider the circuit in which only a resistor and a battery are connected. In this circuit, a current of constant magnitude and a constant direction called a stationary current continues to flow, so it is called the **direct current (DC) circuit**.

Connection of resistances When two or more resistors are

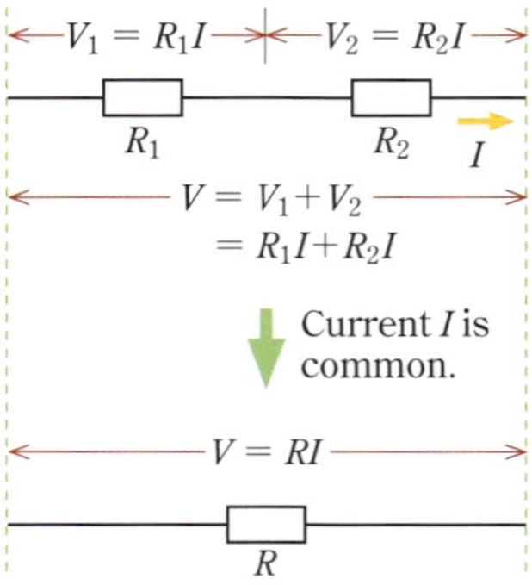

Fig. 19-15 Series connection of resistors. Combined resistance $R = R_1 + R_2$

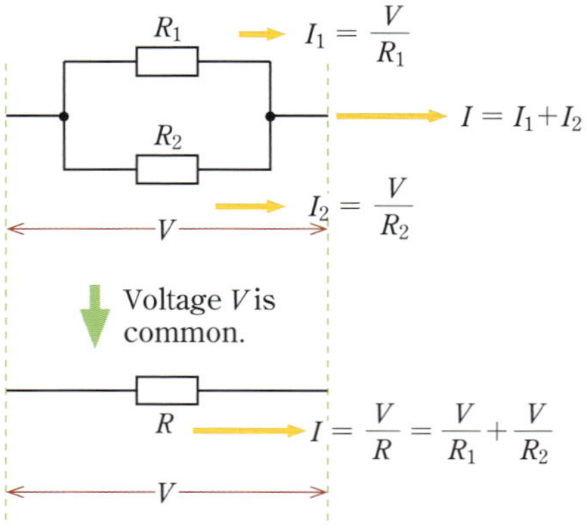

Fig. 19-16 Parallel connection of resistors. Combined resistance $\frac{1}{R} = \frac{1}{R_1} + \frac{1}{R_2}$

connected and regarded as one resistor, the resistance value of this connected resistor is called a combined resistance. Two resistors are connected in series and in parallel.

When two resistors with each resistance R_1 and R_2 are connected in series as shown in Fig. 19-15, the combined resistance R is

$$R = R_1 + R_2 \tag{19.13}$$

and when these resistors are connected in parallel as shown in Fig. 19-16, the combined resistance R is

$$\frac{1}{R} = \frac{1}{R_1} + \frac{1}{R_2}, \qquad R = \frac{R_1 R_2}{R_1 + R_2}. \tag{19.14}$$

Question 2 Show the following : When three or more resistors with each resistance $R_1, R_2, R_3, \cdots$ are connected in series, the combined resistance R is

$$R = R_1 + R_2 + R_3 + \cdots \tag{19.15}$$

and when three or more resistances $R_1, R_2, R_3, \cdots$ are connected in parallel, the combined resistance R is

$$\frac{1}{R} = \frac{1}{R_1} + \frac{1}{R_2} + \frac{1}{R_3} + \cdots. \tag{19.16}$$

Question 3 Find the current which flows through the battery shown in Fig. 19-17.

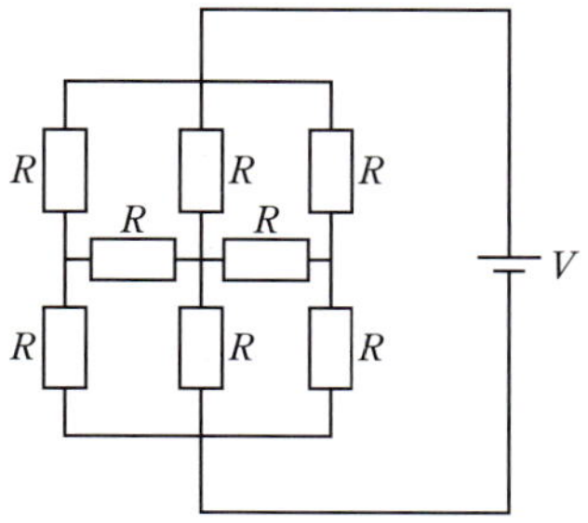

Fig. 19-17

Kirchhoff's laws To find the current which flows through a complex circuit, use of Kirchhoff's two laws mentioned below leads us to the solution.

The first law The sum of the currents flowing into any junction in the circuit is equal to the sum of the currents flowing out from that junction.

The second law In the potential differences in a complete travel around any closed path in the circuit, if the potential increase due to the power supply and resistance is represented as a positive amount, and the drop of the potential is represented as a negative amount, then the sum of the potential differences is always zero.

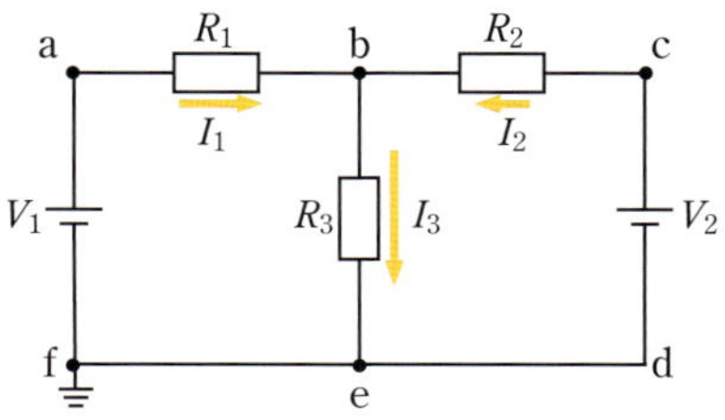

Fig. 19-18

The first law is derived from the charge conservation law. For example, at the junction b in Fig. 19-18, the inflow current is I_1 and I_2, and the outflow current is I_3. Hence,

$$I_1 + I_2 = I_3. \tag{19.17}$$

The second law states the fact when a complete travel around any closed path in the circuit is considered, the potentials at the starting and end points are equal. In using the second law, it is convenient to use the rule of change in potential at the power supply and resistor as shown in Fig. 19-19. For example, if we apply the second law to the closed path fabcdef, we have

$$V_1 - R_1 I_1 + R_2 I_2 - V_2 = 0. \tag{19.18}$$

Kirchhoff's second law is expressed alternatively as follows :

The second law When the potential differences around any closed

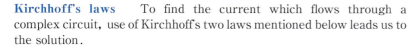

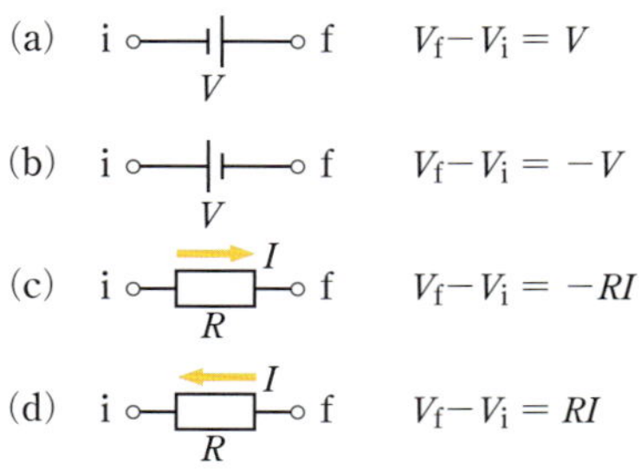

Fig. 19-19 The difference in potential $V_f - V_i$ between point f and point I

path in the circuit is considered, if the current and electromotive force in the direction along the path are assumed to be positive amounts, and the current and electromotive force in the opposite direction of the traveling direction are assumed to be negative amounts, then the sum of electromotive forces in the circuit is equal to the sum of "resistance"×"current" at each resistor.

In this way, equation (19.18) is derived in the form

$$V_1 - V_2 = R_1 I_1 - R_2 I_2. \tag{19.18$'$}$$

Question 4 Show that if we apply the second law to the two paths fabef and dcbed in Fig. 19-18, we will have

$$V_1 = R_1 I_1 + R_3 I_3 \quad \text{and} \quad V_2 = R_2 I_2 + R_3 I_3. \tag{19.19}$$

In addition, find from equations (19.17) and (19.19) the currents I_1, I_2, and I_3 flowing through the circuit shown in Fig. 19-17.

19.4 Current and work

Learning objective To understand the work rate of the power source when the electric wire of resistance R is connected to the power source with the electromotive force V, and the derivation of formula of Joule's heat that gives the amount of generated heat.

Work rate (power) A current flows through the circuit when connected to an electric power source, for example to a battery. When the charge ΔQ is transferred from the negative electrode to the positive electrode of the battery with an electromotive force V against the electric force, the work that the battery performs is $V\,\Delta Q$. When current I flows through the circuit, charge $\Delta Q = I\,\Delta t$ is transferred in time Δt, so work $\Delta W = VI\,\Delta t$ is performed by the power source. The work performed by the power source per unit time

$$P = \frac{\Delta W}{\Delta t} = VI \tag{19.20}$$

is called the **work rate (power)** of power source. The unit of work rate (power) is watt (symbol W).

Unit of power W = J/s
Unit of electric power W = J/s

The work performed by an electric power source is transformed into various types of work that the current I performs when it flows through the circuit with a potential difference of V. The work rate (power P) of current in this case is also given by

$$P = VI. \tag{19.21}$$

The work rate (power) of current (work performed per unit time) is called the **electric power**. The unit of electric power is of course watt. The fact that equation (19.20) and equation (19.21) are identical indicates the energy conservation law, meaning that the work in the process of generating electromotive force in the battery is equal to the work performed by the current flowing through the circuit.

Fig. 19-20 An electric heater

Joule's heat When current flows through a conductor with electric resistance, the temperature of the conductor rises. The mechanism of electric heaters and incandescent lamps is based on this property. When the battery is connected to both ends of the wire, the work of the electric force of the electric field generated in the wire by the electromotive force of the battery is not used to accelerate free electrons. In the wire, free electrons move at a constant average velocity while repeatedly colliding with thermally oscillating positive ions. In other words, the chemical energy of the battery is converted into the energy of the thermally oscillating positive ions in the wire, becoming the internal energy in the wire, and the temperature of the wire rises.

Consider the situation in which current I flows through a wire of resistor R connected to a power supply of electromotive force V. The amount of heat generated per unit time in the wire is equal to the power $P = IV$ of current, so the amount of heat Q generated during time t can be expressed, for example, as

$$Q = VIt = RI^2t = \frac{V^2}{R}t. \tag{19.22}$$

Here Ohm's law $V = IR$ is used. Since it was Joule who discovered that the heat generated by the current is proportional to the square of the current, the heat generated by the current is referred to as Joule's heat. **Joule's heat** corresponds to the frictional heat in mechanics.

The work performed by the current is called the **electric energy**, and its practical unit **kilowatt-hour** (symbol kWh), which is the work performed by 1 kW electric power in one hour, is often used :

$$1\ \mathrm{kWh} = 1000\ \mathrm{W} \times 3600\ \mathrm{s} = 3.6 \times 10^6\ \mathrm{J} \tag{19.23}$$

Practical unit of electric energy kWh
$1\ \mathrm{kWh} = 1000\ \mathrm{W} \times 3600\ \mathrm{s}$
$= 3.6 \times 10^6\ \mathrm{J}$

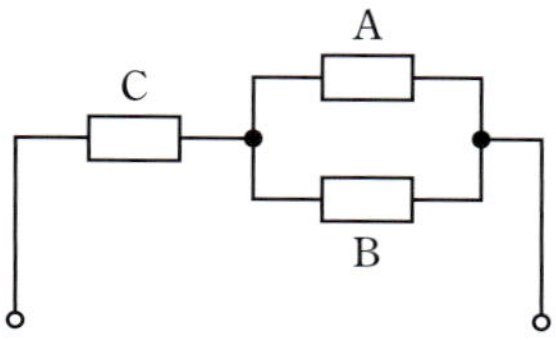

Fig. 19-21

Question 5 The resistances of the three resistors in the circuit in Fig. 19-21 are assumed to be equal. How many times is the energy consumed by the resistor A as large as the energy consumed by the entire circuit?

Question 6 A resistor in the circuit is generating heat at a rate of 1 W. How will the heat generation ratio change when the voltage applied to the resistor is doubled?

19.5 *CR* circuit

Learning objectives To learn the charging of capacitors and to become able to explain the description, "The time constant of the *CR* circuit is *CR*".

In a ***CR* circuit** [Fig. 19-22 (a)] consisting of a capacitor with capacity C, a resistor with resistance R, and a battery with electromotive force V, we charge the capacitor. Let us find, as a function of time t since we turn the switch S on [Fig. 19-22 (b)], the charge $Q(t)$ on the electrode plates and the current $I(t)$ flowing through the resistor.

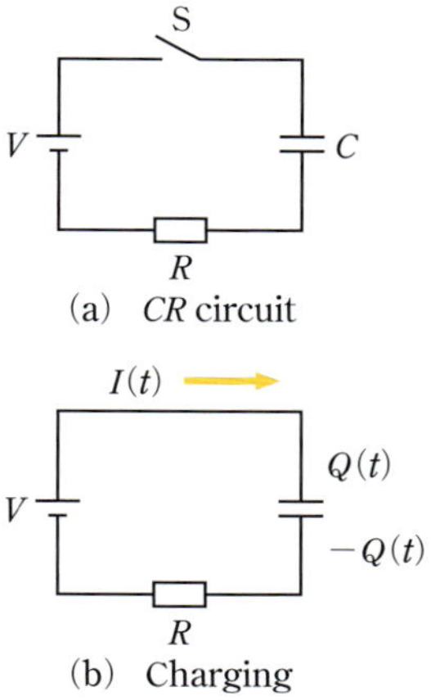

Fig. 19-22

Kirchhoff's first law holds for *CR* circuits as well. To apply Kirchhoff's second law, the potential difference $\dfrac{Q}{C}$ between the plates of capacitor (capacity C) included in the closed circuit must be taken into consideration (see Fig. 19-23).

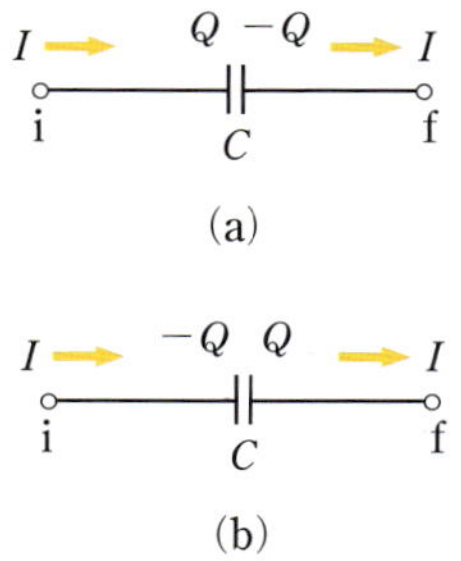

Fig. 19-23 Potential difference $V_f - V_i$ between the point f and the point i

(a) $V_f - V_i = -\dfrac{Q}{C}$, $\Delta Q = I\,\Delta t$

(b) $V_f - V_i = \dfrac{Q}{C}$, $\Delta Q = -I\,\Delta t$

Since the electromotive force V of the battery is equal to the sum of the voltage drop RI at the electric resistance R and the potential difference $\frac{Q}{C}$ between the electrode plates, the relation

$$V = RI + \frac{Q}{C} \tag{19.24}$$

holds. From $\Delta Q = I\,\Delta t$ of the relationship between current and charge we obtain the relationship between current and charge :

$$I = \frac{\mathrm{d}Q}{\mathrm{d}t} \tag{19.25}$$

Substituting equation (19.25) into equation (19.24), gives the following differential equation :

$$\frac{\mathrm{d}Q}{\mathrm{d}t} + \frac{1}{CR}Q = \frac{V}{R}. \tag{19.26}$$

The general solution of equation (19.26) is the sum of the general solution of the homogeneous equation obtained by setting $V = 0$,

$$Q(t) = c\,\mathrm{e}^{-t/CR} \qquad (c \text{ is an arbitrary constant}) \tag{19.27}$$

and the particular solution $Q(t) = CV$

$$Q(t) = c\,\mathrm{e}^{-t/CR} + CV. \tag{19.28}$$

Since at $t = 0$, $Q = 0$, so $c + CV = 0$, with the result that the arbitrary constant $c = -CV$. Therefore, the charge $Q(t)$ on the electrode plate when the time t has elapsed since the switch is turned on is

$$Q(t) = CV(1 - \mathrm{e}^{-t/CR}) \tag{19.29}$$

[Fig. 19-24 (a)]. At $t = CR$, $Q \approx 0.63CV$, $t = 2CR$, $Q \approx 0.86CV$, and $t = 3CR$, $Q \approx 0.95CV$. Therefore, CR is a measure of the time it takes for the charge to reach the final value CV. For this reason, CR is referred to as the **time constant** of the CR circuit.

Differentiation of equation (19.29) with respect to t and use of equation (19.25) leads to

$$I(t) = \frac{V}{R}\mathrm{e}^{-t/CR} \tag{19.30}$$

for the current [Fig. 19-24 (b)].

If the battery is removed after completion of charging and the conducting wire is connected, the capacitor will discharge. The time constant for discharge is also CR.

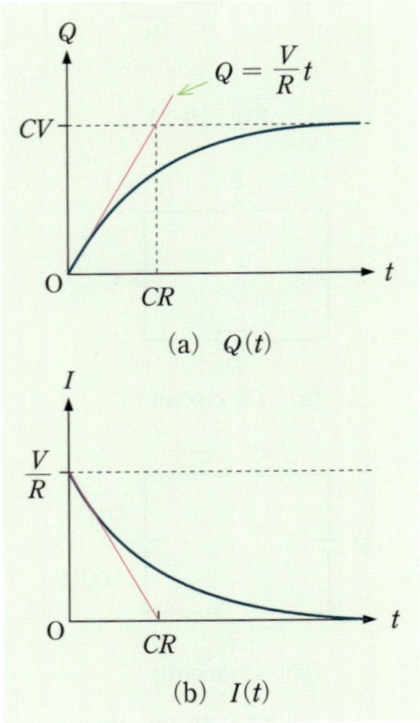

(a) $Q(t)$

(b) $I(t)$

Fig. 19-24

Question 7 Find the time constant of the CR circuit composed of a 100 pF capacitor and a 10 kΩ resistor.

Problems for exercise 19

A

1. Calculate the electric resistance at 20 °C of a 10 m copper wire with a cross-sectional area of 2.0 mm^2.
2. There is a rectangular parallelepiped of carbon. Its size is 1 cm×1 cm×25 cm. Assuming the electric resistivity of carbon is 3×10^{-5} Ω·m, calculate the electric resistance between two square faces.
3. Explain why good conductors of electricity are also good conductors of heat.
4. The circuit in Fig. 1 is a device called a potentiometer, and AB is a homogeneous resistance wire with uniform thickness. When the switch S was moved to the side 1 and the contact point C was moved, the needle of the galvanometer G indicated the 0 position when the AC length was L_1. When the switch was put on the side of 2 and the same operation was performed, the needle of G indicated the 0 position when the AC length was L_2. Show that there is a relationship of $V_1 : V_2 = L_1 : L_2$ between the electromotive forces of V_1 and V_2 of the two batteries.

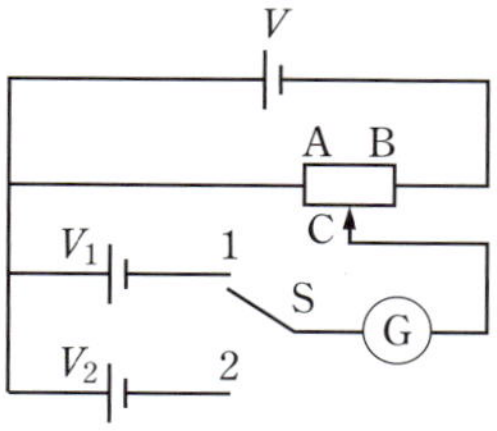

Fig. 1

5. Four 100 Ω resistors are connected as shown in Fig. 2. Find the combined resistance between A and B and between A and C.

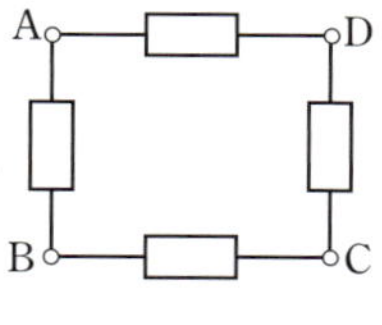

Fig. 2

6. Calculate the current I flowing through the circuit in Fig. 3.

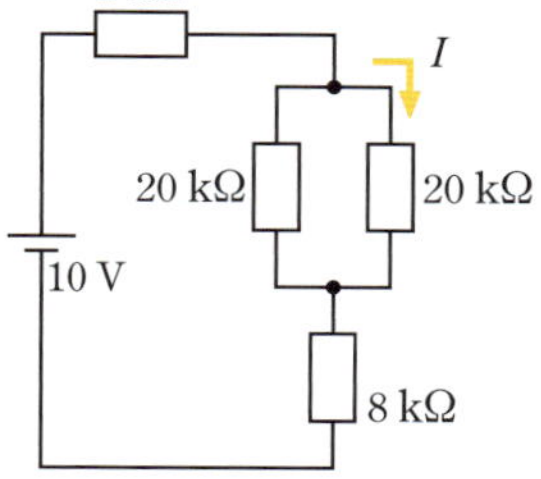

Fig. 3

7. Calculate the combined resistance in Fig. 4.

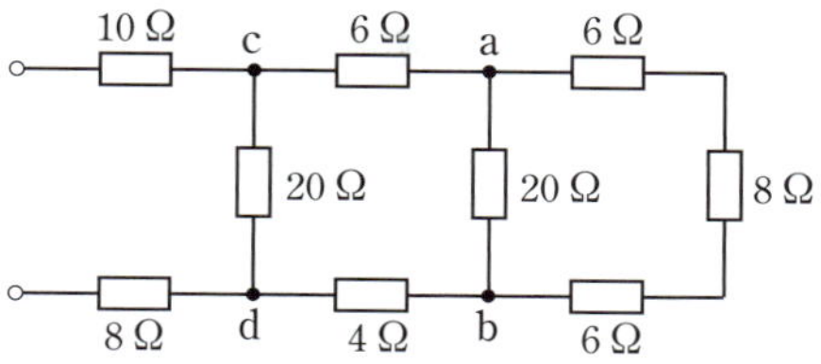

Fig. 4

8. The resistance of a 100 W bulb for 100 V is expected to be 100 Ω, but the measured resistance of the bulb at room temperature was less than 100 Ω. Explain the reason.
9. A 100 W bulb and a 500 W electric heater were connected in parallel. A power line of 100 V was connected to these through a 0.10 Ω wire. Calculate the voltage drop of the wire.
10. Which resistance is greater, 100 W bulb or 60 W bulb? If the filament lengths are the same, which filament is thicker?
11. In the circuit in Fig. 5, all light bulbs have a resistance of 2 Ω, and the electromotive force of the battery is 6 V. Which of the following ways will increase the power consumption of light bulb 3?
 a) To decrease the resistance of bulb 3.
 b) To increase the resistance of bulb 3.
 c) To decrease the electromotive force.
 d) To add one more resistor at C.

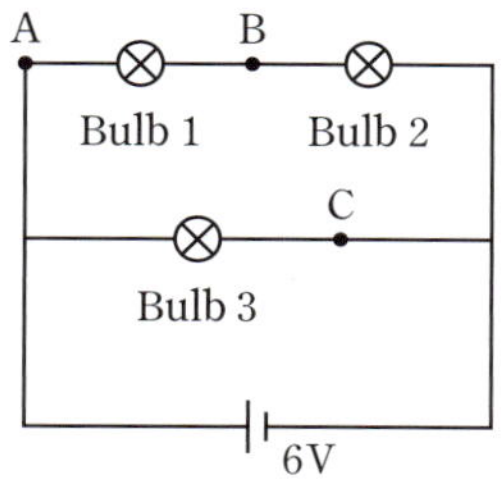

Fig. 5

12. **Wheatstone bridge** In order to find the unknown resistance value R of resistor R, is used a circuit shown in Fig. 6, in which two resistors R_1 and R_2 with known resistances R_1 and R_2, a rheostat (variable resistor) R_3 with resistance R_3, a battery with voltage V and a galvanometer G, and a switch S are connected. For measurement the switch S is closed and the R_3 value of the rheostat is adjusted so that the galvanometer needle indicates the zero position. Show that the unknow resistance value R is given by the equation

$$R = \frac{R_1 R_3}{R_2}.$$

This circuit is referred to as the Wheatstone bridge.

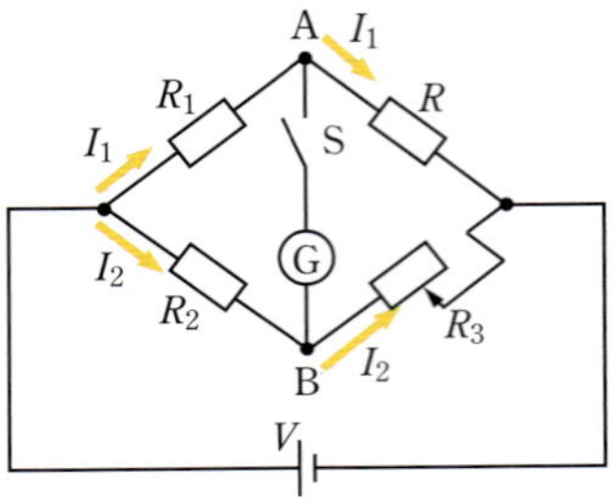

Fig. 6

13. When a dryer is connected to a 100 V power line, a current of 8 A flows.
  (1) How much power is used?
  (2) If it takes 2600 J to evaporate 1 g of water, how long does it take to dry a wet laundry containing 0.5 kg of water?

B

1. Calculate the electric capacity C and electric resistance R of a parallel-plate capacitor filled with thin mica. Assume the following : area of the electrode plate $A = 10^2\ \mathrm{cm}^2$, thickness of mica $d = 10^{-2}$ cm, relative dielectric constant of mica $\varepsilon_r = 6$, and electric conductivity $\sigma = 6\times10^{-15}\ \Omega^{-1}\cdot\mathrm{cm}^{-1}$. When this capacitor was charged to Q_0 and $-Q_0$ at time $t = 0$, it began to discharge from mica. Express the state of discharge using R, C and t.

2. The probability that a free electron in the wire will continue to run for t seconds without collision after colliding once with a thermally oscillating positive ion is $P(t) = \mathrm{e}^{-t/\tau}$. The probability of a collision between time t and $t+\mathrm{d}t$ is $-\dfrac{\mathrm{d}P}{\mathrm{d}t}\mathrm{d}t = P(t)\dfrac{\mathrm{d}t}{\tau}$, so the average time between two collisions is
$$\int_0^\infty \frac{t\,\mathrm{e}^{-t/\tau}}{\tau}\mathrm{d}t = \tau.$$
Show that the average distance that the free electron whose average velocity immediately after the collision is $\mathbf{0}$ travels between two successive collisions is $\dfrac{eE\tau^2}{m}$, and the drift velocity is $\boldsymbol{v} = -\dfrac{e\boldsymbol{E}\tau}{m}$.

3. (1) Since the Earth is negatively charged, an electric field of about 100 V/m is created downward near the Earth surface. The electric resistivity of air on the surface is about $3\times10^{13}\ \Omega\cdot\mathrm{cm}$. Show that the current density flowing down to the Earth due to the atmospheric electric field is $j = 3\times10^{-12}\ \mathrm{A/m^2}$, and that the current flowing into the entire Earth is 1500 A in total.
  (2) Show that the charge density σ induced on the Earth surface by the atmospheric electric field ($E = 100$ V/m) is $\sigma = -10^{-9}\ \mathrm{C/m^2}$, so the charge on the Earth surface is -500000 C. Under the above-mentioned condition, the charge of the Earth would disappear in about 6 minutes. What charges the Earth by supplying a negative charge is the lightning stroke.

Electric Current and Magnetic Field

Magnetism is well known as an electromagnetic phenomenon from old days together with frictional electricity.

Electric charges generate electric fields around their positions and the electric fields exert the electric forces upon the other electric charges. Similarly, magnetic charges of the magnets produce magnetic fields around their positions and the magnetic fields exert magnetic forces upon other magnetic charges. In addition to the magnetic charges electric currents as well as moving charges generate magnetic fields. On the other hand, magnetic fields act not only on magnetic charges but also on electric currents as well as on moving charges by magnetic forces. In total, there occur nine (3 times 3) kinds of interactions.

In this chapter we mainly learn the relationship between the static magnetic field which is time-independent and the stationary electric current.

Fig. 20-1 Hard disc drive for personal computers.
Motors are used for rotating magnetic disks and for the controller of the reading device.

What is important in the magnetic interaction is "magnetic force exerted on the current-carrying conduction wire placed in the magnetic field", which is applied to motors. Motors are installed not only in the home electric appliances such as refrigerators, air conditioners etc. but also in laptop computers and so on.

Now, as we will learn in the Chapter 22, when we consider the electromagnetism inside the matter based on the fundamental laws given by Maxwell equations, there appear two kinds of magnetic fields, $\boldsymbol{B}$ and $\boldsymbol{H}$. Most textbooks and physics encyclopedia call $\boldsymbol{B}$ as "magnetic flux density of magnetic field" and $\boldsymbol{H}$ as "strength of magnetic field". In this book, we denote "magnetic flux density of the magnetic field $\boldsymbol{B}$" as "magnetic field $\boldsymbol{B}$" and "strength of magnetic field $\boldsymbol{H}$" as "magnetic field $\boldsymbol{H}$". The reason for this is that we can describe electromagnetism in the vacuum only in terms of the electric field $\boldsymbol{E}$ and the magnetic field $\boldsymbol{B}$. Therefore, it is considered to be instructive to use the name "magnetic flux density $\boldsymbol{B}$" due to the historical custom, after having realized the relationship between magnetic field $\boldsymbol{B}$ and the electric current[*1].

*1 $\boldsymbol{B}$ is called "magnetic field" in the physics education at US high schools and universities.

20.1 Gauss' law of the magnetic field $\boldsymbol{B}$

Learning objective To understand that a magnetic force acts through a mediation of magnetic field $\boldsymbol{B}$, and that the behavior of magnetic field $\boldsymbol{B}$ can be described by a line of magnetic force (or magnetic line of force) which is a closed curve with neither a starting point nor an end point. To understand that the Gauss' law of the magnetic field $\boldsymbol{B}$ is nothing but the law which shows the absence of magnetic monopoles.

From the Iron Age, it was known that loadstones produced from various places in the world attract irons. The ancient Chinese found that compass needles show the north and south directions. There exist magnetic poles called the north pole (N-pole) and the south pole (S-pole) which attract irons most strongly at the both ends of the magnet. A magnetic pole of the compass needle which points to the north direction is a north pole, while the pole pointing to the south direction is a south pole. When a bar magnet is cut into two pieces, there appear a N-pole and a S-pole in a pair at the cross section, and we cannot extract the magnetic monopole (separated magnetic pole). Thus, the electromagnetism is constructed such that there are no magnetic monopoles.

Magnet's forces attracting irons and turning a compass needle in the north-south direction are called **magnetic forces**. The intensity of a magnetic pole is called the **magnetic charge**[*2]. The intensities of the different kinds of the magnetic poles at the both ends are equal. The magnetic force F acting between the magnetic poles is proportional to the product of magnetic charges Q_m and Q_m' and inversely proportional to the square of the distance r (Fig. 20-2).

$$F = k\frac{Q_m Q_m'}{r^2}. \tag{20.1}$$

The repulsive force acts between the same kinds of magnetic poles and the attracting force acts between the different kinds of magnetic poles.

*2 Although magnetic monopoles do not exist, we introduce magnetic charges for practical convenience when we consider the magnetic action of the magnet. This is because then we can use an analogy of static electricity with the static magnetic interaction by making a correspondence between electric charges and magnetic charges. The real entity which carries the magnetic property of matter is mainly microscopic magnetic dipole moment acquired by electron. The magnetic charge corresponds to the polarization charge appearing on the surface of dielectric substance.

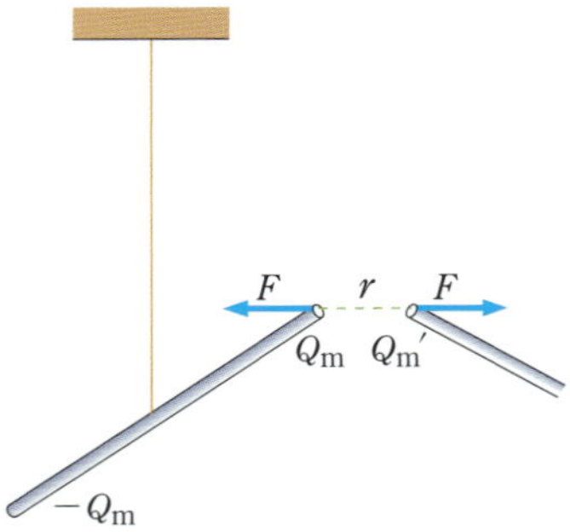

Fig. 20-2 $F = k\dfrac{Q_m Q_m'}{r^2}$ (for the case $Q_m Q_m' > 0$)

We set the sign of the magnetic charge of the N-pole to be positive and that of the S-pole to be negative. The proportional constant k in equation (20.1) will be explained in Section **20.6**.

Magnetic field and lines of magnetic force Magnetic forces are exerted on magnets not only by other magnets but also by neighboring electric currents. In electromagnetism it is considered that the magnetic forces do not directly act on each other between magnetic poles or between electric currents and magnetic poles, but it is considered that the magnetic field $\boldsymbol{B}$ generated by magnets and electric currents around them, will act magnetic forces upon magnets and electric currents.

In Chapter 16, we have mentioned that a point charge q at the position $\boldsymbol{r}$ is exerted an electric force $\boldsymbol{F} = q\boldsymbol{E}(\boldsymbol{r})$ by the electric field $\boldsymbol{E}(\boldsymbol{r})$. Hence, for a while, let us take the magnetic force of magnetic field $\boldsymbol{B}(\boldsymbol{r})$ acting on the magnetic charge Q_m of magnetic pole at the point $\boldsymbol{r}$ to be :

$$\boldsymbol{F} = Q_m \boldsymbol{B} \qquad \text{(in vacuum)}. \tag{20.2}$$

The unit of the magnetic field $\boldsymbol{B}$ is **tesla** (symbol T). The definition of tesla will be given in Section **20.3** using the magnetic force acting on a charged particle moving in the magnetic field.

To illustrate the appearance of the magnetic field $\boldsymbol{B}$, just as in the case of lines of electric force for the electric field, we draw a curve, the tangent of which points to the direction of the magnetic field at that point. We call these curves **lines of magnetic force** of the magnetic field $\boldsymbol{B}$ (In some textbooks they are called lines of magnetic flux). The direction of the magnetic field is that for the magnetic force acting on the positive magnetic charge (N-pole), and the line of the magnetic force generated by the magnet around itself leaves from the N-pole and enters into the S-pole of the magnet (Fig. 20-3).

What about the lines of magnetic force inside the magnet? Faraday, who invented the lines of magnetic force, considered that the lines of magnetic force of the magnetic field $\boldsymbol{B}$ generating the phenomena of electromagnetic induction are closed curves which are connected inside and outside of magnets. As we will learn in the next chapter, unless we take this view, we cannot explain the electromagnetic induction phenomenon. Where the closed line of magnetic force leaves is the N-pole, while it enters into magnets at the S-pole (Fig. 20-4).

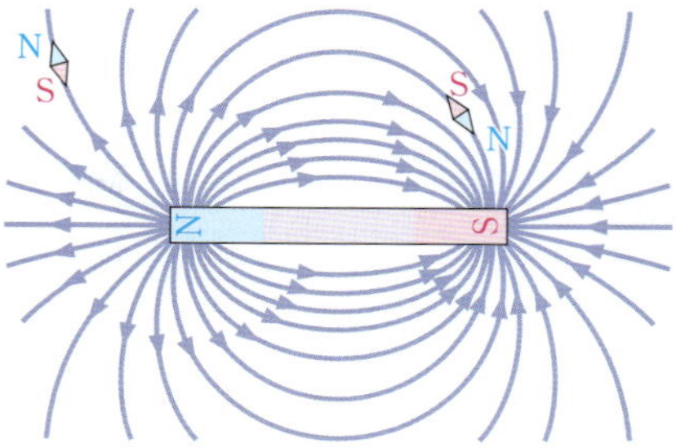

Fig. 20-3 Lines of magnetic force. The appearance of lines of magnetic force of the magnetic fields generated by the magnet, can be seen from an observation that by putting a glass plate on the magnet and scattering iron powder on it and shaking, then the iron powder gets aligned along the lines of magnetic force.

Unit of magnetic field $\boldsymbol{B}$ T
Unit of magnetic charge
N/T = A·m

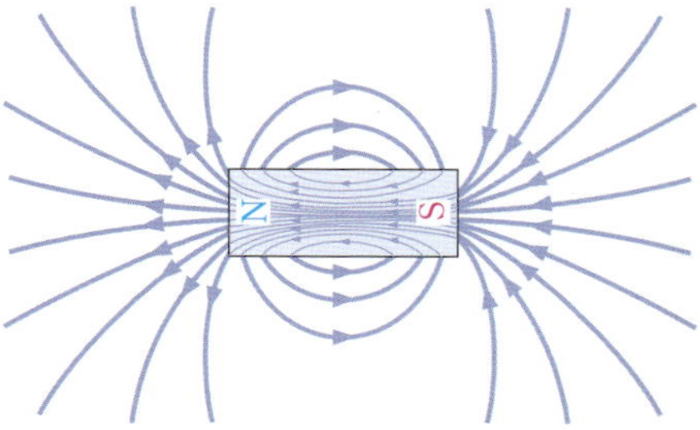

Fig. 20-4 The lines of magnetic force illustrating the magnetic field are closed curves which possess neither a starting point nor an end point.

Magnetic flux As shown in Fig. 20-5, a plane S of area A is placed in the uniform magnetic field $\boldsymbol{B}$. Let the angle between the normal vector $\boldsymbol{n}$ of the plane and the magnetic field $\boldsymbol{B}$ be θ, then we have

$$\Phi_B = BA\cos\theta = B_n A \qquad (B_n = B\cos\theta), \tag{20.3}$$

which we define as a **magnetic flux** passing through the plane S. B_n is a component of the magnetic field $\boldsymbol{B}$ in the direction of the normal vector of the plane S [Fig. 20-5 (c)]. Since the units of the magnetic field $\boldsymbol{B}$ and the area are T and m^2, respectively, and the unit of magnetic flux is $T \cdot m^2$, which we call weber (symbol Wb).

The net magnetic flux Φ_B penetrating an arbitrary curved surface S

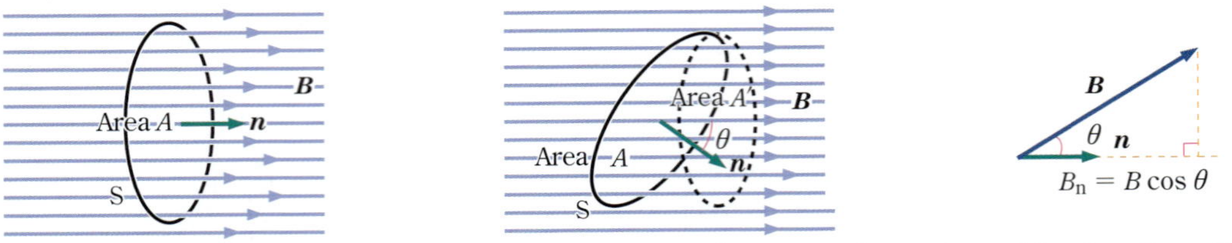

(a) Magnetic flux $\Phi_B = BA$ (b) Magnetic flux $\Phi_B = BA' = BA\cos\theta = B_n A$ (c) $B_n = B\cos\theta$

Fig. 20-5 Magnetic flux Φ_B

from back side to front side, as in the case of the flux of electric line of force, is expressed as

$$\Phi_B = \iint_S B_n \, dA. \quad (20.4)$$

If we draw lines of the magnetic force at the rate of B lines per unit area of the plane S, which is perpendicular to the magnetic field $\boldsymbol{B}$, then the magnetic flux Φ_B is equal to the number of lines of the magnetic force penetrating the surface S.

Gauss' law of the magnetic fields *B* Since there is no magnetic monopole which is either a starting point or an end point of the line of magnetic force, a line of magnetic force of the magnetic field $\boldsymbol{B}$ possesses neither a starting point nor an end point and is a closed curve without a break in the middle. Therefore, the number of lines of magnetic force entering into an arbitrary closed surface S and the number of the lines of magnetic force outgoing from the surface are the same, and so we have

"Magnetic flux Φ_B outgoing from the closed surface S" = 0

$$\iint_S B_n \, dA = 0 \quad \text{(S is a closed surface)}, \quad (20.5)$$

Unit of magnetic flux
Wb = T·m²

which we call **Gauss' law of magnetic fields *B***. This is one of the four Maxwell's laws, as the fundamental laws of the electromagnetism, and is a law which always holds.

20.2 Magnetic field generated by electric currents

Learning objective To understand that it is not only magnets that generate magnetic fields, but magnetic fields are also generated around electric currents. To become able to explain the property of magnetic fields produced by the long linear currents, circular currents, and a current flowing through a long solenoid.

Magnetic field generated by a long linear current It was discovered by Oersted of Denmark in 1820 that when an electric current flows through a long straight conducting wire, magnetic fields are generated around the wire (Fig. 20-6). Placing a magnetic compass needle under the electric current flowing in the south-north direction

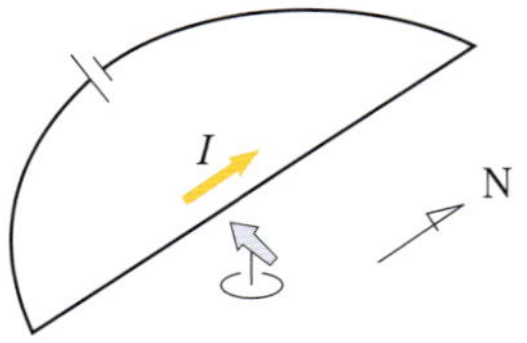

Fig. 20-6 Oersted's experiment. An electric current flowing from south to north deflects the magnetic compass needle under the conducting wire as shown in the figure.

and examining which direction the N-pole points, we find as shown in Fig. 20-6 that the direction of magnetic force acting on the magnetic pole is perpendicular to both directions of the current and the vertical line drawn from the magnetic pole to the current.

If we scatter iron sand on a cardboard placed perpendicular to the current-carrying conducting wire, then the iron sand links up forming concentric circles with the wire at the center, and in this case the line of the magnetic force is a circle having neither a starting point nor an end point (Fig. 20-7). The direction of the magnetic field is the direction to turn of the right-handed screw when it moves in the direction of the current. This is called the **right-handed screw rule**.

When we measure the magnetic field around a long straight current-carrying wire, the strength B of the magnetic field $\boldsymbol{B}$ is proportional to the electric current I and inversely proportional to the distance d from the current. Taking the proportional constant to $\frac{\mu_0}{2\pi}$ we can express B as

$$B = \frac{\mu_0 I}{2\pi d}. \tag{20.6}$$

μ_0 (read "mu zero") in the proportional constant, is given by

$$\mu_0 = 4\pi \times 10^{-7}\ \mathrm{T \cdot m/A}, \tag{20.7}$$

which is called the **magnetic constant***.

Case 1 When we shorted the both electrodes of the size D battery with a slightly thick copper wire, there occurred a current flow of 5 A. The strength B of the magnetic field $\boldsymbol{B}$ at the point with a distance 1 cm from the copper wire is

$$B = \frac{(4\pi \times 10^{-7}\ \mathrm{T \cdot m/A}) \times (5\ \mathrm{A})}{2\pi \times (0.01\ \mathrm{m})} = 10^{-4}\ \mathrm{T}.$$

The strength of the horizontal component of the terrestrial magnetism is 0.3×10^{-4} T. The angle θ which satisfies the equation $\tan\theta = \frac{1}{0.3}$ is 73°, and so the direction of the magnetic compass needle in Fig. 20-6 deflects from the south-north direction by 73°.

Biot-Savart's law Biot and Savart of France, after they had heard about the news of the Oersted's discovery that the magnetic fields are generated around the electric current, performed the experiment making use of ring wires, and wires bent to have the shape of the combination of circular arc and line segment, and they discovered the rule to obtain the magnetic fields produced by the current-carrying wire of an arbitrary shape (see Problems for exercise 20, B5). They found that

The magnetic field $\Delta\boldsymbol{B}$ generated by the infinitesimal element $\Delta\boldsymbol{s}$ of the conducting wire carrying a stationary current I at the point P apart from the element by a distance r (relative position vector $\boldsymbol{r}$) is in magnitude given as

$$\Delta B = \frac{\mu_0 I\ \Delta s \sin\theta}{4\pi r^2} \tag{20.8}$$

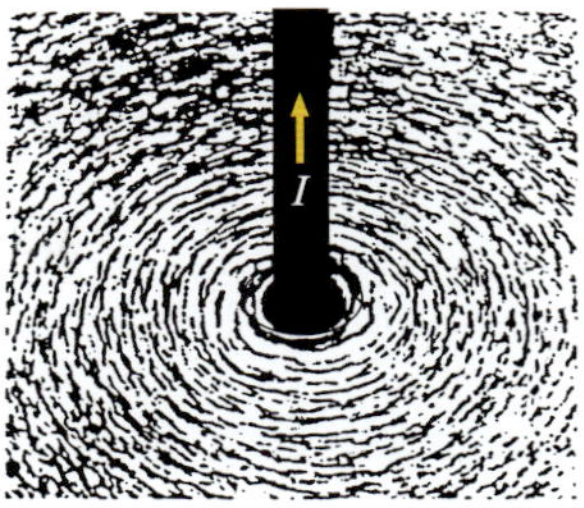

(a)

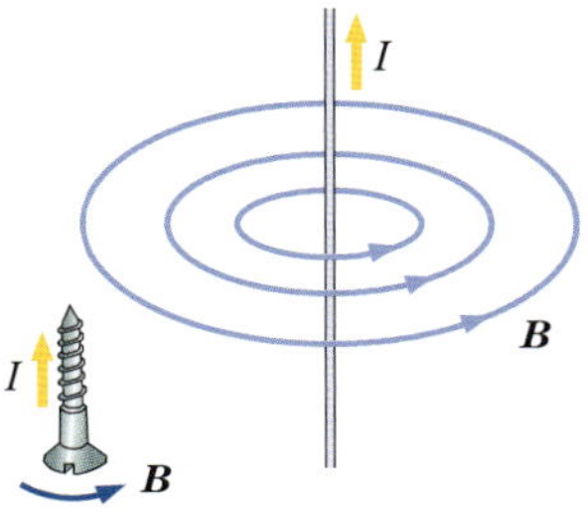

(b)

Fig. 20-7 The lines of magnetic force of the magnetic field generated by a long straight current are concentric circles with the current at the center. When a right-handed screw moves in the direction of the current, the direction to turn of the screw is the direction of the line of magnetic force.

Magnetic constant (permeability of vacuum)
$\mu_0 = 4\pi \times 10^{-7}\ \mathrm{T \cdot m/A}$

* The magnetic constant, μ_0, has been historically called the permeability of vacuum, but it is not a constant representing the property of the vacuum as a magnetic body. μ_0 is a transformation coefficient to match the mechanistic units [m], [kg], and [s], which are independently defined, with the measured numerical values of the electromagnetic unit [A].

In association with the revision of the definition for the unit of the electric current ampere in 2018, the numerical value of the magnetic constant slightly changed. In the calculations of this book, we use the value equation (20.7), for simplicity.

and its direction is perpendicular to both $\Delta \boldsymbol{s}$ and $\boldsymbol{r}$, and $\Delta \boldsymbol{B}$ points to the direction for a right-handed screw to move when turning it from the direction of $\Delta \boldsymbol{s}$ to the direction of $\boldsymbol{r}$.

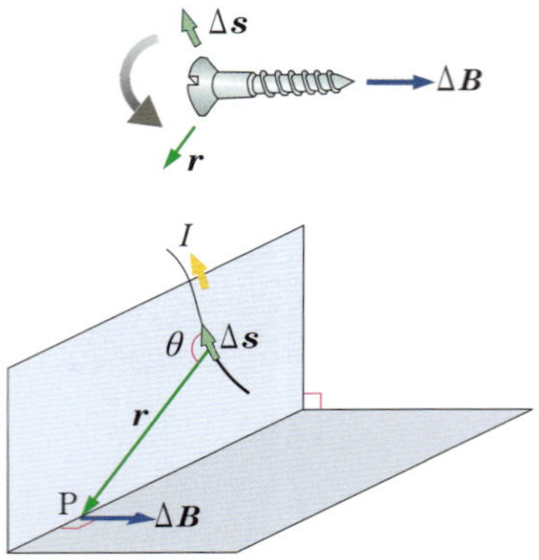

Fig. 20-8 Biot-Savart's law

Here, θ is the angle between $\Delta \boldsymbol{s}$, which is a vector of the length Δs pointing to the current, and $\boldsymbol{r}$ (Fig. 20-8). This is called **Biot-Savart's law**. Using the representation of a vector product, it can be expressed as

$$\Delta \boldsymbol{B} = \frac{\mu_0 I}{4\pi r^3} \frac{\Delta \boldsymbol{s} \times \boldsymbol{r}}{}. \tag{20.8'}$$

The magnetic field $\boldsymbol{B}(\boldsymbol{r})$ generated at the point P (position vector $\boldsymbol{r}$) by the stationary current I flowing through the closed circuit C, is a superposition of the magnetic field $\Delta \boldsymbol{B}$ produced by the infinitesimal element $\Delta \boldsymbol{r}'$ (position vector $\boldsymbol{r}'$) of the conducting wire according to the equation (20.8′) (Notation changed to $\mathrm{d}\boldsymbol{r}'$ instead of $\mathrm{d}\boldsymbol{s}$)* :

$$\boldsymbol{B}(\boldsymbol{r}) = \frac{\mu_0 I}{4\pi} \oint_{\mathrm{C}} \frac{\mathrm{d}\boldsymbol{r}' \times (\boldsymbol{r} - \boldsymbol{r}')}{|\boldsymbol{r} - \boldsymbol{r}'|^3}. \tag{20.9}$$

* When the velocity $\boldsymbol{v}$ of the particle with a charge q at the point $\boldsymbol{r}'$ is small enough in magnitude compared to the speed of light, the magnetic field at $\boldsymbol{r}$, $\boldsymbol{B}(\boldsymbol{r})$, produced by this charged particle is (see Problems for exercise 21, B7) :

$$\boldsymbol{B}(\boldsymbol{r}) = \frac{\mu_0 q}{4\pi} \frac{\boldsymbol{v} \times (\boldsymbol{r} - \boldsymbol{r}')}{|\boldsymbol{r} - \boldsymbol{r}'|^3}.$$

The line of magnetic force of the magnetic field $\boldsymbol{B}$ produced by a linear current is a circle. Moreover, the line of magnetic force of the magnetic field $\boldsymbol{B}$ generated by the stationary current following Biot-Savart's law, is a closed curve having neither an initial point nor an end point (Proof omitted).

Magnetic fields $\boldsymbol{B}$ produced by electric currents satisfy the principle of superposition. That is, the magnetic field $\boldsymbol{B}$ generated when a number of wires are carrying currents, is a vector sum of magnetic fields $\boldsymbol{B}$ due to the current carried by each conducting wire.

Question 1 Compare the strength of the magnetic fields $\boldsymbol{B}$ at points A, B, C and D in Fig. 20-9.

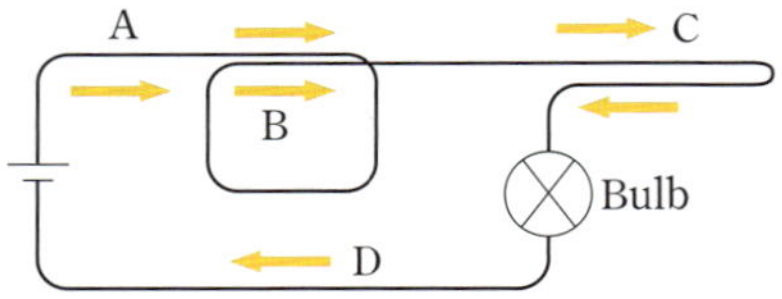

Fig. 20-9

Magnetic field generated by a circular current The magnetic field $\boldsymbol{B}$ due to the electric current flowing in a circular conducting wire (coil) turns out to be as shown in Fig. 20-10. We can understand how it is realized from the fact that the infinitesimal part of the coil generates the magnetic field around it as in the case of linear current and the principle of superposition. The magnetic field penetrating the coil points to the direction in which a right-handed screw moves when it is turned to the direction of the rotating current. The magnetic field appearing far enough from the coil looks like the magnetic field a thin magnet generates at a distant point.

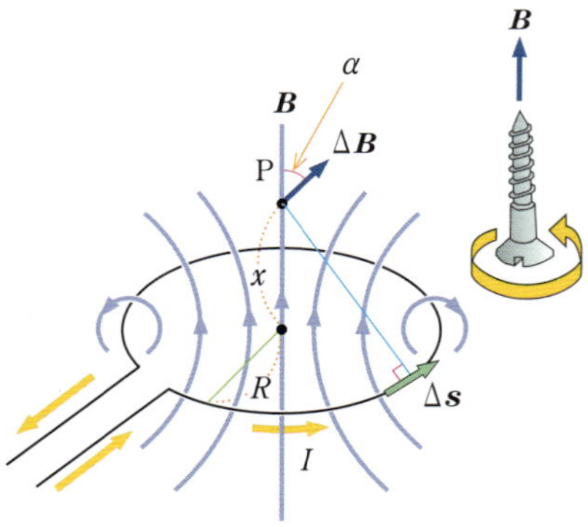

Fig. 20-10 Magnetic field generated by a circular current

The strength of the magnetic field $\boldsymbol{B}$ in the center of a circular coil of one turn (radius R) carrying the current I is given by

$$B = \frac{\mu_0 I}{2R} \quad \text{(at the center of circle of radius } R\text{)}. \tag{20.10}$$

The strength of the magnetic field $\boldsymbol{B}$ due to the coil of N turns is N times this value.

Proof The strength of the magnetic field that the current $I\,\Delta \boldsymbol{s}$ on the infinitesimal element $\Delta \boldsymbol{s}$ of the wire in Fig. 20-11 generates at the center of the circle O, is $\Delta B = \dfrac{\mu_0 I\,\Delta s}{4\pi R^2}$ using Biot-Savart's law. Since the sum of Δs is the circumference of a circle of radius R, that is $\sum \Delta s = 2\pi R$, we have

$$B = \sum \Delta B = \frac{\mu_0 I}{4\pi R^2} \sum \Delta s = \frac{\mu_0 I}{4\pi R^2} \times (2\pi R) = \frac{\mu_0 I}{2R}.$$

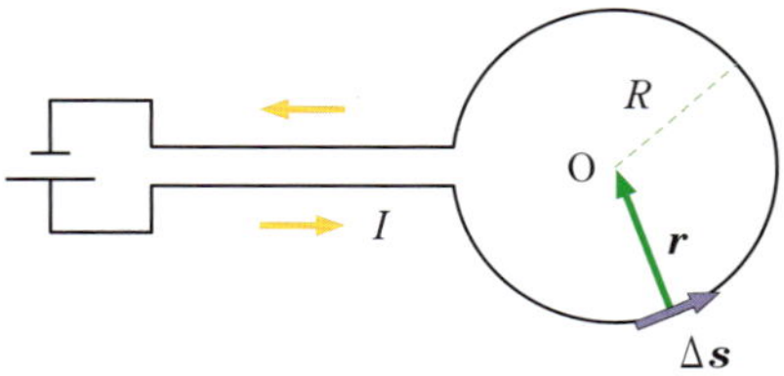

Fig. 20-11 $B = \dfrac{\mu_0 I}{2R}$

Question 2 Find the magnetic field at the point P generated by the current I flowing the conducting wire in Fig. 20-12.

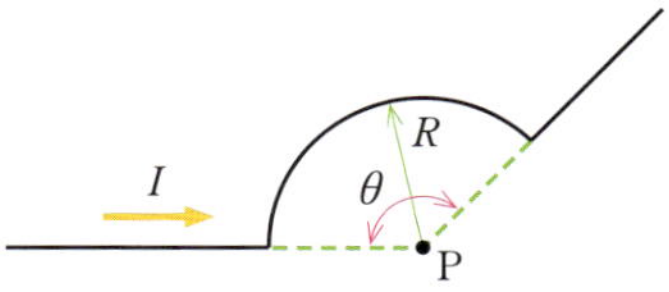

Fig. 20-12

For reference

The strength of the magnetic field $\boldsymbol{B}$ at the point P, on the central axis of the circular current (radius R), apart from the center of the circle with a distance x is

$$B = \frac{\mu_0 I R^2}{2(R^2+x^2)^{3/2}} \quad \text{(on the central axis)}, \qquad (20.11)$$

which can be derived from the fact that ΔB in Fig. 20-10 is $\dfrac{\mu_0 I\,\Delta s}{4\pi(R^2+x^2)}$ and B is the sum of $\Delta B \cos\alpha = \Delta B \dfrac{R}{(R^2+x^2)^{1/2}}$.

Magnetic field due to the current flowing in a long solenoid

Insulated wires densely wound in a cylindrical shape is called a solenoid or a solenoid coil. The magnetic field generated by the current flowing in the solenoid is the superposition of the magnetic fields produced by many circular currents. When we flow an electric current I through a long solenoid with the number of turns per unit length is n and inside of which is a vacuum, we get, inside the solenoid, the magnetic field $\boldsymbol{B}$ which is parallel to the axis of the solenoid and uniform in strength except for the region near both ends, and given by

$$B = \mu_0 n I \quad \text{(inside the solenoid with air core)}, \qquad (20.12)$$

and the direction of the magnetic field points to the direction in which a right-handed screw moves when it is turned to the direction of the current (Fig. 20-13). Furthermore, the magnetic field $\boldsymbol{B}$ outside of the infinitely long solenoid is $\boldsymbol{0}$ everywhere (see Example 1).

A solenoid becomes an electromagnet when it carries an electric current. If an iron core is put into the solenoid, then the iron core is magnetized and becomes a magnet, as a result, magnetic fields $\boldsymbol{B}$ inside as well as outside the solenoid get much stronger. When the inside of a

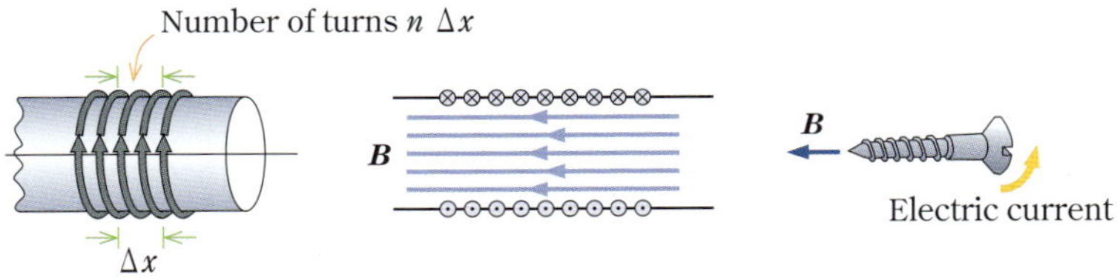

Fig. 20-13 Magnetic field generated by the current flowing in a solenoid. Inside an infinitely long solenoid $B = \mu_0 n I$. The symbol ⊙ indicates that the current flows from the backside of the paper plane to the frontside, while the symbol ⊗ indicates the flow from the frontside to the backside.

long solenoid is filled with a substance, and the strength of the magnetic field $\boldsymbol{B}$ becomes μ_r times the value for the case without the substance, we call μ_r **relative permeability** of the substance. In this case the strength of the magnetic field $\boldsymbol{B}$ inside the long solenoid is as follows :

$$B = \mu_r \mu_0 nI \quad \text{(inside a long solenoid filled with a substance of relative permeability } \mu_r\text{)} \tag{20.13}$$

Proof of equation (20.12) on the central axis A solenoid can be regarded as an aggregation of numerous circular currents. A minute part Δx at the distance between x and $x+\Delta x$ from the point P on the central axis contains $n\,\Delta x$ circular currents. The magnetic field $\Delta \boldsymbol{B}$ at the point P produced by each circular current is given by equation (20.11). Hence the strength of the magnetic field $\boldsymbol{B}$ at the point P on the central axis of the infinitely long solenoid is found to be

$$B = \frac{\mu_0 nIR^2}{2}\int_{-\infty}^{\infty}\frac{\mathrm{d}x}{(R^2+x^2)^{3/2}} = -\int_{\pi}^{0}\frac{\mu_0 nI}{2}\sin\theta\,\mathrm{d}\theta$$

$$= \frac{\mu_0 nI}{2}\cos\theta\Big|_{\pi}^{0} = \mu_0 nI, \tag{20.14}$$

where we have taken the central axis of the solenoid to be x-axis (Fig. 20-14) and we also set

$$x = R\cot\theta = R\frac{\cos\theta}{\sin\theta}, \qquad R = (R^2+x^2)^{1/2}\sin\theta,$$

and use has been made of the relation

$$\mathrm{d}x = -R\frac{\mathrm{d}\theta}{\sin^2\theta}.$$

The strength of the magnetic field $\boldsymbol{B}$ at the center of an end of the long solenoid is $B = \frac{1}{2}\mu_0 nI$. This fact can be derived from the integration of equation (20.14) with respect to θ, not from π to 0, but from $\pi/2$ to 0.

The proof of equation (20.12) other than on the central axis will be done in the Example 1 of the next section.

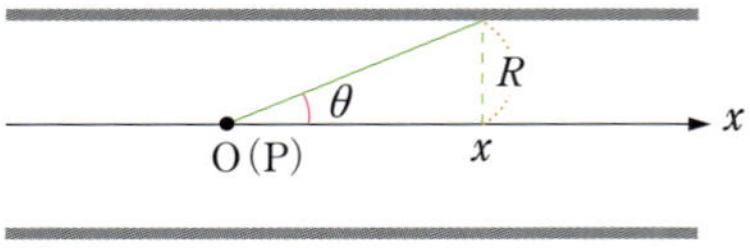

Fig. 20-14

Case 2 When an electric current of 3 A flows through an air core solenoid of length 30 cm, and number of turns 6000, the strength B of the magnetic field $\boldsymbol{B}$ inside the solenoid is to be calculated. Since the number n of turns per one meter is $n = \frac{6000}{0.3\,\text{m}} = 20000/\text{m}$, we obtain

$$B = \mu_0 nI = (4\pi\times10^{-7}\,\text{T}\cdot\text{m/A})\times(20000/\text{m})\times(3\,\text{A}) = 0.075\,\text{T}.$$

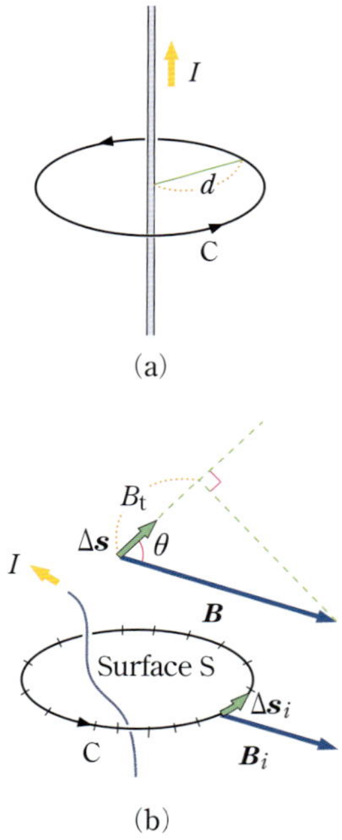

Fig. 20-15 Ampere's law $\oint_C B_t\,\mathrm{d}s = \mu_0 I$

Ampere's law The strength B of the magnetic field $\boldsymbol{B}$ at a point with a distance d from the current I is $B = \frac{\mu_0 I}{2\pi d}$ everywhere. As shown in Fig. 20-15 (a), let us consider the path C going around the circle of radius d with the center the current passes through. The component of the magnetic field $\boldsymbol{B}$ in the direction of the tangent of the circle $B_t = B = \frac{\mu_0 I}{2\pi d}$ times the length of the path C, $2\pi d$ is the product of the magnetic constant μ_0 and the current I passing through the path, that is, $\mu_0 I$, and is independent of the radius of the path d.

Let us consider the case where the path is not a circle and/or the current is not linear. As shown in Fig. 20-15 (b), we divide the path C of the closed curve with the direction designated into small parts. Then we sum up the products of the length of each small parts Δs_i and the component of the magnetic field $\boldsymbol{B}_i$ at the point in the tangential direction of the closed curve B_{it}, that is $B_{it}\,\Delta s_i$. The line integral defined as the limit of $\sum_i B_{it}\,\Delta s_i$ when $\Delta s_i \to 0$, is independent of the shape of the path, and is equal to μ_0 times the sum of the currents penetrating the closed curve C, I, as follows :

$$\oint_C B_t\,ds = \mu_0 I. \tag{20.15}$$

The sign of the current is taken to be positive if the current flows in the direction in which a right-handed screw moves when it is turned to the direction of the closed curve C. This is called **Ampere's law**.

Ampere's law A magnetic field arises around the electric current where the line of magnetic force goes around the current dextrally. The integral of the tangential component of the magnetic field $\boldsymbol{B}$ along the closed curve is equal to μ_0 times the current penetrating the closed curve.

This law corresponds to Ampere-Maxwell's law, which is one of Maxwell's four laws, the fundamental laws of electromagnetism, for the case without time-dependence of electric fields*.

* Historically, Ampere's law was derived from Biot-Savart's law, while Biot-Savart's law can be derived from Ampere's law together with Gauss' law of magnetic fields $\boldsymbol{B}$.

There are some cases where magnetic fields are easy to calculate using Ampere's law.

Example 1 By making use of Ampere's law, show that the magnetic field $\boldsymbol{B}$ generated by the current I flowing in an infinitely long air-core solenoid is given as follows :

$$B = \mu_0 nI \quad \text{(inside the solenoid)} \tag{20.16a}$$

$$B = 0 \quad \text{(outside the solenoid)} \tag{20.16b}$$

Solution The magnetic field $\boldsymbol{B}$ generated by the current flowing in an infinitely long solenoid is invariant under the translation in the direction of the central axis, and hence the direction of the magnetic field $\boldsymbol{B}$ is parallel to the central axis of the solenoid.

Let us consider a rectangle PQRS, one side of which, PQ, is on the central axis as shown in Fig. 20-16, and use the fact that the strength B of the magnetic field on the central axis is $\mu_0 nI$. On the line segments QR, SP perpendicular to the central axis, $B_t = 0$. Let the strength of the magnetic field on the line segment RS be B, and note that the directed segment $\overrightarrow{\mathrm{RS}}$ is opposite to the magnetic field $\boldsymbol{B}$, and hence $B_t \cdot \overline{\mathrm{RS}} = -B \cdot \overline{\mathrm{RS}}$. Making use of this fact we get for the left-hand side of equation (20.15) as follows :

$$\oint_C B_t\,ds = \mu_0 nI \cdot \overline{\mathrm{PQ}} - B \cdot \overline{\mathrm{RS}} \quad (\overline{\mathrm{PQ}} = \overline{\mathrm{RS}}). \tag{20.17}$$

In the case of Fig. 20-16 (a), since the current

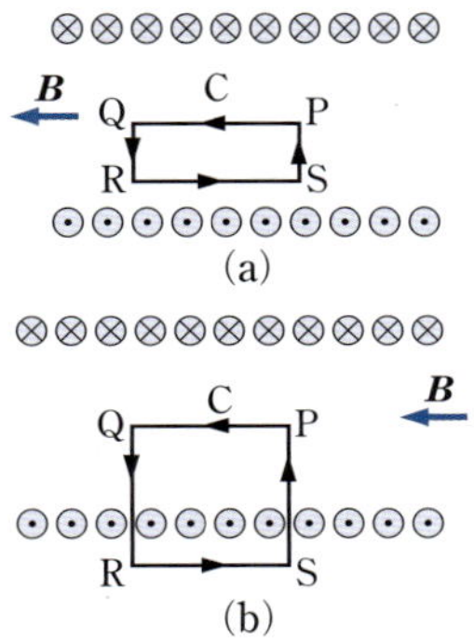

Fig. 20-16 Solenoid

passing through the rectangle PQRS is 0, the right-hand side of equation (20.15) is 0. Therefore, Ampere's law in this case becomes $\mu_0 nI \cdot \overline{PQ} - B \cdot \overline{RS} = 0$, and so we get

$$B = \mu_0 nI \quad \text{(inside the solenoid)}. \qquad (20.18)$$

Namely, inside an infinitely long solenoid, the magnetic field $\boldsymbol{B}$ is constant, and the magnitude is $\mu_0 nI$.

In the case of Fig. 20-16 (b), since the current passing through the rectangle PQRS is $nI \cdot \overline{PQ}$, the right-hand side of equation (20.15) is $\mu_0 nI \cdot \overline{PQ}$. Therefore, Ampere's law in this case becomes $\mu_0 nI \cdot \overline{PQ} - B \cdot \overline{RS} = \mu_0 nI \cdot \overline{PQ}$, and so we get

$$B = 0 \quad \text{(outside the solenoid)}. \qquad (20.19)$$

Namely, outside an infinitely long solenoid, the magnetic field $\boldsymbol{B}$ is $\boldsymbol{0}$ everywhere.

20.3 Force acting on a charged particle (Lorentz force)

Learning objective To become able to explain the direction of a magnetic force acting on a moving charged particle. To understand that the magnetic force acts on a moving charged particle transversely, and hence a charged particle performs a uniform circular motion in the uniform magnetic field.

Force acting on a charged particle As shown in Fig. 20-17, when we let a horseshoe magnet approach the electron beam, like nipping the beam, which is moving from the negative electrode to the positive electrode in the discharge tube, the course of the beam is deflected. This is because the magnetic field generated by the magnet exerts a force on the electron. By inspecting how the electron beam is deflected, it can be realized that the electron is exerted a force in the direction perpendicular to both the direction of motion (direction of the velocity $\boldsymbol{v}$) and the direction of the magnetic field $\boldsymbol{B}$. Furthermore, the magnetic field does not exert a force on the charge at rest.

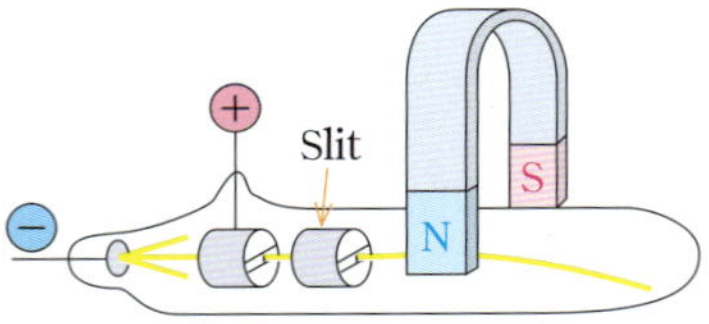

Fig. 20-17 Bringing a magnet closer to the electron beam of a discharge tube.

The magnitude F of the magnetic force $\boldsymbol{F}$ acting on a charged particle of charge q moving with a velocity $\boldsymbol{v}$ through the magnetic field $\boldsymbol{B}$, with taking the angle between the magnetic field $\boldsymbol{B}$ and the velocity $\boldsymbol{v}$ to θ, can be written as

$$F = qvB \sin\theta. \qquad (20.20)$$

When the charged particle moves perpendicular to the magnetic field, the magnetic force is maximum, while the charged particle moves parallel to the magnetic field, no magnetic force acts on the particle.

The direction of the magnetic force acting on the positive charge moving through the magnetic field $\boldsymbol{B}$ is the direction for a right-handed screw to move when turning it from the direction of the motion of the charge (the direction of velocity $\boldsymbol{v}$) to the direction of the magnetic field $\boldsymbol{B}$ [Fig. 20-18 (a)]. In the case where a negative charged particle such as an electron moves in the magnetic field, the magnetic force acts in the opposite direction (in the direction of a right-handed screw moves when it is turned from $q\boldsymbol{v}$ to $\boldsymbol{B}$) [Fig. 10-18 (b)]. Making use of the notation of vector product, the magnetic force can be written as

$$\boldsymbol{F} = q\boldsymbol{v} \times \boldsymbol{B}. \qquad (20.20')$$

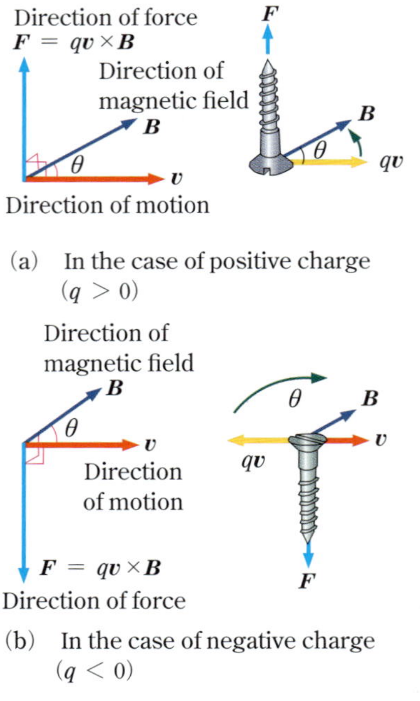

Fig. 20-18 Magnetic force $\boldsymbol{F} = q\boldsymbol{v} \times \boldsymbol{B}$ ($F = qvB\sin\theta$)

In addition, for the case of the electron where $q = -e$, equation (20.20′) becomes $\boldsymbol{F} = -e\boldsymbol{v} \times \boldsymbol{B}$.

Question 3 In a uniform magnetic field $\boldsymbol{B}$ directed from the frontside to the backside of the paper plane in Fig. 20-19, an electron moves upward with a velocity $\boldsymbol{v}$ along the paper plane. Which one, ①, ②, ③, ④, or ⑤, is the direction of the magnetic force acting on the electron?

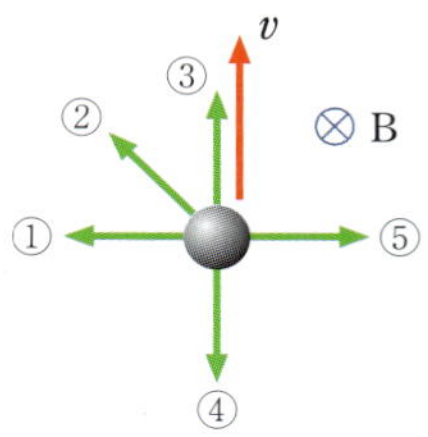

Fig. 20-19

Using the magnetic force of equation (20.20′) acting on a charged particle at $\boldsymbol{r}$ with a charge q and velocity $\boldsymbol{v}$, we define the magnetic field $\boldsymbol{B}$ (conventional name is magnetic flux density) at $\boldsymbol{r}$, which is denoted by $\boldsymbol{B}(\boldsymbol{r})$. From the equation (20.20), tesla T, the unit of magnetic field $\boldsymbol{B}$, is N/C (m/s) = N/(A·m). Namely, it turns out

$$\mathrm{T} = \mathrm{N/(A \cdot m)}. \tag{20.21}$$

Unit of magnetic field $\boldsymbol{B}$
T = N/(A·m)

In the case where an electric field $\boldsymbol{E}$ also exists in addition to the magnetic field $\boldsymbol{B}$, an electric force $q\boldsymbol{E}$ acts on the charged particle too. When a charged particle with a charge q is in motion with velocity $\boldsymbol{v}$ through the electric field $\boldsymbol{E}$ as well as the magnetic field $\boldsymbol{B}$, the force $\boldsymbol{F}$ due to the electric and magnetic fields acting on the particle is

$$\boldsymbol{F} = q\boldsymbol{E} + q\boldsymbol{v} \times \boldsymbol{B}. \tag{20.22}$$

The electromagnetic force acting on the charged particle is called the **Lorentz force** (the magnetic force only is also called the Lorentz force sometimes).

Question 4 In the apparatus shown in Fig. 20-20, an electron is in motion with a constant velocity $\boldsymbol{v}$. When the electric field $\boldsymbol{E}$ and magnetic field $\boldsymbol{B}$ were applied, the electron kept the straight motion as before. Find the speed of the electron. Assume that $\boldsymbol{E} \perp \boldsymbol{B}$, $\boldsymbol{E} \perp \boldsymbol{v}$, $\boldsymbol{B} \perp \boldsymbol{v}$.

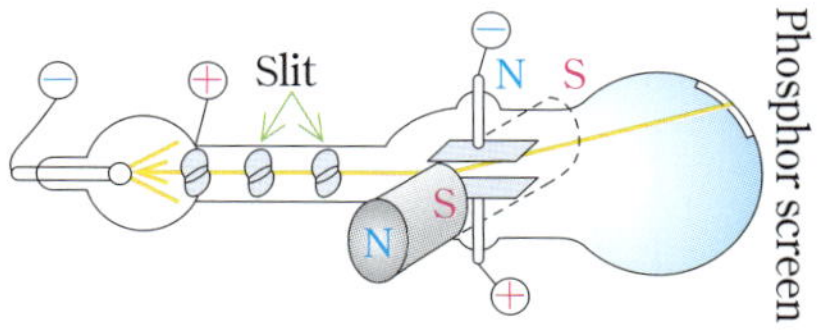

Fig. 20-20

Cyclotron motion The magnetic force acting on a charged particle moving through the uniform magnetic field does not do work, as it exerts the force in the direction perpendicular to the direction of motion. Thus, due to the magnetic force, the direction of the motion of the charged particle is changed but the speed is not changed. Therefore, for a charged particle moving in the uniform magnetic field and to the direction vertical to the magnetic field, the charged particle receives the magnetic force, constant in magnitude and perpendicular to the direction of motion. As a result, the charged particle performs a uniform circular motion (Fig. 20-21). For a radius r and a speed v of the uniform circular motion, the centripetal acceleration is $\dfrac{v^2}{r}$. On the other hand, the magnetic force exerted on the charged particle with a charge q and a mass m by the magnetic field $\boldsymbol{B}$ is qvB in magnitude. Hence, we have for the equation of motion :

$$\frac{mv^2}{r} = qvB. \tag{20.23}$$

Consequently, the speed v of the circular motion and the period $T = \dfrac{2\pi r}{v}$ are as follows :

$$v = \frac{qBr}{m}, \qquad T = \frac{2\pi m}{qB}. \tag{20.24}$$

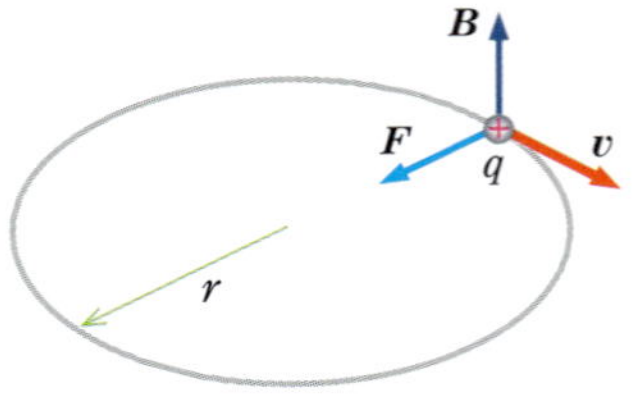

Fig. 20-21 Uniform circular motion in the uniform magnetic field (the case $q > 0$)

$$\frac{mv^2}{r} = F = qvB$$

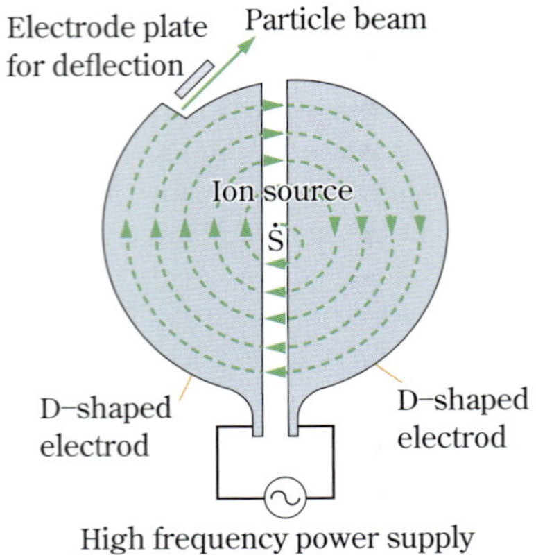

Fig. 20-22 Motion of the ion in the cyclotron. High-frequency electric field with cyclotron frequency is applied on the two D-shaped electrodes placed perpendicular to the uniform magnetic field. The ion emitted from the ion source S is accelerated by the electric field between the electrodes, and keeps accelerated every time it comes to the gap between the electrodes. So, the radius of the circular motion is increasing more and more. Finally, it is extracted from the deflecting electrode plane to outside as a beam.

Then, we find the number of rotations of the circular motion per unit time, $f = \frac{1}{T}$ is

$$f = \frac{qB}{2\pi m}. \tag{20.25}$$

Note that the number of rotations of the circular motion of the charged particle per unit time, f, is independent of the speed v and the radius r. The cyclotron, using this fact, accelerates the ions in the magnetic field with the alternating electric fields of the frequency $f = \frac{qB}{2\pi m}$, which is called the **cyclotron frequency** (Fig. 20-22).

Since the magnetic force does not act in the direction of magnetic field, the motion of the charged particle in the uniform magnetic field, in general, is a spiral motion which is a superposition of the uniform linear motion in the direction of the magnetic field and the uniform circular motion on a plane perpendicular to the magnetic field (Fig. 20-23).

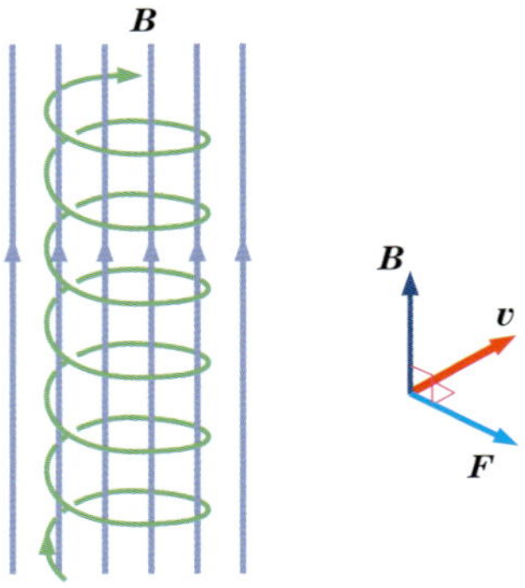

Fig. 20-23 Motion of a charged particle in a uniform magnetic field (for the case $q > 0$). The charged particle moves twining around the line of magnetic force.

Question 5 A particle of mass m, charge q, and velocity $\boldsymbol{v}$ entered the uniform magnetic field $\boldsymbol{B}$ and, after drawing a semicircle, collided with the wall at a distance d from the entrance (slit) (Fig. 20-24). If the particle of the same mass and the same speed, but a double charge enters from the entrance, where on the wall does the particle collide with?

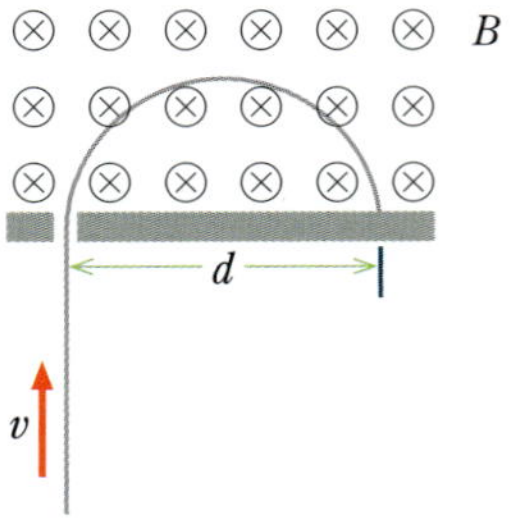

Fig. 20-24

Hall effect The magnetic field acts on a charged particle moving through the conductor with a force transverse to the direction of motion. Since this phenomenon was discovered by Hall in 1879, it is called the **Hall effect**. From the Hall effect, we identify the sign of the charge of the charged particle moving inside the conductor.

When an electric potential difference is applied both ends of the metal or semiconductor, the current flows from the high potential side to the low potential side. In the magnetic action of the current, however, it is unknown whether positively charged particle is moving toward the direction of the electric field $\boldsymbol{E}$ or negatively charged particle is moving toward the opposite direction of the electric field. On the other hand, if we apply the magnetic field $\boldsymbol{B}$ perpendicular to the electric field, then the magnetic force $\boldsymbol{F}$ is generated in the direction perpendicular to both electric $\boldsymbol{E}$ and magnetic $\boldsymbol{B}$ fields. Consequently, the moving direction of the current-carrying charged particle shifts to the transverse direction, and hence the charged particles accumulate on the side surface and as a result on the opposite side there appear an equal amount of charges with opposite sign due to the charge conservation law. The electric field $\boldsymbol{E}_{\mathrm{H}}$ in the transverse direction generated by the accumulated positive and negative charges is called the Hall electric field. Depending on the sign, plus or minus, of the current-carrying charge, the Hall electric field $\boldsymbol{E}_{\mathrm{H}}$ points to one direction or opposite direction (Fig. 20-25). Hence, by measuring the direction and the strength of the Hall electric field $\boldsymbol{E}_{\mathrm{H}}$, we can determine the sign of the charge and density of the current-carrying charged particle.

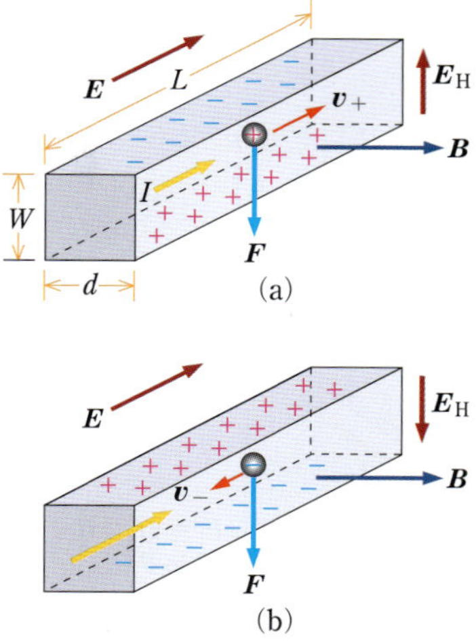

Fig. 20-25 Hall effect. (a) In the case where the charge of the current-carrying charged particle is positive, positive charges accumulate on the lower surface of the conductor. (b) In the case where the charge of the current-carrying charged particle is negative, negative charges accumulate on the lower surface of the conductor.

20.4 Force acting on the current

Learning objective To understand that the magnetic field exerts a magnetic force also on the electric current, and to become able to explain Fleming's left-hand rule concerning the direction of the magnetic force acting on the current. To understand the correspondence between the bar magnet and the current flowing in the coil concerning magnetic phenomena. To become able to explain the operating principle of motors.

As shown in Fig. 20-26, wire which is hung between the poles of a magnet perpendicularly to the magnetic field, and having the current flow, swings in the direction perpendicular to both the magnetic field and the current. Let us consider the case where the wire made from a substance except for the ferromagnet such as iron on which magnetic fields exert the magnetic force even in the absence of the current. The strength F of the magnetic force which acts on the current-carrying wire in the magnetic field $\boldsymbol{B}$ is proportional to the intensity I of the current, the strength B of the magnetic field, and the length L of the wire in the magnetic field. In the case of the wire kept perpendicular to the magnetic field $\boldsymbol{B}$, we find

$$F = IBL. \tag{20.26}$$

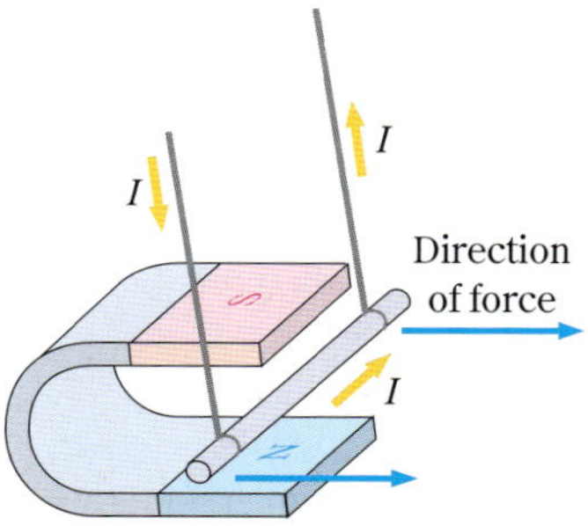

Fig. 20-26 A magnetic force acts on the current in the magnetic field

Performing the experiment with various directions of the wire, we find that the magnetic force acting on the wire is the strongest when the current is perpendicular to the magnetic field, and the magnetic force becomes zero when the current is parallel to the magnetic field. When the angle between the current and the magnetic field is θ, the magnitude F of the magnetic force acting on the wire of length L in the magnetic field $\boldsymbol{B}$ is given by

$$F = IBL \sin\theta \tag{20.27}$$

(Fig. 20-27).

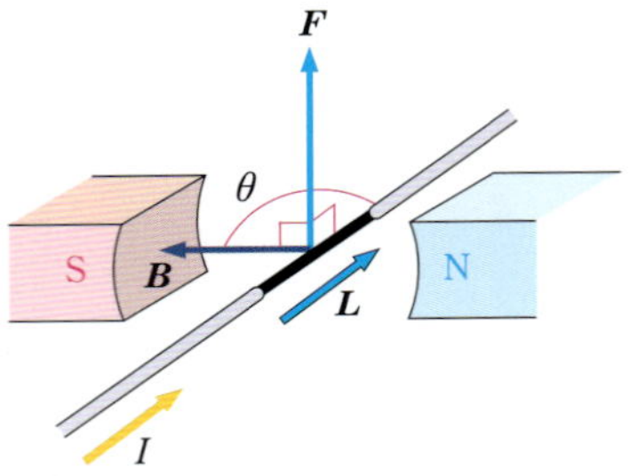

Fig. 20-27 $\boldsymbol{F} = I\boldsymbol{L}\times\boldsymbol{B}$ $(F = IBL\sin\theta)$

If you point the index (first) finger of your left hand in the direction of magnetic field $\boldsymbol{B}$, the middle (second) finger to the direction of the current I, and the thumb to the direction perpendicular to both the index and the middle fingers, then the direction of the force F acting on the current is the direction of the thumb (Fig. 20-28). This is called **Fleming's left-hand rule**. This rule may be memorized as left-hand's FBI rule.

Making use of the notation of vector products, these results can be expressed as (Fig. 20-29)

$$\boldsymbol{F} = I\boldsymbol{L}\times\boldsymbol{B}, \tag{20.27'}$$

where $I\boldsymbol{L}$ is a vector pointing to the direction of the current with the length IL.

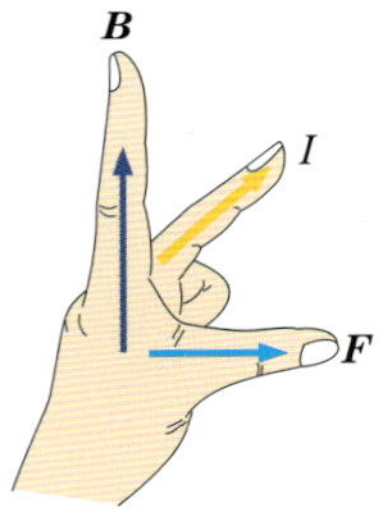

Fig. 20-28 Fleming's left-hand rule

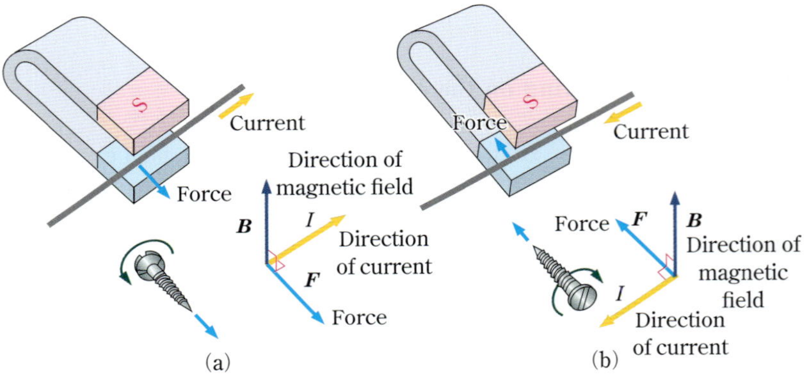

Fig. 20-29 The direction of the magnetic force acting on the current in the magnetic field

Case 3 When we make the current of 10 A flow in the copper wire kept perpendicular to the earth's magnetic field of 4.6×10^{-5} T, the strength F of the magnetic field acting on 1 m of the copper wire is

$$F = ILB = (10\ \text{A})\times(1\ \text{m})\times(4.6\times10^{-5}\ \text{T}\ (=\text{N/A}\cdot\text{m}))$$
$$= 4.6\times10^{-4}\ \text{N}.$$

If the area of the cross section of the copper wire is 1 mm^2 ($= 10^{-6}\ \text{m}^2$), then its mass is about 8 g and the gravity acting on it is 8×10^{-2} N.

In the current-carrying conducting wire, through the space of positively charged ions, the free electron with the negative charge $(-e)$ is moving in the opposite direction of the current. The conducting wire as a whole is not charged, and hence electric force acting on the wire is zero.

For reference Derivation of equation (20.27′) from equation (20.20′)

The magnetic force acting on the current carried by the conducting wire (20.27′) can be derived from the magnetic force acting on the charged particle (20.20′) as follows. Let the density of the free electron in the wire be n, and the area of cross section of the wire be A, then the number of the free electrons contained in the part of the wire of length L is nAL. Let the average velocity of the free electrons in the conducting wire placed in the magnetic field $\boldsymbol{B}$ be $\boldsymbol{v}$, then the free electron in the wire receives the following force on the average from the magnetic field $\boldsymbol{B}$

$$\boldsymbol{f} = -e\boldsymbol{v}\times\boldsymbol{B}. \tag{20.28}$$

The magnetic force $\boldsymbol{F}$ acting on the part of the wire of length L is the sum of the magnetic force $-e\boldsymbol{v}\times\boldsymbol{B}$ over the nAL free electrons, thus we have

$$\boldsymbol{F} = nAL(-e\boldsymbol{v}\times\boldsymbol{B}). \tag{20.29}$$

According to equation (19.4), we note $I = envA$, and thus denoting the vector pointing to the direction of the current (in the direction $-\boldsymbol{v}$) and with the length L by $\boldsymbol{L}$, equation (20.29) becomes

$$\boldsymbol{F} = I\boldsymbol{L}\times\boldsymbol{B}, \tag{20.30}$$

and hence equation (20.27′) has been derived. Thus it has turned out that the magnetic force acting on the free electron in the conducting wire is the total sum of the magnetic force acting on the free electrons in the wire. Note that this holds for the case where the wire is not made from the substance such as iron, which receives the magnetic force even in the absence of the current.

Magnetic force acting on the current-carrying coil in the magnetic field As shown in Fig. 20-30, we let the current I flow in the rectangular coil ABCD which can rotate around the axis OO′ perpendicular to the uniform magnetic field $\boldsymbol{B}$. According to Fleming's left-hand rule, the magnetic force acts on the AB part of the wire in the direction from top to bottom, and on the CD part from bottom to top on the paper plane. The two magnetic forces are the same in magnitude, IBa, and

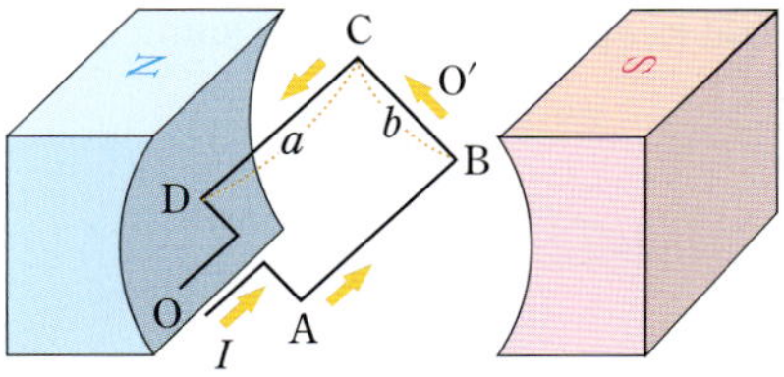

(a)

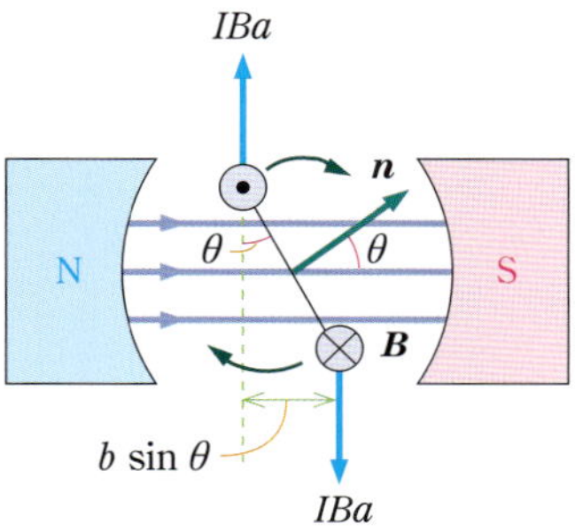

(b) ⊙ mark shows that current flows from the back of the paper to the front, and ⊗ shows that current flows from the front to the back of the paper.

Fig. 20-30 Magnetic force acting on the coil in the magnetic field

opposite in the direction. Since their lines of action are apart from each other at a distance $b\sin\theta$, which leads to the moment of force (torque) N that is a couple of forces with the magnitude $(IBa)\times(b\sin\theta) = IabB\sin\theta$, rotating the coil in the direction such that the plane of the coil becomes perpendicular to the magnetic field. Note that θ is the angle between the normal vector $\boldsymbol{n}$ of the coil plane and the magnetic field $\boldsymbol{B}$. As the area of the coil is $A = ab$, the moment of the magnetic force having action on the coil is expressed as

$$N = IAB\sin\theta. \qquad (20.31)$$

Using the notation of the vector product we can rewritten the above moment of force as

$$\boldsymbol{N} = IA\boldsymbol{n}\times\boldsymbol{B}. \qquad (20.31')$$

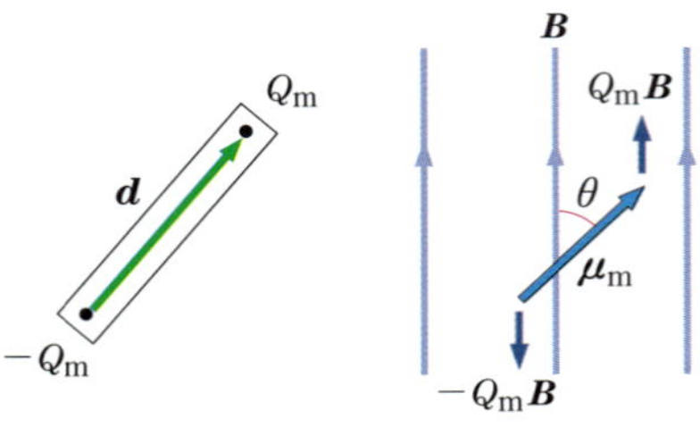

Fig. 20-31 Magnetic dipole. (a) Magnetic moment of magnet $\boldsymbol{\mu}_m = Q_m\boldsymbol{d}$. (b) Moment of magnetic force acting on the magnet $\boldsymbol{N} = \boldsymbol{\mu}_m \times\boldsymbol{B}$ ($N = \mu_m B\sin\theta$)

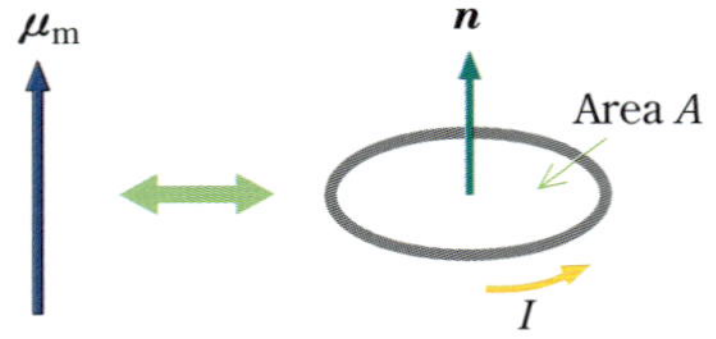

Fig. 20-32 $\boldsymbol{\mu}_m = AI\boldsymbol{n}$ (The length of the normal vector $\boldsymbol{n}$ is 1.)

Equation (20.31) holds for the case of the coil carrying the current I and surrounding an arbitrary plane figure of area A. The direction of the normal vector $\boldsymbol{n}$ is the direction in which the right-handed screw moves when the screw is turned along the direction of the current.

Let us consider the magnetic dipole with the magnetic charges of the magnetic poles, Q_m and $-Q_m$ having the length d as shown in Fig. 20-31. The magnetic moment is given by

$$\boldsymbol{\mu}_m = Q_m\boldsymbol{d}. \qquad (20.32)$$

In the magnetic field $\boldsymbol{B}$, the moment of magnetic force acting on this magnet is

$$\boldsymbol{N} = \boldsymbol{\mu}_m\times\boldsymbol{B} \qquad (N = \mu_m B\sin\theta) \text{ (in vacuum)}. \qquad (20.33)$$

Comparing the above equation with equation (20.31), it turns out that the circuit of the area A carrying the current flow I around the normal vector $\boldsymbol{n}$ is exerted the same couple of moment as the magnet having the magnetic moment, perpendicular to the coil plane, and given by

$$\boldsymbol{\mu}_m = AI\boldsymbol{n} \qquad (\mu_m = AI). \qquad (20.34)$$

Hence, we call $AI\boldsymbol{n}$ or AI **magnetic moment** of the current-carrying coil (Fig. 20-32) [see the paragraph of "electric dipole" on p. 232].

Direct-current motor As shown in Fig. 20-33, in order for the direction of the current flowing in the coil to change every time the coil rotates through 180 degrees, we attach the split ring commutator and make it touch with the brush. Then, the direction of the current in the coil is reversed every time the coil plane becomes perpendicular to the magnetic field. Thus, the coil keeps rotation in the same direction. This

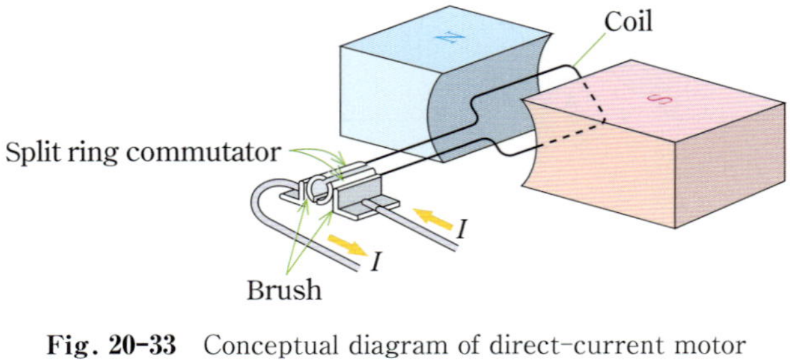

Fig. 20-33 Conceptual diagram of direct-current motor

is the principle of the direct-current motor.

In the case of a one-turn coil, the moment of the magnetic force making the coil rotate around the axis, $IAB \sin\theta$, is maximum (IAB) when the coil plane is parallel to the magnetic field, and is 0 when the coil plane is perpendicular to the magnetic field. Under such situation, the coil does not rotate in a uniform manner. By attaching the heavy flywheel, we can make the rotation of the coli uniform, but the actual motors, by combining many coils pointing to various directions, are devised to rotate with a constant angular frequency.

20.5 Force exerted between the currents

Learning objective To understand that the force is exerted between two currents by the mediation of magnetic fields.

As is the case where the force is exerted between two electric charges as well as between two magnetic poles, the force acts between the currents carried by the two conducting wires. This is because a current flowing through the wire generates a magnetic field around it and that magnetic field acts on the current flowing in another conducting wire.

As shown in Fig. 20-34, we make the two currents I_1, I_2 flow through two long wires a, b, respectively, by keeping the two wires straight and parallel. Assume that the two wires are apart at a distance d. The strength of the magnetic field $\boldsymbol{B}_1$ generated at the position of the wire b by the current I_1 flowing through the wire a, with the use of the equation (20.6), is written as

$$B_1 = \frac{\mu_0 I_1}{2\pi d}, \tag{20.35}$$

where the direction of $\boldsymbol{B}_1$ is perpendicular to the flow of the current I_2, and in the direction the right-handed screw rotates when it is moved to the direction of the current I_1. The magnitude $F_{2\leftarrow1}$ of the magnetic force $\boldsymbol{F}_{2\leftarrow1}$ exerted by this magnetic field $\boldsymbol{B}_1$ on the part of length L in the current I_2 is, noting that $F_{2\leftarrow1} = B_1 I_2 L$ from equation (20.26), written as

$$F_{2\leftarrow1} = B_1 I_2 L = \frac{\mu_0 I_1 I_2 L}{2\pi d}. \tag{20.36}$$

Since the direction of the force is to the direction of the right-handed screw proceeds when it is turned from the direction of the flow of the current I_2 to the direction of $\boldsymbol{B}_1$, the current I_2 flowing in the same direction of I_1 receives an attractive force $\boldsymbol{F}_{2\leftarrow1}$ directed to the current I_1 [Fig. 20-34 (a)]. Similarly, I_2 acts on I_1 with an attractive force $\boldsymbol{F}_{1\leftarrow2}$ directed to I_1. Therefore, an attractive force works for the two parallel currents flowing in the same direction. We also note that for the two parallel currents flowing in the opposite direction, a repulsive force works [Fig. 20-34 (b)].

To summarize what is mentioned above,

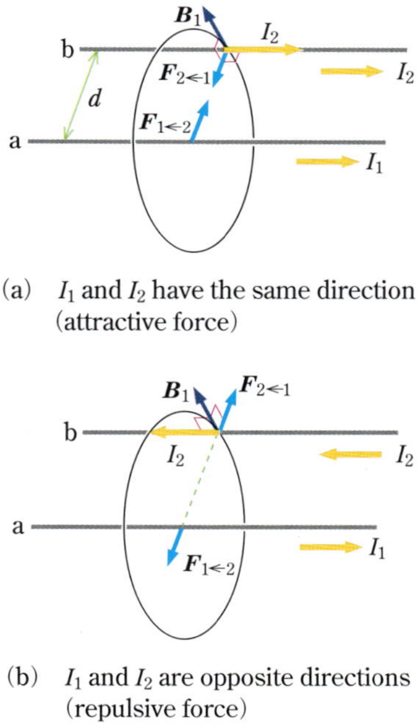

Fig. 20-34 Force acting between the parallel currents

The magnitude of the magnetic force exerted between the parallel direct-currents, is inversely proportional to the distance d between the wires and proportional to the product of the currents $I_1 I_2$. The strength F of the magnetic force, acting on a part of length L of the wire, is

$$F = \frac{\mu_0 I_1 I_2 L}{2\pi d} \qquad (\mu_0 = 4\pi\times10^{-7}\ \mathrm{N/A^2})^{*1} \qquad (20.37)$$

When the directions of the currents are the same the force is attractive, and when the directions are opposite, the force is repulsive.

*1 Since $\mathrm{T} = \mathrm{N/(A\cdot m)}$,
$\mu_0 = 4\pi\times10^{-7}\ \mathrm{T\cdot m/A}$
$= 4\pi\times10^{-7}\ \mathrm{N/A^2}$

The force exerted between the currents was discovered by Ampere in 1820.

Case 4 Suppose that the distance between the two parallel wires is 10 cm and the current of 100 A flows in each wire in the opposite direction. The strength F of the force acting on the 10 m of the wire, is

$$F = \frac{(4\pi\times10^{-7}\ \mathrm{N/A^2})\times(100\ \mathrm{A})\times(100\ \mathrm{A})\times(10\ \mathrm{m})}{2\pi\times(0.1\ \mathrm{m})} = 0.2\ \mathrm{N}$$

and is a repulsive force. The magnitude of this force is nearly equal to the gravity acting on an object of 20 g.

Case 5 When the current is made flow in a helical spring, the spring shrinks due to the attractive force between the currents in the neighboring coils flowing in the same direction.

Unit of the electromagnetism Making use of the law of the force exerted between the parallel direct-currents (20.37), we are able to know the intensity of the current by measuring the strength of the force acting between the parallel currents. In the International System of Units until 2018, 1 ampere (symbol A) used to be defined as the current such that, when it flows in the two parallel wires separated by 1 m from each other in a vacuum, there appears a force of the strength 2×10^{-7} N per 1 m between the parallel wires*2.

Unit of electric current A

*2 From 2018, the unit of the electric current, A, has been set by defining the elementary electric charge e exactly as $1.602\,176\,634\times10^{-19}$ C. Thus, the value of the magnetic constant μ_0 slightly deviates from $4\pi\times10^{-7}\ \mathrm{N/A^2}$.

If the 4 **base units,** in addition to the unit of length m, unit of mass kg, and unit of time s, the unit of electric current A, are once set up, then the units of other quantities concerning the electromagnetism are determined by the combination of these base units as derived units. This system of units is called **MKSA International System of Units**.

20.6 Magnetic field in the presence of magnetic bodies

Learning objective To understand the magnetization $\boldsymbol{M}$ which is a macroscopic physical quantity, the sum of magnetic moments of the atom per unit volume of matter. To understand the characteristics of three kinds of matter ; diamagnetic body, paramagnetic body, and ferromagnet.

To understand that the magnetic field $\boldsymbol{H}$ is the sum of $\boldsymbol{H}^{(c)}$, which is $\frac{1}{\mu_0}$ times the magnetic field $\boldsymbol{B}$ generated by the conduction current, and $\boldsymbol{H}^{(m)}$ computed from magnetic charges using Coulomb's law. To understand that when the long solenoid is filled up with a magnetic body of relative permeability μ_r, the strength of the magnetic field $\boldsymbol{B}$ becomes μ_r times that of the original solenoid.

Magnetization M When a solenoid carries the current it becomes an electromagnet. If the iron core is put into the solenoid, the strength of the electromagnet becomes much increased. This is because the iron core is magnetized and it changes to a magnet. All substances get magnetized (have magnetic properties) in the magnetic field, though there are differences in strength. A substance, when we focus on the magnetic property, is called a **magnetic body**.

All matter is a collection of atoms. Inside an atom, the electrons revolve around the nucleus and rotate on its own axis. In the microscopic world such as atoms, the Newtonian mechanics does not hold, but its world is governed by the quantum mechanics. So, it probably should be said that the electrons perform revolution-like motion and rotation-like motion. When the electron, a charged particle, moves in the atom, a **microscopic current flows** in the atom, as a result many atoms become minute magnets (magnetic dipoles) having tiny magnetic moments. Moreover, the electron due to the rotation-like motion called the **spin**, possesses its own proper magnetic moment whose magnitude is about

$$\mu_B = \frac{eh}{4\pi m} = 9.27\times10^{-24}\ \mathrm{A\cdot m^2}, \tag{20.38}$$

where m is the electron mass, h is the Planck constant.

Although the magnetic moment of each atom is extremely small, when the matter is put in the magnetic field, the directions of the magnetic moment of the atom become aligned in the same direction of the magnetic field (ferromagnet, paramagnet), or in the opposite direction of that of the magnetic field (diamagnet), and the body acquires a macroscopic magnetic moment and gets magnetized. The vector sum of the magnetic moments of the atoms per unit volume of the magnetic body is defined as magnetization M of the magnetic body. That is, let $\boldsymbol{\mu}_j$ be the magnetic moment of the j-th atom, then the magnetization $\boldsymbol{M}$, as the macroscopic field, is defined as (unit of magnetization is A/m)*

$$\boldsymbol{M} = \sum_{j\ (\text{unit volume})} \boldsymbol{\mu}_j. \tag{20.39}$$

Fig. 20-35 The coil part of the superconducting solenoid magnet of the ATLAS detector at the European Organization for Nuclear Research (CERN). By applying the magnetic field, we bend the trajectory of the charged particle.

Unit of magnetization A/m

* The unit of the magnetic moment is the unit of AI, $\mathrm{A\cdot m^2}$.

Equivalent magnet In Section **20.4** we have shown that the magnetic property of the current I flowing in the closed circuit of area A with the normal vector $\boldsymbol{n}$ of the circuit plane, is equivalent to the magnetic property of the magnetic dipole with the magnetic moment

$$\boldsymbol{\mu}_m = AI\boldsymbol{n}. \tag{20.40}$$

Therefore, as shown in Fig. 20-36, if the solenoid of the cross-section area A, length L, and number of turns per unit length n, carries the current I, the magnetic property of this solenoid of volume AL and number of turns nL is the same as that of cylindrical magnet of the same size of the solenoid with the magnetic moment $nLAI$. This can be seen from equation (20.40). Namely, it is equivalent to the magnet of the magnetic moment nI per unit volume.

The magnetization of the magnetic body $\boldsymbol{M}$ is the magnetic moment per unit volume. Hence, the cylindrical magnet of magnetization $\boldsymbol{M}$ and the solenoid of the same shape carrying the current per unit length

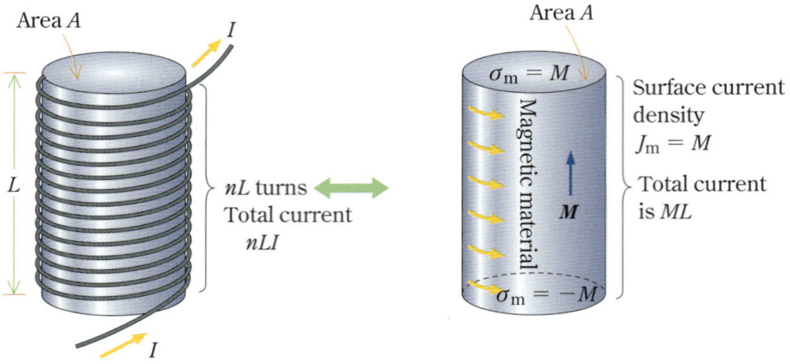

Fig. 20-36 Solenoid and cylindrical magnet ($M = nI$). The magnetic fields $\boldsymbol{B}$ they generate outside are the same.

$nI = |\boldsymbol{M}| = M$ generate the same magnetic field $\boldsymbol{B}$ in the external space. Namely, from the macroscopic point of view, the magnetic field $\boldsymbol{B}$ generated by the microscopic current of the magnetic body with the magnetization $\boldsymbol{M}$, is the same as the magnetic field $\boldsymbol{B}$ produced by the surface current flowing around the side surface of the magnetic body with the surface current density,

$$J_m = M \tag{20.41}$$

as is derived following Biot-Savart's law. This equivalent macroscopic surface current is called the **magnetization current**.

In the same way as the polarization charge per unit area $\sigma_P = \pm P$ appears on the surface of the dielectric substance with polarization $\boldsymbol{P}$, on the surface of the magnetic body with magnetization $\boldsymbol{M}$, there appears the magnetic charge of surface density given by

$$\sigma_m = \pm M. \tag{20.42}$$

Fig. 20-37 The aurora photographed from the International Space Station

Magnetic field $\boldsymbol{H}$ In the electromagnetism the object of which is the macroscopic world, the microscopic magnetic field $\boldsymbol{B}$ heavily changing at the atomic scale is treated as the macroscopic magnetic field $\boldsymbol{B}$ averaged in the region which is as small as invisible for human eyes but large enough compared to the atom. The magnetic field $\boldsymbol{B}$ is the magnetic field generated by all the currents involved. For the electric current I that creates the macroscopic static magnetic field $\boldsymbol{B}$, there exist two kinds of currents ; a conduction current I_0, which is a flow of the free electrons through the discharge tube or the conductor and is measurable by the macroscopic detector of currents, and a magnetization current I' which is the microscopic current in the magnetic body viewed in a macroscopic scale ($I = I_0 + I'$). Therefore, the static magnetic field $\boldsymbol{B}$ is the sum of the magnetic field $\boldsymbol{B}^{(c)}$ produced by the conduction current I_0 and the magnetic field $\boldsymbol{B}^{(m)}$ produced by the magnetization current I', that is, $\boldsymbol{B} = \boldsymbol{B}^{(c)} + \boldsymbol{B}^{(m)}$. Here we note that both magnetic fields $\boldsymbol{B}^{(c)}$ and $\boldsymbol{B}^{(m)}$ are subject to Biot-Savart's law.

The magnetic field $\boldsymbol{B}^{(m)}$ generated by the magnetization current can be proved to be μ_0 times the sum of the magnetic field produced by the magnetic charge Q_m appearing on the surface of the magnetic body,

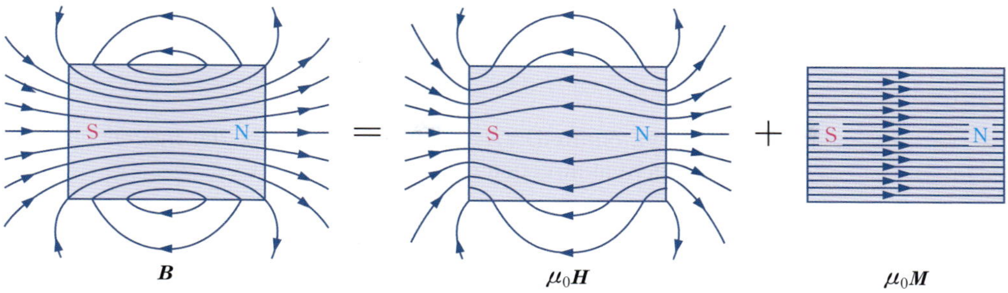

Fig. 20-38 $\boldsymbol{B}$, $\mu_0\boldsymbol{H}$ and $\mu_0\boldsymbol{M}$ generated by the bar of the ferromagnet uniformly magnetized in the direction of the axis.

according to Coulomb' law for the magnetic field, given by

$$\boldsymbol{H}^{(\mathrm{m})}(\boldsymbol{r}) = \frac{Q_{\mathrm{m}}}{4\pi r^2}\frac{\boldsymbol{r}}{r} \quad \text{(in the case } Q_{\mathrm{m}} \text{ at the origin)} \tag{20.43}$$

and the magnetization $\boldsymbol{M}$. That is, we find

$$\boldsymbol{B}^{(\mathrm{m})} = \mu_0(\boldsymbol{H}^{(\mathrm{m})}+\boldsymbol{M}). \tag{20.44}$$

Although we do not give the proof here, you will understand the above relation by looking at Fig. 20-38, which shows the magnetic field $\boldsymbol{B}$ produced by the bar magnet uniformly magnetized in the direction of the axis, and $\mu_0\boldsymbol{H}$ as well as $\mu_0\boldsymbol{M}$. The lines of magnetic force representing the magnetic field $\boldsymbol{H}^{(\mathrm{m})}$ possesses the starting point at the N-pole and the end point at the S-pole, while the lines representing the magnetization $\boldsymbol{M}$ starts from the S-pole and ends at the N-pole.

Now we define the magnetic field $\boldsymbol{H}^{(\mathrm{c})}$ due to the conduction current as

$$\boldsymbol{B}^{(\mathrm{c})} = \mu_0\boldsymbol{H}^{(\mathrm{c})}, \tag{20.45}$$

and we also define the macroscopic magnetic field $\boldsymbol{H}$, which is called, for the historical reason, as "strength of the magnetic field" given by

$$\boldsymbol{H} = \boldsymbol{H}^{(\mathrm{c})}+\boldsymbol{H}^{(\mathrm{m})}. \tag{20.46}$$

The unit of the magnetic field $\boldsymbol{H}$ is the same as the unit of magnetization $\boldsymbol{M}$, and is A/m.

Unit of magnetic field $\boldsymbol{H}$ A/m

In this way, the magnetic field $\boldsymbol{B} = \boldsymbol{B}^{(\mathrm{c})}+\boldsymbol{B}^{(\mathrm{m})} = \mu_0(\boldsymbol{H}^{(\mathrm{c})}+\boldsymbol{H}^{(\mathrm{m})}+\boldsymbol{M})$ can be rewritten as the sum of magnetic field $\boldsymbol{H}$ and the magnetization $\boldsymbol{M}$, that is

$$\boldsymbol{B} = \mu_0(\boldsymbol{H}+\boldsymbol{M}). \tag{20.47}$$

The magnetic field $\boldsymbol{H}$ is the sum of the magnetic field $\boldsymbol{H}^{(\mathrm{c})}$ which is calculated from the conduction current using Biot-Savart's law with $\mu_0 = 1$ and $\boldsymbol{H}^{(\mathrm{m})}$ that is computed from the magnetic charge using Coulomb's law (20.43).

Since $\boldsymbol{M} = \boldsymbol{0}$ outside of the magnet, we note that $\boldsymbol{B} = \mu_0\boldsymbol{H}$, and the $\boldsymbol{B}$ and $\boldsymbol{H}$ are the same except for the proportional constant μ_0.

Ampere's law for the magnetic field $\boldsymbol{H}$ The current I on the right-hand side of Ampere's law (20.15) is the sum of the conduction current I_0 and the magnetization current I', that is, $I = I_0+I'$, while $\boldsymbol{B}$ on the left-hand side is $\mu_0(\boldsymbol{H}+\boldsymbol{M})$. Since it can be proved that the contribution from $\boldsymbol{M}$ on the left-hand side is equal to I' on the right-hand side, we can derive Ampere's law for the magnetic field $\boldsymbol{H}$:

$$\oint_{\mathrm{C}} H_{\mathrm{t}}\,\mathrm{d}s = I_0\,, \tag{20.48}$$

(Proof omitted). H_{t} is the component of the magnetic field $\boldsymbol{H}$ in the tangential direction of the closed curve C.

Coulomb's law for the magnetic force The magnetic field $\boldsymbol{B}$ acts on the magnetic charge Q_{m} in a vacuum with a magnetic force $\boldsymbol{F} = Q_{\mathrm{m}}\boldsymbol{B} = \mu_0 Q_{\mathrm{m}}\boldsymbol{H}$ (Section **20.1**). Therefore, using the equation (20.43) we can derive **Coulomb's law for the magnetic force** where two magnetic charges Q_{m} and Q_{m}' separated at distance r in vacuum are subjected to a force given as

$$F = \frac{\mu_0 Q_{\mathrm{m}} Q_{\mathrm{m}}'}{4\pi r^2} \quad \text{(in vacuum)}. \tag{20.49}$$

Magnetic susceptibility In most of the magnetic bodies, except for the permanent magnet which is magnetized in the absence of applied magnetic fields, the magnetization $\boldsymbol{M}$ is in the same or opposite direction of the magnetic field which generates the magnetization, and the magnitude is proportional to the strength of the magnetic field causing the magnetization. Thus, we denote the relation of proportionality between $\boldsymbol{M}$ and $\boldsymbol{H}$ as

$$\boldsymbol{M} = \chi_{\mathrm{m}}\boldsymbol{H} \tag{20.50}$$

and the proportional constant χ_{m}, which is determined by the substance, is called the **magnetic susceptibility***. The magnetic susceptibility χ_{m} is a dimensionless constant. By substituting equation (20.50) into equation (20.47) we get

* Proportionality relation between $\boldsymbol{M}$ and $\boldsymbol{B}$ can be expressed $\boldsymbol{M} = \dfrac{\chi_{\mathrm{m}}}{\mu_0}\boldsymbol{B}$. This χ_{m} is sometimes called the magnetic susceptibility.

$$\boldsymbol{B} = \mu_0(1+\chi_{\mathrm{m}})\boldsymbol{H}. \tag{20.51}$$

If we define the **relative permeability** μ_{r} as

$$1+\chi_{\mathrm{m}} = \mu_{\mathrm{r}}\,, \tag{20.52}$$

equation (20.51) can be rewritten as

$$\boldsymbol{B} = \mu_{\mathrm{r}}\mu_0\boldsymbol{H}. \tag{20.53}$$

We call $\mu = \mu_{\mathrm{r}}\mu_0$ the **magnetic permeability**.

As in the case where a long solenoid is filled up inside with a magnetic body, if the magnetic charges appearing at both ends are located far away, so that we can ignore $\boldsymbol{H}^{(\mathrm{m})}$ and determine the magnetic field $\boldsymbol{H}^{(\mathrm{c})}(=\boldsymbol{H})$ from the current carried by the coil, the magnetic field in the substance $\boldsymbol{B}$ can be found using $\boldsymbol{B} = \mu_{\mathrm{r}}\mu_0\boldsymbol{H}$.

Diamagnetic body In the glass, antimony, bismuth, gold, silver, copper, ordinary organic matter, most salts, water, many gasses except for oxygen are, when applied magnetic fields, the magnetization $\boldsymbol{M}$ induced by electromagnetic induction we learn in the next chapter points to the direction opposite to the magnetic field and the magnetization is weak. This kind of matter with $\chi_{\mathrm{m}} < 0$ and $1 \gg |\chi_{\mathrm{m}}| > 0$ is called

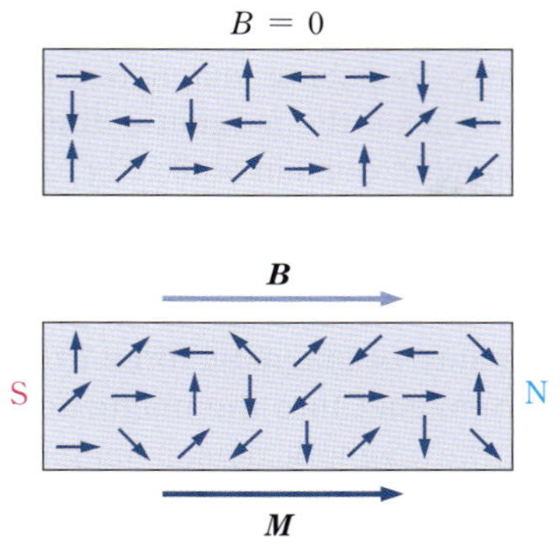

Fig. 20-39 Magnetization of paramagnetic body
(a) The molecules of the paramagnetic body in the absence of applied magnetic fields are tiny magnetic dipoles, but the magnetic dipoles point to various directions due to thermal motion.
(b) When the magnetic field is applied, part of the magnetic dipoles point to the magnetic field.

a **diamagnetic body**. At room temperature, χ_m of the solid diamagnetic body is around -10^{-5} (for the case of copper -9.7×10^{-6}).

Paramagnetic body In the elements such as platinum, aluminum, chrome, manganese, transition metals and their compounds, gasses like oxygen and nitrogen monoxide, the magnetization $\boldsymbol{M}$ induced in the magnetic field points to the direction of the magnetic field and the magnitude of the magnetization $\boldsymbol{M}$ is small. This kind of matter with $\chi_m > 0$ and $1 \gg \chi_m > 0$ is called a **paramagnetic body**. At room temperature, χ_m of the solid paramagnetic body is in the range $10^{-5}-10^{-2}$ (in the case of aluminum 2.1×10^{-5}, manganese 8.3×10^{-4}).

Question 6 As shown in Fig. 20-40, a small test body, which is not magnetized, is hung on the symmetrical axis of the non-uniform but strong magnetic field and then the electric current is made flow through the electromagnet. Show that the substance attracted with a weak force to the region of strong magnetic field [Fig (a)] is the paramagnetic body, while the substance repelled by a weak repulsive force from the region of strong magnetic field [Fig (b)] is the diamagnetic body.

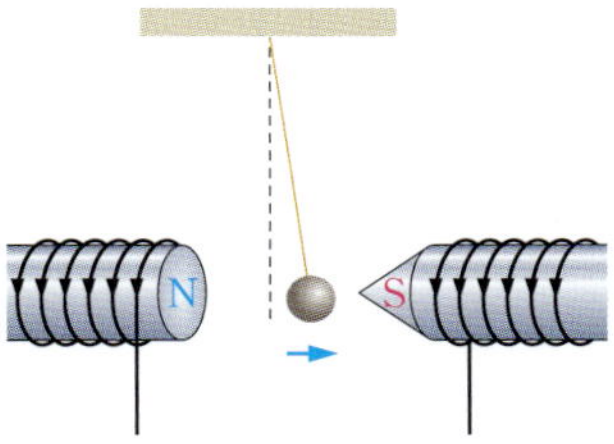

(a) Ball of paramagnetic body

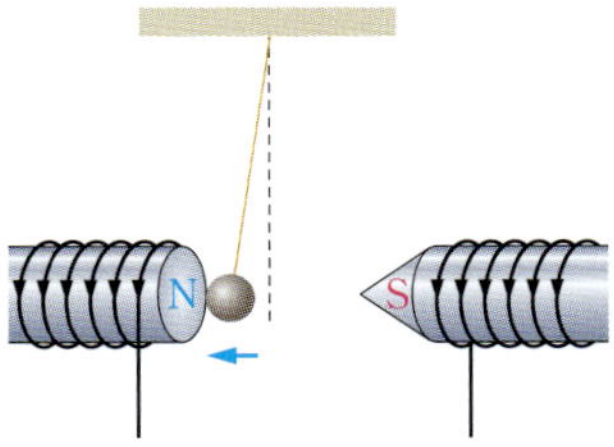

(b) Ball of diamagnetic body

Fig. 20-40 Balls of paramagnetic and diamagnetic bodies hung in the strong but not uniform magnetic field. The magnetic field is stronger near the S-pole with a sharp end than near the N-pole.

Ferromagnet Iron, cobalt, nickel, and alloys and compounds including these metals as main components as well as compounds of rare earth elements and so on, become permanent magnets or have a big magnetization in the magnetic field. Because of these properties they are called **ferromagnets**.

The characteristic of ferromagnets such as iron which becomes permanent magnet is that, even in the absence of the external magnetic field, the magnetic moments of the atom spontaneously point to the same direction due to the so-called exchange force between the atoms originating from the quantum mechanical effects. If we place an actual ferromagnet at the location without magnetic fields, it is not in such a state, but it is divided into a number of differently directed ferromagnetic domains. In each domain, the magnetic moment points

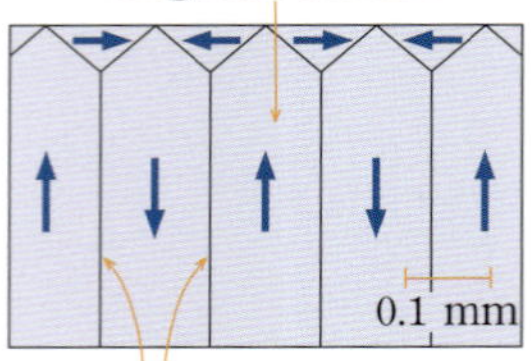

Fig. 20-41 Magnetic domain. Arrows show the directions of magnetization.

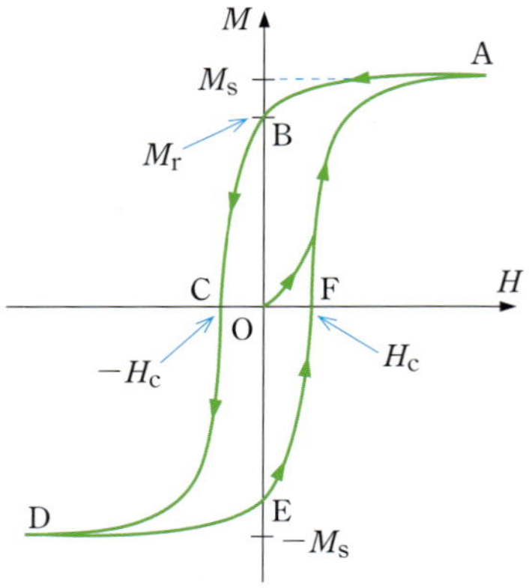

Fig. 20-42 Magnetization curve of ferromagnet

to a certain direction, but it is quite often that the ferromagnet, as a whole, possesses nearly vanishing magnetization. This small zone (about 10^{-7}–10^{-2} cm) is called the **magnetic domain**, the boundary of which is called the magnetic domain wall (Fig. 20-41). The cause for the magnetic domain to be created is to lower the internal energy of the ferromagnet. Although the internal energy is raised at the domain wall, the internal energy, as a whole, decreases by making the magnetic domain to prevent the lines of magnetic force from going outside.

When we put a ferromagnet into the magnetic field generated by a long air-core solenoid, the magnetic domain magnetized in the direction of the magnetic field grows and the direction of the magnetization in the magnetic domain rotates pointing to the direction of the magnetic field.

In this way, the magnetization M of the ferromagnet increases with the strength of the magnetic field H (intensity of the current flowing through the solenoid) (path O $\rightarrow$ A in Fig. 20-42). However, because even in the case where all the magnetic moments of the molecules point to the magnetic field only the finite magnetization M_s is obtained, the magnetization hardly increases at extremely high H (point A in Fig. 20-42). This is called a saturated state of magnetization (Actually, in the case of the saturated iron, on the average, 2.2 electrons per one atom direct the proper magnetic moment to the magnetic field).

After the saturated state of the magnetization, when the magnetic field is weakened, the magnetization M decreases along the path A $\rightarrow$ B. Even when the magnetic field H becomes 0, the magnetization does not vanish but remains a finite value of magnetization M_r ($M_r = \overline{OB}$). This is called the **residual magnetization**. Permanent magnets make use of this residual magnetization. When the direction of the magnetic field (direction of the current) is reversed ($H < 0$), the magnetization still remains in the region of path B $\rightarrow$ C in Fig. 20-42. When $H = -H_c = -\overline{OC}$, we first arrive at $M = 0$. This H_c is called the **coercive force**. Increasing the magnetic field in the opposite direction, the magnetization of opposite direction is saturated at D. The curve representing the relation between the magnetization M and magnetic field H is called the **magnetization curve** or the **hysteresis loop**.

As described above, the phenomenon that the strength of the magnetization M is related to the history of the past magnetization is called the **magnetic hysteresis**. In the transformer, the iron core is magnetized periodically, the area surrounded by the magnetization curve multiplied by μ_0 ($\mu_0 \oint H\,dM$) is lost as a heat by the iron core per unit volume every one cycle. This is called the **hysteresis loss**. The material of small hysteresis loss has been developed and used for the iron core of the transformer.

The magnitude M_s of the magnetization at a saturated state depends upon the temperature. With the increase in temperature, M_s decreases, and at a certain temperature called the **Curie temperature** (constant determined by the material) becomes $M_s = 0$. At temperatures above the Curie temperature, the ferromagnetism disappears, and the magnetic body shows paramagnetism. This is because when the temperature increases the thermal motion becomes intense and the thermal motion overcomes the exchange force. The Curie temperature of the iron is 1043 K.

The ferromagnet is roughly classified into two groups ; (1) The ferromagnet which does not get magnetized easily, when the magnetic field applied, and the residual magnetization of which is large when the magnetic field is removed. (2) The ferromagnet which shows very large magnetization even when the tiny magnetic field is applied, and possesses small residual magnetization as well as small coercive force. The former is called the hard magnetic material, and the latter is called the soft magnetic material. The hard magnetic material is suitable for making the permanent magnet. While, the soft magnetic material is used for the transformer or iron core of the choke coil.

The residual magnetization of the ferromagnet can be regarded as if this material keeps the memory at the time when it was exposed to the magnetic field. This magnetic memory is applied for the magnetic tape and magnetic disc. For this purpose, the material which has a certain amount of coercive force and a magnetization curve of an angulated shape, is desirable. This is called the semi-hard magnetic material or recording material.

Fig. 20-43 Magnetic tape for the data storage of the computer

Case 6 | The case where the current I is made flow through an infinitely long solenoid inside of which is filled up with a substance of relative permeability μ_r In the case of the infinitely long solenoid, the magnetic field $\boldsymbol{H}$ produced by the polarized magnetic charges can be ignored. Hence the strength of the magnetic field $\boldsymbol{H}$ is equal to that generated by the conduction current I flowing in the coil according to Biot-Savart's law :

$$H = nI \quad \text{(inside the solenoid)}, \tag{20.54a}$$

$$H = 0 \quad \text{(outside the solenoid)}. \tag{20.54b}$$

Therefore, using equation (20.53) we have

$$B = \mu_r\mu_0 H = \mu_r\mu_0 nI \quad \text{(inside the solenoid)}, \tag{20.55a}$$

$$B = 0 \quad \text{(outside the solenoid)}. \tag{20.55b}$$

Namely, the magnetic field $\boldsymbol{B}$ is μ_r times the magnetic field of the solenoid with a vacuum core. The directions of the magnetic fields $\boldsymbol{H}$ and $\boldsymbol{B}$ are obtained from the direction of the current and the right-handed screw rule.

Case 7 | The case where the current I is made flow through a toroid (N turns) inside of which is filled up with a substance of relative permeability μ_r (Fig. 20-44) We apply Ampere's law to the magnetic field $\boldsymbol{H}$ for the closed curve C which is the circle of radius R with the center at the central axis of the toroid. Since $H_t = H$ from the symmetry, equation (20.48) becomes inside the toroid, $2\pi RH = NI$, and outside the toroid, $2\pi RH = 0$. Then we have

$$H = \frac{NI}{2\pi R}, \quad B = \mu_r\mu_0 H = \frac{\mu_r\mu_0 NI}{2\pi R} \quad \text{(inside the toroid)} \tag{20.56a}$$

$$H = 0, \quad B = 0 \quad \text{(outside the toroid)} \tag{20.56b}$$

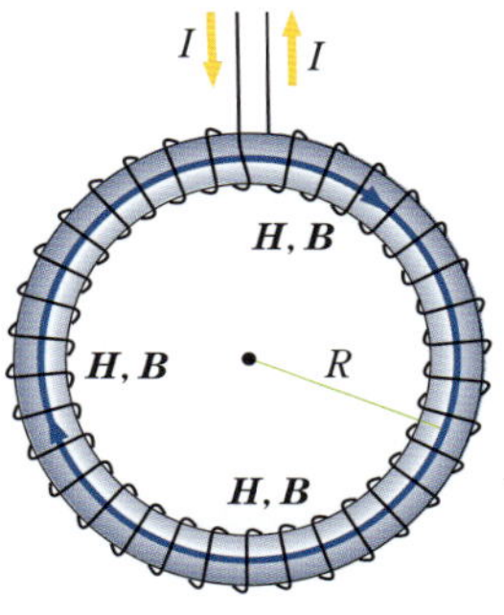

Fig. 20-44 Toroid

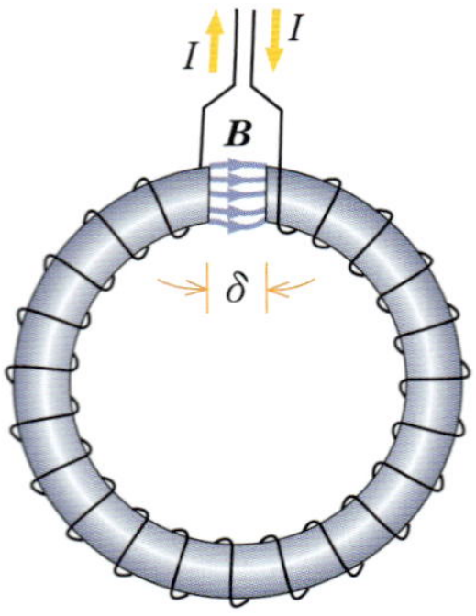

Fig. 20-45 C-shaped electromagnet

Case 8 | C-shaped electromagnet In a toroid filled up inside with soft iron of the relative permeability μ_r, we make a short gap of length δ in the iron core (Fig. 20-45). Since the line of magnetic force of the magnetic field $\boldsymbol{B}$ possesses neither the starting point nor the end point, even though we have a small gap in the iron core, the line of magnetic force is mostly a circle, and the strength B of the magnetic field $\boldsymbol{B}$ is approximately constant inside the iron core as well as in the gap. The strength of the magnetic field $\boldsymbol{H}$ is at the gap $H_0 \approx \dfrac{B}{\mu_0}$, inside the iron core $H_1 \approx \dfrac{B}{\mu_r \mu_0}$. Along the central axis of the iron core of length $L = 2\pi R$, we apply Ampere's law (20.48), and find

$$NI = H_0\delta + H_1 L \approx B\left(\frac{\delta}{\mu_0} + \frac{L}{\mu_r \mu_0}\right). \tag{20.57}$$

Thus, we have

$$B \approx \frac{\mu_r \mu_0 NI}{L + \delta\mu_r}. \tag{20.58}$$

In the case where the gap δ is narrow, $\delta \ll \dfrac{L}{\mu_r}$, the strength of the magnetic field $\boldsymbol{B}$ (20.58) is μ_r times the magnetic field $\dfrac{\mu_0 NI}{L}$ of the air core toroid of length L with the same number N of turns, and so the strength can be raised to thousands times the air core case. This is the principle of electromagnet.

Problems for exercise 20

A

1. At which of the two points, A or B in Fig. 1, is the magnetic field stronger? Judge from the appearance of lines of magnetic force. What kind of force acts on the magnetic compass needle placed between the magnetic poles?

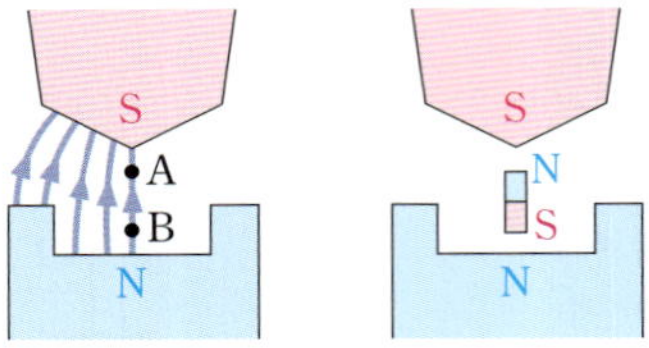

Fig. 1

2. Suppose the Earth is a big magnet. Which pole of the Earth, the north pole or the south pole, does the N-pole of this big magnet correspond to? If we regard the Earth magnetism as due to the circular current flowing inside the Earth, on what plane and in which direction does the circular current flow?
3. The current of 10 A is flowing in an infinitely long conducting wire. Find the strength of the magnetic field B at the point with a distance 1 cm from the wire.
4. One of the two parallel straight wires separated by 10 cm, carries the current of 4 A, while another wire carries the current of 6 A, flowing in the opposite direction. Find the strength of the magnetic field B in the middle point between the two wires.
5. Find the strength of the magnetic force due to the Earth's magnetic field ($B \sim 4.5 \times 10^{-5}$ T) acting on the electron (charge $-e = -1.6 \times 10^{-19}$ C, mass $m = 9.1 \times 10^{-31}$ kg) flying from outer space with a speed of 1/10 of the speed of light ($v \sim 3 \times 10^7$ m/s) to Japanese sky. Find also the gravity of the Earth acting on the electron. Calculate the radius r of the spiral motion this electron performs around the line of magnetic force originating from the geomagnetism.
6. Examining the trajectory of the electron moving in the substance in the presence of the uniform magnetic field and in the absence of the electric field, we get the curve given in Fig. 2 (The electron reduces the speed in the substance).

(1) Which is the direction of electron's motion, A → B or B → A?

(2) Which is the direction of the magnetic field, front side → back side or back side → front side of the paper plane?

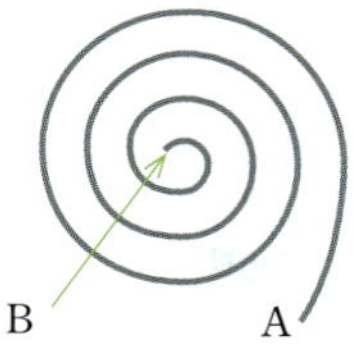

Fig. 2

7. As shown in Fig. 3 (see also Fig. 20.30), the current is flowing between the magnetic poles of the two magnets. Draw the direction of the force acting on the current and that of the force acting on the magnetic poles.

Fig. 3

8. When the current of 20 A is made flow in the wire perpendicular to the magnetic field of 3×10^{-5} T, find the magnetic force acting on 1 m of the wire.

9. For the Earth's magnetic field at Tochigi prefecture, the angle between the magnetic field and vertically downward direction is 40.5° (inclination = 49.5°), and its horizontal component is 3.0×10^{-5} T (The magnitude of the magnetic field is 4.61×10^{-5} T). When the current of 10 A flows through the horizontal wire which points to the south-north direction, find the magnetic force acting on the length 2 m of the wire.

10. What sort of the magnetic force acts between the circular current shown in Fig. 4 and the linear current passing through the center of the circle and perpendicular to the circle?

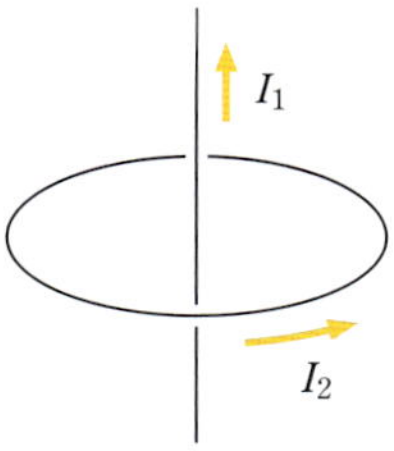

Fig. 4

11. The circular wire of one turn with the radius 10 cm is connected to the battery of 6 V with the resistance 100 Ω. Find the strength of the magnetic field B at the center of the circle.

12. The solenoid, which is made by winding the cylinder of length 30 cm with the conducting wire 1200 turns, carries the current of 1 A. What is the strength of magnetic field B inside the solenoid?

13. The conducting wire of 1 km is uniformly wound around the tube of length 1 m and circumference 0.2 m. Calculate the necessary current to have the magnetic field of 0.1 T at the central part inside the tube.

14. The solenoid with the number of turns per 1 m is 4000, possesses an iron core with relative permeability $\mu_r = 1000$. When we flow the current of 1 A through the solenoid, find the strength of the magnetic field $\boldsymbol{H}$ and that of the magnetic field $\boldsymbol{B}$.

B

1. We want to construct a cyclotron accelerating the proton to get a kinetic energy of 2 MeV. Assume that the strength of the magnetic field is 0.3 T. The mass of the proton is 1.67×10^{-27} kg, and 1 MeV = 1.6×10^{-13} J.

(1) How much should be the minimum value of the radius of the magnetic pole?

(2) How much should be the frequency of the AC power source for acceleration?

2. Find the magnetic field $\boldsymbol{B}$ at the center c of the circle (a) with radius R and also that of the semicircle (b) in Fig. 5. Is there any contribution from the current flowing through the linear part?

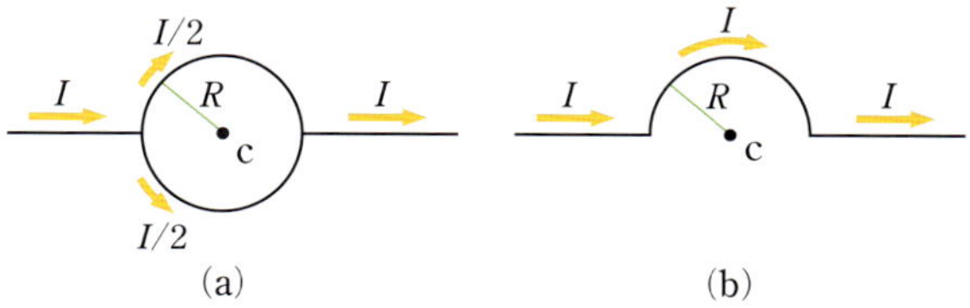

Fig. 5

3. Show that the strength of the magnetic field B generated by an infinitely long linear current is given by equation (20.6) with the use of Biot-Savart's law.

4. Let us consider a long straight coaxial cable with the cross section shown in Fig. 6. The two currents I, $-I$ flow through the two conductors with the same magnitude but in the opposite directions. How does the strength of the magnetic field $\boldsymbol{B}$ change with the distance r from the center?

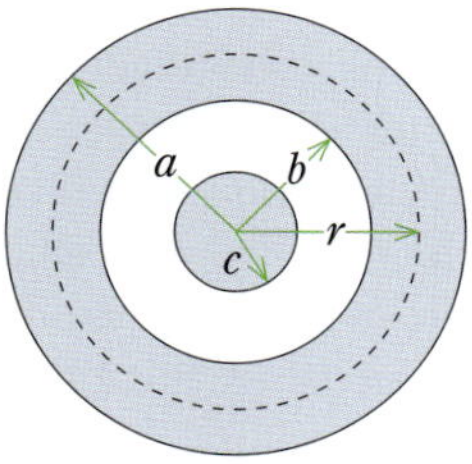

Fig. 6

5. The current flowing in the conducting wire generates the magnetic field. The magnetic field generated by the current flowing in the whole circuit can be observed, but the magnetic field generated only by its minute part cannot be directly observed. Biot and Savart was successful in observing it by an ingenious idea.

Show that the difference of the magnetic fields at the point P produced by the current I flowing in the wire ABC in Fig. 7 and that produced by the current

I flowing in the wire A′B′B″C′ is equal to the magnetic field produced at P by the current I flowing in a part of the wire B′B″.

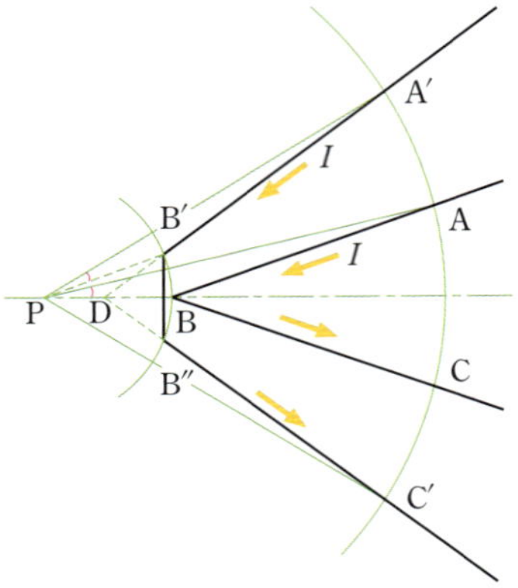

Fig. 7

6. **Magnetic field generated by moving charged particle** ▌ If the velocity $\boldsymbol{v}$ of an object with the charge q is small enough compared to the speed of light, then the magnetic field $\boldsymbol{B}$ generated by this charged object at the position vector $\boldsymbol{r}$ from the charge q is

$$\boldsymbol{B} = \frac{\mu_0 q \boldsymbol{v} \times \boldsymbol{r}}{4\pi r^2}. \quad (1)$$

Of course, this charged object produces the electric field as well.

(1) Compare equation (1) with Biot-Savart's law.

(2) Show that the charged particle 1 with charge q_1, velocity $\boldsymbol{v}_1$ and the charged particle 2 with charge q_2, velocity $\boldsymbol{v}_2$ interact with each other by the following magnetic forces mediated by the magnetic field (1)

$$\boldsymbol{F}_{1\leftarrow 2} = \frac{\mu_0 q_1 q_2}{4\pi} \frac{[\boldsymbol{v}_1 \times \{\boldsymbol{v}_2 \times (\boldsymbol{r}_1 - \boldsymbol{r}_2)\}]}{|\boldsymbol{r}_1 - \boldsymbol{r}_2|^3}$$

$$\boldsymbol{F}_{2\leftarrow 1} = \frac{\mu_0 q_1 q_2}{4\pi} \frac{[\boldsymbol{v}_2 \times \{\boldsymbol{v}_1 \times (\boldsymbol{r}_2 - \boldsymbol{r}_1)\}]}{|\boldsymbol{r}_1 - \boldsymbol{r}_2|^3}$$

Show also that the $\boldsymbol{F}_{2\leftarrow 1}$ and $\boldsymbol{F}_{1\leftarrow 2}$ do not obey the law of 1action and reaction.

Synchrotron radiation

An electric charge at rest produces only the time-independent static electric field in its surroundings. Thus, the electric charge at rest does not emit an electromagnetic wave. The charge in a uniform motion with a constant velocity generates the electric filed as well as the magnetic field around it. However, for an observer moving with the same velocity as this charge, the charge appears at rest. Therefore, this charge also does not emit an electromagnetic wave. What emits an electromagnetic wave is only the charge which is in an accelerated motion. Thus, one of the most powerful methods to produce strong electromagnetic waves is the one to accelerate electrons, which is easy to accelerate because the mass is very small, in a large amount.

The electromagnetic wave emitted when the direction of the electron moving straight with nearly the speed of light, is bent by magnets etc., is called the synchrotron radiation. The synchrotron radiation is emitted in the direction of motion of the electron. The higher the energy of electrons becomes and the larger the direction of motion changes, the more the light is focused and bright including the light with short wavelength such as X ray.

Since the electron is charged, it is generating electric fields in the surroundings. When the electron of high energy is bent in its direction by the magnets, the electric field nearby the electron is shaken off and emitted in the tangential direction as an electromagnetic wave (Fig. 20-A). You may

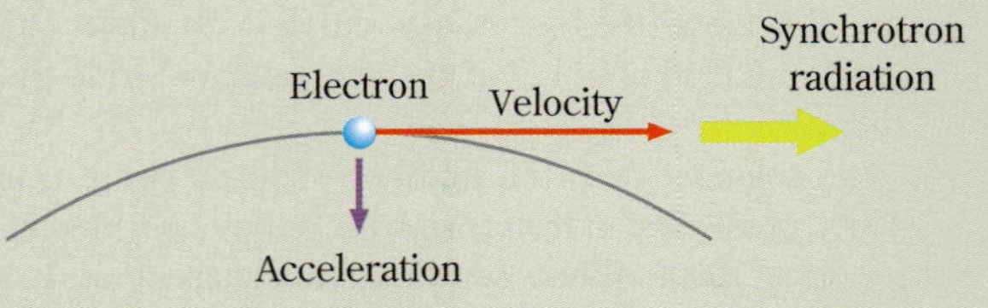

Fig. 20-A Principle of the emission of synchrotron radiation

just think that this is a synchrotron radiation.

As an apparatus to attain strong synchrotron radiations, there has been constructed an electron synchrotron which is a circular accelerator as well as a storage ring by which the electron is accelerated and confined in the vacuum pipe for a long time keeping the circular motion with the magnetic force supplied by the electromagnet. The typical example is the Spring-8 (Fig. 20-B) in the Harima Science Garden City of Hyogo Prefecture.

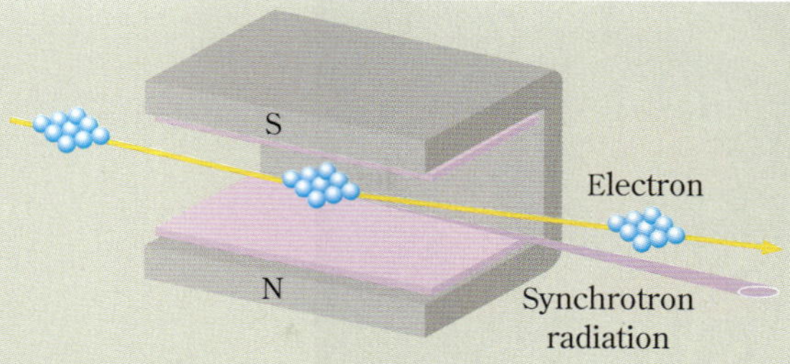

Fig. 20-C Synchrotron radiation from the bending magnet

Fig. 20-B Spring-8

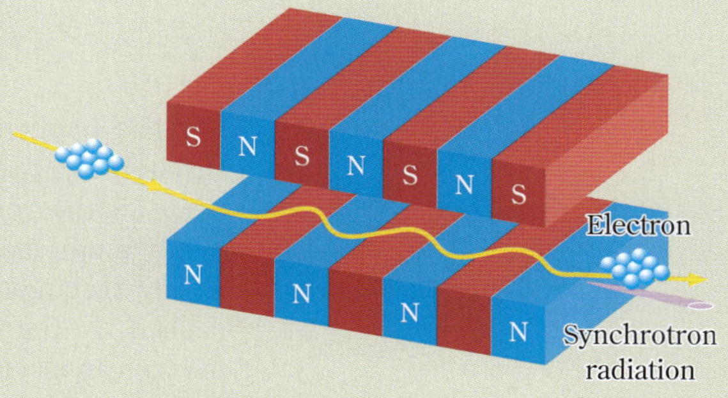

Fig. 20-D Synchrotron radiation from the undulator

In this apparatus, as the type of the magnet which is used to change the moving direction of the electron, in addition to the bending magnet (Fig. 20-C) to confine the electrons in the ring-shaped accelerator, there exist a combination of the magnets in a special form such as the undulator shown in Fig. 20-D which is inserted near the circular orbit to have small oscillations of the orbit. From these two types of magnets we obtain each characteristic synchrotron radiation

The synchrotron radiation has such characteristics that "it is extremely bright and finely focused including the range of the wavelength from infrared to X rays, naturally polarized, and the repetition of the short pulses". Therefore, synchrotron radiation, as an excellent luminous source with the range of the wavelength from ultra-violet to X rays, is used in a wide range of fields in science and technology.

Electromagnetic Induction

What happens if we drop one of two neodymium (Nd) magnets through a copper or aluminum pipe in free fall, and at the same time drop another one in free fall in the air from the same height? The neodymium magnet, irrespective of its orientation, free-falls more slowly through the copper or aluminum pipe than in the air. This is caused by the electromagnetic induction where the moving magnet induces an electric current in the pipe, and then this current generates a magnetic field which interrupts the motion of the magnet.

The electromagnetic induction we will learn in this chapter, is a phenomenon where variation of the magnetic flux (number of lines of magnetic force) penetrating the circuit induces the electromotive force, that is, time-varying magnetic fields give rise to the electric field.

Metal detector is designed to detect metals with a subtle fluctuation of magnetic fields when a human or an object passes through the aperture (head).

Electromagnetic induction is a phenomenon which is widely applied to our familiar electric equipments such as electric transformers. When Faraday gave a talk about the electromagnetic induction, which he just discovered, to science amateurs, the finance minister at that time asked him "What is it useful for?". It is told that Faraday answered to this question "Of course Sir. Soon you will be able to impose a tax on it." The electromagnetic induction discovered in 1831 is today applied to broad areas including the generation of alternating currents by electricity generators and the voltage transformation of alternating currents.

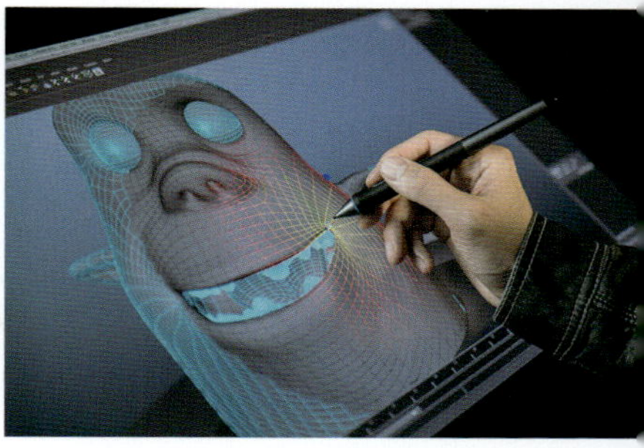

Fig. 21-1 Electronic pen (digital pen) of electromagnetic induction type. Magnetic fields are generated on the surface of the tablet. When the electronic pen moves in the magnetic field, electric currents flow in the built-in coil of the pen, and then a guide signal is emitted from the pen. By receiving the guide signal the tablet reads off the position as well as the movement of the pen, and the repetition of this process enables the screen of the tablet to display the trajectory of the pen smoothly.

21.1 Discovery of electromagnetic induction

Learning objective To understand, through the story about the discovery of electromagnetic induction, what sort of phenomenon the electromagnetic induction is.

The story starts from about 10 years later after Oersted found the magnetic effect of electric currents. Faraday had an impression that just as the electric currents give rise to magnetism, conversely, electric currents are generated from magnetism. If we bring an iron rod close to a magnet, this iron rod gets magnetism and attracts other iron fragments. And so, Faraday wondered if he puts a coil with electric currents flowing close to another coil, then electric current would appear in the second coil.

In 1831, Faraday carried out a series of experiments. As shown in Fig. 21-2, he wound a copper coil A around a half of the soft-iron ring, and wound another coil B around the other half of the ring. Both ends of the coil B was connected to a detector consisting of a conducting wire placed over a magnetic needle. Just at the moment when both ends of the coil A were connected to the battery, the needle twitched and then oscillated gradually moving back to the initial position and stopped. After that, the needle was at rest while a constant current continues to flow, but when the current of coil A was switched off, at that moment the needle twitched again and then after oscillation it stopped at the initial position. What is important here is the following. "The current flows in coil B and the magnetic force acts on the needle only at the moment when the current is switched to pass through the coil A and also at the moment when the current is switched off." The directions of the force acting on the needle are different between the moment of switching on and that of switching off. That is, the directions of the current passing through coil B are opposite in both cases.

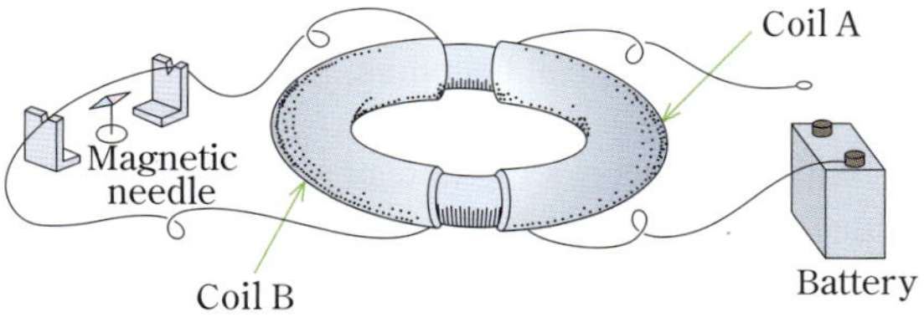

Fig. 21-2 Faraday's experiment of electromagnetic induction (magnetic needle is a galvanometer)

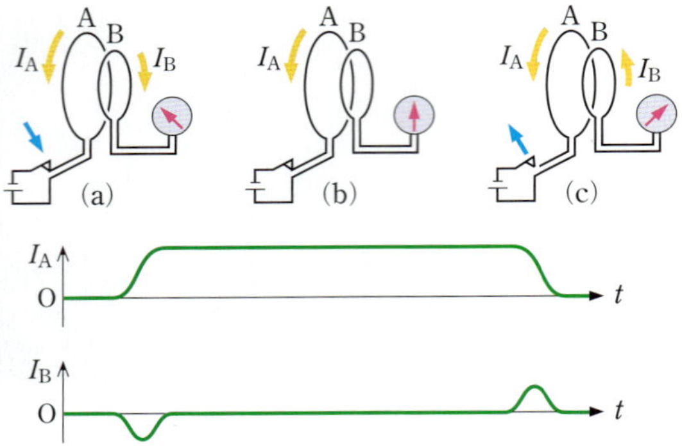

Fig. 21-3 Explanation drawing of Faraday's experiment of electromagnetic induction. (a) Start to flow the current in coil A. (b) A constant current is flowing in coil A. (c) Switch off the current in coil A.

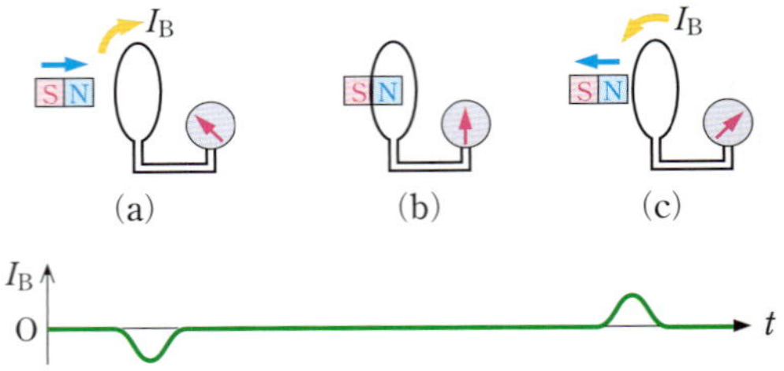

Fig. 21-4 Experiment of electromagnetic induction using a magnet. (a) Move a magnet to the right. (b) Keep a magnet at rest. (c) Move a magnet to the left.

The ring of the soft iron used in this experiment plays a role of enhancing the magnetic effects of the current flowing in the coil A. But, if we remove the ring and in order to show the phenomena easy to understand, we have an illustration in Fig. 21-3. The direction of the current I_B flowing in the coil B at the moment of starting the current flow I_A in the coil A is opposite to the direction of I_A. While, the direction of the current I_B flowing in the coil B at the moment of switching off the current I_A in the coil A, is the same as that of I_A.

What is the cause of generating the electric current in the coil B? When we start flowing the current in the coil A, the magnetic field is generated in the location of the coil B. The current flows in the coil B only at the moment when we switch on or switch off the current in the coil A. And so, Faraday suspected that the cause which gives rise to the current in coil B is the varying magnetic field, and performed an experiment with a use of a magnet as illustrated in Fig. 21-4. obtaining the result shown in the figure. In this way, it has turned out that, when the magnetic field in the vicinity of a coil varies, an electric current is produced in the coil.

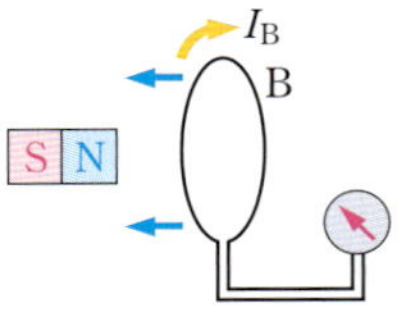

Fig. 21-5 Instead of moving a magnet to the right, the coil B is moved to the left toward the magnet at rest.

Moreover, he found that instead of moving a magnet closer to a coil at rest, putting the coil to approach the magnet at rest, generates, in the same way, a current flow in the coil (see Fig. 21-5). Since the relative motions of the coil and the magnet are the same for the experiment of Fig. 21-4 (a) and that of Fig. 21-5, it is expected that the same current flows in the coil.

Comparing the two cases where we let the magnet approach rapidly or slowly to the coil, the rapid approaching makes a needle of the galvanometer deflect bigger than the slow approaching.

The bigger the area of the coil becomes, the bigger the galvanometer deflects.

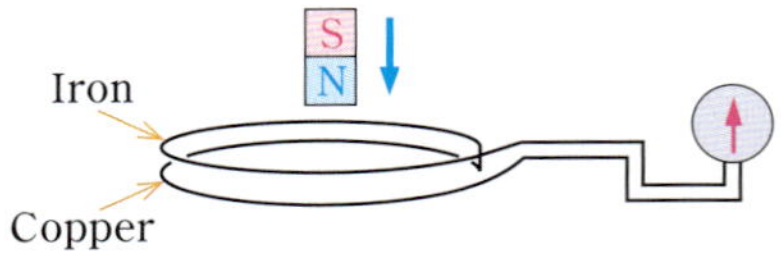

Fig. 21-6 There occurs no electric current flow when a magnet is approached a combined coil with two loops of the same size made of the copper and iron wires and connected in the opposite directions.

The reason why there occurs the flow of an induced current in the coil is that an induced electromotive force is produced in the coil. This fact can be confirmed by the following experiment (Fig. 21-6). Namely, making two coils with the same radius and the same winding number made from iron and copper wires with different resistance, and then connecting the two coils in the opposite direction, there occurs no deflection of the galvanometer when we put the magnet to approach it. The induced electromotive forces generated in the two coils are the same, but the directions are opposite, and so they cancel each other, hence it is confirmed that the total net induced electromotive force of the combined coil is zero.

In this way, Faraday discovered the following fact

The variation of magnetic flux (net number of magnetic lines of force) penetrating a circuit (coil) generates an induced electromotive force in the circuit (coil).

This phenomenon is called the **electromagnetic induction**. The electromagnetic induction was discovered independently by, an American physicist, Henry in 1831.

What about the direction of the electromotive force, or the direction of current flow? In 1834, Lenz summarized the experimental results of Fig. 21-3 and Fig. 21-4 in a form easy to understand as follows :

An induced electromotive force generated by the electromagnetic induction, will appear in such a direction that it opposes the change of the magnetic field which is produced by the induced current and penetrates the circuit.

This is referred to as **Lenz's law**.

21.2 Law of electromagnetic induction

Learning objective To understand the law of electromagnetic induction based on the phenomenon where the variation of the magnetic flux (number of magnetic lines of force) penetrating the coil generates the electromotive force. Then it becomes possible to explain to which direction of the coil electromotive force will appear. To understand that variation of a magnetic field $\boldsymbol{B}$ with time induces an electric field.

With the use of magnetic flux, from the experimental results shown in Figs. 21-2—21-6, it turns out that the following law about magnitude and direction of the induced electromotive force holds.

(1) The induced electromotive force V_i appears only during the time when magnetic flux Φ_B penetrating the circuit is changing, and the magnitude is proportional to the changing speed (time change rate) of Φ_B, $\dfrac{\mathrm{d}\Phi_\mathrm{B}}{\mathrm{d}t}$.

(2) An induced electromotive force is generated in such a direction that it opposes the change of the magnetic flux penetrating the circuit by the magnetic field produced by the induced current (Lenz's law).

The mathematical expression for the law of electromagnetic induction is given as follows :

$$V_\mathrm{i} = -\frac{\mathrm{d}\Phi_\mathrm{B}}{\mathrm{d}t}, \qquad (21.1)$$

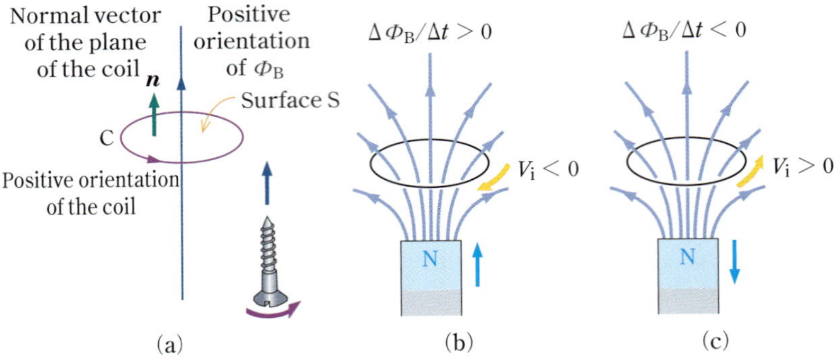

Fig. 21-7 Electromagnetic induction. (a) Positive direction of magnetic flux and positive direction of induced electromotive force. (b) A magnet is moved close to the coil. (c) A magnet is moved away from the coil.

where V_i is the induced electromotive force generated in the coil (a closed curve with the direction indicated) C, and Φ_B is a magnetic flux penetrating the surface S with the boundary C (an orientable surface with the boundary C),

$$\Phi_B = \iint_S B_n \, dA. \tag{21.2}$$

The unit of magnetic flux is Wb, that of induced electromotive force is V, and that of time is s*. The normal vector $\boldsymbol{n}$ of surface S (positive direction of magnetic flux penetrating surface S) points to such a direction that the right-handed screw moves on when it is rotated along the direction of the closed curve C. The sign of the induced electromotive force is taken to be positive as a convention when it is directed to the direction of the closed curve C [Fig. 21-7 (a)]. With this convention, we understand the reason why there is the negative sign of the right-hand side of equation (21.1).

* From equation (21.1) it can be seen that the unit of magnetic flux is Wb = V·s.

There are infinitely many surfaces with the closed curve C as the boundary. Owing to Gauss' law of the magnetic field $\boldsymbol{B}$ (20.5) there is neither starting point nor end point for the lines of magnetic force of the magnetic field $\boldsymbol{B}$. Therefore, any choice of the surface gives the same number of lines of magnetic force penetrating the surface. That is, magnetic flux is invariant under the choice of the surface (Fig. 21-8).

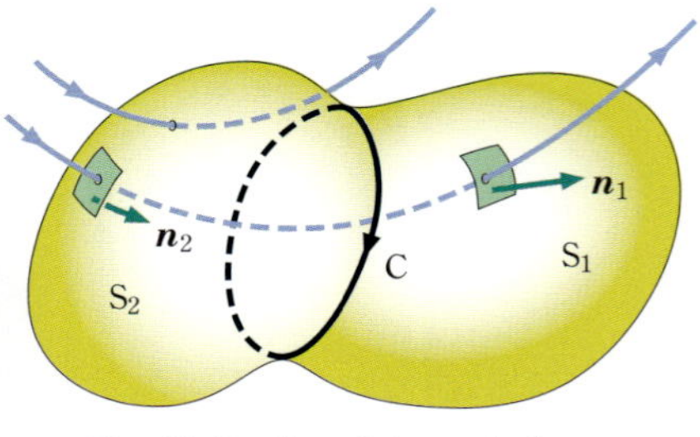

Fig. 21-8 Gauss' law of the magnetic field $\boldsymbol{B}$ is applied to the region surrounded by the two surfaces S_1 and S_2 with the boundary C. Since $\boldsymbol{n}_2$ is an inward normal of the closed surface, we have

$$\iint_{S_1} B_n \, dA - \iint_{S_2} B_n \, dA = 0.$$

When the magnetic flux Φ_B penetrating the coil of N windings changes with time, equation (21.1) can be applied to the electromotive force induced in the coil per one winding, and so the total induced electromotive force occurred in the whole coil is given by

$$V_i = -N \frac{d\Phi_B}{dt}. \tag{21.3}$$

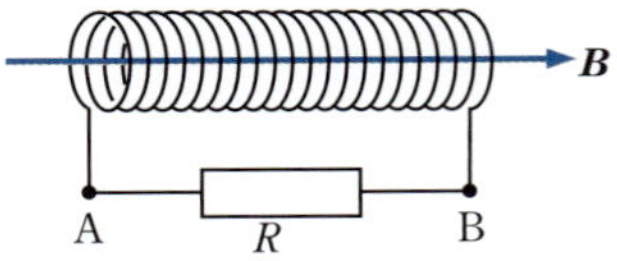

Fig. 21-9

Case 1 The magnetic flux penetrating a coil of 2000 turns in the direction of arrow as shown in Fig. 21-9 is decreasing in a rate of 1.0×10^{-3} Wb per second. The electromotive force generated in the coil is found to be

$$V_i = -N \frac{d\Phi_B}{dt} = \frac{2000\times1.0\times10^{-3}\ \text{Wb}(=\text{V}\cdot\text{s})}{1\ \text{s}} = 2.0\ \text{V},$$

and with the direction A → coil → B. Hence the electric current I flowing through the resistor of 100 Ω connected to this coil is

$$I = \frac{V_i}{R} = \frac{2.0\ \text{V}}{100\ \Omega\ (= \text{V/A})} = 2.0\times 10^{-2}\ \text{A},$$

where the direction of the current is B → A.

Question 1 When a short bar magnet was dropped through a coil as shown in Fig. 21-10 (a), there occurred an induced electromotive force V_i as shown in Fig. 21-10 (b).

(1) Explain the reason why the height of the crest is smaller than the depth of the trough.

(2) What is the relation between the area of the crest and that of the trough?

Use the following equation,

$$\int_{-\infty}^{\infty} V_i\,dt = -\int_{-\infty}^{\infty} \frac{d\Phi_B}{dt}\,dt = -\Phi_B(\infty)+\Phi_B(-\infty).$$

Question 2 Three rings of conducting wires are aligned as shown in Fig. 21-11. The middle ring carrying an electric current is approaching the ring A at rest, while retreating from the ring B at rest. What sort of electric currents are flowing in the ring A and B? Or are they carrying no current?

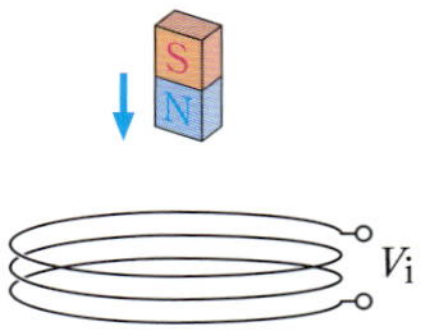

(a)

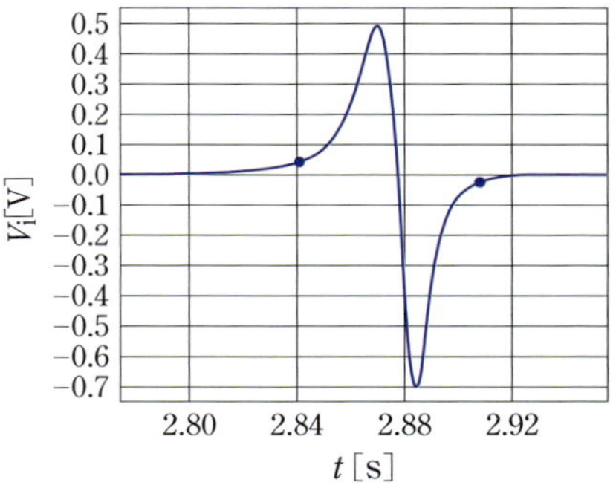

(b)

Fig. 21-10 Electromotive force due to a fall of a short bar magnet in the coil

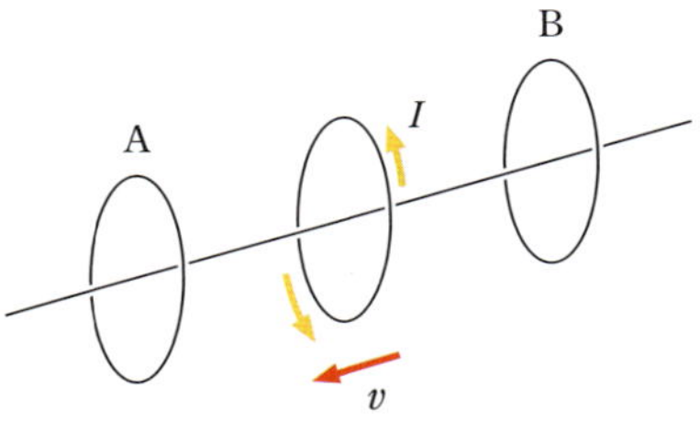

Fig. 21-11

The reason why electric currents flow in a coil due to the electromagnetic induction is that a free electron (charge $-e$) in the coil receives either an electric force $-e\boldsymbol{E}$ or a magnetic force $-e\boldsymbol{v}_e\times\boldsymbol{B}$ or both forces. The velocity $\boldsymbol{v}_e$ of a free electron is the sum of the velocity of the coil $\boldsymbol{v}$ and the drift velocity $\boldsymbol{v}_d$ with respect to the coil, $\boldsymbol{v}_e = \boldsymbol{v}+\boldsymbol{v}_d$ (Fig. 21-12). When an electric charge goes around a coil (closed circuit C) once, the work done by electric and magnetic forces exerted on a unit positive charge is the induced electromotive force V_i generated in the coil. This work is a line integral of the tangential component of $\boldsymbol{E}+(\boldsymbol{v}+\boldsymbol{v}_d)\times\boldsymbol{B}$ along the closed curve C (its component in the direction of an infinitesimal element of the circuit, $d\boldsymbol{s}$, in Fig. 21-12). Now, $\boldsymbol{v}_d$ is parallel to $d\boldsymbol{s}$, and so $(\boldsymbol{v}_d\times\boldsymbol{B})$ is perpendicular to $d\boldsymbol{s}$, hence it turns out that $(\boldsymbol{v}_d\times\boldsymbol{B})\cdot d\boldsymbol{s} = 0$. Therefore, we can express the induced electromotive force V_i due to electromagnetic induction in the general case as

$$V_i = \oint_C [\boldsymbol{E}+(\boldsymbol{v}\times\boldsymbol{B})]\cdot d\boldsymbol{s} \qquad \text{(induced electromotive force)}, \tag{21.5}$$

where, $\boldsymbol{v}$ is the velocity of the coil.

The variation of the magnetic flux penetrating the circuit occurs in the following cases,

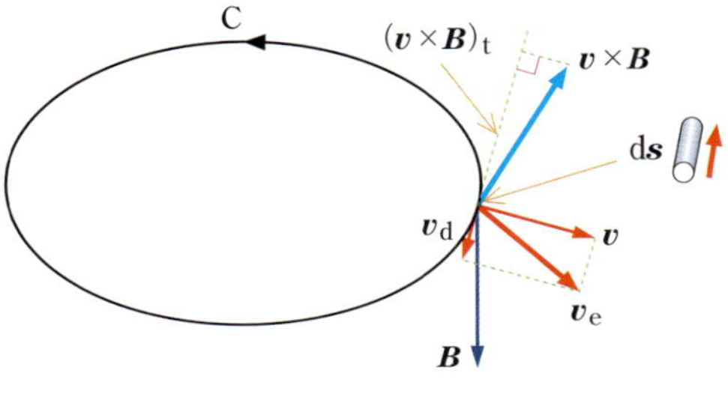

Fig. 21-12 A circuit moving with a velocity $\boldsymbol{v}$ in the magnetic field $\boldsymbol{B}$

(1) The case where the circuit is at rest, while the magnetic field changes with time,
(2) The case where the magnetic field does not change with time, while the circuit moves,
(3) The case where the magnetic field changes with time and the circuit moves,

the law of electromagnetic induction (21.1) holds in any case. Electric currents flow through the coil in the electromagnetic induction, that is because either an electric or a magnetic force exerts on free electrons in the conducting wire. In the case of (1), it is due to the electric force $-e\boldsymbol{E}$ exerted by an induced electric field $\boldsymbol{E}$ arising from the time-varying magnetic field $\boldsymbol{B}$. In the case of (2), it is due to the magnetic force acting on the electron in the moving coil.

Time-varying magnetic fields generate electric fields First of all, let us consider **the case where the circuit is at rest and the magnetic field changes with time**. The experiment illustrated in Fig. 21-4 (a) is an example of this case. As the circuit is at rest, we set $\boldsymbol{v} = \boldsymbol{0}$ in equation (21.5). Hence, the induced electromagnetic force is found to be

$$V_{\mathrm{i}} = \oint_{\mathrm{C}} \boldsymbol{E}\cdot \mathrm{d}\boldsymbol{s} = \oint_{\mathrm{C}} E_{\mathrm{t}}\, \mathrm{d}s. \tag{21.6}$$

The law of electromagnetic induction (21.1) becomes

$$\frac{\mathrm{d}\Phi_{\mathrm{B}}}{\mathrm{d}t} = -\oint_{\mathrm{C}} E_{\mathrm{t}}\, \mathrm{d}s \tag{21.7}$$

In the case of the experiment shown in Fig. 21-4 (a), though the coil is at rest and the there is no charged object, an induced electromotive force arises. Even in the presence of a charged object, because the Coulomb force is a conservative force, the electrostatic field satisfies the following relation,

$$\oint_{\mathrm{C}} \boldsymbol{E}\cdot \mathrm{d}\boldsymbol{s} = \oint_{\mathrm{C}} E_{\mathrm{t}}\, \mathrm{d}s = 0 \qquad \text{(for the case of an electrostatic field).} \tag{21.8}$$

[(see equation (16.59)]. Therefore, the electromotive force giving the current flow in the coil is generated because an electric field other than the electrostatic field is induced in the coil. That is,

In the case where the magnetic field changes with time, an electric field is generated due to the electromagnetic induction.

In the case where the magnetic field changes with time at a certain point, an induced electric field is generated irrespective of whether there is a conducting object or not at that point. In general, we find

Electric field $\boldsymbol{E}$ = "electric field generated by electric charge" +"electric field due to electromagnetic induction".

A line of electric force of an electric field (electrostatic field), produced by an electric charge at rest, is a curve which has a positive charge at the initial point and a negative charge at the end point, and it is by no means a closed curve without initial and end points. On the other hand, a line of electric field generated in the surroundings by changing the magnetic flux penetrating the coil with time is a closed curve as shown in Fig. 21-13 (a). When we let a positive charge go

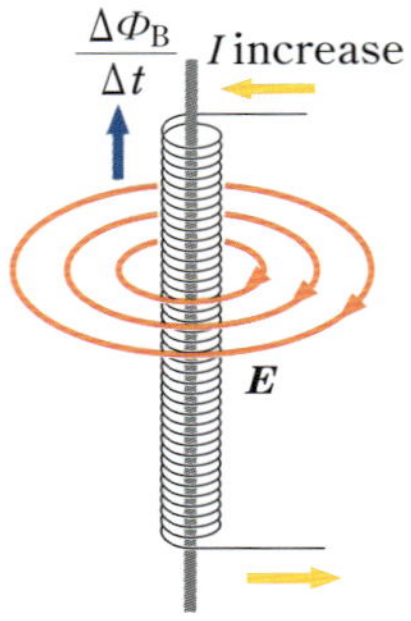

(a) Induced electric field

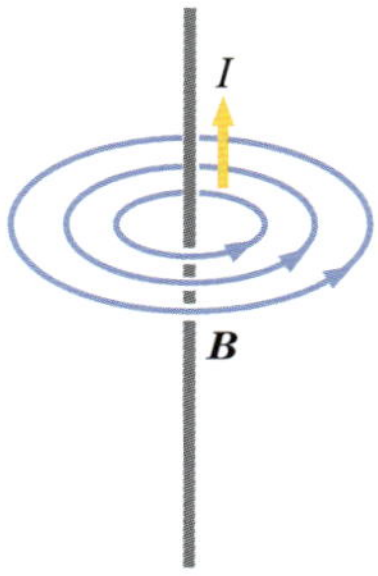

(b) Magnetic field generated by electric current

Fig. 21-13 (a) Induced electric field (b) Magnetic field generated by electric current

around once along the line of electric force, then induced electric field will do a work on the charge. Hence, through the coil placed along the closed line of electric force electric currents flow in the direction of the line of the electric force. The appearance of the induced electric field looks like that of the magnetic field generated by an electric current [Fig. 21-3 (b)].

For the electrostatic field, a line integral of the tangential component E_t of the electric field $\boldsymbol{E}$ along the closed curve vanishes. Thus, the electric potential can be defined for the electrostatic field. In the presence of the induced electric field, a line integral of E_t along the closed curve does not vanish, and so we cannot define the electric potential at each point or potential difference between two points. However, the potential difference along a path between the two points on the closed curve can be defined.

Case 2 A long solenoid in Fig. 21-14 carries an alternating current, and the current due to the electromagnetic induction flows through the conducting wire which encircles the solenoid, and as a result, the miniature bulb A, B are lighted up. Let us consider the brightness of the miniature bulb A, B when the switch S is turned on. Inserting an alternating-current source with the same electromotive force as the one generated by the electromagnetic induction at the point C on the wire, and consider the brightness of the miniature bulb A, B using the relation, "induced electromotive force" = "electric resistance" × "electric current". It can be seen that the miniature bulb A becomes bright, while the miniature bulb B goes out.

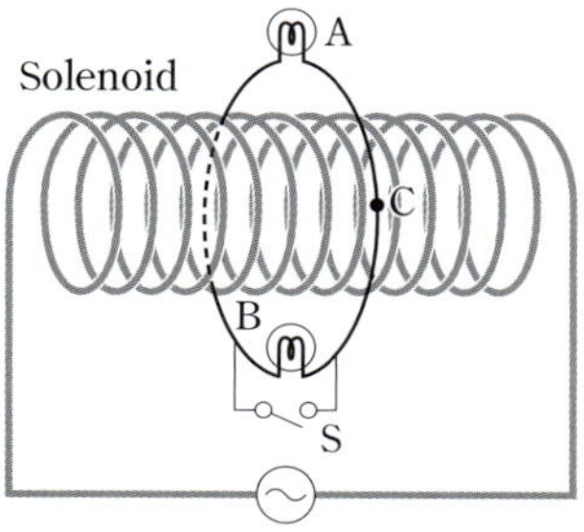

Fig. 21-14

Example 1 Let us consider a long solenoid of n turns per unit length. When an electric current I starts to flow through this solenoid as depicted in Fig. 21-15, show that the direction of the electric field $\boldsymbol{E}$ at the point with a distance of r from the central axis is as shown in the figure. Show also that the intensity of the electric field is given by

$$E = \frac{1}{2}\mu_0 nr \frac{\mathrm{d}I}{\mathrm{d}t}. \qquad (21.9)$$

The magnetic field outside the solenoid vanishes. Is an induced electric field also vanishing outside the solenoid?

Solution The fact that there is neither initial point nor end point for the lines of force of the induced electric field and the rotational symmetry around the central axis, it turns out that the lines of electric force are concentric circles with a central axis as the center. In the case where this circle is inside the solenoid, by taking the circle of a radius r in the figure for the closed curve C in equation (21.7) we have

$$V_i = \oint_C E_t \, \mathrm{d}s = 2\pi rE = -\frac{\mathrm{d}\Phi_B}{\mathrm{d}t}$$
$$= -\frac{\mathrm{d}(\mu_0 nI\pi r^2)}{\mathrm{d}t}, \qquad (21.10)$$

and hence the intensity of the induced electric field E is found to be

$$E = \frac{1}{2}\mu_0 nr \frac{\mathrm{d}I}{\mathrm{d}t}. \qquad (21.11)$$

The negative sign of the right-hand side of equation (21.10) indicates that the direction of the electric field is opposite to that of the electric current.

Outside of the solenoid we find

$$2\pi rE = -\frac{\mathrm{d}\Phi_B}{\mathrm{d}t} = -\frac{\mathrm{d}(\mu_0 nI\pi R^2)}{\mathrm{d}t},$$

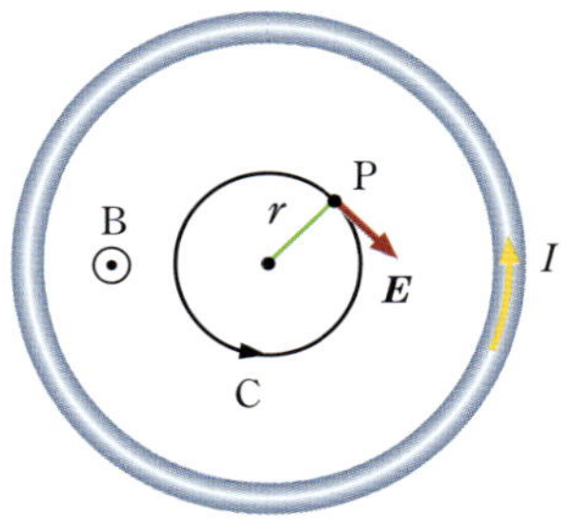

Fig. 21-15

and we obtain for the intensity of the induced electric field as

$$E = \frac{1}{2r}\mu_0 n R^2 \frac{\mathrm{d}I}{\mathrm{d}t}, \qquad (21.12)$$

and the direction of the electric field is opposite to that of the electric current. Namely, there exists no magnetic field outside the solenoid, but there appears an induced electric field outside the solenoid, provided the magnetic field inside the solenoid varies with time.

21.3 Electromagnetic induction with constant magnetic field and moving coil

Learning objective To become able to explain the operating principle of the generator.

The experiment shown in Fig. 21-5, is an example where a magnetic field does not change with time, while a circuit moves. In this case, as the magnetic field does not change, no induced electric field appears. The electromotive force generated in the coil moving with a velocity $\boldsymbol{v}$ is caused by the magnetic force acting on the charged particle given as

$$V_{\mathrm{i}} = \oint_{\mathrm{C}} (\boldsymbol{v}\times\boldsymbol{B})\cdot \mathrm{d}\boldsymbol{s}. \qquad (21.13)$$

It can be proved that the variation of the magnetic flux originates from that of the surface S and is given by (Proof omitted)

$$\frac{\mathrm{d}\Phi_{\mathrm{B}}}{\mathrm{d}t} = -\oint_{\mathrm{C}} (\boldsymbol{v}\times\boldsymbol{B})\cdot \mathrm{d}\boldsymbol{s}. \qquad (21.14)$$

Therefore, the law of electromagnetic induction (21.1) holds also in the case where the magnetic field does not vary while the coil moves.

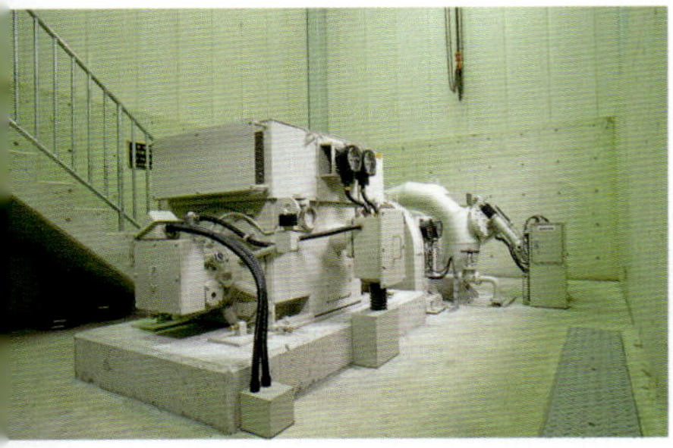

Fig. 21-16 Alternating current 3-phase generator at Yatogawa-Daisan Power Plant (Fig. 5-13)

Electromotive force generated in the rotating coil in the magnetic field-AC generator In the homogeneous magnetic field $\boldsymbol{B}$, we let a rectangular conducting wire (coil) rotate with a constant angular velocity ω around the axis OO′ perpendicular to the magnetic field [Fig. 21-17 (a)]. Let the angle between the normal vector $\boldsymbol{n}$ and the magnetic field $\boldsymbol{B}$ be $\theta = \omega t$. If we take the area of the rectangle surrounded by the coil to be A, the magnetic flux penetrating the coil Φ_{B} is found to be

$$\Phi_{\mathrm{B}} = BA\cos\theta = BA\cos\omega t. \qquad (21.15)$$

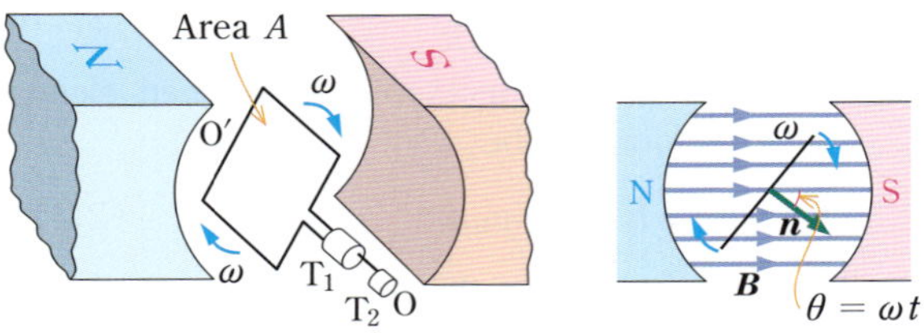

(a) A coil that rotates at an angular velocity ω in a uniform magnetic field

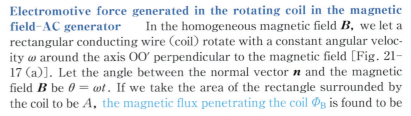

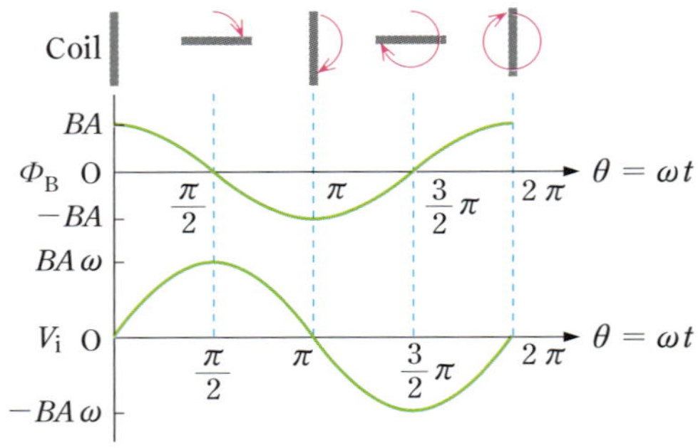

(b) $\Phi_{\mathrm{B}} = BA\cos\omega t$, $V_{\mathrm{i}} = BA\,\omega\sin\omega t$

Fig. 21-17 AC generator

The induced electromotive force V_i produced in the coil by electromagnetic induction is

$$V_i = -\frac{d\Phi_B}{dt} = BA\omega \sin \omega t. \tag{21.16}$$

The induced electromotive force points to the direction from T_1 through the coil to T_2. The electromotive force becomes maximum when the magnetic field $\boldsymbol{B}$ is parallel to the surface of the coil, while the electromotive force vanishes when the magnetic field is perpendicular to the coil surface. When the coil is rotated through 180 degrees, the direction of the electromotive force becomes opposite. As shown in Fig. 21-17 (b), the electromotive force is the alternating current (AC) electromotive force (alternating voltage) with

$$\text{period } T = \frac{2\pi}{\omega}, \qquad \text{frequency } f = \frac{\omega}{2\pi}. \tag{21.17}$$

This is the principle of the AC generator. The current generated by AC electromotive force is the alternating current.

For reference Relation between the experiment in Fig. 21-4 (a) and that in Fig. 21-5

In the experiment of Fig. 21-4 (a) the current flowing in the circuit is caused by the electric force due to the induced electric field, while in the experiment of Fig. 21-5 the current in the circuit originates from the magnetic force due to the magnetic field. So, at first glance, they have no connection. However, in these experiments shown in Figs. 21-4 (a) and 21.5, the magnet and the coil are approaching each other, and the relative motions of the magnet and coil are the same, so it is expected that the same current should flow in the two cases.

In Figs. 21-18 (a) and (b), the relative motions of the magnet and the conducting wire are the same. In the case of Fig. 21-18 (a), the magnetic force acting on the charged particle in the conducting wire with a charge q moving to the right through the magnetic field $\boldsymbol{B}$ at a speed v is $f = qvB$ in magnitude. In the case of Fig. 21-18 (b), the induced electric field $\boldsymbol{E}$ generated by the magnetic field originating from the magnet moving to the left with a speed v acts an electric force $q\boldsymbol{E}$ on the charged particle with a charge q in the conducting wire which is at rest. Then from the condition that both forces are the same, $qvB = qE$, it turns out that the intensity of the induced electric field is $E = vB$. That is, between the magnetic poles of the magnet moving at a speed v, an electric field $\boldsymbol{E}$ in the direction depicted in Fig. 21-18 (b) with a magnitude

$$E = vB \tag{21.18}$$

appears*. This induced electric field due to the electromagnetic induction is produced in the same way whether there is a conducting wire or not.

* The electromagnetic fields $\boldsymbol{E}, \boldsymbol{B}$ observed in the frame fixed to the magnet, and the electromagnetic fields $\boldsymbol{E}', \boldsymbol{B}'$ observed in the frame fixed to the conducting wire are related by equation (23.14) in the theory of relativity. For the case of Fig. 21-18, $\boldsymbol{E}' \approx \boldsymbol{v}\times\boldsymbol{B}$, $\boldsymbol{B}' \approx \boldsymbol{B}$.

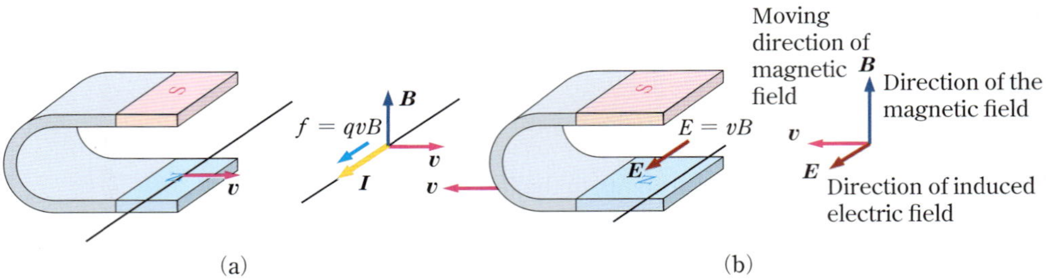

Fig. 21-18 The case (a) where a conducting wire moves to the right with a speed v between the magnetic poles of the magnet, and the case (b) where the conducting wire is at rest, while the magnet moves to the left with a speed v. In both cases, the relative motions are the same.
(a) When we move a conducting wire to the right in the magnetic field, an induced electromotive force is generated in the conducting wire. This electromotive force is due to the magnetic force $f = qvB$ which acts on the charged particle in the conducting wire with a charge q ($q > 0$). (b) When we move a magnet to the left, an electric force acting on the charge q is $f = qE = qvB$, and hence the induced electric field $E = vB$ is generated. As this induced electric field appears even in the absence of the wire, we draw the symbols for the wire and electric field apart from each other.

Question 3 By looking at Fig. 21-18 (a), check the following **Fleming's right hand rule**: "If you point the thumb of your right hand in the direction of the velocity $\boldsymbol{v}$ of the conducting wire and point the index (first) finger in the direction of magnetic field $\boldsymbol{B}$, then direction of the middle (second) finger, which is perpendicular to both the thumb and the index finger, corresponds to the direction of the electromotive force generated in the conducting wire (direction of electric current I flowing through the conducting wire)". (Considering that an external force $\boldsymbol{F}$ is acting in the direction of the velocity $\boldsymbol{v}$ of the conducting wire, we can call this rule right-hand's FBI rule.)

21.4 Self-induction and mutual induction

Learning objective To understand that in the electromagnetic induction, there exist the mutual induction, where two coils are involved, and the self-induction, where the change of the current flowing through a coil gives rise to an induction electromotive force in the direction which opposes the change of the current. To understand the role of the self-induction against the electric current flowing through the circuit. To understand that a magnetic field is accompanied by the energy of the magnetic field.

Self-induction If the electric current I flowing through the coil changes, then as the magnetic flux penetrating the coil also changes, thus there appears in the coil an induced electromotive force in the direction which opposes the change of the current. This phenomenon of the electromagnetic induction is called the **self-induction**. The electromotive force due to the self-induction is sometimes called the **counter electromotive force**, because it is generated in the direction which opposes the change of the electric current (opposite direction) which is originally the cause for the current.

The magnetic field produced by the electric current flowing in the coil is proportional to the electric current, and hence the magnetic flux Φ_B

penetrating the coil of one turn is proportional to the electric current I. Therefore, the total magnetic flux penetrating the coil of N turns, $N\Phi_B$, can be expressed as $N\Phi_B = LI$* (L is a coefficient). If the electric current flowing through the coil changes from I to $I+\Delta I$ by ΔI within a time Δt, then the variation of the total magnetic flux is $N\,\Delta\Phi_B = L\,\Delta I$. Thus, from equation (21.3) we have

* If the coil is so long that the magnetic flux Φ_B penetrating the coil is not constant, use the average value of Φ_B for the $N\Phi_B$.

When an electric current I flowing in a closed circuit changes, in the direction which opposes this change, an induced electromotive force V_i due to the self-induction arises, where

$$V_i = -L\frac{dI}{dt}. \qquad (21.19)$$

This proportional constant $L = \dfrac{N\Phi_B}{I}$ is called the **inductance** or the **self-inductance**. L is a constant determined by the shape of the closed circuit and its number of turns as well as magnetic bodies in the surroundings. The unit is Wb/A = T·m²/A = V·s/A, which we call Henry (symbol H) in connection with the physicist, Henry, who discovered the self-induction. The electromotive force due to the self-induction appears in the direction opposing the change in the electric current, L is always positive,

Unit of inductance
H = Wb/A = T·m²/A
= V·s/A

$$L > 0. \qquad (21.20)$$

As shown in Fig. 21-19, a coil L, an electric resistor R and a battery with an electromotive force V are connected in series. When the switch is turned on, the electric current starts to flow, but the counter electromotive force is generated due to the self-induction and opposes the electric current to reach in a moment to the value $\dfrac{V}{R}$ which is derived from Ohm's law $V = RI$. When we switch off the circuit carrying the current $I = \dfrac{V}{R}$, the current does not become zero at the moment of switch-off, because of the self-induction.

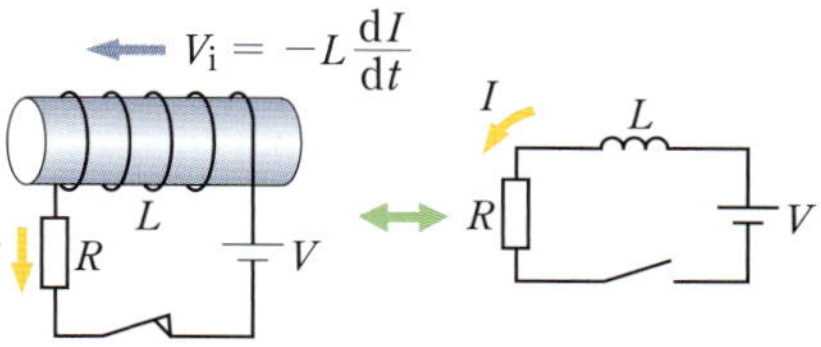

Fig. 21-19 Self-induction

Example 2 | *LR* circuit Find the electric current when time t has elapsed after the switch of the circuit in Fig. 21-19 is turned on.

Solution The electromotive force of the circuit is the sum of the electromotive force of the battery V and that of self-induction $-L\dfrac{dI}{dt}$. Thus, Ohm's law leads to

$$V - L\frac{dI}{dt} = RI. \qquad (21.21)$$

The general solution to this differential equation, is a sum of a special solution to equation (21.21), $I = \dfrac{V}{R}$, and general solution to the equation (21.21) where we set $V = 0$, namely, the general solution to the differential equation $\dfrac{dI}{dt} = -\dfrac{R}{L}I$, $I = ce^{-Rt/L}$,

$$I = \frac{V}{R} + ce^{-Rt/L} \qquad (c \text{ is an arbitrary constant}). \qquad (21.22)$$

At the moment $t = 0$ when the switch is turned on, $I = 0$. Hence, we have $c = -\dfrac{V}{R}$. Thus the current $I(t)$ when time t has elapsed after the switch is turned on, is given by

$$I(t) = \frac{V}{R}(1 - e^{-Rt/L}), \qquad (21.23)$$

which is increasing with time and after the time of the order several times of $\dfrac{L}{R}$ passes, it approaches the final value $\dfrac{V}{R}$ (Fig. 21-20). $\dfrac{L}{R}$ is

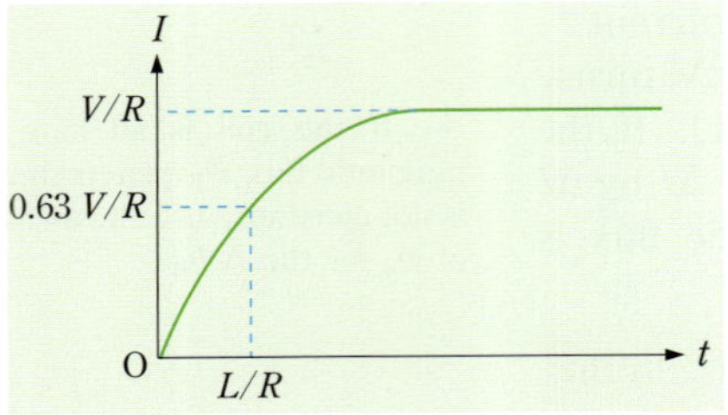

Fig. 21-20 The electric current which flows the circuit in Fig. 21-19, where the switch is turned on at $t = 0$.

called the **time constant** of the LR circuit.

When we short both ends of the battery of the LR circuit which carries a current $I_0 = \frac{V}{R}$, the current does not vanish immediately, but decreases as

$$I = I_0\, e^{-Rt/L}. \qquad (21.24)$$

Example 3 (1) Calculate the self-inductance of a long air core solenoid (Fig. 21-21).

(2) The self-inductance of the solenoid wound around the iron core of relative permeability μ_r is the self-inductance of the air core solenoid multiplied by μ_r. Find the self-inductance of the solenoid consisting of an iron core with cross sectional area 8 cm^2, length 10 cm, and relative permeability $\mu_r = 1000$, around which the conducting wire wound uniformly 1000 times.

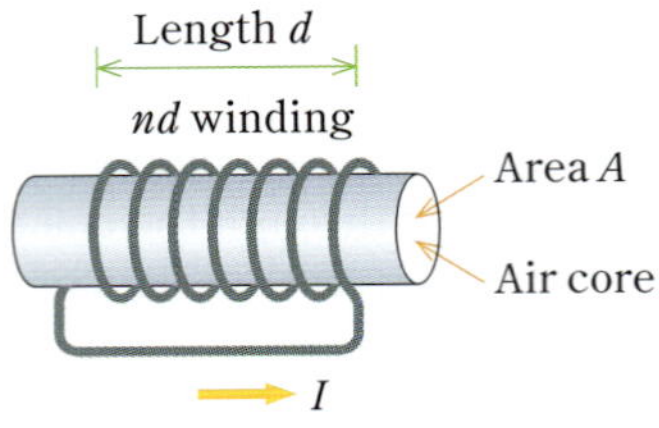

Fig. 21-21 Air core solenoid

Solution (1) If the solenoid is long enough, the intensity B of the magnetic field is taken to be $\mu_0 nI$, then the magnetic flux Φ_B penetrating one turn of the coil of area A is given by

$$\Phi_B = BA = \mu_0 nIA. \qquad (21.25)$$

Hence, the self-inductance L of the solenoid with the total number of turns $N = nd$ is

$$L = \frac{N\Phi_B}{I} = \frac{nd\Phi_B}{I} = \mu_0 n^2 Ad \qquad \text{(air core)}. \qquad (21.26)$$

(2) The self-inductance for the case with an iron core is equation (21.26) times μ_r, that is

$$L = \mu_r \mu_0 n^2 Ad \qquad \text{(with iron core)}. \qquad (21.27)$$

Therefore, we get the following numerical value :

$$L = 1000 \times (4\pi \times 10^{-7}\ \mathrm{T{\cdot}m/A}) \times \left(\frac{1000}{0.1\ \mathrm{m}}\right)^2 \times (8 \times 10^{-4}\ \mathrm{m^2}) \times (0.1\ \mathrm{m})$$
$$= 10\ \mathrm{T{\cdot}m^2/A} = 10\ \mathrm{H}.$$

Energy of magnetic field In order to increase the electric current flowing through the coil, the electric power source must act on free electrons and do the work to let them pass through the coil against the counter electromotive force $-L\frac{\Delta I}{\Delta t}$. Here in this case, the work done by the power source is stored in the current-carrying coil as energy. When the current is increased from I to $I + \Delta I$ for an infinitesimal time Δt, the amount of charge moved is $\Delta Q = I\,\Delta t$, so the necessary work ΔW for increasing the current by ΔI is given as

$$\Delta W = L\frac{\Delta I}{\Delta t} I\,\Delta t = LI\,\Delta I. \qquad (21.28)$$

Hence, the necessary work for increasing the current from 0 to I is found to be

$$W = \int_0^I LI\,\mathrm{d}I = \frac{1}{2}LI^2. \qquad (21.29)$$

In a coil with a self-inductance L carrying a current I, the magnetic energy equal to the amount of the above work as given by

$$U = \frac{1}{2}LI^2, \tag{21.30}$$

is stored.

Energy stored in long solenoids When the current I is flowing in an air core solenoid of long length (length d, cross section A, number of turns per unit length n), the intensity B of the magnetic field inside the solenoid is

$$B = \mu_0 nI,$$

and the self-inductance L of the solenoid is [equation (21.26)]

$$L = \mu_0 n^2 Ad.$$

Therefore, the magnetic energy U stored in the solenoid can be rewritten as

$$U = \frac{1}{2}LI^2 = \frac{1}{2\mu_0}B^2(Ad). \tag{21.31}$$

Since Ad is the internal volume of the solenoid, equation (21.31) indicates that inside the solenoid with intensity B of the magnetic field $\boldsymbol{B}$ there exists an **energy of the magnetic field** per unit volume given by

$$u_B = \frac{1}{2\mu_0}B^2 \quad \text{(in vacuum)}. \tag{21.32}$$

In the case where the inside of a long solenoid is filled with an iron core of relative permeability μ_r, we have $L = \mu_r\mu_0 n^2 Ad$, $B = \mu_r\mu_0 nI$ and equation (21.32) becomes

$$u_B = \frac{1}{2\mu_r\mu_0}B^2 \quad \text{(in substance with relative permeability } \mu_r\text{)}. \tag{21.33}$$

Fig. 21-22 Installation of solenoid in p.287.

Mutual induction In the case where two coils are placed close to each other or wound around the same iron core, through the varying magnetic field, each coil undergoes a phenomenon of electromagnetic induction mutually with another coil. As shown in Fig. 21-23, we place the two coils L_1, L_2 close to each other, and then flow a current I_1 through the first coil L_1, hence there appears a magnetic field, and its magnetic flux $\Phi_{2\leftarrow1}$ passes through one turn of the second coil L_2. The magnetic flux $\Phi_{2\leftarrow1}$ is proportional to the current I_1 flowing in the coil L_1. Hence, the total magnetic flux $N_2\Phi_{2\leftarrow1}$ passing through the coil L_2 of N_2 turns can be expressed as $N_2\Phi_{2\leftarrow1} = M_{21}I_1$* ($M_{21}$ is a proportional constant). When the current I_1 flowing in the coil L_1 changes from I_1 to $I_1+\Delta I_1$ for a time Δt, the change of the total magnetic flux passing through the coil L_2 is $N_2\,\Delta\Phi_{2\leftarrow1} = M_{21}\,\Delta I_1$. Therefore, the change of the current I_1 flowing in the coil L_1 generates an induced electromotive force in the coil L_2 given by

$$V_{2\leftarrow1} = -M_{21}\frac{dI_1}{dt}, \tag{21.34}$$

which produces a current in the second coil L_2 opposing the change of the magnetic flux $\Phi_{2\leftarrow1}$. In this way, the phenomenon where the

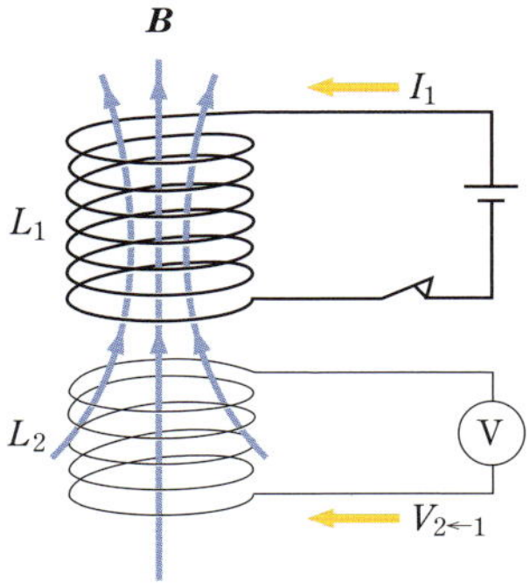

Fig. 21-23 Mutual induction

* If the coil L_2 is so long that the magnetic flux $\Phi_{2\leftarrow1}$ passing through the coil L_2 is not constant, use the average value $\Phi_{2\leftarrow1}$ for $N_2\Phi_{2\leftarrow1}$.

change in a current of one closed circuit generates an induced electromotive force in another closed circuit is called the **mutual induction**. The proportional constant M_{21} given as

$$M_{21} = \frac{N_2 \Phi_{2\leftarrow 1}}{I_1} \tag{21.35}$$

is called the **mutual inductance** of the two coils. M_{21} is determined by the shape and number of turns of each coil L_1, L_2 as well as the relative position and magnetic bodies in the surroundings. The unit of mutual inductance is also Henry (symbol H). What Faraday found in the experiment in Fig. 21-2 was in fact a mutual induction.

Unit of mutual inductance H

Similarly, the change of the current I_2 flowing in the second coil L_2 generates an induced electromotive force in the first coil L_1, $V_{1\leftarrow 2}$, which is given by

$$V_{1\leftarrow 2} = -M_{12}\frac{\mathrm{d}I_2}{\mathrm{d}t}, \tag{21.36}$$

where, the mutual inductance M_{12} reads

$$M_{12} = \frac{N_1 \Phi_{1\leftarrow 2}}{I_2}. \tag{21.37}$$

Here $\Phi_{1\leftarrow 2}$ is the magnetic flux passing through one turn of the first coil L_1 by the current I_2. There exists a relation between M_{12} and M_{21} given as

$$M_{12} = M_{21}, \tag{21.38}$$

which is called the **reciprocity theorem of mutual inductance**.

21.5 Alternating current

Learning objective To become able to explain how to express the alternating voltage and alternating current, especially effective value, and also the definition of impedance Z and phase shift ϕ. To understand how resistors, capacitors, and coils contribute to Z and ϕ. To become able to write down the equations obeyed by currents flowing through the RLC circuit. To understand the principle of transformer.

Alternating current An electric current such as the one obtained from a battery in which the direction of flow does not change with time is called the direct current (DC). On the contrary, for the coil rotating with a constant angular velocity ω in a homogeneous magnetic field there appears an oscillating electromotive force with time as (see Section **21.3**)

$$V(t) = V_{\mathrm{m}} \sin \omega t = \sqrt{2}\, V_{\mathrm{e}} \sin \omega t. \tag{21.39}$$

The electromotive force in which the direction of the current flow is constantly changing is called the **alternating current** (AC) **electromotive force** or **alternating voltage**.

When an alternating voltage $V(t)$ is applied to both ends of a resistor R, we find from Ohm's law,

$$V(t) = RI(t), \tag{21.40}$$

which leads to the **alternating current** (AC) given as (Fig. 21-24)

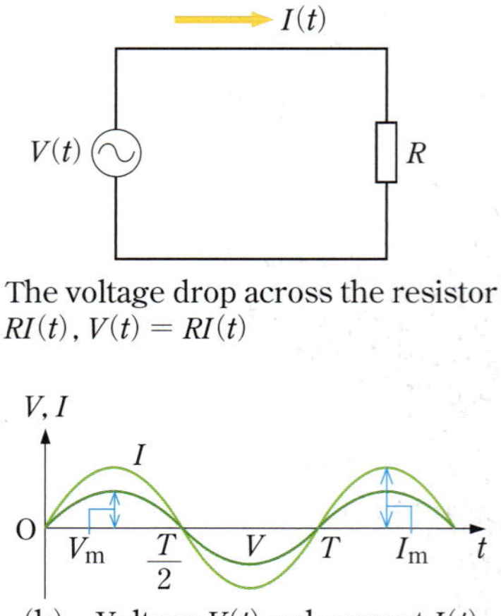

(a) The voltage drop across the resistor is $RI(t)$, $V(t) = RI(t)$

(b) Voltage $V(t)$ and current $I(t)$

Fig. 21-24 Circuit only with alternating voltage and resistor

$$I(t) = I_m \sin \omega t = \sqrt{2} I_e \sin \omega t \tag{21.41}$$

$$V_m = RI_m, \qquad V_e = RI_e. \tag{21.42}$$

V_m and I_m are the maximum values of oscillating voltage and current, respectively. V_e and I_e are **effective values** of voltage and current, respectively. The effective values are maximum values times $\frac{1}{\sqrt{2}}$,

$$V_e = \frac{V_m}{\sqrt{2}}, \qquad I_e = \frac{I_m}{\sqrt{2}}. \tag{21.43}$$

The values displayed at an AC voltmeter or at an AC ammeter are effective values. For an electric power of 100 V for family use, the effective value of voltage V_e is 100 V, the maximum value V_m is 141 V.

In the case of the alternating current, electric power consumed as a Joule heat reads

$$V(t)I(t) = V_m I_m \sin^2 \omega t = 2V_e I_e \sin^2 \omega t = V_e I_e (1 - \cos 2\omega t) \tag{21.44}$$

which varies with time. Taking the average with respect to time we have

$$\langle P \rangle = V_e I_e = RI_e^2 = \frac{V_e^2}{R}. \tag{21.45}$$

Thus, if we use effective values V_e and I_e instead of V and I, the formula of Joule heat (19.22) and Ohm's law (19.5) hold also for an alternating current. Here we have used the time average of $\cos 2\omega t$, which is $\langle \cos 2\omega t \rangle = 0$*.

* The time averages of the square of alternating voltage $V(t)$ and the square of alternating current $I(t)$ are $\langle V(t)^2 \rangle = V_e^2$ and $\langle I(t)^2 \rangle = I_e^2$.

An alternating voltage and an alternating current oscillate with time. Period T satisfies $\omega T = 2\pi$, and we find $T = \frac{2\pi}{\omega}$. We call ω the **angular frequency** of the alternating current.

$$f = \frac{1}{T} = \frac{\omega}{2\pi} \tag{21.46}$$

is a number of oscillations per unit time, which we call the **frequency** of the alternating current. ωt in the sine function is called a **phase** of alternating voltage. Here we assume that the phase is 0 at $t = 0$. The unit of frequency is 1/s, which we call Hertz (symbol Hz). The frequency of electric power which is supplied by electricity company is 50 Hz in East Japan and 60 Hz in West Japan.

Unit of frequency Hz = 1/s

If the alternating voltage $V(t) = \sqrt{2} V_e \sin \omega t$ is applied to the circuit containing a coil and a capacitor, we have the flow of the following alternating current

$$I(t) = \sqrt{2} I_e \sin(\omega t - \phi) \tag{21.47}$$

which has the same angular frequency but different phase. The angle ϕ which represents the delay of the phase of the current compared to that of the voltage is called the **phase shift**. The ratio of V_e to I_e, $\frac{V_e}{I_e}$, is called the **impedance**, and denoted as Z.

$$V_e = ZI_e. \tag{21.48}$$

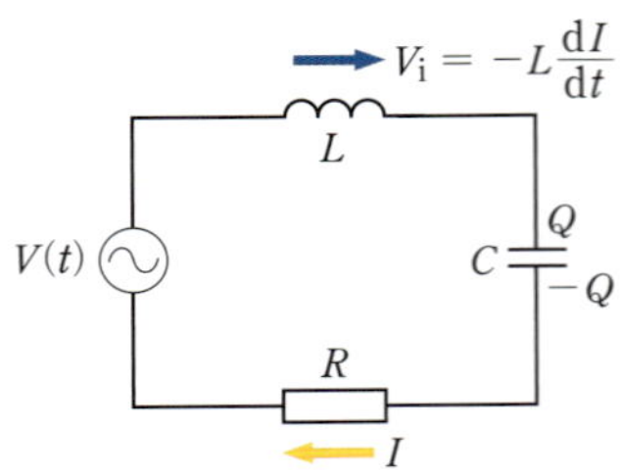

Fig. 21-25 *RLC* series circuit

Impedance in the AC circuit corresponds to the resistance in the DC circuit, and the unit is Ohm (Ω).

***RLC* series circuit** As shown in Fig. 21-25, AC circuit, in which a resistor of electric resistance R, a coil of self-inductance L, and a capacitor of electric capacity C are connected in series and applied by the AC electromotive force given as

$$V(t) = \sqrt{2}\,V_e \sin \omega t, \tag{21.49}$$

at both ends, is called the ***RLC* series circuit**. In this case the following relation holds :

"voltage drop at the electric resistance RI"+"voltage drop at the capacitor $\frac{Q}{C}$" = "electromotive force contained in the circuit $V(t)$"+ "induced electromotive force generated in the coil $-L\frac{dI}{dt}$"

$$RI + \frac{Q}{C} = \sqrt{2}\,V_e \sin \omega t - L\frac{dI}{dt}. \tag{21.50}$$

Differentiating the above equation with respect to t and using $I = \frac{dQ}{dt}$ we get

$$L\frac{d^2 I}{dt^2} + R\frac{dI}{dt} + \frac{1}{C}I = \sqrt{2}\,\omega V_e \cos \omega t. \tag{21.51}$$

This equation has the same form as the equation of motion for a mass point with mass m on which a periodically changing force $F = F_0 \cos \omega t$, a drag force proportional to the velocity $-2m\gamma v$, and a spring force $-kx$ are acting, and is given by

$$m\frac{d^2 x}{dt^2} + 2m\gamma\frac{dx}{dt} + kx = F_0 \cos \omega t. \tag{21.52}$$

[See equation (4.32) in Section **4.2**. External force in equation (4.32) is $F = F_0 \cos \omega_f t$.]

For the solutions to the differential equation (21.52), there are two kinds ; one is a solution which represents the damped oscillation exponentially damping with time, and another is a solution representing a periodic motion with period $T = \frac{2\pi}{\omega}$. If we consider a transient phenomenon occurring in such a case where we switch on and off the circuit, we have to deal with the exponentially changing solution. Here, however, we only consider the solution of forced oscillation which represents a periodic motion with a constant amplitude. In terms of the impedance of the circuit :

$$Z = \left\{R^2 + \left(\omega L - \frac{1}{\omega C}\right)^2\right\}^{1/2} \tag{21.53}$$

and the phase shift ϕ (Fig. 21-6) :

$$\sin \phi = \frac{1}{Z}\left(\omega L - \frac{1}{\omega C}\right), \qquad \cos \phi = \frac{R}{Z} \tag{21.54}$$

the solution to equation (21.51) can be expressed as

$$I(t) = \frac{1}{Z}\sqrt{2}\,V_e \sin(\omega t - \phi) \tag{21.55}$$

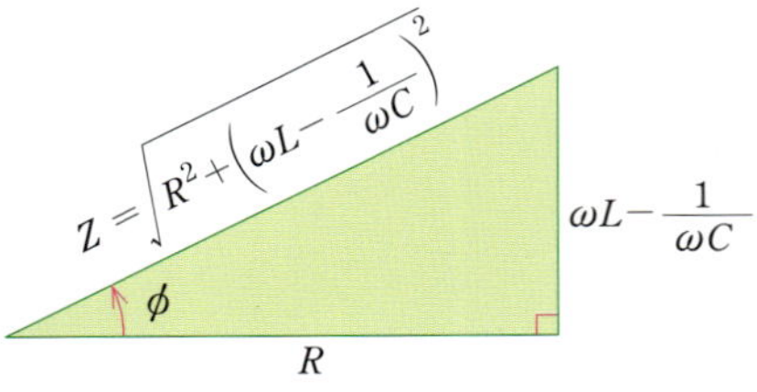

Fig. 21-26 Impedance Z and phase shift ϕ

which can be checked to be a solution by substituting equation (21.55) to equation (21.51). In the case where the circuit contains only a power source and a coil, we have $\phi = 90°$, and so the current lags the voltage in phase by 90°. In the case where there exist only a power source and a capacitor, we find $\phi = -90°$, and hence the current leads the voltage in phase by 90°.

The impedance of the circuit varies depending on the angular frequency ω of the AC electromotive force supplied from outside, and becomes minimum ($Z = R$) at

$$\omega L - \frac{1}{\omega C} = 0$$

$$\therefore \quad \omega = \omega_R = \frac{1}{\sqrt{LC}}. \tag{21.56}$$

At this time, the current becomes maximum, and it is said that the **resonance** occurs. The frequency :

$$f_R = \frac{\omega_R}{2\pi} = \frac{1}{2\pi\sqrt{LC}}$$

is called the **resonance frequency** or characteristic frequency. Making use of this fact, we can extract the component of a specific frequency from the current which is a mixture of various frequencies. In fact, it is used for a **tuned circuit** for a radio or a TV set to select the broadcast with a specific frequency (Fig. 21-27).

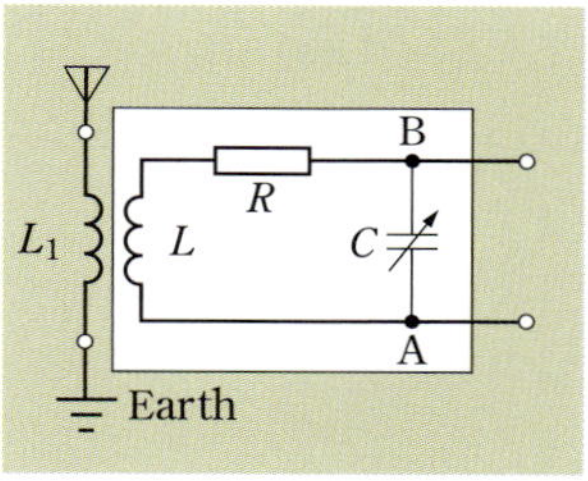

Fig. 21-27 Tuned circuit

The electric power consumed in the RLC circuit is per unit hour on the average as

$$\langle P \rangle = V_e I_e \cos\phi = R I_e^2. \tag{21.57}$$

Therefore, at the electric resistance in the RLC circuit, at the rate of $\langle P \rangle = R I_e^2$ on the average, electric power is consumed as Joule heat. Since the electric power consumed at the DC circuit is $V_e I_e$, the average electric power $\langle P \rangle = V_e I_e \dfrac{R}{Z}$ consumed at the AC circuit contains the factor $\dfrac{R}{Z}$ called the **power factor**. In the derivation of equation (21.57), we have used the relation,
$2\langle \sin(\omega t - \phi) \sin \omega t \rangle = \cos\phi \langle (1 - \cos 2\omega t) \rangle - \sin\phi \langle \sin 2\omega t \rangle = \cos\phi$.

Transformer An apparatus to raise or lower the voltage of alternating currents using the mutual inductance as shown in Fig. 21-28 is a **transformer**. A transformer consists of an iron core of quadrilateral shape and the primary and secondary coils wound around the iron core.

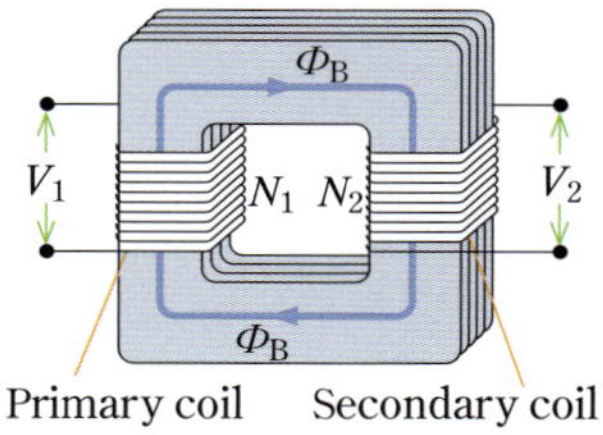

Fig. 21-28

Let the number of turns of the primary and secondary coils be N_1 and N_2, respectively. When the AC voltage V_1 is applied on the primary coil, by the alternating current through the coil, the varying magnetic flux Φ_B is generated in the iron core. The magnetic flux does not leak from the iron core and goes through the secondary coil. Here we only consider an ideal transformer in which the magnetic flux produced in the primary coil entirely goes through the secondary coil. If the magnetic flux changes by $\Delta\Phi_B$ in an infinitesimal time Δt, the counter

Fig. 21-29 A lot of transformers are used in our surroundings for such a purpose as the transmission of electricity from power stations. Here shown is the photograph of a transformer (161 kV/13.2 kV) used in a factory.

electromotive force V_{i1} generated by self-induction in the primary coil and the induced electromotive force V_2 produced in the secondary coil can be written as

$$V_{i1} = -N_1 \frac{\Delta \Phi_B}{\Delta t}, \quad V_2 = -N_2 \frac{\Delta \Phi_B}{\Delta t}. \tag{21.58}$$

Provided we can ignore the resistance in the primary coil, as the counter electromotive force generated in the primary coil is valanced with the AC voltage applied from outside ($V_1 + V_{i1} = 0$), from equation (21.58) there is a relation between the voltage and the number of turns of the primary and the secondary coils as follows :

$$\frac{|V_2|}{|V_1|} = \frac{N_2}{N_1}. \tag{21.59}$$

Therefore, between the effective value of the AC voltage V_{1e} applied on the primary coil and the effective value of the induced electromotive force V_{2e} generated in the secondary coil, we have the following relation :

$$\frac{V_{2e}}{V_{1e}} = \frac{N_2}{N_1}. \tag{21.60}$$

That is, if we take the number of turns of the secondary coil more than that of the primary coil, we would get a higher voltage and if we take the number of turns of the secondary coil less than that of the primary coil, we would get a lower voltage. The relation (21.60) was derived under the condition that there is no leakage of the magnetic flux from the iron core. In this case, the self-inductances for the primary and secondary coils, L_1, L_2 and the mutual inductance M satisfies the relation : $L_1 L_2 = M^2$.

Connecting the resistance R with the secondary coil, we find the secondary coil carries a current I_2 and the electric power is consumed at the resistor. In the ideal transformer where no energy is consumed at the iron core and the coil, because of the law of the energy conservation, the electric power consumed at the secondary coil is equal to the electric power supplied at the primary coil. Hence in this case, denoting the effective values of the currents flowing in the primary and secondary coils by I_{1e} and I_{2e}, respectively, from the energy conservation law we get the following relation :

$$I_{1e} V_{1e} = I_{2e} V_{2e}, \tag{21.61}$$

or

$$\frac{I_{2e}}{I_{1e}} = \frac{N_1}{N_2}. \tag{21.62}$$

Note that in this case we have

$$I_{1e} = \frac{I_{2e} N_2}{N_1} = \frac{V_{2e}}{R}\frac{N_2}{N_1} = \frac{V_{1e}}{R}\left(\frac{N_2}{N_1}\right)^2. \tag{21.63}$$

As can be seen from equation (21.58), if there is no variation in the magnetic flux, no voltage will appear in the transformer. In order to have changes in the magnetic flux, it is necessary for the flowing current to change. Namely, the transformer works with alternating currents. If we let a direct current flow through the transformer, we do not find any current flowing in the secondary coil, moreover, there is a risk that the winding wire in the primary coil becomes heated and the transformer burns.

Problems for exercise 21

A

1. (1) A conducting wire ($R = 20\ \Omega$) surrounding a square with area $0.25\ \mathrm{m}^2$ is placed perpendicular to the magnetic field of $B = 0.30\ \mathrm{T}$. Find the magnetic flux penetrating the square.
   (2) This magnetic field became zero in 0.01 s. Calculate the average induced electromotive force of the induced electric field. Find also the average current.
2. What is the amplitude of an induced electromotive force generated by rotating the coil of area $25\ \mathrm{cm}^2$ at a speed of 100 turns per second in a uniform magnetic field of 0.010 T? Assume that the rotating axis is perpendicular to the magnetic field (see Fig. 21-17).
3. In order to measure the intensity of the magnetic field which is oscillating with a known angular frequency ω, we place a **search coil** with a surface perpendicular to the magnetic field. Suppose that if the cross-sectional area of the coil is A, and the number of turns is N, then the voltage of the both ends of the coil is $V_0 \sin \omega t$. How can you express the change of the magnetic field with time?
4. The current flowing in a solenoid of $L = 0.1\ \mathrm{H}$ is increasing by 100 mA every 0.01 s. What is the magnitude of the induced electromotive force?
5. If we push an electromagnet into a coil with both ends connected, we feel resistance. The more number of turns of the coil becomes, the more the resistance acts. What is the reason for that?
6. (1) Find the self-inductance of the solenoid consisting of an iron core with cross-sectional area $10\ \mathrm{cm}^2$, length 10 cm, and relative permeability $\mu_r = 1000$, around which the conducting wire is wound uniformly 1000 times.
   (2) The current flowing in this solenoid increased from 0 to 10 mA in 0.01 s. What is the magnitude of the average induced electromotive force generated in the coil?
7. A coil of $L = 0.5\ \mathrm{H}$ and a resistance of $R = 100\ \Omega$ are connected in series, and further this circuit is connected with AC power source of $V_e = 100\ \mathrm{V}$, $f = 50\ \mathrm{Hz}$ (Fig. 1). Find the effective value of current I_e and the phase shift ϕ.

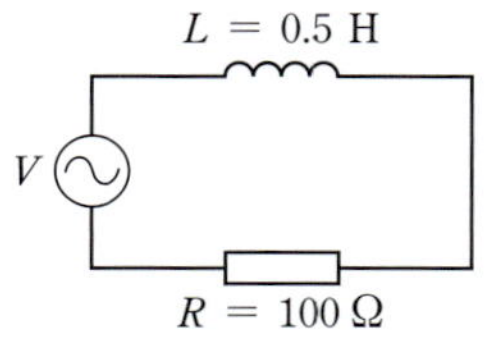

Fig. 1

8. Suppose the inductance of the coil in a radio receiver is 200 μH. In order to receive the radio wave of AM broadcast with a frequency from 500 kHz to 2000 kHz, to what values should we choose the variable range of the electric capacity of the capacitor?

B

1. When we move a conducting wire of 1 m length placed vertical to the magnetic field of $4.6 \times 10^{-5}\ \mathrm{T}$ in the air to the direction perpendicular to both of the magnetic field and the wire with a speed 10 m/s, find the electric potential difference generated between both ends of the wire.
2. We make a huge coil of $L = 40\ \mathrm{H}$ from a superconducting material.
   (1) When we connect it to the DC power source of $V = 7.5\ \mathrm{V}$, find the time for the current I flowing in the coil to reach 1500 A (Superconducting materials have 0 electric resistance).
   (2) We produced with this coil a magnetic field of 1.8 T in a vacuum of volume $27\ \mathrm{m}^3$. What is the energy of the magnetic field U?
3. Find the mutual inductance of the two concentric circular one-turn coils, L_1 and L_2 shown in Fig. 2.

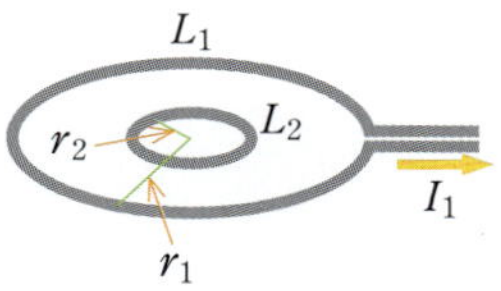

Fig. 2

4. (1) Calculate the energy stored in a space of volume $1\ \mathrm{m}^3$ which is applied by an electric field of 1 MV/m ($= 10^6$ V/m).
   (2) Calculate the energy stored in a space of volume $1\ \mathrm{m}^3$ which is applied by a magnetic field of 10 T.

Maxwell Equation and Electromagnetic Wave

VERA in Ogasawara, Tokyo (Project to make a precise three-dimensional map of the Milky Way of galaxy, in which radio telescopes with 20 m diameter are installed at 4 locations throughout Japan, and the observation is going on day and night). Radio telescope and southern sky taken at Ogasawara Observatory. Scorpio and Sagittarius, aiming at the Scorpio from the opposite bank of the Milky Way, are seen in the picture of the night sky in the summer. The bright stars around the telescope are, Saturn, Spica, and Mars.

Most of the laws in the electromagnetism we have learned so far were known by 180 years ago. Nonetheless, the concept of the electric field and the magnetic field was unknown at that time. Faraday understood the electromagnetic phenomena in terms of the lines of electric and magnetic forces. The physicist who established the electromagnetism by introducing the concept of electric and magnetic fields, organizing the laws of electromagnetism, and selecting the fundamental laws of electric and magnetic fields called Maxwell equations was Maxwell.

He derived the existence of the electromagnetic wave from his equations, and showed that the light is a kind of electromagnetic wave. In the present chapter, we will learn Maxwell's equations and the electromagnetic wave.

We have learned a number of laws about the electromagnetism. Now, let us recollect them.

The main characters of the electromagnetism are the electric field $\boldsymbol{E}$ and the magnetic field $\boldsymbol{B}$ as well as the electric charge q and the electric current I. The electric charge q receives from the electric field $\boldsymbol{E}$ the action of the electric force $q\boldsymbol{E}$, while from the magnetic field (magnetic flux density) $\boldsymbol{B}$, it receives the action of the magnetic force $q\boldsymbol{v}\times\boldsymbol{B}$, and the part with the length L of the conducting wire carrying the current I receives from the magnetic field $\boldsymbol{B}$ the action of the magnetic force $I\boldsymbol{L}\times\boldsymbol{B}$. When dielectric materials or magnetic materials come to the scene, the microscopic charge and current are represented by the macroscopic fields such as the polarization $\boldsymbol{P}$ and the magnetization $\boldsymbol{M}$.

An electric field $\boldsymbol{E}$ is generated around the electric charge, while a magnetic field $\boldsymbol{B}$ is generated around the electric current. The electric field produced by the static charge follows Coulomb's law. The line of electric force of the electric field is a line starting from a positive charge and ending at negative charge and is continuous without a break. It is Gauss' law of the electric field that $\dfrac{Q}{\varepsilon_0}$ lines of electric force emerge from the electric charge Q.

The magnetic field generated around the stationary electric current can be derived by Biot-Savart's law, and satisfies the Gauss' law and Ampere's law for the magnetic field $\boldsymbol{B}$. Gauss' law of the magnetic field $\boldsymbol{B}$ means that the line of magnetic force of the magnetic field $\boldsymbol{B}$ is a closed curve without starting and end points. Hence it shows that there is no isolated magnetic monopole.

The electric field is generated not only around the charge but also around the varying magnetic field. This is the electromagnetic induction. The line of the electric force arising from the electromagnetic induction is a closed curve with neither starting point nor end point.

22.1 Maxwell equation

Learning objective To understand that the magnetic field $\boldsymbol{B}$ is generated when the electric field $\boldsymbol{E}$ changes with time. To become able to explain what physical contents are represented by each of the four equations, usually referred to as Maxwell equations, which are the fundamental laws of electromagnetism.

The fundamental laws of electromagnetism became almost in the present form which is considered to be due to Maxwell's paper with the title "Dynamical theory of electromagnetic fields" published in 1865*. He selected following 4 fundamental laws for the electric and magnetic fields, by organizing the many laws concerning the electromagnetic phenomena that had been by that time obtained.

The first three are the following three laws.

(1) Gauss' law of electric field $\boldsymbol{E}$:

"flux of line of electric force Φ_E emerging from a closed surface S"

$$= \frac{\text{"total electric charge } Q_{\text{in}} \text{ inside a closed surface S"}}{\varepsilon_0} \qquad (16.19)$$

* In the beginning, the Maxwell equations were composed of 20 equations. The physicist who reorganized them into four vector equations was Heaviside and it was in 1884.

$$\iint_S E_n \, dA = \frac{1}{\varepsilon_0} Q_{in} \tag{22.1a}$$

(2) Gauss' law of magnetic field *B* : Separated magnetic monopole does not exist

"magnetic flux Φ_B emerging from a closed surface S" = 0 (20.5)

$$\iint_S B_n \, dA = 0 \tag{22.1b}$$

(3) Faraday's law of electromagnetic induction : Time-varying magnetic field ***B*** generates an electric field ***E***.

"line integral of the tangential component of an electric field ***E*** along a closed curve C (electromotive force)" = − "time-varying rate of a magnetic flux Φ_B penetrating a surface S with a closed curve C as the boundary" (21.7)

$$\oint_C E_t \, ds = -\frac{d\Phi_B}{dt} = -\iint_S \frac{\partial B_n}{\partial t} dA \tag{22.1c}$$

As a candidate for the final law, let us take Ampere's law for the magnetic field ***B*** generated by a steady electric current (20.15)
"line integral of the tangential component of a magnetic field ***B*** along a closed curve C"
= μ_0 × "electric current I penetrating a surface S with a closed curve C as the boundary"

$$\oint_C B_t \, ds = \mu_0 I$$

and apply this law to the experiment shown in Fig. 22-1. When a charged capacitor is discharged, the electric current flows through the conducting wire and generates the magnetic field, but there is no current flow between the capacitor plates. Hence, choosing a surface which intersects the conducting wire for the surface S with the closed curve C as the boundary illustrated in Fig. 22-1, we get I on the right-hand side of Ampere's law given as $I = -\frac{dQ}{dt}$. But, if we choose a surface passing between the two capacitor plates for the surface S, then we have $I = 0$. Because of this ambiguity on the right-hand side, Ampere's law does not hold in the case of time-dependent electric current.

Maxwell, therefore, realized that corresponding to the electromagnetic induction where an electric field is generated by the time-varying magnetic field, **a magnetic field should be induced by the time-varying electric field**. Although there is no current flow between the

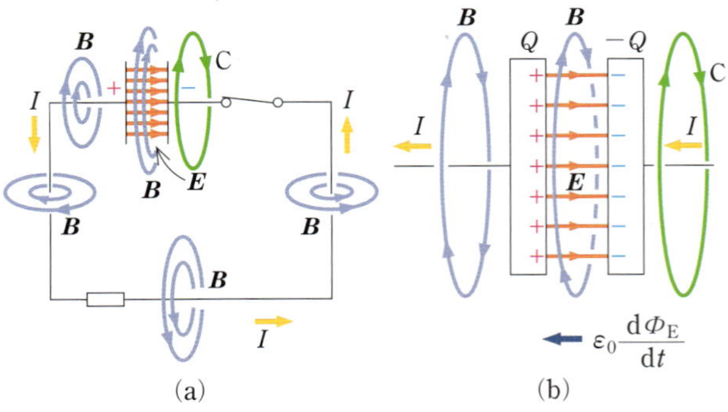

Fig. 22-1 (a) Induced magnetic field arising between the plates of a capacitor in the state of discharge
(b) Magnified figure of the neighborhood of a capacitor. $\frac{d\boldsymbol{E}}{dt}$ between the capacitor plates points to the left.

plates, the electric field there varies with time, and the time-varying rate of the flux of electric line of force Φ_E satisfies the relation $\varepsilon_0 \frac{d\Phi_E}{dt} = -\frac{dQ}{dt}$ [This relation can be derived by differentiating, with respect to t, the relation $\varepsilon_0 \Phi_E = -Q$ that is obtained by applying equation (22.1a) to the closed surface surrounding the right-hand side plate with the charge $-Q$ and by noting that $\boldsymbol{E} = \boldsymbol{0}$ on the right side of the plate. We assume that the space between the plates is a vacuum].

Maxwell thought "When the electric field varies, a magnetic field is generated just like in the case where there is a flow of the current which is equal to ε_0 times the time-varying rate of the flux of electric line of force Φ_E, $\frac{d\Phi_E}{dt}$"*. So he extended Ampere's law (20.15) which says that the magnetic field arises surrounding the current dextrally, and added the following law.

* Maxwell called $\varepsilon_0 \frac{d\Phi_E}{dt}$ the displacement current.

(4) **Ampere-Maxwell's law**: Around an electric current, a magnetic field is generated surrounding the electric current dextrally. When an electric field changes with time a magnetic field is generated.

"line integral of the tangential component of a magnetic field $\boldsymbol{B}$ along a closed curve C"
= $\mu_0 \times$ "electric current I penetrating a surface S with a closed curve C as the boundary"
+ $\varepsilon_0 \mu_0 \times$ (time-varying rate of the flux of electric line of force Φ_E penetrating a surface S with a closed curve C as the boundary)

$$\oint_C B_t \, ds = \mu_0 I + \mu_0 \varepsilon_0 \frac{d\Phi_E}{dt}$$
$$= \mu_0 I + \mu_0 \varepsilon_0 \iint_S \frac{\partial E_n}{\partial t} dA \quad \text{(in vacuum)} \quad (22.1d)$$

Since where the electric current I is disconnected there appears the same amount of displacement current $\varepsilon_0 \frac{d\Phi_E}{dt}$, the right-hand side of equation. (22.1d) does not change by the choice of the surface S with a closed curve C as the boundary. The line of magnetic force of the induction magnetic field generated by the time variation of the electric field is a closed curve as in the case of that of the magnetic field generated by the electric current (Fig. 22-2).

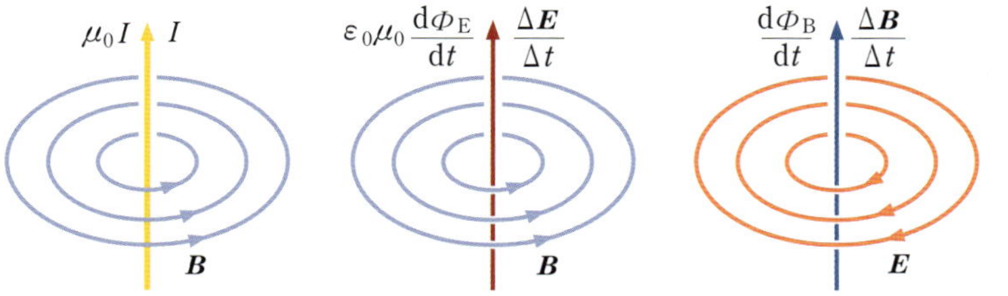

(a) A magnetic field arises around an electric current.
(b) A magnetic field arises when an electric field changes with time.
(c) An electric field arises when a magnetic field changes with time.

Fig. 22-2

Four equations (22.1a)-(22.1d) collected together are called **Maxwell equations**, which are the equations telling us about the followings :

(1) What sort of an electric field $\boldsymbol{E}$ and a magnetic field $\boldsymbol{B}$ do the charge Q and the current I produce?

(2) What sort of an electric field $\boldsymbol{E}$ is generated when a magnetic field $\boldsymbol{B}$ varies?

(3) What sort of a magnetic field $\boldsymbol{B}$ is generated when an electric field $\boldsymbol{E}$ varies?

For example, Coulomb's law obeyed by an electric field generated by a static electric charge, is derived from equation (22.1a) together with equation (22.1c) with right-hand side set to be 0, that is, the equation (16.50) which guarantees the existence of the electric potential (Proof omitted). Furthermore, Biot-Savart's law which a magnetic field generated by a steady electric current obeys, is derived from equation (22.1b) together with equation (22.1d) with right-hand side $\frac{\mathrm{d}\Phi_\mathrm{E}}{\mathrm{d}t}$ set to be 0, that is, Ampere's law (20.15) (Proof omitted). Historically, Gauss' law of the electric field was derived from Coulomb's law, and Gauss' law and Ampere's law for the magnetic field $\boldsymbol{B}$ were derived from Biot-Savart's law. However, in the theoretical system of the electromagnetism, Coulomb's law and Biot-Savart's law are regarded as derived from Maxwell equations.

Electric fields and magnetic fields are closely related with each other as seen from equations (22.1c) and (22.1d), hence they are collectively called **electromagnetic fields**.

Electric fields and magnetic fields are generated by the electric charge and the electric current, respectively. Conversely, the electric field $\boldsymbol{E}$ and the magnetic field $\boldsymbol{B}$ act electric force and magnetic force on the electric charge and the electric current, respectively. Namely we can summarize as follows :

(1) An electric field $\boldsymbol{E}$ exerts the electric force $\boldsymbol{F} = Q\boldsymbol{E}$ on a charged object with an electric charge Q,

(2) A magnetic field $\boldsymbol{B}$ exerts the magnetic force $\boldsymbol{F} = Q\boldsymbol{v}\times\boldsymbol{B}$ on a charged object with a charge Q and velocity $\boldsymbol{v}$,

(3) A magnetic field $\boldsymbol{B}$ exerts the magnetic force $\boldsymbol{F} = I\boldsymbol{L}\times\boldsymbol{B}$ on a part of conductor with a length L in which an electric current I flows.

For reference Maxwell equations in the presence of materials

In the case where there exist a dielectric substance and a magnetic body, by introducing the polarization vector $\boldsymbol{P}$ and the magnetization $\boldsymbol{M}$, respectively given by

$$\boldsymbol{D} = \varepsilon_0\boldsymbol{E}+\boldsymbol{P} \tag{22.2}$$

$$\boldsymbol{B} = \mu_0(\boldsymbol{H}+\boldsymbol{M}) \tag{22.3}$$

we define the electric flux density $\boldsymbol{D}$ and the magnetic field (intensity of magnetic field) $\boldsymbol{H}$.

Then we have from equation (22.1a) and equation (22.1d)

$$\iint_\mathrm{S} D_\mathrm{n}\,\mathrm{d}A = Q_0 \tag{22.1a'}$$

$$\oint_C H_t\,ds = I_0 + \frac{d\psi_E}{dt} = I_0 + \iint_S \frac{\partial D_n}{\partial t}\,dA \qquad (22.1d')^*$$

Q_0 is a free charge freely movable through vacuum or the material inside the closed surface S, while I_0 is a conduction current as a flow of a free charge freely movable through conducting wire or vacuum. The Maxwell equations in the presence of the dielectric substance and the magnetic body are given by a set of (22.1a′), (22.1b), (22.1c), and (22.1d′).

* ψ_E is an electric flux penetrating a surface S bounded with a closed curve C (see Section **18.1**).

For reference Example where a magnetic field is generated by the time-varying electric field

When the two plates (area A) of the parallel-plate capacitor store positive and negative charges $Q = \sigma A$, $-Q = -\sigma A$, respectively, then there appears an electric field with the intensity $E = \dfrac{\sigma}{\varepsilon_0}$ between the plates (see Section **17.2**). For an observer a who is at rest by the side of the parallel-plate capacitor, there exits only this electric field [Fig. 22-3 (a)].

While, for an observer b moving by the side of capacitor with a velocity $\boldsymbol{v}$, the electric current with the current density of $\boldsymbol{j} = \pm\sigma\boldsymbol{v}$ is flowing, and hence b observes, in addition to the electric field, the magnetic field of the intensity

$$B = \mu_0 j = \mu_0 \sigma v = v\mu_0\varepsilon_0 E \qquad (22.4)$$

(The readers only have to understand this issue qualitatively here. see Section **23.5**).

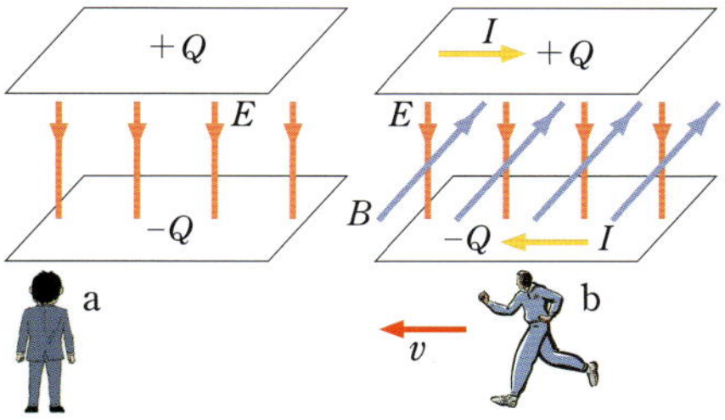

Fig. 22-3 (a) Observing only an electric field (b) Observing, in addition to the electric field, a magnetic field of the intensity $B = v\mu_0\varepsilon_0 E$.

22.2 Electromagnetic wave

Learning objective To become able to explain the reason why electromagnetic waves exist as solutions to the Maxwell equations. To become able to explain that the light is a transverse wave with an argument based on the polarization phenomena. To become able to explain how Hertz generated and detected electromagnetic waves so that he confirmed the electromagnetic wave.

When we make an electric oscillation in the circuit composed of the coil and the antenna (capacitor) shown in Fig. 22-4, oscillating electric fields are generated in the surroundings of the antenna. According to Maxwell's theory, an oscillating magnetic field arises around an oscillating electric field. By the electromagnetic induction, an oscillating electric field arises around an oscillating magnetic field. In this way, oscillating electric fields are generated around the antenna, and the oscillation of electromagnetic fields propagates outward through the space as a wave originating from the antenna, the source of the wave (Fig. 22-5). Namely, the observation that the oscillation of the electric and magnetic fields propagates as a wave leads to the existence of **electromagnetic waves**.

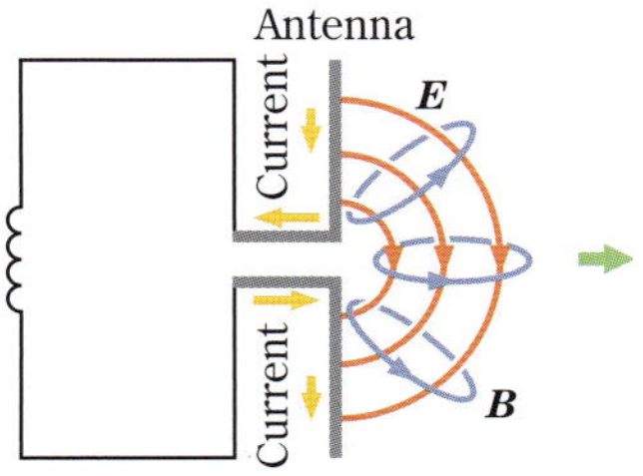

Fig. 22-4 Conceptual diagram of generating electromagnetic waves

Maxwell noticed that from the calculation of the speed of the electromagnetic wave in vacuum, it amounts to 300 000 km per second (3×10^8 m/s), which is equal to the speed of light in the air. The speed

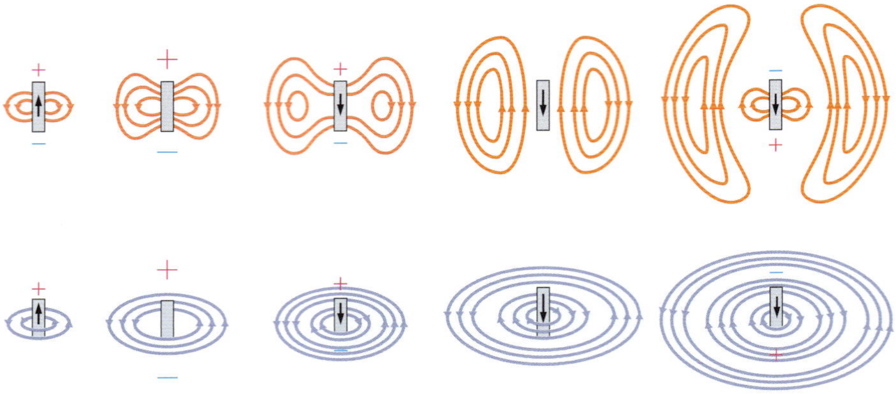

Fig. 22-5 Emission of electromagnetic waves (conceptual diagram). To make it easy to see, we have shown the lines of electric force and the magnetic line of force separately.

of light in the air was measured by Fizeau in 1849, and he had obtained the result of about 300 000 km per second (see Problems for exercise 22, B22). Thus, Maxwell arrived at an amazing conclusion :

Light is a propagation of oscillating electric and magnetic fields, namely, an electromagnetic wave.

Maxwell wrote in his paper of 1864 as follows. "As this speed is very close to the speed of light, we seem to have a strong reason to conclude that light, including the radiating heat (infrared ray) and other kind of radiation (if they were present), follows the laws of the electromagnetism and is an oscillation of electric and magnetic fields propagating as a form of a wave, that is, an electromagnetic wave. …"

Thus, it has turned out that light is a kind of the electromagnetic wave. Since the sound is a propagation of vibration of particles of the air, it does not travel through the vacuum in the absence of the air. On the other hand, as the electromagnetic field exists even in a vacuum, the electromagnetic wave propagates through the vacuum. Light is an electromagnetic wave emitted by the electric current due to the motion of the electron inside the molecule or the atom.

Fig. 22-6 Hiratsuka TV relay station

Electromagnetic waves are called by various names depending on the wavelength. As shown in Table 22-1, in addition to the light (visible light), electromagnetic waves include gamma ray (γ ray), X ray, ultraviolet ray, infrared ray, radio wave and so on. The electromagnetic wave used for the communication with the wavelength longer than 0.1 mm is called a **radio wave**. The wavelength of the AM radio broadcast is 190–560 m and that of the FM radio broadcast is 3.3–3.9 m.

In order to detect the electromagnetic wave, one only has to adjust the resonance frequency of the tuning circuit shown in Fig. 21-27 to match the frequency of the electromagnetic wave and to detect the oscillating current generated in the circuit.

Speed of electromagnetic wave Although the behavior of the oscillation of electric and magnetic fields of an electromagnetic wave in the vicinity of an antenna is complicated, the electromagnetic wave

Table 22-1 various electromagnetic waves

wavelength [m]	frequency [Hz]	Name and frequency		
10^5	3×10^3			
10^4	3×10^4	Very Low Frequency (VLF)	Radio wave	3～30 kHz
10^3	3×10^5	Low Frequency (LF)	Radio wave	30～300 kHz
10^2	3×10^6	Medium Frequency (MF)	Radio wave	300～3000 kHz
10	3×10^7	High Frequency (HF)	Radio wave	3～30 MHz
1	3×10^8	Very High Frequency (VHF)	Radio wave	30～300 MHz
10^{-1}	3×10^9	Ultra High Frequency (UHF)	Micro wave	300～3000 MHz
10^{-2}	3×10^{10}	Super High Frequency (SHF)	Micro wave	3～30 GHz
10^{-3}	3×10^{11}	Extremely High Frequency (EHF)	Micro wave	30～300 GHz
10^{-4}	3×10^{12}	Submillimeter Wave		300～3000 GHz
10^{-5}	3×10^{13}	Infrared ray		
10^{-6}	3×10^{14}	7.7×10^{-7} m	Visible light	
10^{-7}	3×10^{15}	3.8×10^{-7} m	Visible light	
10^{-8}	3×10^{16}	Ultraviolet ray		
10^{-9}	3×10^{17}	Ultraviolet ray		
10^{-10}	3×10^{18}	X ray		
10^{-11}	3×10^{19}	X ray		
10^{-12}	3×10^{20}	γ ray		
10^{-13}	3×10^{21}	γ ray		

emitted from the antenna may be considered to propagate as a plane wave and hence becomes simple, provided we limit the region far enough away from the antenna.

In the case of a plane wave, the state of the electromagnetic fields at some instant, is constant on the plane perpendicular to the direction of propagation of the wave, and varies periodically along the travelling direction. If the electromagnetic wave were a longitudinal wave in which the electric and magnetic fields oscillate parallel to the direction of the propagation, the lines of electric force and magnetic line of force would be created or annihilated in the vacuum in the absence of electric charges (Fig. 22-7). Therefore, the electric and magnetic fields have to be perpendicular to the travelling direction of the wave, hence the electromagnetic wave is a transverse wave.

We can derive that the speed of electromagnetic wave is about 3×10^8 m/s as follows :

(1) As shown in Fig. 21-18, the magnetic field $\boldsymbol{B}$ passing through a point with a velocity $\boldsymbol{v}$ produces an induced electric field $\boldsymbol{E}$ which is perpendicular to both $\boldsymbol{v}$ and $\boldsymbol{B}$, having the magnitude [Fig. 22-8 (a)]

$$E = vB. \tag{22.5}$$

(2) Comparing equation (22.1c) with equation (22.1d) where we set $I = 0$, we see that $\boldsymbol{E}$ corresponds to $\boldsymbol{B}$, and $\boldsymbol{B}$ corresponds to $-\varepsilon_0\mu_0\boldsymbol{E}$ (Note the minus sign). This electric field $\boldsymbol{E}$ moves with the same velocity $\boldsymbol{v}$ as $\boldsymbol{B}$ moves. Hence, corresponding to (1), $\boldsymbol{E}$ generates an induced magnetic field $\boldsymbol{B}$ which is perpendicular to both $\boldsymbol{v}$ and $\boldsymbol{E}$, having the magnitude

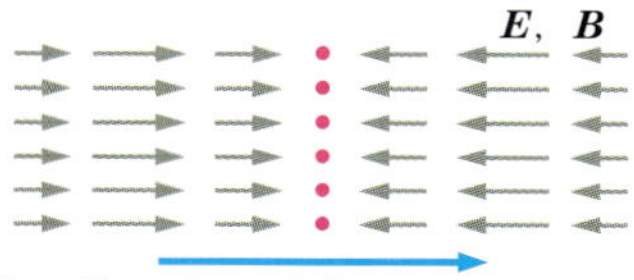

Fig. 22-7 Let us suppose that the electromagnetic wave is a longitudinal wave. If we draw the lines of electric and magnetic forces such that their densities are proportional to the strength of the electric and magnetic fields shown in the figure, there must occur creations and annihilations of the lines everywhere.

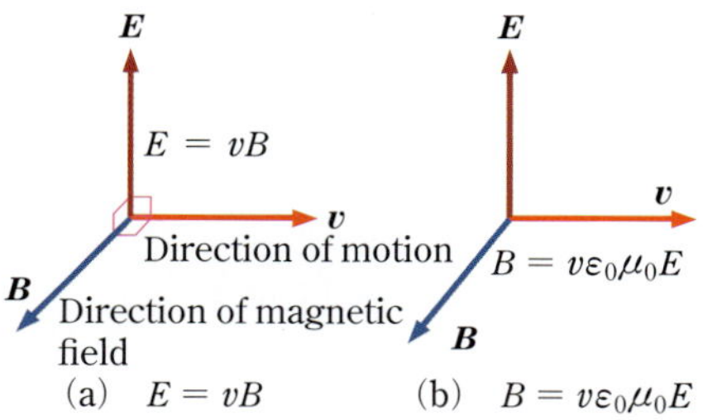

Fig. 22-8

$$B = v\varepsilon_0\mu_0 E. \qquad (22.4)$$

[Fig. 22-8 (b), see the "For reference" at the end of the previous section.]

(3) The condition for equations (22.4) and (22.5) to hold at the same time, that is, the condition for **B** in (2) to be equal to the original **B** is given by $B = v\varepsilon_0\mu_0 E = v^2\varepsilon_0\mu_0 B$. Namely, from $v^2 = \frac{1}{\varepsilon_0\mu_0}$, the speed of the electromagnetic wave in a vacuum, c is obtained as follows :

$$c = \frac{1}{\sqrt{\varepsilon_0\mu_0}} \approx 3\times 10^8\ \mathrm{m/s}, \qquad (22.6)$$

where we have used $\frac{1}{4\pi\varepsilon_0} \approx 9\times 10^9\ \mathrm{N{\cdot}m^2/(A^2{\cdot}s^2)}$ and $\mu_0 = 4\pi\times 10^{-7}\ \mathrm{N/A^2}$.

Summarizing the results derived in the above, we have the following conclusions.

In the case where the oscillation of the electric and magnetic fields propagates as a wave in a vacuum*,

* Make use of equation (22.1d′) instead of equation (22.1d).

(1) The speed is constant and is equal to the speed of light in a vacuum, c.

(2) The direction of oscillation of the electric field and that of the magnetic field are perpendicular to each other

$$\boldsymbol{E} \perp \boldsymbol{B} \qquad (22.7)$$

(3) The propagating direction $\boldsymbol{k}$ of the electromagnetic wave is perpendicular to both direction of oscillation of the electric field and that of the magnetic field

$$\boldsymbol{k} \perp \boldsymbol{E}, \qquad \boldsymbol{k} \perp \boldsymbol{B} \qquad (22.8)$$

$\boldsymbol{k}$ points in the direction to proceed the right-handed screw when it is turned from the direction of $\boldsymbol{E}$ to that of $\boldsymbol{B}$ (in terms of the vector product, the direction of $\boldsymbol{E}\times\boldsymbol{B}$).

[Note that the electromagnetic wave is a transverse wave. (Fig. 22-9)]

(4) $$E = cB \qquad (22.9)$$

The speed of light c_r in a homogeneous material with a relative dielectric constant ε_r and a relative permeability μ_r is given by*

$$c_r = \frac{1}{\sqrt{\varepsilon_r\mu_r\varepsilon_0\mu_0}} = \frac{c}{\sqrt{\varepsilon_r\mu_r}} \qquad (22.10)$$

For a paramagnetic material or a diamagnetic material, $\mu_r \approx 1$, and hence $c_r \approx (\varepsilon_r)^{-1/2}c$.

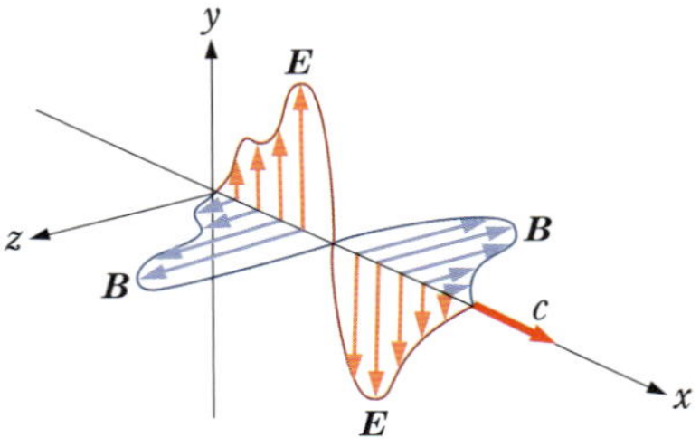

Fig. 22-9 An electromagnetic wave propagating in the direction of +x-axis

Polarized light It can be realized that light is a transverse wave, if we use two polarizing plates (polarizers) called the polaroid. In the sun light which is called a natural light and an incandescent light, the oscillating direction of the magnetic field points to various directions on a plane perpendicular to the travelling direction (Fig. 22-11, The arrow means the direction of the magnetic field).

Polaroid is made by aligning the directions of the needle crystals of an organic compound and then by implanting them into plastic plates.

Fig. 22-10 Polarized light

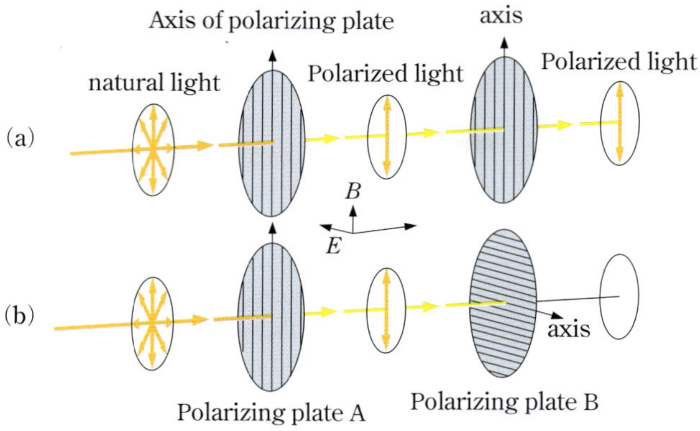

Fig. 22-11 Polarized light by polarizer

When the oscillating direction of the electric field points to the axial direction of polarizer (direction of the needle crystal), the energy of the oscillating electric field is absorbed by the crystal. Thus, the polarizer lets only the light in which the magnetic field oscillates in the axial direction (the electric field oscillates in the direction perpendicular to the axis) pass through. The light in which the oscillating direction is polarized is called the polarized light. As illustrated in Fig. 22-11, we superimpose two polarizing plates and then rotate one of them. When the axes of the two polarizers are aligned, we have most bright light, the brightness gradually decreases during the rotation, and it becomes darkest when rotated by 90 degree. This experimental result can be easily understood if light is a transverse wave, but cannot be understood if it is a longitudinal wave.

Hertz's experiment An important conclusion of Maxwell's theory is that "There exist electromagnetic waves having various wavelengths, and all the electromagnetic waves propagate with a speed of 300 000 km per second in a vacuum". Since the wavelengths of light reside in the limited range of a few millionths meter, we need to generate and detect electromagnetic waves other than light, in order to establish Maxwell's theory. An electromagnetic wave can be generated by an oscillating electric current.

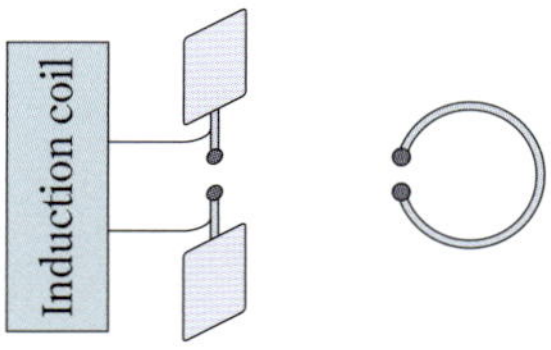

Fig. 22-12 Conceptual diagram of Hertz's experiment. Left-hand side is a generator of electromagnetic waves, and right-hand side is a detector (a wire loop with a small gap)

The electromagnetic wave Maxwell predicted was experimentally proved in 1888, 20 years later after his prediction and also after his death. Hertz constructed two apparatuses which resonate with the same frequency (generator and detector of electromagnetic waves) (Fig. 22-12).

An apparatus with an induction coil is the generator of electromagnetic waves (Fig. 22-13). Switching on and off the electric current of the coil A by oscillating switch S, generates a heavily varying magnetic field in the iron core, which in turn generates an alternating high voltage in the coil B rolled many times as a result of electromagnetic induction. Then molecules in the air are ionized and sparks occur between the terminals. Sparks imply the existence of the oscillating currents which rapidly go back and forth between the terminals. The

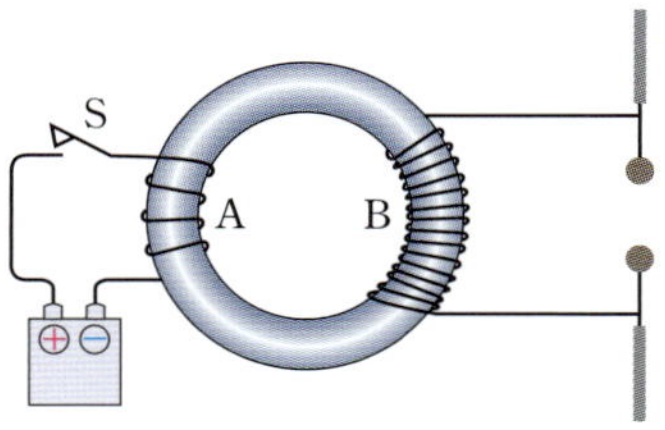

Fig. 22-13 Conceptual diagram of a generator of electromagnetic waves

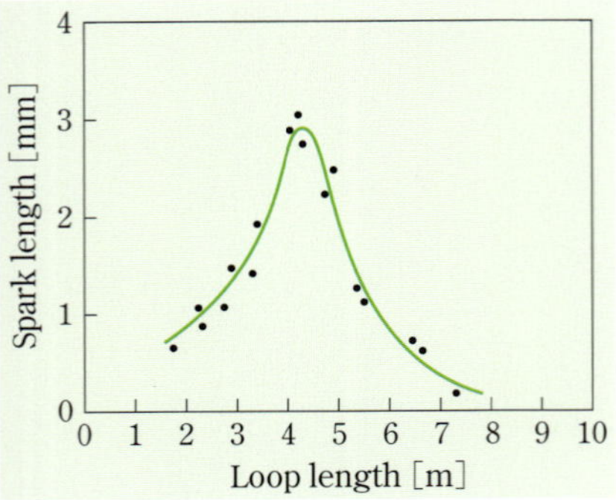

Fig. 22-14 Resonance curve observed by Hertz

frequency can be adjusted by the size and shape of the terminals.

Hertz used a conducting wire loop with a small gap as a detector for the electromagnetic wave, and he discovered that the sparks fly between the terminals of the induction coil and also at the gap of the loop at the same time (Fig. 22-14). This experimental result can be interpreted as follows. The oscillating current between the terminals produces the oscillation of electric and magnetic fields, which propagates through the space as an electromagnetic wave. Then this electromagnetic wave, during the passing through the loop, generates oscillating electric and magnetic fields. Finally, this oscillating electric field gives rise to sparks between the gap of the loop.

The rotation or changing the size of the generator brings the changes in the way of discharges. This shows the fact that the produced electromagnetic wave is polarized as well as the fact that it possesses a specific wavelength.

Moreover, Hertz measured the speed of electromagnetic waves in 1888, and confirmed that it is the same as the speed of light just as Maxwell predicted. Further, Hertz found that this electromagnetic wave undergoes reflection and refraction on the surface of solid materials, and it also shows interference and diffraction phenomena.

22.3 Electromagnetic fields

Learning objective To understand that oscillation of the electric and magnetic fields propagates through the vacuum as an electromagnetic wave, which carries energy as well as momentum, therefore the electric field and the magnetic field actually exist as physical objects.

Energy of electromagnetic wave The energy of electric field exists where there is an electric field, while the energy of the magnetic field exists where there is a magnetic field. The electromagnetic wave, which represents the propagation of the oscillating electric and magnetic fields through space, carries energy of electric and magnetic fields. From equations (17.16) and (21.32) we find that there is the following energy per unit volume in a vacuum where the electromagnetic wave propagates,

$$u = \frac{1}{2}\varepsilon_0 E^2 + \frac{1}{2\mu_0}B^2 = \varepsilon_0 E^2 \qquad \text{(in vacuum)} \tag{22.11}$$

(We have used $E = cB = \dfrac{B}{\sqrt{\varepsilon_0\mu_0}}$). Therefore, the energy S of the electromagnetic field passing through the unit area perpendicular to the travelling direction of the electromagnetic wave per unit time is given by

$$S = c\varepsilon_0 E^2 \qquad \text{(in vacuum)}. \tag{22.12}$$

Now we define the vector $\boldsymbol{S}$ which has the magnitude S and points to the travelling direction of the electromagnetic wave (propagating direction of energy) and represents the state of energy propagation, due to electromagnetic wave, as

$$\boldsymbol{S} = \frac{1}{\mu_0}\boldsymbol{E}\times\boldsymbol{B} \quad \text{(in vacuum)} \tag{22.13}$$

which is called **Poynting's vector**.

Momentum of electromagnetic wave An electromagnetic wave carries energy as well as momentum. In terms of Poynting's vector $\boldsymbol{S}$ and the energy density u of an electromagnetic wave, the momentum density $\boldsymbol{P}$ carried by the electromagnetic wave propagating though vacuum is

$$\boldsymbol{P} = \frac{\boldsymbol{S}}{c^2}, \qquad P = \frac{u}{c} \quad \text{(in vacuum)}. \tag{22.14}$$

With this definition we can show that the sum of the momentum of a charged particle and an electromagnetic field is conserved (Proof omitted).

According to Newton's second law of motion, the time change rate of the momentum of an object is equal to the force acting on the object. If light has a momentum, when it hits an object and is absorbed or reflected, the change of the momentum of light leads to the change of the momentum of the object. Thus, the object receives a pressure from the light. We call this the radiation pressure of light, which was confirmed independently by American physicists Nichols and Hull, and by a Russian physicist Lebedev in 1901–1903 (Fig. 22-15).

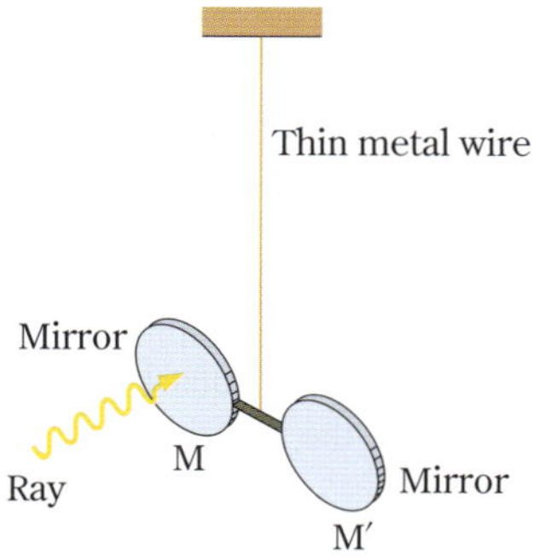

Fig. 22-15 A thin metal wire is twisted by the radiation pressure of light made incident on the mirror M. The measurement of the angle of the twist determines the radiation pressure of light.

Essential nature of electric and magnetic fields In Chapter 16, we have learned how an electric field $\boldsymbol{E}$, which exerts the electric force $\boldsymbol{F} = q\boldsymbol{E}$ on a charged object with the charge q, appears around an electric charge. In Chapter 20, we have learned what sort of the magnetic field is generated around an electric current or a magnet, and further what sort of the magnetic forces are exerted on an electric current and a charged object by the magnetic field. In these chapters, main characters are charge, current and magnet and so on. So, you may have an impression that the electric and magnetic fields are mere supporting characters or they are introduced just for a convenience of calculation.

On the other hand, the essence of the electromagnetic induction we have learned in Section **21.1**, is not that by moving magnets or by changing the currents in the vicinity of a coil we have a current flow in the coil, but that variation of the magnetic field near the coil generates an electric field. In other words, the electromagnetic induction is a phenomenon where electric and magnetic fields are intertwined with each other. When a magnetic field changes there appear an electric field. On the other hand the changes of an electric field generate a magnetic field. When a magnetic field changes, an electric field is generated even in the absence of the coil. And also, if an electric field varies there appears a magnetic field. This statement is confirmed by the fact that an electromagnetic wave exists.

Moreover, the fact that an electromagnetic wave carries energy and

Fig. 22-16 An imaginary figure of surroundings of a super huge black hole. The neighborhood of a black hole is at high temperature, and strong electromagnetic waves fly around. It is considered that active star-formation activities take place, where the materials are pushed out and collide with other materials by the radiation pressure of the electromagnetic wave.

momentum indicates that electric and magnetic fields are not virtual, but actually exist as a physically realistic object.

Then, what is real entity of the electric and magnetic fields? As light and electromagnetic waves propagate through vacuum, the electromagnetic waves, which convey the vibration of electric and magnetic fields, are not connected with so-called materials. We may say that the electromagnetic field is a property of the space where we are present. We will come back again this subject in the next chapter.

Problems for exercise 22

A

1. Find the wavelength of electromagnetic wave of the radio broadcast with frequency 1200 kHz and that of television broadcast with frequency 500 MHz. Assume the speed of light is $c = 3\times 10^8$ m/s.

B

1. The maximum value of the electric field, at the point with the distance 50 km apart from the antenna of height 40 m of the transmitting station of radio broadcast, is 10^{-3} V/m. In this case, how many watts (W) is the energy flow (per 1 m^2)? And what is the maximum value of the magnetic field there?
2. **Fizeau's experiment** When the rotation speed of the cogwheel [number of cogs (teeth) $N = 720$] of the apparatus shown in Fig. 1 is adjusted, the light passing through between the cogs (teeth) and then reflected by the mirror M, is totally blocked by the adjacent cogs (teeth) rotated during that time. Fizeau increased the number of rotations per unit time, n, gradually from 0. Then he found at a rotation speed $n = 12.6$ turn/s, the eyesight of an observer O became darkest for the first time. Find the speed c of light from the result of this experiment.

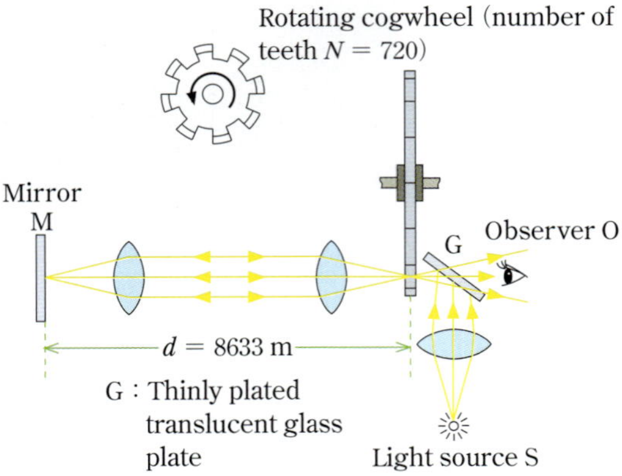

Fig. 1

3. As shown in Fig. 2, we set up the mirror A which rotates around the center O with a high speed and another mirror B which is fixed. The light emitted from the light source S hits and reflects at A, B and then returns to A. At that moment, A is rotated from M to M′, hence the reflected light proceeds to the direction of OS′ deviating from that of OS by an angle β. By measuring this angle β, Foucault determined the speed of light in 1850. Assume that $\overline{PO} = 20$ m, the rotation speed of the rotating mirror was 800 turns/s, and $\beta = 1.34\times 10^{-3}$ rad. What is the speed of light from this experiment? Note that $\beta = 2\theta$.

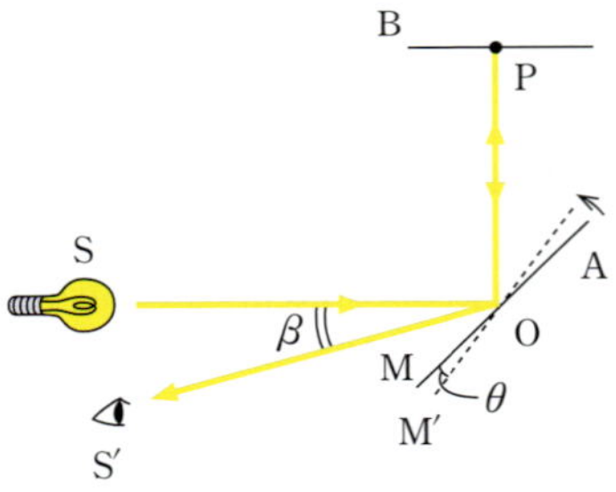

Fig. 2

Energy transfer line from battery to miniature light bulb (resistor)—Poynting's vector—

The current circuit transfers the electric energy from the power source to the circuit element. In this case, through which route is the electric energy transferred? As shown in Fig. 22-A, a battery, a miniature light bulb (resistor) and a switch are connected in series. Unless the current flows through the circuit by switching on, the miniature light bulb does not light up. So, one might imagine that the energy travels through the conducting wire. Well then, is the energy carried through the conducting wire from the positive electrode to the miniature bulb and the direction of electric current? Or is it carried through the wire from the negative electrode to the miniature bulb and the direction of electron current?

The electromagnetism tells us that the flow of the electromagnetic energy is expressed by Poynting's vector

$$\boldsymbol{S} = \frac{1}{\mu_0}\boldsymbol{E}\times\boldsymbol{B}.$$

Thus, let us examine how Poynting's vector behaves in our case. The upper part of the circuit of Fig. 22-A is higher than the lower part in electric potential by 1.5 V. So, there exist electric fields directed from above to below. Since the starting point of the line of electric force is a positive charge and the end point is a negative charge, the conducting wire, the both ends of the resistor, and the electrodes of the battery are charged as shown in Fig. 22-A.

Around the current in Fig. 22-A, magnetic fields appear in a way for the magnetic line of force going around dextrally. Hence, Poynting's vector points to the direction, to which the right-handed screw moves when rotating it from the direction of the electric field $\boldsymbol{E}$ to that of the magnetic field $\boldsymbol{B}$ as drawn with the green lines in Fig. 22-A.

The electromagnetic energy, which is outgoing from the inside of the battery to outside by the flow from the negative to the positive electrodes, originates from the chemical electromotive force of the battery. Most of the electromagnetic energies propagate outside of the conducting wire entering into the resistor from its side surface and turn to the Joule heat. If we focus on the aspect of energy, the electric current is a type of flow totally different from the water current.

Finally, we add one comment. If we put the conducting wire at a superconducting state, there exist no electric fields inside the conducting wire, and hence all the electromagnetic energies propagate outside the conducting wire.

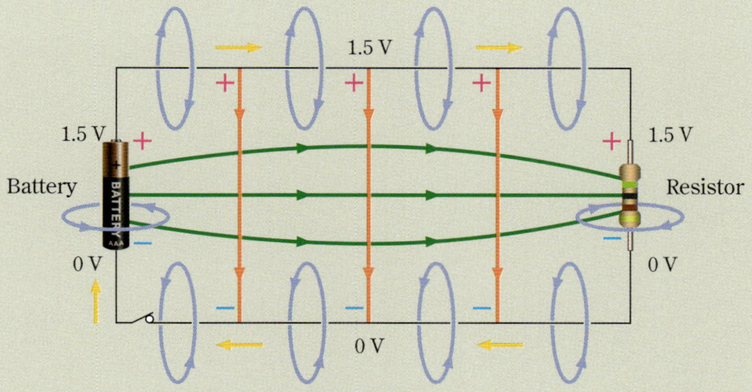

Fig. 22-A

Theory of Relativity

The phenomenon in which the vibration propagates in the air is the propagation of the sound wave, while the phenomenon where the vibration propagates through the Earth's crust is the earthquake. The electromagnetic wave is the propagation of the oscillation of the electric and magnetic fields. Here we wonder if the oscillation of the electric and magnetic fields is the oscillation of the unknown matter which is the medium of the electromagnetic wave. In 19th century, physicists thought that it was unreasonable for the light wave to propagate through the vacuum where no matter exists, and that the universe must be filled everywhere with the matter called "ether" which plays the role of medium through which the light wave propagates.

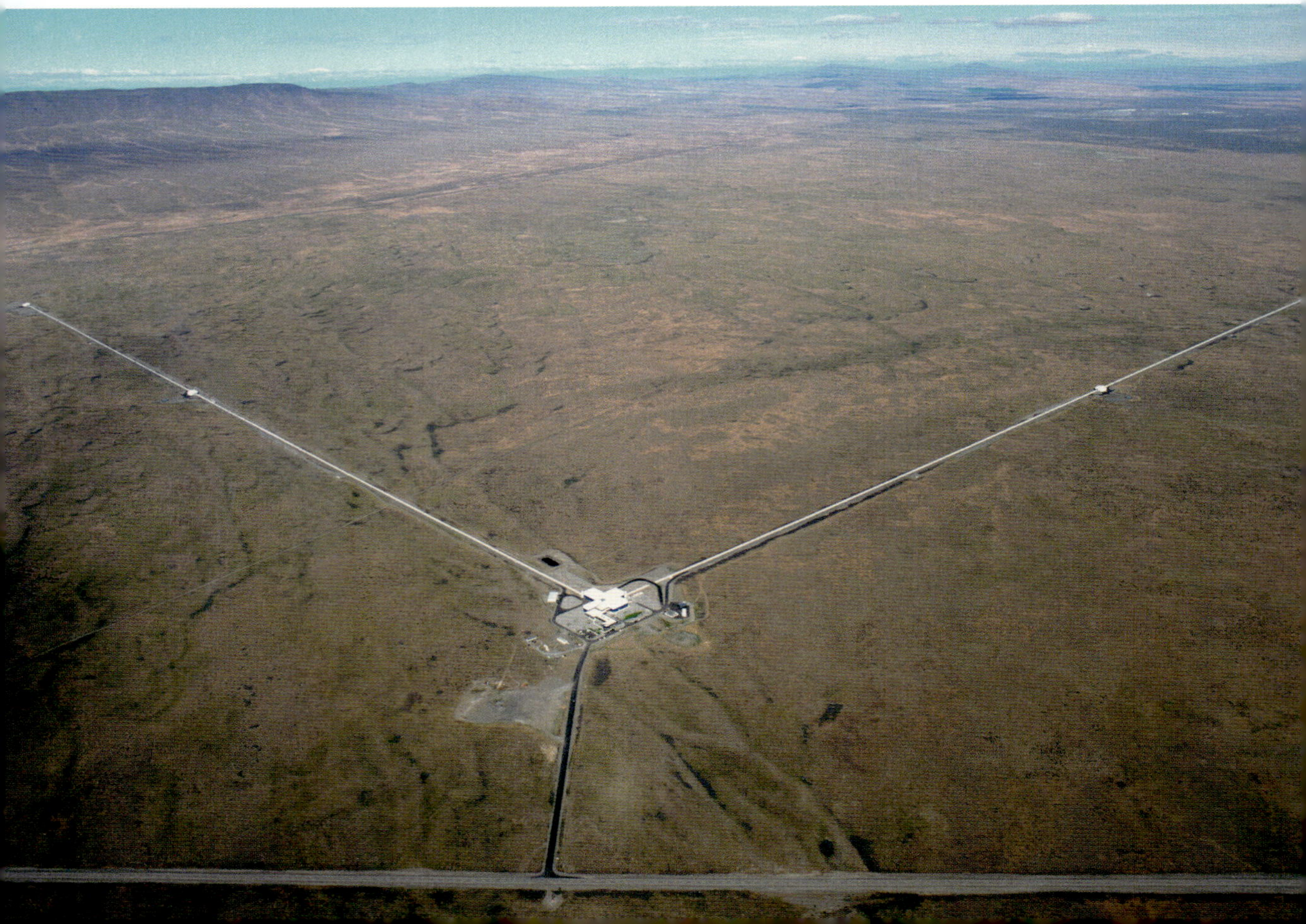

LIGO (Laser Interferometer Gravitational-Wave Observatory)
The gravitational waves (wave in which the oscillated distortion of the space-time propagates with the speed of light) emitted when two black holes, 1.3 billion light years apart from the Earth, collided and merged, were observed in 2016 for the first time. Photograph taken in 2008.

Since the light from the fixed star arrives at the Earth, the ether must fill the whole universe uniformly. The light (electromagnetic wave) is a transverse wave, therefore the medium which can transmit the transverse wave is only a solid that can be twisted. Moreover, the speed of light is so high that the ether has to be very hard (twisting elasticity is so large) and the density must be very small. However, it is not conceivable that the ether resists against the motion of the ordinary object. If the ether ever exists, it must have such a curious property.

Now, as the wave propagates at a constant speed with respect to the medium, the observer moving at the velocity $\boldsymbol{u}$ with respect to the medium, is supposed to observe the velocity of the wave front moving at the velocity $\boldsymbol{c}$ with respect to the medium, as $\boldsymbol{c}'$ where $\boldsymbol{c}' = \boldsymbol{c} - \boldsymbol{u}$. Therefore, only when we measure the speed of the light (light wave) in the reference frame fixed to the ether, the speed of light should be constant irrespective of the direction. While, when we measure the speed of light in the reference frame moving relatively to the ether, the speeds of light are different depending on the direction. Now, let us start with the story about the experiment of Michelson-Morley which tried to detect the effects of relative motion of the Earth to the ether, with the above expectation.

23.1 The Michelson-Morley Experiment

Learning objective For the light which propagates as a wave in space, there exists no matter as the medium. To understand the principle and the result of the Michelson-Morley experiment which showed this fact.

Principle of the Michelson-Morley experiment Let us take a motion of the ship traveling in the river as an analogy of the light propagating in the ether. The ship which moves on the static water with the speed c, when going down the river with the flow u, will have a speed $c+u$ with respect to the bank of the river. While, when it goes up the river, it will have a speed $c-u$. Hence, the round-trip time t_1 from a certain point to another point downstream with a distance L is given by

$$t_1 = \frac{L}{c+u} + \frac{L}{c-u} = \frac{2cL}{c^2-u^2} \tag{23.1}$$

[see Fig. 23-1 (a)]. Next, let us take the time as t_2 for this ship to go

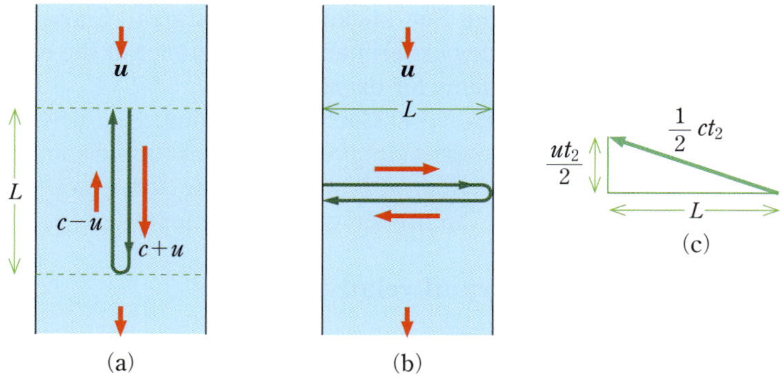

Fig. 23-1 Principle of the Michelson-Morley experiment. $\boldsymbol{u}$ is the velocity of the flow of the river. c is the speed of the ship with respect to the static water.

and back across the river with the width L. As the river flows with the speed u, if the ship goes by aiming at the point $\frac{ut_2}{2}$ upstream on the opposite bank, then it moves in the direction vertical to the bank of the river. Therefore, as $2\sqrt{L^2+\left(\frac{ut_2}{2}\right)^2} = ct_2$ we obtain

$$t_2 = \frac{2L}{\sqrt{c^2-u^2}} \neq t_1 \tag{23.2}$$

[see Fig. 23-1 (b), (c)]. In this way, when the river is flowing, the time for a ship to travel the same distance depends on the direction to go.

The Earth rotates its own axis and revolves around the Sun. Since we do not think that the ether moves with the Earth together, when we investigate if there is any difference in the time for the light to go a round trip depending on the direction, we should be able to see whether there is any medium for the light or not.

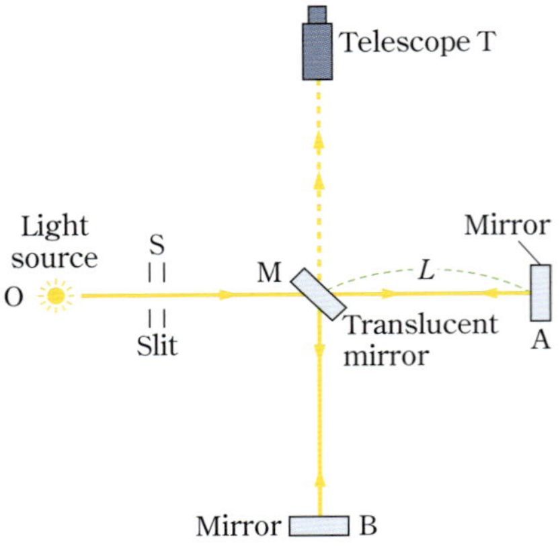

Fig. 23-2 Conceptual diagram of the Michelson-Morley experiment.

The experiment of Michelson and Morley In 1887, Michelson and Morley conducted an experiment using the apparatus shown in Fig. 23-2. They transform the light from the light source O into the thin light ray by the slit S, and then split it by the translucent mirror M to the 2 light rays MA and MB which are perpendicular to each other and subsequently let them reflect by the mirrors A and B. The two reflected light rays either reflect or transmit the translucent mirror M and enter the telescope T, and interfere each other forming the interference fringes there. Since this apparatus is floating above the mercury, we can rotate the whole apparatus on the horizontal plane. Even when the apparatus is rotated by 90 degrees, there appeared no shift in the interference pattern.

As the Earth performs rotation and revolution, it is regarded as moving with respect to the inertial frame. Therefore, the effects of the Earth's motion with respect to the ether, should turn into phase difference of lights corresponding to the difference between t_1 given by equation (23.1), and t_2 given by equation (23.2). When the apparatus is rotated by 90 degrees, t_1 and t_2 are exchanged, so the change in the phase difference will give rise to a shift of the interference fringes, and then this shift must be observed. As the result of the experiment, in any season of the year when the observation is carried out, no shift in the interference fringes was detected. It is not conceivable that the ether revolves around the Sun sticking to the Earth. Consequently, from the Michelson-Morley experiment, the existence of the ether was rejected (see B1 of Problems for exercise 23).

It has turned out that the electric and magnetic fields which propagate light (electromagnetic wave, in general) in the vacuum, are not the properties associated with a matter like the ether, but the property of the space or the property of the vacuum.

23.2 Special theory of relativity

Learning objective To understand what principle Einstein's special theory of relativity is based on.

The inertial frame is the coordinate system where the law of inertia holds. In Section **9.1**, we have introduced Galilei's principle of relativity in which we consider "The coordinate system moving with a constant speed with respect to an inertial frame is another inertial frame. For all the inertial frames in mechanics, the acceleration, the mass and the force are the same, and the law of motion, $m\boldsymbol{a} = \boldsymbol{F}$ in the same form holds".

Then, what about the laws of electromagnetism? In Section **21.1**, we have learned that in the two inertial frames moving with the constant relative velocity, the law of electromagnetic induction holds in the same form. In other words, it is likely that the fundamental laws of the electromagnetism of the same form hold in both of the inertial frames. And in Section **22.2**, we learned that the speed of light can be calculated from the Maxwell equations. If the Maxwell equations hold in the same form both in the two inertial frames, then the speed of light should be of the same value in both reference frames. This coincides with the result of the Michelson-Morley experiment.

In 1905 Einstein proposed the **special theory of relativity** with the following two postulates as the fundamental principle* :

(1) The coordinate system moving with a constant speed with respect to an inertial frame is also an inertial frame, and the fundamental laws of physics hold in the same form in all inertial frames (**Einstein's principle of relativity**).
(2) In all inertial frames, the speed of light is a constant irrespective of the traveling direction (**Principle of the constancy of the speed of light**).

However, the constancy of the speed of light is in contradiction with the transformation law of the velocity in Newtonian mechanics (See the following column). Einstein noticed that the assumption in the Newtonian mechanics that the time can be measured using the common clock in all inertial frames is the cause of this contradiction. And he thought that the contradiction will be resolved if we think that the timing of the clock differs in the different inertial frames. The transformation law of the position coordinates and the times in different inertial frames is the Lorentz transformation explained in the following column.

Fig. 23-3 For a person who is surfing, the wave moving with the same speed as that of the person looks as if the wave is at rest. If we can run after the light with the same speed of light, does light wave look at rest? The theory developed from this kind of question Einstein asked is the special theory of relativity.

* The theory based on the principle of general relativity which states that the fundamental laws of physics hold in the same form in all coordinate systems is the general theory of relativity proposed by Einstein in 1916. The general theory of relativity has its fundamental basis on the principle of equivalence which says that the force of inertia gives rise to the same effect as the gravitational force.

For reference Transformation law of the velocity in Newtonian mechanics

As shown in Fig. 23-4, in Newtonian mechanics, there is a relation between the two coordinate systems, system S and system S′, moving in a relative motion with the velocity $\boldsymbol{u}$ in the direction of x-axis :

$$x' = x - ut, \quad y' = y, \quad z' = z, \quad t' = t. \tag{23.3}$$

Since it is considered that we can use a common clock for the measurement of the time in the two coordinate systems, we set $t' = t$. From equation (23.3) we can derive the relation between the veloci-

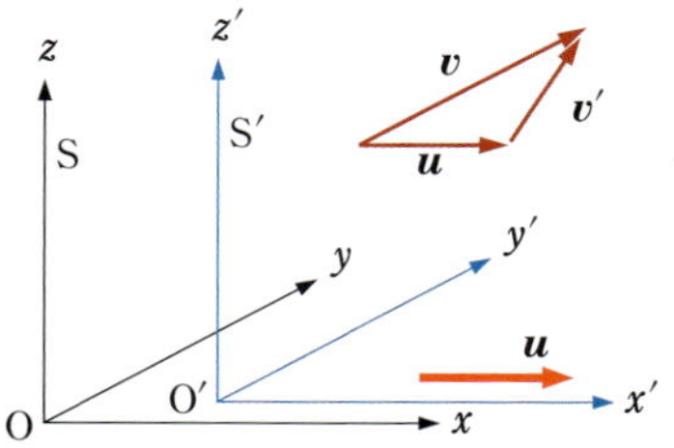

Fig. 23-4 Two coordinate systems ; system S (O-x, y, z system) and system S′ (O′-x', y', z' system). The system S′ is in a uniform linear motion with respect to the system S along the x-axis with a constant velocity $\boldsymbol{u}$. In Newtonian mechanics, $\boldsymbol{v}' = \boldsymbol{v} - \boldsymbol{u}$.

ties of the objects $\boldsymbol{v} = (v_x, v_y, v_z)$ and $\boldsymbol{v}' = (v_x', v_y', v_z')$ in the two coordinates :

$$v_x' = \frac{\mathrm{d}x'}{\mathrm{d}t'} = \frac{\mathrm{d}}{\mathrm{d}t}(x-ut) = v_x - u, \qquad v_y' = v_y, \qquad v_z' = v_z. \tag{23.4}$$

So, we get the relation $\boldsymbol{v}' = \boldsymbol{v} - \boldsymbol{u}$ (Fig. 23-4)

For reference　Lorentz transformation

However, if we apply equation (23.4) to the velocity of light, it contradicts with the principle of the constancy of the speed of light. Equations (23.3) and (23.4) hold for the phenomena we experience in the daily life, but do not hold in the case of the light as well as the objects moving as fast as light. The time does not elapse independently of the space, but they are related with each other. Hence, the relation valid in these cases is called the Lorentz transformation. The transformation in 4 dimensions is given by

$$x' = \frac{x-ut}{\sqrt{1-\dfrac{u^2}{c^2}}}, \qquad y' = y, \qquad z' = z, \qquad t' = \frac{t - \dfrac{ux}{c^2}}{\sqrt{1-\dfrac{u^2}{c^2}}} \tag{23.5}$$

together with the relation between the velocity $\boldsymbol{v}$ at the system S and the velocity $\boldsymbol{v}'$ at the system S′ which is derived from equation (23.5) :

$$v_x' = \frac{v_x - u}{1-\dfrac{uv_x}{c^2}}, \qquad v_y' = \frac{\sqrt{1-\dfrac{u^2}{c^2}}}{1-\dfrac{uv_x}{c^2}} v_y, \qquad v_z' = \frac{\sqrt{1-\dfrac{u^2}{c^2}}}{1-\dfrac{uv_x}{c^2}} v_z \tag{23.6}$$

[see the B4 of Exercise 23 for the derivation of equation (23.6)].

The transformation laws in the theory of relativity, equations (23.5) and (23.6), are different from those in the Newtonian mechanics, equations (23.3) and (23.4). However, when $u \ll c$, the former approximately coincides with the latter. Therefore, in the case where u is small, we can apply the Newtonian mechanics.

In the case where $\boldsymbol{v} = (c, 0, 0)$, we get $\boldsymbol{v}' = (c, 0, 0)$, and hence the speed of light turns out to be the same for the both coordinate systems. Moreover, we can prove that we cannot go over the speed of light in any method combining two velocities lower than the speed of light, by using the relativistic composition rule of velocities, equation (23.6).

23.3 The time dilation of the moving clock and the length contraction of the moving rod

Learning objective According to the special theory of relativity, the clock moving with high speed looks running slowly, while the rod moving with high speed looks contracted. These facts contrary to the experience of the daily life are predicted there. It will become possible to explain these facts from the principle of the constancy of the speed of light.

Time dilation of the moving clock We line up many clocks along the x-axis of a certain coordinate system (S-system), for example, along the straight railway track (Fig. 23-6). For the observer in the S-system, these clocks are adjusted to show the same time t. At the origin O′ of the coordinate system (S′-system), which is linearly moving with a constant velocity $\boldsymbol{u}$ with respect to the S-system in the direction of positive x-axis, for example at the window of a high-speed train, a clock is placed and shows the time t'. We adjust the clocks such that, when the origins O and O′ of the two coordinate systems coincide, all the clocks of the S-system show $t=0$, and the clock at the origin O′ of the S′-system shows $t'=0$ [Fig. 23-6 (a)]. The time t later in the S-system, the origin O′ of the S′-system moves for a distance ut, namely it is located at the position $x=ut$ [Fig. 23-6 (b)]. At that moment, the clock placed at the origin O′ of the S′-system shows

$$t'=t\sqrt{1-\frac{u^2}{c^2}}\,. \tag{23.7}$$

Fig. 23-5 Quasi-Zenith Satellite System "Michibiki". The Global Positioning System, GPS, is the system with a number of satellites going around the Earth at a height of 20 000 km with the period of half a day. By receiving the signals from at least 4 satellites, the position is measured. The clock of the satellite orbiting the Earth with the high speed is adjusted taking account of the theory of relativity.

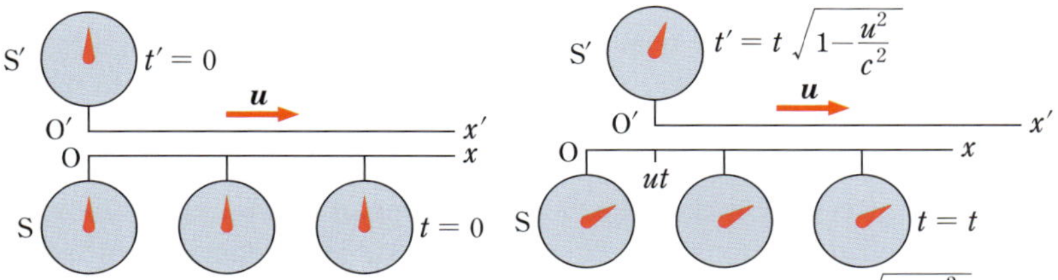

(a) $t=0,\ t'=0$　　(b) $t=t,\ t'=t\sqrt{1-\frac{u^2}{c^2}}$

Fig. 23-6 Time dilation of moving clock

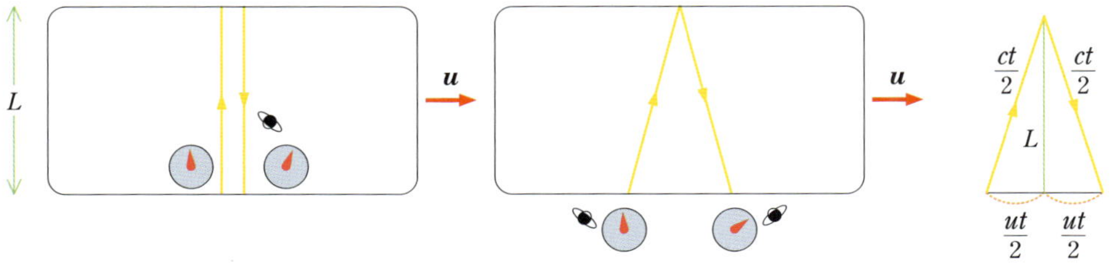

(a) When looking inside a car, one observer measures two times $t'=0$ and $t'=\frac{2L}{c}$ with one clock.

(b) When viewed from the ground, one observer measures time $t=0$, and another observer measures time $t=\frac{2L}{\sqrt{c^2-u^2}}$ with another clock.

(c) $t=\frac{2L}{\sqrt{c^2-u^2}}$ because $\left(\frac{ct}{2}\right)^2=\left(\frac{ut}{2}\right)^2+L^2$

Fig. 23-7 The light going a round-trip across the train moving with a high speed

This can be shown as follows.

For the means to measure the time, we use the distance for the light to pass through, and let us consider the light going across and back inside a train which is running with a high speed. Let us take the width of the train to L. When observed inside the train, the distance the light travelled is $2L$ and hence the round-trip time is $t' = \frac{2L}{c}$ [Fig. 23-7 (a)]. While, in the case where we observe this event on the ground, as the train is moving with a speed u, the distance the light travelled is longer than $2L$ [Fig. 23-7 (b)]. If we apply the Pythagoras' theorem to the triangle depicted in Fig. 23-7 (c), it turns out that the round-trip time of the light is $t = \frac{2L}{c\sqrt{1-\frac{u^2}{c^2}}}$. Since the speed c of light is common for both observers, if we compare the one moving clock with the two clocks at rest in the two locations on the ground, the running time (t') of the moving clock is slower than the running time (t) of the clock at rest, by a factor $\sqrt{1-\frac{u^2}{c^2}}$. That is, equation (23.7) was derived.

Proper time The time counted by the clock fixed at a moving object is called the **proper time** of the object. The running rate of the proper time of the object moving with the speed u with respect to an inertial frame, as we have shown above, is the running rate of the clock at rest in the inertial frame times $\sqrt{1-\frac{u^2}{c^2}}$.

The conclusion of the theory of relativity stating that the clock of the moving object runs slow was verified in the experiments. Unstable nuclei or elementary particles, when measured by the clock fixed to themselves (proper time), decay in a constant mean lifetime. For example, the mean lifetime of the elementary particle called a muon is 2.2×10^{-6} s when it is at rest. On the other hand, the mean lifetime while moving with a speed u, compared to the case at rest, is extended by a factor $\frac{1}{\sqrt{1-\frac{u^2}{c^2}}}$, which was confirmed by experiments.

Length contraction of moving rod (Lorentz contraction) The watch of the driver of the train running with a high speed u, when observed from the ground, appears to count the time passing between two points with a distance L_0 on the ground not as $\frac{L_0}{u}$ but $L_0\frac{\sqrt{1-\frac{u^2}{c^2}}}{u}$ [Fig. 23-6 (b)]. Therefore, for the train driver, the distance (= velocity × passing time) between the two points on the ground appears not as L_0 but contracted as $L_0\sqrt{1-\frac{u^2}{c^2}}$. In other words, the rod with a length L_0 at rest, when measured by an observer moving linearly in the direction along the rod with a speed u, appears to be contracted in length to the value :

$$L = \sqrt{1-\frac{u^2}{c^2}}\, L_0. \tag{23.8}$$

Compared to the length L_0 when the rod is at rest, it appears contracted in the direction of motion. This is called the **Lorentz contraction**. Note that for the moving object, the length in the direction perpendicular to the direction of motion appears to be the same as in the case it is at rest.

23.4 Theory of relativity and mechanics

Learning objective To understand the meaning of the equation $E = mc^2$.

Conservation law of momentum and momentum When the two objects receive the action only from the internal forces, the sum of the "momenta", defined as "mass" times "velocity", is conserved. This is derived in the Newtonian mechanics (Section **7.2**). In order for this conservation law of momentum to hold in the theory of relativity, we only have to define the momentum $\boldsymbol{p}$ for the object with a rest mass m_0 and the velocity $\boldsymbol{v}$ as

$$\boldsymbol{p} = \frac{m_0 \boldsymbol{v}}{\sqrt{1-\dfrac{v^2}{c^2}}} \tag{23.9}$$

(proof omitted).

What equation (23.9) implies is that the mass m of the object moving with a speed v becomes bigger than the mass when it is at rest (**rest mass**), m_0, and is given as

$$m = \frac{m_0}{\sqrt{1-\dfrac{v^2}{c^2}}} \tag{23.10}$$

Then, the momentum in the theory of relativity, $\boldsymbol{p}$, can be expressed as $m\boldsymbol{v}$. The speed of the mass point varies with the time, the mass m also changes with the time. The law of motion in the theory of relativity is not "mass"×"acceleration" = "force", but "time changing rate of momentum" = "force", namely,

$$\frac{\mathrm{d}\boldsymbol{p}}{\mathrm{d}t} = \boldsymbol{F}. \tag{23.11}$$

Fig. 23-8 The Superconducting Ring Cyclotron. In the Superconducting Ring Cyclotron, superconducting magnets producing the magnetic fields are placed on the circumference of the ring at the separated sectors, by which the variation of the magnetic field is enhanced and the strong power of the convergence is obtained. With this machine we can accelerate light ions to 440 MeV per nucleon, and accelerate heavy ions to 350 MeV per nucleon. Uranium can be accelerated to a speed of about 70 % of the velocity of light.

Mass and energy When an external force $\boldsymbol{F}$ acts on an object, that object is accelerated. In the Newtonian mechanics, the kinetic energy of the object increases for the amount of the work done by the external force acting on the object. While in the theory of relativity, "mass"× (speed of light in vacuum)^2 of the object, that is, mc^2 increases for the amount of the work done by the external force (see B5 of the Problems for exercise 23). Thus, Einstein regarded mc^2 as the energy E which the object with the rest mass m_0 acquires at the speed u. This is nothing but the statement "Mass is a form of energy" given by the famous equation of energy :

$$E = mc^2 = \frac{m_0 c^2}{\sqrt{1-\frac{u^2}{c^2}}}. \qquad (23.12)$$

In the case where the speed of the object u is much smaller than the speed of light c, and the Newtonian mechanics can be applied, the equation of the energy of the object, Eq. (23.12) reduces to

$$E \approx m_0 c^2 + \frac{1}{2} m_0 u^2. \qquad (23.13)$$

The second term on the right-hand side $\frac{1}{2} m_0 u^2$ is the kinetic energy in the Newtonian mechanics. The first term on the right-hand side, $m_0 c^2$ which can be interpreted as the energy of the object at rest is called the **rest energy**.

Mass as a form of energy implies that if the mass decreases in a nuclear reaction by Δm, then the nuclear energy of the amount $\Delta m \cdot c^2$ turns into another form of energy. As the speed of light in the vacuum is $c = 3\times10^8$ m/s, the object of mass 1 kg possesses a huge amount of a potential nuclear energy given as :

$$(1\ \text{kg})\times(3\times10^8\ \text{m/s})^2 = 9\times10^{16}\ \text{J}.$$

The energy emitted by the decrease of the mass in the nuclear fission reaction is practically realized as a nuclear power generation. Further, the energy emitted from the Sun is arising from the decrease of the mass due to the nuclear fusion in the core of the Sun. In the nuclear fusion process inside the Sun, when the 1 kg of hydrogen nuclei is fused, about 7 g of the mass disappears turning into another form of energy.

23.5 Electromagnetic fields and coordinate systems

Learning objective To understand that in the Lorentz transformation, electric fields and magnetic fields mix with each other.

The velocities are different in the two inertial frames in a relative motion. Similarly, in the two inertial frames, electric currents are different and also electric and magnetic fields are different. For example, in the case of Fig. 23-9, the observer S′ who is at rest in the train (S′-system) observes only the electric field $\boldsymbol{E}'$ due to the electric charge of the charged body. On the other hand, the observer S

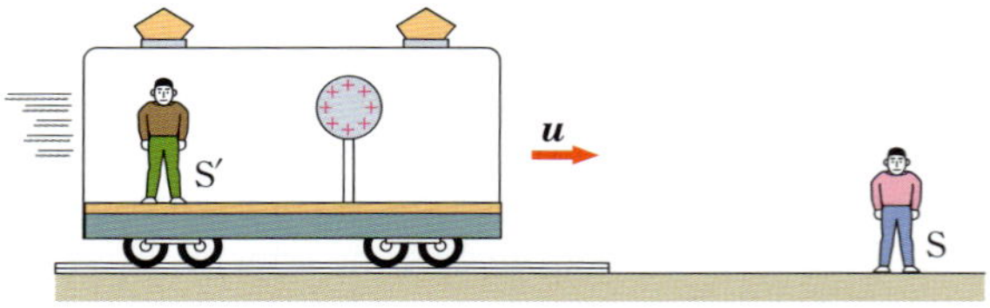

Fig. 23-9 The observer S′ in the train only observes the electric field $\boldsymbol{E}'$. The observer S on the ground observes the electric field $\boldsymbol{E}$ and the magnetic field $\boldsymbol{B}$.

standing on the ground (S-system) observes the electric field $\boldsymbol{E}$ as well as the magnetic field $\boldsymbol{B}$ due to the electric charge of the charged body moving with a velocity $\boldsymbol{u}$.

In general, the electric field $\boldsymbol{E}$ and the magnetic field (magnetic flux density) $\boldsymbol{B}$ in the S-system are related with the electric field $\boldsymbol{E}'$ and the magnetic field (magnetic flux density) $\boldsymbol{B}'$ in the S′-system by the Lorentz transformation (23.5) as

$$E_x' = E_x, \quad E_y' = \gamma(\boldsymbol{E}+\boldsymbol{u}\times\boldsymbol{B})_y, \quad E_z' = \gamma(\boldsymbol{E}+\boldsymbol{u}\times\boldsymbol{B})_z$$
$$B_x' = B_x, \quad B_y' = \gamma\left(\boldsymbol{B}-\frac{\boldsymbol{u}\times\boldsymbol{E}}{c^2}\right)_y, \quad B_z' = \gamma\left(\boldsymbol{B}-\frac{\boldsymbol{u}\times\boldsymbol{E}}{c^2}\right)_z \qquad (23.14)$$

where $\gamma = \dfrac{1}{\sqrt{1-\dfrac{u^2}{c^2}}}$ (proof omitted).

The electric charge of the charged particle has the same value when observed in any inertial frame, and does not change depending on the velocity or acceleration of the charged particle. Since the S-system and the S′-system are inertial frames, the equations of motion for the charged particle with a charge q in the electric field $\boldsymbol{E}$ and the magnetic field $\boldsymbol{B}$ have the same form :

$$\frac{\mathrm{d}\boldsymbol{p}}{\mathrm{d}t} = q(\boldsymbol{E}+\boldsymbol{v}\times\boldsymbol{B}) \quad \text{(S-system)}$$
$$\frac{\mathrm{d}\boldsymbol{p}'}{\mathrm{d}t'} = q(\boldsymbol{E}'+\boldsymbol{v}'\times\boldsymbol{B}') \quad \text{(S′-system)}. \qquad (23.15)$$

Problems for exercise 23

A

1. A train with the total length 500 m is running with a speed of $0.6c$. How many meters does the observer on the ground measure the length L of the train to be?
2. In order to increase one's own weight by 1 %, how fast does that person run?
3. How much amount of energy in Joule does mass of 1 g correspond to?

B

1. In the Michelson-Morley experiment (Fig. 23-2), the sodium (Na) D-line with the wavelength $\lambda = 6\times10^{-7}$ m was used, and the distance L between the translucent mirror M and the mirror A, B was about 1 m. However, as the light travelled the round-trip between M and A, B for 10 times, so we should take the distance practically as $L = 10$ m. If the ether were present, show that the expected phase shift by rotating the apparatus by 90 degrees, is $\dfrac{4\pi L\left(\dfrac{u}{c}\right)^2}{\lambda}$ and estimate its value (Assume that u is of the order of the speed of the Earth's revolution 30 km/s).
2. The two trains running with a speed $0.6c$ are approaching from the opposite directions. When observed by the observer in one of the trains, find the relative velocity of the other train.
3. When the twin brothers A, B were at the age of 20, one of them, A, started the round-trip to the α star at the constellation Centaurs, which is the nearest fixed star from the Earth (The distance from the Earth is 4.4 light years), by a spacecraft with a speed $0.99c$. How old are they when A is back?
4. Derive the relativistic transformation law of the velocities (23.6) from the Lorentz transformation (23.5).
5. When we define the energy E of the object by equation (23.12), show that the following relation between the work and the energy hold.

$$\int_{P_1}^{P_2} \boldsymbol{F}\cdot\mathrm{d}\boldsymbol{r} = \int_{P_1}^{P_2} \frac{\mathrm{d}}{\mathrm{d}t}(m\boldsymbol{v})\cdot\mathrm{d}\boldsymbol{r}$$
$$= m_0\int_{v_1}^{v_2} \mathrm{d}\left(\frac{\boldsymbol{v}}{\sqrt{1-\dfrac{v^2}{c^2}}}\right)\cdot\boldsymbol{v}$$

$$= m_0 \int_{v_1}^{v_2} \mathrm{d}\left(\frac{c^2}{\sqrt{1-\frac{v^2}{c^2}}}\right)$$

$$= \frac{m_0 c^2}{\sqrt{1-\frac{v_2^2}{c^2}}} - \frac{m_0 c^2}{\sqrt{1-\frac{v_1^2}{c^2}}}.$$

6. Show that the relativistic energy $E = m_0 \frac{c^2}{\sqrt{1-\frac{u^2}{c^2}}}$ and the relativistic momentum $\boldsymbol{p} = m_0 \frac{\boldsymbol{u}}{\sqrt{1-\frac{u^2}{c^2}}}$ satisfy the relation $E^2 = (pc)^2 + (m_0 c^2)^2$.

7. The electromagnetic field produced by the electric charge q which is at rest at the point $\boldsymbol{r}'$ is now seen by the observer moving with respect to the charge with a velocity $-\boldsymbol{v}$. When the velocity is much smaller than the speed of light, show that the magnetic field $\boldsymbol{B}(\boldsymbol{r})$ which is produced by the charged particle at the position $\boldsymbol{r}$ is given by

$$\boldsymbol{B}(\boldsymbol{r}) = \frac{\mu_0 q}{4\pi} \frac{\boldsymbol{v} \times (\boldsymbol{r} - \boldsymbol{r}')}{|\boldsymbol{r} - \boldsymbol{r}'|^3}$$

Atomic Physics

There are two prefixes in English : "macro" and "micro". The term macro implies "long", "large", and "gigantic" etc., while the term micro originates from mikros in Greek meaning to be small and the term micro means such as "minute", "related to microscope", or "one millionth". One millionth of a meter is the size of a cell, and in the discussion of atomic world a prefix "nano" which means "minimal" or "one billionth" is used. The diameter of hydrogen atom is 0.1 nm (0.1 nm $= 10^{-10}$ m).

Light and electrons exhibit both wave and particle natures. In this chapter, we will study atomic physics focusing on the duality of waves and particles, which is a feature of the microscopic world.

Doctor Kōsuke Morita (see p. vi) and GARIS-II. RIKEN has already developed GARIS-II to be used for the synthesis of 119th element and beyond, and will continue to search for new elements over 119th.

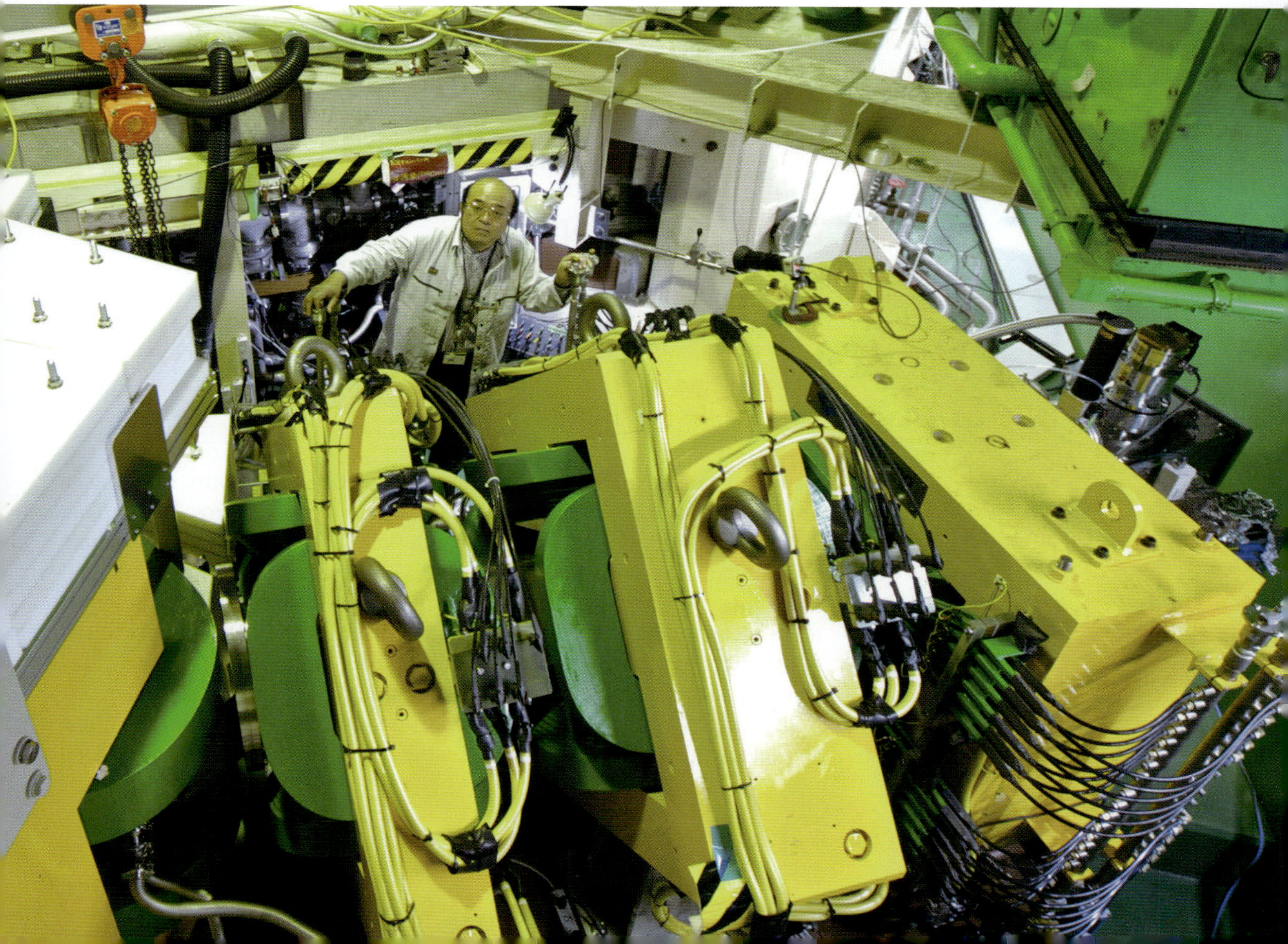

Subjects attracting the interests of many people are mostly the phenomena around us, which are so-called macro-phenomena. Yet, why is it required for us to know the physics of invisible microscopic world such as atoms? The best answer to this query is found by looking back on the history of physics.

Physics originated from exploring the laws of phenomena that can be seen through the eyes and/or touched with the hand. For example, parabolic motions of the stones visible to the naked eye, motion of the celestial bodies, heat felt with our hands, visible light, sounds we can hear. Such phenomena were the subjects that attracted the interests of physicists.

With the advances of physics research, however, to understand the nature of visible light, it became necessary to accept the existence of electric and magnetic fields that are not directly visible. As a result, the use of electromagnetic waves of various wavelength became possible, and the innovation of methods in information exchange among people around the world has greatly advanced.

Also, it became clear that in order to have a deep understanding of the laws of the world that we see and touch, such as the thermal phenomena and the properties of matter both of which we experience in our daily life, we have to know about the worlds of molecule and atom in which electrons are actively playing a role. Neither of these small worlds can be seen with the naked eye or touched with the hand.

As a result, it has been gradually clarified what the inside of the atom is and what rules govern there. For example, to find the answer to the question why substances can be classified into conductors that conduct electricity and insulators that do not, it is required to understand the structure of atomic world and quantum mechanics governing the atomic world.

The first step to understand the quantum mechanics governing the atomic world is to understand the duality of substance.

24.1 Structure of atom

Learning objective To understand the atomic structure, namely that an atom is formed by bonding of a nucleus and electron (s), and that the atomic radius is about 10^{-10} m and the radius of heavy nuclei is about 10^{-14} m, so the radius of atomic nucleus is less than about 1/10000 of the atomic radius.

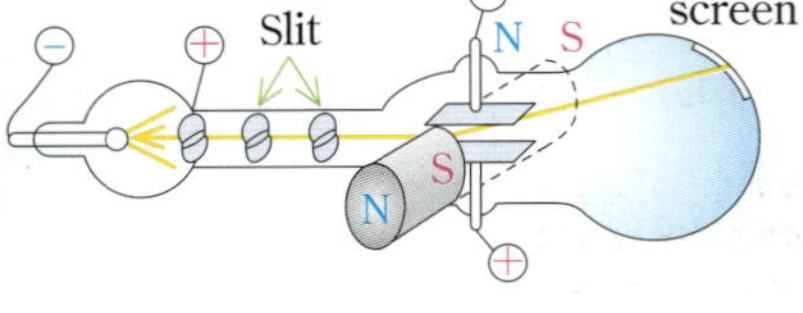

Fig. 24-1 An experiment conducted by Thomson. The point at which the electrons collide with the fluorescent screen correspond to the point where a particle with a fixed mass and a negative charge moves in the electric and magnetic fields according to Newton's equation of motion.

Discovery of electron It has long been known that pure substances are classified into chemical elements and compounds. To explain quantitative equations such as law of definite proportion, law of multiple proportion, law of gaseous reaction etc., the idea of atoms and molecules was born at the beginning of the 19th century. The term atom means "indivisible". But Thomson discovered in 1897 as a result of experiment using the apparatus (discharge tube) shown in Fig. 24-1 that constituent elements exist in the atom.

In this experiment, negatively charged particles jump from the negative electrode (cathode) of the heated metal to the positive electrode (anode) and move in the electric and magnetic fields under the action of electric and magnetic forces and collide with the fluorescent screen on the back side of the positive electrode to generate bright spots. Thomson discovered that the particle has a fixed mass and a fixed charge and follows the same trajectory as the particle moving according to Newton's equation of motion, and that the particle has a small mass of about 1/2000 times that of a hydrogen atom. Even if the cathode metal was changed to another metal, the same particle appeared, so it was found that this particle is a constituent particle common to various substances, and was named **electron**. Electron is the lightest constituent particle of substance, and it moves easily in substance, so it plays a leading role in physical phenomena and chemical changes that various substances exhibit.

Atomic model It was found that in the atom, an electron with a mass of about 1/1840 of the mass of a hydrogen atom and a negative charge exists. In what form does the positively charged substance with most of the mass of the atom exist in the atom?

Geiger and Marsden under the supervision of Rutherford conducted in 1909 an experiment in which a beam of helium nuclei called α (alpha) particle collided with a thin gold foil. They found that many α particles passed the gold foil but some came back in the opposite direction (Fig. 24-2). α particles do not bounce back in the opposite direction even if they collide with light electrons with a mass of only about 1/7000 of their mass. If the positive charge of a gold atom is uniformly distributed within the atomic radius of about 10^{-10} m, since the charge density of the positive charge is small, the repulsive force between the α particle and the positive charge is weak. Because the magnitude of the potential energy is several thousandths of the kinetic energy of the α particle, the α particle is not repelled backward by the repulsive force with the positive charge of the gold atom

The repulsive force acting between the charged particles with the same sign is inversely proportional to the square of the distance, so it becomes extremely large at short distances. If an α particle collides with a gold atom and its traveling direction is bent more than 90° by the electric force, the positively charged part of the gold atom must be gathered in a very small part inside the atom (Fig. 24-3). Rutherford called this small part the **atomic nucleus**. In order for the α particle to head-on collide with the gold nucleus and come back, the electric potential energy on the surface of the gold nucleus must be greater than the initial kinetic energy of the α particle, so the size of the gold nucleus must be less than about 10^{-14} m (see Problems for exercise 24, B1).

In this way, Rutherford discovered in 1911 the atomic model described as follows. An atom with a radius of about 10^{-10} m consists of an atomic nucleus with a radius of about 10^{-14} m and Z electrons surrounding the nucleus and each electron has a negative charge $-e$. The nucleus has a positive charge Ze equal to the atomic number times the elementary charge e and most atomic mass.

The force that combines the nucleus and electrons to form an atom is the Coulomb attraction that acts between the positive charge of nucleus and the negative charge of the electron, but the motion law of the microscopic world is not Newtonian mechanics but quantum mechanics.

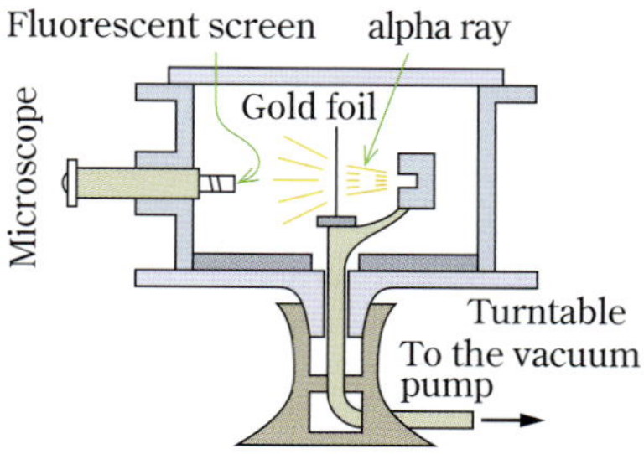

(a) Conceptual diagram of device

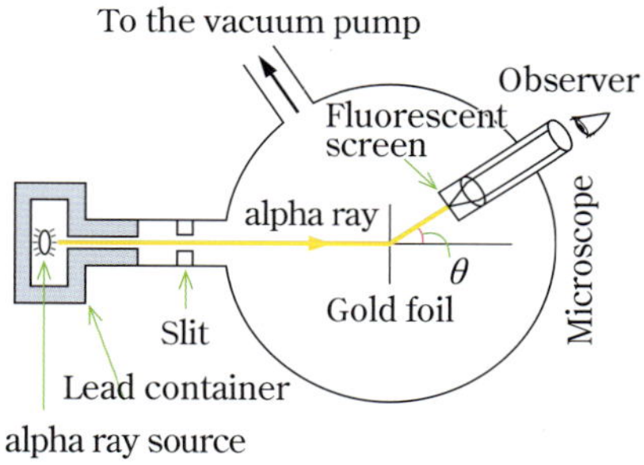

(b) Conceptual diagram of experiment

Fig. 24-2 Conceptual diagram of experiments conducted by Geiger and Marsden

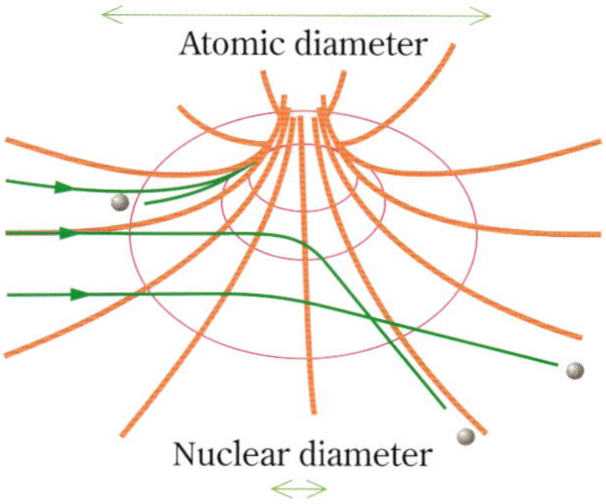

Fig. 24-3 Potential distribution of electric field created by a nuclear positive charge, and change in traveling direction of α particles.

24.2 Duality of light

Learning objective To understand that light propagates in space as electromagnetic waves, but light of frequency ν and wavelength λ is emitted or absorbed by substances as particles (photons) with energy $E = h\nu$ and momentum $p = \frac{h}{\lambda}$. To understand that light interference fringes reflect the magnitude of the probability of photons colliding with the screen.

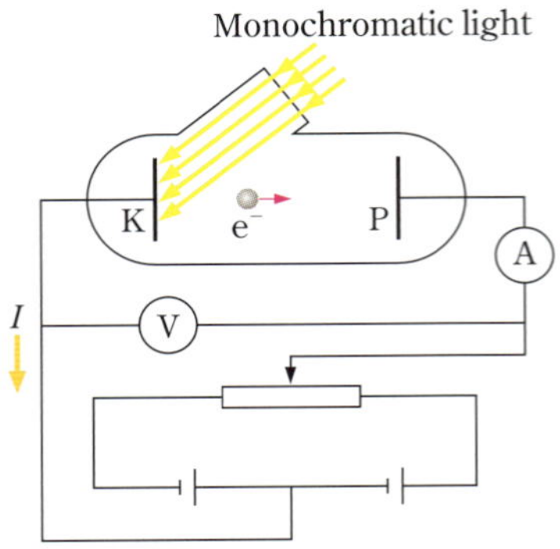

Fig. 24-4 Conceptual diagram of experiment of photoelectric effect. As the potential V of the positive electrode P relative to the negative electrode K is lowered, the current I decreases. When $V = V_0 < 0$, if I takes $I = 0$, then $K_{max} = eV_0$.

* In this chapter the symbol ν (nu) is used as the symbol of the frequency of light.

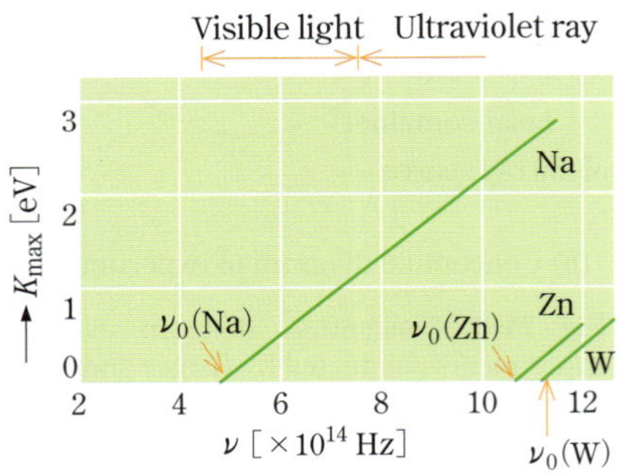

Fig. 24-5 Relationship between frequency ν of monochromatic light and maximum energy K_{max} of electrons. The unit of vertical axis is eV, 1 eV $= 1.6 \times 10^{-19}$ J.

A study of electromagnetism revealed that light is a kind of electromagnetic wave and that the vibration of the electromagnetic field is transmitted as a wave in space. The wave nature of light is confirmed by diffraction and interference, and the wavelength can be determined based on the diffraction and interference phenomena. However, there exist phenomena that cannot be explained even if light is considered as a wave. Those phenomena are the photoelectric effect and Compton scattering of X-rays.

The phenomenon that electrons are emitted when short-wavelength visible light or ultraviolet light is irradiated onto a metal is the **photoelectric effect** (Fig. 24-4). This phenomenon has the following characteristics*.

(1) No matter how strong light is irradiated, if the light frequency ν is lower than a fixed value (threshold frequency) ν_0 specific to the metal to be irradiated, electrons are not emitted.

However,

(2) if the irradiated light frequency is higher than the threshold frequency ν_0, electrons are emitted. The emitted electrons have kinetic energy of various intensities, but the kinetic energy K_{max} of the fastest electron is not related to the irradiated light intensity and is determined only by the light frequency ν (Fig. 24-5), and expressed as

$$K_{max} = h\nu - h\nu_0 \tag{24.1}$$

where h is the Planck constant $h = 6.626 \times 10^{-34}$ J·s which appears in equation (14.7) of Planck's law.

In 1900, Planck showed that to theoretically derive Planck's law for radiation of light (generally electromagnetic waves), it is necessary to introduce the hypothesis, "The energy E of the electromagnetic wave of frequency ν is restricted to an integer multiple of $h\nu$":

$$E = nh\nu \quad (n = 0, 1, 2, \cdots). \tag{24.2}$$

In 1905, Einstein, considering "Light (generally electromagnetic waves) of frequency ν is a flow of particles with energy of magnitude

$$E = h\nu \tag{24.3}$$

and in the photoelectric effect, when this particle of light collides with an electron in the metal, its entire energy $h\nu$ is absorbed by the electron at once", succeeded to explain the experimental results of photoelectric effect. The particle of light was named the **photon**.

We learned in Section **22.3** that the electromagnetic field also has momentum and when the energy density of the electromagnetic wave is u, the magnitude P of the momentum density of the electromagnetic wave is $\frac{u}{c}$. Therefore, the photon having the energy of magnitude $E = h\nu$ has the momentum

$$p = \frac{E}{c} = \frac{h\nu}{c} = \frac{h}{\lambda} \qquad (24.4)$$

which is directed in the traveling direction of light ($c = \lambda\nu$).

In the case of scattering of wave on the water surface by a pile in a pond, the wavelengths of the incident and scattered waves are the same. Compton, however, discovered in 1923 that when a substance is irradiated with X-rays that are electromagnetic waves, the X-rays scattered by the substance have a component with a wavelength λ' longer than λ in addition to the wavelength λ that is the same as the incident wave (Fig. 24-6). Scattering of X-rays in which the wavelength changes is called the **Compton scattering**. He considered X-rays with frequency ν and wavelength λ as a flow of photons with energy $E = h\nu$ and momentum $p = \frac{h}{\lambda}$, and explained this phenomenon as the elastic collision of X-ray photons with electrons (Fig. 24-7). Scattering in which change in wavelength does not occur is explained as scattering of X-ray photons by the nucleus.

Question 1 Derive the formula of equation (24.5)

$$\Delta\lambda = \lambda' - \lambda = \frac{h}{m_e c}(1 - \cos\phi) = 2.43\times10^{-12}(1 - \cos\phi)\ \text{m} \qquad (24.5)$$

for the wavelength λ' of scattered X-rays in the Compton scattering. ϕ is the scattering angle of photon and m_e is the mass of electron at rest.

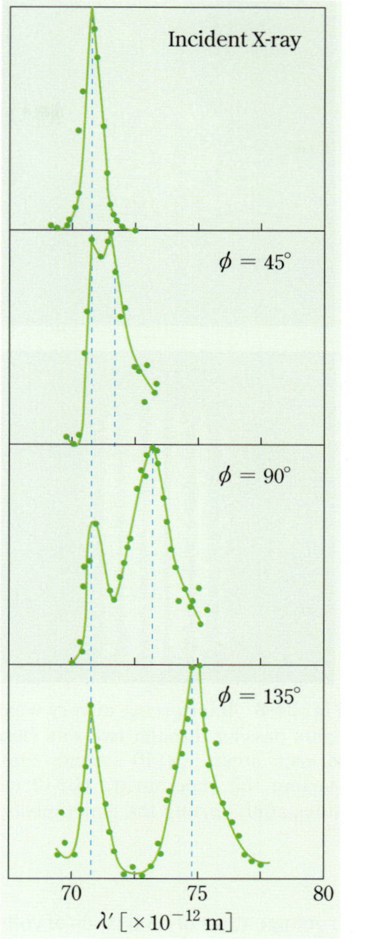

Fig. 24-6 Distribution of scattering angle ϕ and wavelength λ' of scattered X-ray. Scattering of the incident X-ray of wavelength $\lambda = 7.1\times 10^{-11}$ m by graphite. The vertical axis corresponds to the intensity of scattered X-ray.

As described above, light exhibits both the wave nature and the particle nature. In other words, light is transmitted through space as a wave of frequency ν and the wavelength λ, but when it is emitted or absorbed by materials it behaves as a collection of photons (particles of light) with energy E and momentum p.

$$E = h\nu, \qquad p = \frac{h}{\lambda}. \qquad (24.6)$$

Such a behavior is referred to as the **duality of light**. Light has the wave nature and the particle nature but between them the intimate relation of equation (24.6) holds.

The reality of the duality of light is clarified by the formation of interference fringes representing the wave nature of light on the highly light-sensitive detection surface. The photo of Fig. 24-8 shows the interference phenomena when light from a very weak light source passes through two gaps (slits). The bright spot generated when the light collides with the detection surface indicates that the photon has collided. That is, it can be seen that light collides as photons (particles) when it collides with the detection surface (fluorescent substance).

Since the number of photons reached in 10 seconds after starting the experiment is small, the arrival positions of photons seem to have no regularity [Fig. 24-8 (a)]. However, it can be seen that a great number of photons reach in 10 minutes after starting the experiment

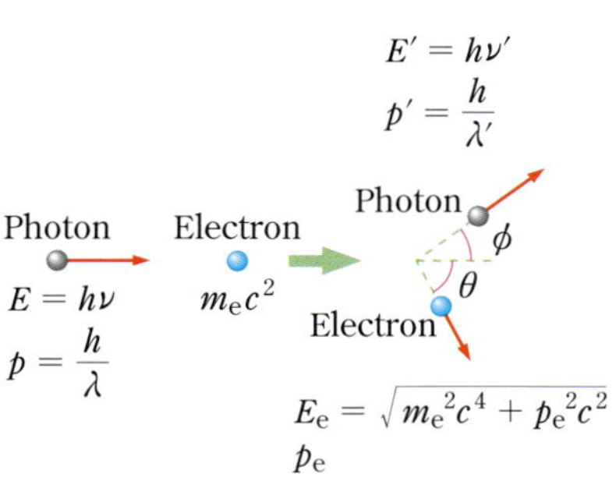

Fig. 24-7 Compton scattering due to an electron in the atom

(a)

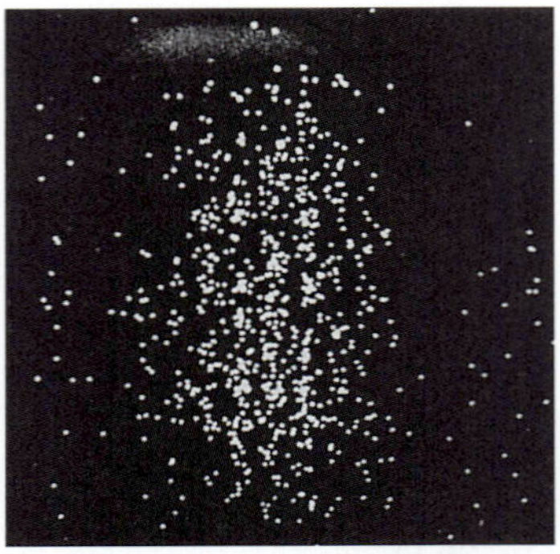

(b)

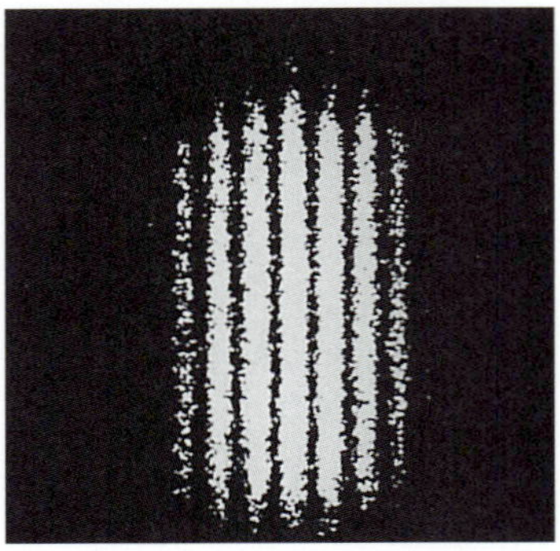

Fig. 24-8 Interference of very weak lights passing through two slits close to each other. (a) 10 seconds after starting the experiment. (b) 10 minutes after starting the experiment.

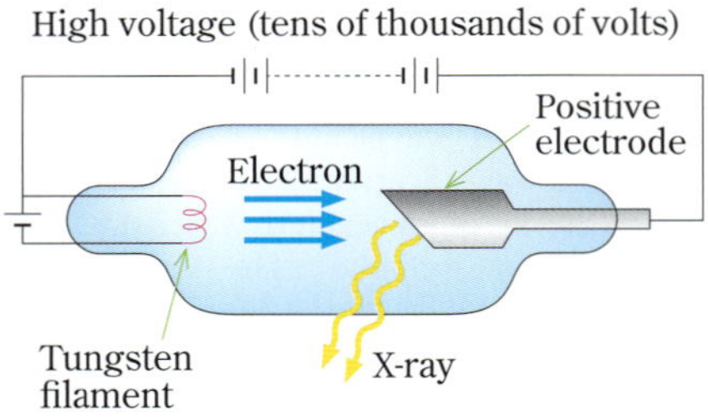

Fig. 24-9 Conceptual diagram of X-ray generation apparatus

leading to the result that a large number of photons reached the bright part of the bright and dark stripes generated by the light wave interference, and few photons reached the dark part [Fig. 24-8 (b)]. As shown, in the behavior of a collection of lots of photons, interference fringes that represent the characteristic nature of waves appear. This is the reality of the duality of light. The motion law that photons follow is quantum electromagnetism.

For reference Electron volt

In atomic physics, as a practical unit of energy, is used the work done by an electric field when a charged particle with an elementary electric charge e passes through a potential difference of 1 V, and the unit is called electron volt and expressed 1 eV.

$$1\,\mathrm{eV} \approx 1.602\times10^{-19}\,\mathrm{J} \tag{24.7}$$

1 keV = 10^3 eV and 1 MeV = 10^6 eV are also used.

For reference X-rays

X-rays were discovered in 1895 in an occasion when Röntgen, conducting an experiment on a discharge tube, noticed that an unused photographic plate placed near the discharge tube was exposed to light. X-rays are electromagnetic waves with a wavelength of about 10^{-9}–10^{-12} m, which is shorter than that of light. They do not penetrate metal or bone but penetrate paper or glass, and have an ionizing ability. When penetrating into crystals, they show diffraction and interference phenomena. They have the dual nature like light, and satisfy the two relations of equation (24.6).

A conceptual diagram of X-ray generation apparatus is given in Fig. 24-9. When electrons (charge $-e$) emitted from a heated filament are accelerated with a high voltage V, the electrons collide with the metal plate of the positive electrode, and all or part of the kinetic energy eV changes into X-rays. The relationship between the wavelength and intensity of X-ray generated in this way is shown in Fig. 24-10. The spectrum of X-rays consists of smooth curve parts (continuous X-rays) and sharp peaks (characteristic X-rays). Characteristic X-rays are X-rays of a wavelength specific to the metal atom of the positive electrode, corresponding to the line spectrum of the light emitted by the atom (see Section **24.5**).

Continuous X-rays have the shortest wavelength λ_0 determined by the acceleration voltage V of electron. λ_0 corresponds to the case where the kinetic energy eV of the electron colliding with the positive electrode changes wholly into the energy of generated X-ray photons. Using equation (24.6) and the relation $\nu\lambda = c$, we have

$$eV = \frac{ch}{\lambda_0} \tag{24.8}$$

$$\lambda_0 = \frac{ch}{eV} = 1.240\times10^{-10}\times\frac{10^4}{V\,[\mathrm{V}]}\,\mathrm{m}. \tag{24.9}$$

Here V [V] is the numerical part of acceleration voltage V measured

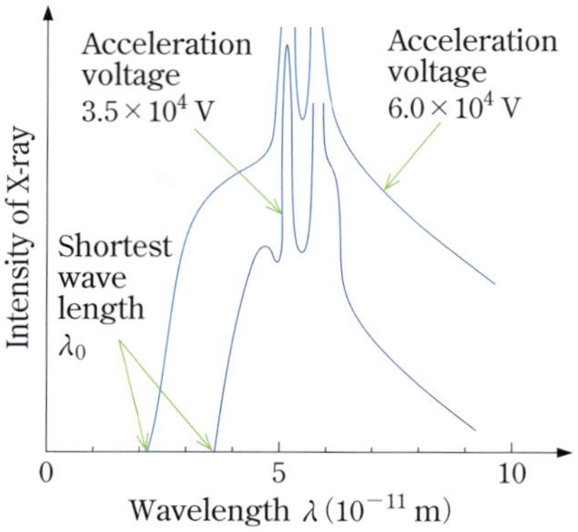

Fig. 24-10 X-ray spectrum (the positive electrode is Pd.)

in V unit. In the case of X-rays generated by an X-ray generator with an acceleration voltage of 4.0×10^4 V, the shortest wavelength λ_0 is 3.1×10^{-11} m.

24.3 Duality of electron

Learning objective To become able to explain based on concrete examples that electrons have both the particle nature and the wave nature, and to understand that electrons travel in space as waves and when emitted or absorbed by substances, they behave as particles.

In the previous section, it was explained that light that was thought to be waves behaves like particles. Then, let us consider whether electrons that behave like particles also exhibit the wave nature, or it just exhibits the particle nature.

The photos of Fig. 24-12 show the record of electrons that reached the detection surface placed at the merging position of the two electron streams passing through the two slits set in the electron-stream path (Fig. 24-11). Looking at Fig. 24-12 (e), we can read the bright and dark stripes indicating the interference phenomenon that is characteric-tic of waves. That is, these photos demonstrate that two electron waves ψ_1 and ψ_2 passing through two slits overlap each other to become $\psi_1+\psi_2$, and the two waves ψ_1 and ψ_2 (if they have the same sign) reinforce or (if they have the different sigh) weaken each other on the detection surface and as a result the distribution of intensity $|\psi_1+\psi_2|^2$ of the electron wave on the detection surface forms bright and dark stripes. The experiments were conducted in a situation where the simultaneous presence of two or more electrons in the experimental apparatus is rare, so the bright and dark stripes are not generated by the interaction of two or more electrons. In other words, these photos show that "one" electron that generates one bright spot passes through both slits simultaneously.

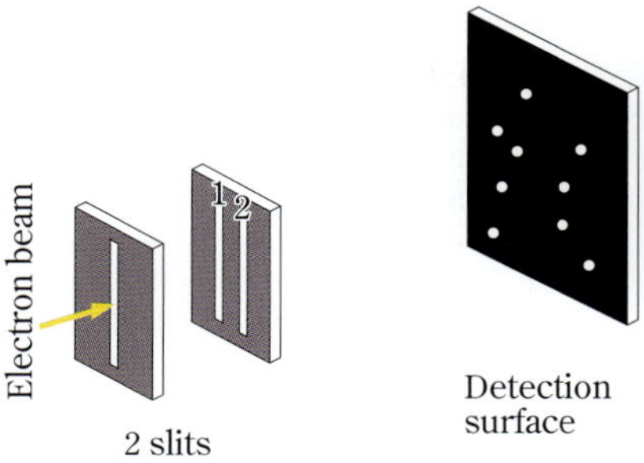

Fig. 24-11 Electron beam and two slits 1 and 2 (conceptual diagram)

When we sequentially look at Figs. 24-12 (a) to (e) which record the forming process of the bright and dark stripes, we notice that the brightness of the bright and dark stripes does not increase continuously, but electrons as "particles" collide with the detection surface (fluorescent film) one by one to generate bright spots. We also notice

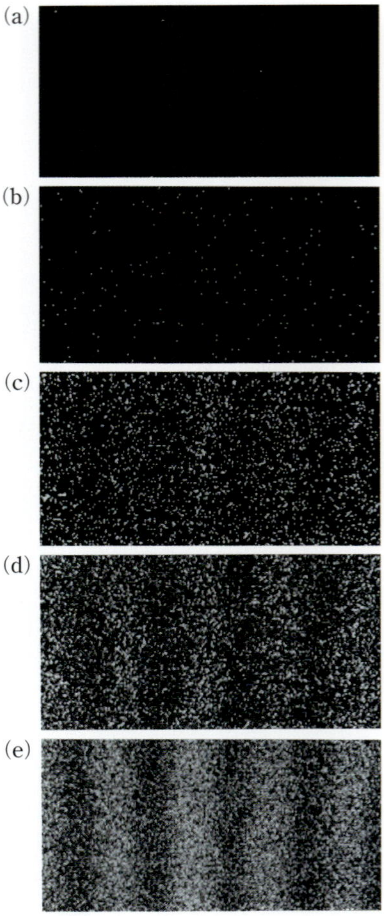

Fig. 24-12 Forming process of interference fringes observed using an electron microscope. Electrons pass through the two slits and reach the detector at interval one by one. When the electrons reach the fluorescent film on the detector surface, they are detected, recorded in a recording device, and displayed on a monitor. This figure shows that the electrons arrive one by one on the detection surface, and as a result, formation process of interference fringes is in order of photographs of a to e.

that, bright and dark stripes are gradually formed as a result of the difference in the probability of collision depending on the location.

Fig. 24-8 for photons and Fig. 24-12 for electrons are quite similar. Also in the case of electrons, the wave phenomenon called interference fringes appears as a difference in the spatial distribution of the probability to find the particles (electrons).

In this way, it has been shown that similarly as for the case of light, electrons have the duality of particles and waves. The mechanics that electrons follow is **quantum mechanics**. The equation that determines the wave function $\psi(x, y, z, t)$ representing how electrons travel through space as a wave is the Schrödinger equation :

$$-\frac{\hbar^2}{2m}\left(\frac{\partial^2\psi}{\partial x^2}+\frac{\partial^2\psi}{\partial y^2}+\frac{\partial^2\psi}{\partial z^2}\right)+V\psi = i\hbar\frac{\partial\psi}{\partial t}. \qquad (24.10)$$

$|\psi(x, y, z, t)|^2$ represents the probability that an electron of mass m is detected at a point (x, y, z) at t, and $\hbar$ means $\frac{h}{2\pi}$, which is read h bar.

In the case of light, the momentum p of photon of the light of wavelength λ was $\frac{h}{\lambda}$. When the electron beam with mass m and velocity $\boldsymbol{v}$ exhibits wave properties, its wavelength λ is the same as in the case of light,

$$\lambda = \frac{h}{p} = \frac{h}{mv}. \qquad (24.11)$$

This wavelength is called **de Broglie wavelength** after de Broglie who proposed the relation of equation (24.11).

Protons and neutrons, similarly as electrons and photons, have been confirmed to exhibit both the wave nature and the particle nature.

Case 1 | Davisson-Germer's experiment In 1927, when Davisson and Germer irradiated an electron beam perpendicularly on the surface of a single crystal of nickel [Figure 24-13 (a)], they found that the intensity of electrons scattered by the surface becomes stronger in a specific direction [Fig. 24-13 (b)], and that the direction (scattering angle θ) changes with the acceleration voltage of the electron beam.

When the electron (mass m) is accelerated between electrodes of the potential difference V, the work eV of the electric field becomes the kinetic energy of the electron, so the de Broglie wavelength of the electron wave whose kinetic energy is

$$\frac{1}{2}mv^2 = \frac{p^2}{2m} = \frac{h^2}{2m\lambda^2} = eV \qquad (24.12)$$

the de Broglie wavelength of the electron waves becomes

$$\lambda = \frac{h}{\sqrt{2meV}} = \sqrt{\frac{150.41}{V[\mathrm{V}]}}\times 10^{-10}\ \mathrm{m} \qquad (24.13)$$

(mass of electron $m = 9.109\times 10^{-31}$ kg, electric charge $-e = 1.602 \times 10^{-19}$ C). V[V] is the numerical part of the potential difference V measured in the unit of V.

The condition that determines the angle θ at which the electron wave of wavelength λ is strongly scattered by the crystal surface with interatomic spacing d is

$$d \sin\theta = n\lambda \qquad (n = 1, 2, 3, \cdots) \tag{24.14}$$

[21.13 (c)]. Davisson and Germer succeeded in determining the wavelength λ of the electron wave from the measurement results of the angle θ [Fig. 24-13 (b)] at which the reflected electron beam intensity is maximal [Fig. 24-13 (b)] and the atomic distance d. Their experimental values and the theoretical values of (24.13) showed a good agreement (see Problems for exercise 24, B4).

Question 2 Find the wavelength λ of the electron beam obtained by accelerating an electron with zero kinetic energy at a voltage of 100 V.

An example of devices that use the wave nature of electron is an **electron microscope**. It is the wave diffraction phenomenon that prevents observation of microscopic objects with a microscope. The shorter the wavelength, the more difficult the wave diffraction. Since the wavelength of the electron wave becomes extremely short as the acceleration voltage of the electron increases, the arrangement of molecules and atoms can also be observed with an electron microscope.

24.4 Uncertainty principle

Learning objective To understand the uncertainty principle stating that it is impossible to measure both the position and momentum of minute particles such as electrons accurately at the same time.

When waves are split into two and then they merge, they overlap and interfere. Particles do not have such a property. The nature of waves and the nature of particles are incompatible in our daily experience. Why can electrons have the duality? Let us consider the reason. In order to observe how electrons move in space as waves or particles, it is necessary to irradiate the electron path with light and scatter the light with electrons.

When we want to measure the position of an object precisely, it is necessary to apply a finely focused light beam to the object. However, optical studies have shown that the width of light beam of wavelength λ can only be narrowed to about $\frac{\lambda}{2\pi}$. Hence, there is an uncertainty Δx of about $\frac{\lambda}{2\pi}$ in the measured value of the object position obtained using the light of wavelength λ. On the other hand, due to the particle nature of light, the intensity of the light applied to electrons cannot be reduced to less than one photon. Since the momentum of one photon is $\frac{h}{\lambda}$, when the light of wavelength λ is applied, the momentum of the object changes, and in the measured value of momentum, the uncertainty Δp of about $\frac{h}{\lambda}$ exists.

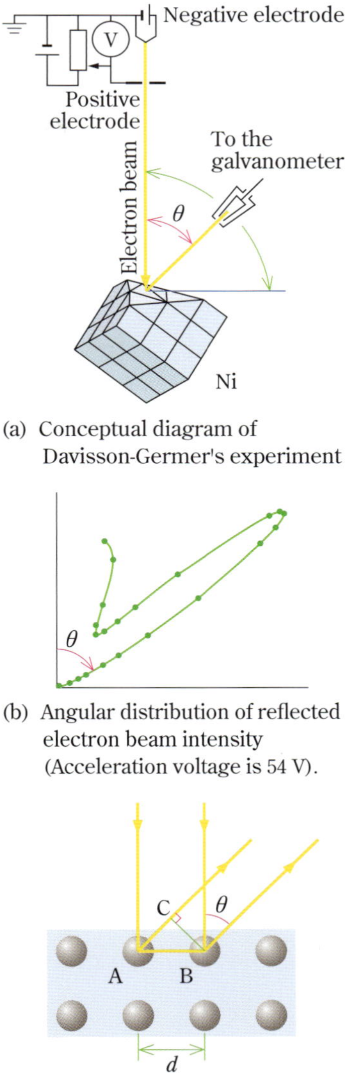

(a) Conceptual diagram of Davisson-Germer's experiment

(b) Angular distribution of reflected electron beam intensity (Acceleration voltage is 54 V).

(c) Condition for electron waves to be strongly scattered $\overline{AC} = d\sin\theta = n\lambda$

Fig. 24-13

If we attempt to determine the position x of an electron accurately using light of a short wavelength, the uncertainty Δp of the measured value of momentum becomes large, and if we attempt to accurately determine the momentum p of the electron using light of a long wavelength, then the uncertainty Δx of the measured value of position increases. Consequently, we are led to Heisenberg's **uncertainty principle** :

$$\text{“uncertainty } \Delta x \text{ of measured value of position”} \times \text{“uncertainty } \Delta p \text{ of measured value of momentum”} \geqq \frac{h}{4\pi}. \quad (24.15)$$

The uncertainty principle says that it is impossible to measure both the "position" and "momentum" of minute objects like electrons accurately at the same time. This principle holds not only for electrons but also protons and neutrons. Because "momentum" = "mass"×"velocity", when light is irradiated, the motion of electrons with a small mass is most disturbed.

For example, in the case of Fig. 24-12, if to identify which of the two slits the electrons pass through we irradiate the electrons with light of a wavelength shorter than the slit interval, the motion of electrons is greatly disturbed and more electrons go to the dark part of the stripes, so the bright and dark stripes disappear. If we want to investigate the behavior of the particle nature, the behavior of the wave nature disappears. Therefore, detection of the wave nature and the particle nature of electrons at the same time is impossible.

As described above, in principle we cannot accurately measure both the position and velocity of an electron in a very narrow space such as an atom, so it is theoretically impossible for us to consider in detail the situation where electrons move in a circular orbit in the atom (see Problems for exercise 24, B2).

24.5 Stationary state of the atom and the line spectrum of light

Learning objective To understand that the energy values that an atom can take are discrete and therefore, when the light emitted from the atom is dispersed, a line spectrum is generated.

How should we consider the motional state of electrons in an atom? According to quantum mechanics which is the mechanics of the atomic world, electrons move as waves in the atom. There are two types of waves. One is a traveling wave like a wave spreading on the water surface. The other is a standing wave that occurs in strings when we play a guitar or piano string. The frequency ν of the standing wave can only take the discrete values (see Section **12.5**). For example, if we hit one of the piano keys, no sound other than the fundamental frequency and its harmonics is emitted.

Electron waves in an atom are standing waves that oscillate at a discrete frequency, just like the waves that occur in a string, In the world of quantum mechanics, energy is h times the wave frequency ν, so the energy of an atom ($E = h\nu$) takes only the discrete values E_1, E_2, E_3, $\cdots$. Each of these discrete energy states is called the **stationary state**. The stationary state with the lowest energy is called the ground state, and other stationary states are called excited states.

An atom in the high-energy stationary state E_n is unstable and emits photons to shift to the low-energy stationary state E_m (it does a transition from E_n to E_m). At this time, the excess energy $E_n - E_m$ changes into the energy of photon (Fig. 24-14). Since the energy of photon is $h\nu$, the frequency ν of the light emitted by the atom in the transition is limited to the discrete value :

$$\nu = \frac{E_n - E_m}{h}. \tag{24.16}$$

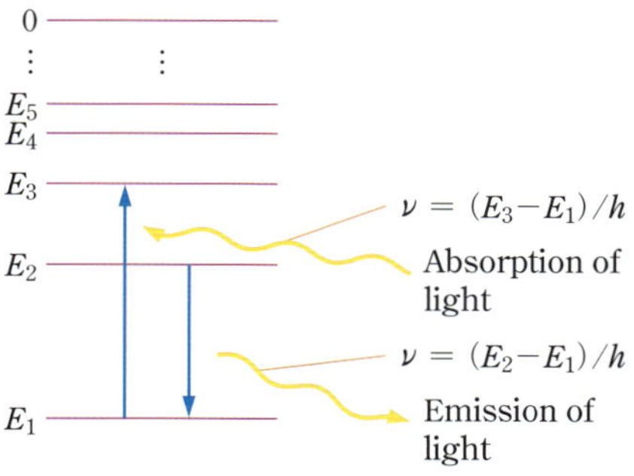

Fig. 24-14 Energy levels of atom and the emission-absorption of light.

Through the experience with neon signs we know that the frequency of light emitted by gas atoms is limited to discrete values. The gas in the discharge tube emits a specific color of light. For example, neon emits red and argon emits purple. When an atom is heated to a high temperature, or stimulated by electric sparks or atomic collisions, the atom emits light, which is split into many lines when it is dispersed by a diffraction grating (Fig. 24-15). These lines called line spectra correspond to the lights of discrete frequency.

On the contrary, when an atom in the low energy stationary state E_m absorbs one photon of light of frequency $\nu = \dfrac{E_n - E_m}{h}$, it shifts (it does a transition) to the high energy stationary state E_n.

Similarly as the case of gas molecules sealed in a tube, when an immense number of constituent particles move randomly colliding with each other or with walls in a closed space at an absolute temperature T, the probability that a constituent particle has energy E is, according to the Boltzmann distribution we discussed in Section **14.3**, proportional to

$$e^{-E/kT}. \tag{24.17}$$

Therefore, at room temperature most atoms and molecules are in the ground state of the lowest energy.

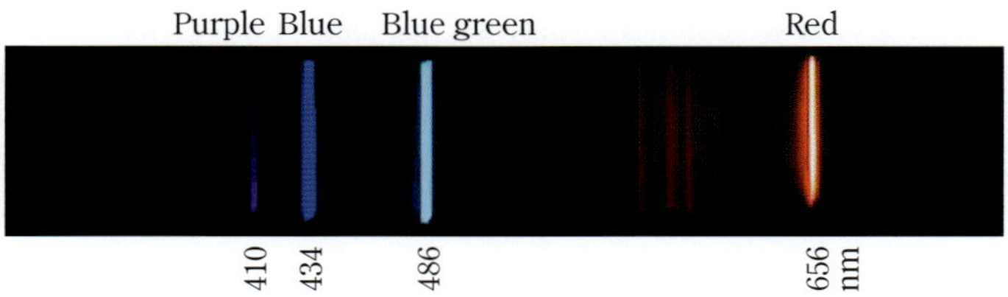

Fig. 24-15 Part of the line spectra of hydrogen atom. The number given below the figure is the wavelength. The frequency ν of light emitted by hydrogen atoms is only the discrete value that satisfies the condition

$$\nu = (3.29 \times 10^{15}\ \mathrm{s}^{-1})\left(\frac{1}{m^2} - \frac{1}{n^2}\right).$$

m and n are positive integers with $n > m$. The spectra of this figure correspond to the case $m = 2$, called Balmer series.

24.6 Periodic law of elements

Learning objective To understand how the periodic law of elements can be explained by atomic physics. To understand the atomic orbitals and the quantum numbers.

At the center of the atom is a nucleus composed of positively charged protons and uncharged neutrons. The number of protons contained in the nucleus is called the **atomic number**. The atomic number is also the number of electrons contained in the atom, which, as will be explained, determines the chemical nature of the atom.

Quantum mechanics can be applicable to calculate the energy of a stationary state of an atom. A standing wave of an atom can be considered as a collection of standing waves that correspond to the individual electrons contained in the atom. In other words, each electron behaves as a standing wave independent of other electrons, and it can be considered that the energy of each electron takes only discrete specific values of energy. The energy levels of electron vary from atom to atom, but they can be qualitatively shown as in Fig. 24-16. Each Z electron in the atom of atomic number Z occupies one of these energy states (stationary states represented by a standing wave). These states are referred to as **atomic orbitals** from the comparison with the orbitals of electrons in atoms.

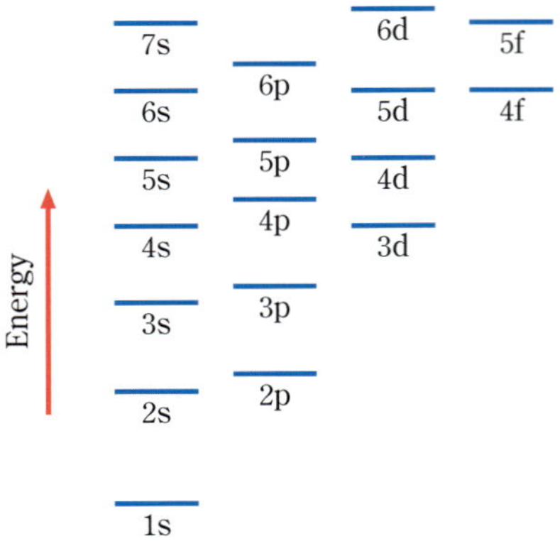

Fig. 24-16 Electronic energy levels in heavy atoms

Such symbols as 1s, 2s, 2p, ⋯ are the names of state. The number specifying the state is called the **quantum number**. The numbers 1, 2, 3, 4, ⋯ are the number of node surfaces (amplitude is 0) of a standing wave minus 1, and are called the **principal quantum numbers**. The symbols s, p, d, f, ⋯ indicate respectively the number 0, 1, 2, 3, ⋯ of the azimuthal quantum number (**quantum number of orbital angular momentum**) l representing the magnitude of the angular momentum when electrons revolve around the nucleus. In the state of orbital angular momentum quantum number l, there are $2l+1$ states corresponding to the difference in the direction of the rotation axis of the revolution. Electrons perform a rotational motion called spin (see Section **20.6**), and one state of rotational motion is occupied by two electrons with different spin orientations.

Accordingly, if we consider spin, 1s, 2s, 3s, ⋯ have two states, 2p, 3p, ⋯ have six states, 3d, 4d, ⋯ have 10 states, and 4f, 5f, ⋯ have 14 states each. On account of the restriction imposed by **Pauli's principle** that only one electron can occupy one state, in the ground state of the atom with atomic number Z, Z electrons occupy the levels shown in Fig. 24-16 from the 1s state of the lowest energy to the Z-th state.

When elements are arranged in atomic number order, there appears the so-called **periodic law of elements** telling that elements with similar chemical properties appear at regular intervals. A table in which elements are arranged based on the periodic law is called the **periodic table**. Looking at the electron configuration in Fig. 24-17, we can see that the electron configuration corresponds to the periodic table of elements. In other words, the number of electrons that occupy the group of state in the final filling stage determines the chemical

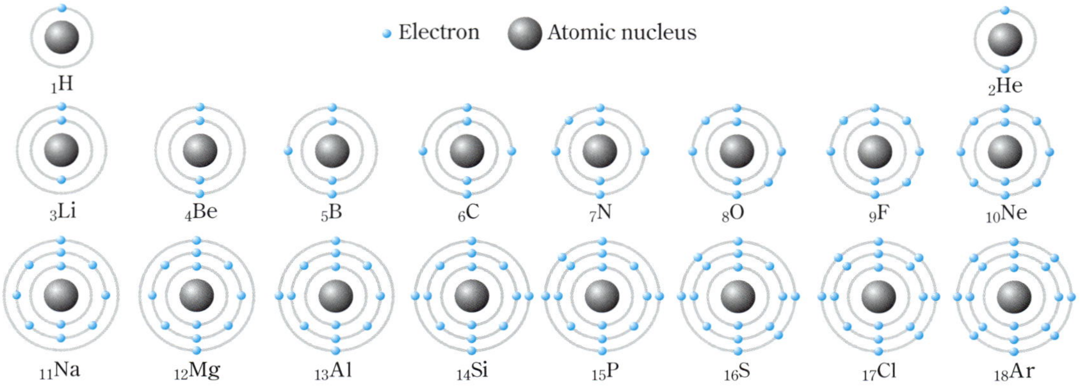

Fig. 24-17 Electron configuration of atoms in the ground state. In reality, electrons exist near the shell as a wave.

properties of the element. Since the higher the energy, the farther the electrons are from the nucleus, the group of state is called the shell, and the state occupied in the final stage is called the outermost shell. The number of electrons in the outermost shell corresponds to the atomic valence that determines the chemical properties of the atom, so the electrons in the outermost shell are also called valence electrons.

The atoms with the filled outermost shell are atoms of inert gases such as helium He, neon Ne, and argon Ar atoms. Atoms such as hydrogen H, lithium Li and sodium Na emit only one electron in the outermost shell and tend to become monovalent positive ions, while atoms such as fluorine F and chlorine Cl have only one vacancy in the outermost shell and they easily put an electron into the vacant seat to become monovalent negative ions.

In the ground state of atom, the lowest energy state to the Z-th state are occupied by electrons. Instead, in the excited state, there are vacancies in the low energy state and at the same time electrons exist in the high energy state. At room temperature, most atoms in matter are in the ground state, but some atoms are in the excited state due to thermal motion. To excite gas atoms, it is sufficient to heat the gas or to apply a voltage between the electrodes of the discharge tube to accelerate the electron to collide with the gas atoms.

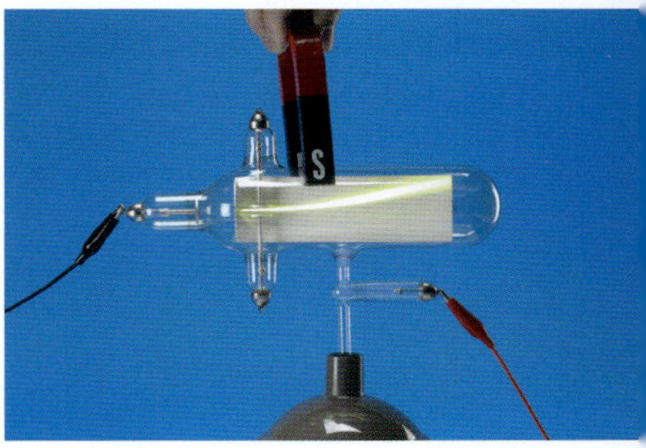

Fig. 24-18 A discharge tube

24.7 Metals, insulators, and semiconductors

Learning objective To understand what causes the difference in the electrical conductivity of metals, insulators, and semiconductors. To understand the mechanism of electric conduction in p-type and n-type semiconductors.

Band Let us consider the electrons inside a solid in which atoms are tightly packed. When one atom is present alone, the energy of the electron can only take the discrete values shown at the left end of Fig. 24-19.

But, if two atoms are brought close each other, the electrons of one

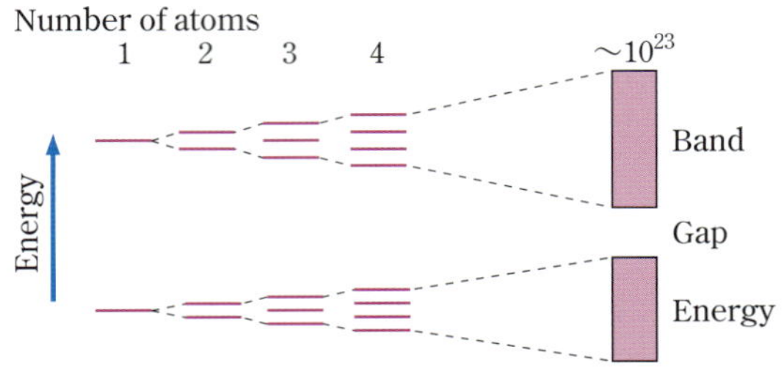

Fig. 24-19 Formation of energy band

atom interact with the electrons of the other atom, so the frequency of the stationary wave of electrons changes. Thus, in the situation of two atoms in close proximity, the electron energy level is the second from the left in Fig. 24-19. This phenomenon resembles that in Fig. 24-20 in which when two pendulums are oscillated simultaneously, the energy exchange occurs between the pendulums and the two types of oscillation with slightly different frequencies are performed.

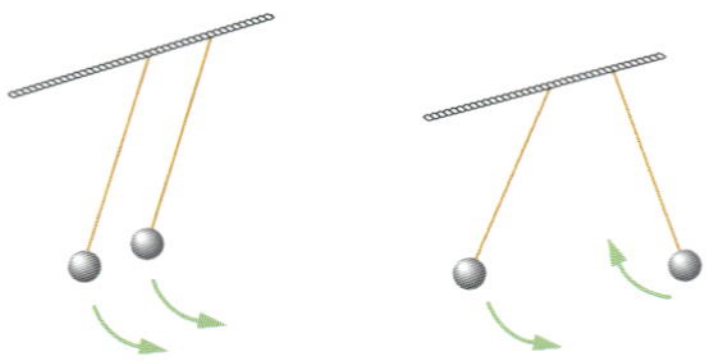

Fig. 24-20 The same two pendulums are suspended from one string. The frequencies of the two types of oscillation are slightly different.

As the number of atoms in proximity increases to 3, 4, ⋯ the energy levels become 3rd, 4th, ⋯ from the left in Fig. 24-19. Therefore, when a large number of atoms gather to form a crystal, the energy values which electrons can take have a width around the energy level of the atom, as shown at the right side of Fig. 24-19. A range of energy having this width is called an **energy band** or **band**. On the other hand, the range of energy that electrons cannot take is called the **energy gap** or **gap**.

If the number of atoms constituting a crystal is N, nN electrons can enter into the band corresponding to the group of energy level in which n electrons enter in the case of single atom. Electrons occupy in order from the low energy band. The band where free electrons (also called conduction electrons) move freely in matter is called the **conduction band**.

Fig. 24-21 Silicon single crystal with eleven nines (99.999999999 %) purity and uniform crystal structure

Metals In the case of metals, the conduction band of the band containing free electrons is only partially occupied by electrons [Fig. 24-22 (a)], so even with a small voltage, electrons are continuously accelerated to a high energy state, and the current flows.

Insulators In the case of insulators in which no current flows even with an applied voltage, the conduction band is completely filled with electrons up to a band called the valence band, and the conduction band having no electron is across the gap [Fig. 20-22 (a)]. When we attempt to accelerate electrons in the valence band by applying a voltage to move electrons to the conduction band, the electrons must jump over

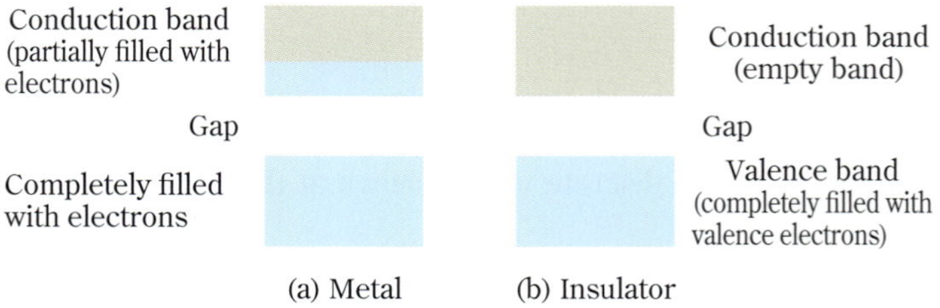

Fig. 24-22 Metal, insulator, semiconductor

the gap to reach the conduction band above the gap. Since the energy difference between the conduction band and the valence band is larger than the energy obtained from the electric field, the electrons cannot move to the conduction band. For this reason, even if a voltage is applied to insulators, electrons are not accelerated and no current flows.

Semiconductors In semiconductors with an electrical resistivity between metals and insulators, the valence band is completely filled with electrons like insulators and across over the gap there is a conduction band with no electron [Fig. 24-22 (b)]. The difference from insulators is that the gap is narrow, so when the temperature is high, the thermal motion of electrons excites a small amount of electrons from the valence band to the conduction band. Therefore, the electric resistivity of semiconductors is much larger than that of metals, but decreases rapidly with the increase in temperature.

Typical semiconductor materials include group 14 elements of germanium Ge and silicon Si, 1 : 1 compounds between group 13 and group 15 elements like InSb, InAs, GaAs etc., and 1 : 1 compound CdSe between group 12 and group 16 elements.

Silicon is an element having four valence electrons like carbon, and each silicon atom generates four valence electrons, shares eight electrons with surrounding silicon atoms, and bonds together in a mechanism called the covalent bond.

In the case of silicon, the gap between the electron-filled valence band and the empty conduction band is narrow (1.17 eV). Therefore, when their valence electrons forming covalent bond receive thermal energy, they jump the gap to reach the conduction band to become free electrons [Fig. 24-23 (a)]. In this case, a cavity is created after an electron escapes from the covalent bond, and in turn a nearby electron gets into this cavity, and the electron of other atoms gets into the cavity thus generated. In this way, the cavities generated by the escape of electrons move in the crystal, similarly as bubbles move in the water. Thus, when a voltage is applied, a situation occurs in which the positively charged cavities move in the direction of the electric field, that is, in the opposite direction to the movement of electrons in the reverse direction of the electric field (Fig. 24-24). These cavities are called **holes**. Accordingly, in this situation, electric conduction occurs in both free electrons and holes. Such materials are called **intrinsic semiconductors**. Because there are few free electrons, the electric resistivi-

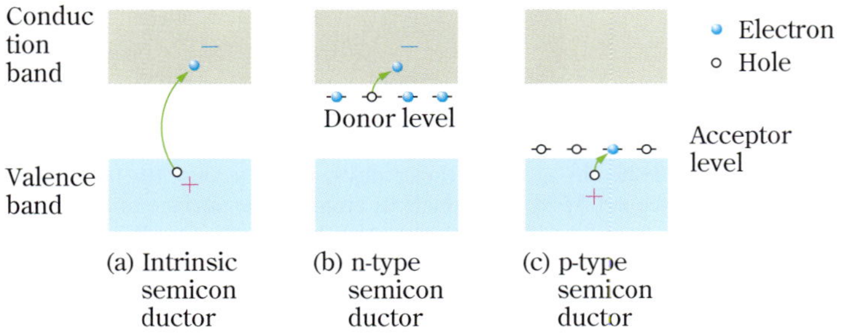

Fig. 24-23 Energy band of semiconductor

ty of intrinsic semiconductors is much higher than metals but much lower than insulators.

Among materials called semiconductors, materials that are important for application are those obtained by doping impurities into silicon Si. When the crystal of silicon is doped with phosphorus P, arsenic As, antimony Sb, or bismuth Bi, namely, elements having five valence electrons as impurities, these impurity atoms enter the crystal lattice point and emit four electrons to be covalently bonded to the surrounding four silicon atoms. Consequently, one valence electron of the impurity atom is left for each. This electron should enter in the conduction band above the valence band, but it is at an impurity level slightly below the conduction band because it is pulled by the electric force of the impurity atom that has become positive ion. However, with acquirement of a small amount of energy, it leaves the impurity atom, jumps to the conduction band, and becomes a free electron that can move around in the crystal. When a voltage is applied, free electrons move and current flows. Such materials are called **n-type semiconductors**. The impurity level is called the donor level [Fig. 24-23 (b)].

When a crystal of silicon is doped with impurities such as boron B, aluminum Al, gallium Ga, or indium In, etc., which have only three valence electrons, one valence electron is lacking for each to form a covalent bond with the surrounding atoms. As a result, it may seem that there are as many cavities in the valence band as the number of impurity atoms. But if an electron is put into this cavity, the impurity atom has a negative charge, so the energy of the electron entering the cavity is slightly higher than the energy of the electron in the valence band. As a result, in the energy level of the semiconductor crystal containing this impurity, an impurity level that is not occupied at low temperatures exists immediately above the valence band. This is called the acceptor level [Fig. 24-23 (c)].

When electrons in the valence band get energy, they jump to the free acceptor level and cavities are created in the valence band. This cavity

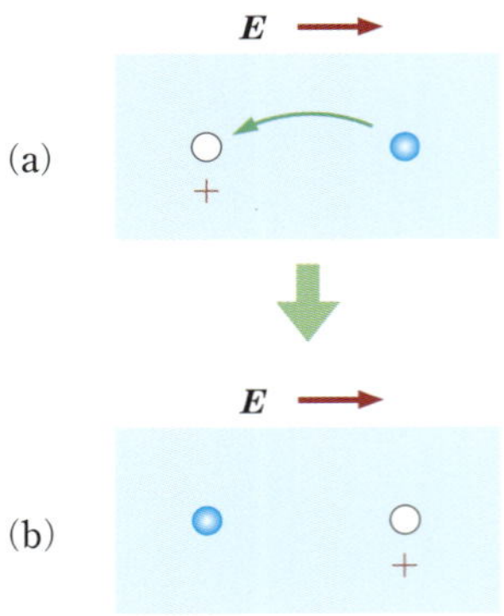

Fig. 24-24 The transfer of electrons in the reverse direction of the electric field ***E*** can be regarded as the movement of holes in the direction of the electric field ***E***.

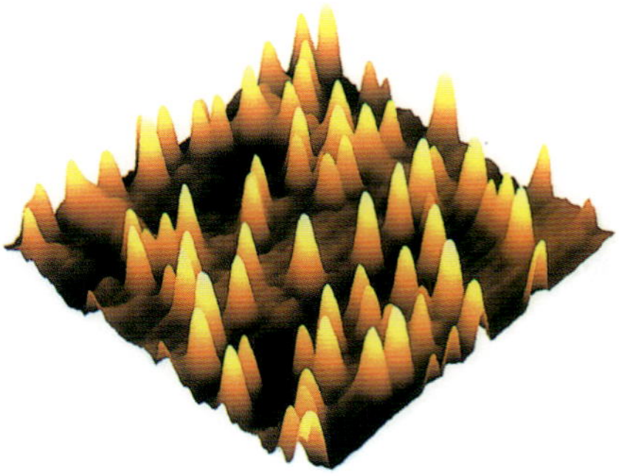

Fig. 24-25 An atomic force microscopy image of InAs quantum dots on InP (001) surface. By carefully supplying the group 13 element In and the group 15 element As to the crystal plane of InP under the condition of accurately in one atomic layer, an extremely fine three-dimensional structure of InAs can be formed. Crystals of group 13-15 compound exhibit semiconductor properties.

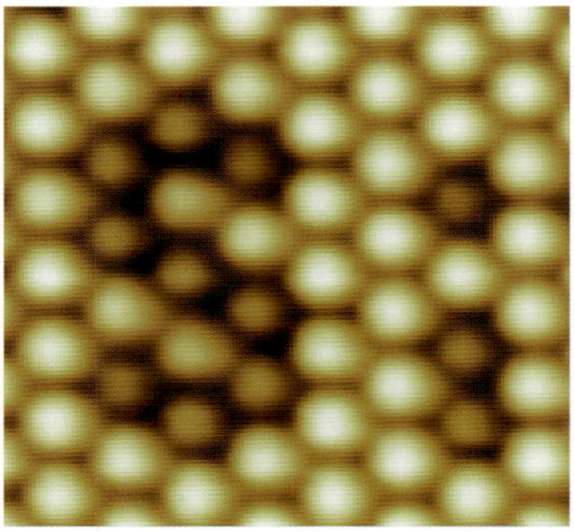

Fig. 24-26 In a study these days, a very thin tip of an atomic force microscope (AFM) can touch an atom on the surface to apply a force, and exchange the position with the atom held at the tip. With this technology, precise injection of impurities is possible. In this figure, a silicon (Si) atom held at the tip of the probe was exchanged and buried on the surface where tin (Sn) atoms are regularly arranged, and the character "Si" is formed.

is the hole described above. Then, when a voltage is applied, a current flows due to the movement of holes. Such materials are called **p-type semiconductors**. The names of n-type and p-type depend on whether the charge carried by the carrier is negative or positive.

Thus, it can be understood that the electric conduction of solids is well explained by introducing the concept of energy band.

The characteristics of semiconductors are significantly influenced by impurities contained. Based on this fact, first, high-purity silicon is produced, and then a fixed amount of a certain kind of impurity is dissolved (referred to as doping) in it to fabricate p-type and n-type semiconductors having the desired characteristics at a desired site.

24.8 Application of semiconductors

Learning objective To become aware of how semiconductors are applied.

pn junction diode A structure in which part of the silicon crystal is p-type and the other part is n-type so that the p-type semiconductor and the n-type semiconductor are in contact is called a **pn junction**, and a pn junction diode with two attached electrodes is called a pn junction diode [Fig. 24-28 (a)]. The pn junction diode has a rectification function allowing the current to flow in only one direction under specified conditions (Fig. 24-29).

Let us consider first what happens when a p-type semiconductor and an n-type semiconductor are joined. When they are joined, free electrons diffuse from the n-type part near the junction into the p-type part, and holes from the p-type part near the junction diffuse into the n-type part and both are combined and annihilated each other. Thus, the layer near the junction has no carrier (no free electrons nor holes). This layer is called the depletion layer. As a result, within the depletion layer, positive charges appear in the n-type part near the junction while negative charges appear in the p-type part [Fig. 24-28 (a)]. These charges prevent the carriers in the p-type and n-type parts from further diffusion.

When the n-type electrode is connected to the positive electrode of the battery and the p-type electrode is connected to the negative electrode, the electrons in the n-type part and the holes in the p-type

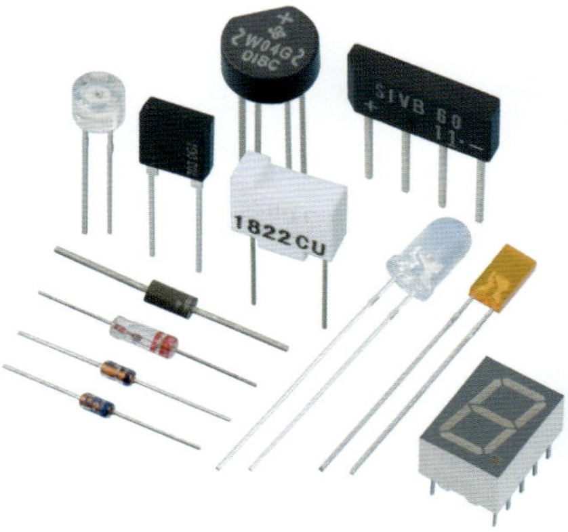

Fig. 24-27 General-purpose rectifier diodes

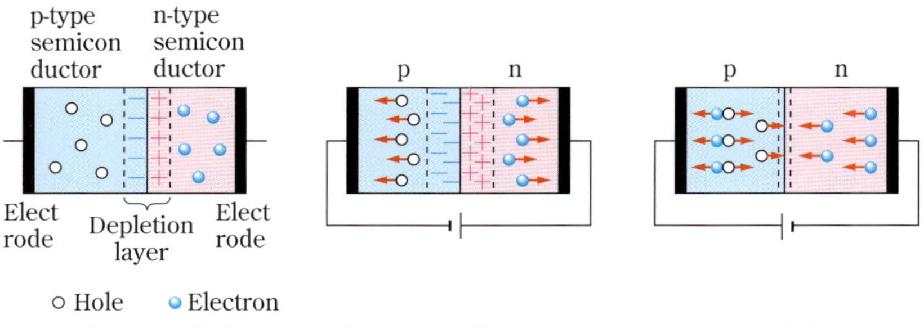

Fig. 24-28 pn junction diode

part are also drawn toward the respective electrodes. Accordingly the depletion layer spreads and carriers cannot move through the junction, so almost no current flows [Fig. 24-28 (b)].

Conversely, when the p-type electrode is connected to the positive electrode of the battery and the n-type electrode is connected to the negative electrode, the holes in the p-type part move to the n-type part and the electrons in the n-type part move to the p-type part. As a result, the depletion layer narrows, and when a voltage of more than a certain level (about 0.6 V or more) is applied, carriers can flow each other across the depletion layer and current flows. At this time, free electrons flow toward the n-type part from the electrode attached to the n-type part to replenish electrons. Also, from the inside of the p-type part, electrons are directed toward the electrode attached to it, which can be viewed as holes being supplied to the p-type part from the electrode. Therefore, in this case, the current continues to flow [Fig. 24-28 (c)].

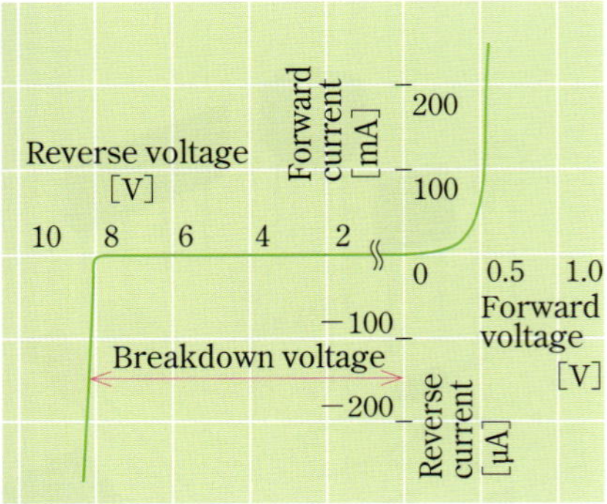

Fig. 24-29 Characteristics of pn-junction diode

Thus, in pn-junction diodes, current flows only when the potential of the p-type part becomes positive with respect to the n-type part, and the current does not flow in the opposite case (Fig. 24-29). This is called the **rectification** of diode, and the former situation is called the forward direction, while the latter situation is called the reverse direction. When the reverse voltage exceeds a certain level, the current starts to flow rapidly. This voltage is called the breakdown voltage.

Transistor A typical example of semiconductor application is a transistor used in electronic circuits. A transistor is a semiconductor circuit element with three terminals and has the amplification and switching functions. The invention of the transistor opened the way to reduce the size and power consumption of electronic devices. In what follows, bipolar transistors and MOS field effect transistors will be introduced.

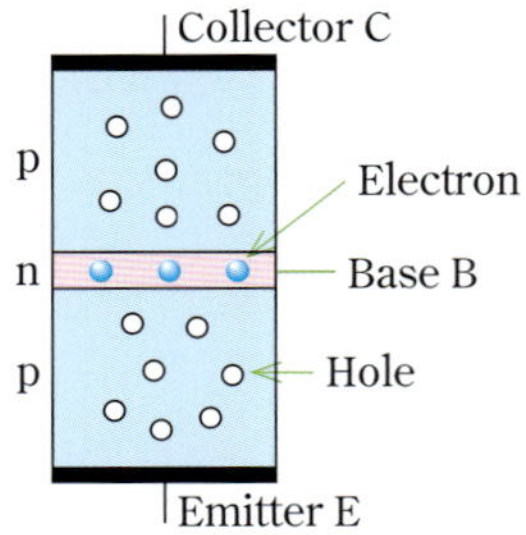

Fig. 24-30 Bipolar transistor

In Fig. 24-30 is shown a bipolar transistor which has an electrode-attached p-type, n-type and p-type parts of silicon crystal. The central n-type part is called the base and is made very thin. Even if a voltage is applied between the p-type emitter and collector, no current flows, because one of the two pn junctions with the base is reversed. But, for example, when a forward voltage is applied between emitter E and base B, a current flows between E and B and a large amount of holes are injected into the base. Accordingly, the base is as if it were a p-type, so the diffused holes flow into the collector and a large amount of collector current flows. This characteristic is used for voltage amplification, because the collector current changes greatly with a small change in the base voltage. This type of transistor operates on both electrons and holes, so it is called the bipolar transistor.

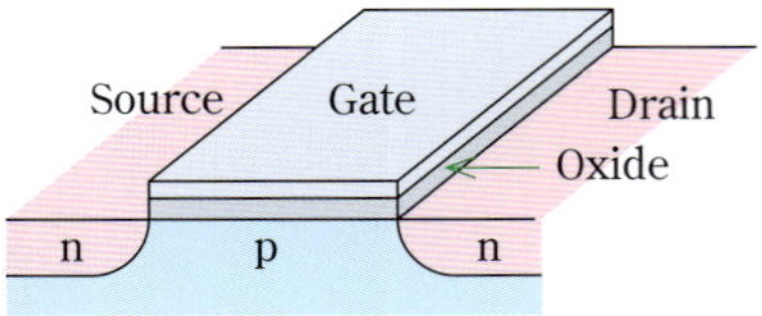

Fig. 24-31 MOS field effect transistor

Fig. 24-31 shows an example of what is called a MOS field effect transistor. This is prepared first by oxidizing the surface of a p-type silicon substrate to form a thin SiO_2 film on it, then forming a metal film (gate electrode) on the oxide film, and then opening two cavities in the oxide film on both sides to make two high density n-type doped electrodes (source and drain). In this state, even if a voltage is applied between the source and the drain, no current flows because one of the two pn junctions becomes a reverse junction. However, when a positive voltage is applied to the gate electrode, holes are removed from the portion under the gate called the channel and it turns into n-typed and the source and the drain are connected through the n-type channel, and a current called a drain current flows. A slight change in the gate voltage causes a significantly great change in the drain current, so this property is applied to voltage amplification. In addition, the drain current can be turned on and off. A field effect transistor is called a unipolar transistor because it works with either electrons or holes. Since field effect transistors can be made extremely small in size, they are used in the manufacture of super-LSIs, such as semiconductor memories and CPUs.

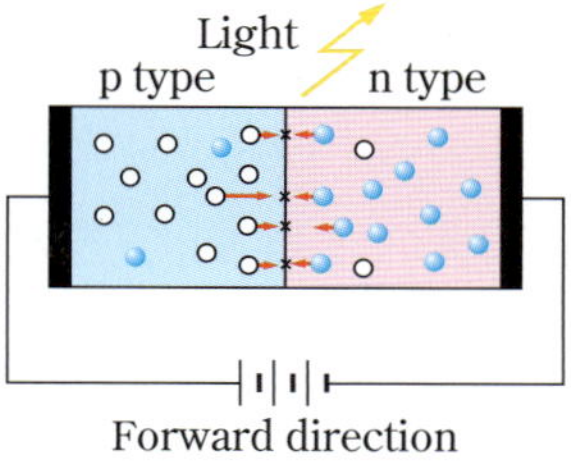

Fig. 24-32 Light emission of light emitting diode

Light emitting diode A light emitting diode (LED) is a diode device with a pn junction as shown in Fig. 24-32 in which semiconductors that emit light easily, such as gallium arsenide (GaAs) or gallium phosphide (GaP), are used in place of silicon. When a forward voltage is applied to this pn junction diode, electrons and holes are combined to neutralize near the junction surface. This process is a process in which electrons in the high energy (E_n) conduction band are brought to empty sites in the low energy (E_m) valence band. In this process the difference in electron energy $E_n - E_m$ is emitted not in the form of heat of crystal but as photons from the junction. The emission color changes depending on the semiconductor material.

Fig. 24-33 A conventional signal (upper) and the one using light emitting diode (lower)

Solar battery A solar battery is an element that uses semiconductors to directly convert the energy of sunlight into electric energy (Fig. 24-34). When the photon with energy larger than the energy gap is irradiated on the surface near the pn junction so that electron-hole pairs are formed, the electric field of the depletion layer causes the electrons to move to the n-type part and the holes to the p-type part. This generates a photovoltaic power that charges the p-type part positively and the n-type part negatively.

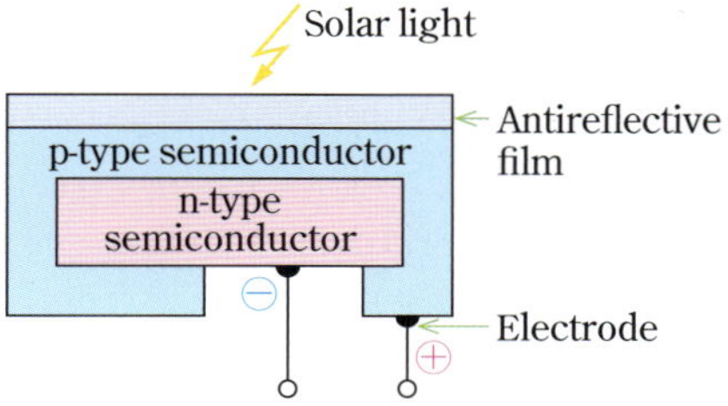

Fig. 24-34 Solar battery

24.9 Laser

Learning objective To understand the overview of the principle of laser which produces a thin, powerful monochromatic light beam.

The light of an incandescent lamp utilizes blackbody radiation of a tungsten filament heated to high temperatures, and the light of a fluorescent lamp utilizes light emission by atomic ions excited by the impact of collisions with accelerated electrons. In both cases, since individual atoms emit light independently of the other atoms (referred to as **spontaneous emission**), so light is emitted in all directions, and the phases of the electromagnetic wave generated by the spontaneous emission of different atoms are independent of each other. Therefore, even in the case of monochromatic light (light with a specific frequency) emitted from one light source, it is not a sine wave whose phase keeps changing uniformly in time but the uniform phase state lasts only for a short time (about 10^{-9} second) and wave phases soon become non-uniform. In natural light, the length of this sine wave is about several tens of centimeters, which is called the coherence length. For this reason, interference does not occur when the optical path difference is several tens cm or more. In addition, the light from the two light sources does not interfere even at the same frequency.

The term laser is an abbreviated expression of English acronym made from Light Amplification by Stimulated Emission of Radiation and in the early stages it only referred to the visible light and those of its surrounding frequency domain and sometime later it became a generic term for light of any frequency. A laser is a device that generates a narrow, powerful beam of monochromatic light in which vibrations of the light waves emitted from each atom are in phase and which does not diffuse over long distances.

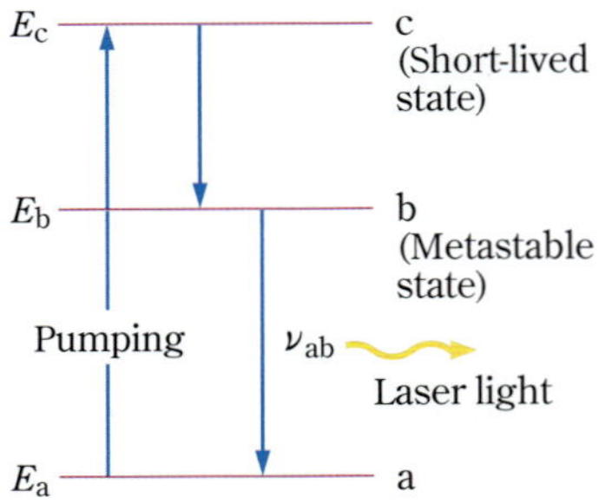

Fig. 24-35 Pumping

When an atom with an energy level structure as shown in Fig. 24-35 is in the excited state b, it emits light of frequency $\nu_{ab} = \dfrac{E_b - E_a}{h}$ and makes a transition to the ground state a. This transition occurs even in the absence of light (spontaneous emission) around the atom, but when light of frequency ν_{ab} is incident on the atom, the atom is induced by this light to emit light of the same frequency ν_{ab} in the same direction and with the same phase as that of the incident light. This phenomenon is called the **stimulated emission**. The light of stimulated emission and the incident light cause enhanced interference and become strong light. If there are a large number of atoms in the excited state b, stronger stimulated emission occurs and the energy of light increases (in contrast to electrons that can exist only in one state, any number of photons can exist in one state). In other words, light amplification occurs.

However, in the natural state, the number of atoms in the ground state a is much larger than the number of atoms in the excited state b. Therefore, light of frequency ν_{ab} is absorbed by the atoms in the ground state, and amplification of light fails to occur.

In order to emit highly strong light with the stimulated emission of radiation, it is necessary to create the state called population inversion in which the population of atom in the excited state b is denser than that of the ground state a. For this purpose, atoms in the ground state a are first excited to the excited state c by being irradiated with an electron beam or an intense light of a specific frequency. This operation is similar to pumping water to a high position with a pump, so it is called pumping. Atoms in the short-lived excited state c immediately make a transition to the long-lived metastable state b. Accordingly, the population of atoms in the excited state b increases remarkably and the population inversion between states a and b is attained. The material in this state amplifies light of a specific wavelength, so it works as an optical amplifier.

When a material in the light-amplifying state (state of population inversion) is placed between two reflectors facing each other, the resonant frequency light which becomes a standing wave between the two reflectors is amplified by stimulated radiation to becomes an oscillation state, and part of it is emitted as narrow monochromatic light

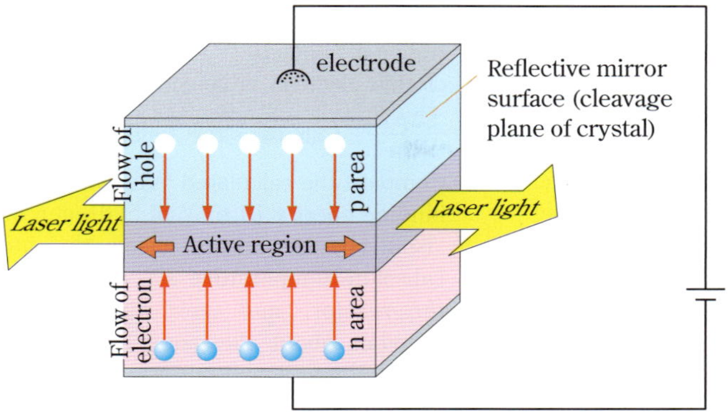

Fig. 24-36 Fundamental structure of semiconductor laser (conceptual diagram)

to the outside. This is the principle of laser oscillators.

There are many substances that exhibit laser action, and gases, liquids, solids, semiconductors, etc. are used as optical media for oscillation. Laser light, born in 1960, is used for optical communication via optical fibers, CD reading, laser printers and many other devices.

The overwhelming majority of lasers currently manufactured are semiconductor lasers. The semiconductor laser has a structure in which a thin active region for amplifying light is sandwiched between a p-type semiconductor and an n-type semiconductor (Fig. 24-36). When a voltage in the direction shown in the figure is applied to this structure, electrons from the n-type region and holes from the p-type region flow into the active region. As a result, high energy electrons and low energy holes increase in the active region. This is the method to create the state of population inversion by pumping in a semiconductor laser. When electrons enter holes, extra energy is emitted as light. In semiconductor lasers, the cleavage plane obtained by cutting the crystal in the easily-cleaving direction becomes the reflective mirror.

The laser beam travels as a narrow beam, and can be focused onto a spot as small as the wavelength using a lens with a short focal length. At this time, the energy density of light at the focal point becomes very high. All materials can be ionized by increasing the electric field strength of the laser light. To eliminate the risk of blindness, every possible precaution must be taken to prevent the laser light from directly hitting the eyes.

Fig. 24-37 Four laser beams emitted from the Subaru Telescope (at right in the photo) and others. The purpose is to create an artificial star with lasers to correct the blurring of star image due to atmospheric fluctuations.

Problems for exercise 24

A

1. Find the energy is one photon of orange light with a wavelength of 0.6 μm.

2. The laser emitted 10 J of energy in one pulse of 5×10^{-11} s.
   (1) Find the length of this pulse in vacuum.
   (2) Find the energy per unit volume of the beam when the cross-sectional area of this beam is $2\times10^{-6}\ \text{m}^2$.
   (3) Find the electric field strength in the beam.
   (4) When the wavelength of this beam is 6.9×10^{-7} m, how many photons are contained in one pulse?

3. Calculate the speed v of electron in the electron beam when the de Broglie wavelength is the size of atom (about 10^{-10} m). Compare this speed with the speed of light in vacuum $c = 3\times10^8$ m/s. Find the approximate value of the kinetic energy E of this

electron in eV unit. Assume the mass of electron $m = 9.11 \times 10^{-31}$ kg.

4. Find the de Broglie wavelength of neutron with a speed of 1.0×10^4 m/s. Assume the mass of neutron $m = 1.67 \times 10^{-27}$ kg.

5. If the following particles have the same kinetic energy, which particle has the longest de Broglie wavelength ; electron, positron, α particle (nucleus of helium atom)?

B

1. Calculate the electric potential energy $U(r) = \dfrac{ZZ'e^2}{4\pi\varepsilon_0 r}$ for gold nucleus ($Z = 79$) and helium nucleus ($Z' = 2$) at $r = 10^{-10}$ m and 10^{-14} m, and answer in eV unit. If a gold atom is approximated by a point charge, up to how much distance from the gold atom can an a particle having a kinetic energy of 4.79 MeV approach? Using this result and based on the experiment in which the α particle approaches the gold nucleus and its path is greatly bent, we can estimate the radius of gold nucleus.

2. Assume that the position of the electron is determined with the accuracy ($\Delta x = 0.5 \times 10^{-10}$ m) of the size of hydrogen atom. Calculate the uncertainty Δp of the momentum of electron and the uncertainty $\Delta v = \dfrac{\Delta p}{m}$ of the speed under this assumption. When the electron is confined in a region of about 10^{-10} m in length, the kinetic energy of this electron is considered to be approximately $\dfrac{(\Delta p)^2}{2m}$. Calculate the kinetic energy in eV unit. Compare the calculated value with the ground state binding energy 13.6 eV of hydrogen atom.

3. Neutrons generated inside the atomic reactor (absolute temperature T) collide with the atoms in the reactor, and their kinetic energy for each becomes about the same as the thermal energy $\dfrac{3}{2}kT$ of the atom. These neutrons are called thermal neutrons. Find the de Broglie wavelength λ and speed v of the thermal neutron at $T = 600$ K. Assume the Boltzmann constant $k = 1.38 \times 10^{-23}$ J/K.

4. In Davisson-Germer's experiment shown in Fig. 24-13, find the angle θ at which the intensity of the reflected wave of the electron beam from Ni is maximized (in the case of $n = 1$) when the acceleration voltage is 54 V and also 181 V. Let the lattice distance d be $d = 2.17 \times 10^{-10}$ m.

5. A high power laser emits a light pulse of 2000 J. Find the momentum of this pulse.

Tunnel effect and Esaki diode

Dr. Leo Esaki (1925-) discovered in 1958 that a sandwich-type circuit element (Esaki diode) in which a very thin insulator (actually a semiconductor) was inserted between a p-type semiconductor and an n-type semiconductor exhibits a tunneling effect. Dr. Esaki was awarded the Nobel Prize in Physics in 1973 for this discovery.

When the glass window is closed, the air flow between the outside and the room is blocked, but if the glass plate is thin, sounds are transmitted through the glass. This represents the difference between the particle nature and the wave nature. Free electrons as particles cannot enter into the thin insulator of Esaki diode but they can enter as electron waves. The amplitude of the electron waves decays as they travel the insulator, but alter they pass through the insulator, they again travel as free electrons. Since free electrons seem to be passing through tunnels and obstacles (insulators), this phenomenon is called the tunnel effect.

The tunnel effect is a phenomenon unique to quantum mechanics and arising from the duality of particle nature and wave nature.

Nuclei and Elementary Particles

From ancient times people undertook smelting of metals and producing the bronze or iron from mineral ore. However, they could not create gold in any kind of processing. Chemistry was born from these kinds of attempt called "alchemy". Chemical reaction is nothing but a rearrangement of atoms between molecules and it is impossible to convert one element to another. This is because the nucleus in the center of the atom cannot be changed through the chemical reaction. On the other hand, the nucleus is a composite object consisting of the proton and the neutron, and when the nucleus is collided with another nucleus of a large energy, there occurs a rearrangement of protons and neutrons between the nuclei, and as a result, the nucleus changes.

In this chapter, we learn about nuclei and their constituents, elementary particles.

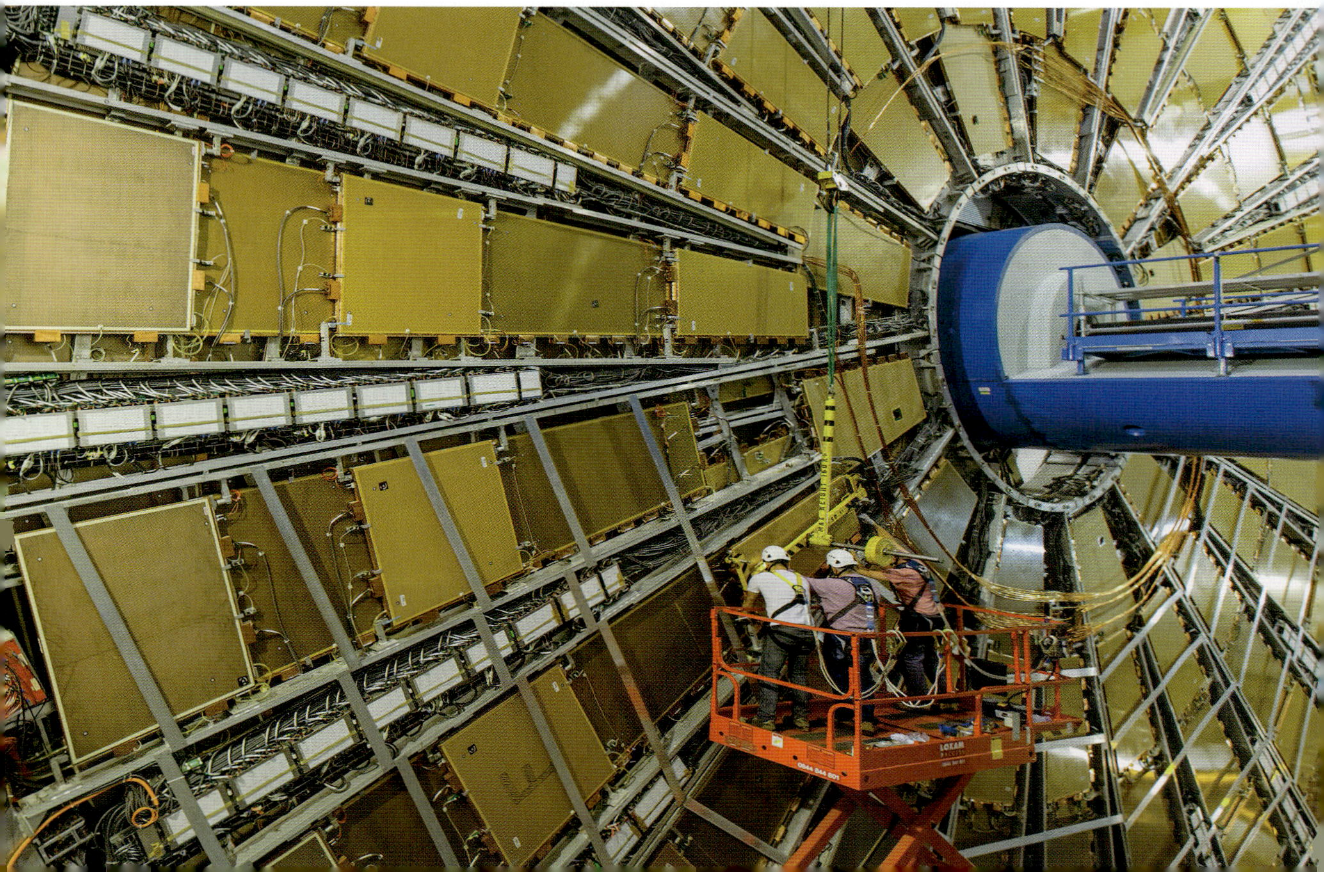

ATLAS detector at CERN in Geneva. Discovery of unknown particles is expected (see p.287, 311).

25.1 Composition of Nucleus

Learning objective To understand that the nucleus is composed of protons and neutrons which are bound together by the nuclear force.

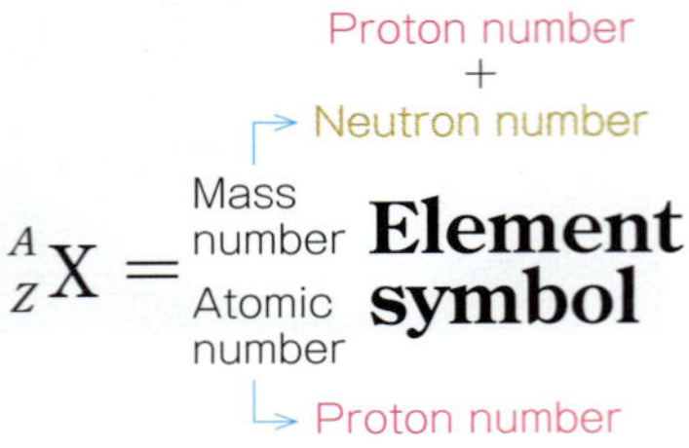

Fig. 25-1 $^{9}_{4}\mathrm{Be}$ represents a nucleus of beryllium with atomic number 4 and mass number 9.

As in the case of the atom, which is not the smallest building block of matter that is indivisible, but consists of the nucleus and electrons, the nucleus located in the center of the atom is not the smallest and indivisible unit of matter. The evidence that indicates the presence of structure for the nucleus comes from the fact that in many elements, the atomic weight which is proportional to the mass of the atom is nearly integral multiple of the atomic weight of the hydrogen which is the lightest atom. For example, the atomic weights of hydrogen, carbon, nitrogen, oxygen are 1, 12, 14, 16, respectively. In some cases, the atomic weights deviate from the integral multiples, but these are due to the effects of isotopes which will be explained later. As an example, the atomic weight of chlorine is 35.5, but this is because it is a mixture of the isotopes of the atomic weight 35 and 37 with the ratio about 3 to 1. The mass of the atom comes nearly from that of the nucleus, hence the mass of the nucleus is approximately an integral multiple of mass of the hydrogen nucleus. This integral value is called the **mass number** of the nucleus. The nucleus of element X with atomic number Z and mass number A is denoted by $^{A}_{Z}\mathrm{X}$. The atom with the nucleus $^{A}_{Z}\mathrm{X}$ is also denoted by the same symbol $^{A}_{Z}\mathrm{X}$ (Fig. 25-1).

Toward the end of the 19th century, the radioactivity was discovered, and from the fact that the radioactive element decays into another element, it was confirmed that the nucleus is not the smallest and indivisible unit of material structure.

The nucleus can be transformed artificially. In 1919, Rutherford, by colliding the nucleus of the helium $^{4}_{2}\mathrm{He}$, called the α (alpha) particle, on the target of the nitrogen nucleus, has shown that the reaction :

$$^{14}_{7}\mathrm{N}+^{4}_{2}\mathrm{He} \longrightarrow ^{1}_{1}\mathrm{H}+^{17}_{8}\mathrm{O} \qquad (25.1)$$

can take place, and thus he proved that nucleus can be transformed artificially. The hydrogen nucleus $^{1}_{1}\mathrm{H}$ can be emitted by the collision of various nuclei and it is the lightest nucleus with mass number 1, and hence it is considered to be the constituent particle of the nucleus and is called the **proton** (the symbol is p).

In 1932, Chadwick, a British physicist, confirmed that the radiation emitted in the collision of α particle on the beryllium target $^{9}_{4}\mathrm{Be}$ is identified as a neutral particle with mass nearly the same as proton' s. He named this particle the **neutron** (the symbol is n). This reaction can be expressed as

$$^{4}_{2}\mathrm{He}+^{9}_{4}\mathrm{Be} \longrightarrow ^{1}_{0}\mathrm{n}+^{12}_{6}\mathrm{C}. \qquad (25.2)$$

As a result of the discovery of the neutron, it has turned out that the nucleus is composed of protons and neutrons. The proton and the neutron are collectively referred to as **nucleon**. The masses of the proton and the neutron are roughly the same*.

$$m_{\mathrm{p}} = 1.6726\times10^{-27}\ \mathrm{kg}$$
$$m_{\mathrm{n}} = 1.6749\times10^{-27}\ \mathrm{kg}$$

The atomic number Z is the number of protons contained in the

* $\frac{1}{12}$ of the mass of one carbon atom with mass number 12, $^{12}_{6}\mathrm{C}$ is called the atomic mass unit (the symbol u), and is used for the practical unit of the atom and the nucleus.

$1\ \mathrm{u} = 1.660\,538\,78\times10^{-27}\ \mathrm{kg}$

nucleus of the atom, and is also the number of electrons existing in the neutral atom. The sum of the number of the protons Z and that of the neutrons N contained in the nucleus, $A = Z+N$, is the mass number of the nucleus. Now, the atoms or the nuclei with the same atomic number but different mass numbers are called **isotope** for each other. The chemical properties of the isotopes are the same. For reference, the mass of the electron (the symbol e^-) is shown as follows :

$$m_e = 9.109 \times 10^{-31}\ \mathrm{kg}.$$

In the nuclear reaction where rearrangements of protons and neutrons between the nuclei take place, the sum of the numbers of protons as well as the sum of the numbers of neutrons do not change. This conservation law, of course, holds for the reactions (25.1) and (25.2).

At early period of the nuclear research, the alpha particle from the radioactive elements which emit alpha particles through the decay, was used as an energetic projectile nucleus which collides with the target nucleus. Later on, Cockcroft and Walton, Lawrence, Van de Graaff and others invented various types of nuclear accelerator, and then it has become possible to accelerate many kinds of nuclei from hydrogen to heavy nucleus using electric fields ending up with large energies.

The nucleus is approximately spherical in shape, and the volume is roughly proportional to the mass number with the radius r is 10^{-15}–10^{-14} m.

Nuclear force The force which acts between nucleons binding nucleons together and becomes the cause to form a nucleus is called the **nuclear force**. The nuclear force is not an electrical force. The Coulomb force does not exert on neutrons which are uncharged, and the electric force between protons is a repulsive force. The nuclear force is not a gravitational force too. The strength of the gravitational force between protons is only about $1/10^{36}$ of that of the electric force.

At short distance like a distance between nucleons inside the nucleus, the nuclear force must be an attractive force which is much stronger than the Coulomb repulsive force between protons. The proton-proton collision, for the inter-proton distance bigger than about 2×10^{-15} m, can be explained as a scattering due to Coulomb repulsive force. Hence the nuclear force is a force of extremely short distance with the range about 2×10^{-15} m. For a distance less than 5×10^{-16} m, the nuclear force is a strong repulsive force (Fig. 25-2). The nuclear forces acting between proton and proton, neutron and neutron, proton and neutron are approximately the same strength.

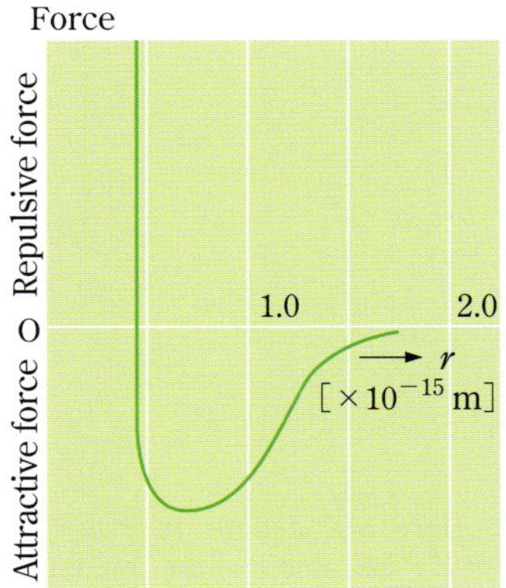

Fig. 25-2 Nuclear force. Distance r between nucleons and the strength of the nuclear force

Since the nuclear force is a short-range force, the nucleon interacts only with the neighboring nucleons. According to the Pauli principle that two protons or two neutrons cannot exist at the same state, the proton tends to attract the neutron nearby, and the neutron in turn tends to attract the proton in the neighborhood. Hence, there exist approximately the same number of protons and neutrons inside the nucleus. The Coulomb repulsive force acts between the two protons in addition to the nuclear force. Because this repulsive force is a long-range force, it acts on any proton inside the nucleus. Therefore, when the number of the protons inside the nucleus increases, the inter-proton distance tends to become larger than inter-neutron distance. Thus there exists a tendency for the ratio of number of neutrons N to number of protons Z inside the nucleus, $\frac{N}{Z}$ to increase.

In 1935, Hideki Yukawa proposed the meson theory for the nuclear force in which he claims that the nuclear force originates from an exchange of an elementary particle called "**pi meson**" between nucleons just like playing catch with a ball. According to the Yukawa theory, the range of the nuclear force, d, is inversely proportional to the mass of the pi meson, m_π, and given by $d = \frac{h}{2\pi m_\pi c}$. Yukawa combined this relation with the range of the nuclear force $d \sim 2\times10^{-15}$ m, and predicted that the mass of the pi meson m_π is 250 times the mass of the electron m_e. In 1947, the pi meson was discovered. The fact that the measured mass was $m_\pi \approx 270 m_e$ is a great achievement of theoretical physics.

25.2 Binding energy of nucleus

Learning objective To understand that when the protons and neutrons are bound together to form a nucleus, the binding energy amounts to the mass defect of the nucleus. To understand that the binding energy per nucleon varies from one nucleus to another, hence there exist nuclei which exhibit the nuclear fission or the nuclear fusion.

The mass of a nucleus is nearly proportional to its mass number A and approximately equal to the sum of the masses of the constituent nucleons. When the mass of the nucleus is precisely measured, it turns out that the mass of the nucleus is smaller than the sum of masses of the constituent nucleons. That is, the mass of the nucleus with mass number A and atomic number Z, i.e. ${}^{A}_{Z}\mathrm{X}$ and the mass denoted as $m({}^{A}_{Z}\mathrm{X})$ is smaller than the sum of Z times proton mass m_p and $(A-Z)$ times neutron mass m_n. This mass difference :

$$\Delta m = Zm_p + (A-Z)m_n - m({}^{A}_{Z}\mathrm{X}) \qquad (25.3)$$

is called the **mass defect** of the nucleus.

When the nucleons get together forming a nucleus, compared to the unbound state, the energy of the bound state is lower by the amount of the potential energy (negative value) of nuclear force. According to the theory of relativity, the mass m is equivalent to the energy $E = mc^2$, and hence this amount of energy decrease ΔE leads to the decrease of mass by $\Delta m = \frac{\Delta E}{c^2}$, namely mass defect*. In order to break up the nucleus we have to provide the nucleons in the nucleus with $\Delta E = \Delta m \cdot c^2$ from outside the nucleus, and so we refer to this energy ΔE as the **binding energy** of the nucleus. We have shown in Fig. 25-3 the binding energy per nucleon, $\frac{\Delta E}{A}$, which is the binding energy of the nucleus divided by the mass number. The nucleus with a large value of this quantity is more stable than the one with a smaller value.

* Here we denote $m \to m - \Delta m$, $E \to E - \Delta E$ due to the binding of nucleons.

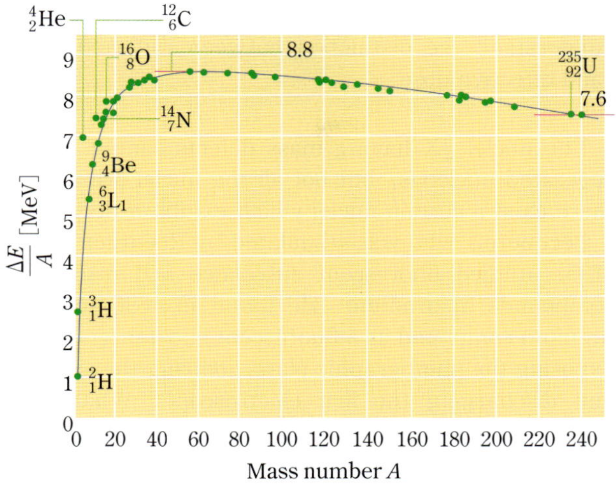

Fig. 25-3 Average binding energy per nucleon $\frac{\Delta E}{A}$ and mass number A

From Fig. 25-3, it can be seen that the nuclei with the mass number A around 60 possess the largest value of $\frac{\Delta E}{A}$ (about 8.8 MeV), and hence they are most stable*. When the mass number increases greater than about 60, the proton number also increases, therefore the nucleus becomes unstable because of the Coulomb repulsive force between nucleons and $\frac{\Delta E}{A}$ will decrease. If the mass number decreases less than about 60, then the number of the other nucleons on which the nuclear force acts decreases, thus $\frac{\Delta E}{A}$ will also decrease.

* The nucleus with the maximal value of $\frac{\Delta E}{A}$ is the nickel nucleus ${}^{62}_{28}$Ni, while the nucleus with the minimal value of the mass per nucleon is the iron nucleus ${}^{56}_{26}$Fe.

From these facts, there is a possibility for two light nuclei fuse together to become one nucleus. This is called the **nuclear fusion**. On the other hand, there is a possibility for a very heavy nucleus to break up into 2 nuclei with about half mass numbers of the heavy nucleus. We call this process the **nuclear fission**.

Moreover, as we will learn in the next section, the number of stable nuclei is not so large, but about 270 kinds. The stable nucleus with the largest atomic and mass numbers is ${}^{208}_{82}$Pb, and nuclei with atomic number as well as mass number greater than this nucleus are all unstable.

In the reaction and decay of nucleus, the sum of the masses of the nuclei involved will change before and after the reaction. The energy absorbed or emitted through the reaction or decay due to the change of mass is called nuclear energy. Considering the nuclear energy, the conservation law of energy holds also for the reaction and decay of nuclei.

25.3 Nuclear decay and radiation

Learning objective To understand that there exist unstable nuclei which undergo α decay or β decay among the nuclei with mass defects. To understand the laws of nuclear decay of unstable nuclei and the half-life. To understand property of radiation, and to understand the distinction between radioactivity, absorbed dose, and effective dose.

The nuclei ${}^{A}_{Z}\mathrm{X}$ having mass defects cannot be decomposed into A separated nucleons, but not all the nuclei are stable. This is because there exist a certain nuclei which exhibit α decay or β decay.

Becquerel, who was stimulated by Roentgen's discovery of X ray in 1895, studied fluorescent substance and discovered radiation from compounds of uranium. He found in 1896 something emitted from uranium compound as a fluorescent substance and it penetrates matter exposing the photographic plate, ionizing the air to make it conductive and discharging the leaf electroscope. What was emitted in this phenomenon is called radiation, while the ability or action for emitting radiation is called the **radioactivity**.

Mr. and Mrs. Curie in order to make sure if substances other than uranium U exhibit the same property, performed the chemical analysis of pitchblende of uranium ore separating it into components. As a result of this analysis they discovered in 1898 the elements which emit radiation much more strongly than uranium, that is, radium Ra and polonium Po.

Fig. 25-4 Measuring the radioactivity by a Geiger counter

As for the radiation emitted by the natural radioactive material, it has turned out that there are 3 kinds ; α (alpha) ray. β (beta) ray, and γ (gamma) ray. The **α ray** is positively charged and shielded by one thin sheet of paper. The **β ray** is bended significantly by magnetic fields. It is negatively charged and shielded by thin aluminum plate. The **γ ray** is not bended by magnetic fields, and it is necessary to have a lead plate with about 10 cm in thickness for shielding. The real entity of α ray is the nucleus of helium ${}^{4}_{2}\mathrm{He}$, β ray is identified as a high-speed electron, and γ ray is an electromagnetic wave with short wavelength in its reality.

The phenomenon in which nuclei decay by emitting radiation is called radioactive decay, and the decays emitting α ray, β ray, γ ray are called α decay, β decay, γ decay, respectively. In the α decay where a helium nucleus ${}^{4}_{2}\mathrm{He}$ is emitted, the original nucleus is changed into the nucleus with the atomic number decreased by 2 and the mass number decreased by 4. The β decay where an electron e^- and a neutrino ν^0 are emitted, leads to the nucleus with the mass number unchanged and the atomic number increased by 1. In the γ decay, both mass and atomic numbers do not change and the nucleus transits to the lower energy state. These decays occur in the case where the sum of the masses of the decay products is smaller than the mass of the decaying nucleus.

Nucleus having radioactivity is called the **radioactive isotope** (radioisotope for short)

For reference β decay of neutron

Neutron has a larger mass than the proton. The difference between the neutron and the proton mass is larger than the mass of electron ($m_\mathrm{n} - m_\mathrm{p} > m_\mathrm{e}$). Therefore neutron is unstable and in an average life-time about 15 minutes by emitting an electron e^- and a neutrino ν^0 it undergoes β decay and becomes proton.

$$n^0 \longrightarrow p^+ + e^- + \nu^0 \qquad (25.4)$$

The difference between the rest energy of the initial and final states $(m_n - m_p - m_e)c^2$ (about 0.78 MeV) turns to the kinetic energy of the decay products. Neutron is unstable when it is alone and undergoes β decay, but exists as a stable state inside the nucleus due to the binding energy (because the mass becomes effectively small).

Neutrino is an electrically neutral particle and possesses an extremely small mass (less than 1/250 000 of electron mass). The force which causes the β decay is called the weak force.

Law of decay and half-life We cannot predict precisely when a radioisotope decays. It may decay one second later, or it may decay ten thousand years later. In this way, the decay phenomenon occurs randomly but it follows the law of probability.

The probability for a radioisotope to decay in a unit of time is determined only by the kind of isotope. The time, $T_{1/2}$, that elapses before the radioisotope which was contained in the substance having a large amount of the radioisotope reduces to a half amount, is characteristic to each radioisotope, and it has nothing to do with the time passed from its creation until present, temperature, pressure, and chemical bonding state. This time $T_{1/2}$ is called the **half-life** of the radioisotope. The amount of the radioisotope, N, decreases with time t as shown in Fig. 25-5. Some examples of the half-life are listed in Table 25-1.

Suppose that there exist N_0 radioisotopes at time $t = 0$. The remaining number of the radioisotopes at time t, $N(t)$ is given by

$$N(t) = N_0\left(\frac{1}{2}\right)^{t/T_{1/2}} = N_0 e^{-\lambda t}. \qquad (25.5)$$

This is called the **law of decay**, and λ is referred to as the decay constant. There is a following relationship :

$$\lambda T_{1/2} = \log_e 2 \approx 0.693. \qquad (25.6)$$

The larger λ is (The smaller $T_{1/2}$ is), the faster the decaying speed is. The smaller λ is (The larger $T_{1/2}$ is), the slower the decaying speed is. The number of decays is proportional to the number of radioisotopes remained without decay up to that time [Precisely speaking, it is proportional to $\lambda N(t)$].

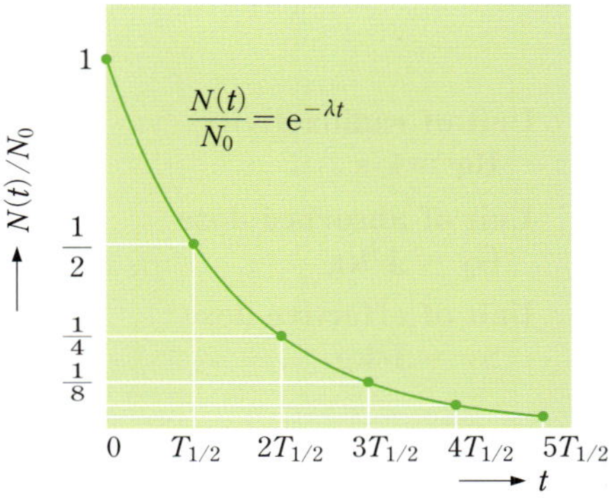

Fig. 25-5 Time t and the number of radioisotopes which remain without decay

Table 25-1 Half-life of radioactive isotope

Nucleus	Type of decay	Half-life
$^{14}_{6}C$	β	5.70×10^3 years
$^{32}_{15}P$	β	14.268 days
$^{45}_{20}Ca$	β	162.61 days
$^{60}_{27}Co$	β	5.2713 years
$^{90}_{38}Sr$	β	28.79 years
$^{131}_{53}I$	β	8.0252 days
$^{137}_{55}Cs$	β	30.08 years
$^{226}_{88}Ra$	α	1.600×10^3 years
$^{238}_{92}U$	α	4.468×10^9 years

Radiation Nowadays, in addition to α ray, β ray and γ ray arising from radioactive decays, X ray, neutron, high-speed ion, and stream of electrons or elementary particles are also called radiations. The radiation when passing through a matter, strikes electrons out from the atom inside the matter and ionizes the atom. This effect is called the **ionization effect**. The strength of ionization effect varies depending on the kind of radiation as well as on the energy involved. The radiation particle with an electric charge, if it goes slower then the time acting an electric force on the each atom in the surroundings for longer time, has a strong ionization effect. α ray runs slowly and possesses an electric charge $2e$, so, it has the strongest ionization effects. Next comes β ray with high-speed and charged $-e$ concerning the ionization effect. γ

ray, which is electrically neutral and ionizes atoms through photo-electric effects and Compton scatterings, possesses the smallest ionization effect. Due to the ionization effect, the radiation loses its energy, and it soon stops inside thick matters. The penetrating power through material is larger for the radiation having smaller ionization effects, and it decreases in the order : γ ray, β ray, and α ray.

Unit of radioactivity and radiation dose The intensity of the radiation of a material is expressed as how many pieces of radiation that material can emit per second, or in other words, how many numbers of unstable nuclei in the material decay per second. The unit of the radioactivity in the International System of Units (SI) is becquerel (Symbol Bq) which stands for the intensity of the radioactivity where one nucleus decays per second.

Unit of radioactivity
Bq = 1/s
Unit of absorbed dose
Gy = J/kg
Unit of effective dose
Sv = J/kg

The effects which are given to the material by the radioactive source even with the same intensity, strongly depend upon the kind of radiation emitted and also upon the energy. Therefore "intensity" of radiation is expressed by the effects given to the substance. When the radiation irradiated is absorbed by the substance of 1 kg at the rate of 1 J, it is called the **absorbed dose** of 1 Gray (Symbol Gy).

Effects of radiation given to the living things is not determined only by the absorbed radiation dose. The effect of radiation on human body varies in degree even for the same absorbed dose depending on the kind of radiation and on the tissue or the organ having an exposure to the radiation. The radiation weighting factor expressing the difference due to the kinds of radiation multiplied by the tissue weighting factor expressing the difference arising from the tissue or the organ is equal to the **effective dose** representing the effects on human body. The unit for the effective dose is called the sievert (Symbol Sv). In the case where the human body is exposed to β ray, γ ray, and X ray uniformly, it turns out that "effective dose" = "absorbed dose"*.

* This equality means that the numerical parts are equal. The unit of the left-hand side is Sv, while that of the right-hand side is Gy.

When the human body is exposed to the radiation except for β ray, γ ray, and X ray uniformly, "effective dose" = "radiation weighting factor"×"absorbed dose". The radiation weighting factor for the proton is 2, that for the α particle and also the heavier ion is 20, that for the neutron varies from 2.5 to 20 depending on the energy.

Human being is exposed to the radiation originating from the Galaxy, the radiation emitted by natural radioisotopes such as the radon Rn contained in the earth or in the air, as well as the radiation emitted from natural radioisotopes like the potassium ^{40}K contained in foods or in human bodies. The exposure to these background radiations called natural radiation per year is estimated to be 2.4mSv for the world average. The intensity of the natural radiation differs a big amount depending on the location due to differences in geological condition, height, and the strength of geomagnetism.

As a quantity which expresses the intensity of the environmental radiation, there exists the **space dose rate**. The intensity of the radiation passing through a point in space can be expressed by the effect of the radiation on the human body if that human were present at that point. As for the unit for this quantity, μSv/h (micro sievert per hour) is used. In the case where one stays at the location with space dose rate

1 μSv/h keeping exposed for a year, the effective dose amounts to about 9 mSv.

25.4 Nuclear energy

Learning objective To become able to explain what the nuclear energy is, and to understand that the nuclear energy is the source of the solar energy and also the energy source of the nuclear power generation.

Solar energy Outside of the atmosphere of the Earth, the solar radiation energy received by the 1 m^2 area facing straight to the Sun is 1.37 kJ. This is called the solar constant. From this fact, it is known that the Sun emits the energy of 3.85×10^{26} J per second.

The source of the energy the Sun emits is the nuclear energy which is released when the hydrogen nuclei undergo nuclear fusion and become a helium nucleus in the center of the Sun at the temperature 1.57×10^{7} K. Although this nuclear fusion takes place through several processes, it can be summarized finally as

$$p^{+}+p^{+}+p^{+}+p^{+}+e^{-}+e^{-} \longrightarrow {}^{4}_{2}He^{++}+\nu^{0}+\nu^{0}+26.7\ \text{MeV}.$$

When the hydrogen undergoes nuclear fusion in this process, the energy of 6.4×10^{14} J is released per 1 kg of hydrogen. Therefore, in the Sun, about 6.0×10^{11} kg hydrogen undergoes nuclear fusion per second, and an enormous number of neutrinos are produced. These solar neutrinos are detected at the Super-Kamiokande detector installed in the underground of Kamioka in Gifu prefecture.

In order to generate nuclear fusion, two nuclei have to come close against the Coulomb repulsion and touch together. In the center of the Sun which is at high temperature, there exist a certain number of nuclei having extremely high energy of thermal motion, and they collide each other leading to the nuclear fusion. This kind of reaction is called the thermonuclear fusion reaction.

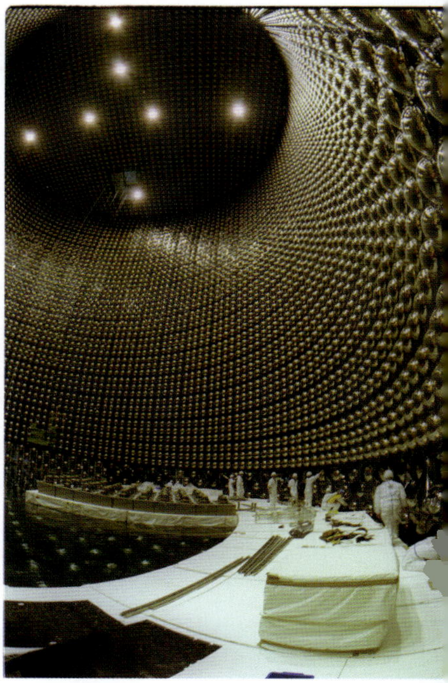

Fig. 25-6 Super-Kamiokande detector. A huge cylindrical water tank with diameter 39.2 m, and height 41.4 m stores 50,000 tons of pure water, and 11146 photomultiplier tubes with diameter 50 cm are attached around its whole wall.

Nuclear fission Uranium nuclei of atomic number 92 ${}^{238}_{92}U$, ${}^{235}_{92}U$ are unstable and α decay with a long half-life. These uranium nuclei are energetically possible to have nuclear fission and to decay into two nuclei with masses about a half of the parent and accompanied by a certain number of neutrons. However, despite the gravitational potential of the water in a lake at the top of the mountain is larger than that at the foot of the mountain, the water is not coming to the foot of the mountain unless we dig a tunnel through the slope of the mountain. The uranium nucleus naturally α decays quite slowly, but does not have a nuclear fission in nature. Now, the neutron is electrically neutral, so it can approach the uranium nucleus closely without having repulsion by positive charges in the nucleus. Thus, by hitting and stimulating the uranium nucleus with a neutron we deform it in an oval shape to generate a trigger for a splitting, then the uranium nucleus undergoes nuclear fission.

The nuclear fission of the uranium nucleus was discovered by Hahn and Strassmann in 1938. They detected an isotope of $_{56}Ba$ in the decay product, and confirmed the nuclear fission. When $^{235}_{92}U$ is hit by a thermal neutron (slow neutron) and induced a nuclear fission, a nucleus of A around 95 and another nucleus of A around 140 together with 2–3 neutrons are produced, then about 200 MeV nuclear energy is transformed into the kinetic energy of the fission product. This energy is so large that it cannot be compared to the one obtained in the chemical reaction. For example, the produced energy in the combustion of carbon $C+O_2 \longrightarrow CO_2$ is about 4 eV per one carbon atom.

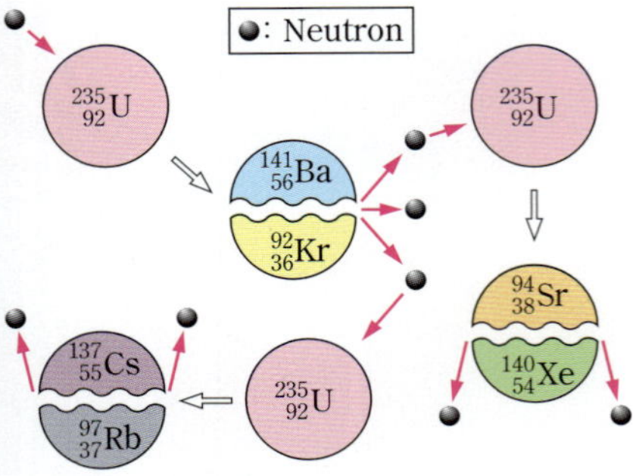

Fig. 25-7 Chain reaction of nuclear fission of $^{235}_{92}U$. Typical three examples for the fission products are shown.

If the neutrons emitted from the nuclear fission are absorbed by another uranium nucleus, a new nuclear fission is triggered. A multiple number of neutrons are produced for a one nuclear fission, hence it is possible for nuclear fissions to occur one after another (Fig. 25-7). This is called the **chain reaction**. In order for the chain reaction to take place, the neutrons emitted have to be used without escaping to the outside. For that purpose, the nuclei undergoing the nuclear fission must exist together with a certain amount. The minimum amount of the uranium nuclei necessary for the chain reaction to occur is called the **critical mass**. If a cluster of uranium is less than the critical mass, then the neutrons escape to the outside before triggering the next fission and the chain reaction does not occur.

There are three isotopes for the natural uranium. They are $^{238}_{92}U$ (abundance ratio 99.274 %), $^{235}_{92}U$ (abundance ratio 0.720 %), and $^{234}_{92}U$ (abundance ratio 0.005 %). Among these isotopes, the one which undergoes nuclear fission by the thermal neutrons is only $^{235}_{92}U$ with abundance ration 0.72 %. $^{235}_{92}U$ absorbs thermal neutrons and goes through nuclear fission, and then emits, on the average, 2.5 pieces of fast neutrons (with the energy larger than thermal neutrons). For the natural uranium, most abundant $^{238}_{92}U$ absorbs the neutrons and the chain reactions do not continue. When fast neutrons hit light water (ordinary water, H_2O), heavy water (D_2O), graphite (C), and so on, the fast neutrons, colliding these light nuclei and giving them kinetic energies, become thermal neutrons[*1, *2]. These materials are called moderators. The thermal neutrons are not absorbed by $^{238}_{92}U$, and hence the chain reaction is maintained.

*1 D stands for the nucleus of deuterium or deuteron ^2_1H.
*2 The neutron which reaches a state of thermal equilibrium by the collision with the atoms nearby and the kinetic energy becomes around $\frac{3}{2}kT$ is called the thermal neutron (see Problems for exercise 24, B3).

When we control the chain reaction and keep it continue to occur with a constant pace, we refer to this state as the **critical state**. The apparatus to realize the critical state is the **nuclear reactor**. The nuclear power generation is an electric power generation by the thermal engine in which the high-temperature heat reservoir is the interior of the reactor where nuclear energy is transformed into thermal energy, while the low-temperature heat reservoir is sea water or river water. We show the conceptual diagram for the nuclear power generation in Fig. 25-8. The uranium compound as a fuel is stuffed into the metal tube, and it is called the fuel rod. In order to control the reaction, the control rod is inserted and removed through the fuel rod. The control rod is made out of cadmium Cd or boron B which absorb the neutrons very well. In the reactor where the light water is used as the moderator, the absorption of neutrons by the light water is so large that the chain reaction does not occur if the natural uranium which contains $^{235}_{92}U$ only 0.72 % is used as a fuel. Thus, we use the enriched uranium

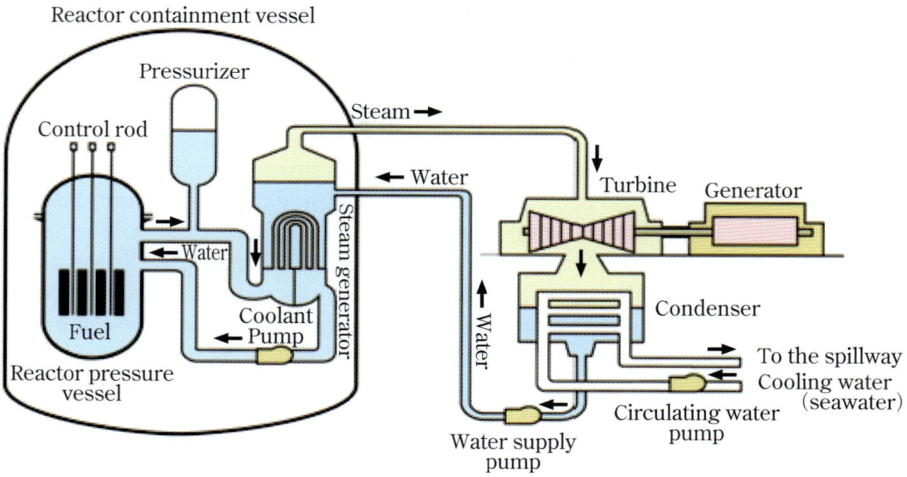

Fig. 25-8 Conceptual diagram for the Pressurized light Water Reactor (PWR) for power generation

Nuclear reactors used in Japan are two types ; pressurized light water reactor shown here and Boiling light Water Reactor (BWR) in which the boiled water vapor by nuclear fuel in pressure vessel is directly sent to the turbine. The light water reactor is a reactor using ordinary water (light water) as a working substance for the thermal engine. The reactor of Fukushima Daiichi nuclear power plant which had an accident at the occasion of the great east Japan earthquake is a boiling light water reactor. In the pressurized light water reactor, the pressure of about 160 atmosphere is applied to the water filling the pressure vessel, so water of about 320 °C is not boiled.

as a fuel which is obtained by enriching ${}^{235}_{92}\mathrm{U}$ to a few %. In 1942, Fermi and his collaborators showed that the chain reaction occurred in the reactor using the natural uranium and graphite.

Furthermore, by absorbing neutrons ${}^{238}_{92}\mathrm{U}$ becomes ${}^{239}_{92}\mathrm{U}$, and the latter then β decays twice turning into ${}^{239}_{94}\mathrm{Pu}$ which undergoes the nuclear fission by thermal neutrons. The critical mass of the ${}^{239}_{94}\mathrm{Pu}$ is much smaller than that of ${}^{235}_{92}\mathrm{U}$.

25.5 Elementary particles

Learning objective To learn the overview of the world of elementary particles.

After the neutron was discovered in 1932, it turned out that materials are ultimately composed of the electron, the proton and the neutron. Since 1930's, these particles together with the light quantum, the photon, have become generically called **elementary particles** as the meaning of fundamental particles of the structure of matter. There are other members in the group of elementary particles. For example, as the typical particles which were discovered after they had been predicted, there exist the positron Dirac predicted, the neutrino Pauli predicted, and the pi meson Yukawa predicted. In addition to these particles, there are many elementary particles which were discovered without the predictions.

There are a number of characteristics or features in the elementary particles. The first feature is the fact that each elementary particle has the definite mass and charge. Therefore, we cannot distinguish the two elementary particles of the same kind since they are completely identical. Elementary particles are classified into two groups ; The **fermion** which follows Pauli's principle saying that only one of the same kind of elementary particles can exist at the same state. While, for the **boson,** any number of particles can exist at the same state. The electron, proton and neutron are fermions, while the photon is a boson. The nuclei, atoms and molecules possess the same property.

The second feature is the fact that there exists an **anti-particle** with the same mass, but opposite sign of the electric charge for each elementary particle. The anti-particle of the electron (symbol e^-) with a minus charge $-e$ and a mass m_e is the positron (e^+) with a positive charge e and a mass m_e. The anti-particle of the proton is called an anti-proton and that of the neutron is called an anti-neutron. If the particle and anti-particle (for example, electron and positron) are collided together, then they are annihilated turning into an energy. It may also happen that the energy is transformed into producing a pair of particle and anti-particle.

The third feature is that the elementary particle changes. For example, the neutron is unstable when it is alone, and decays into a proton and a neutrino, but it is not composed of the proton and the neutrino. When the neutron decays, the neutron is annihilated and at the same time a proton and a neutrino are produced. In this way, the elementary particle has a property to change.

Fig. 25-9 KEKB accelerator at the High Energy Accelerator Research Organization (KEK).
Colliding-beam accelerator which performs experiments of particle physics by accumulating high-energy electrons (8 billion electron-volts, 8 GeV) and positrons (3.5 billion electron-volts, 3.5 GeV) into each of two rings and by colliding them at the intersection (IR) of the two rings.

Since around 1950, it has become possible to accelerate protons and neutrons to high-energies by using huge accelerators. When we let a high-energy proton collide on another proton at rest, various new particles are produced. The kinetic energy of the colliding proton turns into masses of the produced particles. Among the produced elementary particles there exist the pi meson which was predicted by Hideki Yukawa in 1935. In this way, the number of new elementary particles that get out from the submicroscopic world by big accelerators eventually became more than 100 kinds, and then it was realized that not all of these particles are regarded as fundamental particles of the material structure. In the present day, strongly interacting particles such as the proton, neutron, pi meson (pion) and others are composed of more fundamental particles called **quarks**. The quark was proposed by Gell-Mann and Zweig in 1964.

By letting high-energy electrons accelerated by accelerators with short de Broglie's wavelengths collide on protons or neutrons and probing inside the nucleon, it has turned out that the nucleon possesses the extension in size with radius about 8×10^{-16} m and contains three smaller particles. These are quarks. The quarks contained in the nucleon are u quark $\left(\text{up quark, electric charge } \frac{2}{3}e\right)$ and d quark $\left(\text{down quark, electric charge } -\frac{1}{3}e\right)$, and the proton is a composite state made up of 2 u quarks and one d quark, while the neutron is a composite state made up of one u quark and 2 d quarks (Fig. 25-10).

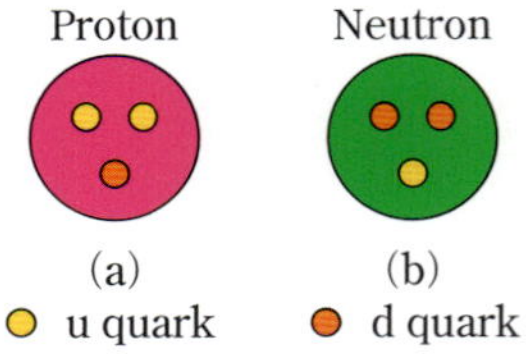

Fig. 25-10 Nucleons in the quark model. (a) proton p is composed of uud, (b) neutron n is composed of udd.

Even if we let two high-energy protons undergo a head-on collision for the purpose of striking quarks out of the proton, quarks do not come out. It is considered that we cannot isolate one quark alone.

At present, in total, six types of quarks ; u quark, d quark, s quark $\left(\text{strange quark, electric charge } -\frac{1}{3}e\right)$, c quark $\left(\text{charm quark, electric charge } \frac{2}{3}e\right)$, b quark $\left(\text{bottom quark, electric charge } -\frac{1}{3}e\right)$, t quark $\left(\text{top quark, electric charge } \frac{2}{3}e\right)$ have been discovered.

Interactions and classification of elementary particles There exist various forces in nature. Among them, some of the forces are fundamental, but some others are not. Gravity (universal gravitation) and electromagnetic force are fundamental forces, but the frictional force is a complex force originating from the electromagnetic force acting between molecules, and is not fundamental. The nuclear force is different from the gravity and the electromagnetic force, but it is other kind of force called the **strong force**. In the β decay of the nuclei, the neutrino and electron or positron are produced, and the force which causes this decay is called the **weak force**.

Currently, 4 types of forces ; gravity, electromagnetic force, strong force, and weak force are considered to be fundamental forces in nature. In the past, electric force and magnetic force were regarded to be totally unrelated, but afterward it was realized that they are inextricably linked, and they were unified and became the electromagnetic force. Similarly, the research by Weinberg and Salam has elucidated that the electromagnetic force and the weak force are closely related and it is suitable to unify the two forces and call it **electroweak force**. Moreover, the attempt to unify the strong force with electroweak force is the grand unified theory. Note that masses of elementary particles are extremely so small that we can neglect the gravity.

The particle which receives the interaction of the strong force is called a **hadron**. The hadron is composed of quarks. Hadrons receive all the interactions ; strong, electromagnetic and weak forces. On the other hand, the electron and the neutrino do not receive the interaction of the strong force. The electron receives electromagnetic and weak forces, while the neutrino only receives the weak force. The particles which does not receive the interaction of strong force such as electrons and neutrinos are referred to as the **lepton**. At present, as for the lepton, there are 3 types of charged particles (electron e^-, muon μ^-, tau τ^-) and 3 types of neutrinos (electron neutrino ν_e, muon neutrino ν_μ, tau neutrino ν_τ), namely, in total, 6 kinds of leptons have been found together with their anti-particles (e^+, μ^+, τ^+, $\bar{\nu}_e$, $\bar{\nu}_\mu$, $\bar{\nu}_\tau$). It is intriguing that both quarks and leptons exist 6 kinds for each.

The electron, muon and tau particle have exactly the same property except for the different masses. In addition, e and ν_e, μ and ν_μ, τ and ν_τ as in the following

$$\pi^+ \longrightarrow \mu^+ + \nu_\mu, \qquad \mu^+ \longrightarrow e^+ + \nu_e + \bar{\nu}_\mu, \qquad \tau^- \longrightarrow e^- + \bar{\nu}_e + \nu_\tau$$

always appear in pairs. Thus we note that in fundamental particles

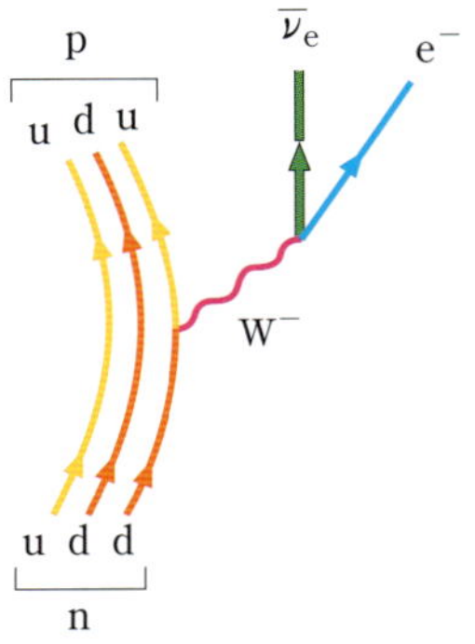

Fig. 25-11 The β decay of the neutron is mediated by W boson.

there are three generations : (u, d, ν_e, e^-), (c, s, ν_μ, μ^-), (t, b, ν_τ, τ^-).

The photon γ is a particle which mediates the electromagnetic force, and belongs to the group called the **gauge particle** mediating the fundamental forces. The gauge particles which mediate the weak force are W boson (W^+ and W^-) and Z boson (Z^0), and they were found in 1983 (Fig. 25-11). The masses of these particles are about 100 times heavier than the proton mass, hence the range of the force they mediate is so short, less than 10^{-17} m, that it is indeed a weak force. If the energy of the colliding particles becomes extremely large and the de Broglie's wavelength is of order 10^{-17} m, then the weak force grows as strong as the electromagnetic force.

The gauge particle mediating the strong force is called the **gluon**, which makes quarks strongly bound by the mediating force and confines the quarks inside the hadrons. The gauge particle mediating gravity is called the **graviton**.

In addition to the three groups, hadrons (quarks), leptons, gauge particles, there exists Higgs particle. According to the standard model of elementary particles, the universe is filled with the Higgs field, which gives the particles therein masses by the interaction. The excited state of the Higgs field is the Higgs particle*. The Higgs particle was discovered in 2012 in the experiment carried out at the accelerator constructed by European organization for nuclear research (CERN). The accelerator is the world largest colliding-beam circular machine (27 km circumference) called the LHC (Large Hadron Collider), where two proton beams were accelerated to 3.5~4 TeV for each and had undergone head-on collisions.

* An excited state of the electromagnetic field is a photon. In the vacuum, the average value of the electromagnetic field, which is a vector field, is **0**. The average value of the Higgs field, which is a scalar field, possesses a certain magnitude and it is regarded as the origin of the mass of elementary particles.

Now, have all the fundamental particles in nature been discovered so far? According to the recent results of observational cosmology, which tries to understand the history of the universe based on the results of the observation of the universe, there exist the tremendous amounts of substance called dark matters which have neither strong nor electromagnetic interactions, but only have the gravitational interaction. The total abundance of the dark matter is several times more than the ordinary matter (atom). Thus, detection of the elementary particles composing the dark matter is now intensively going on.

Problems for exercise 25

A

1. What are the proton number and neutron number each for $^{208}_{82}$Pb, $^{235}_{92}$U?
2. Which nucleus possesses the largest binding energy per nucleon among ^{4}He, ^{12}C, ^{16}O, ^{56}Fe, ^{238}U?
   ① ^{4}He ② ^{12}C ③ ^{16}O ④ ^{56}Fe ⑤ ^{238}U
3. How many gram does the 1 gram of radioactive sodium (Na) which β decays with half-life 15 hours become after 45 hours?
4. If the nucleus produced by the α or β decays is unstable, it continues to decay until it becomes stable. This sequence of nuclei is called the **decay series**. In the uranium-radium series where $^{238}_{92}$U decays via radium $^{226}_{88}$Ra to a stable isotope of lead $^{206}_{82}$Pb, how many times do we have the α decays and the β decays, each?
5. In order to determine the age of the ancient relic, there is a method in which we measure the radioactivity of the ^{14}C contained in the sample and compare it with the radioactivity of the ^{14}C contained in the organic matter at present. Cosmic rays coming to the Earth from the Sun or from the outer space collide with atmospheric molecules producing the

neutron which in turn collides with the nitrogen in the atmosphere giving ^{14}C in the reaction $n+{}^{14}N \longrightarrow {}^{14}C+{}^{1}H$. The carbon ^{14}C is coupled with the oxygen forming the radioactive carbon dioxide $^{14}CO_2$ which is absorbed into the living organism. As a result, the radioactivity of the carbon in all the living organism is about 15.3 counts per minute for one gram (about 1/4 Bq). When a living organism dies, ^{14}C is no longer absorbed into the living organism, and the radioactivity of the relic will decrease. The half-life of ^{14}C is 5700 years. Now, the radioactivity of the charcoal contained in an ancient relic 1/4 of that of new charcoal. How old is this charcoal?

6. In the case where the absorbed dose of the water due to irradiation of radioactive ray is 10 Gy, how much is the temperature increase of water in degree Celsius?

Note that the specific heat capacity of water is 4.2×10^3 J/(kg·K).

① 0.0024 ② 0.024 ③ 0.24
④ 0.42 ⑤ 4.2

B

1. Suppose that the radius of the nucleus with mass number A is $r = 1.2\times10^{-15}A^{1/3}$ m. What is the density of the nucleus in the unit g/cm^3? Take the mass of the nucleon to be 1.67×10^{-27} kg. Note that the mass of the Sun is 2.0×10^{30} kg and the radius is 7.0×10^8 m. If the density of the Sun becomes equal to that of the nucleus, how many meters is the radius R of the Sun?

2. When the nucleus X (mass M) is decomposed into the nucleus Y (mass m) and α particle (mass m_α), calculate the kinetic energy of α particle.

Dr. Masatoshi Koshiba and neutrino astronomy

For the reason "pioneering contributions to astrophysics, in particular for the detection of cosmic neutrinos." Dr. Masatoshi Koshiba (1926—), who succeeded in the observation of neutrinos from supernova for the first time ever, was awarded Nobel prize in physics 2002.

The hypothesis of the grand unified theory for particle physics predicts that the proton, the fundamental particle of material structure, is not stable, and although it is a very small probability the proton is expected to decay. Therefore, in order to detect the proton decay, Dr. Koshiba in 1978 proposed the construction of the Kamiokande detector which stores water of 3000 ton in the tank surrounded by the photomultiplier tubes at the 1000 m underground of Kamioka mine in Gifu prefecture. The experiment was started in 1983. The last three letters, NDE, of Kamiokande means the initial of Nucleon Decay Experiment. Although the proton decay, which is the original aim, was not found in this experiment, the neutrinos from supernova was observed. In fact, the neutrinos emitted by the explosion of supernova 160 000 years ago at Large Magellanic Cloud, which is an accompanying nebula of the galaxy, was observed for 11 events at 7 : 35 a.m., February 23rd, in 1987 (Greenwich Mean Time). This observation opened a new field of science called the neutrino astrophysics. The supernova explosion is an emergence of the shock wave occurred at the last stage of the stellar evolution where the star turns to a neutron star due to the gravitational contraction. At that time 99 % of the energy are emitted to the outer space as neutrinos.

The apparatus in which the performance was enhanced 10 to 100 times that of the Kamiokande detector was constructed at the 1000 m underground of the Kamioka mine, 900 m apart from the Kamiokande detector. This is the Super-Kamiokande detector.

The primary purpose for constructing the Kamiokande detector was to detect the proton decay predicted by the grand unified theory. So far the proton decay has not been observed yet, but the Kamiokande and Super-Kamiokande detectors achieved great results for the observation of solar neutrinos referred at the Section **25.4**.

Dr. Takaaki Kajita who has shown that the neutrinos possess tiny masses

The Nobel prize in physics 2015 was awarded to Dr. Takaaki Kajita (1959-) for "The discovery of neutrino oscillation indicating that the neutrinos possess mass" which was the great achievement using Super-Kamiokande detector.

The pi meson produced at the collision of the cosmic rays (The main component is proton), coming from the outer space, with the atmosphere is unstable. It decays in two steps : pi meson → muon+muon neutrino, muon neutrino → electron +muon neutrino+electron neutrino (see p.378). Therefore the ratio of muon neutrinos ν_μ to electron neutrino ν_e produced at the collision of cosmic rays with atmospheric air should be 2 : 1 ($\nu_\mu/\nu_e = 2$).

When observed, the neutrinos produced up in the air and falling from the sky to the detector show the ratio $\nu_\mu/\nu_e = 2$ as well as the absolute number for each as expected. However, the neutrinos produced at the atmosphere of the opposite side of the Earth and penetrating through the Earth show the ratio ν_μ/ν_e nearly one. Namely the amount of ν_μ is anomalously smaller than expected. This phenomenon implies that the muon neutrinos turns into the tau neutrinos (not detected by the detector) during the passing through the Earth.

From the study of the beta decay of the nuclei, it was known that the mass of the electron neutrino is less than one two-hundred fifty thousandth (1/250 000) of the electron mass. Hence, in the standard model of elementary particles, the masses of three kinds of neutrinos had been regarded all zero. And if the neutrinos are massive and different in their masses, it is expected that they transform into another kind of neutrinos during the flight. Thus, this phenomenon had been named the neutrino oscillation.

The discovery of the neutrino oscillation at the Super-Kamiokande detector was reported by Dr. Kajita at the international conference held in 1998 at Takayama in Gifu prefecture. The discovery of the neutrino oscillation means the discovery that the neutrino possess mass and the existence of the phenomena beyond the standard model of elementary particles.

Dr. Shin-ichiro Tomonaga and Dr. Hideki Yukawa

The scientists who developed the foundation of theoretical physics in Japan, especially theoretical research of elementary particles, are Dr. Shin-ichiro Tomonaga (1906-1979) and Dr. Hideki Yukawa (1907-1981), and further Dr, Yoshio Nishina who watched the former two warmly. Dr. Tomonaga formulated the quantum electromagnetism which is the relativistic quantum theory of electromagnetism, and proposed the renormalization theory resolving its theoretical difficulties. Dr. Yukawa elucidated the nuclear force with which the proton and the neutron get together to form nuclei, by introducing a new elementary particle called pi meson (pion). Dr. Yukawa was awarded the Nobel prize in physics 1949, and Dr. Tomonaga was awarded the Nobel prize in physics 1965 together with Dr. Schwinger and Dr. Feynman.

In order to know their thoughts and personalities, it would be a good way to read the sentences included in the collected works of Shin-ichiro Tomonaga (Misuzu shobo) and those of Hideki Yukawa (Iwanami shoten). Let me briefly introduce the characteristics of their sentences.

The sentence of Dr. Tomonaga is as follow : "After the world war I is over, almost nothing easy to study remains in the research about the planetary electron (electron inside atom), the researchers started to care about what is going on inside of the nucleus. Since we cannot affect the nucleus from outside easily, we need a big progress in the

Fig. 25-A Dr. Shin-ichiro Tomonaga

Fig. 25-B Dr. Hideki Yukawa

technology of experiments for this research. Just as we need a special qualification or money offering to visit the sanctum of a temple, a special technology is necessary to enter into the inside of the nucleus, and it costs money". From the above citation you will notice that he has careful consideration to let the readers understand exactly what he means by writing logically and using a metaphor.

On the other hand, Dr. Yukawa learned "Four Books and Five Classics of Confucianism" in his childhood at his home, thus he had a detailed knowledge about Chinese classics. The expression of his sentence is like "Science is considered too narrowly in Japan. To make our attitudes toward daily life rational or reasonable is a progress of the science in a broad sense." or "For younger people, it may be much better to jump into a new attempt without studying old things. I was so in the past". Like these citations, his sentences are concise, clear and get to the point. Moreover, putting a Japanese poem of his own into the sentence and make it rich in content even in the simplicity.

Snow is close high on Mt. Hiei looking cold and clean day after day
My heart is at the bottom of the loneliness, while my way never ends

When the heaven and the earth were divided, it is said to have been created
Now the atom (nucleus) is again broken down to pieces

The first poem was composed, while he was young, expressing his state of mind when he did not get much progress in his research and going home looking at Mt. Hiei. The second was made when he visited European organization for nuclear research (CERN) and having a look at the huge accelerator for the experiment of nuclei collisions.

Appendix Mathematical Formulas

A.1 Properties of trigonometric functions

$\sin\theta = \dfrac{y}{r} \qquad \cos\theta = \dfrac{x}{r}$

$\tan\theta = \dfrac{y}{x} \qquad \cot\theta = \dfrac{x}{y}$

$\sin^2\theta + \cos^2\theta = 1$

$\tan\theta = \dfrac{\sin\theta}{\cos\theta} \qquad \cot\theta = \dfrac{\cos\theta}{\sin\theta}$

$\sin 2\theta = 2\sin\theta\cos\theta$

$\cos 2\theta = \cos^2\theta - \sin^2\theta = 1 - 2\sin^2\theta$
$= 2\cos^2\theta - 1$

$\sin^2\theta = \dfrac{1}{2}(1-\cos 2\theta)$

$\cos^2\theta = \dfrac{1}{2}(1+\cos 2\theta)$

$\sin(\alpha\pm\beta) = \sin\alpha\cos\beta \pm \cos\alpha\sin\beta$
(Double-sign corresponds)

$\cos(\alpha\pm\beta) = \cos\alpha\cos\beta \mp \sin\alpha\sin\beta$
(Double-sign corresponds)

$a\sin\theta + b\cos\theta = \sqrt{a^2+b^2}\sin(\theta+\alpha)$

where $\sin\alpha = \dfrac{b}{\sqrt{a^2+b^2}}$

$\cos\alpha = \dfrac{a}{\sqrt{a^2+b^2}}$

In the following formulas the unit of θ is radian (rad).

$\sin\left(\dfrac{\pi}{2}-\theta\right) = \cos\theta \qquad \cos\left(\dfrac{\pi}{2}-\theta\right) = \sin\theta$

$\sin n\pi = 0$ (n is an integer.)

$\lim_{\theta\to 0}\dfrac{\sin\theta}{\theta} = 1$, if $|\theta| \ll 1$ $\sin\theta \approx \theta$

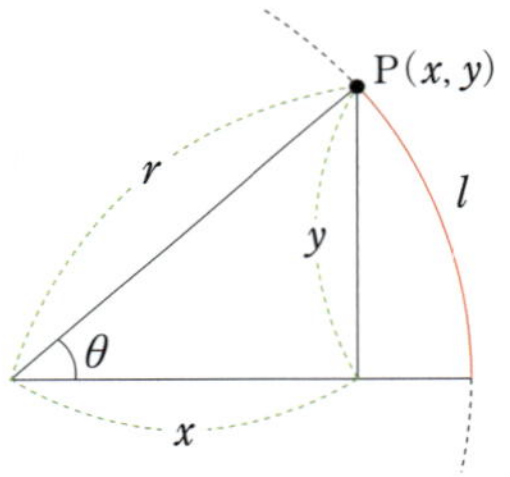

Fig. A.1 $\theta = \dfrac{l}{r}$ [rad]

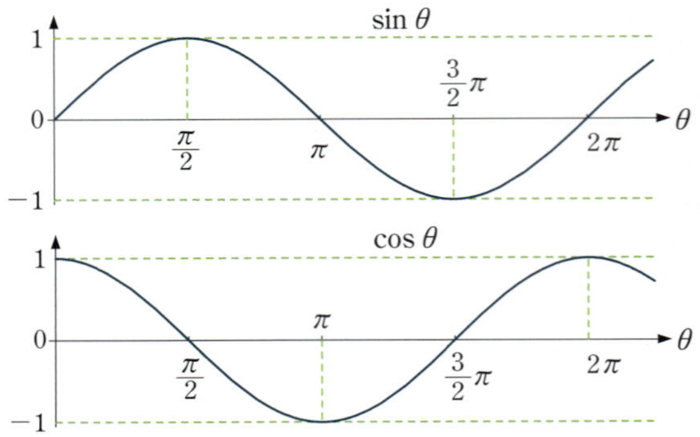

Fig. A.2

Table A.1

Degree (°)	0	30	45	about 57	60	90	180	270	360
Radian (rad)	0	$\dfrac{\pi}{6}$	$\dfrac{\pi}{4}$	1	$\dfrac{\pi}{3}$	$\dfrac{\pi}{2}$	π	$\dfrac{3\pi}{2}$	2π

Table A.2

θ [rad]	0	$\dfrac{\pi}{6}$	$\dfrac{\pi}{4}$	$\dfrac{\pi}{3}$	$\dfrac{\pi}{2}$	$\dfrac{2}{3}\pi$	$\dfrac{3}{4}\pi$	$\dfrac{5}{6}\pi$	π
$\sin\theta$	0	$\dfrac{1}{2}$	$\dfrac{1}{\sqrt{2}}$	$\dfrac{\sqrt{3}}{2}$	1	$\dfrac{\sqrt{3}}{2}$	$\dfrac{1}{\sqrt{2}}$	$\dfrac{1}{2}$	0
$\cos\theta$	1	$\dfrac{\sqrt{3}}{2}$	$\dfrac{1}{\sqrt{2}}$	$\dfrac{1}{2}$	0	$-\dfrac{1}{2}$	$-\dfrac{1}{\sqrt{2}}$	$-\dfrac{\sqrt{3}}{2}$	-1
$\tan\theta$	0	$\dfrac{1}{\sqrt{3}}$	1	$\sqrt{3}$	—	$-\sqrt{3}$	-1	$-\dfrac{1}{\sqrt{3}}$	0

A.2 Exponential function

The constant e defined by the condition

$$\lim_{x\to 0}(1+x)^{1/x} = \mathrm{e}$$

is an irrational number, and its value is 2.718281⋯.

Properties of the exponential function e^x with base e

$$\mathrm{e}^x\mathrm{e}^y = \mathrm{e}^{x+y} \qquad \frac{\mathrm{e}^x}{\mathrm{e}^y} = \mathrm{e}^{x-y}$$

$$\mathrm{e}^0 = 1 \qquad \mathrm{e}^1 = \mathrm{e}$$

$$\lim_{x\to-\infty}\mathrm{e}^x = 0 \qquad \lim_{x\to\infty}\mathrm{e}^x = \infty$$

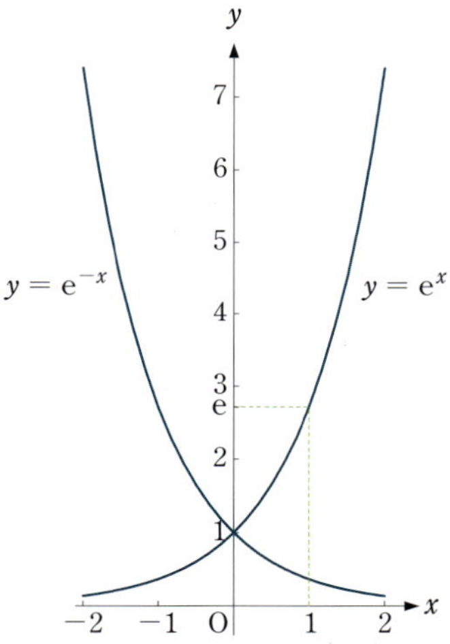

Fig. A.3 $y = \mathrm{e}^x,\ y = \mathrm{e}^{-x}$

A.3 Properties of the natural logarithm (logarithm with base e, $\log_{\mathrm{e}} x$) $\log x$

$$\log(xy) = \log x + \log y$$

$$\log\frac{x}{y} = \log x - \log y$$

$$\log x^n = n\log x \qquad \log\frac{1}{x^n} = -n\log x$$

$$\log \mathrm{e} = 1 \quad \text{and} \quad \log 1 = 0$$

The domain of $y = \log x$ is a set of whole positive real numbers :

$$\lim_{x\to+0}\log x = -\infty \qquad \lim_{x\to\infty}\log x = \infty.$$

Since $y = \log x$ is the inverse function of $y = \mathrm{e}^x$, $y = \log x$ and $x = \mathrm{e}^y$ are equivalent.
In pocket computers, the symbol $\ln x$ is used.

A.4 Primitive function and derivative function (C is an arbitrary constant ; a, b, d, n are constants.)

$f(t)+C = \int\frac{\mathrm{d}f}{\mathrm{d}t}\,\mathrm{d}t$	$\frac{\mathrm{d}f}{\mathrm{d}t}$
at^n+C	ant^{n-1}
$a\sin t+C$	$a\cos t$
$a\sin(bt+d)+C$	$ab\cos(bt+d)$
$a\cos t+C$	$-a\sin t$
$a\cos(bt+d)+C$	$-ab\sin(bt+d)$
$a\,\mathrm{e}^t+C$	$a\,\mathrm{e}^t$
$a\,\mathrm{e}^{bt}+C$	$ab\,\mathrm{e}^{bt}$
$\frac{1}{a}\log\lvert at+b\rvert+C$	$\frac{1}{at+b}$

A.5 Formulas of Vectors

Scalar product

$$\boldsymbol{A}\cdot\boldsymbol{A} = |\boldsymbol{A}|^2 = A^2 = A_x{}^2+A_y{}^2+A_z{}^2$$

$$\boldsymbol{A}\cdot\boldsymbol{B} = \boldsymbol{B}\cdot\boldsymbol{A} = AB\cos\theta = A_xB_x+A_yB_y+A_zB_z$$

$$\boldsymbol{A}\cdot(\boldsymbol{B}+\boldsymbol{C}) = \boldsymbol{A}\cdot\boldsymbol{B}+\boldsymbol{A}\cdot\boldsymbol{C} \quad \textbf{(Distribution rule)}$$

Vector product

$$\boldsymbol{A}\times\boldsymbol{B} = AB\sin\theta\cdot\boldsymbol{n} = (A_yB_z-A_zB_y)\boldsymbol{i}+(A_zB_x-A_xB_z)\boldsymbol{j}+(A_xB_y-A_yB_x)\boldsymbol{k}$$

where $\boldsymbol{n}$ is perpendicular to both $\boldsymbol{A}$ and $\boldsymbol{B}$. When we point the thumb of our right hand in the direction of $\boldsymbol{A}$ and the index finger in the direction of $\boldsymbol{B}$, then $\boldsymbol{n}$ is a unit vector which points to the direction of the middle finger. $\boldsymbol{i}, \boldsymbol{j},$ and $\boldsymbol{k}$ are unit vectors in the direction of x-axis, y-axis and z-axis, respectively. θ is the angle between $\boldsymbol{A}$ and $\boldsymbol{B}$.

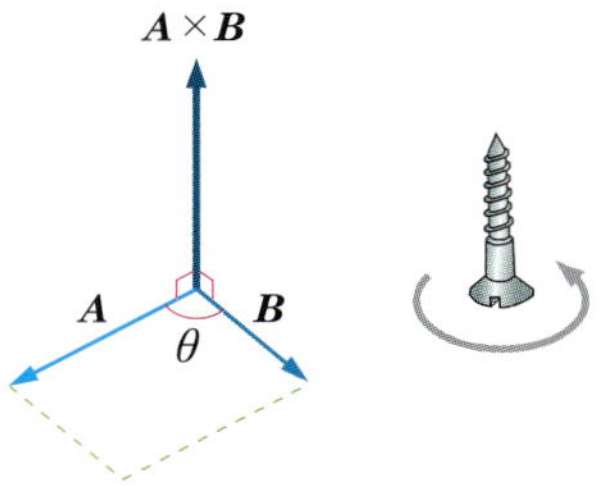

Fig. A.4 Vector product $\boldsymbol{A}\times\boldsymbol{B}$

$$\boldsymbol{A}\times\boldsymbol{B} = -\boldsymbol{B}\times\boldsymbol{A}$$
$$\boldsymbol{A}\times\boldsymbol{A} = \boldsymbol{0}$$
$$\boldsymbol{A}\cdot(\boldsymbol{B}\times\boldsymbol{C}) = \boldsymbol{B}\cdot(\boldsymbol{C}\times\boldsymbol{A}) = \boldsymbol{C}\cdot(\boldsymbol{A}\times\boldsymbol{B})$$
$$(\boldsymbol{A}\times\boldsymbol{B})\times\boldsymbol{C} = (\boldsymbol{A}\cdot\boldsymbol{C})\boldsymbol{B}-(\boldsymbol{B}\cdot\boldsymbol{C})\boldsymbol{A}$$
$$[\boldsymbol{A}\cdot\{\boldsymbol{B}\times(\boldsymbol{C}\times\boldsymbol{D})\}] = (\boldsymbol{A}\times\boldsymbol{B})\cdot(\boldsymbol{C}\times\boldsymbol{D}) = (\boldsymbol{A}\cdot\boldsymbol{C})(\boldsymbol{B}\cdot\boldsymbol{D}) - (\boldsymbol{A}\cdot\boldsymbol{D})(\boldsymbol{B}\cdot\boldsymbol{C})$$

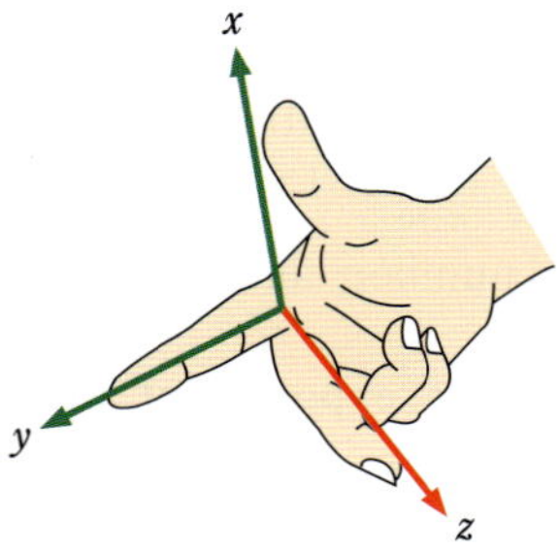

Fig. A.5 The right-handed system is a rectangular coordinate system where if we point the thumb of the right hand in the direction of $+x$-axis, and the index finger in the direction of $+y$-axis, then $+z$-axis points to the direction of the middle finger.

Answers to Questions and Problems for exercises

Chapter 1

Questions

1. ①-⑤, ②-④, ③-①, ④-⑥, ⑤-②, ⑥-③
2. 10 m/s, 20 m/s, 30 m/s. 5 m, 20 m, 45 m.
3. $t = \sqrt{\dfrac{2x}{g}} = \sqrt{\dfrac{2\times 4.9\ \text{cm}}{9.8\ \text{m/s}^2}} = \sqrt{\dfrac{1}{100}}\ \text{s} = 0.1\ \text{s}$
4. Use $1° = \dfrac{\pi}{180}$ rad.
5. $\dfrac{\theta}{2\pi}$ times the circle area πr^2.
6. (1) Section 3→4 with the smallest radius of curvature

   (2) Sections of straight lines 2→3 and 4→1

Problems for exercise 1

A

1. (1) km/h = 1000 m/3600 s = $\dfrac{1}{3.6}$ m/s,

   (2) omitted.
2. $\dfrac{552.6\ \text{km}}{4.2\ \text{h}} = 132\ \text{km/h} = 37\ \text{m/s}$
3.

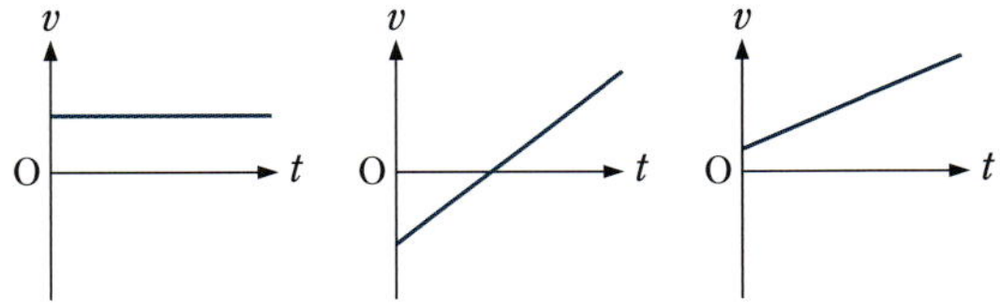

Fig. S-1

4. omitted.
5. $x = \dfrac{ad-bc}{a-b},\ t = \dfrac{d-c}{a-b}$
6. $t = \dfrac{v}{a} = \dfrac{55\ \text{m/s}}{0.25\ \text{m/s}^2} = 220\ \text{s}$
7. For the ball at the lowest position, $t = \dfrac{11}{30}$ s and $x = 0.65$ m, and the second lowest ball, $t = \dfrac{10}{30}$ s and $x = 0.54$ m. Calculate g using the formula $g = \dfrac{2x}{t^2}$. In both cases, a value $g = 9.7\ \text{m/s}^2$ is obtained.
8. $-bV_0,\ 2bV_0(bt-1)$
9. (1) $2at+b,\ 2a$

   (2) $ab\cos(bt+c),\ -ab^2\sin(bt+c)$

   (3) $-ab\sin(bt+c),\ -ab^2\cos(bt+c)$

   (4) $\dfrac{ab}{bt+c},\ -\dfrac{ab^2}{(bt+c)^2}$

   (5) $am\,e^{mt} - bn\,e^{-nt},\ am^2 e^{mt} + bn^2 e^{-nt}$
10. $10\sqrt{3}$ m/s, 10 m/s
11. $\omega = \dfrac{2\pi\ \text{rad}}{24\times 60\times(60\ \text{s})} = 7.3\times 10^{-5}\ \text{rad/s}$
12. $\dfrac{v^2}{r} \approx g \quad \therefore \quad v \approx \sqrt{rg} = \sqrt{(0.5\ \text{m})\times(9.8\ \text{m/s}^2)}$
    $= \sqrt{4.9}\ \text{m/s} = 2.2\ \text{m/s}$
13. (1) $f = \dfrac{1}{T} = 0.1\ \text{s}^{-1},\ \omega = 2\pi f = 0.63\ \text{s}^{-1}$

    (2) $v = r\omega = (4\ \text{m})\times(0.2\pi\ \text{s}^{-1}) = 2.5\ \text{m/s}$

    (3) $a = r\omega^2 = (4\ \text{m})\times(0.2\pi\ \text{s}^{-1})^2$
    $= 1.58\ \text{m/s}^2$, 0.16 times

B

1. $\omega = \dfrac{d\theta}{dt} = 0.01\ \text{s}^{-1}$ (rad can be omitted)

   $a = v\omega = (20\ \text{m/s})\times(0.01\ \text{s}^{-1}) = 0.2\ \text{m/s}^2$
2. When $a_0 > 0$, downward-convex parabola, and when $a_0 < 0$, upward-convex parabola. The tangential slope at the origin is v_0, so when $v_0 > 0$, it is positive, when $v_0 = 0$, it is zero, and when $v_0 < 0$, it is negative.
3. $v > 0$ indicates the motion in the $+x$ direction and $v < 0$ indicates the motion in the $-x$ direction. $a > 0$ is the accelerated motion in the $+x$ direction (1), or decelerated motion in the $-x$ direction (3). $a < 0$ is the decelerated motion in the $+x$ direction (2), or accelerated motion in the $-x$ direction (4).

Chapter 2

Questions

1. See Fig. S-2.

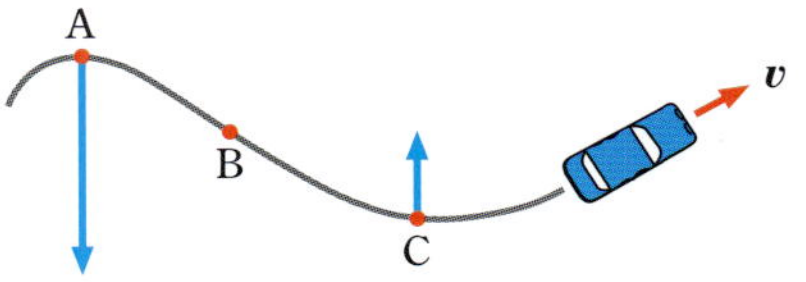

Fig. S-2

2. The adult moves forward, because the frictional force that the ground pushes the adult is stronger than the frictional force that pushes the infant.
3. It does not move.
4. $F_{m\leftarrow 2m} = \dfrac{F}{3}$
5. $\boldsymbol{F}_{\text{string}\leftarrow\text{A}} + \boldsymbol{F}_{\text{string}\leftarrow\text{B}} = m_{\text{string}}\boldsymbol{a}_{\text{string}} \approx \boldsymbol{0}$.

   $\therefore\ \boldsymbol{F}_{\text{A}\leftarrow\text{string}} = -\boldsymbol{F}_{\text{string}\leftarrow\text{A}} \approx \boldsymbol{F}_{\text{string}\leftarrow\text{B}} = -\boldsymbol{F}_{\text{B}\leftarrow\text{string}}$
6. 20 N

Problems for exercise 2

A

1. $F = ma = (30\ \mathrm{kg}) \times (4\ \mathrm{m/s^2}) = 120\ \mathrm{N}$
2. $a = \dfrac{(0\ \mathrm{m/s}) - (30\ \mathrm{m/s})}{6\ \mathrm{s}} = -5\ \mathrm{m/s^2}$,
 $F = (20\ \mathrm{kg}) \times (-5\ \mathrm{m/s^2}) = -100\ \mathrm{N}$
3. $a = \dfrac{F}{m} = \dfrac{12\ \mathrm{N}}{2\ \mathrm{kg}} = 6\ \mathrm{m/s^2}$
4. $a = \dfrac{F}{m} = \dfrac{20\ \mathrm{N}}{2\ \mathrm{kg}} = 10\ \mathrm{m/s^2}$,
 $v = at = (10\ \mathrm{m/s^2}) \times (3\ \mathrm{s}) = 30\ \mathrm{m/s}$
5. (1) $a = \dfrac{(30\ \mathrm{m/s}) - (20\ \mathrm{m/s})}{5\ \mathrm{s}} = 2\ \mathrm{m/s^2}$
 (2) $F = (1000\ \mathrm{kg}) \times (2\ \mathrm{m/s^2}) = 2000\ \mathrm{N}$
6. At the switching point P, since the moving direction of the wheel changes by a finite angle θ in a short time t, the acceleration $\dfrac{v\theta}{t}$ of the wheel is large even if the speed v is small.
7. (a)
 In the case of (a), $a = \dfrac{F}{m} = \dfrac{0.98\ \mathrm{N}}{0.4\ \mathrm{kg}} = 2.5\ \mathrm{m/s^2}$.
 In the case of (b), $a = \dfrac{0.98\ \mathrm{N}}{(0.4+0.1)\ \mathrm{kg}} = 2.0\ \mathrm{m/s^2}$.
8. Take the elevator and the person as one object.
 $(M+m)a = T-(M+m)g$. $a = \dfrac{T}{M+m} - g$

B

1. We assume that the friction between the desk and the object B is negligible. The equations of motion for the objects A and B, $m_A a = m_A g - S$ and $m_B a = S$ hold. Eliminating S from these equations, we have for the acceleration $a = \dfrac{m_A}{m_A+m_B} g$ and for the tension $S = m_B a = \dfrac{m_A m_B}{m_A+m_B} g < m_A g$. With the increase of m_B, $\dfrac{m_A m_B}{m_A+m_B} g$ increases and gradually approaches $m_A g$.
2. $F = G\dfrac{m^2}{r^2} = \{6.7\times10^{-11}\ \mathrm{m^3/(kg\cdot s^2)}\}$
 $\times \dfrac{(1\ \mathrm{kg})^2}{(0.05\ \mathrm{m})^2} = 2.7\times10^{-8}\ \mathrm{N}$
3. The forces acting on the sleigh and the passenger are the force $\boldsymbol{F}$ from a sleigh-pulling person, gravity $\boldsymbol{W}$, normal force $\boldsymbol{N}$, and maximum frictional force $\boldsymbol{F}_{max}$. Conditions of balance for these forces are,
 in perpendicular direction :
 $$W = N + F(\sin 30^\circ) = N + \frac{F}{2}$$
 $$\therefore\ N = W - \frac{F}{2}$$
 in horizontal direction : $F(\cos 30^\circ) = \dfrac{\sqrt{3}}{2}F = \mu N$

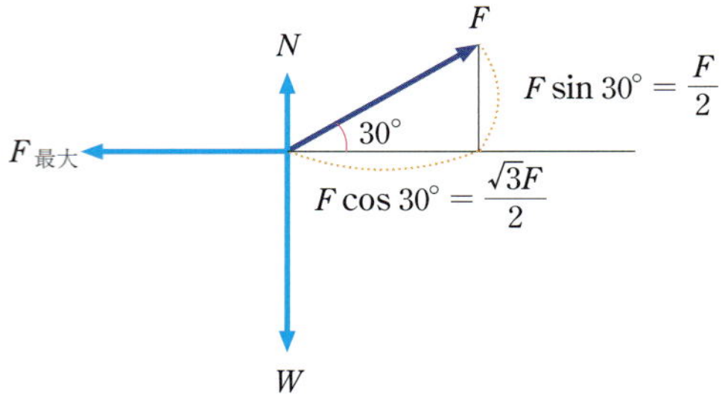

Fig. S-3

$$= 0.25\left(W - \frac{F}{2}\right)$$
$$\therefore\ F = \frac{0.5W}{\sqrt{3}+0.25} = 0.25\times(60\ \mathrm{kgf}) = 15\ \mathrm{kgf}$$

4. $F > 0.30\times(25\ \mathrm{kg})\times(9.8\ \mathrm{m/s^2}) + 0.15\times(5\ \mathrm{kg})\times(9.8\ \mathrm{m/s^2}) = 81\ \mathrm{N}$

Chapter 3

Questions

1. omitted
2. The amount of momentum change is the same irrespective of how to land. Therefore, the magnitude of force which the ground exerts on our body is inversely proportional to the acting time duration of the force.
3. $\dfrac{1}{2}\times(24\ \mathrm{m/s})\times(20\ \mathrm{s}) + (24\ \mathrm{m/s})\times(100\ \mathrm{s}) + \dfrac{1}{2}\times(24\ \mathrm{m/s})\times(30\ \mathrm{s}) = 3000\ \mathrm{m}$
4. Find the area of the trapezoid.
5. Use $0 = v_0 - bt_1$ and $x = \dfrac{1}{2}v_0 t_1$.
6. ①. Uniformly accelerated motion of acceleration $g\sin\theta$ in section AB, uniform motion with zero acceleration in section BC, and uniformly accelerated motion of $-\mu' g$ after passing the point C.
7. Set the right-hand side of the third equation of equation (3.31) to 0 and find the solution that is not $t = 0$.
8. Use $\sin 2(90^\circ - \theta_0) = \sin(180^\circ - 2\theta_0) = \sin 2\theta_0$. Since $\sin(90^\circ - \theta_0) \neq \sin\theta_0$, the residence times in the air are different.
9. $R = \dfrac{{v_0}^2}{g} > 100\ \mathrm{m}$
 $\therefore\ v_0 > \sqrt{g\times(100\ \mathrm{m})} = 31\ \mathrm{m/s}$
10. $R < \dfrac{{v_0}^2}{g} = \dfrac{(10\ \mathrm{m/s})^2}{10\ \mathrm{m/s^2}} = 10\ \mathrm{m}$
11. $v_t = \dfrac{mg}{b} = \dfrac{mg}{6\pi\eta R} \propto m$
12. $m\dfrac{d^2x}{dt^2} = mg - \dfrac{1}{2}C\rho A v^2$. Since the acceleration immediately after the start of the fall is g, the fall

velocity increases, but when the object reaches the terminal velocity $v_t = \sqrt{\dfrac{2mg}{C\rho A}}$, it performs a motion of constant velocity.

13. $v(t) = \dfrac{mg}{b}(1-e^{-\frac{bt}{m}}) \approx \dfrac{mg}{b}\left\{1-\left(1-\dfrac{bt}{m}\right)\right\} = gt$

14. $x = \int_0^t \dfrac{dx}{dt}dt = \dfrac{mg}{b}\int_0^t (1-e^{-\frac{bt}{m}})\,dt$
$= \dfrac{mg}{b}\left[t+\dfrac{m}{b}e^{-\frac{bt}{m}}\right]_0^t$
$= \dfrac{mg}{b}\left(t+\dfrac{m}{b}e^{-\frac{bt}{m}}-\dfrac{m}{b}\right)$

Problems for exercise 3

A

1. (1) $\dfrac{1}{3}at^3+\dfrac{1}{2}bt^2+ct+d\log|t|-\dfrac{e}{t}+C$
(2) $-\dfrac{a}{\omega}\cos(\omega t+b)+C$
(3) $\dfrac{a}{\omega}\sin(\omega t+b)+C$

2. $122.5\ \text{m} = \dfrac{1}{2}\times(9.8\ \text{m/s}^2)t^2$
$\therefore\ t = 5\ \text{s},\ v = gt = (9.8\ \text{m/s}^2)\times(5\ \text{s}) = 49\ \text{m/s}$

3. omitted

4. (1) $50\ \text{km/h} = \dfrac{50\times(1000\ \text{m})}{3600\ \text{s}} = 13.9\ \text{m/s}$,
$(13.9\ \text{m/s})\times(0.5\ \text{s}) = 6.9\ \text{m}$
(2) $100\ \text{km/h} = \dfrac{100000\ \text{m}}{3600\ \text{s}} = 27.8\ \text{m/s}$. From the third relation of equation (3.26), $x = \dfrac{{v_0}^2}{2b} = \dfrac{(27.8\ \text{m/s})^2}{2\times(7\ \text{m/s}^2)} = 55\ \text{m}$

5. (1) $v = (20\ \text{m/s})-(10\ \text{m/s}^2)t$
(2) $x-x_0 = (20\ \text{m/s})t-(5\ \text{m/s}^2)t^2$
$= -(5\ \text{m/s}^2)(t-2\ \text{s})^2+20\ \text{m}$.
At $t = 2$ s, $x-x_0$ takes the maximum value 20 m. At $t = 5$ s, $x-x_0 = -25$ m. The travel distance is (20 m) + (45 m) = 65 m, and the displacement is −25 m.

6. (1) $v = (9.8\ \text{m/s}^2)\times(3.0\ \text{s}) = 29.4\ \text{m/s}$
(2) $h = \dfrac{1}{2}\times(9.8\ \text{m/s}^2)\times(3.0\ \text{s})^2 = 44.1\ \text{m}$
(3) $\dfrac{44.1\ \text{m}}{3.0\ \text{s}} = 14.7\ \text{m/s}$

7. (1) Because the height of the highest points are the same, they are the same.
(2) Because the height of the highest points are the same, they are the same.
(3) The residence times in the air are the same. From the distance traveled in the same time, the magnitudes of initial speed are a → b → c. in the ascending order.
(4) The ascending order is a → b → c.

8. $x = v_0t,\ z = -\dfrac{1}{2}gt^2+z_0 = -\dfrac{1}{2}\dfrac{gx^2}{{v_0}^2}+z_0$.
At $x = 12$ m, $z = (2.5\ \text{m})-(0.54\ \text{m}) = 2.0\ \text{m}$.
∴ The ball goes over the net.
Substituting $t = \sqrt{\dfrac{2z_0}{g}}$ into $x = x_0t$ leads to $x = v_0\sqrt{\dfrac{2z_0}{g}} = 26\ \text{m}$.

B

1. (1) $a_1 = \dfrac{v}{t_1},\ -a_2 = -\dfrac{v}{t_3-t_2}$
(2) $x_1 = \dfrac{1}{2}vt_1,\ x_2 = v(t_2-t_1)+x_1$
$x_3 = \dfrac{1}{2}v(t_3-t_2)+x_2$
(3) omitted

2. (1) $a = \dfrac{F}{m} = \dfrac{20\ \text{N}}{10\ \text{kg}} = 2\ \text{m/s}^2$
(2) $a = \dfrac{10\ \text{N}}{10\ \text{kg}} = 1\ \text{m/s}^2$,
$x = \dfrac{1}{2}at^2 = \dfrac{1}{2}\times(1\ \text{m/s}^2)\times(10\ \text{s})^2 = 50\ \text{m}$,
$v = at = (1\ \text{m/s}^2)\times(10\ \text{s}) = 10\ \text{m/s}$
(3) $a = -\dfrac{20\ \text{N}}{10\ \text{kg}} = -2\ \text{m/s}^2$, $t = -\dfrac{v_0}{a} = \dfrac{20\ \text{m/s}}{2\ \text{m/s}^2} = 10\ \text{s}$, $x = \dfrac{1}{2}v_0t = \dfrac{1}{2}\times(20\ \text{m/s})\times(10\ \text{s}) = 100\ \text{m}$
(4) $a = \dfrac{(40\ \text{m/s})-(20\ \text{m/s})}{5\ \text{s}} = 4\ \text{m/s}^2$, $F = ma = (10\ \text{kg})\times(4\ \text{m/s}^2) = 40\ \text{N}$

3. From the third relation of equation (3.26), $b = \dfrac{{v_0}^2}{2d} = \dfrac{(200\ \text{m/s})^2}{2d} = \dfrac{20000\ \text{m}^2/\text{s}^2}{d} \leqq 6\times(9.8\ \text{m/s}^2)$
$\therefore\ d = \dfrac{20000\ \text{m}^2/\text{s}^2}{6\times(9.8\ \text{m/s}^2)} = 340\ \text{m}$

4. (1) $v_0 = \sqrt{2gh} = \sqrt{2\times(9.8\ \text{m/s}^2)\times(3.0\ \text{m})} \approx 7.7\ \text{m/s}$
(2) The deceleration time t and the acceleration $-a$ satisfy the relations $0.6\ \text{m} = \dfrac{1}{2}at^2$ and $at = 7.7\ \text{m/s}$. Therefore,
$\therefore\ a = \dfrac{(7.7\ \text{m/s})^2}{1.2\ \text{m}} = 49\ \text{m/s}^2 = 5.0g$
$\therefore\ F = ma+mg = 6\ mg = 6\times(40\ \text{kg})\times(9.8\ \text{m/s}^2) = 2.4\times10^3\ \text{N}$

5. $m\dfrac{dv}{dt} = mg-bv^2 \quad \therefore\ \dfrac{dv}{v^2-{v_t}^2} = -\dfrac{b}{m}dt$.
Integration of both sides gives
$$\int\frac{dv}{v^2-{v_t}^2} = \frac{1}{2v_t}\log\left|\frac{v-v_t}{v+v_t}\right| = -\frac{b}{m}\int dt = -\frac{bt}{m}+c$$

$v_t = \sqrt{\dfrac{mb}{b}}$. At $t=0$, $v=0$, so the arbitrary constant $c=0$.

$$\therefore \quad \log\left|\frac{v-v_t}{v+v_t}\right| = -\frac{2bv_t t}{m} = -\frac{2gt}{v_t}$$

$$\to v = v_t \frac{1-e^{-2gt/v_t}}{1+e^{-2gt/v_t}}$$

6. Use $m\dfrac{d^2x}{dt^2} = -mg\sin\theta$ and $m\dfrac{d^2y}{dt^2} = -mg\cos\theta$. T is obtained by setting $y=0$. Substituting T into the relation for x gives

$$X = \frac{2v_0{}^2 \sin\alpha\cos(\alpha+\theta)}{g\cos^2\theta} = \frac{v_0{}^2}{g\cos^2\theta}\{\sin(2\alpha+\theta) - \sin\theta\},$$

X takes the maximum when $2\alpha+\theta = \dfrac{\pi}{2}$ provided

$$\alpha+\theta = \frac{\pi}{4}+\frac{\theta}{2} > \frac{\pi}{4}$$

$$\therefore \quad \alpha = \frac{\pi}{4}-\frac{\theta}{2} < \frac{\pi}{4}.$$

7. $\dfrac{dv_x}{\dfrac{g}{\beta}-v_x} = \beta\,dt \qquad \therefore \quad \log\left|\dfrac{g}{\beta}-v_x\right| = -\beta t + C,$

$$C = \log\left|\frac{g}{\beta}-v_{x0}\right|$$

Therefore, from $\dfrac{\dfrac{g}{\beta}-v_x}{\dfrac{g}{\beta}-v_{x0}} = e^{-\beta t}$, v_x is determined.

$$\frac{dv_y}{v_y} = -\beta\,dt \qquad \therefore \quad \log v_y = -\beta t + C,\ C = \log v_{y0}$$

$$\therefore \quad \frac{v_y}{v_{y0}} = e^{-\beta t}$$

Using equation (3.18) and from v_x and v_y, x and y are obtained, similarly as for the case of Question 14 of this chapter.

Chapter 4

Questions

1. The restoring forces of the two springs acting on the weight are $-k_1x$ and $-k_2x$, so the equation of motion of the weight is

$$m\frac{d^2x}{dt^2} = -k_1x-k_2x = -(k_1+k_2)x$$

This equation corresponds to the case where k of the equation of motion of the spring pendulum is k_1+k_2, Thus the frequency f is

$$f = \frac{1}{2\pi}\sqrt{\frac{k_1+k_2}{m}}$$

2. $T = 2\pi\sqrt{\dfrac{2\text{ m}}{9.8\text{ m/s}^2}} = 2.8\text{ s}$

3. omitted.

4. ③ [Note that the tangential component of the trajectory of the acceleration is $-g\sin\theta$ and the magnitude of the centripetal acceleration is $\dfrac{v^2}{L}$.]

Problems for exercise 4

A

1. (1) 0 (2) 0 (3) simple harmonic oscillation of frequency $f = \dfrac{1}{2\pi}\sqrt{\dfrac{k}{m}}$

2. $T = 2\pi\sqrt{\dfrac{m}{k}}$. $k = \dfrac{4\pi^2 m}{T^2} = \dfrac{4\pi^2\times(2\text{ kg})}{(2\text{ s})^2} = 20$ kg/s^2

3. (1) $k = \dfrac{mg}{x_0} = \dfrac{(1\text{ kg})\times(9.8\text{ m/s}^2)}{0.1\text{ m}} = 98\text{ kg/s}^2$

(2) $T = 2\pi\sqrt{\dfrac{m}{k}} = 2\pi\sqrt{\dfrac{1.0\text{ kg}}{98\text{ kg/s}^2}} = 0.63\text{ s}$

(3) $a_{max} = A\omega^2 = \dfrac{Ak}{m} = \dfrac{(0.05\text{ m})\times(98\text{ kg/s}^2)}{1\text{ kg}} = 5\text{ m/s}^2$. $\dfrac{a_{max}}{g} = \dfrac{5\text{ m/s}^2}{9.8\text{ m/s}^2} = 0.5$

4. $\sqrt{\dfrac{1}{0.17}} = 2.4$ [times]

5. It does not change.

B

1. angular frequency

$$\omega = \sqrt{\frac{k}{m}} = \sqrt{\frac{5.0\times10^4\text{ N/m}}{500\text{ kg}}} = 10\text{ s}^{-1}$$

(1) $f = \dfrac{\omega}{2\pi} = \dfrac{10}{2\pi}\text{ s}^{-1} = 1.6\text{ s}^{-1}$,

$T = \dfrac{1}{f} = \dfrac{1}{1.6\text{ s}^{-1}} = 0.63\text{ s}$

(2) $v_{max} = A\omega = (1.0\text{ cm})\times(10\text{ s}^{-1}) = 10\text{ cm/s} = 0.1\text{ m/s}$

(3) $a_{max} = A\omega^2 = (1\text{ cm})\times(10\text{ s}^{-1})^2 = 1.0\text{ m/s}^2$

2. In the equation of motion of the pendulum $mL\dfrac{d^2\theta}{dt^2} = -mg\sin\theta$, the mass on the left-hand side is proportional to the inertia of an object and is called the **inertial mass**, while the mass on the right-hand side is proportional to the gravity and is called the **gravitational mass**. Newton's experiment shows "inertial mass" = "gravitational mass".

3. (1) In the case of Fig. 2 (a), the stretch of the spring is $\dfrac{x}{2}$ and the restoring force acting on the weight is $k\dfrac{x}{2}$. In the case of Fig. 2 (b), the stretch of each spring is x and the restoring force acting on the weight is $2kx$.

(2) In the case of Fig. 2 (a), the period is $2\pi\sqrt{\dfrac{2M}{k}}$ and in the case of Fig. 2 (b), the period is $2\pi\sqrt{\dfrac{M}{2k}}$.

The period of the former is two times as long as the latter.

4. The solution is $x = \dfrac{mg}{k} + A\cos(\omega t + \theta_0)$, where A and θ_0 are arbitrary constants. Since this solution satisfies the equation of motion and contains two arbitrary constants, it is a general solution.

Chapter 5

Questions

1. In A → B and B → A, if a process follows the same path in reverse, in the same minute section, $\boldsymbol{F}_i$ is the same but the direction of $\Delta \boldsymbol{s}_i$ is reverse. Thus, $W^{\text{Con}}{}_{\text{A}\to\text{B}} = -W^{\text{Con}}{}_{\text{B}\to\text{A}}$.
2. Since $F = -\dfrac{\Delta U}{\Delta x}$, if the U-x graph is descending to the right, then the force is facing right $F > 0$). If it is ascending to the right, the force is facing to the left ($F < 0$), and if it is horizontal, $F = 0$. In the range $0\ \text{m} < x < 5\ \text{m}$, $F = 0\ \text{N}$. In the range $5\ \text{m} < x < 10\ \text{m}$, $F = 6\ \text{N}$. In the range $10\ \text{m} < x < 15\ \text{m}$, $F = 0\ \text{N}$. In the range $15\ \text{m} < x < 20\ \text{m}$, $F = -3\ \text{N}$, and in the range $20\ \text{m} < x < 25\ \text{m}$, $F = 1\ \text{N}$.
3. omitted.

Problems for exercise 5

A

1. (1) $mgh = 1.6 \times 10^3\ \text{J}$
   (2) In mechanics, 0 J.
   (3) $-mgh = -1.6 \times 10^3\ \text{J}$
2. 144 km/h = 40 m/s. $\dfrac{1}{2} \times (0.15\ \text{kg}) \times (40\ \text{m/s})^2 = 120\ \text{J}$. Work is also 120 J.
3. 0
4. The same speed.
5. b. The mechanical energy at the highest point in the air is the sum of the kinetic energy and the potential energy of gravity. Therefore, the highest point in the air is lower than the point A.
6. When the string is vertical, $mgL = \dfrac{1}{2}mv^2$.
   $S = mg + \dfrac{mv^2}{L} = 3\,mg$.
7. Since the two points A and B are at the same height, $0.5 + 0.5\cos\theta = \dfrac{\sqrt{3}}{2}$. $\cos\theta = \sqrt{3} - 1 = 0.73$. $\theta = 43°$
8. $P = Fv = mgv = (50\ \text{kg}) \times (9.8\ \text{m/s}^2) \times (2\ \text{m/s}) = 980\ \text{W}$
9. $P \geqq mgv = (1000\ \text{kg}) \times (9.8\ \text{m/s}^2) \times \dfrac{10\ \text{m}}{60\ \text{s}} = 1633\ \text{W}$
10. $\dfrac{4.6 \times 10^7\ \text{W}}{(65\ \text{m}^3/\text{s}) \times (10^3\ \text{kg/m}^3) \times (9.8\ \text{m/s}^2) \times (77\ \text{m})} = 0.94 \quad \therefore\ 94\ \%$
11. (1) $(40\ \text{kg}) \times (9.8\ \text{m/s}^2) \times (3000\ \text{m}) = 1.2 \times 10^6\ \text{J}$
    (2) $\dfrac{1.2 \times 10^6\ \text{J}}{(3.8 \times 10^7\ \text{J/kg}) \times 0.20} = 0.16\ \text{kg}$
12. The work that the gravity performs is $10\,mgh = 10 \times (3.0\ \text{kg}) \times (9.8\ \text{m/s}^2) \times (3.0\ \text{m}) = 882\ \text{J}$. The amount of heat required to raise the temperature by T is $(500\ \text{g}) \times (4.2\ \text{J/(g·°C)})\ T = 2100\ T\ (\text{J/°C})$,
    $\therefore\ T = \dfrac{882\ \text{J}}{2100\ \text{J/°C}} = 0.42\ \text{°C}$.
13. The energy of the ball is quadrupled. Thus, the initial velocity is doubled and the upward moving distance is quadrupled. If the ball is launched in the horizontal direction, it travels the twice longer distance during the fall.

B

1. $10.8\ \text{km/h} = 3\ \text{m/s}$, $h = (3\ \text{m/s}) \times (120\ \text{s}) \times 0.087 = 31.3\ \text{m}$,
   $P = \dfrac{(75\ \text{kg}) \times (9.8\ \text{m/s}^2) \times (31.3\ \text{m})}{120\ \text{s}} = 192\ \text{W}$
2. 0
3. Let the speed at the highest position on the circumference of circle of radius r be v. Then, $\dfrac{1}{2}mv^2 = mg(d-r)$. The condition that the weight is kept stretched at the highest position is that the tension of the thread $S \geqq 0$. $S = \dfrac{mv^2}{r} - mg = \dfrac{2mg(d-r)}{r} - mg \geqq 0$. $2d \geqq 3r = 3(L-d)$.
   From $5d \geqq 3L$, $d \geqq \dfrac{3}{5}L$.
4. (1) $\dfrac{(8.0 \times 10^3\ \text{W}) \times (1\ \text{m})}{22.4\ \text{m/s}} = 3.6 \times 10^2\ \text{J}$
   (2) The frictional force is $3.6 \times 10^2\ \text{N}$, and the gravity is $1.0 \times 10^4\ \text{N}$. The ratio is 0.036. See Case 2 of Section 8.1.
   (3) $\dfrac{(3.6 \times 10^2\ \text{J/m}) \times (1.7 \times 10^4\ \text{m})}{3.3 \times 10^7\ \text{J}} = 0.19$
   $\therefore$ 19 %

Chapter 6

Question

1. Since the area ΔS that the line segment connecting the point O and the object covers in time Δt is $\Delta S = \dfrac{dv\,\Delta t}{2}$, the areal velocity $\dfrac{\mathrm{d}S}{\mathrm{d}t} = \dfrac{vd}{2}$. The angular momentum $L = mvd = 2m\dfrac{\mathrm{d}S}{\mathrm{d}t}$.

Problems for exercise 6

A

1. Because $\dfrac{a^3}{T^2}$ = constant, when T is 70-fold, then, a is $70^{2/3} = 17$-fold.
2. One-stage rockets move on an elliptical orbit whose

one of the focal points is the center of the earth, so they always collide with the earth.

3. Duration of vernal equinox → autumn equinox is 186 days, and that of autumn equinox → vernal equinox is 179 days. The perihelion is in the period of autumn equinox → vernal equinox.

B

1. Assume the length of the string is l. $L = mlv = ml\dfrac{d(l\theta)}{dt}$. $N = -mgl\sin\theta$. $\dfrac{dL}{dt} = N$.

$\therefore \quad l\dfrac{d^2\theta}{dt^2} = -g\sin\theta$.

2. What is considered is the motion due to the central force, so the motion is planar. This plane is taken to be the xy plane.

(1) Since the equation of motion is given by $m\dfrac{d^2x}{dt^2} = -kx$ and $m\dfrac{d^2y}{dt^2} = -ky$, the motion is a composite of simple oscillations in the x and y directions. If we choose the coordinate axes so that the point where r is the largest is on the x axis, then, $y = 0$ when $|x|$ is the maximum.

$\therefore \quad x = A\cos(\omega t+\alpha)$, $y = B\sin(\omega t+\alpha)$.

$$\left(\frac{x}{A}\right)^2+\left(\frac{y}{B}\right)^2 = 1$$

(2) Because the motion is due to the central force.

(3) $T = \dfrac{2\pi}{\omega} = 2\pi\sqrt{\dfrac{m}{k}}$ is independent of the amplitudes A and B.

Chapter 7

Questions

1. The gravity W acting on the log is the sum of $W_A = 80$ kgf and $W_B = 70$ kgf, so $W = 150$ kgf. The mass of the log is 150 kg. Assuming that the distance from the end A to the center of gravity is x, $xW_A = (6\,\text{m}-x)W_B$, $80x = 70(6\,\text{m}-x)$, $x = 2.8$ m.
2. Fragment spheres of the fireworks ball which are generated around the center of gravity without gravity fall freely as a whole.
3. (a) The right coin stops and the left coin moves to the left.
(b) The right coin stops and the leftmost coin moves to the left. The case (a) happened consecutively.
4. omitted.
5. omitted.
6. $\dfrac{d\boldsymbol{L}_G}{dt} = \boldsymbol{R}\times M\boldsymbol{A} = \boldsymbol{R}\times\boldsymbol{F} = \boldsymbol{N}_G$

Problems for exercise 7

A

1. A and B start to move at the constant speeds of $-\dfrac{mv}{M_A}$ and $\dfrac{mv}{M_B+m}$, respectively.
2. The space craft moves 3 m till it reaches the center of gravity.
3. It is possible (when a long train passes the hill, the center of gravity of the train is lower than the top of the hill in the meantime.)
4. $m_A\boldsymbol{v}_A+m_B\boldsymbol{v}_B = (m_A+m_B)\boldsymbol{v}'$

$\therefore \quad \boldsymbol{v}' = \dfrac{m_A\boldsymbol{v}_A+m_B\boldsymbol{v}_B}{m_A+m_B}$

5. (1) $mV = (m+M)v$

$\therefore \quad v = \dfrac{mV}{m+M} = 0.87$ m/s

(2) $h = \dfrac{v^2}{2g} = \dfrac{(0.87\text{ m/s})^2}{2\times(9.8\text{ m/s}^2)} = 0.039\text{ m} = 3.9$ cm

B

1. We use an approximation that the mass of human body is concentrated near the center of gravity G. When we stand up near the lowest point of the swinging path, the line of action of the force acting on a human body by the swing and the line of action of the gravity pass through the center O of rotation, so the angular momentum mvr does not change when we stand up. Therefore, when we stand up and the radius r of rotation decreases, the speed v increases. When we bend the knees near the highest point, the speed is almost zero and does not change. Repeating this operation will increase the speed. The angular momentum of the human body is not conserved except at the lowest point.
2. The total angular momentum of human body and swivel chair is conserved at every stage (it is zero).
3. If the components of the galaxy lose energy, they tend to approach each other by the universal gravity. They can approach close to the direction parallel to the rotation axis, but cannot approach the rotation axis because of the angular momentum conservation law. As a result, the galaxy becomes disk-like.
4. We select the center of gravity as the origin and set $\boldsymbol{R} = \boldsymbol{0}$ in equation (1). Since $\boldsymbol{N} = \boldsymbol{0}$, the moment of gravity around the center of gravity is $\boldsymbol{0}$.
5. $\boldsymbol{r}_i\times\boldsymbol{F}_{i\leftarrow j}+\boldsymbol{r}_j\times\boldsymbol{F}_{j\leftarrow i} = (\boldsymbol{r}_i-\boldsymbol{r}_j)\times\boldsymbol{F}_{i\leftarrow j} = \boldsymbol{0}$

Chapter 8

Questions

1. (b) Since $\sqrt{6}R \ll L$, $\dfrac{1}{2}MR^2 < \dfrac{1}{12}ML^2$.
2. $\dfrac{1}{3}ML^2 = \dfrac{1}{12}ML^2+M\left(\dfrac{L}{2}\right)^2$
3. Since $V = R\omega$, $\dfrac{1}{2}MV^2 : \dfrac{1}{2}I_G\omega^2 = MR^2 : I_G$
4. The energy $\dfrac{1}{2}MV^2$ of center of gravity motion is

$\dfrac{1}{1+\dfrac{I_G}{MR^2}}$ times larger.

Problems for exercise 8

A

1. $F\times(15\text{ cm}) = (3\text{ kgf})\times(2.5\text{ cm})$
 $\therefore\ F = 0.5\text{ kgf}$
2. (1) Length of the rope $L = \sqrt{h^2+l^2} = \sqrt{(4.0\text{ m})^2+(3.0\text{ m})^2} = 5.0\text{ m}$. Distance between the hinge and the tension S is $d = l\times\dfrac{h}{L} = (3.0\text{ m})\times\dfrac{4}{5} = 2.4\text{ m}$. From the condition that the sum of the moment of force around the hinge $= 0$,
 $$(2.4\text{ m})S = (1.8\text{ m})W = (1.8\text{ m})\times(40\text{ kgf})$$
 $$\therefore\ S = 30\text{ kgf}$$
 (2) From the condition of balance of the force acting on the bar,
 $N = \dfrac{3}{5}S = 18\text{ kgf},\ F = W-\dfrac{4}{5}S = 16\text{ kgf}$
3. From the balance of the moments of force around the lower end of spinal column $= 0$,
 $T(\sin 12^\circ)\left(\dfrac{2L}{3}\right)-0.4W(\cos\theta)\left(\dfrac{L}{2}\right)-(0.2W+Mg)\times(\cos\theta)L = 0,$
 $T\sin 12^\circ = \left(0.6W+\dfrac{3}{2}Mg\right)\cos\theta$
 $\therefore\ T = 2.5W+6.2Mg = 2.7\times10^2\text{ kgf}$
4. $M = \rho\pi R^2h = (8\text{ g/cm}^3)\times\pi\times(1\text{ m})^2\times(1\text{ m}) = 8\pi\times10^3\text{ kg}$. $I = \dfrac{1}{2}MR^2 = \dfrac{1}{2}\times(8\pi\times10^3\text{ kg})\times(1\text{ m})^2 = 4\pi\times10^3\text{ kg·m}^2$. $\omega = \dfrac{2\pi\times600}{60\text{ s}} = 20\pi/\text{s}$.
 $K = \dfrac{1}{2}I\omega^2 = 8\pi^3\times10^5\text{ J} = 2.5\times10^7\text{ J}$
5. $I = \dfrac{3ML^2}{3} = (200\text{ kg})\times(5.0\text{ m})^2 = 5\times10^3\text{ kg·m}^2$. $\omega = \dfrac{2\pi\times300}{60\text{ s}} = 10\pi\text{ s}^{-1}$. $K = \dfrac{1}{2}I\omega^2 = \dfrac{5}{2}\pi^2\times10^5\text{ J} = 2.5\times10^6\text{ J}$
6. (a) whose I is smaller than that of (b).
7. To increase the moment of inertia. Even if the balance is lost, the angular velocity is small, so there is a time margin for restoring the balance.
8. (1) To the right.
 (2) To increase the moment of inertia.
9. When an object moves by the distance d with acceleration A, the speed $V = \sqrt{2Ad}$. Because the ratio of acceleration is $1 : \dfrac{5}{7} = 7 : 5$, so the ratio of speed V is $\sqrt{7} : \sqrt{5}$.
10. The first is the can containing liquid beer for which the $\dfrac{I_G}{MR^2}$ value is the smallest. The next is the can in which beer is frozen.

B

1. Because the resistant force from the floor passes through the sphere center of the hemisphere, it becomes unstable when the center of gravity of the weight and the hemisphere moves out of the hemisphere.
 $\therefore\ mh > M\left(\dfrac{3R}{8}\right)$ $\quad\therefore$ It is unstable when $h > \dfrac{3MR}{8m}$.
2. There are four equations : condition for the balance of force in the vertical direction $N_1+N_2-Mg = 0$, condition for the balance of moment of force around center of gravity $N_2l_2-N_1l_1-Fh = 0$, equation of motion in the horizontal direction $MA = F$, and $F \leqq \mu N_2$ (for rear-wheel drive vehicles). A takes the maximum value at $F = \mu N_2$, or $N_2 = Mg$ (since $N_1 \geqq 0$). $\therefore\ MA_{max} = \mu Mg,\ A_{max} = \mu g$.
 $N_2l_2-N_1l_1-Fh = N_2l_2-\mu N_2h$. $\therefore\ l_2 = \mu h$.
 In this solution, effects of wheel rotation are neglected.
3. $d = \dfrac{1}{2}\sqrt{a^2+b^2},\ \dfrac{I}{Md} = \dfrac{2}{3}\sqrt{a^2+b^2},$
 $T = 2\pi\left(\dfrac{2}{3g}\sqrt{a^2+b^2}\right)^{1/2}$
4. Let the force to hit the ball be F. The equation of motion of the center of gravity is $MA = F$, and the equation of rotational motion around the center of gravity is $I_G\alpha = \dfrac{2}{5}MR^2\alpha = \dfrac{2}{5}RF$. The acceleration at the point of contact with the floor is $A-R\alpha = 0$.
5. Since the speed of the thread-winding reel at the point in contact with the floor is zero, the law of rotational motion around the contact point P is $I_P\alpha = N$. In the case of $\boldsymbol{F}_1$, the thread-winding reel moves to the right because $N < 0$, and in the case of $\boldsymbol{F}_2$, the reel does not move because $N = 0$, and in the case of $\boldsymbol{F}_3$, $N > 0$ and the thread-winding real moves to the left.

Chapter 9

Questions

1. (1) If a person sees this phenomenon on the ground, $m\boldsymbol{a}_0 = \boldsymbol{S}+m\boldsymbol{g}$. If the person sees it in the train, $\boldsymbol{S}+m\boldsymbol{g}-m\boldsymbol{a}_0 = \boldsymbol{0}$. (2) $\tan\theta = \dfrac{a_0}{g}$. (3) The inclination of the string is in the opposite direction of the apparent gravitation. It faces to the direction of the string which is hanging the weight.
2. $ma = mg-N$. $N = mg-ma = 50\times\{(9.8\text{ m/s}^2)-(1\text{ m/s}^2)\} = 440\text{ N} = 45\text{ kgf}$. It is because the

supporting force does not act.

Problems for exercise

A

1. 20 gf = 0.20 N, $\dfrac{0.20\ \mathrm{N}}{0.10\ \mathrm{kg}} = 2.0\ \mathrm{m/s^2}$ upward.
2. (1) It holds. (2) It does not hold. (3) It does not hold.
3. $\dfrac{v^2}{r} = g\tan\theta$. $\tan\theta = \dfrac{(30\ \mathrm{m/s})^2}{(800\ \mathrm{m})\times(9.8\ \mathrm{m/s^2})} =$ 0.11. $\theta = 6.5°$
4. In an inertial frame which is consistent with a non-inertial frame fixed to the mast (because no relative velocity exists at the base of the mast) at the moment when the lead ball starts to fall without initial velocity, the tip of the mast with height h is moving to the east at speed ωh, so the ball falls to the east from the base point of the mast (0.8 cm to the east if h is 50 m).

B

1. A person on a ground interprets the motion of the ball as a linear motion at a constant speed, since no force acts on it from the moment the string is broken. A person riding on a merry-go-round may consider it as a result of action of the centrifugal force and the Coriolis force. Immediately after the breaking of the rope, $v' = 0$ and only the centrifugal force acts. With the increase in v', the Coriolis force becomes significant.
2. Consider the inertial frame that was at rest with respect to the launching site at the moment of launching and use the hint.
3. Coordinate axes of the inertial frame at the north pole are determined by the direction of the star, and at the North Pole, the Earth surface rotates with a period of 24 hours relative to the inertial system. Because the oscillating surface of the pendulum is at rest with respect to the inertial frame, the oscillating surface of the pendulum rotates in a 24-hour cycle. See Fig. 9-9.

Chapter 10

Question

1. The normal component and the tangential component of the force F are $F\cos\theta$ and $F\sin\theta$, respectively. Divide these by the cross-sectional area $\dfrac{A}{\cos\theta}$.

Problems for exercise

A

1. One person
2. $\Delta L = \dfrac{1}{E}\dfrac{FL}{A} = \dfrac{(9.8\ \mathrm{N})\times(1\ \mathrm{m})}{\pi\times(10^{-4}\ \mathrm{m})^2\times(2\times10^{11}\ \mathrm{N/m^2})}$ $= 1.6\times10^{-3}\ \mathrm{m} = 1.6\ \mathrm{mm}$
3. (1) $\theta \approx \dfrac{1\ \mathrm{cm}}{30\ \mathrm{cm}} = 0.033\ \mathrm{rad}$

   (2) $\tau = \dfrac{0.98\ \mathrm{N}}{(0.3\ \mathrm{m})^2} = 11\ \mathrm{N/m^2}$

   (3) $G = \dfrac{\tau}{\theta} = \dfrac{11\ \mathrm{N/m^2}}{0.033} = 3.3\times10^2\ \mathrm{N/m^2}$
4. (1) $(1.2\times10^8\ \mathrm{N/m^2})\times(6\times10^{-4}\ \mathrm{m^2}) = 7\times10^4\ \mathrm{N}$ $= 7\times10^3\ \mathrm{kgf}$

   (2) $\dfrac{\Delta L}{L} = \dfrac{1}{E}\dfrac{F}{A} = \dfrac{1.2\times10^8\ \mathrm{Pa}}{1.7\times10^{10}\ \mathrm{Pa}} = 7\times10^{-3}$.

   It stretches by 0.7%.
5. $\Delta p = \rho g h = (10^3\ \mathrm{kg/m^3})\times(9.8\ \mathrm{m/s^2})\times(10^4\ \mathrm{m})$ $= 1\times10^8\ \mathrm{Pa}$, $\dfrac{\Delta V}{V} \approx 3\dfrac{\Delta D}{D} = -\dfrac{\Delta p}{k} = \dfrac{-1\times10^8\ \mathrm{Pa}}{1.7\times10^{11}\ \mathrm{Pa}}$ $= -6\times10^{-4}$, $\Delta D = \dfrac{1}{3}\times(20\ \mathrm{cm})\times(-6\times10^{-4}) =$ $-4\times10^{-3}\ \mathrm{cm} = -0.04\ \mathrm{mm}$

B

1. When the bar is stretched by x, the force $F = \dfrac{AEx}{L}$. Therefore, $W = \dfrac{AE}{2L}(\Delta L)^2$.
2. The tension F is applied to the two opposing faces of the elastic cube having a side length L, and at the same time the tension F' is applied to the other two pairs of opposing faces so that deformation in these two directions does not occur. The stretched length in the direction of the applied tension F is $\Delta L = \dfrac{F}{EL} - 2\sigma\dfrac{F'}{EL}$. The condition that the other two directions do not stretch is $(1-\sigma)\dfrac{F'}{EL} - \sigma\dfrac{F}{EL} = 0$.

   $\therefore\ F' = \dfrac{\sigma}{1-\sigma}F$. $EL\,\Delta L = \left(1-\dfrac{2\sigma^2}{1-\sigma}\right)F =$ $\dfrac{1-\sigma-2\sigma^2}{1-\sigma}F = \dfrac{(1-2\sigma)(1+\sigma)}{1-\sigma}F$

   $\therefore\ \dfrac{F}{L\,\Delta L} = \dfrac{1-\sigma}{(1+\sigma)(1-2\sigma)}E = k + \dfrac{4}{3}G$
3. The side length of the cube shrinks by $\dfrac{Lp}{E}$ due to the longitudinal pressure, but it extends by $\dfrac{2\sigma Lp}{E}$ through Poisson's ratio due to the pressure in the two lateral directions. Therefore, the change in volume is $\Delta V = \left\{L-(1-2\sigma)\dfrac{pL}{E}\right\}^3 - L^3 \approx -3(1-2\sigma)\dfrac{pV}{E}$

   $\therefore\ k = -\dfrac{pV}{\Delta V} = \dfrac{E}{3(1-2\sigma)}\quad (V = L^3)$
4. When the stress is applied as shown in Fig. 2 (a), as seen from Fig. 2 (b), the tangential stress shown in Fig. 2 (c) is generated.

   $\tan\left(\dfrac{\pi}{4}-\dfrac{\theta}{2}\right) \approx 1-\theta = \dfrac{\mathrm{\overline{A'H'}}}{\mathrm{\overline{A'E'}}} = \dfrac{1-(1+\sigma)\dfrac{\tau}{E}}{1+(1+\sigma)\dfrac{\tau}{E}}$

   $\approx 1-2(1+\sigma)\dfrac{\tau}{E}\qquad \therefore\ G = \dfrac{\tau}{\theta} = \dfrac{E}{2(1+\sigma)}$

Chapter 11
Questions

1. Due to the balance of pressure, the air pressures in the nostril, trachea and lungs are equal to the pressure of seawater that presses the skin.
2. Since the lift in equation (11.14) balances with the gravity Mg, $\frac{[\text{area of wing } A]}{[\text{mass } M]}$ is inversely proportional to (velocity v)^2.

Problems for exercise 11
A

1. $\frac{10^3\times9.8\text{ N}}{4\times3\times1.01\times10^5\text{ Pa}}=0.008\text{ m}^2=80\text{ cm}^2$.
2. Buoyancy is 1.29 kgf, and gravity is (0.178 kgf)+(0.200 kgf) = 0.378 kgf. (1.29 kgf)−(0.38 kgf) = 0.91 kgf $\quad\therefore\quad$ 0.91 kg
3. Bernoulli's law is $p_0+\rho gh=p_0+\frac{1}{2}\rho v^2$
$\therefore\quad v=\sqrt{2\times(9.8\text{ m/s}^2)\times(0.5\text{ m})}=3.1\text{ m/s}$
4. (1) $v=(0.2\text{ m/s})\times\frac{(3\text{ cm})^2}{(1\text{ cm})^2}=1.8\text{ m/s}$
(2) $p_A-p_B=\frac{1}{2}\rho v_B{}^2-\frac{1}{2}\rho v_A{}^2=\frac{1}{2}\times(10^3\text{ kg/m}^3)\times\{(1.8\text{ m/s})^2-(0.2\text{ m/s})^2\}=1.6\times10^3\text{ Pa}=1.6\times10^{-2}\text{ atm}$
5. The sphere must be pulled from both sides at a stronger force than that the air pressure acting on a circle of radius 20 cm, namely $\pi(20\text{ cm})^2\times(1.033\text{ kgf/cm}^2)=1298\text{ kgf}$.
6. Because the velocity relative to the air is small, the pressure (inertial resistance) decreases.
7. Under the non-slip condition, the rotating ball rotates the air around it. According to Bernoulli's law, for the observer in Fig. 1, the air pressure of the fast-flow side below the ball is lower than the air pressure of the slow-flow side above the ball. As a result, air exerts the downward force on the ball due to the pressure difference between the upper and lower pressures.

B

1. When the ship bottom is in close contact with the seabed, the seabed does not apply sufficient pressure to raise the ship bottom.
2. $p_A=\frac{1}{2}\rho_{air}u^2+p_B$, $\frac{1}{2}\rho_{air}u^2=p_A-p_B=\rho_0gh$
$\therefore\quad u=\sqrt{\frac{2\rho_0gh}{\rho_{air}}}$
3. Assume the depth at time t be h ($h=H$ at $t=0$).
$A\,dh=-vS\,dt=-\sqrt{2gh}S\,dt$, $\frac{A}{S}\frac{dh}{\sqrt{h}}=-\sqrt{2g}\,dt$, $\frac{2A\sqrt{h}}{S}=-\sqrt{2g}t+c$ (constant). At $t=T$, $h=0$, so $c=\sqrt{2g}T$.
$\frac{2A\sqrt{h}}{S}=\sqrt{2g}(T-t)\qquad\therefore\quad T=\sqrt{\frac{2H}{g}}\left(\frac{A}{S}\right)$
4. Since the amount of drainage per unit time is $S\sqrt{2gh}$, the water supply will be larger if the amount of water in the tank is reduced. Let the depth at the time of full water be H. The lowest water level is the one at which the falling speed $\frac{dh}{dt}=-\sqrt{2gh}\frac{S}{A}$ of water surface by drainage is equal to the rising speed $\frac{H}{2T}=\frac{\frac{S}{A}\sqrt{gH}}{2\sqrt{2}}$ of the water surface by water supply. Thus, the level is never less than $\frac{h}{H}=\frac{1}{16}$.
5. Because the inertial resistance is large in seawater, both iron balls will immediately fall at the terminal velocity $v_t=\sqrt{\frac{2mg}{C\rho A}}$ (Chapter 3, Question 12). Since the mass $m\propto$ (radius)^3 and the cross-sectional area $A\propto$ (radius)^2, the larger iron ball has a higher terminal velocity and falls to the seabed first.
6. Because the resistance of water increases with speed.

Chapter 12
Questions

1. $\mu=\frac{0.3\text{ kg}}{4\text{ m}}=0.075\text{ kg/m}$,
$v=\sqrt{\frac{9.8\text{ kg·m/s}^2}{0.075\text{ kg/m}}}=11\text{ m/s}$
2. $\frac{\sin 45^\circ}{\sin\theta_t}=1.41\qquad\therefore\quad\sin\theta_t=\frac{1}{2},\ \theta_t=30^\circ$
3. Referring to the case of $n=1$ in Fig. 12-20, draw a diagram for the case of $n>1$.
4. $0.1\text{ N/m}^2\approx10^{-6}$ atm

Problems for exercise 12
A

1. (1) 3 cm (2) 5 Hz (3) 20 cm
(4) 100 cm/s = 1 m/s
2. (1) $\sin\theta=\frac{V}{v}$
(2) $v=\frac{V}{\sin\theta}=\frac{340\text{ m/s}}{0.5}=680\text{ m/s}$
3. The velocity of the medium is not zero. The energy of wave is changed into the kinetic energy of the medium.
4. $|f-440\text{ Hz}|=6\text{ Hz}$. $\quad\therefore\quad f=446$ Hz or 434 Hz. With the decrease in tension, the frequency decreases. From the decrease in beat frequency, we can conclude $f=446$ Hz.
5. (340 m/s)×(3.0 s) = 1020 m
6. $v=72$ km/h = 20 m/s. Before passing-by : $f'=f\frac{V+v}{V-v}=(500\text{ Hz})\times\frac{360\text{ m/s}}{320\text{ m/s}}=563$ Hz. After

passing-by : $f' = f\dfrac{V-v}{V+v} = (500\,\text{Hz})\times\dfrac{320\,\text{m/s}}{360\,\text{m/s}} = 444\,\text{Hz}$.

B

1. omitted.
2. Differentiate equation (12.5) with respect to t under the condition of constant x.
3. The twist is propagated at the speed $v = \sqrt{\dfrac{G}{\rho}}$ of the transverse wave.
4. (1) $20\log_{10}\left(\dfrac{28\,\text{Pa}}{2\times10^{-5}\,\text{Pa}}\right) = 123\,[\text{dB}]$

   (2) $1\,\text{atm} = 1.01\times10^5\,\text{Pa}$

   $\therefore\ P = \dfrac{28\,\text{Pa}}{1.01\times10^5\,\text{Pa}} = 2.8\times10^{-4}\,\text{atm}$

Chapter 13

Question

1. $R = \dfrac{(1.5-1)^2}{(1.5+1)^2} = 0.04$

Problems for exercise 13

A

1. If we draw on the surface of the glass a ray line connecting A (eye) and B so that the law of refraction is satisfied, it looks like B is in the direction of the ray entering the eye. B looks above the actual position.
2. $\sin\theta_c = \dfrac{V_1}{V_2} = \dfrac{340\,\text{m/s}}{1500\,\text{m/s}} = 0.23 \quad \therefore\ \theta_c = 13°$
3. $\sin\theta_c = \dfrac{1}{n} = \dfrac{1}{2.42} \quad \therefore\ \theta_c = 24.4°$
4. $\sin\theta = \dfrac{\lambda}{D} = \dfrac{5\times10^{-7}\,\text{m}}{10^{-5}\,\text{m}} = 5\times10^{-2}$. Width $\approx 2l\theta = 2\times(1\,\text{m})\times(5\times10^{-2}) = 0.1\,\text{m} = 10\,\text{cm}$
5. $d = \dfrac{\lambda}{\sin\theta} = \dfrac{0.5\times10^{-6}\,\text{m}}{0.5} = 10^{-6}\,\text{m}$. $\dfrac{1}{10^{-6}\,\text{m}} = 10^6/\text{m} = 10^4/\text{cm}$. $\therefore\ 10^4$ slits in 1 cm.

B

1. The sun rises from the direction of the critical angle and sets in the direction of the critical angle.
2. $\theta \approx \dfrac{0.61\lambda}{R} = \dfrac{0.61\times5.00\times10^{-7}\,\text{m}}{10^{-4}\,\text{m}} = 3.1\times10^{-3}$.

   Diameter $D \approx 2l\theta = 6\times10^{-3}\,\text{m} = 6\,\text{mm}$.
3. Since sounds of various frequencies are mixed, destructive interference and constructive interference coexist, and also since the amplitudes of sound waves of the same frequency emitted from left and right speakers are not equal, the interference is not so significant. In addition, reflected sounds disturb the interference.

Chapter 14

Question

1. (1) If we consider the oxygen molecule as a cube whose one side length is d, the density is $\dfrac{32\,\text{g}}{6.02\times10^{23}\times d^3} = 1.12\,\text{g/cm}^3$. Thus,

   $d = \left(\dfrac{32\,\text{g}}{(1.12\,\text{g/cm}^3)\times(6.02\times10^{23})}\right)^{1/3} = 3.6\times10^{-8}\,\text{cm} \approx 4\times10^{-10}\,\text{m}$

   (2) $\rho = \dfrac{32\,\text{g}}{22.4\times10^3\,\text{cm}^3} = 1.43\times10^{-3}\,\text{g/cm}^3$.

   $\left(\dfrac{1.12}{1.43\times10^{-3}}\right)^{1/3} = 9.2$

Problems for exercise 14

A

1. We consider the case where the temperature lowers slowly. When the temperature lowers, the temperature of water near the lake surface lowers. When the water temperature near the lake surface is 3.98 °C or more, the cooled water sinks to the bottom because the density is higher than the water near the bottom. Eventually, the temperature of the whole lake water reaches about 3.98 °C, the temperature of maximum density. Since the density of water near the surface is lower than that of the bottom, the movement of water toward the bottom does not occur, and the water near the surface cools down more, and the lake surface starts to freeze at 0 °C.
2. Surface temperatures of red stars are lower than those of blue stars.
3. $(0.8\,\text{J/(g·K)})\times(3\,\text{g/cm}^3)\times(4.5\times10^5\,\text{cm})^3\times(100\,\text{K}) = 2.2\times10^{19}\,\text{J} = 6\times10^{12}\,\text{kWh} = 6$ 兆 kWh
4. $\Delta V = [(9.5-0.4)\times10^{-4}\,\text{K}^{-1}]\times(60\,\text{L})\times(30\,\text{K}) = 1.6\,\text{L}$
5. $H = \dfrac{kA\,\Delta T}{L} = \dfrac{[0.15\,\text{J/(m·s·K)}]\times(20\,\text{m}^2)\times(25\,\text{K})}{0.05\,\text{m}} = 1500\,\text{J/s}$
6. $L = \dfrac{1}{\sqrt{2}\pi Nd^2} = \dfrac{1}{\sqrt{2}\pi(p/kT)d^2} = \dfrac{kT}{\sqrt{2}\pi pd^2}$

B

1. Since the straight line showing $p = 1$ atm passes only the solid phase and the gas phase, solid dry ice sublimes into gaseous carbon dioxide.
2. Consider a cube whose one side length is L. $\Delta V = (L+\Delta L)^3 - L^3 \approx 3L^2\,\Delta L = 3L^3\alpha\,\Delta T = 3V\alpha\,\Delta T = \beta V\,\Delta T$. $\therefore\ \beta = 3\alpha$
3. $a\sigma A(T_1{}^4 - T_2{}^4)\,t = 0.7\times[5.67\times10^{-8}\,\text{W/(m}^2\text{·K}^4)]\times(1.2\,\text{m}^2)\times[(309\,\text{K})^4-(293\,\text{K})^4]\times(1\,\text{s}) = 83\,\text{J}$
4. At critical point, isotherms are stationary (the first and second derivatives are 0). Thus from equation (14.36) we have

   $$\left(\frac{\partial p}{\partial V}\right)_{T=一定} = -\frac{nRT}{(V-nb)^2} + \frac{2an^2}{V^3} = 0$$

   and

$$\left(\frac{\partial^2 p}{\partial V^2}\right)_{T=一定} = \frac{2nRT}{(V-nb)^3} - \frac{6an^2}{V^4} = 0.$$

From the above two equations we have $V_c = 3nb$ and $T_c = \frac{8a}{27bR}$. Substituting these relations into equation (14.36), we have $p_c = \frac{a}{27b^2}$.

Chapter 15

Questions

1. $TV^{0.4} = \text{constant}$. $\quad\therefore\; TV^{0.4} = T_0 V_0^{\,0.4}$.
$\frac{V}{V_0} = \left(\frac{T_0}{T}\right)^{2.5} = \left(\frac{283\ \text{K}}{373\ \text{K}}\right)^{2.5} = 0.50 \quad \therefore\; 50\ \%$

2. Because in constant pressure change, the volume expands as the temperature of the gas rises, and the gas performs work to the outside.

3. Since $\frac{W}{Q_H} > \frac{T_H - T_L}{T_H} = \frac{30\ \text{K}}{298\ \text{K}} = 0.1$,
$W > (1\ \text{J}) \times 0.1 = 0.10\ \text{J}$.

Problems for exercise 15

A

1. (1) $mgh = mC\,\Delta T$, $C = 1.0\ \text{cal/(g·°C)} = 4.2\ \text{J/(g·°C)} = 4.2\times10^3\ \text{J/(kg·°C)}$, $\Delta T = \frac{gh}{C} = \frac{(9.8\ \text{m/s}^2)\times(50\ \text{m})}{4.2\times10^3\ \text{J/(kg·°C)}} = 0.12\ \text{°C}$

(2) $P = 0.20\times\frac{mgh}{t}$

$$= \frac{0.20\times(4\times10^5\times10^3\ \text{kg})\times(9.8\ \text{m/s}^2)\times(50\ \text{m})}{60\ \text{s}}$$

$= 6.5\times10^8\ \text{W}$

2. (1) B→C, D→A (2) A→B, C→D
(3) C→D, D→A (4) A→B, C→D
(5) $p_2(V_B - V_A)$, $C_p(T_2 - T_1)$

3. $TV^{0.4} = T_0 V_0^{\,0.4}$. $T = T_0\left(\frac{V_0}{V}\right)^{0.4} = (300\ \text{K}) \times 20^{0.4} = 994\ \text{K} = 721\ \text{°C}$

4. $T = T_0\left(\frac{V_0}{V}\right)^{0.4} = (293\ \text{K})\times\left(\frac{1}{2}\right)^{0.4} = 222\ \text{K} = -51\ \text{°C}$

5. $\eta = \frac{(673\ \text{K}) - (323\ \text{K})}{673\ \text{K}} = 0.52.\ 52\ \%$

B

1. While ice melts, the temperature remains constant at $T = 273$ K, and the amount of heat absorbed (heat of fusion) is $Q = (80\ \text{cal/g})\times(10^3\ \text{g})\times(4.2\ \text{J/cal}) = 3.4\times10^5\ \text{J}$. Therefore,

$$S_B - S_A = \int_A^B \frac{dQ}{T} = \frac{1}{T}\int_A^B dQ = \frac{Q}{T} = \frac{3.4\times10^5\ \text{J}}{273\ \text{K}}$$
$$= 1.2\times10^3\ \text{J/K}$$

2. In order that a state is transferred from the initial state to the final state in a reversible change, it is sufficient that the amount of heat Q is reversibly given from the heat reservoir at temperature T. Hence, the entropy increases by $\frac{Q}{T}$.

3. (1) $dQ = nC_V\,dT$

$$\therefore\; S_B - S_A = \int_A^B \frac{dQ}{T} = nC_V\int_{T_A}^{T_B}\frac{dT}{T} = nC_V \log\frac{T_B}{T_A}$$

(2) In the isochoric change from the reference state $A(p_A, V_A, T_A)$ to the state $B(p_B, V_A, T)$, $\Delta S = nC_V \log\frac{T}{T_A}$. In the isothermal change to the state (p, V, T), from equation (15.48) $\Delta S = nR\log\frac{V}{V_A}$.

$$\therefore\; \Delta S = nC_V\log T + nR\log V + \text{constant}$$
$$= nC_p\log T - nR\log p + \text{constant}.$$

4. Use $\Delta H = \Delta U + p\,\Delta V + V\,\Delta p = \Delta Q + V\,\Delta p$.

5. $p = (\text{constant})\times V^{-\gamma}$, so $\frac{dp}{dV} = -\frac{\gamma p}{V}$.

$$c = \sqrt{\frac{\gamma p}{nM/V}} = \sqrt{\frac{\gamma pV}{nM}} = \sqrt{\frac{\gamma RT}{M}}.$$

In the isothermal change, $\frac{dp}{dV} = -\frac{p}{V}$.

6. (a) If the Thomson statement does not hold and whole amount of heat from the high-temperature heat reservoir can be converted to work, it becomes possible to operate the refrigerator using this work, and the heat is transferred from the low-temperature heat reservoir to the high-temperature heat reservoir without causing changes elsewhere. Therefore, the Clausius statement does not hold.

(b) If the Clausius statement does not hold and whole amount of heat can be transferred from the low-temperature heat reservoir to the high temperature heat reservoir without causing changes elsewhere, the heat released from the heat engine to the low-temperature heat reservoir can be returned to the high-temperature heat reservoir. Thus, the whole amount of heat from the high-temperature heat reservoir can be transformed into work, and the Thomson statement does not hold.

Therefore, if the Thomson statement holds, the Clausius statement also holds, and if the Clausius statement holds, the Thomson statement also holds (the contraposition is true).

7. $\Delta Q = T\,\Delta S$. $\eta = 1 - \frac{Q_L}{Q_H}$

$$= 1 - \frac{(300\ \text{K})\times(150\ \text{J/K})}{(700\ \text{K})\times(100\ \text{J/K}) + (400\ \text{K})\times(50\ \text{J/K})} =$$

$0.5. \quad \therefore\; 50\ \%$.

8. If the work done to the outside is not 0, the isothermal circulation process in one direction of the reversible process becomes a circulation process that receives heat from one heat reservoir and performs a positive work to the outside. This violates the second

law of thermodynamics (Thomson statement). When the substance undergoes the isothermal reversible circulation process of A → B → C → D → E → C → A at temperature T, the work performed by the system to the outside when the system performs an isothermal reversible circulation process is zero, thus

$$\int_{V_l}^{V_c} p\,dV + \int_{V_c}^{V_g} p\,dV + \int_{V_g}^{V_c} p_0\,dV + \int_{V_c}^{V_l} p_0\,dV = 0$$

$$\therefore \quad \int_{V_c}^{V_g} (p-p_0)\,dV = \int_{V_l}^{V_c} (p_0-p)\,dV$$

Chapter 16

Questions

1. Let the force from the upper right charge $-q$ be $\boldsymbol{F}_1$, and the force from the lower left charge $-q$ be $\boldsymbol{F}_2$, and the force from the lower right charge q be $\boldsymbol{F}_3$ (Fig. S-4). Since the length of the diagonal is $\sqrt{2}L$, the magnitudes of the three forces are

$$F_1 = F_2 = \frac{q^2}{4\pi\varepsilon_0 L^2}$$

and

$$F_3 = \frac{q^2}{4\pi\varepsilon_0(\sqrt{2}L)^2} = \frac{q^2}{8\pi\varepsilon_0 L^2}.$$

The direction of the resultant force $\boldsymbol{F}$ of these three forces is towards the center of the square. Since the components of force $\boldsymbol{F}_1$ and $\boldsymbol{F}_2$ in the direction of resultant force are respectively $F_1 \cos 45° = F_2 \cos 45° = \frac{F_1}{\sqrt{2}} = \frac{q^2}{4\pi\varepsilon_0\sqrt{2}L^2}$, the magnitude of the resulting force $\boldsymbol{F}$ is

$$F = F_1 \cos 45° + F_2 \cos 45° - F_3 = \frac{(2\sqrt{2}-1)q^2}{8\pi\varepsilon_0 L^2}.$$

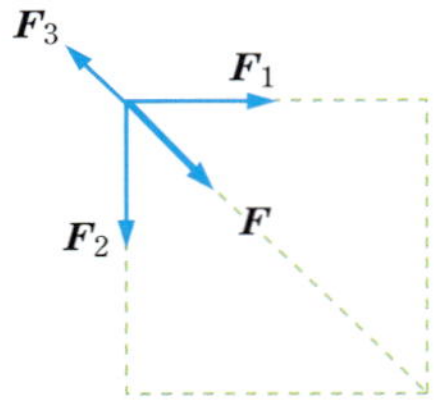

Fig. S-4

2. omitted.
3. omitted.
4. The direction of the electric field is leftward, and the electric field at point a is stronger.
5. The electric field at point P where the equipotential lines are closely spaced is stronger.

Problems for exercise 16

A

1. Similarity : The strength is inversely proportional to r^2. The strength is proportional to the product of the quantities (mass or charge) representing the strength of the force source of two objects.

Difference : The mass is always positive, and the attractive force acts between positive masses. On the other hand, there are positive and negative charges, and the repulsive force acts between charges of the same sign. Since attractive forces act between the masses, objects with large masses can exist. Repulsive forces act between charges of the same sign, and attractive forces act between charges of the opposite sign, so there is no object with a large charge.

2. $\dfrac{4}{x^2} = \dfrac{10}{(2\,\mathrm{m}-x)^2}$ $\qquad \sqrt{10}\,x = 2(2\,\mathrm{m}-x)$

$\therefore \quad x = \dfrac{4\,\mathrm{m}}{\sqrt{10}+2} = 0.77\,\mathrm{m}$

3. (1) $E = \dfrac{F}{Q} = \dfrac{6.0\times10^{-4}\,\mathrm{N}}{3.0\times10^{-6}\,\mathrm{C}} = 2.0\times10^2\,\mathrm{N/C}$

(2) $F = QE = (-6.0\times10^{-6}\,\mathrm{C})\times(2.0\times10^2\,\mathrm{N/C}) = -1.2\times10^{-3}\,\mathrm{N}$. The direction of the force is opposite to the initial force.

4. (1) The point at which the electric field $E = 0$ lies in the range $0 < x < 9.0\,\mathrm{cm}$ on the x-axis. $\dfrac{4.0}{x^2} = \dfrac{1.0}{(9.0\,\mathrm{cm}-x)^2}$. $\quad \therefore\ 2.0\times(9.0\,\mathrm{cm}-x) = 1.0x$.

From $18\,\mathrm{cm} = 3.0x$, $x = 6.0\,\mathrm{cm}$.

(2) When $x = 15\,\mathrm{cm}$,

$$E_x = (9.0\times10^9\,\mathrm{N\cdot m^2/C^2})\times\left(\frac{4.0\times10^{-6}\,\mathrm{C}}{(0.15\,\mathrm{m})^2} + \frac{1.0\times10^{-6}\,\mathrm{C}}{[(0.15\,\mathrm{m})-(0.09\,\mathrm{m})]^2}\right)$$

$$= 9.0\times10^9\times4.6\times10^{-4}\,\mathrm{N/C} = 4.1\times10^6\,\mathrm{N/C}$$

$$\boldsymbol{E} = (4.1\times10^6\,\mathrm{N/C}, 0, 0).$$

5. $eE = mg$, $E = \dfrac{(9.1\times10^{-31}\,\mathrm{kg})\times(9.8\,\mathrm{m/s^2})}{1.6\times10^{-19}\,\mathrm{C}} = 5.6\times10^{-11}\,\mathrm{N/C}$

$ma = eE$, $a = \dfrac{(1.6\times10^{-19}\,\mathrm{C})\times(10000\,\mathrm{N/C})}{9.1\times10^{-31}\,\mathrm{kg}} = 1.8\times10^{15}\,\mathrm{m/s^2}$

6. Since the universal gravity also follows Gauss' law, the universal gravity is zero inside the spherical shell [see Case 4 (1)]. Therefore, it becomes a weightless state like in a spacecraft.
7. This is the case of $\lambda = 0$ in equation (16.27).
8. Outside the parallel plates $\dfrac{\sigma}{\varepsilon_0}$, and between the parallel plates 0 (Fig. S-5).

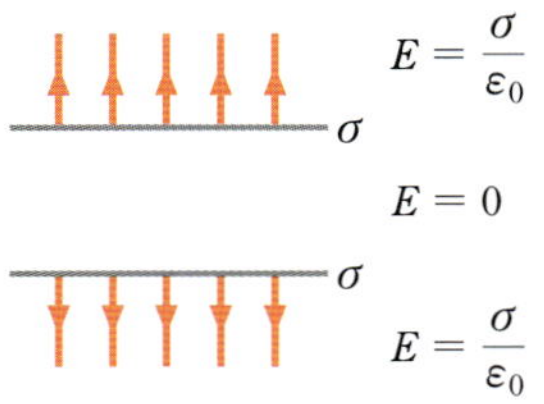

Fig. S-5

The upward electric field $E = \frac{\sigma}{2\varepsilon_0}$ created by the charge of the lower surface acts on the charge σ on the unit area of the upper surface with the upward repulsive force $\frac{\sigma^2}{2\varepsilon_0}$.

9. Using Gauss' law, $E_{1n}A_1 + E_{2n}A_2 = 0$.

10. (1) Directions of the electric field are all facing to the left, and directions of the electric force are all facing to the right.
(2) Since the intensity of the electric field is proportional to the density of the equipotential lines, it weakens in the order of (b), (a), and (c).

11. (a) $V_A = \frac{Q}{2\pi\varepsilon_0 d} - \frac{Q}{8\pi\varepsilon_0 d} = \frac{3Q}{8\pi\varepsilon_0 d}$,

$V_B = \frac{Q}{4\pi\varepsilon_0 d} - \frac{Q}{4\pi\varepsilon_0 d} = 0, \quad \therefore \quad V_A - V_B = \frac{3Q}{8\pi\varepsilon_0 d}$

(b) $V_A = \frac{Q}{8\pi\varepsilon_0 d}$, $V_B = \frac{Q}{4\pi\varepsilon_0 d}$,

$\therefore \quad V_A - V_B = -\frac{Q}{8\pi\varepsilon_0 d}$

12. (1) $E = \frac{V}{d} = \frac{24\ \text{V}}{0.08\ \text{m}} = 300\ \text{V/m} = 300\ \text{N/C}$
(2) $F = QE = (3\times10^{-6}\ \text{C})\times(300\ \text{N/C})$
$= 9\times10^{-4}\ \text{N}$ Downward
(3) $W = FL = (9\times10^{-4}\ \text{N})\times(0.04\ \text{m})$
$= 3.6\times10^{-5}\ \text{N}\cdot\text{m} = 3.6\times10^{-5}\ \text{J}$
(4) $V_B - V_A = EL = (300\ \text{V/m})\times(0.04\ \text{m})$
$= 12\ \text{V}$

B

1. Since the central angle of the semicircle is π (rad), the charge ΔQ in the portion of central angle $\Delta\theta$ (rad) of Fig. 7 is $\Delta Q = (10\ \mu\text{C})\times\frac{\Delta\theta}{\pi}$. The electric field strength at the center O due to this charge is $\Delta E = \frac{\Delta Q}{4\pi\varepsilon_0 r^2}$. The direction of the electric field $\boldsymbol{E}$ at the center O is, due to the symmetry, facing to the direction of the arrow pointing to the right, so the strength is

$$E = \Sigma\Delta E\cos\theta = \Sigma\frac{\cos\theta}{4\pi\varepsilon_0 r^2}\Delta Q$$
$$= \frac{1}{4\pi\varepsilon_0}\frac{10\ \mu\text{C}}{\pi r^2}\int_{-\pi/2}^{\pi/2} d\theta\cos\theta$$
$$= \frac{1}{4\pi\varepsilon_0}\times\frac{2\times10\ \mu\text{C}}{\pi r^2}$$
$$= \frac{(9.0\times10^9\ \text{N}\cdot\text{m}^2/\text{C}^2)\times2\times(10\times10^{-6}\ \text{C})}{\pi(0.1\ \text{m})^2}$$
$$= 5.7\times10^6\ \text{N/C}$$

2. (1) Since the charge $q(r)$ inside the spherical surface of radius r is $q(r) = \frac{4\pi}{3}\rho r^3$, $E(r) = \frac{q(r)}{4\pi\varepsilon_0 r^2} = \frac{\rho r}{3\varepsilon_0}$. $\boldsymbol{E}(\boldsymbol{r}) = E(r)\frac{\boldsymbol{r}}{r} = \frac{\rho\boldsymbol{r}}{3\varepsilon_0}$

(2) Since the electric force acting on the negative charge $-q$ is the force $\boldsymbol{F} = -q\boldsymbol{E} = -\frac{q\rho\boldsymbol{r}}{3\varepsilon_0}$, facing to the center of the sphere, the center of the sphere is the stable balanced point.
(3) If the mass point is displaced from the balanced point, the restoring force $-\frac{q\rho\boldsymbol{r}}{3\varepsilon_0}$ proportional to the displacement $\boldsymbol{r}$ works. The motion is the simple harmonic oscillation with frequency $\omega = \left(\frac{q\rho}{3\varepsilon_0 m}\right)^{1/2}$.

3. (1) The electric field inside a sphere of radius R uniformly charged with charge density ρ is, if we assume the position vector from the center is $\boldsymbol{r}$, according to the result of the above Problem 2 (1), $\boldsymbol{E} = \frac{\rho\boldsymbol{r}}{3\varepsilon_0}$. Let the position vectors from the center of positive and negative charges be $\boldsymbol{r}'$ and $\boldsymbol{r}$, respectively. Then, the electric field in the sphere becomes uniform or $\frac{\rho}{3\varepsilon_0}(\boldsymbol{r}'-\boldsymbol{r}) = -\frac{\rho}{3\varepsilon_0}\boldsymbol{\delta}$ where $\boldsymbol{\delta} = \boldsymbol{r}'-\boldsymbol{r}$ is the vector showing deviation.
(2) Since the thickness of the surface charge is $\delta\cos\theta$, the magnitude of the surface charge density is $\rho\delta\cos\theta$.

4. Let the electric fields created by the charges of the two plates be $\boldsymbol{E}_1$ and $\boldsymbol{E}_2$, then $\boldsymbol{E} = \boldsymbol{E}_1 + \boldsymbol{E}_2$. $\boldsymbol{E}_1$ and $\boldsymbol{E}_2$ are perpendicular to the plates, and $E_1 = \frac{\rho}{\varepsilon_0}$ and $E_2 = \frac{\rho}{2\varepsilon_0}$. As for the direction, see Fig. S-6.

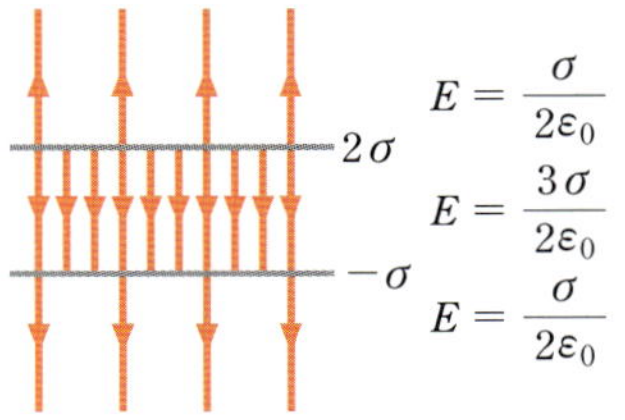

Fig. S-6

5. The universal gravitation is inversely proportional to the square of the distance r as is the Coulomb force, so Gauss' law holds, and the result of Case 4 (2) can be used. That is, the universal gravitation exerted by the object A is the same as when mass m_A is concentrated at the center of A. The law of action and reaction says that the magnitude of the universal gravitation of the object B received by the object A is equal to the magnitude of the universal gravitation which the object B exerts on the mass m_A at the center of the object A. This is, according to Gauss' law, equal to the universal gravitation exerted on the mass m_A at the center of A when the mass m_B is at the

center of B. $\therefore \quad F = \dfrac{Gm_A m_B}{r^2}$.

6. $E_x = -\dfrac{\partial V(r)}{\partial x} = -\dfrac{\partial r}{\partial x}\dfrac{dV(r)}{dr}$

$= -\dfrac{x}{r}\left(-\dfrac{q}{4\pi\varepsilon_0 r^2}\right) = \dfrac{q}{4\pi\varepsilon_0 r^2}\dfrac{x}{r}.$

The following is omitted.

7. The electric potential at point (x, y, z) is, from equation (16.44)

$V(x, y, z)$

$= \dfrac{1}{4\pi\varepsilon_0}\left(\dfrac{q}{\sqrt{(x-a)^2+y^2+z^2}} - \dfrac{q}{\sqrt{(x+a)^2+y^2+z^2}}\right)$

$\dfrac{1}{\sqrt{(x\mp a)^2+y^2+z^2}} \approx \dfrac{1}{\sqrt{x^2+y^2+z^2\mp 2ax}}$

$= \dfrac{1}{\sqrt{x^2+y^2+z^2}}\left(1\mp\dfrac{2ax}{x^2+y^2+z^2}\right)^{-1/2}$

$\approx \dfrac{1}{\sqrt{x^2+y^2+z^2}}\left(1\pm\dfrac{ax}{x^2+y^2+z^2}\right)$

Approximation is made in the above calculation. Then,

$V(x, y, z) = \dfrac{px}{4\pi\varepsilon_0(x^2+y^2+z^2)^{3/2}} = \dfrac{p\cos\theta}{4\pi\varepsilon_0 r^2}$

$(p = 2aq)$

$E_x(x, y, 0) = \dfrac{p}{4\pi\varepsilon_0}\left\{\dfrac{3x^2}{(x^2+y^2)^{5/2}} - \dfrac{1}{(x^2+y^2)^{3/2}}\right\}$

$= \dfrac{p}{4\pi\varepsilon_0 r^3}(3\cos^2\theta - 1)$

$E_y(x, y, 0) = \dfrac{3pxy}{4\pi\varepsilon_0(x^2+y^2)^{5/2}}$

$= \dfrac{3p}{4\pi\varepsilon_0 r^3}\sin\theta\cos\theta$

$E_z(x, y, 0) = 0$

Chapter 17

Questions

1. omitted

Problems for exercise 17

A

1. Since $\boldsymbol{E} = \boldsymbol{0}$ inside the conductor, if we consider the closed surface S surrounding the cavity inside the conductor and use Gauss' law of the electric field, the total electric quantity inside S is 0 (Fig. S-7). Use this fact and the charge conservation law.

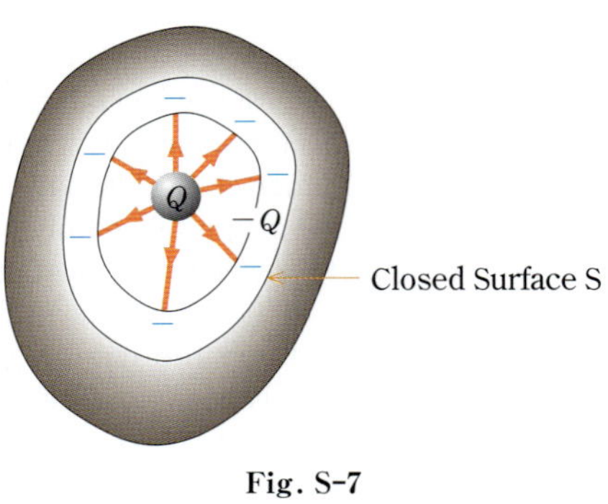

Fig. S-7

2. Inside the empty can, the electric field is 0. Outside the can, an attractive force acts on the cork ball with electrostatically induced charges of opposite sign.
3. (1) Electric potential energy of the charge on the outer surface of a metal spherical shell.
(2) Because the electric potential energy is lower when the charge is outside the metal sphere shell than when the charge is inside the metal sphere shell. There is no electric field inside the metal spherical shell.
(3) Because the hair is charged and the electric repulsion works,
4. Since the electric field [Fig. 3 (b)] when the plane equidistant from the two charges Q and $-Q$ in Fig. 3 (a) is replaced with the conductor plane is the same as the right half of Fig. 3 (a), the electric force acting on the charge Q in Fig. 3 (b) has the same magnitude as the electric force acting on the charge Q on the right side of Fig. 3 (a).
5. $C = \dfrac{\varepsilon_0 A}{d} = \dfrac{(8.85\times10^{-12}\ \mathrm{F/m})\times(0.05\ \mathrm{m})^2}{10^{-3}\ \mathrm{m}}$
$= 2.2\times10^{-11}\ \mathrm{F}$
6. This is a parallel connection of three parallel plate capacitors of electric capacity $\dfrac{\varepsilon_0 A}{d}$. Thus, $\dfrac{3\varepsilon_0 A}{d}$.
7. Total electric capacity $C = 15\ \mu\mathrm{F}$.
$V_c = \dfrac{Q_c}{C_c} = \dfrac{CV}{C_c} = \dfrac{(15\times10^{-6}\ \mathrm{F})\times(10\ \mathrm{V})}{20\times10^{-6}\ \mathrm{F}} = 7.5\ \mathrm{V}$
8. $\dfrac{1}{2}CV^2 = 0.5\times(20\times10^{-6}\ \mathrm{F})\times(200\ \mathrm{V})^2 = 0.4\ \mathrm{J}$

B

1. Prepare two capacitors connected in series, and connect this with five capacitors in parallel.
2. 1 μF (the total electric capacity of the rightmost three capacitors is 2 μF.)
3. This is a parallel connection of a spherical capacitor and an isolated spherical conductor with radius b.
$$C = 4\pi\varepsilon_0 b + \frac{4\pi\varepsilon_0 ab}{b-a} = \frac{4\pi\varepsilon_0 b^2}{b-a}$$
4. $U = \dfrac{1}{2}\varepsilon_0 E^2\times\text{volume} = \dfrac{1}{2}\times(8.85\times10^{-12}\ \mathrm{F/m})\times(10^6\ \mathrm{V/m})^2\times(1\ \mathrm{m}^3) = 4.4\ \mathrm{J}$
5. (1) 50 V (2) $2\times\dfrac{1}{2}CV^2 = (100\times10^{-6}\ \mathrm{F})\times(50\ \mathrm{V})^2 = 0.25\ \mathrm{J}$
(3) The energy the capacitor A initially has is 0.5 J. The difference of 0.25 J is the heat generated in the wire.
6. Since the electric field in the conductor is

$\boldsymbol{E}_1+\boldsymbol{E}_2=\boldsymbol{0}$, $E_2=\dfrac{\sigma}{2\varepsilon_0}$.

Chapter 18

Questions

1. The attraction between the charged object and the nearby charge generated by dielectric polarization is stronger than the repulsion between the charged object and the distant charge. Accordingly, the paper is attracted to the charged object.

Problems for exercise 18

A

1. $Q=CV=(10^{-6}\,\text{F})\times(100\,\text{V})=10^{-4}\,\text{C}$
2. The charges on the electrode plates are $C_1V_1=(C_1+C_2)V_2$, $C_1=\varepsilon_r C_2$ $\quad\therefore\quad \varepsilon_r=\dfrac{V_2}{V_1-V_2}$
3. $C=\dfrac{\varepsilon_r\varepsilon_0 A}{d}=\dfrac{3.5\times(8.85\times10^{-12}\,\text{F/m})\times(1\,\text{m}^2)}{0.0001\,\text{m}}=3.1\times10^{-7}\,\text{F}=0.31\,\mu\text{F}$
4. (1) $C=\dfrac{\varepsilon_r\varepsilon_0 A}{d}=\dfrac{8\times(9\times10^{-12}\,\text{F/m})\times(10^{-4}\,\text{m}^2)}{10^{-8}\,\text{m}}=7\times10^{-7}\,\text{F}=0.7\,\mu\text{F}$

(2) $U=\dfrac{1}{2}CV^2=0.5\times(7\times10^{-7}\,\text{F})\times(0.1\,\text{V})^2=4\times10^{-9}\,\text{J}$

(3) $E=\dfrac{V}{d}=\dfrac{0.1\,\text{V}}{10^{-8}\,\text{m}}=10^7\,\text{V/m}$. $\sigma=\varepsilon_0 E=(10^{-11}\,\text{F/m})\times(10^7\,\text{V/m})=10^{-4}\,\text{C/m}^2$.
$Q=\sigma A=(10^{-4}\,\text{C/m}^2)\times(10^{-4}\,\text{m}^2)=10^{-8}\,\text{C}$

B

1. This is a series connection of the capacitors with electric capacities $\dfrac{\varepsilon_1\varepsilon_0 A}{d_1}$ and $\dfrac{\varepsilon_2\varepsilon_0 A}{d_2}$. Therefore,

$$C=\frac{\varepsilon_0 A}{\dfrac{d_1}{\varepsilon_1}+\dfrac{d_2}{\varepsilon_2}}$$

2. $D=\varepsilon_1\varepsilon_0 E=\sigma_{\text{free}}$. Thus, $E=\dfrac{\sigma_{\text{free}}}{\varepsilon_1\varepsilon_0}$.

Chapter 19

Questions

1. omitted.
2. The combined resistance can be derived in the same way as connecting two resistors.
3. Three resistors of resistance $2R$ are connected in parallel. The current flowing the circuit is the same as when the combined resistance $\dfrac{2}{3}R$ is connected to the battery. Thus, $I=\dfrac{3V}{2R}$.
4. $I_1=\dfrac{(R_2+R_3)V_1-R_3V_2}{R_1R_2+R_2R_3+R_3R_1}$, $I_2=\dfrac{(R_1+R_3)V_2-R_3V_1}{R_1R_2+R_2R_3+R_3R_1}$, $I_3=\dfrac{R_2V_1+R_1V_2}{R_1R_2+R_2R_3+R_3R_1}$
5. As $P=RI^2$, when the resistance value R is the same, the power consumption is proportional to the square of the current. Since the ratio of the magnitudes of the currents flowing through the resistors A, B and C is 1 : 1 : 2, the power consumed by A is $\dfrac{1}{6}$.
6. Since $P=\dfrac{V^2}{R}$, the heat generating rate is quadrupled. $\quad\therefore\quad$ 4 W.
7. $CR=100\,\text{pF}\times10\,\text{k}\Omega=10^{-10}\,\text{F}\times10^4\,\Omega=10^{-6}\,\text{s}$

Problems for exercise 19

A

1. $R=\dfrac{\rho L}{A}=\dfrac{(1.68\times10^{-8}\,\Omega\cdot\text{m})\times(10\,\text{m})}{2.0\times10^{-6}\,\text{m}^2}=8.4\times10^{-2}\,\Omega$
2. $R=\dfrac{(3\times10^{-5}\,\Omega\cdot\text{m})\times(0.25\,\text{m})}{(0.01\,\text{m})^2}=8\times10^{-2}\,\Omega$
3. Free electrons transmit both electricity and heat.
4. Since the electrical resistance between A and C is proportional to the wire length $\overline{\text{AC}}$, the potential difference between A and C is also proportional to the length $\overline{\text{AC}}$.
5. Between AB : parallel connection of 100 Ω and 300 Ω, so 75 Ω.
Between AC : parallel connection of 200 Ω and 200 Ω, so 100 Ω.
6. Since the resistance of the circuit is 2 kΩ+10 kΩ+8 kΩ = 20 kΩ, the current flowing through the battery is 10 V÷20 kΩ = 0.5 mA. $\quad\therefore\quad I=0.25$ mA.
7. The combined resistance of the rightmost four resistors is 10 Ω. The combined resistance of the seven resistors from the right end is also 10 Ω. Thus, the combined resistance of the circuit is 28 Ω.
8. The resistance of the tungsten filament of the bulb increases with the rise in temperature.
9. The electrical resistances of the bulb and the heater are 100 Ω and 20 Ω, respectively. The combined resistance is 16.67 Ω, so the current is $\dfrac{100\,\text{V}}{(16.67\,\Omega)+(0.10\,\Omega)}=6.0\,\text{A}$, and the voltage drop is $RI=(0.10\,\Omega)\times(6.0\,\text{A})=0.6\,\text{V}$.
10. Since $P=\dfrac{V^2}{R}$, R is larger for the bulb of smaller P, namely 60 W bulb. The thicker wire has the smaller resistance, so the wire of 100 W bulb is thicker.

11. a)
12. The current does not flow through the galvanometer. The current I_1 flows through R_1 and R while the current I_2 flows through R_2 and R_3. In addition, the points A and B are equipotential. $\therefore$ $R_1 I_1 = R_2 I_2$, $RI_1 = R_3 I_2$, $\therefore$ $\frac{R_1}{R_2} = \frac{R}{R_3}$
13. (1) $P = VI = (100\ \mathrm{V})\times(8\ \mathrm{A}) = 800\ \mathrm{W}$
(2) $\frac{(500\ \mathrm{g})\times(2600\ \mathrm{J/g})}{800\ \mathrm{W}} = 1.6\times10^3\ \mathrm{s} = 27\ \mathrm{min}$

B

1. $C = \frac{\varepsilon_r\varepsilon_0 A}{d} = \frac{6\times(8.9\times10^{-12}\ \mathrm{F/m})\times(10^{-2}\ \mathrm{m^2})}{10^{-4}\ \mathrm{m}}$
$= 5.3\times10^{-9}\ \mathrm{F}$.
$R = \frac{d}{\sigma A} = \frac{10^{-4}\ \mathrm{m}}{(6\times10^{-15}\ \Omega^{-1}\cdot\mathrm{m}^{-1})\times(10^{-2}\ \mathrm{m^2})} = 1.7\times10^{12}\ \Omega$. $CR = 9\times10^3\ \mathrm{s}$. $Q(t) = Q_0\,\mathrm{e}^{-t/CR}$.
2. Since the magnitude of acceleration is $\frac{eE}{m}$, the magnitude of mean displacement d is the expected value of $\frac{eEt^2}{2m}$,
$$d = \int_0^\infty \frac{eE}{2m} t^2 \mathrm{e}^{-t/\tau}\frac{1}{\tau}\,\mathrm{d}t = \frac{eE\tau^2}{m}$$
In addition, the drift velocity $\boldsymbol{v}$ is "mean displacement" divided by "mean time",
$$\boldsymbol{v} = -\frac{e\boldsymbol{E}\tau^2}{m\tau} = -\frac{e\boldsymbol{E}\tau}{m}$$
3. (1) $i = \frac{E}{\rho} = \frac{100\ \mathrm{V/m}}{3\times10^{13}\ \Omega\cdot\mathrm{m}} = 3\times10^{-12}\ \mathrm{A/m^2}$
$I = 4\pi(6.4\times10^6\ \mathrm{m})^2\times(3\times10^{-12}\ \mathrm{A/m^2}) = 1500\ \mathrm{A}$
(2) $\sigma = \varepsilon_0 E = -(9\times10^{-12}\ \mathrm{F/m})\times(100\ \mathrm{V/m}) \approx -10^{-9}\ \mathrm{C/m^2}$. $Q = 4\pi(6.4\times10^6\ \mathrm{m})^2\times(-10^{-9}\ \mathrm{C/m^2}) = -5\times10^5\ \mathrm{C}$

Chapter 20

Questions

1. The ratio of the strength of the magnetic field at A, B, C, and D is 1 : 2 : 0 : 1.
2. $B = \frac{\mu_0 I}{4\pi R^2}\sum\Delta s = \frac{\mu_0 I}{4\pi R^2}\times(R\theta) = \frac{\mu_0 I\theta}{4\pi R}$. The direction is from the frontside of the paper plane to the backside. The current flowing in the straight-line part does not contribute to the magnetic field at the point P.
3. ⑤
4. Electric force eE = magnetic force evB. Thus we get $v = \frac{E}{B}$.
5. From the first equation of equation (20.24) $d = \frac{2mv}{qB}$. When q becomes doubled, the particle collides with the wall at the point $\frac{d}{2}$.
6. For the paramagnetic body an N-pole (S-pole) is induced in the S-pole (N-pole) side of the magnet. Since the magnetic force from the S-pole with a sharp end of the magnet is stronger than that from the N-pole, the test body swings to the right. In the case of the diamagnetic body, the S-pole induced in the side of the S-pole with a sharp end of the magnet receives a stronger repulsive force, and so the test body swings to the left.

Problems for exercise 20

A

1. The magnetic field at the point A with a high density of the lines of the magnetic force is stronger than that at the point B. The magnetic compass needle is attracted to the direction of the S-pole (upward).
2. South pole. From east to west on the place including the equator of the Earth.
3. $B = \frac{\mu_0 I}{2\pi d} = \frac{(2\times10^{-7}\ \mathrm{T\cdot m/A})\times(10\ \mathrm{A})}{10^{-2}\ \mathrm{m}} = 2\times10^{-4}\ \mathrm{T}$
4. $B = \frac{\mu_0 I_1}{2\pi d} + \frac{\mu_0 I_2}{2\pi d}$
$= \frac{(4\pi\times10^{-7}\ \mathrm{T\cdot m/A})\times[(4+6)\mathrm{A}]}{2\pi\times(0.05\ \mathrm{m})} = 4\times10^{-5}\ \mathrm{T}$
5. Let the angle between the Earth magnetic field and the velocity of the electron be θ.
$F = qvB\sin\theta = (1.6\times10^{-19}\ \mathrm{C})\times(3\times10^7\ \mathrm{m/s})\times(4.5\times10^{-5}\ \mathrm{T})\sin\theta = 2\times10^{-16}\sin\theta\ \mathrm{N}$.
The gravity is $mg = (9.1\times10^{-31}\ \mathrm{kg})\times(9.8\ \mathrm{m/s^2}) = 10^{-29}\ \mathrm{N}$, and hence if $\theta \neq 0$ it can be ignored compared to the magnetic force. From the equation of motion $\frac{m(v\sin\theta)^2}{r} = q(v\sin\theta)B$, we get
$r = \frac{mv\sin\theta}{eB} = \frac{(9.1\times10^{-31}\ \mathrm{kg})(3\times10^7\ \mathrm{m/s})\sin\theta}{(1.6\times10^{-19}\ \mathrm{C})\times(4.5\times10^{-5}\ \mathrm{T})}$
$= (3.8\sin\theta)\ \mathrm{m}$
6. (1) A → B (Since the electron reduces the speed, the radius of curvature decreases.)
(2) Backside → frontside.
7. See Fig. S-8 (The magnetic pole receives the force from the other magnetic pole as well as the force from the current)

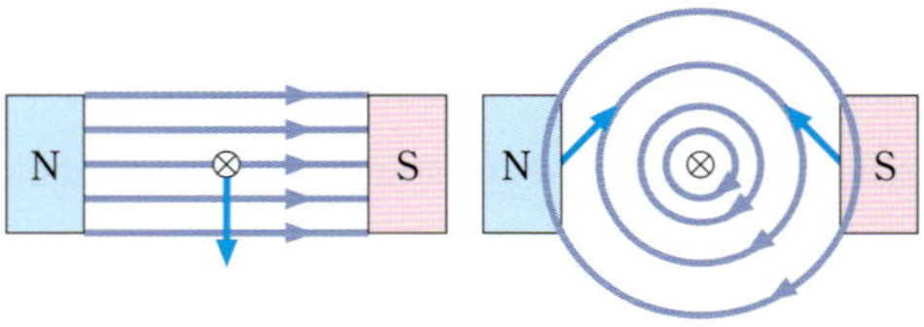

Fig. S-8

8. $F = ILB = (20\ \mathrm{A})\times(1\ \mathrm{m})\times(3\times10^{-5}\ \mathrm{T}) = 6\times10^{-4}\ \mathrm{N}$

9. $F = ILB\sin\theta = (10\ \mathrm{A})\times(2\ \mathrm{m})\times(4.61\times10^{-5}\ \mathrm{T})\times\sin 49.5^\circ = 7.0\times10^{-4}\ \mathrm{N}$

10. The two currents do not interact with each other.

11. $I = \dfrac{V}{R} = \dfrac{6\ \mathrm{V}}{100\ \Omega} = 0.06\ \mathrm{A}$. $B = \dfrac{\mu_0 I}{2r} = \dfrac{(2\pi\times10^{-7}\ \mathrm{T\cdot m/A})\times(0.06\ \mathrm{A})}{0.1\ \mathrm{m}} = 3.8\times10^{-7}\ \mathrm{T}$

12. $B = \mu_0 nI = (4\pi\times10^{-7}\ \mathrm{T\cdot m/A})\times\dfrac{1200}{0.3\ \mathrm{m}}\times(1\ \mathrm{A}) = 5.0\times10^{-3}\ \mathrm{T}$

13. $n = \dfrac{10^3\ \mathrm{m}}{(0.2\ \mathrm{m})\times(1\ \mathrm{m})} = 5\times10^3\ \mathrm{m^{-1}}$ $\quad B = \mu_0 nI = (4\pi\times10^{-7}\ \mathrm{T\cdot m/A})\times(5\times10^3\ \mathrm{m^{-1}})I = 0.1\ \mathrm{T}$,
$\therefore\ I = 16\ \mathrm{A}$

14. $H = nI = (4000\ \mathrm{m^{-1}})\times(1\ \mathrm{A}) = 4000\ \mathrm{A/m}$
$B = \mu_0\mu_r H = (4\pi\times10^{-7}\ \mathrm{T\cdot m/A})\times1000\times(4000\ \mathrm{A/m}) = 5.0\ \mathrm{T}$

B

1. (1) The speed v of the proton of 2 MeV is $v = \sqrt{\dfrac{2E}{m}} = \sqrt{\dfrac{2\times2\times10^6\times1.6\times10^{-19}\ \mathrm{J}}{1.67\times10^{-27}\ \mathrm{kg}}} = 2.0\times10^7\ \mathrm{m/s}$.
$r = \dfrac{mv}{eB} = \dfrac{(1.67\times10^{-27}\ \mathrm{kg})\times(2.0\times10^7\ \mathrm{m/s})}{(1.6\times10^{-19}\ \mathrm{C})\times(0.3\ \mathrm{T})} = 0.70\ \mathrm{m}$
(2) $f = \dfrac{eB}{2\pi m} = \dfrac{(1.6\times10^{-19}\ \mathrm{C})\times(0.3\ \mathrm{T})}{2\pi\times(1.67\times10^{-27}\ \mathrm{kg})} = 4.6\times10^6\ \mathrm{Hz}$

2. (a) 0 (b) $\dfrac{\mu_0 I}{4R}$, in the direction from the frontside of the paper plane to the backside. No contribution from the straight-line part.

3. The magnetic field $\Delta\boldsymbol{B}$ generated by the current element $I\,\Delta x$ in Fig. S-9 at the point P has the direction from the frontside of the paper plane to the backside and the magnitude is given by $\Delta B = \dfrac{\mu_0 I\,\Delta x\sin\theta}{4\pi r^2}$.
Since $x = -d\cot\theta$,
$\mathrm{d}x = \dfrac{d}{\sin^2\theta}\mathrm{d}\theta$.

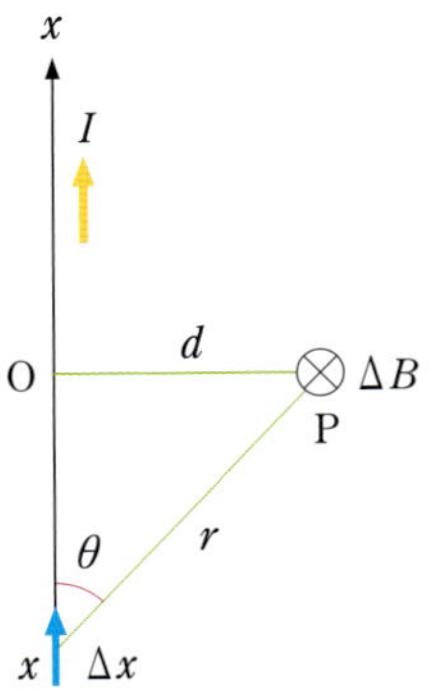

Fig. S-9

$$B = \int_{-\infty}^{\infty}\frac{\mu_0 I\sin\theta\,\mathrm{d}x}{4\pi r^2} = \int_0^{\pi}\frac{\mathrm{d}\theta\,\mu_0 Id}{4\pi r^2\sin\theta} = \frac{\mu_0 I}{4\pi d}\int_0^{\pi}\sin\theta\,\mathrm{d}\theta = -\frac{\mu_0 I\cos\theta}{4\pi d}\Big|_0^{\pi} = \frac{\mu_0 I}{2\pi d}$$

4. From the symmetry, the lines of the magnetic force of $\boldsymbol{B}$ are concentric circles with the central axis as the center. The direction of $\boldsymbol{B}$ follows the right-handed screw rule. The magnitude of $\boldsymbol{B}$ on the circumference with the distance r from the axis, $B(r)$, is obtained from Ampere's law with the current penetrating the circle being set to $I(r)$, and is given as
$B(r) = \dfrac{\mu_0 I(r)}{2\pi r}$.

$$B(r) = \begin{cases} \dfrac{\mu_0 Ir}{2\pi c^2} & (r \leqq c) \\ \dfrac{\mu_0 I}{2\pi r} & (c \leqq r \leqq b) \\ \dfrac{\mu_0 I(a^2-r^2)}{2\pi r(a^2-b^2)} & (b \leqq r \leqq a) \\ 0 & (a \leqq r) \end{cases}$$

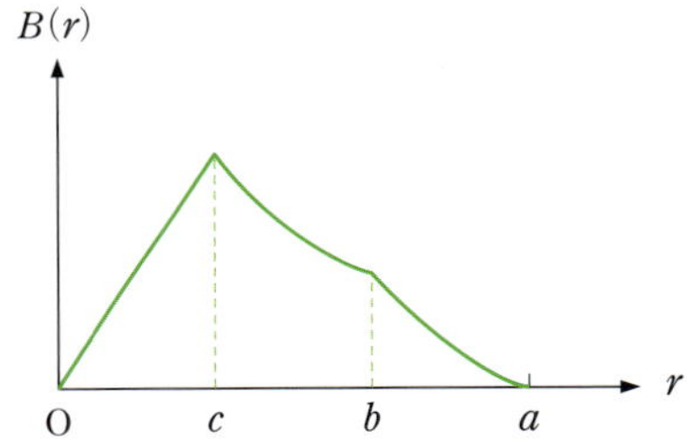

Fig. S-10

5. The magnetic field generated at the point P by the current I flowing in the part AB and that generated by the current I flowing in the part A′B′ are the same.

6. (1) Since the current is $I = evnA$, equation (1) is the equation obtained by replacing $\boldsymbol{v}$ appearing in Biot-Savart's law $I\,\mathrm{d}\boldsymbol{s} = (evnA)\,\mathrm{d}\boldsymbol{s} = (en)(A\,\mathrm{d}s)\boldsymbol{v}$ = "the charge inside the wire of length ds", by $q\boldsymbol{v}$.
(2) Since $\boldsymbol{v}_1\times\{\boldsymbol{v}_2\times(\boldsymbol{r}_1-\boldsymbol{r}_2)\}$
$= \boldsymbol{v}_2\{\boldsymbol{v}_1\cdot(\boldsymbol{r}_1-\boldsymbol{r}_2)\}-(\boldsymbol{v}_1\cdot\boldsymbol{v}_2)(\boldsymbol{r}_1-\boldsymbol{r}_2)$
$\boldsymbol{v}_2\times\{\boldsymbol{v}_1\times(\boldsymbol{r}_2-\boldsymbol{r}_1)\}$
$= \boldsymbol{v}_1\{\boldsymbol{v}_2\cdot(\boldsymbol{r}_2-\boldsymbol{r}_1)\}-(\boldsymbol{v}_1\cdot\boldsymbol{v}_2)(\boldsymbol{r}_2-\boldsymbol{r}_1)$,
because of the first term of each equation, $\boldsymbol{F}_{1\leftarrow2} \neq -\boldsymbol{F}_{2\leftarrow1}$.

Chapter 21

Questions

1. (1) Since the falling speed of the magnet increases with time, the changing rate of the magnetic flux increases.
(2) $\int_{-\infty}^{\infty} V_\mathrm{i}\,\mathrm{d}t$ = "area of crest" − "area of trough" $= -\Phi_\mathrm{B}(\infty)+\Phi_\mathrm{B}(-\infty) = 0$.
See the figure for $\boldsymbol{B}$ of Fig. 20-38.

2. The current flows through the ring A in the

opposite direction of the current in the middle ring, while the current flows through the ring B in the same direction.

3. See Fig. S-11.

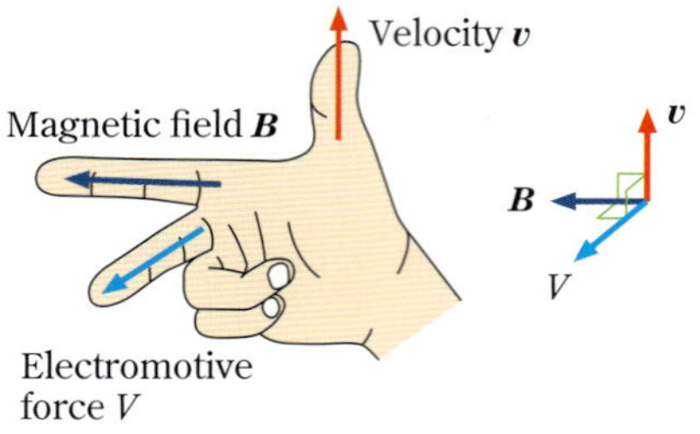

Fig. S-11

Problems for exercise 21

A

1. (1) $\Phi_B = BA\cos\theta = (0.30\ \mathrm{T})\times(0.25\ \mathrm{m}^2)\times 1 = 7.5\times10^{-2}\ \mathrm{Wb}$

(2) $V_i = \dfrac{\Delta\Phi_B}{\Delta t} = \dfrac{7.5\times10^{-2}\ \mathrm{Wb}}{0.01\ \mathrm{s}} = 7.5\ \mathrm{V}$

$\langle I\rangle = \dfrac{V_i}{R} = \dfrac{7.5\ \mathrm{V}}{20\ \Omega} = 0.38\ \mathrm{A}$

2. $\omega BA = (2\pi\times100\ \mathrm{s}^{-1})\times(0.01\ \mathrm{T})\times(25\times10^{-4}\ \mathrm{m}^2) = 1.6\times10^{-2}\ \mathrm{V}$

3. $V_0\sin\omega t = -NA\dfrac{\mathrm{d}B}{\mathrm{d}t}$, $\quad\therefore\quad B = \dfrac{V_0}{NA\omega}\cos\omega t$

4. $V_i = L\dfrac{\Delta I}{\Delta t} = \dfrac{(0.1\ \mathrm{H})\times(100\times10^{-3}\ \mathrm{A})}{0.01\ \mathrm{s}} = 1\ \mathrm{V}$

5. This is because when the number of the turns becomes more, the magnetic field which is generated in the coil due to the current of mutual induction and pushes back the electromagnet, becomes stronger.

6. (1) $\mu_r\mu_0 n^2 Ad = 1000\times(4\pi\times10^{-7}\ \mathrm{T\cdot m/A})\times(10^4\ \mathrm{m}^{-1})^2\times(10\times10^{-4}\ \mathrm{m}^2)\times(0.10\ \mathrm{m}) = 4\pi\ \mathrm{H} = 13\ \mathrm{H}$

(2) $L\dfrac{\Delta I}{\Delta t} = \dfrac{(13\ \mathrm{H})\times(10\times10^{-3}\ \mathrm{A})}{0.01\ \mathrm{s}} = 13\ \mathrm{V}$

7. $Z = \sqrt{R^2+\omega^2L^2} = \sqrt{R^2+(2\pi fL)^2}$
$= \sqrt{(100\ \Omega^2)+(50\pi\ \Omega)^2} = 186\ \Omega$

$I_e = \dfrac{V_e}{Z} = \dfrac{100\ \mathrm{V}}{186\ \Omega} = 0.54\ \mathrm{A}$

$\tan\phi = \dfrac{2\pi fL}{R} = \dfrac{50\pi\ \Omega}{100\ \Omega} = 1.57 \qquad \therefore\quad \phi = 57.5°$

8. $C = \dfrac{1}{(2\pi f)^2 L}$, $C = (5.0\times10^{-10}\text{–}3.1\times10^{-11})\ \mathrm{F}$

B

1. $V_i = vBL = (10\ \mathrm{m/s})\times(4.6\times10^{-5}\ \mathrm{T})\times(1\ \mathrm{m}) = 4.6\times10^{-4}\ \mathrm{V}$

2. (1) Since the electromotive force of the DC power source, V, and the electromotive force of the self-induction, $L\dfrac{\mathrm{d}I}{\mathrm{d}t}$, are balanced, we have $V = L\dfrac{\mathrm{d}I}{\mathrm{d}t}$. The solution to this equation is $I = \dfrac{V}{L}t$ ($I=0$ at $t=0$). The time to reach $I = 1500\ \mathrm{A}$, $t = \dfrac{(1500\ \mathrm{A})\times(40\ \mathrm{H})}{7.5\ \mathrm{V}} = 8000\ \mathrm{s}$.

(2) $U = \dfrac{B^2}{2\mu_0}\times$"volume" $= \dfrac{1}{8\pi}\{10^7\ \mathrm{A/(T\cdot m)}\}\times(1.8\ \mathrm{T})^2\times(27\ \mathrm{m}^3) = 3.5\times10^7\ \mathrm{J}$.

3. The current I_1 flowing in L_1 generates $B = \dfrac{\mu_0 I_1}{2r_1}$ at the center of the circle. Hence

$\therefore\quad \Phi_{2\leftarrow1} = \dfrac{\pi r_2^{\,2}\mu_0 I_1}{2r_1}$,

$\therefore\quad M_{12} = M_{21} = \dfrac{\mu_0\pi r_2^{\,2}}{2r_1}$.

4. (1) $u_E = \dfrac{1}{2}\varepsilon_0E^2 = \dfrac{1}{2}(8.85\times10^{-12}\ \mathrm{F/m})\times(10^6\ \mathrm{V/m})^2 = 4.4\ \mathrm{J/m^3}$

(2) $u_B = \dfrac{1}{2\mu_0}B^2 = \dfrac{1}{8\pi}\{10^7\ \mathrm{A/(T\cdot m)}\}\times(10\ \mathrm{T})^2 = 4.0\times10^7\ \mathrm{J/m^3}$

Chapter 22

Problems for exercise 22

A

1. $\lambda = \dfrac{c}{f} = \dfrac{3\times10^8\ \mathrm{m/s}}{1200\times10^3/\mathrm{s}} = 250\ \mathrm{m}$, $\dfrac{3\times10^8\ \mathrm{m/s}}{500\times10^6/\mathrm{s}} = 0.60\ \mathrm{m}$.

B

1. The strength of the electric field changes as $E(t) = (10^{-3}\ \mathrm{V/m})\cos\omega t$, and hence the time average of the energy flow $S = c\varepsilon_0E^2(t)$ is $\langle S\rangle = \dfrac{1}{2}c\varepsilon_0\times(10^{-3}\ \mathrm{V/m})^2 = \dfrac{1}{2}\times(3\times10^8\ \mathrm{m/s})\times\{8.9\times10^{-12}\ \mathrm{C^2/(N\cdot m^2)}\}\times(10^{-6}\ \mathrm{V^2/m^2}) = 1.3\times10^{-9}\ \mathrm{J/(m^2\cdot s)} = 1.3\times10^{-9}\ \mathrm{W/m^2}$.

$B = \dfrac{E}{c} = \dfrac{10^{-3}\ \mathrm{V/m}}{3\times10^8\ \mathrm{m/s}} = 3\times10^{-12}\ \mathrm{T}$.

2. The time for the light to travel the distance $2d$ is $\dfrac{1}{2nN}$, and therefore $c = \dfrac{2d}{\frac{1}{2nN}} = 4dnN = 3.13\times10^8\ \mathrm{m/s}$.

3. The round-trip time is $t = \dfrac{\theta}{2\pi\times(800\ \mathrm{s}^{-1})} = \dfrac{1.34\times10^{-3}}{3200\pi\ \mathrm{s}^{-1}} = 1.33\times10^{-7}\ \mathrm{s}$. $\quad c = \dfrac{2\times(20\ \mathrm{m})}{1.33\times10^{-7}\ \mathrm{s}} = 3.00\times10^8\ \mathrm{m/s}$.

Chapter 23

Problems for exercise 23

A

1. $(500\ \text{m})\times\sqrt{1-0.6^2}=400\ \text{m}$
2. $\dfrac{m}{m_0}=\left(1-\dfrac{u^2}{c^2}\right)^{-1/2}=1.01,\quad \therefore\quad u\approx 0.14c$
3. $(10^{-3}\ \text{kg})\times(3.0\times10^8\ \text{m/s})^2=9\times10^{13}\ \text{J}$

B

1. The effect of the Earth's motion with respect to the ether leads to the phase difference of lights, with wavelength λ and frequency f, corresponding to the difference between t_1 and t_2 given by $2\pi f(t_1-t_2)$ $\approx \dfrac{2\pi L\left(\dfrac{u}{c}\right)^2}{\lambda}$. When the apparatus is rotated by 90 degrees, t_1 and t_2 are exchanged, and hence the change in the phase difference $\approx \dfrac{4\pi L\left(\dfrac{u}{c}\right)^2}{\lambda}$ should be observed as the shift of the interference fringes.
$$\frac{4\pi L\left(\dfrac{u}{c}\right)^2}{\lambda}=\frac{4\pi\times(10\ \text{m})\times\left(\dfrac{30}{300000}\right)^2}{6\times10^{-7}\ \text{m}}=2\ \text{rad}$$
2. If we put $v_x=-u=0.6c$ in the first equation of equation (23.6) we get $v=\dfrac{0.6c+0.6c}{1+0.6^2}=\dfrac{15}{17}c$.
3. When observed by B who was staying in the Earth, the time for A to have made a round-trip between the Earth and the α star is $\dfrac{4.4\times2}{0.99}=8.9$ years. Hence B is 28.9 years old. A's clock, as observed by B, runs slow at the speed of $\sqrt{1-0.99^2}=0.141$ times the speed of B's clock. Therefore, A has stayed in the spacecraft for 8.9×0.141 years $=1.25$ years. Hence, A is 21.25 years old. On the contrary, when observed by A who has made a space travel, B looks younger than A (twin paradox). However, this kind of argument does not make sense, because A has undergone the accelerated motion at the takeoff and landing on the Earth as well as at the turning back from the α star, which corresponds to the motion in the non-inertial frame.
4. If we put $\gamma=\dfrac{1}{\sqrt{1-\dfrac{u^2}{c^2}}}$ in equation (23.5), we get $dx'=\gamma(dx-u\,dt)$, $dy'=dy$, $dz'=dz$, $dt'=\gamma\left(dt-\dfrac{u}{c^2}dx\right)$. Therefore using $v_x'=\dfrac{dx'}{dt'}$, $\dfrac{dx}{dt}=v_x$ we can derive equation (23.6).
5. Use $\boldsymbol{v}\cdot d\left(\dfrac{\boldsymbol{v}}{\sqrt{1-\dfrac{v^2}{c^2}}}\right)=\dfrac{\boldsymbol{v}\cdot d\boldsymbol{v}}{\sqrt{1-\dfrac{v^2}{c^2}}}$
$$+\frac{v^2(\boldsymbol{v}\cdot d\boldsymbol{v})}{c^2\left(1-\dfrac{v^2}{c^2}\right)^{3/2}}=\frac{\boldsymbol{v}\cdot d\boldsymbol{v}}{\left(1-\dfrac{v^2}{c^2}\right)^{3/2}}=d\left(\frac{c^2}{\sqrt{1-\dfrac{v^2}{c^2}}}\right)$$
6. omitted
7. If we put the Coulomb electric field $\boldsymbol{E}(\boldsymbol{r})=\dfrac{q}{4\pi\varepsilon_0}\dfrac{(\boldsymbol{r}-\boldsymbol{r}')}{|\boldsymbol{r}-\boldsymbol{r}'|^3}$ into $\boldsymbol{E}$ on the right-hand side of equation (23.14) where $\boldsymbol{u}$ is replaced by $-\boldsymbol{v}$, we get
$$\boldsymbol{B}'(\boldsymbol{r})\approx\boldsymbol{v}\times\boldsymbol{E}(\boldsymbol{r})/c^2=\frac{q\mu_0}{4\pi}\frac{\boldsymbol{v}\times(\boldsymbol{r}-\boldsymbol{r}')}{|\boldsymbol{r}-\boldsymbol{r}'|^3}$$

Chapter 24

Questions

1. We can derive from the energy conservation law,
$$\frac{ch}{\lambda}+m_ec^2=\frac{ch}{\lambda'}+\sqrt{m_e{}^2c^4+p_e{}^2c^2},$$
and from the momentum conservation law,
$$\frac{h}{\lambda}=\frac{h}{\lambda'}\cos\phi+p_e\cos\theta,\quad \frac{h}{\lambda'}\sin\phi=p_e\sin\theta.$$
Using $\sin^2\theta+\cos^2\theta=1$ to eliminate θ of the scattering angle of electrons, we have
$$p_e{}^2=\left(\frac{h}{\lambda}-\frac{h}{\lambda'}\right)^2+\frac{2h^2}{\lambda\lambda'}(1-\cos\phi).$$
By combining this equation with the first equation to eliminate the momentum p_e of the recoil electron, equation (24.5) is derived.
2. 1.23×10^{-10} m

Problems for exercise 24

A

1. $\nu=\dfrac{c}{\lambda}=\dfrac{3\times10^8\ \text{m/s}}{0.6\times10^{-6}\ \text{m}}=5\times10^{14}\ \text{Hz}$.
$E=h\nu=(6.6\times10^{-34}\ \text{J·s})\times(5\times10^{14}\ \text{s}^{-1})=3.3\times10^{-19}\ \text{J}$.
2. (1) $(5\times10^{-11}\ \text{s})\times(3.0\times10^8\ \text{m/s})=1.5\times10^{-2}\ \text{m}=1.5\ \text{cm}$.
(2) $\dfrac{10\ \text{J}}{(1.5\times10^{-2}\ \text{m})\times(2\times10^{-6}\ \text{m}^2)}=3.3\times10^8\ \text{J/m}^3$
(3) $w=\varepsilon_0E^2$
$$\therefore\quad E=\sqrt{\frac{w}{\varepsilon_0}}=\left(\frac{3.3\times10^8\ \text{J/m}^3}{8.85\times10^{-12}\ \text{F/m}}\right)^{1/2}=6\times10^9\ \text{V/m}.$$
(4) $h\nu=\dfrac{hc}{\lambda}=\dfrac{(3\times10^8\ \text{m/s})\times(6.6\times10^{-34}\ \text{J·s})}{6.9\times10^{-7}\ \text{m}}=2.9\times10^{-19}\ \text{J}$.
Number of photons $n=\dfrac{10\ \text{J}}{2.9\times10^{-19}\ \text{J}}=3.4\times10^{19}$.
3. Since the momentum $p=mv=\dfrac{h}{\lambda}$, $v=\dfrac{h}{m\lambda}=$
$$\frac{6.6\times10^{-34}\ \text{J·s}}{(9.11\times10^{-31}\ \text{kg})\times(10^{-10}\ \text{m})}=7\times10^6\ \text{m/s}.$$

$\frac{v}{c} = \frac{7\times10^6\ \mathrm{m/s}}{3\times10^8\ \mathrm{m/s}} = \frac{1}{40}$ $E = \frac{1}{2}mv^2 = \frac{1}{2}\times(9.11\times10^{-31}\ \mathrm{kg})\times(7\times10^6\ \mathrm{m/s})^2 = 2.2\times10^{-17}\ \mathrm{J} = 1.4\times10^2\ \mathrm{eV}$

4. $\lambda = \frac{h}{mv} = \frac{6.63\times10^{-34}\ \mathrm{J\cdot s}}{(1.67\times10^{-27}\ \mathrm{kg})\times(1.0\times10^4\ \mathrm{m/s})} = 4.0\times10^{-11}\ \mathrm{m}$

5. Since $K = \frac{1}{2}mv^2 = \frac{p^2}{2m} = \frac{h^2}{2m\lambda^2}$, if the kinetic energy K is the same, the de Broglie wavelength λ is longer as the mass m is smaller. The de Broglie wavelength of electron with the smallest mass is the longest.

B

1. $\frac{ZZ'e^2}{4\pi\varepsilon_0} = 79\times2\times(9\times10^9\ \mathrm{N\cdot m^2/C^2})\times(1.6\times10^{-19}\ \mathrm{C})^2 = 3.64\times10^{-26}\ \mathrm{J\cdot m} = 2.27\times10^{-7}\ \mathrm{eV\cdot m}$. When $r = 10^{-10}$ m, $U = 2.27\times10^3$ eV. When $r = 10^{-14}$ m, $U = 2.27\times10^7\ \mathrm{eV} = 22.7$ MeV. From $\frac{1}{2}mv^2 = \frac{ZZ'}{4\pi\varepsilon_0 r}$, $r = \frac{2.27\times10^{-7}\ \mathrm{eV\cdot m}}{4.79\times10^6\ \mathrm{eV}} = 4.7\times10^{-14}$ m.

2. $\Delta p \gtrsim \frac{h}{2\pi(\Delta x)} = \frac{6.63\times10^{-34}\ \mathrm{J\cdot s}}{2\pi(0.5\times10^{-10}\ \mathrm{m})} = 2\times10^{-24}$ kg·m/s.

$\Delta v = \frac{\Delta p}{m} \gtrsim \frac{2\times10^{-24}\ \mathrm{kg\cdot m/s}}{9.1\times10^{-31}\ \mathrm{kg}} = 2\times10^6$ m/s.

$\frac{(\Delta p)^2}{2m} \gtrsim 2\times10^{-18}$ J. Since $13.6\ \mathrm{eV} = 2.2\times10^{-18}$ J, the magnitudes are similar. Hence, it is impossible to make Δx of an electron with an energy of 13.6 eV much smaller than the radius of hydrogen atom.

3. $p = \sqrt{3mkT} = \frac{h}{\lambda}$

$$\lambda = \frac{h}{\sqrt{3mkT}} = \frac{6.63\times10^{-34}\ \mathrm{J\cdot s}}{\sqrt{3\times(1.67\times10^{-27}\ \mathrm{kg})\times(1.38\times10^{-23}\ \mathrm{J/K})\times(600\ \mathrm{K})}} = 1.0\times10^{-10}\ \mathrm{m},$$

$$v = \frac{p}{m} = \sqrt{\frac{3kT}{m}} = \sqrt{\frac{3(1.38\times10^{-23}\ \mathrm{J/K})\times(600\ \mathrm{K})}{1.67\times10^{-27}\ \mathrm{kg}}} = 3.9\times10^3\ \mathrm{m/s}$$

4. $\lambda = \sqrt{\frac{150.4}{54}}\times(10^{-10}\ \mathrm{m}) = 1.67\times10^{-10}$ m. $\sin\theta = \frac{\lambda}{d} = \frac{1.67\times10^{-10}\ \mathrm{m}}{2.17\times10^{-10}\ \mathrm{m}} = 0.77$. $\therefore\ \theta = 50°$.

When $V = 181$ V, since $\lambda = 0.91\times10^{-10}$ m, $\theta = 25°$.

5. $P = \frac{E}{c} = \frac{2000\ \mathrm{J}}{3\times10^8\ \mathrm{m/s}} = 6.7\times10^{-6}$ kg·m/s

Chapter 25

Problems for exercise 25

A

1. 82, 126 and 92, 143
2. ④ ^{56}Fe
3. $\left(\frac{1}{2}\right)^{45/15} = \left(\frac{1}{2}\right)^3 = \frac{1}{8}$ [g]
4. The mass number does not change in the β decay, and it reduces by 4 in each α decay. Therefore, the number of times of the α decays is $\frac{238-206}{4} = 8$ times. The atomic number increases by 1 in the β decay, and it reduces by 2 in each α decay. Therefore, the number of the times of β decays is $-(92-82-2\times8) = 6$ times.
5. From $\frac{1}{4} = \left(\frac{1}{2}\right)^2 = \left(\frac{1}{2}\right)^{t/T_{1/2}}$, we have $2 = \frac{t}{T_{1/2}}$. Hence, we have $t = 2T_{1/2} = 2\times5700\ \mathrm{y} = 11400$ y.
6. ① $(10\ \mathrm{J/kg})\div[4.2\times10^3\ \mathrm{J/(kg\cdot K)}] = 0.0024$ K

B

1. $$\rho \approx \frac{m_pA}{\frac{4\pi(1.2\times10^{-15}A^{1/3}\ \mathrm{m})^3}{3}} = \frac{3\times(1.67\times10^{-27}\ \mathrm{kg})}{4\pi(1.2\times10^{-15}\ \mathrm{m})^3} = 2.3\times10^{17}\ \mathrm{kg/m^3} = 2.3\times10^{14}\ \mathrm{g/cm^3}.$$ $$r = \left(\frac{3M_S}{4\pi\rho}\right)^{1/3} = \left(\frac{M_S}{m_p}\right)^{1/3}\times(1.2\times10^{-15}\ \mathrm{m}) = \left(\frac{2.0\times10^{30}}{1.67\times10^{-27}}\right)^{1/3}\times(1.2\times10^{-15}\ \mathrm{m}) = 1.3\times10^4\ \mathrm{m} = 13\ \mathrm{km}$$

2. Suppose that X at rest decays into Y (velocity $\boldsymbol{v}$) and α (velocity $\boldsymbol{v}'$). From the momentum conservation law, we have $m\boldsymbol{v}+m_\alpha\boldsymbol{v}' = \boldsymbol{0}$. From the energy conservation law, we get $Mc^2 = mc^2+\frac{1}{2}mv^2+m_\alpha c^2+\frac{1}{2}m_\alpha v'^2$.

Hence, if we write $E \equiv (M-m-m_\alpha)c^2$, we obtain $E = \frac{1}{2}mv^2+\frac{1}{2}m_\alpha v'^2 = \frac{m_\alpha{}^2v'^2}{2m}+\frac{1}{2}m_\alpha v'^2$.

Thus, we get $\frac{1}{2}m_\alpha v'^2 = \frac{Em}{m+m_\alpha}$.

Photo Credits

Book cover and jacket-front side：Erich Lessing/PPS
Book cover and jacket-back side：Alamy/PPS

A photograph of the Earth at the top left of the first page of each chapter：NASA

p.1：RIKEN

Chapter 0

p.2：NAOJ
Fig. 0.1：NAOJ
Fig. 0.2：NASA
Fig. 0.3：gezzeg/123RF
Fig. 0.6：AIST
Fig. 0.8：RIKEN Quantum Metrology Laboratory
Fig. 0.9：AIST
Fig. 0.10：AIST
Fig. 0.11：Eizo Ohno (Hokkaido University)

Chapter 1

p.11：Alamy/PPS
Fig. 1.8：photolibrary
Fig. 1.18：Beautifulblossom-Fotolia.com
Fig. 1.21：Arrows-Fotolia.com
Fig. 1.28：lamax-Fotolia.com
Fig. 1.A (left)：sudowoodo/123RF
Fig. 1.A (right)：oni-Fotolia.com

Chapter 2

p.27：Alamy/PPS
Fig. 2.1：williammanning/123RF
Fig. 2.3：eintracht/123RF
Fig. 2.14：NASA
Fig. 2.26：rafaelbenari/123RF

Chapter 3

p.40：Alamy/PPS
Fig. 3.4：swimwitdafishes/123RF
Fig. 3.14：Joggie Botma-Fotolia.com
Fig. 3.17：Shuji Ukon

Chapter 4

p.57：Alamy/PPS
Fig. 4.1：addricky/123RF
Fig. 4.3：Tamio Sasagawa http://www.mars.dti.ne.jp/~stamio
Fig. 4.7：flynt/123RF
Fig. 4.8：http://www.uchiyama.info/
Fig. 4.12：smuay/123RF
Fig. 4.13：Honshu-Shikoku Bridge Expressway Company Limited.

Chapter 5

p.68：AGE/PPS
Fig. 5.5：stephanscherhag/123RF
Fig. 5.8：michaklootwijk/123RF
Fig. 5.13：Public Enterprise Bureau of Shimane Prefecture
Fig. 5.19：photolibrary
Fig. 5.20：toliknik/123RF

Chapter 6

p.83：Alamy/PPS
Fig. 6.7：seventysix/123RF

Chapter 7

p.90：Alamy/PPS
Fig. 7.9：JAXA/NASA
Fig. 7.11：thierry burot-Fotolia.com
Fig. 7.16：olga_besnard/123RF

Chapter 8

p.102：Masami Goto/PPS
Fig. 8.1：kaowenhua/123RF
Fig. 8.3：actionsports/123RF
Fig. 8.9：photolibrary

Chapter 9

p.119：Science Source/PPS
Fig. 9.4：nd3000-Fotolia.com
Fig. 9.5：Robert Ford-Fotolia.com

Chapter 10

p.126：Alamy/PPS
Fig. 10.4：serezniy/123RF
Fig. 10.9：macor/123RF

Chapter 11
p.133 : Rex/PPS
Fig. 11.1 : photolibrary
Fig. 11.5 : vanbeets/123RF
Fig. 11.6 : cylonphoto/123RF
Fig. 11.14 : lello4d/123RF
Fig. 11.19 : http://www.tohnic.co.jp/

Chapter 12
p.144 : HIP/PPS
Fig. 12.7 : katyphotography-Fotolia.com
Fig. 12.10 : NNP
Fig. 12.12 : Greg Brave-Fotolia.com
Fig. 12.21 : whitestone/123RF
Fig. 12.22 : bradengunem/123RF

Chapter 13
p.166 : Alamy/PPS
Fig. 13.1 : NNP
Fig. 13.3 : ziggy-Fotolia.com
Fig. 13.6 (top) : Olivier Le Moal-Fotolia.com
Fig. 13.6 (bottom) : nobasuke-Fotolia.com
Fig. 13.8 : whitetag/123RF
Fig. 13.9 : anaken2012/123RF

Chapter 14
p.173 : Alamy/PPS
Fig. 14.1 : zhengzaishanchu-Fotolia.com
Fig. 14.2 : Kawaguchi Liquefaction Chemical Corporation
Fig. 14.4 : hanapon1002-Fotolia.com
Fig. 14.6 : Hoda Bogdan-Fotolia.com
Fig. 14.8 : missisya-Fotolia.com
Fig. 14.9 : JAXA
Fig. 14.10 : Denkikogyo Co., Ltd.

Chapter 15
p.190 : Alamy/PPS
Fig. 15.4 : photolibrary
Fig. 15.9 : tsubakiya_k-Fotolia.com
Fig. 15.10 : photolibrary
Fig. 15.15 : Toshiba Energy Systems & Solutions Corporation
Fig. 15.23 : JAXA/NASA

Chapter 16
p.214 : SPL/PPS
Fig. 16.1 : pacoayala/123RF

Chapter 17
p.237 : Science Source/PPS
Fig. 17.9 : Lennard-Fotolia.com
Fig. 17.13 : Panasonic Corporation
Fig. 17.16 : TH

Chapter 18
p.248 : SPL/PPS
Fig. 18.1 : TDK Corporation
Fig. 18.7 : Kojundo Chemical Laboratory Co., Ltd.
https://www.kojundo.co.jp/en_index.html

Chapter 19
p.255 : Alamy/PPS
Fig. 19.2 : Shikoku Electric Power CO., Inc.
Fig. 19.11 : BillionPhotos.com-Fotolia.com
Fig. 19.12 : Murata Manufacturing Co., Ltd.
http://www.murata.com
Fig. 19.14 : Yokohama Physics Circle Annex Tenjin's Page
http://www2.hamajima.co.jp/~tenjin/tenjin.htm
Fig. 19.20 : unaikatsuhiro-Fotolia.com

Chapter 20
p.269 : Science Source/PPS
Fig. 20.1 : photolibrary
Fig. 20.35 : TOSHIBA CORPORATION/ATLAS JAPAN
Fig. 20.37 : JAXA/NASA
Fig. 20.43 : claudiodivizia/123RF
Fig. 20.B : RIKEN

Chapter 21
p.298 : SPL/PPS
Fig. 21.1 : Wacom
https://www.wacom.com
Fig. 21.16 : Public Enterprise Bureau of Shimane Prefecture
Fig. 21.22 : TOSHIBA CORPORATION/ATLAS JAPAN
Fig. 21.29 : Mitsubishi Electric Corporation

Chapter 22
p.318 : NAOJ
Fig. 22.6 : photolibrary

Fig. 22.10 : Yasushi Shibata of Himagine Laboratory
Fig. 22.16 : JAXA

Chapter 23

p.332 : SPL/PPS
Fig. 23.3 : Mike Thomas-Fotolia.com
Fig. 23.5 : JAXA
Fig. 23.8 : RIKEN

Chapter 24

p.343 : RIKEN
Fig. 24.12 : Akira Tonomura (Hitachi, Ltd.)
Fig. 24.15 : Toshio Ito
Fig. 24.18 : NNP
Fig. 24.21 : Shin-Etsu Chemical., Ltd.
Fig. 24.25 : Arao Nakamura/Yoshikazu Takeda (Nagoya University)
Fig. 24.26 : Yoshiaki SUGIMOTO
Fig. 24.27 : Murata Manufacturing Co., Ltd. http://www.murata.com
Fig. 24.33 : paylessimages/123RF
Fig. 24.37 : NAOJ/Dan Birchall

Chapter 25

p.365 : SPL/PPS
Fig. 25.4 : sergeyussr/123RF
Fig. 25.6 : Kamioka Observatory, ICRR (Institute for Cosmic Ray Research), The University of Tokyo
Fig. 25.9 : High Energy Accelerator Research Organization (KEK)
Fig. 25.A : SPL/PPS
Fig. 25.B : SPL/PPS

INDEX

Greek alphabet terms

α decay（α 崩壊） 370
α ray（α 線） 370
β decay（β 崩壊） 370
β ray（β 線） 370
γ decay（γ 崩壊） 370
γ ray（γ 線） 370
Ω（unit：オーム） 258

A

A（unit：アンペア） 8, 256, 286
absolute temperature（絶対温度） 181
absorbed dose（吸収線量） 372
AC electromotive force（交流起電力） 314
AC generator（交流発電機） 306
acceleration（加速度） 14, 18
acceptor level（アクセプター準位） 359
adiabatic change（断熱変化） 193
adiabatic free expansion（断熱自由膨張） 194
alternating current（交流電流，交流） 312
alternating voltage（交流電圧） 312
amount of heat（熱量） 176
Ampere-Maxwell's law（アンペール-マクスウェルの法則） 287
Ampere's law（アンペールの法則） 276
Ampere's law for the magnetic field ***H***（磁場 ***H*** のアンペールの法則） 289
amplitude（振幅） 60, 146
angle of incidence（入射角） 153
angle of reflection（反射角） 153
angular frequency（角振動数，角周波数） 60, 313
angular momentum（角運動量） 85
angular momentum conservation law（角運動量保存則） 86, 99
angular momentum of the system of particles（質点系の角運動量） 98
angular velocity（角速度） 22
anti-particle（反粒子） 376
antinodes（腹） 157
apparent force（見かけの力） 122
Archimedes' principle（アルキメデスの原理） 136
areal velocity（面積速度） 86
atom（原子） 344
atomic mass unit（原子質量単位） 366
atomic nucleus（原子核） 345
atomic number（原子番号） 354
atomic orbital（原子軌道） 354
atomic weight（原子量） 366
Avogadro constant（アボガドロ定数） 181
Avogadro law（アボガドロの法則） 181

B

balance of rigid body（剛体のつり合い） 105
band（バンド） 356
base units（基本単位） 286
beat（うなり） 164
Becquerel, A. H.（ベクレル） 370
Bernoulli's law（ベルヌーイの法則） 137, 138
binding energy（結合エネルギー） 368
Biot-Savart's law（ビオ-サバールの法則） 273
Boltzmann constant（ボルツマン定数） 183
Boltzmann distribution（ボルツマン分布） 184
Borda's pendulum（ボルダの振り子） 110
boson（ボース粒子） 376
boundary condition（境界条件） 47
boundary layer（境界層） 141
Boyle-Charles' law（ボイル-シャルルの法則） 182
Boyle's law（ボイルの法則） 181
Bq（unit：ベクレル） 372
bulk modulus（体積弾性率） 129
buoyancy（浮力） 136

C

C（unit：クーロン） 216
cal（unit：カロリー） 78, 176
capacitor（キャパシター，コンデンサー） 241
Carnot, N. L. S.（カルノー） 199
Carnot's cycle（カルノー・サイクル） 200
Carnot's principle（カルノーの原理） 202
causality（因果律） 47
Cavendish, H.（キャベンディッシュ） 33
Celsius temperature scale（セルシウス温度目盛） 174
center of gravity（重心） 91
center of mass（質量中心） 91
central force（中心力） 85
centrifugal force（遠心力） 122
centripetal acceleration（向心加速度） 23, 30
centripetal force（向心力） 30
Chadwick, J.（チャドウイック） 366
chain reaction（連鎖反応） 374
characteristic frequency（固有振動数，固有周波数） 157, 315
characteristic vibration（固有振動） 157
charge conservation law（電荷保存則） 216
Charles' law（シャルルの法則） 181
chemical energy（化学エネルギー） 78
circuit（回路） 262
circuit element（回路素子） 262
Clausius, R. J. E.（クラウジウス） 186
Clausius' inequality（クラウジウスの不等式） 207
closed tube（閉管） 160
Cockcroft, S. J. D.（コッククロフト） 367
coefficient of cubical expansion（体膨張率） 177
coefficient of kinetic friction（動摩擦係数） 37
coefficient of linear expansion（線膨張率） 177
coefficient of static friction（静止摩擦係数） 37
coercive force（保磁力） 292
component of force（分力） 30
compressibility（圧縮率） 129
compression wave（疎密波） 145
Compton scattering（コンプトン散乱） 347
Compton, A. H.（コンプトン） 347
condenser（コンデンサー） 241
conduction band（伝導帯） 356
conduction electron（伝導電子） 238
conductor（導体） 238
conservation law of mechanical energy（力学的エネルギー保存則） 76, 77
conservative force（保存力） 72
convection（対流） 178
Coriolis force（コリオリの力） 123
Coulomb energy（クーロン・エネルギー） 228
Coulomb potential（クーロン・ポテンシャル） 228
Coulomb, C. A.（クーロン） 216
Coulomb's force（クーロン力） 218
Coulomb's law（クーロンの法則） 216, 219

Coulomb's law for the magnetic force
（磁気力のクーロンの法則） 290
counter electromotive force
（逆起電力） 308
couple of forces（偶力） 88
CR circuit（*CR* 回路） 265
crest（山） 146
critical angle（臨界角） 169
critical damping（臨界減衰） 64
critical mass（臨界量，臨界質量） 374
critical point（臨界点） 175
critical pressure（臨界圧） 187
critical state（臨界状態） 374
critical temperature（臨界温度） 187
Curie temperature（キュリー温度） 292
cycle（サイクル） 200
cyclic process（循環過程） 200
cyclotron frequency
（サイクロトロン周波数） 280
cyclotron motion
（サイクロトロン運動） 279

D

damped oscillation（減衰振動） 64
dark matter（ダークマター） 36
Davisson, C. J.（デビソン） 350
Davisson-Germer's experiment
（デビソン-ガーマーの実験） 350
dB（unit：デシベル） 160
de Broglie wavelength
（ド・ブロイ波長） 350
de Broglie, L.（ド・ブロイ） 350
decay constant（崩壊定数） 371
decay series（崩壊系列） 378
definite integral（定積分） 41
derived units（組立単位） 8
diamagnetic body（反磁性体） 290
dielectric polarization（誘電分極） 250
dielectric substance（誘電体） 249
differential equation（微分方程式） 37
diffraction（回折） 154, 170
diffraction grating（回折格子） 171
dimension（次元，ディメンション） 9
Dirac, P. A. M（ディラック） 375
direct current circuit（直流回路） 262
direct-current motor（直流モーター） 284
dispersion（分散） 169
displacement（変位） 12, 18
donor level（ドナー準位） 357
Doppler effect（ドップラー効果） 160
drift velocity（ドリフト速度） 257
duality（二重性） 347, 350
duality of electron
（電子の二重性） 349
duality of light（光の二重性） 347
Dulong, P. L.（デュロン） 196
Dulong-Petit's law
（デュロン-プティの法則） 196

E

effective dose（実効線量） 372
effective value（実効値） 313
efficiency of heat engine
（熱機関の効率） 199
Einstein, A.（アインシュタイン） 335, 346
Einstein's principle of relativity
（アインシュタインの相対性原理） 335
elastic body（弾性体） 127
elastic collision（弾性衝突） 96
elastic constant（弾性定数） 58, 128
elastic deformation（弾性変形） 126
elastic force（弾力） 58
elastic limit（弾性限界） 127
elastic potential energy
（弾性ポテンシャルエネルギー） 74
elastic wave（弾性波） 151
elasticity（弾性） 126, 127
electric capacity（電気容量） 242
electric charge（電荷） 217
electric conductivity（電気伝導率） 259
electric current（電流） 256
electric current density（電流密度） 259
electric dipole（電気双極子） 232
electric dipole moment
（電気双極子モーメント） 232
electric energy（電力量） 265
electric field（電場，電界） 218, 230
electric flux density（電束密度） 252
electric force（電気力） 216
electric potential（電位） 227
electric power（電力） 264
electric resistance（電気抵抗） 258
electric resistivity（電気抵抗率） 259
electric shielding（静電遮蔽） 240
electric susceptibility（電気感受率） 251
electrical constant（電気定数） 217
electromagnetic field（電磁場） 328
electromagnetic induction
（電磁誘導） 300
electromagnetic wave（電磁波） 323
electromotive force（起電力） 257
electron（電子） 345
electron microscope（電子顕微鏡） 351
electrostatic field（静電場） 219, 248
electrostatic induction
（静電誘導） 238
electrostatic tension（静電張力） 247
electroweak force（電弱力） 377
elementary charge
（電気素量，素電荷） 216
elementary particle（素粒子） 376
enegy conservation law
（エネルギー保存則） 79
energy（エネルギー） 68, 339
energy band（エネルギーバンド） 356
energy gap（エネルギーギャップ） 356
energy level（エネルギー準位） 356
energy of electric field
（電場のエネルギー） 245, 254
energy of electromagnetic wave
（電磁波のエネルギー） 328
energy of magnetic field
（磁場のエネルギー） 310
entropy（エントロピー） 205, 206
equation of continuity（連続方程式） 137
equation of motion for the center of gravity（重心の運動方程式） 93, 103
equation of motion of the car
（自動車の運動方程式） 117
equation of state（状態方程式） 191
equation of state of the ideal gas
（理想気体の状態方程式） 182
equipotential line（等電位線） 231
equipotential surface（等電位面） 231
equivalent magnet（等価磁石） 287
Esaki, L.（江崎玲於奈） 354
escape velocity（脱出速度） 76
ether（エーテル） 333
eV（unit：電子ボルト） 229, 348
exponent（指数） 9
external force（外力） 32, 91

F

F（unit：ファラド） 242
Faraday constant（ファラデー定数） 217
Faraday, M.（ファラデー） 236, 299
fermion（フェルミ粒子） 376
ferroelectrics（強誘電体） 252
ferromagnet（強磁性体） 291
field（場） 219
first law of motion
（運動の第 1 法則） 28
first law of thermodynamics
（熱力学の第 1 法則） 191
first order phase transition
（1 次相転移） 174
fixed end（固定端） 154
Fizeau, A. H. L.（フィゾー） 330
Fizeau's experiment
（フィゾーの実験） 330
Fleming's left-hand rule
（フレミングの左手の法則） 282

Fleming's right-hand rule
（フレミングの右手の法則） 308
fluid（流体） 133
flux of electric line of force
（電気力線束） 222
force（力） 32
Force exerted between the currents
（電流の間に作用する力） 285
force in macroscopic scale
（巨視的に見た力） 36
force of inertia（慣性力） 119, 120, 122
forced vibration（強制振動） 65
Foucault, J. B. L.（フーコー） 125
Franklin, B.（フランクリン） 236, 240
Fraunhofer diffraction
（フラウンホーファーの回折） 170
free electric charge（自由電荷） 237
free electron（自由電子） 238
free end（自由端） 154
frequency（振動数，周波数）
60, 146, 313
Fresnel diffraction（フレネルの回折）
170
frictional force（摩擦力） 36
fundamental theorem of differential
and integral calculus
（微分積分学の基本定理） 42
fundamental vibration（基本振動）
157

G

Galilean principle of relativity
（ガリレオの相対性原理） 121
Galileo Galilei（ガリレオ） 55, 62
gap（ギャップ） 356
gas constant（気体定数） 182
gauge particle（ゲージ粒子） 378
Gauss' law（ガウスの法則）
222, 252, 272, 319
Gauss' law of electric field
（電場のガウスの法則） 222
Gauss' law of electric field ***E***
（電場 ***E*** のガウスの法則） 319
Gauss' law of electric flux density
（電束密度のガウスの法則） 252
Gauss' law of magnetic field ***B***
（磁場 ***B*** のガウスの法則） 272, 320
Geiger, H. W.（ガイガー） 345
Gell-Mann, M.（ゲルマン） 376
general solution（一般解） 47
Germer, L. H.（ガーマー） 350
Gibbs free energy
（ギブズの自由エネルギー） 212
gluon（グルーオン） 378
gravitational acceleration
（重力加速度） 15
gravitational constant（重力定数） 33
graviton（重力子） 378
gravity（重力） 34
group velocity（群速度） 164
Gy（unit：グレイ） 372

H

H（unit：ヘンリー） 309, 312
hadron（ハドロン） 377
Hagen-Poiseuille's law
（ハーゲン-ポアズイユの法則） 141
half-life（半減期） 371
Hall effect（ホール効果） 280
heat（熱） 176
heat capacity（熱容量） 176
heat conduction（熱伝導） 177
heat engine（熱機関） 199
heat of transition（転移熱） 174
Heisenberg, W. K.（ハイゼンベルク）
352
Helmholtz free energy
（ヘルムホルツの自由エネルギー）
211
Helmholtz, H. L. F.（ヘルムホルツ）
192
Henry, J.（ヘンリー） 301
Hertz, G. L.（ヘルツ） 327
Hertz's experiment（ヘルツの実験）
327
high temperature heat source
（高温熱源） 199
hole（ホール） 357
Hooke's law（フックの法則） 58, 127
hydrostatic pressure（静水圧） 134
hysteresis loop
（ヒステリシスループ） 292
hysteresis loss（ヒステリシス損失）
292
Hz（unit：ヘルツ） 60, 146, 313

I

ideal gas（理想気体） 182
impedance（インピーダンス） 313
impulse（力積） 43
incompressible fluid（非圧縮性流体）
137
indefinite integral（不定積分） 37
inductance（インダクタンス） 309
inelastic collision（非弾性衝突） 97
inertia（慣性） 28
inertial frame（慣性系） 119, 120
inertial resistance（慣性抵抗） 49, 141
initial condition（初期条件） 47
inner product（内積） 69
instantaneous acceleration
（瞬間加速度） 19
instantaneous velocity（瞬間速度）
13, 17
insulator（絶縁体，不導体） 248, 356
interference（干渉） 152
internal energy（内部エネルギー）
176, 185
internal energy of the ideal gas
（理想気体の内部エネルギー） 185
internal force（内力） 32, 93
intrinsic semiconductor
（真性半導体） 358
ionization（電離作用） 372
irreversible process（不可逆変化）
198
isobaric change（定圧変化） 192
isochoric change（定積変化） 193
isochronism（等時性） 60, 62
isochronism of pendulum
（振り子の等時性） 62
isolated spherical conductor
（孤立導体球） 243
isotherm（等温曲線） 187
isothermal change（等温変化） 193
isotope（同位体） 367

J

J（unit：ジュール） 69, 78, 176
Joule, J. P.（ジュール） 78, 192
Joule's experiment（ジュールの実験）
78
Joule's heat（ジュール熱） 265

K

K（unit：ケルビン） 181
Kajita, T.（梶田隆章） 380
Kamerlingh-Onnes, H.
（カマリング・オネス） 261
Kepler, J.（ケプラー） 88
Kepler's laws（ケプラーの法則） 88
kg（unit：キログラム） 8
kgf（unit：重力キログラム） 34
kgw（unit：キログラム重） 34
kinetic energy（運動エネルギー） 71
kinetic friction（動摩擦力） 37
kinetic theory of gases
（気体の分子運動論） 182
Kirchhoff's laws
（キルヒホッフの法則） 263
Koshiba, M.（小柴昌俊） 379
kWh（unit：キロワット時） 265

L

laminar flow（層流） 142
laser（レーザー） 362
law of action and reaction
（作用反作用の法則） 31
law of black body radiation
（黒体放射の法則） 178
law of constant areal velocity
（面積速度一定の法則） 88
law of decay（崩壊の法則） 371
law of electromagnetic induction

(電磁誘導の法則) 301, 320
law of equation of motion for the center of gravity of the rigid body (剛体の重心運動の法則) 103
law of equipartition of energy (エネルギー等分配法則) 184
law of force (力の法則) 32
law of inertia (慣性の法則) 28
law of motion (運動の法則) 29
law of reflection (反射の法則) 153
law of refraction (屈折の法則) 153
law of rotational motion (回転運動の法則) 85, 104
law of rotational motion of rigid body (剛体の回転運動の法則) 107
law of universal gravitation (万有引力の法則) 33
Lawrence, E. O. (ローレンス) 367
length (長さ) 8
Lenz, E. K. (レンツ) 301
Lenz's law (レンツの法則) 301
lepton (レプトン) 377
lift (揚力) 139
lift constant (揚力定数) 139
light (光) 166
light emitting diode (発光ダイオード) 361
line spectrum (線スペクトル) 353
lines of electric flux (電束線) 252
lines of electric force (電気力線) 220
lines of magnetic force (磁力線) 271
longitudinal wave (縦波) 145
Lorentz contraction (ローレンツ収縮) 339
Lorentz force (ローレンツ力) 215, 279
Lorentz transformation (ローレンツ変換) 335
low-temperature heat reservoir (低温熱源) 199
LR circuit (*LR* 回路) 309

M

m (unit：メートル) 9
Mach number (マッハ数) 165
macro (マクロ) 343
macroscopic electric field (巨視的な電場) 238
magnet (磁石) 287
magnetic body (磁性体) 287
magnetic charge (磁荷) 270
magnetic constant (磁気定数) 273
magnetic dipole (磁気双極子) 284
magnetic domain (磁区) 292
magnetic field (磁場，磁界) 215, 271
magnetic flux (磁束) 271
magnetic force (磁気力) 270
magnetic hysteresis (磁気ヒステリシス) 292
magnetic moment (磁気モーメント) 284
magnetic permeability (透磁率) 290
magnetic susceptibility (磁化率，磁気感受率) 290
magnetization (磁化) 287
magnetization current (磁化電流) 288
magnetization curve (磁化曲線) 292
Magnus effect (マグヌス効果) 143
Marsden, E. (マースデン) 345
mass (質量) 8, 34, 339
mass defect (質量欠損) 368
mass number (質量数) 366
mass point (質点) 12, 90
maximum frictional force (最大摩擦力) 37
Maxwell distribution (マクスウェル分布) 184
Maxwell, J. C. (マクスウェル) 184, 319
Maxwell's equations (マクスウェルの方程式) 215, 224, 321
Maxwell's rule (マクスウェルの規則) 188
Mayer's relation (マイヤーの関係式) 195
mean acceleration (平均加速度) 14, 17
mean free path (平均自由行程) 185
mean velocity (平均速度) 12, 18
mechanical energy (力学的エネルギー) 76
mechanical similarity law (力学的相似性) 142
medium (媒質) 144
metal (金属) 356
Meyer, J. R. (マイヤー) 192
Michelson, A. A. (マイケルソン) 333
Michelson-Morley experiment (マイケルソン-モーリーの実験) 333
micro (ミクロ) 343
microscopic current (微視的な電流) 287
MKS system of units (MKS 単位系) 8
MKSA International System of Units (MKSA 国際単位系) 286
MKSA system of units (MKSA 単位系) 8
modulus of elasticity (伸び弾性率) 129
molar heat capacity (モル熱容量) 176
Molar heat capacity at constant pressure (定圧モル熱容量) 195
molar heat capacity at constant volume (定積モル熱容量) 195
molar heat capacity of the ideal gas (理想気体のモル熱容量) 195
moment (モーメント) 84
moment of external force (外力のモーメント) 98
moment of force (力のモーメント) 84
moment of inertia (慣性モーメント) 107, 108
moments of gravity (重力のモーメント) 91
momentum (運動量) 42
momentum conservation law (運動量保存則) 96
momentum of electromagnetic wave (電磁波の運動量) 329
Morley, E. W. (モーリー) 333
Mr. and Mrs. Curie (キュリー夫妻) 370
mutual inductance (相互インダクタンス) 312
mutual induction (相互誘導) 311

N

N (unit：ニュートン) 29
n-type semiconductor (n 型半導体) 358
nano (ナノ) 343
neutrino (ニュートリノ) 370
neutron (中性子) 366
Newton, I. (ニュートン) 24
Newton's equation of motion (ニュートンの運動方程式) 29
node (節) 157
non-inertial frame (非慣性系) 120
nonslip condition (滑りなしの条件) 140
normal distribution (正規分布) 10
normal force (垂直抗力) 36
normal stress (法線応力) 127
nuclear decay (原子核の崩壊) 369
nuclear energy (核エネルギー) 340, 373
nuclear fission (核分裂) 369, 373
nuclear force (核力) 367
nuclear fusion (核融合) 369
nuclear reactor (原子炉) 374
nucleon (核子) 366
nucleus (原子核) 365

O

Oersted, H. C. (エルステッド) 272
Ohm's law (オームの法則) 258
open end correction (開口端補正) 160
open tube (開管) 160
optical fiber (光ファイバー) 169
Otto cycle (オットー・サイクル) 203
overdamping (過減衰) 64

P

p-type semiconductor（p 型半導体） 359
Pa（unit：パスカル） 127
parabolic motion（放物運動） 47
parallel-axis theorem（平行軸の定理） 110
parallel-plate capacitor（平行板キャパシター） 242
paramagnetic body（常磁性体） 291
particular solution（特殊解） 47
Pascal's principle（パスカルの原理） 135
Pauli, W.（パウリ） 375
Pauli's principle（パウリ原理） 354
perfect fluid（完全流体） 134
perfectly inelastic collision（完全非弾性衝突） 100
period（周期） 24, 147
periodic law of elements（元素の周期律） 354
periodic table（周期表） 354
permeability of vacuum（真空の透磁率） 272
permittivity of vacuum（真空の誘電率） 217
perpetual motion machine（永久機関） 194
perpetual motion machine of the first kind（第 1 種の永久機関） 195
perpetual motion machine of the second kind（第 2 種の永久機関） 195
phase（位相） 60, 147, 313
phase（相） 174
phase diagram（相図） 175
phase of the reflected wave（反射波の位相） 154
phase shift（位相のずれ） 313
phase transition（相転移） 175
phase velocity（位相速度） 164
phenomenological force（現象論的な力） 36
photoelectric effect（光電効果） 346
photon（光子，フォトン） 346, 375, 376
physical quantity（物理量） 7
pi meson（パイ中間子） 368
planar motion of rigid body（剛体の平面運動） 111
Planck constant（プランク定数） 178
Planck, M. K. E. L.（プランク） 189
Planck's law（プランクの法則） 178
plane wave（平面波） 153
plasticity（塑性） 127
pn junction（pn 接合） 359
pn junction diode（pn 接合ダイオード） 359
point charge（点電荷） 217
Poisson ratio（ポアッソン比） 129
polar coordinates（極座標） 21
polarization（分極） 250
polarized charge（分極電荷） 250
polarized light（偏光） 326
position（位置） 12
position vector（位置ベクトル） 17
positron（陽電子） 376
potential difference（電位差） 228
potential energy（位置エネルギー，ポテンシャルエネルギー） 72
potential energy of gravity（重力ポテンシャルエネルギー） 73
potential energy of the universal gravitation（万有引力ポテンシャルエネルギー） 74
power（仕事率，パワー） 70, 264
power factor（力率） 315
power in wave（波の強さ） 148
Poynting's vector（ポインティングのベクトル） 329, 331
precession（歳差運動） 115
prefix（接頭語） 8
pressure（圧力） 127
primitive function（原始関数） 37
principal quantum number（主量子数） 354
principle of increase of entropy（エントロピー増大の原理） 207
principle of superposition of electric field（電場の重ね合わせの原理） 220
principle of superposition of electric force（電気力の重ね合わせの原理） 218
principle of superposition of waves（波の重ね合わせの原理） 152
principle of the constancy of the speed of light（光速一定の原理） 335
proper time（固有時） 338
proton（陽子） 366
pulse（パルス） 146

Q

quantity of state（状態量） 191
quantum mechanics（量子力学） 350
quantum number（量子数） 354
quantum number of orbital angular momentum（軌道角運動量量子数） 354
quark（クォーク） 376

R

rad（unit：ラジアン） 22
radiation（放射線） 370
radio wave（電波） 324
radioactive isotope（放射性同位体） 370
radioactivity（放射能） 370
radioisotope（ラジオアイソトープ） 370
reciprocity theorem of mutual inductance（相互インダクタンスの相反定理） 312
rectification（整流作用） 359
reduced mass（換算質量） 97
reflection rate（反射率） 168
refracted angle（屈折角） 153
refraction（屈折） 153
refractive index（屈折率） 153
relation between impulse and momentum change（力積と運動量変化の関係） 43
relationship between work and kinetic energy（仕事と運動エネルギーの関係） 71
relative dielectric constant（比誘電率） 251
relative permeability（比透磁率） 276, 290
relative position vector（相対位置ベクトル） 97
relative refractive index（相対屈折率） 153
relative velocity（相対速度） 19, 98
residual magnetization（残留磁化） 292
resistance（抵抗） 258
resistor（抵抗器） 258
resonance（共振，共鳴） 65, 315
resonance frequency（共振周波数） 315
rest energy（静止エネルギー） 340
rest mass（静止質量） 339
resultant force（合力） 30
reversible engine（可逆機関） 202
reversible process（可逆変化） 198
Reynolds number（レイノルズ数） 142
right-handed screw rule（右ねじの規則） 273
rigid body（剛体） 91
rigid pendulum（剛体振り子） 108
rigidity（剛性率） 130
RLC circuit（*RLC* 回路） 315
Roentgen, W. C.（レントゲン） 370
rotating frame（回転座標系） 124
Rutherford, E.（ラザフォード） 345

S

s (unit：秒) 8
Salam, A. (サラム) 377
scalar product (スカラー積) 69, 383
Schrödinger equation
(シュレーディンガー方程式) 350
search coil (さぐりコイル) 317
second law of motion
(運動の第 2 法則) 29
second law of thermodynamics
(熱力学の第 2 法則) 198
self-inductance
(自己インダクタンス) 309
self-induction (自己誘導) 308
semiconductor (半導体) 357
semiconductor laser
(半導体レーザー) 363
separable differential equation
(変数分離形の微分方程式) 51
shear modulus (ずれ弾性率) 130
shear strain (ずれ変形) 129
shock wave (衝撃波) 165
SI (International System of Units)
(国際単位系) 8
significant figures (有効数字) 10
simple harmonic oscillation (単振動) 58
simple pendulum (単振り子) 61
sine wave (正弦波) 147
solar battery (太陽電池) 361
solar constant (太陽定数) 179, 372
solar energy (太陽エネルギー) 372
solenoid (ソレノイド) 275
sound pressure (音圧) 160
sound pressure level (音圧レベル) 160
sound wave (音波) 158
space dose rate (空間線量率) 372
special theory of relativity
(特殊相対性理論) 335
specific heat (比熱容量) 176
spectrum (スペクトル) 169
speed of electromagnetic wave
(電磁波の速さ) 324
speed of light (光の速さ) 168
speed of light in a homogeneous
material (物質の中での光の速さ) 326
speed of sound wave (音波の速さ) 158
speed of wave (波の速さ) 147
spherical capacitor
(球形キャパタシー) 243
spherical wave (球面波) 153
spin (スピン) 287
spontaneous emission (自発放射) 361
spring constant (ばね定数) 58
standard deviation (標準偏差) 10
standard uncertainty
(標準不確かさ) 10
standing wave (定在波，定常波) 156
state of thermal equilibrium
(熱平衡状態) 173
state variable (状態変数) 191
static friction (静止摩擦力) 36
stationary electric current
(定常電流) 257
stationary state (定常状態) 353
statistical mechanics (統計力学) 184
steady flow (定常流) 136
Stefan-Boltzmann's law
(シュテファン-ボルツマンの法則) 179
stimulated emission (誘導放射) 362
Stokes' law (ストークスの法則) 49, 141
strain (ひずみ) 128
stream-tube (流管) 137
streamline (流線) 137
streamline shape (流線形) 142
stress (応力) 127
strong force (強い力) 377
Super-Kamiokande detector
(スーパーカミオカンデ検出器) 373
superconductivity (超伝導) 261
Sv (unit：シーベルト) 372
synchrotron radiation (放射光) 296
system of particles (質点系) 90
system of two particles (2 質点系) 97

T

T (unit：テスラ) 271
tangential stress (接線応力) 127
temperature (温度) 174
temperature coefficient of electric
resistivity
(電気抵抗率の温度係数) 259
tension (張力) 127
terminal velocity (終端速度) 50
thermal equilibrium (熱平衡) 174
thermal expansion (熱膨張) 177
thermal radiation (熱放射) 178
thermodynamic temperature
(熱力学温度) 205
thermodynamics (熱力学) 190
third law of motion
(運動の第 3 法則) 31
Thomson, J. J. (トムソン) 344
time (時間) 8
time constant (時定数) 266, 310
time dilation of the moving clock
(動いている時計の遅れ) 337
Tomonaga, S. (朝永振一郎) 380
torque (トルク) 84
Torricelli's law (トリチェリの法則) 137
total momentum of the system of
particles (質点系の全運動量) 94
total reflection (全反射) 169
transformer (変圧器) 315
transistor (トランジスター) 360
transverse wave (横波) 145
traveling waves (進行波) 157
triple point (三重点) 175
trough (谷) 146
tube of electric force (電気力管) 233
tuned circuit (同調回路) 315
tunnel effect (トンネル効果) 364
turbulent flow (乱流) 142
two-body problem (2 体問題) 97
Tycho Brahe (ティコ・ブラーエ) 88

U

ultrasonic wave (超音波) 158
uncertainty (不確かさ) 10
uncertainty principle
(不確定性関係) 352
uniform circular motion
(等速円運動) 22
uniform electric field (一様な電場) 221
uniformly accelerated linear motion
(等加速度直線運動) 45
unit (単位) 8
universal gravitation (万有引力) 33

V

V (unit：ボルト) 228, 257
v-t graph (v-t グラフ) 13
valence electron (価電子) 358
Van de Graaff, R. J.
(バン・デ・グラーフ) 245, 367
van der Waals, J. D.
(ファン・デル・ワールス) 187
van der Waals' equation of state
(ファン・デル・ワールスの
状態方程式) 187
vector (ベクトル) 16
vector product (ベクトル積) 87, 383
velocity (速度) 13, 18, 17
Venturi tube (ベンチュリ管) 137
viscosity (粘性，粘度) 134, 140
viscosity coefficient (粘性係数) 140
viscous drag (粘性抵抗) 49, 141
viscous force (粘性力) 140
voltage (電圧) 258
voltage drop (電圧降下) 258

W

W (unit：ワット) 70, 264
Walton, E. T. S. (ウォルトン) 367
wave (波，波動) 144
wave equation (波動方程式) 148

wave form（波形） 146
wave front（波面） 153
wave function（波動関数） 350
wave number（波数） 163
wave packet（波束） 163
wave source（波源） 144
wavelength（波長） 147
Wb（unit：ウェーバ） 272
weak force（弱い力） 377
Weinberg, S.（ワインバーグ） 377
Wheatstone bridge
（ホイートストーン・ブリッジ） 267
Wien's displacement law
（ウィーンの変位則） 179
work（仕事） 69
working substance（作業物質） 199

X

x-t graph（x-t グラフ） 12

Y

Young's modulus（ヤング率） 129
Yukawa, H.（湯川秀樹） 368, 376, 380

Z

zero vector（零ベクトル） 17
zeroth law of thermodynamics
（熱力学の第 0 法則） 174
Zweig, G.（ツバイク） 376

【著者略歴】

原　康夫
1934 年　神奈川県鎌倉にて出生
1957 年　東京大学理学部物理学科卒業
1962 年　東京大学大学院修了（理学博士）
1962 年　東京教育大学理学部助手
1966 年　東京教育大学理学部助教授
1975 年　筑波大学物理学系教授
1997 年　帝京平成大学教授
2004 年　工学院大学エクステンションセンター客員教授
この間，カリフォルニア工科大学研究員，
シカゴ大学研究員，プリンストン高級研究所員．
1977 年　仁科記念賞受賞
現在　筑波大学名誉教授

訳者　川泉 文男　元名古屋大学教授
植松 恒夫　京都大学名誉教授

Basics of Physics

2019 年 12 月 5 日　第 1 版　第 1 刷　印刷
2019 年 12 月 15 日　第 1 版　第 1 刷　発行

著　者　原　康夫
訳　者　川泉文男
植松恒夫
発行者　発田和子
発行所　株式会社 学術図書出版社
〒113-0033　東京都文京区本郷 5-4-6
TEL 03-3811-0889　振替 00110-4-28454
印刷　三美印刷（株）

定価はカバーに表示してあります．

本書の一部または全部を無断で複写（コピー）・複製・転載することは，著作権法で認められた場合を除き，著作者および出版社の権利の侵害となります．あらかじめ，小社に許諾を求めてください．

© 2019　Printed in Japan

ISBN978-4-7806-0775-8

Prefixes of the metric system

Factor	Symbol	name	Factor	Symbol	name
10	da	deca	10^{-1}	d	deci
10^2	h	hecto	10^{-2}	c	centi
10^3	k	kilo	10^{-3}	m	milli
10^6	M	mega	10^{-6}	µ	micro
10^9	G	giga	10^{-9}	n	nano
10^{12}	T	tera	10^{-12}	p	pico
10^{15}	P	peta	10^{-15}	f	femto
10^{18}	E	exa	10^{-18}	a	atto
10^{21}	Z	zetta	10^{-21}	z	zepto
10^{24}	Y	yotta	10^{-24}	y	yocto

The Greek letters

Uppercase	Lowercase	English equivalent	Greek alphabet name
A	α	a, ā	alpha
B	β	b	beta
Γ	γ	g	gamma
Δ	δ	d	delta
E	ε, ϵ	e	epsilon
Z	ζ	z	zeta
H	η	ē	eta
Θ	θ, ϑ	th	theta
I	ι	i, ī	iota
K	κ	k	kappa
Λ	λ	l	lambda
M	μ	m	mu
N	ν	n	nu
Ξ	ξ	x	xi
O	o	o	omicron
Π	π	p	pi
P	ρ	r	rho
Σ	σ, ς	s	sigma
T	τ	t	tau
Υ	υ	u, y	upsilon
Φ	ϕ, φ	ph (f)	phi
X	χ	ch	chi, khi
Ψ	ψ	ps	psi
Ω	ω	ō	omega